SOLIDWORKS 2018
Reference Guide

A comprehensive reference guide
with over 260 standalone tutorials

David C. Planchard
CSWP & SolidWorks Accredited Educator

SDC Publications

SDC Publications

P.O. Box 1334

Mission, KS 66222

913-262-2664

www.SDCpublications.com

Publisher: Stephen Schroff

Examination Copies

Books received as examination copies are for review purposes only and may not be made available for student use. Resale of examination copies is prohibited.

Electronic Files

Any electronic files associated with this book are licensed to the original user only. These files may not be transferred to any other party.

Trademarks

SOLIDWORKS®, eDrawings®, SOLIDWORKS Simulation®, SOLIDWORKS Flow Simulation, and SOLIDWORKS Sustainability are a registered trademark of Dassault Systèmes SOLIDWORKS Corporation in the United States and other countries; certain images of the models in this publication courtesy of Dassault Systèmes SOLIDWORKS Corporation.

Microsoft Windows®, Microsoft Office® and its family of products are registered trademarks of the Microsoft Corporation. Other software applications and parts described in this book are trademarks or registered trademarks of their respective owners.

The publisher and the author make no representations or warranties with respect to the accuracy or completeness of the contents of this work and specifically disclaim all warranties, including without limitation warranties of fitness for a particular purpose. No warranty may be created or extended by sales or promotional materials. Dimensions of parts are modified for illustration purposes. Every effort is made to provide an accurate text. The author and the manufacturers shall not be held liable for any parts, components, assemblies or drawings developed or designed with this book or any responsibility for inaccuracies that appear in the book. Web and company information was valid at the time of this printing.

The Y14 ASME Engineering Drawing and Related Documentation Publications utilized in this text are as follows: ASME Y14.1 1995, ASME Y14.2M-1992 (R1998), ASME Y14.3M-1994 (R1999), ASME Y14.41-2003, ASME Y14.5-1982, ASME Y14.5-1999, and ASME B4.2. Note: By permission of The American Society of Mechanical Engineers, Codes and Standards, New York, NY, USA. All rights reserved.

Additional information references the American Welding Society, AWS 2.4:1997 Standard Symbols for Welding, Braising, and Non-Destructive Examinations, Miami, Florida, USA.

ISBN-13: 978-1-63057-150-4

ISBN-10: 1-63057-150-4

Printed and bound in the United States of America.

INTRODUCTION

SOLIDWORKS® 2018 Reference Guide with video instruction is a comprehensive text written to assist beginner to intermediate users of SOLIDWORKS. SOLIDWORKS is an immense software package, and no one book can cover all topics for all users. The book provides a centralized reference location to address many of the System and Document properties, FeatureManager, PropertyManager, ConfigurationManager, and RenderManager along with 2D and 3D sketch tools, sketch entities, 3D feature tools, Sheet Metal, Motion Study, SOLIDWORKS Simulation, PhotoView 360, Pack and Go, 3D PDFs, Intelligent Modeling techniques, 3D printing and more.

Chapter 1 provides a basic overview of the concepts and terminology used throughout this book using SOLIDWORKS® 2018 software. If you are completely new to SOLIDWORKS, you should read Chapter 1 in detail and complete Lesson 1, Lesson 2 and Lesson 3 in the SOLIDWORKS Tutorials.

If you are familiar with an earlier release of SOLIDWORKS, you still might want to skim Chapter 1 to be acquainted with some of the commands, menus and features that you have not used, or you can simply jump to any section in any chapter.

Each chapter provides detailed PropertyManager information on key topics with individual stand-alone short tutorials to reinforce and demonstrate the functionality and ease of the SOLIDWORKS tool or feature. The book provides access to over 260 models, their solutions and additional support materials. Learn by doing, not just by reading.

Formulate the skills to create, modify and edit sketches and solid features. Learn the techniques to reuse features, parts and assemblies through symmetry, patterns, copied components, design tables and configurations.

The book is designed to complement the Online Tutorials and Online Help contained in SOLIDWORKS 2018. The goal is to illustrate how multiple design situations and systematic steps combine to produce successful designs.

Redeem the code on the inside cover of the book. View the provided videos and models to enhance the user experience.

- Start a SOLIDWORKS session.

- Understand the SOLIDWORKS Interface.

- Create 2D Sketches, Sketch Planes and use various Sketch tools.

- Create 3D Features and apply Design Intent.

- Create an Assembly.

- Create fundamental Drawings Part 1 & Part 2.

The author developed the tutorials by combining his own industry experience with the knowledge of engineers, department managers, professors, vendors and manufacturers. These professionals are directly involved with SOLIDWORKS every day. Their responsibilities go far beyond the creation of just a 3D model.

About the Book

You will find a wealth of information in this book. Short stand-alone step-by step tutorials are written for each topic with the new and intermediate user in mind.

The following conventions are used throughout this book:

1. The term document is used to refer to a SOLIDWORKS part, drawing or assembly file.

2. The list of items across the top of the SOLIDWORKS interface is the Main menu. Each item in the Main menu has a pull-down menu. When you need to select a series of commands from these menus, the following format is used: Click **Insert ➢ Reference Geometry ➢ Plane** from the Main menu bar. The Plane PropertyManager is displayed.

3. Screen shots in the book were made using SOLIDWORKS 2018 SP0 running Windows® 10.

4. The book is organized into chapters. You can read any chapter without reading the entire book. Each chapter has short stand-alone step-by-step tutorials to practice and reinforce the subject matter and objectives. Learn by doing, not just by reading.

> SOLIDWORKS 2018
Name
BottomUpAssemblyModeling
Bracket
Center of Mass
COM
CSWSA FEA Model Folder 2018
Design Library
Detached Drawings
Drawings
eDrawing
Flex Feature
Freeform Feature
Intelligent Modeling
Lofted Feature
LOGO
Measure
MotionStudy
MY-TEMPLATES
MY-TOOLBOX
Pack and Go
PhotoView 360
SheetMetal
SMC
SOLIDWORKS Explorer
Solutions
Split Line

5. The ANSI overall drafting standard and Third Angle projection is used as the default setting in this text. IPS (inch, pound, second) and MMGS (millimeter, gram, second) unit systems are used.

6. Redeem the code on the inside cover of the book. View the provided videos and models to enhance the user experience. Download and copy the SOLIDWORKS files and folders to your hard drive. The book provides access to over 250 models, their solutions and additional support materials.

7. Compare your results with the tutorial documents in the Summary folders. All models for the stand-alone tutorials are included.

Instructor's information contains over 45 classroom presentations, along with helpful hints, what's new, sample quizzes, avi files of assemblies, projects and all initial and final SOLIDWORKS models.

About the Author

David Planchard is the founder of D&M Education LLC. Before starting D&M Education, he spent over 27 years in industry and academia holding various engineering, marketing, and teaching positions. He holds five U.S. patents. He has published and authored numerous papers on Machine Design, Product Design, Mechanics of Materials, and Solid Modeling. He is an active member of the SOLIDWORKS Users Group and the American Society of Engineering Education (ASEE). David holds a BSME, MSM with the following professional certifications: CCAI, CCNP, CSDA, CSWSA-FEA, CSWP, CSWP-DRWT and SOLIDWORKS Accredited Educator. David is a SOLIDWORKS Solution Partner, an Adjunct Faculty member and the SAE advisor at Worcester Polytechnic Institute in the Mechanical Engineering department. In 2012, David's senior Major Qualifying Project team (senior capstone) won first place in the Mechanical Engineering department at WPI. In 2014, 2015 and 2016, David's senior Major Qualifying Project teams won the Provost award in Mechanical Engineering for design excellence.

David Planchard is the author of the following books:

- **SOLIDWORKS® 2018 Reference Guide with video instruction**, 2017, 2016, 2015, 2014, 2013, 2012, 2011, 2010, and 2009

- **Engineering Design with SOLIDWORKS® 2018 with video instruction**, 2017, 2016, 2015, 2014, 2013, 2012, 2011, 2010, 2009, 2008, 2007, 2006, 2005, 2004, and 2003

- **Engineering Graphics with SOLIDWORKS® 2018 with video instruction**, 2017, 2016, 2015, 2014, 2013, 2012, and 2011

- **SOLIDWORKS® 2018 Quick Start with video instruction**

- **SOLIDWORKS® 2017 in 5 Hours with video instruction**, 2016, 2015, and 2014

- **SOLIDWORKS® 2018 Tutorial with video instruction**, 2017, 2016, 2015, 2014, 2013, 2012, 2011, 2010, 2009, 2008, 2007, 2006, 2005, 2004, and 2003

- **Drawing and Detailing with SOLIDWORKS® 2014**, 2012, 2010, 2009, 2008, 2007, 2006, 2005, 2004, 2003, and 2002

- **Official Certified SOLIDWORKS® Professional (CSWP) Certification Guide with video instruction, Version 4: 2015 - 2017**, Version 3: 2012 - 2014, Version 2: 2012 - 2013, Version 1: 2010 - 2010

- **Official Guide to Certified SOLIDWORKS® Associate Exams: CSWA, CSDA, CSWSA-FEA Version 3: 2015 - 2017**, Version 2: 2012 - 2015, Version 1: 2012 - 2013

- **Assembly Modeling with SOLIDWORKS® 2012**, 2010, 2008, 2006, 2005-2004, 2003, and 2001Plus

- **Applications in Sheet Metal Using Pro/SHEETMETAL & Pro/ENGINEER**

Acknowledgements

Writing this book was a substantial effort that would not have been possible without the help and support of my loving family and of my professional colleagues. I would like to thank Professor John M. Sullivan Jr., Professor Jack Hall and the community of scholars at Worcester Polytechnic Institute who have enhanced my life, my knowledge and helped to shape the approach and content to this text.

The author is greatly indebted to my colleagues from Dassault Systèmes SOLIDWORKS Corporation for their help and continuous support: Avelino Rochino and Mike Puckett.

Thanks also to Professor Richard L. Roberts of Wentworth Institute of Technology, Professor Dennis Hance of Wright State University, Professor Jason Durfess of Eastern Washington University and Professor Aaron Schellenberg of Brigham Young University - Idaho who provided vision and invaluable suggestions.

SOLIDWORKS certification has enhanced my skills and knowledge and that of my students. Thank you to Ian Matthew Jutras (CSWE) who is a technical contributor and the creator of the videos and Stephanie Planchard, technical procedure consultant.

Contact the Author

We realize that keeping software application books current is imperative to our customers. We value the hundreds of professors, students, designers, and engineers that have provided us input to enhance the book. Please contact me directly with any comments, questions or suggestions on this book or any of our other SOLIDWORKS books at dplanchard@msn.com or planchard@wpi.edu.

Note to Instructors

Please contact the publisher www.SDCpublications.com for classroom support materials (.ppt presentations, labs and more) and the Instructor's Guide with model solutions and tips that support the usage of this text in a classroom environment.

Trademarks, Disclaimer and Copyrighted Material

SOLIDWORKS®, eDrawings®, SOLIDWORKS Simulation®, SOLIDWORKS Flow Simulation, and SOLIDWORKS Sustainability are a registered trademark of Dassault Systèmes SOLIDWORKS Corporation in the United States and other countries; certain images of the models in this publication courtesy of Dassault Systèmes SOLIDWORKS Corporation.

Microsoft Windows®, Microsoft Office® and its family of products are registered trademarks of the Microsoft Corporation. Other software applications and parts described in this book are trademarks or registered trademarks of their respective owners.

The publisher and the author make no representations or warranties with respect to the accuracy or completeness of the contents of this work and specifically disclaim all warranties, including without limitation warranties of fitness for a particular purpose. No warranty may be created or extended by sales or promotional materials. Dimensions of parts are modified for illustration purposes. Every effort is made to provide an accurate text. The author and the manufacturers shall not be held liable for any parts, components, assemblies or drawings developed or designed with this book or any responsibility for inaccuracies that appear in the book. Web and company information was valid at the time of this printing.

The Y14 ASME Engineering Drawing and Related Documentation Publications utilized in this text are as follows: ASME Y14.1 1995, ASME Y14.2M-1992 (R1998), ASME Y14.3M-1994 (R1999), ASME Y14.41-2003, ASME Y14.5-1982, ASME Y14.5-1999, and ASME B4.2. Note: By permission of The American Society of Mechanical Engineers, Codes and Standards, New York, NY, USA. All rights reserved.

Additional information references the American Welding Society, AWS 2.4:1997 Standard Symbols for Welding, Braising, and Non-Destructive Examinations, Miami, Florida, USA.

References

- SOLIDWORKS Help Topics and What's New, SOLIDWORKS Corporation, 2018.

- Beers & Johnson, <u>Vector Mechanics for Engineers</u>, 6th ed. McGraw Hill, Boston, MA.

- Gradin, Hartley, <u>Fundamentals of the Finite Element Method</u>, Macmillan, NY 1986.

- Hibbler, R.C, <u>Engineering Mechanics Statics and Dynamics</u>, 8th ed, Prentice Hall.

- Jensen & Helsel, <u>Engineering Drawing and Design</u>, Glencoe, 1990.

- Lockhart & Johnson, <u>Engineering Design Communications</u>, Addison Wesley, 1999.

- Olivo C., Payne, Olivo, T, <u>Basic Blueprint Reading and Sketching</u>, Delmar 1988.

- Walker, James, <u>Machining Fundamentals</u>, Goodheart Wilcox, 1999.

- 80/20 Product Manual, 80/20, Inc., Columbia City, IN, 2012.

- Ticona Designing with Plastics - The Fundamentals, Summit, NJ, 2009.

- SMC Corporation of America, Product Manuals, Indiana, USA, 2012.

- Emerson-EPT Bearing Product Manuals and Gear Product Manuals, Emerson Power Transmission Corporation, Ithaca, NY, 2009.

- Emhart - A Black and Decker Company, On-line catalog, Hartford, CT, 2012.

During the initial SOLIDWORKS installation, you are requested to select either the ISO or ANSI drafting standard. ISO is typically a European drafting standard and uses First Angle Projection. The book is written using the ANSI (US) overall drafting standard and Third Angle Projection for drawings.

Screen shots in the book were made using SOLIDWORKS 2018 SP0 running Windows® 10.

Redeem the code on the inside cover of the book. View the provided videos and models to enhance the user experience. All templates, logos and model documents along with additional support materials are available.

TABLE OF CONTENTS

Redeem the code on the inside cover of the book. View the provided videos and models to enhance the user experience. The book provides access to over 260 models, their solutions and additional support materials.

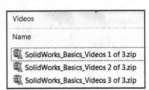

| Videos |
| Name |
| SolidWorks_Basics_Videos 1 of 3.zip |
| SolidWorks_Basics_Videos 2 of 3.zip |
| SolidWorks_Basics_Videos 3 of 3.zip |

Command Syntax

The following command syntax is used throughout the text. Commands that require you to perform an action are displayed in **Bold** text.

Format:	Convention:	Example:
Bold	• All commands actions. • Selected icon button. • Selected geometry: line, circle. • Value entries.	• Click **Options** ⚙ from the Menu bar toolbar. • Click the **Extruded Boss/Base** feature. • Click **Corner Rectangle** ⬜ from the Consolidated Sketch toolbar. • Click the **centerpoint**. • Enter **3.0** for Radius.
Capitalized	• Filenames. • First letter in a feature name.	• Save the **Flashlight** assembly. • Click the **Fillet** ◳ feature.

Windows Terminology in SOLIDWORKS

The mouse buttons provide an integral role in executing SOLIDWORKS commands. The mouse buttons execute commands, select geometry, display Shortcut menus and provide information feedback. A summary of mouse button terminology is displayed below:

Item:	Description:
Click	Press and release the left mouse button.
Double-click	Double press and release the left mouse button.
Click inside	Press the left mouse button. Wait a second, and then press the left mouse button inside the text box. Use this technique to modify Feature names in the FeatureManager design tree.
Drag/Drop	Point to an object, press and hold the left mouse button down. Move the mouse pointer to a new location. Release the left mouse button.
Right-click	Press and release the right mouse button. A Shortcut menu is displayed. Use the left mouse button to select a menu command.
Tool Tip	Position the mouse pointer over an Icon (button). The tool name is displayed below the mouse pointer.

A mouse with a center wheel provides additional functionality in SOLIDWORKS. Roll the center wheel downward to enlarge the model in the Graphics window. Hold the center wheel down. Drag the mouse in the Graphics window to rotate the model.

Visit SOLIDWORKS website: http://www.SOLIDWORKS.com/sw/support/hardware.html to view their supported operating systems and hardware requirements.

The book is designed to expose the new user to numerous tools and procedures. It may not always use the simplest and most direct process.

Hardware & System Requirements
Research graphics cards hardware, system requirements, and other related topics.

SolidWorks System Requirements
Hardware and system requirements for SolidWorks 3D CAD products.

Data Management System Requirements
Hardware and system requirements for SolidWorks Product Data Management (PDM) products.

SolidWorks Composer System Requirements
Hardware and system requirements for SolidWorks Composer and other 3DVIA related products.

SolidWorks Electrical System Requirements
Hardware and system requirements for SolidWorks Electrical products.

Graphics Card Drivers
Find graphics card drivers for your system to ensure system performance and stability.

Anti-Virus
The following Anti-Virus applications have been tested with SolidWorks 3D CAD products.

Hardware Benchmarks
Applications and references that can help determine hardware performance.

The book does not cover starting a SOLIDWORKS session in detail for the first time. A default SOLIDWORKS installation presents you with several options. For additional information for an Education Edition, visit the following site: http://www.SOLIDWORKS.com/sw/engineering-education-software.htm

The Instructor's information contains over 45 classroom presentations, along with helpful hints, What's new, sample quizzes, avi files of assemblies, projects, and all initial and final SOLIDWORKS model files.

CHAPTER 1: QUICK START

Chapter Objective

Chapter 1 provides a basic overview of the concepts and terminology used throughout the book using SOLIDWORKS® 2018 software. If you are completely new to SOLIDWORKS, you should read Chapter 1 in detail and complete Lesson 1, Lesson 2 and Lesson 3 in the SOLIDWORKS Tutorials under the Getting Started category.

Click the What's New Examples button in the SOLIDWORKS Tutorials section to learn what's new for 2018.

Chapter 1 introduces many of the basic operations of SOLIDWORKS such as starting a SOLIDWORKS session, using the User Interface along with opening and closing files, creating a part, assembly, applying design intent and a multi-view drawing with a Bill of Materials (BOM) and more.

On the completion of this chapter, you will be able to:

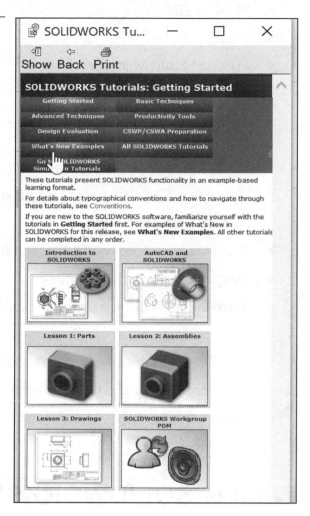

- Utilize the SOLIDWORKS Welcome dialog box.

- Start a SOLIDWORKS session.

- Recognize and use the SOLIDWORKS User Interface (UI) and CommandManager:

    - Menu Bar toolbar, Menu Bar menu, Drop-down menu, Right-Click, Pop-up toolbar, Pop-up menus, Consolidated fly-out tool buttons, System feedback, Confirmation Corner, FeatureManager, Fly-out FeatureManager, Orientation dialog box, Heads-up View toolbar and more.

- Understand basic concepts in SOLIDWORKS.

- Identify the FeatureManager Design Tree:

  - Right-click.

  - Show or Hide.

  - Filter.

  - Fly-out FeatureManager.

- Comprehend the Task Pane:

  - SOLIDWORKS Resources, Design Library, File Explorer, Search, View Palette, Appearances, Scenes and Decals, Custom Properties and Forum.

- Understand Motion Study:

  - Create two 3D parts: Axle and Flatbar.

- Create a 3D assembly:

  - Download provided model files and folders.

  - Open the AirCylinder assembly.

  - Insert components and apply Standard mates to the assembly.

  - Save the assembly.

- Create a 2D ANSI drafting standard assembly drawing:

  - Insert four standard orthographic views: Front, Top, Right and Isometric.

  - Insert a simple Bill of Materials (BOM).

  - Save the drawing.

What is SOLIDWORKS?

SOLIDWORKS® is a mechanical design automation software package used to build parts, assemblies and drawings that takes advantage of the familiar Microsoft® Windows graphical user interface.

SOLIDWORKS is an easy to learn design and analysis tool (SOLIDWORKS Simulation, SOLIDWORKS Motion, SOLIDWORKS Flow Simulation, SOLIDWORKS Sustainability, etc.), which makes it possible for designers to quickly sketch 2D and 3D concepts, create 3D parts and assemblies and detail 2D drawings.

Model dimensions in SOLIDWORKS are associative between parts, assemblies and drawings. Reference dimensions are one-way associative from the part to the drawing or from the part to the assembly.

This book is written for the beginner user with three or more months of experience to the intermediate user and assumes that you have some working knowledge of an earlier version of SOLIDWORKS.

Start a SOLIDWORKS 2018 Session

Start a SOLIDWORKS session and familiarize yourself with the SOLIDWORKS User Interface. As you read and perform the tasks in this chapter, you will obtain a sense of how to use the book and the structure. Actual input commands or required actions in the chapter are displayed in bold.

This book does not cover starting a SOLIDWORKS session in detail for the first time. A default SOLIDWORKS installation presents you with several options. For additional information, visit http://www.SOLIDWORKS.com.

Redeem the code on the inside cover of the book. View the provided videos and models to enhance the user experience.

Activity: Start a SOLIDWORKS Session.

Start a SOLIDWORKS session.

1)   Type **SOLIDWORKS** in the Search window.

2)   Click **All Programs.**

3)   Click the **SOLIDWORKS 2018** application (or if available, **double-click** the SOLIDWORKS icon on the Desktop). When you open the SOLIDWORKS software, view the Welcome dialog box. The Welcome dialog box provides a convenient means to open documents, view folders, access SOLIDWORKS resources, and stay updated on SOLIDWORKS news.

4)   **View** your options. Do not open a document at this time.

You can also click **Welcome to SOLIDWORKS** (Standard toolbar), **Help** > **Welcome** to SOLIDWORKS, or **Welcome to SOLIDWORKS** on the SOLIDWORKS Resources tab in the Task Pane to open the Welcome dialog box.

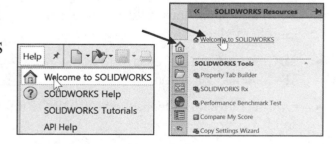

Home Tab

The Home tab lets you open new and existing documents, view recent documents and folders, and access SOLIDWORKS resources.

Sections in the Home tab include *New*, *Recent Documents*, *Recent Folders*, and *Resources*.

Recent Tab

The Recent tab lets you view a longer list of recent documents and folders. Sections in the Recent tab include *Documents* and *Folders*.

The Documents section includes thumbnails of documents that you have opened recently. Click a thumbnail to open the document, or hover over a thumbnail to see the document location and access additional information about the document. When you hover over a thumbnail, the full path and last saved date of the document appears.

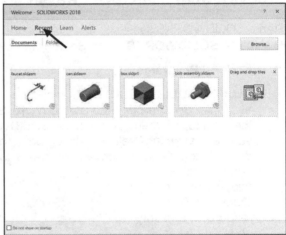

Learn Tab

The Learn tab lets you access instructional resources to help you learn more about the SOLIDWORKS software.

Sections in the Learn tab include:

- **Introducing SOLIDWORKS**. Opens the Introducing SOLIDWORKS book.

- **Tutorials**. Opens the step-by-step tutorials in the software.

- **MySolidWorks Training**. Opens the Training section at MySolidWorks.com.

- **Samples**. Opens local folders containing sample models.

- **3DContentCentral**. Opens 3DContentCentral.com.

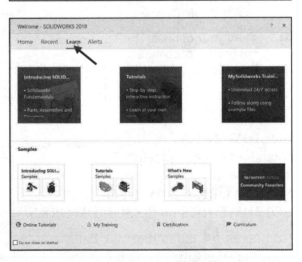

- **Online Tutorials**. Opens the SOLIDWORKS Tutorials (videos) section at solidworks.com.

- **My Training**. Opens the My Training section at MySolidWorks.com.

- **Certification**. Opens the SOLIDWORKS Certification Program section at solidworks.com.

- **Curriculum**. Opens the Curriculum section at solidworks.com.

When you install the software, if you do not install the Help Files or Example Files, the Tutorials and Samples links are unavailable.

Alerts Tab

The Alerts tab keeps you updated with SOLIDWORKS news.

Sections in the Alerts tab include:

Critical. The Critical section includes important messages that used to appear in a dialog box. If a critical alert exists, the Welcome dialog box opens to the Critical section automatically on startup, even if you selected Do not show at startup in the dialog box. Alerts are displayed until you select Do not show this message again.

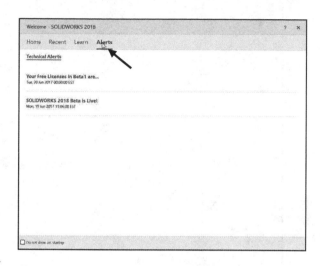

The Critical section does not appear if there are no critical alerts to display.

Troubleshooting. The Troubleshooting section includes troubleshooting messages and recovered documents that used to be on the SOLIDWORKS Recovery tab in the Task Pane.

If the software has a technical problem and an associated troubleshooting message exists, the Welcome dialog box opens to the Troubleshooting section automatically on startup, even if you selected Do not show at startup in the dialog box.

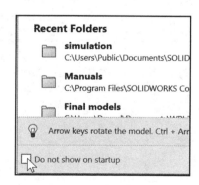

Technical Alerts. The Technical Alerts section opens the contents of the SOLIDWORKS Support Bulletins RSS feed at solidworks.com.

Close the Welcome dialog box.

5) Click **Close** ⊠ from the Welcome dialog box. The SOLIDWORKS Graphics window is displayed.

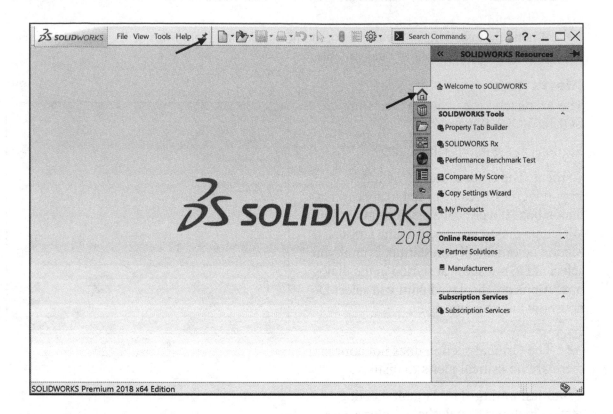

If you do not see this screen, click the SOLIDWORKS Resources ⌂ icon on the right side of the Graphics window located in the Task Pane.

6) **Hover** the mouse pointer over the SOLIDWORKS icon.

7) If needed, **Pin** the Menu Bar toolbar. View your options from the Menu bar menu: **File**, **View Tools** and **Help**.

Menu Bar toolbar

The SOLIDWORKS (UI) is designed to make maximum use of the Graphics window. The Menu Bar toolbar contains a set of the most frequently used tool buttons from the Standard toolbar.

The following default tools are available:

- **New** □ - Creates a new document; **Open** ⌐ - Opens an existing document; **Save** ⊟ - Saves an active document; **Print** 🖶 - Prints an active document; **Undo** ↺ - Reverses the last action; **Select** ⌖ - Selects Sketch entities, components and more; **Rebuild** ❚ - Rebuilds the active part, assembly or drawing; **File Properties** ▤ - Shows the summary information on the active document; and **Options** ⚙▾ - Changes system options and Add-Ins for SOLIDWORKS.

Menu Bar menu

Click SOLIDWORKS in the Menu Bar toolbar to display the Menu Bar menu. SOLIDWORKS provides a context-sensitive menu structure. The menu titles remain the same for all three types of documents, but the menu items change depending on which type of document is active.

Example: The Insert menu includes features in part documents, mates in assembly documents, and drawing views in drawing documents. The display of the menu is also dependent on the workflow customization that you have selected. The default menu items for an active document are *File, Edit, View, Insert, Tools, Window, Help* and *Pin*.

The Pin ⚲ option displays the Menu bar toolbar and the Menu bar menu as illustrated. Throughout the book, the Menu bar menu and the Menu bar toolbar are referred to as the Menu bar.

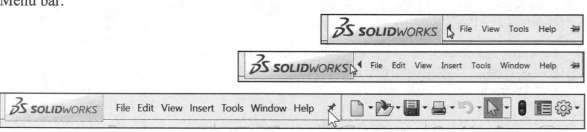

Drop-down menu

SOLIDWORKS takes advantage of the familiar Microsoft®
Windows user interface. Communicate with SOLIDWORKS
through drop-down menus, Context sensitive toolbars,
Consolidated toolbars or the CommandManager tabs.

☀ A command is an instruction that informs
SOLIDWORKS to perform a task.

To close a SOLIDWORKS drop-down menu, press the Esc
key. You can also click any other part of the SOLIDWORKS
Graphics window or click another drop-down menu.

Create a New Part Document

In the next section create a new part document.

Activity: Create a new Part Document.

A part is a 3D model, which consists of features. What are
features?

- Features are geometry building blocks.

- Most features either add or remove material.

- Some features do not affect material (Cosmetic Thread).

- Features are created either from 2D or 3D sketched profiles
  or from edges and faces of existing geometry.

- Features are individual shapes that combined with other
  features make up a part or assembly. Some features, such as
  bosses and cuts, originate as sketches. Other features, such as
  shells and fillets, modify a feature's geometry.

- Features are displayed in the FeatureManager as illustrated
  (Boss-Extrude1, Cut-Extrude1, Cut-Extrude2, Mirror1,
  Cut-Extrude3 and CirPattern1).

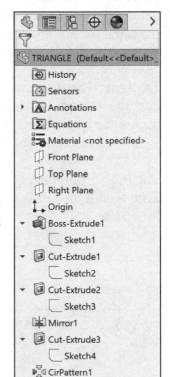

☀ The first sketch of a part is called the Base Sketch. The
Base sketch is the foundation for the 3D model. In this book, we
focus on 2D sketches and 3D features.

There are two modes in the New SOLIDWORKS Document dialog box: *Novice* and *Advanced*. The *Novice* option is the default option with three templates. The *Advanced* mode contains access to additional templates and tabs that you create in system options. Use the *Advanced* mode in this book.

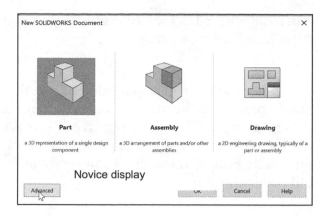

Create a new part.

8)   Click **New** ⬜ from the Menu bar. The New SOLIDWORKS Document dialog box is displayed.

Select the Advanced mode.

9)   Click the **Advanced** button as illustrated. The Advanced mode is set.

10) Click the **Templates** tab.

11) Click **Part**. Part is the default template from the New SOLIDWORKS Document dialog box.

12) Click **OK** from the New SOLIDWORKS Document dialog box.

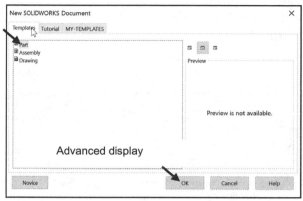

🔆 Illustrations may vary depending on your SOLIDWORKS version and operating system.

The Advanced mode remains selected for all new documents in the current SOLIDWORKS session. When you exit SOLIDWORKS, the Advanced mode setting is saved.

The default SOLIDWORKS installation contains two tabs in the New SOLIDWORKS Document dialog box: *Templates* and *Tutorial*. The *Templates* tab corresponds to the default SOLIDWORKS templates. The *Tutorial* tab corresponds to the templates utilized in the SOLIDWORKS Tutorials.

Part1 is displayed in the FeatureManager and is the name of the document. Part1 is the default part window name.

The Part Origin ⌐ is displayed in blue in the center of the Graphics window. The Origin represents the intersection of the three default reference planes: *Front Plane*, *Top Plane* and *Right Plane*. The positive X-axis is horizontal and points to the right of the Origin in the Front view. The positive Y-axis is vertical and points upward in the Front view. The FeatureManager contains a list of features, reference geometry, and settings utilized in the part.

Edit the document units directly from the Graphics window as illustrated.

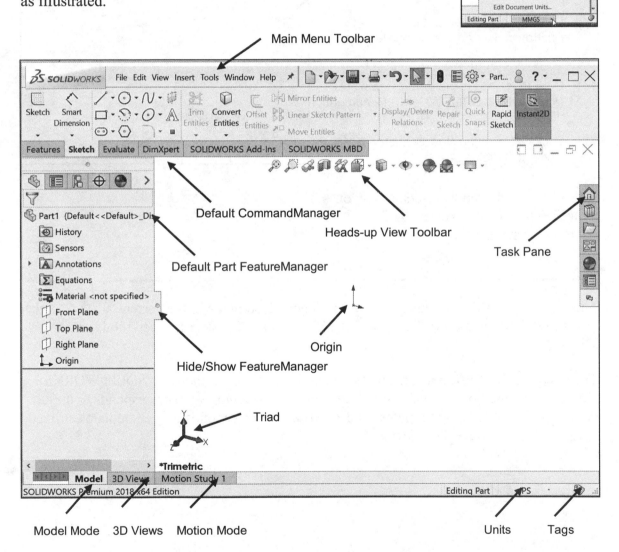

View the Default Sketch Planes.

13) Click the **Front Plane** from the FeatureManager.

14) Click the **Top Plane** from the FeatureManager.

15) Click the **Right Plane** from the FeatureManager.

16) Click the **Origin** from the FeatureManager. The Origin is the intersection of the Front, Top and Right Planes.

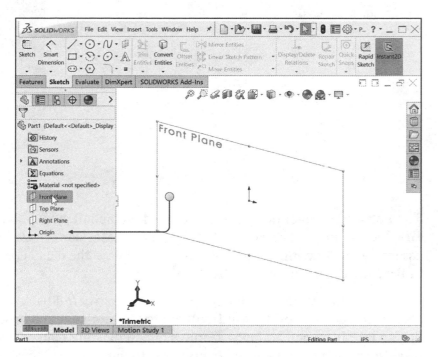

Redeem the code on the inside cover of the book. Download the **SOLIDWORKS 2018** folder to a local hard drive. Open the Bracket part. Review the features and sketches in the Bracket FeatureManager. Work directly from your hard drive.

Activity: Download the SOLIDWORKS 2018 folder. Open a Part.

Download the SOLIDWORKS 2018 folder. Open an existing SOLIDWORKS part.

17) **Download** the SOLIDWORKS 2018 folder to a local hard drive.

18) Click **Open** from the Menu bar menu.

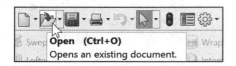

19) Browse to the **SOLIDWORKS 2018\Bracket** folder.

20) Double-click the **Bracket** part. The Bracket part is displayed in the Graphics window.

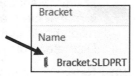

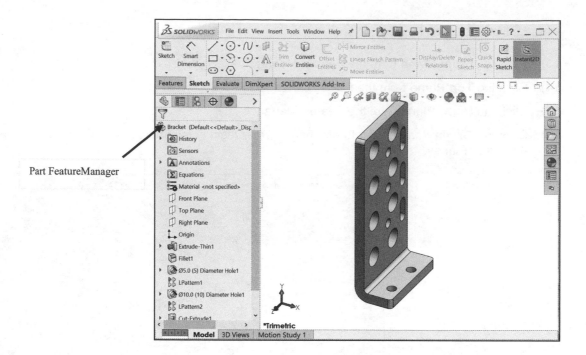

The FeatureManager design tree is located on the left side of the SOLIDWORKS Graphics window. The FeatureManager provides a summarized view of the active part, assembly, or drawing document. The tree displays the details on how the part, assembly or drawing document was created.

Use the FeatureManager rollback bar to temporarily roll back to an earlier state, to absorbed features, roll forward, roll to previous, or roll to the end of the FeatureManager design tree. You can add new features or edit existing features while the model is in the rolled-back state. You can save models with the rollback bar placed anywhere.

In the next section, review the features in the Bracket FeatureManager using the Rollback bar.

Activity: Use the FeatureManager Rollback Bar option.

Apply the FeatureManager Rollback Bar. Revert to an earlier state in the model.

21) Place the **mouse pointer** over the rollback bar in the FeatureManager design tree as illustrated. The pointer changes to a hand 🖑. Note the provided information on the feature. This is called Dynamic Reference Visualization.

22) Drag the **rollback bar** up the FeatureManager design tree until it is above the features you want rolled back, in this case 10.0 (10) Diameter Hole1.

23) **Release** the mouse button.

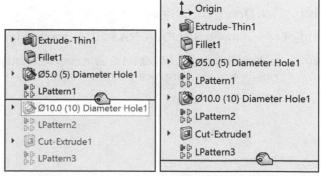

View the first feature in the Bracket Part.

24) Drag the **rollback bar** up the FeatureManager above Fillet1. View the results in the Graphics window.

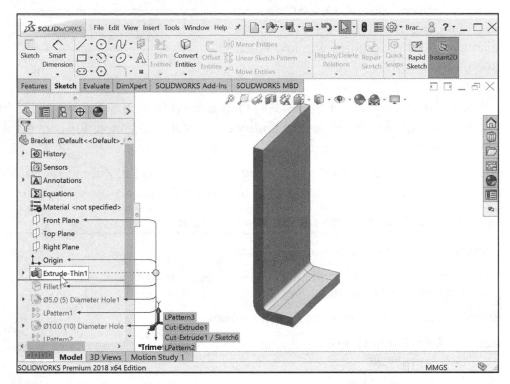

Return to the original Bracket Part FeatureManager.

25) Right-click **Extrude-Thin1** in the FeatureManager. The Pop-up Context toolbar is displayed.

26) Click **Roll to End**. View the results in the Graphics window.

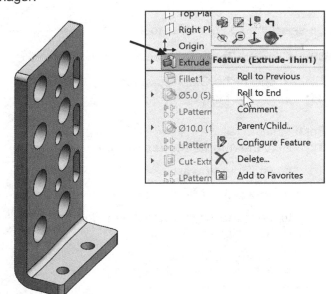

Heads-up View toolbar

SOLIDWORKS provides the user with numerous view options. One of the most useful tools is the Heads-up View toolbar displayed in the Graphics window when a document is active.

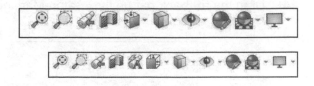

Dynamic Annotation Views : Only available with SOLIDWORKS MBD (Model Based Definition). Provides the ability to control how annotations are displayed when you rotate models.

In the next section, apply the following tools: Zoom to Fit, Zoom to Area, Zoom out, Rotate and select various view orientations from the Heads-up View toolbar.

Activity: Utilize the Heads-up View toolbar.

Zoom to Fit the model in the Graphics window.

27) Click the **Zoom to Fit** icon. The tool fits the model to the Graphics window.

Zoom to Area on the model in the Graphics window.

28) Click the **Zoom to Area** icon. The Zoom to Area icon is displayed.

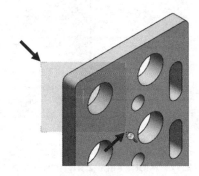

Zoom in on the top left hole.

29) **Window-select** the top left corner as illustrated. View the results.

De-select the Zoom to Area tool.

30) Click the **Zoom to Area** icon.

Fit the model to the Graphics window.

31) Press the **f** key.

Rotate the model.

32) Hold the **middle mouse button** down. Drag **upward** , **downward** , to the **left** and to the **right** to rotate the model in the Graphics window.

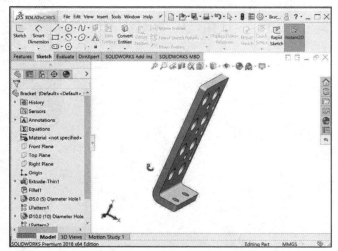

Display a few Standard Views.

33) Click **inside** the Graphics window.

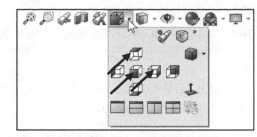

34) Click **Front** 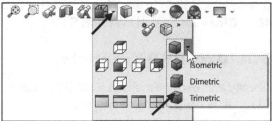 from the drop-down Heads-up view toolbar. The model is displayed in the Front view.

35) Click **Right** from the drop-down Heads-up view toolbar. The model is displayed in the Right view.

36) Click **Top** from the drop-down Heads-up view toolbar. The model is displayed in the Top view.

Display a Trimetric view of the Bracket model.

37) Click **Trimetric** from the drop-down Heads-up view toolbar as illustrated. Note your options. View the results in the Graphics window.

SOLIDWORKS Help

Help in SOLIDWORKS is context-sensitive and in HTML format. Help is accessed in many ways, including Help buttons in all dialog boxes and PropertyManager (or press F1) and Help ⑦ tool on the Standard toolbar for SOLIDWORKS Help.

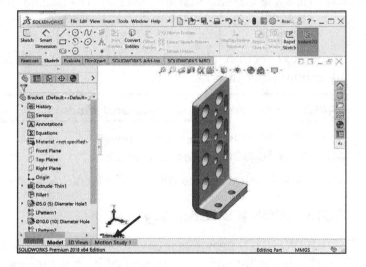

38) Click **Help** from the Menu bar.

39) Click **SOLIDWORKS Help**. The SOLIDWORKS Help Home Page is displayed by default. View your options.

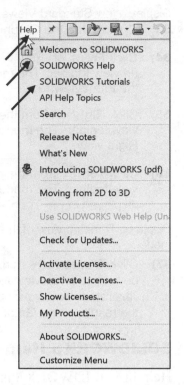

🔆 SOLIDWORKS Web Help is active by default under Help in the Main menu.

Close Help. Return to the SOLIDWORKS Graphics window.

40) **Close** ❎ SOLIDWORKS Home.

SOLIDWORKS Tutorials

Display and explore the SOLIDWORKS tutorials.

41) Click **Help** from the Menu bar.

42) Click **SOLIDWORKS Tutorials**. The SOLIDWORKS Tutorials are displayed. The SOLIDWORKS Tutorials are presented by category.

43) Click the **Getting Started** category. The Getting Started category provides lessons on parts, assemblies, and drawings.

In the next section, close all models, tutorials and view the additional User Interface tools.

Activity: Close all Tutorials and Models.

Close SOLIDWORKS Tutorials and models.

44) **Close** ❎ SOLIDWORKS Tutorials.

45) Click **Window**, **Close All** from the Menu bar menu.

SOLIDWORKS Icon Style

SOLIDWORKS provides a new icon style. It also allows vector-based scaling for superior support of high resolution high pixel density displays. The icon style standardized the perspective of icons. It removes non-essential details, and emphasizes primary elements. Consistent visual styling applies to all icons.

Additional User Interface Tools

The book utilizes additional areas of the SOLIDWORKS User Interface. Explore an overview of these tools in the next section.

Right-click

Right-click in the Graphics window on a model, or in the FeatureManager on a feature or sketch to display the Context-sensitive toolbar. If you are in the middle of a command, this toolbar displays a list of options specifically related to that command.

Right-click an empty space in the Graphics window of a part or assembly, and a selection context toolbar above the shortcut menu is displayed. This provides easy access to the most commonly used selection tools.

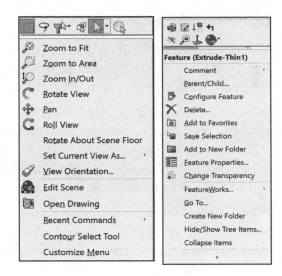

Consolidated toolbar

Similar commands are grouped together in the CommandManager. For example, variations of the Rectangle sketch tool are grouped in a single fly-out button as illustrated.

If you select the Consolidated toolbar button without expanding:

For some commands such as Sketch, the most commonly used command is performed. This command is the first listed and the command shown on the button.

For commands such as rectangle, where you may want to repeatedly create the same variant of the rectangle, the last used command is performed. This is the highlighted command when the Consolidated toolbar is expanded.

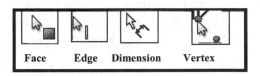

System feedback icon

SOLIDWORKS provides system feedback by attaching a symbol to the mouse pointer cursor.

The system feedback symbol indicates what you are selecting or what the system is expecting you to select.

As you move the mouse pointer across your model, system feedback is displayed in the form of a symbol, riding next to the cursor as illustrated. This is a valuable feature in SOLIDWORKS.

Confirmation Corner

When numerous SOLIDWORKS commands are active, a symbol or a set of symbols is displayed in the upper right hand corner of the Graphics window. This area is called the Confirmation Corner.

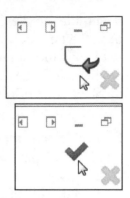

When a sketch is active, the confirmation corner box displays two symbols. The first symbol is the sketch tool icon. The second symbol is a large red X. These two symbols supply a visual reminder that you are in an active sketch. Click the sketch symbol icon to exit the sketch and to save any changes that you made.

When other commands are active, the confirmation corner box provides a green check mark and a large red X. Use the green check mark to execute the current command. Use the large red X to cancel the command.

Confirm changes you make in sketches and tools by using the D keyboard shortcut to move the OK and Cancel buttons to the pointer location in the Graphics window.

Heads-up View toolbar

SOLIDWORKS provides the user with numerous view options from the Standard Views, View and Heads-up View toolbar.

The Heads-up View toolbar is a transparent toolbar that is displayed in the Graphics window when a document is active.

You can hide, move or modify the Heads-up View toolbar. To modify the Heads-up View toolbar, right-click on a tool and select or deselect the tools that you want to display.

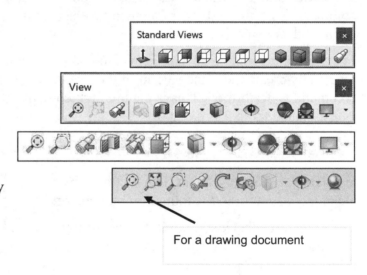

For a drawing document

The following views are available. Note: available views are document dependent.

- *Zoom to Fit* : Zooms the model to fit the Graphics window.

- *Zoom to Area* : Zooms to the areas you select with a bounding box.

- *Previous View* : Displays the previous view.

- *Section View* 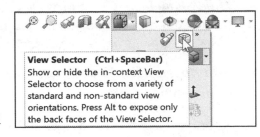 : Displays a cutaway of a part or assembly, using one or more cross section planes.

- *Dynamic Annotation Views* : Only available with SOLIDWORKS MBD. Provides the ability to control how annotations are displayed when you rotate models.

The Orientation dialog has an option to display a view cube (in-context View Selector) with a live model preview. This helps the user to understand how each standard view orientates the model. With the view cube, you can access additional standard views. The views are easy to understand and they can be accessed simply by selecting a face on the cube.

To activate the Orientation dialog box, press (Ctrl + spacebar) or click the View Orientation icon from the Heads up View toolbar. The active model is displayed in the View Selector in an Isometric orientation (default view).

Click the View Selector icon in the Orientation dialog box to show or hide the in-context View Selector.

Press **Ctrl + spacebar** to activate the View Selector.

Press the **spacebar** to activate the Orientation dialog box.

- *View Orientation box* 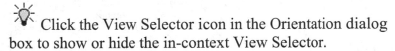 : Provides the ability to select a view orientation or the number of viewports. The available options are *Top, Left, Front, Right, Back, Bottom, Single view, Two view - Horizontal, Two view - Vertical, Four view*. Click the drop-down arrow to access Axonometric views: Isometric, Dimetric and Trimetric.

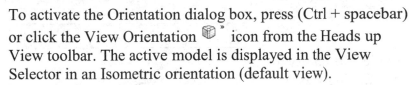

- *Display Style* 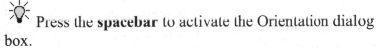 : Provides the ability to display the style for the active view. The available options are *Wireframe, Hidden Lines Visible, Hidden Lines Removed, Shaded, Shaded With Edges*.

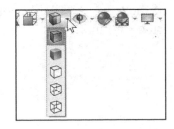
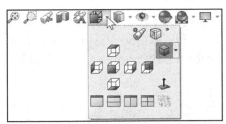

- *Hide/Show Items*  : Provides the ability to select items to hide or show in the Graphics window. The available items are document dependent. Note the View Center of Mass ✛ icon.

- *Edit Appearance* 🌑: Provides the ability to edit the appearance of entities of the model.

- *Apply Scene* 🖼 ˅: Provides the ability to apply a scene to an active part or assembly document. View the available options.

- *View Setting*🖵 ˅: Provides the ability to select the following settings: *RealView Graphics*, *Shadows In Shaded Mode, Ambient Occlusion, Perspective* and *Cartoon*.

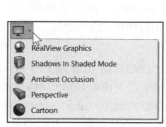

- *Rotate view* ↻ : Provides the ability to rotate a drawing view. Input Drawing view angle and select the ability to update and rotate center marks with view.

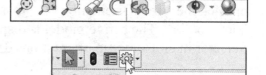

- *3D Drawing View* 📐: Provides the ability to dynamically manipulate the drawing view in 3D to make a selection.

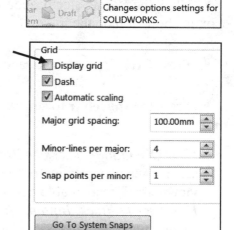

🔆 To display a grid for a part, click Options ⚙ ˅, Document Properties tab. Click Grid/Snaps, check the Display grid box.

🔆 Add a custom view to the Heads-up View toolbar. Press the space key. The Orientation dialog box is displayed. Click the New View 🔧 tool. The Name View dialog box is displayed. Enter a new named view. Click OK.

🔆 Use commands to display information about the triad or to change the position and orientation of the triad. Available commands depend on the triad's context.

Triad

SOLIDWORKS CommandManager

The SOLIDWORKS CommandManager is a Context-sensitive toolbar that automatically updates based on the toolbar you want to access. By default, it has toolbars embedded in it based on your active document type. When you click a tab below the CommandManager, it updates to display that toolbar. For example, if you click the Sketch tab, the Sketch toolbar is displayed.

For commercial users, SOLIDWORKS Model Based Definition (MBD) is a separate application. For education users, SOLIDWORKS MBD is included in the SOLIDWORKS Education Edition as an Add In.

Below is an illustrated CommandManager for a default Part document.

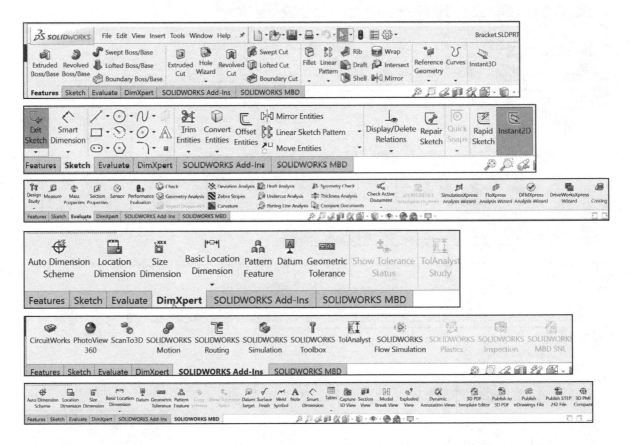

Button sizes. You can set sizes for buttons from the Toolbars tab of the Customize dialog box. To facilitate element selection on touch interfaces such as tablets, you can set up the larger Size buttons and text from the Options menu (Standard toolbar).

The SOLIDWORKS CommandManager is a Context-sensitive toolbar that automatically updates based on the toolbar you want to access. By default, it has toolbars embedded in it based on your active document type.

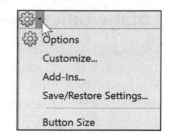

For commercial users, SOLIDWORKS Model Based Definition (MBD) is a separate application. For education users, SOLIDWORKS MBD is included in the SOLIDWORKS Education Edition as an Add In.

Below is an illustrated CommandManager for a default Drawing document.

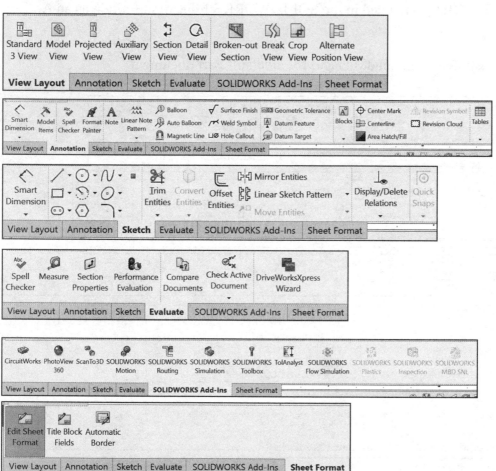

To add a custom tab to your CommandManager, right-click on a tab and click Customize CommandManager from the drop-down menu. The Customize dialog box is displayed. You can also select to add a blank tab as illustrated and populate it with custom tools from the Customize dialog box.

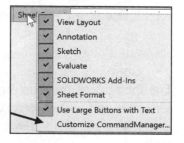

The SOLIDWORKS CommandManager is a Context-sensitive toolbar that automatically updates based on the toolbar you want to access. By default, it has toolbars embedded in it based on your active document type.

💡 For commercial users, SOLIDWORKS Model Based Definition (MBD) is a separate application. For education users, SOLIDWORKS MBD is included in the SOLIDWORKS Education Edition as an Add In.

Below is an illustrated CommandManager for a default Assembly document.

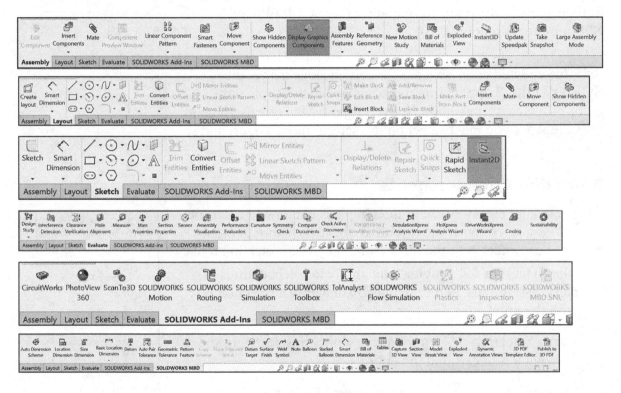

By default, the illustrated options are selected in the Customize box for the CommandManager. Right-click on an existing tab and click Customize CommandManager to view your options.

💡 You can set the number of mouse gestures to 2, 3, 4, 6, 8, or 12 gestures. If you set the number to 2 gestures, you can orient them vertically or horizontally.

Float the CommandManager. Drag the Features, Sketch or any CommandManager tab. Drag the CommandManager anywhere on or outside the SOLIDWORKS window.

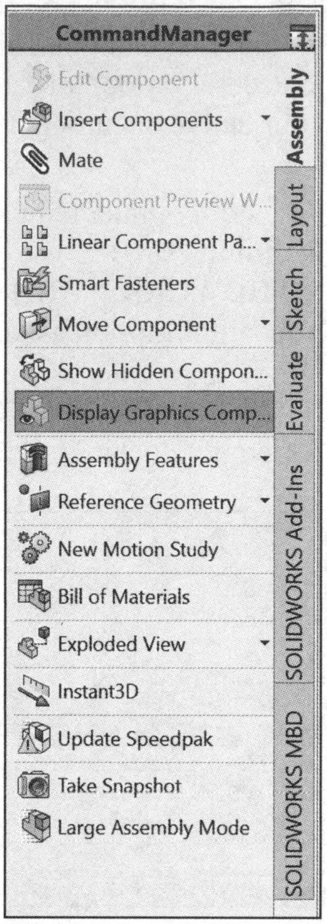

To dock the CommandManager, perform one of the following:

While dragging the CommandManager in the SOLIDWORKS window, move the pointer over a docking icon - Dock above, Dock left, Dock right and click the needed command.

Double-click the floating CommandManager to revert the CommandManager to the last docking position.

Screen shots in the book were made using SOLIDWORKS 2018 SP0 running Windows® 10.

An updated color scheme for certain icons makes the SOLIDWORKS application more accessible to people with color blindness. Icons in the active PropertyManager use blue to indicate what you must select on the screen: faces, edges, and so on.

Selection Enhancements

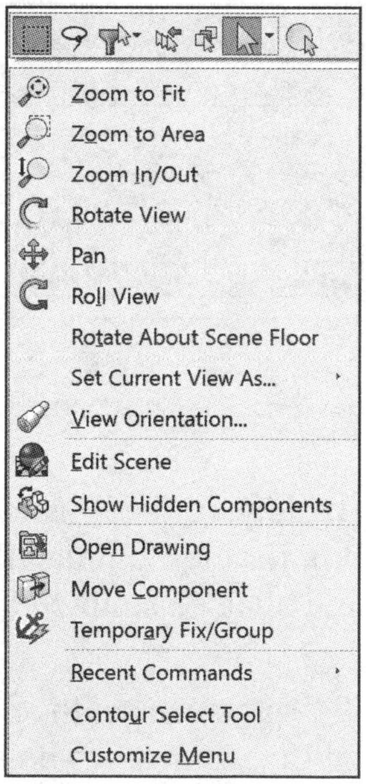

Right-click an empty space in the Graphics window of a part or assembly; a selection context toolbar above the shortcut menu provides easy access to the most commonly used selection tools.

- **Box Selection**. Provides the ability to select entities in parts, assemblies, and drawings by dragging a selection box with the pointer.
- **Lasso Selection**. Provides the ability to select entities by drawing a lasso around the entities.
- **Selection Filters**. Displays a list of selection filter commands.
- **Previous Selection**. Displays the previous selection.
- **Select Other**. Displays the Select Other dialog box.
- **Select**. Displays a list of selection commands.
- **Magnified Selection**. Displays the magnifying glass, which gives you a magnified view of a section of a model.

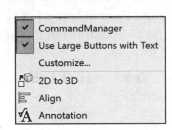

Save space in the CommandManager, right-click in the CommandManager and un-check the Use Large Buttons with Text box. This eliminates the text associated with the tool.

DimXpert provides the ability to graphically check if the model is fully dimensioned and toleranced. DimXpert automatically recognizes manufacturing features. Manufacturing features are not SOLIDWORKS features. Manufacturing features are defined in 1.1.12 of the ASME Y14.5M-1994 Dimensioning and Tolerancing standard. See SOLIDWORKS Help for additional information.

FeatureManager Design Tree

The FeatureManager consists of five default tabs:

- *FeatureManager design tree* tab.

- *PropertyManager* tab.

- *ConfigurationManager* tab.

- *DimXpertManager* tab.

- *DisplayManager* tab.

Select the Hide FeatureManager Tree Area

arrows as illustrated to enlarge the Graphics window for modeling.

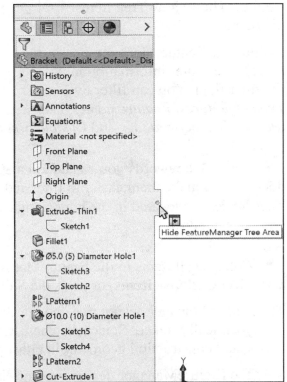

The Sensors tool located in the FeatureManager monitors selected properties in a part or assembly and alerts you when values deviate from the specified limits. There are five sensor types: Simulation Data, Mass properties, Dimensions, Measurement and Costing Data.

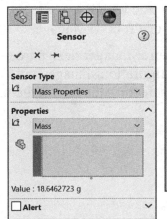

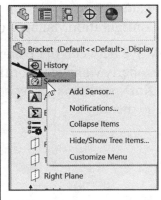

Various commands provide the ability to control what is displayed in the FeatureManager design tree. They are:

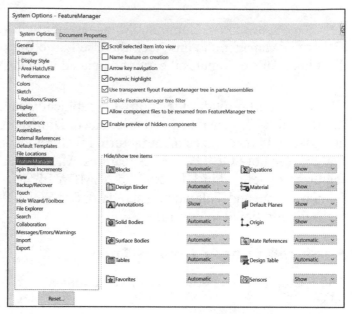

1. Show or Hide FeatureManager items.

💡 Click **Options** ⚙ from the Menu bar. Click **FeatureManager** from the System Options tab. **Customize** your FeatureManager from the Hide/Show Tree Items dialog box.

2. Filter the FeatureManager design tree. Enter information in the filter field. You can filter by *Type of features, Feature names, Sketches, Folders, Mates, User-defined tags* and *Custom properties*.

💡 Tags are keywords you can add to a SOLIDWORKS document to make them easier to filter and to search. The Tags 🏷 icon is located in the bottom right corner of the Graphics window.

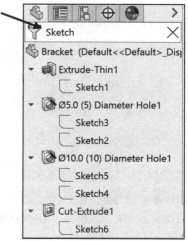

💡 Collapse all items in the FeatureManager, **right-click** and select **Collapse items**, or press the **Shift + C** keys.

The FeatureManager design tree and the Graphics window are dynamically linked. Select sketches, features, drawing views, and construction geometry in either pane.

Split the FeatureManager design tree and either display two FeatureManager instances, or combine the FeatureManager design tree with the ConfigurationManager or PropertyManager.

Move between the FeatureManager design tree, PropertyManager, ConfigurationManager and DimXpertManager by selecting the tabs at the top of the menu.

Split line

The ConfigurationManager is located to the right of the FeatureManager. Use the ConfigurationManager to create, select and view multiple configurations of parts and assemblies.

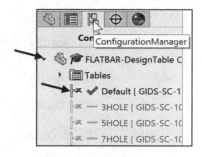

The icons in the ConfigurationManager denote whether the configuration was created manually or with a design table.

The DimXpertManager tab provides the ability to insert dimensions and tolerances manually or automatically. The DimXpertManager provides the following selections: **Auto Dimension Scheme** ⊕, **Basic Location Dimension** ⊢⊣,

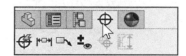

Basic Size Dimension ▢ **Show Tolerance Status** ⁺⊚, **Copy Scheme** ⊕ and **TolAnalyst Study** ⊓.

Fly-out FeatureManager

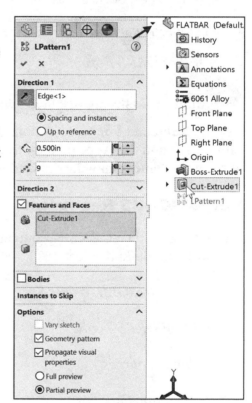

The fly-out FeatureManager design tree provides the ability to view and select items in the PropertyManager and the FeatureManager design tree at the same time.

Throughout the book, you will select commands and command options from the drop-down menu, fly-out FeatureManager, Context toolbar or from a SOLIDWORKS toolbar.

⚡ Another method for accessing a command is to use the accelerator key. Accelerator keys are special key strokes, which activate the drop-down menu options. Some commands in the menu bar and items in the drop-down menus have an underlined character.

Pressing the Alt or Ctrl key followed by the corresponding key to the underlined character activates that command or option.

⚡ Illustrations may vary depending on your SOLIDWORKS version and operating system.

Task Pane

The Task Pane is displayed when a SOLIDWORKS session starts. You can show, hide, and reorder tabs in the Task Pane. You can also set a tab as the default so it appears when you open the Task Pane, pin or unpin to the default location.

The Task Pane contains the following default tabs:

- *SOLIDWORKS Resources* ⌂.

- *Design Library* ▦ .

- *File Explorer* 🗀 .

- *View Palette* ▦.

- *Appearances, Scenes and Decals* ⬤.

- *Custom Properties* ▥ .

- *SOLIDWORKS Forum* 🗨 .

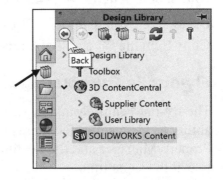

 Additional tabs are displayed with Add-Ins.

Use the **Back** and **Forward** buttons in the Design Library tab and the Appearances, Scenes, and Decals tab of the Task Pane to navigate in folders.

SOLIDWORKS Resources

The basic SOLIDWORKS Resources ⌂ menu displays the following default selections:

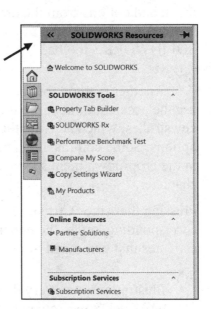

- *Welcome to SOLIDWORKS.*

- *SOLIDWORKS Tools.*

- *Online Resources.*

- *Subscription Services.*

Other user interfaces are available during the initial software installation selection: *Machine Design*, *Mold Design, Consumer Products Design, etc.*

Design Library

The Design Library 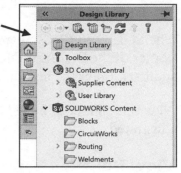 contains reusable parts, assemblies, and other elements including library features.

The Design Library tab contains four default selections. Each default selection contains additional sub categories.

The default selections are:

- *Design Library.*

- *Toolbox.*

- *3D ContentCentral (Internet access required).*

- *SOLIDWORKS Content (Internet access required).*

Activate the SOLIDWORKS Toolbox. Click Tools, Add-Ins.., from the Main menu. Check the SOLIDWORKS Toolbox Library and SOLIDWORKS Toolbox Utilities box from the Add-ins dialog box or click SOLIDWORKS Toolbox from the SOLIDWORKS Add-Ins tab.

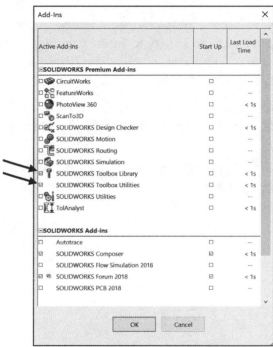

To access the Design Library folders in a non-network environment, click Add File

Location 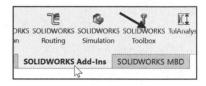 and browse to the needed path. Paths may vary depending on your SOLIDWORKS version and window setup. In a network environment, contact your IT department for system details.

File Explorer

File Explorer duplicates Windows Explorer from your local computer and displays:

- *Recent Documents.*

- *Samples.*

- *Open in SOLIDWORKS.*

- *Desktop.*

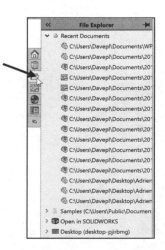

Search

The SOLIDWORKS Search box is displayed in the upper right corner of the SOLIDWORKS Graphics window (Menu Bar toolbar). Enter the text or key words to search.

New search modes have been added to SOLIDWORKS Search as illustrated.

View Palette

The View Palette tool located in the Task Pane provides the ability to insert drawing views of an active document, or click the Browse button to locate the desired document.

Click and drag the view from the View Palette into an active drawing sheet to create a drawing view.

The selected model is FLATBAR-DesignTables in the illustration.

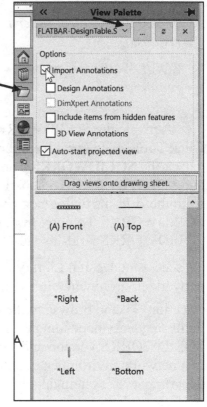

Appearances, Scenes, and Decals

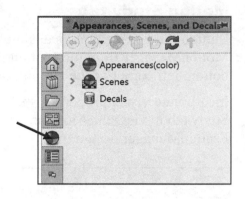

Appearances, Scenes, and Decals ● provide a simplified way to display models in a photo-realistic setting using a library of Appearances, Scenes, and Decals.

An appearance defines the visual properties of a model, including color and texture. Appearances do not affect physical properties, which are defined by materials.

Scenes provide a visual backdrop behind a model. In SOLIDWORKS they provide reflections on the model. PhotoView 360 is an Add-in. Drag and drop a selected appearance, scene or decal on a feature, surface, part or assembly.

Custom Properties

The Custom Properties ▤ tool provides the ability to enter custom and configuration specific properties directly into SOLIDWORKS files.

SOLIDWORKS Forum

Click the SOLIDWORKS Forum 🗩 icon to search directly within the Task Pane. An internet connection is required. You are required to register and to log in for postings and discussions.

User Interface for Scaling High Resolution Screens

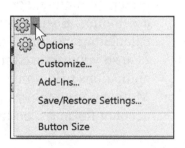

The SOLIDWORKS software supports high-resolution, high-pixel density displays. All aspects of the user interface respond to the Microsoft Windows® display scaling setting. In dialog boxes, PropertyManagers, and the FeatureManager design tree, the SOLIDWORKS software uses your display scaling setting to display buttons and icons at an appropriate size. Icons that are associated with text are scaled to a size appropriate for the text. In addition, for toolbars, you can display Small, Medium, or Large buttons. Click the **Options drop-down arrow** from the Standard Menu bar, and click Button size to size the icons.

Motion Study tab

Motion Studies are graphical simulations of motion for an assembly. Access the MotionManager from the Motion Study tab. The Motion Study tab is located in the bottom left corner of the Graphics window.

Incorporate visual properties such as lighting and camera perspective. Click the Motion Study tab to view the MotionManager. Click the Model tab to return to the FeatureManager design tree.

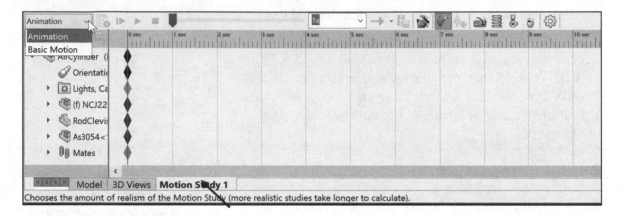

The MotionManager displays a timeline-based interface and provides the following selections from the drop-down menu as illustrated:

- *Animation:* Apply Animation to animate the motion of an assembly. Add a motor and insert positions of assembly components at various times using set key points. Use the Animation option to create animations for motion that do **not** require accounting for mass or gravity.

- *Basic Motion:* Apply Basic Motion for approximating the effects of motors, springs, collisions and gravity on assemblies. Basic Motion takes mass into account in calculating motion. Basic Motion computation is relatively fast, so you can use this for creating presentation animations using physics-based simulations. Use the Basic Motion option to create simulations of motion that account for mass, collisions or gravity.

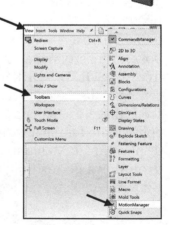

If the Motion Study tab is not displayed in the Graphics window, click **View ➢ Toolbars ➢ MotionManager** from the Menu bar.

3D Views tab

SOLIDWORKS MBD (Model Based Definition) lets you create models without the need for drawings giving you an integrated manufacturing solution. MBD helps companies define, organize, and publish 3D product and manufacturing information (PMI), including 3D model data in industry standard file formats.

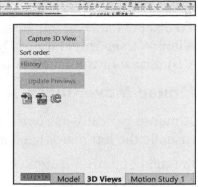

Create 3D drawing views of your parts and assemblies that contain the model settings needed for review and manufacturing. This lets users navigate back to those settings as they evaluate the design.

Use the tools in the SOLIDWORKS MBD CommandManager to set up your model with selected configurations, including explodes and abbreviated views, annotations, display states, zoom level, view orientation and section views. Capture those settings so that you and other users can return to them at any time using the 3D view palctte.

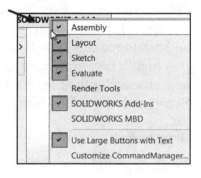

To access the 3D View palette, click the 3D Views tab at the bottom of the SOLIDWORKS window or the SOLIDWORKS MBD tab in the CommandManager. The Capture 3D View button opens the Capture 3D View PropertyManager, where you specify the 3D view name, and the configuration, display state and annotation view to capture. See SOLIDWORKS help for additional information.

Dynamic Reference Visualization (Parent/Child)

Dynamic Reference Visualization provides the ability to view the parent/child relationships between items in the FeatureManager design tree. When you hover over a feature with references in the FeatureManager design tree, arrows display showing the relationships. If a reference cannot be shown because a feature is not expanded, the arrow points to the feature that contains the reference and the actual reference appears in a text box to the right of the arrow.

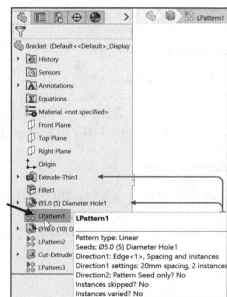

Use Dynamic reference visualization for a part, assembly and ever mates.

To display the Dynamic Reference Visualization, click **View ➤ User Interface ➤ Dynamic Reference Visualization** from the Main menu bar.

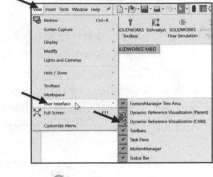

Mouse Movements

A mouse typically has two buttons: a primary button (usually the left button) and a secondary button (usually the right button). Most mice also include a scroll wheel between the buttons to help you scroll through documents and to Zoom in, Zoom out and rotate models in SOLIDWORKS. It is highly recommended that you use a mouse with at least a Primary, Scroll and Secondary button.

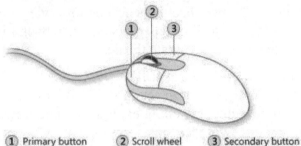

① Primary button   ② Scroll wheel   ③ Secondary button

Single-click

To click an item, point to the item on the screen, and then press and release the primary button (usually the left button). Clicking is most often used to select (mark) an item or open a menu. This is sometimes called single-clicking or left-clicking.

Double-click

To double-click an item, point to the item on the screen, and then click twice quickly. If the two clicks are spaced too far apart, they might be interpreted as two individual clicks rather than as one double-click. Double-clicking is most often used to open items on your desktop. For example, you can start a program or open a folder by double-clicking its icon on the desktop.

Right-click

To right-click an item, point to the item on the screen, and then press and release the secondary button (usually the right button). Right-clicking an item usually displays a list of things you can do with the item. Right-click in the open Graphics window or on a command in SOLIDWORKS, and additional pop-up context is displayed.

Scroll wheel

Use the scroll wheel to zoom-in or to zoom-out of the Graphics window in SOLIDWORKS. To zoom-in, roll the wheel backward (toward you). To zoom-out, roll the wheel forward (away from you).

Create the Axle Part

Tutorial: Axle 1-1

If needed, start a SOLIDWORKS session.

1) Double-click the **SOLIDWORKS** icon from the Desktop.

2) Close the **Welcome** dialog box.

3) If needed, **pin** 📌 the Menu bar menu as illustrated. Use both the Menu bar menu and the Menu bar toolbar in this book.

Create a New Part Document.

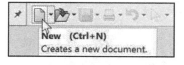

4) Click **New** 🗋 from the Menu bar or click **File**, **New** from the Menu bar menu. The New SOLIDWORKS Document dialog box is displayed. Advanced mode is used in this book.

5) Click the **Templates** tab.

6) Double-click **Part** from the New SOLIDWORKS Document dialog box. Part 1 is displayed in the FeatureManager.

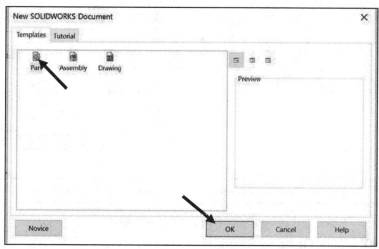

☼ The first system default Part filename is Part1. The system attaches the .sldprt suffix to the created part. The second created part in the same session increments to the filename Part2.

Display the origin.

7) Click **View Origins** from the Heads-up toolbar drop down menu. The origin is displayed in the Graphics window.

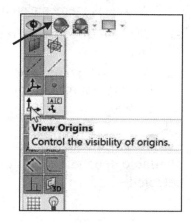

Directional input refers by default to the global coordinate system (X, Y and Z), which is based on Plane1 with its origin located at the origin of the part or assembly. Plane1 (Front) is the first plane that appears in the FeatureManager design tree and can have a different name. The reference triad shows the global X-, Y- and Z-directions.

The three default ⊥ reference planes, displayed in the FeatureManager design tree, represent infinite 2D planes in 3D space.

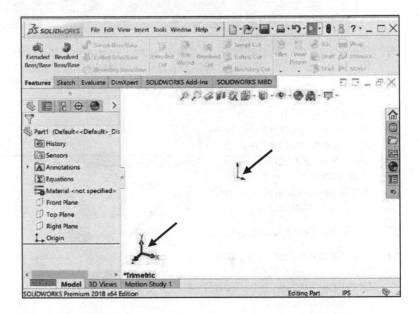

Activity: Set Document Properties for the Axle.

Set Document Properties. Set drafting standard.

8) Click **Options** ⚙ from the Menu bar toolbar. The System Options General dialog box is displayed.

9) Click the **Document Properties** tab.

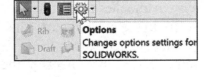

10) Select **ANSI** from the Overall drafting standard drop-down menu.

💡 Various detailing options are available depending on the selected standard.

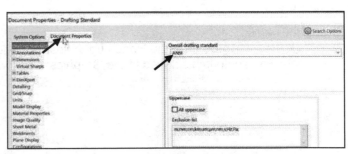

The Overall drafting standard determines the display of dimension text, arrows, symbols, and spacing. Units are the measurement of physical quantities. Millimeter dimensioning and decimal inch dimensioning are the two most common unit types specified for engineering parts and drawings.

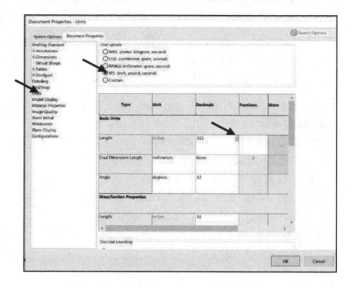

Set Document Properties. Set units and precision.

11) Click the **Units** folder.

12) Click **IPS** (inch, pound, second) for Unit system.

13) Select **.123**, (three decimal places) for Length basic units.

14) Click **OK** from the Document Properties - Units dialog box. The Part FeatureManager is displayed.

Save the model. Enter name.

15) Click **Save** 💾 from the Menu bar.

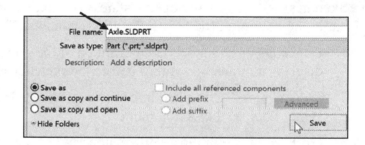

16) Create a new folder named **SOLIDWORKS 2018** in a convenient location on your hard drive.

17) Enter **Axle** for the File name in the SOLIDWORKS 2018.

18) Click **Save** from the Save As dialog box.

The origin ⌐ represents the intersection of the Front, Top and Right planes.

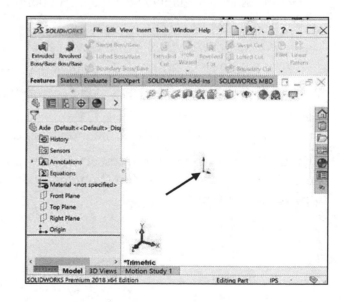

Redeem the code on the inside cover of the book. View the provided videos on creating 2D Sketches, Sketch Planes and using various Sketch tools along with 3D Features and Design Intent.

2D Sketching - Identify the Correct Sketch Plane

Most SOLIDWORKS features start with a 2D sketch. Sketches are the foundation for creating features. SOLIDWORKS provides the ability to create either 2D or 3D sketches.

A 2D sketch is limited to a flat 2D sketch plane located on a reference plane, face or a created plane. 3D sketches are very useful in creating sketch geometry that does not lie on an existing or easily defined plane.

Does it matter what plane you start the base 2D sketch on? Yes. When you create a new part or assembly, the three default planes are aligned with specific views. The plane you select for your first sketch determines the orientation of the part. Selecting the correct plane to start your model is very important.

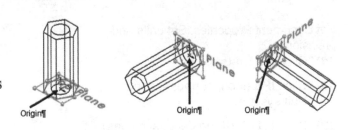

Sketch States

Sketches can exist in any of five states. The state of the sketch is displayed in the status bar at the bottom of the SOLIDWORKS window. These are the five sketch states:

1. *Under Defined.* Inadequate definition of the sketch, (blue). The FeatureManager displays a minus (-) symbol before the sketch name.

2. *Fully Defined.* Complete information, (black). The FeatureManager displays no symbol before the sketch name.

3. *Over Defined.* Duplicate dimensions and or relations, (orange-red). The FeatureManager displays a (+) symbol before the sketch name. The What's Wrong dialog box is displayed.

4. *Invalid Solution Found.* Your sketch is solved but results in invalid geometry, such as a zero length line, zero radius arc or a self-intersecting spline (yellow).

5. *No Solution Found.* Indicates sketch geometry that cannot be resolved, (brown).

☼ Color indicates the state of the individual Sketch entities.

In SOLIDWORKS, it is not necessary to fully dimension or define sketches before you use them to create features. You should fully define sketches before you consider the part finished for manufacturing.

🔆 The name used to describe a 2D or 3D profile is called a sketch.

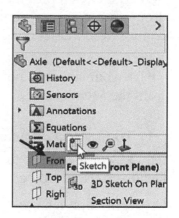

Create a 2D Sketch on the Front Plane.
19) Right-click **Front Plane** from the FeatureManager design tree. This is your Sketch plane for the first feature.

20) Click **Sketch** ⬚ from the Context toolbar. The Sketch toolbar is displayed. Remember, the toolbar is document dependent.

21) Click **Circle** ⊙ from the Sketch toolbar. The Sketch opens on the Front Plane in the Front view by default. The Circle PropertyManager is displayed. The Circle tool uses a Consolidated Circle PropertyManager.

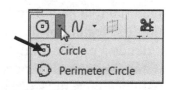

22) Drag the **mouse pointer** into the Graphics window. The cursor displays the Circle feedback symbol. The center point of the

circle is positioned at the origin. The part origin ⤬ is displayed in the center of the Graphics window. The origin represents the intersection of the three default reference planes. They are Front Plane, Top Plane and Right Plane. The positive X-axis is horizontal and points to the right of the origin in the Front view. The positive Y-axis is vertical and points upward in the Front view.

23) Click the **origin** ⤓ from the Graphics window. This is the first point of the circle. The red dot feedback indicates the origin point location. The mouse pointer displays the Coincident to point feedback symbol.

Sketch the circle.
24) Drag the mouse pointer to the **right** of the origin.

25) Click a **position** to create the circle.

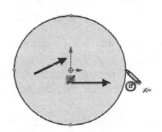

🔆 Control the Sketch relation display. Click **View ➤ Hide/Show ➤ Sketch Relations** from the Menu bar.

Insert a dimension. Apply the Smart Dimension tool.
26) Right-click the **Smart Dimension** ✦ tool from the Context toolbar. The mouse cursor displays the dimension icon.

27) Click the **circumference** of the circle.

28) Click a **position** to locate the dimension in the Graphics window. The Dimension PropertyManager is displayed.

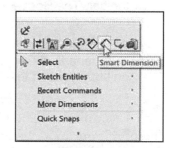

29) Enter **.188in** in the Modify dialog box. The circular sketch is centered at the origin.

30) Click **OK** ✔ from the Modify dialog box.

31) Click **OK** ✔ from the Dimensions PropertyManager.

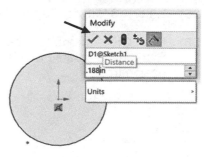
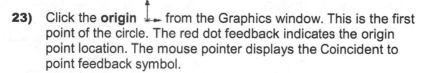

SOLIDWORKS provides the ability to select a unit drop-down menu to modify units in a sketch or feature from the set document properties.

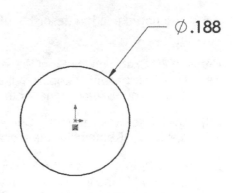

☀ If your sketch is not correct, select **UNDO** ↺ from the Menu bar.

☀ To fit your sketch to the Graphics window, press the f key or the Zoom to Fit 🔍 tool from the Heads-up View toolbar.

Create the first feature. Create an Extruded Boss/Base feature. The Extruded Boss/Base feature adds material to a part. The Extruded Boss/Base feature is the first feature of the AXLE part. An extrusion extends a profile along a path normal to the profile plane for some distance. The movement along that path becomes the solid model. The 2D circle is sketched on the Front Plane.

An Extruded Boss/Base feature is a feature in SOLIDWORKS that utilizes a sketched profile and extends the profile perpendicular (⊥) to the Sketch plane. The Base feature is the first feature that is created. Keep the Base feature simple.

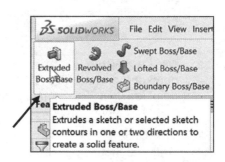

Create the Base Feature.
32)   Click the **Features** tab from the CommandManager. The Features toolbar is displayed. Remember, the toolbar display is document dependent.

33)   Click **Extruded Boss/Base** 🗔 from the Features toolbar. The Boss-Extrude PropertyManager is displayed.

34)   Select **Mid Plane** for End Condition in Direction 1. The Mid Plane End Condition extrudes the sketch equally on both sides of the Sketch plane.

☀ Use different End Condition options to affect the Design Intent of your model and design symmetry.

35)   Enter **1.375in** for Depth. The Depth defines the distance.

36)   Click **OK** ✔ from the Boss-Extrude PropertyManager. Boss-Extrude1 is created and is displayed in the FeatureManager design tree.

Fit the model to the Graphics window.

37) Press the **f** key.

38) **Save** the part. View the created FeatureManager.

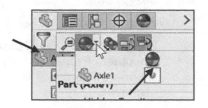

Modify the color (appearance) of the Axle part.

39) Right-click the **Axle** icon from the FeatureManager design tree.

40) Click the **Appearances** drop-down menu as illustrated.

41) Click the **Edit color box** as illustrated. The Color PropertyManager is displayed.

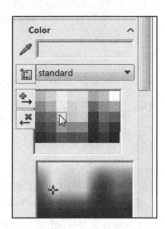

42) Select a **color** from the Color box as illustrated. Note: The Appearances/Scenes tab in the Task Pane is used to add additional patterns and textures.

43) Click **OK** from the Color PropertyManager. The Axle is displayed with the selected color.

Display an Isometric view and save the part.

44) Press the **space bar** to display the Orientation dialog box.

45) Click the **Isometric view** icon.

46) **Save** the part. You created a 2D Sketch on the Front Plane and inserted a geometric relation and dimension. You applied the Extruded Boss/Base feature with a Mid Plane End Condition option. You also applied an Appearance (color) to the part.

Create the Flatbar Part

Tutorial: Flatbar 1-2

Create a new part named Flatbar. Create a 2D Sketch on the Front Plane. Apply geometric relations and dimensions. Insert the needed features.

1) Click **New** from the Menu bar. The New SOLIDWORKS Document dialog box is displayed. Part is the default template.

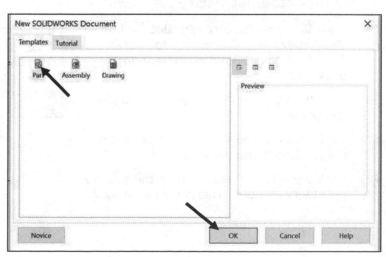

2) Double-click the **Part** icon. Note: The Advanced mode should be selected.

3) Click **Save** from the Menu bar.

4) Enter **Flatbar** in the SOLIDWORKS 2018 folder that you created on your hard drive.

5) Click **Save**. Flatbar is displayed in the FeatureManager design tree.

Set Document Properties. Set drafting standard, units and precision.

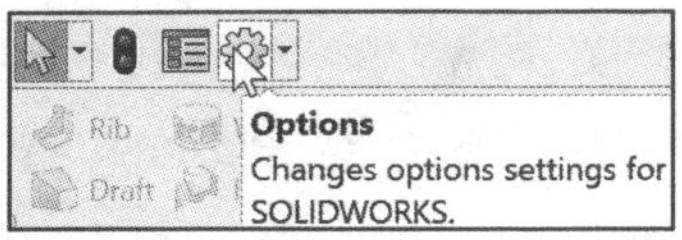

6) Click **Options** ➤ **Document Properties** tab. The Document Properties - Drafting Standard dialog box is displayed.

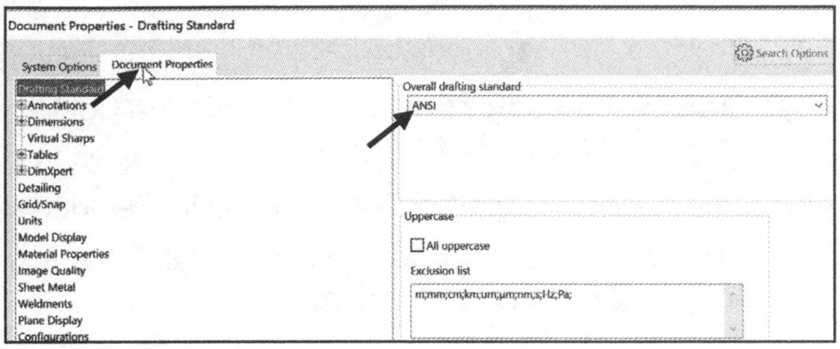

7) Select **ANSI** from the Overall drafting standard drop-down menu.

8) Click **Units**.

9) Click **IPS** for Unit system.

10) Select **.123** (three decimal places) for Length Basic Units.

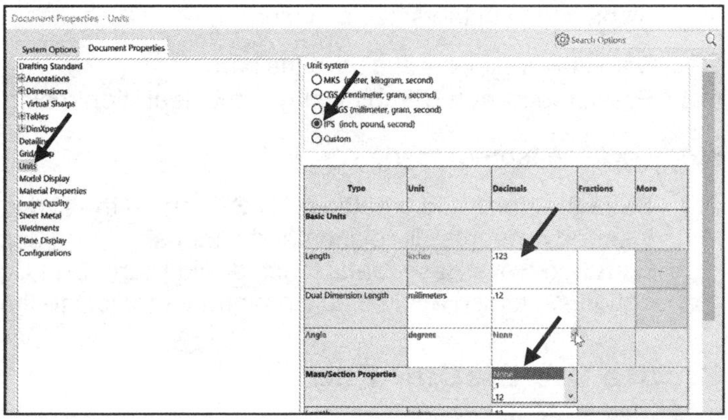

11) Select **None** for Angular unit decimal places.

12) Click **OK** from the Document Properties - Unit dialog box.

Create a 2D Sketch. This is your Base Sketch.

13) Right-click **Front Plane** from the FeatureManager design tree. This is your Sketch plane.

14) Click **Sketch** from the Context toolbar. The Sketch toolbar is displayed.

The Consolidated Slot Sketch toolbar is great to create a slot in three easy steps.

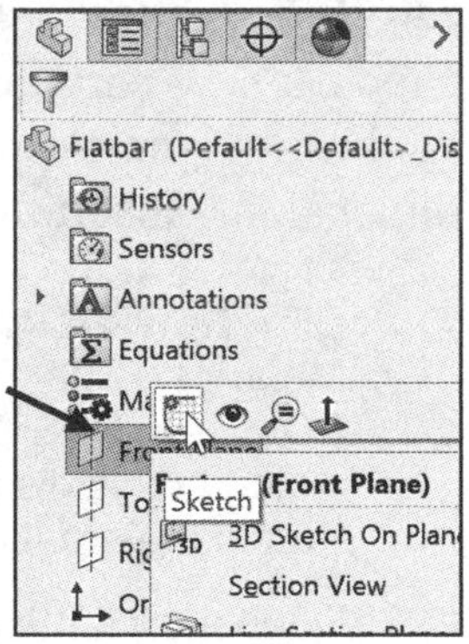

15) Click **Centerpoint Straight Slot** from the Consolidated Slot Sketch toolbar as illustrated. The Slot PropertyManager is displayed.

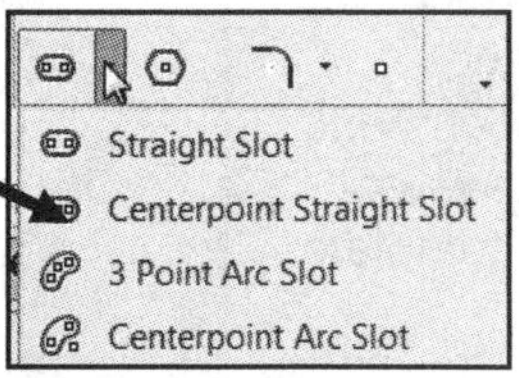

16) Click the **Origin** as illustrated. This is your first point.

17) Click a position directly to the **right** of the Origin. This is your second point.

18) Click a position **directly above the second point** as illustrated. This is your third point.

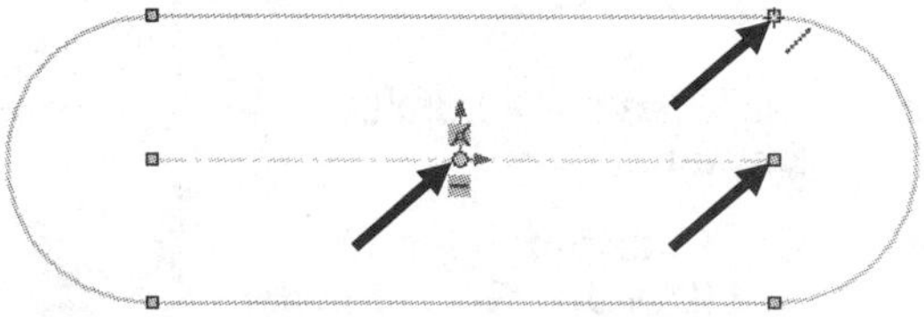

Dimension the 2D sketch.

19) Click **Smart Dimension** 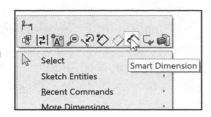 from the Sketch toolbar. Note: You can also Right-click the Smart Dimension ✎ tool from the Pop-up Context toolbar.

20) **Dimension** the Flatbar as illustrated. The sketch is fully defined. The sketch is displayed in black.

Create the first feature for the Flatbar. Create an Extruded Boss/Base feature. The Extruded Boss/Base feature adds material to a part. Extrude the sketch to create the first feature.

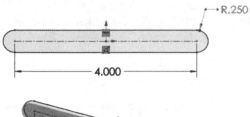

21) Click the **Features** tab from the CommandManager. The Features toolbar is displayed.

22) Click **Extruded Boss/Base** 📦 from the Features toolbar. Blind is the default End Condition in Direction 1.

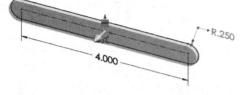

23) Enter **.060in** for Depth in Direction 1. Note the direction of the extrude feature.

24) Click **OK** ✔ from the Boss-Extrude PropertyManager. Boss-Extrude1 is displayed in the FeatureManager design tree.

🔆 Right-click Select to deselect a tool, and to select needed geometry and sketch entities.

25) **Expand** Boss-Extrude1 from the FeatureManager design tree. Sketch1 is fully defined.

26) **Fit** the model to the Graphics window.

27) **Save** the model.

🔆 The Instant3D tool provides the ability to drag geometry and dimension manipulator points to resize and create features directly in the Graphics window. Activate the Instant3D tool. Click the face. Select a manipulator point. Click and drag. Click a location along the ruler for the required dimension. The rule increments are set in System Options.

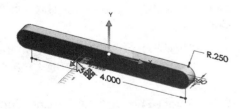

The RapidSketch tool provides the ability to select planar faces or planes, and with any sketch tool active, start to sketch. As you move to each planar face, a plane is created and a sketch is opened.

Utilize the Extruded Cut feature to create the first hole. Insert a new sketch for the Cut-Extrude1 feature.

28) Right-click the **front face** of Boss-Extrude1 for the Sketch plane. This is your Sketch plane.

29) Click **Sketch** from the Context toolbar. The Sketch toolbar is displayed.

30) Display a **Front view** from the Heads-up View toolbar.

31) Click the **Circle** Sketch tool. The Circle PropertyManager is displayed.

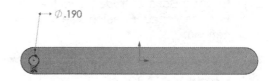

Wake-up the center point of the slot arc.

32) Place the mouse pointer **on the left arc**. Do not click. The center point of the slot arc is displayed.

The process of placing the mouse pointer over an existing arc to locate its center point is called "wake up."

33) Click the **center point** of the arc.

34) Click a **position to the right** of the center point to create the circle.

Add a dimension.

35) Right-click **Smart Dimension** from the Context toolbar.

36) **Dimension** the circle as illustrated. The sketch is fully defined.

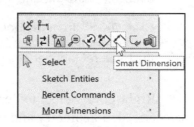

Insert an Extruded Cut feature to create the first hole (seed feature) in the Flatbar part. Think about Design Intent. Why would you select Through All for the End Condition?

37) Click **Extruded Cut** from the Features toolbar. The Cut-Extrude PropertyManager is displayed.

38) Select **Through All** for End Condition in Direction1.

39) Click **OK** from the Cut-Extrude PropertyManager. Cut-Extrude1 is displayed in the FeatureManager.

40) **Expand** Cut-Extrude1 from the FeatureManager. Sketch2 is fully defined.

41) Click **Cut-Extrude1** in the FeatureManager.

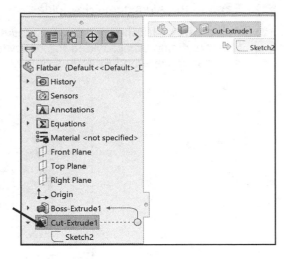

The blue Cut-Extrude1 icon in the FeatureManager indicates that the feature is selected.

Create a Linear Pattern feature. Use a linear pattern to create multiple instances of one or more features that you can space uniformly along one or two linear paths. Utilize the Linear Pattern feature to create additional holes in the Flatbar part.

42) Click **Linear Pattern** [icon] from the Features toolbar. The Linear Pattern PropertyManager is displayed.

Display an Isometric view.
43) Press the **space bar** to display the Orientation dialog box.

44) Click the **Isometric** view [icon] icon.

45) Click the **top edge** of Boss-Extrude1 for Direction 1. Edge<1> is displayed in the Pattern Direction box for Direction 1. The direction arrow points to the right. If required, click the Reverse Direction button.

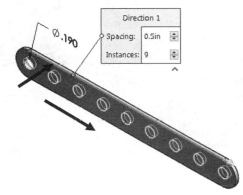

46) Enter **0.5in** for Spacing.

47) Enter **9** for Number of Instances. Instances are the number of occurrences of a feature. Note: Cut-Extrude1 is displayed in the Features to Pattern box.

48) Click **OK** [icon] from the Linear Pattern PropertyManager. LPattern1 is displayed in the FeatureManager.

Apply material properties to the Flatbar part.
49) Right-click **Material** from the Flatbar FeatureManager design tree.

50) Click **Edit Material**. The Material dialog box is displayed. View your options.

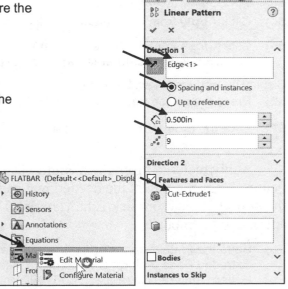

51) **Expand** the Aluminum Alloy folder.

52) Select **6061 Aluminum Alloy** for material. View the Physical Properties of the material. The Material dialog box provides the ability to apply Appearance, CrossHatch, Custom, Application Data and Favorites options.

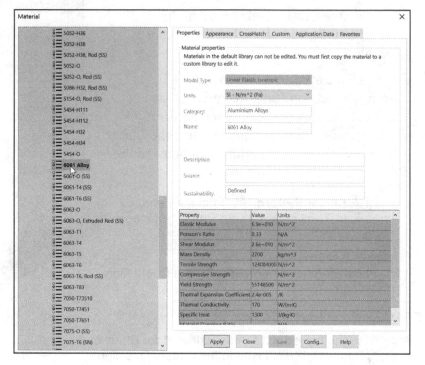

 If needed, select the English (IPS) units category.

53) Click each **tab** and **explore** the options.

Apply the Material to the part.
54) Click **Apply** from the Materials dialog box.

55) Click **Close** from the Materials dialog box. 6061 Alloy is displayed in the FeatureManager design tree.

Display an Isometric view. Save the model.
56) Press the **space bar** to display the Orientation dialog box.

57) Click the **Isometric** view 📦 icon.

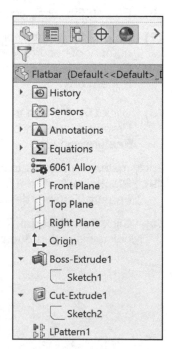

 Apply material properties. Right-click Material from the FeatureManager. Select the needed material if available from the pop-up toolbar or click Edit Material. The material is applied to the model.

58) **Save** the Flatbar part. You completed the Flatbar part using the Extruded Base feature, Extruded Cut feature and the Linear Pattern feature. You applied 6061 Alloy material to the part.

Design Intent is how your part reacts as parameters are modified. Example: If you have a hole in a part that must always be .125≤ from an edge, you would dimension to the edge rather than to another point on the sketch. As the part size is modified, the hole location remains .125≤ from the edge.

Create an Assembly

An assembly is a document that contains two or more parts. An assembly inserted into another assembly is called a sub-assembly. A part or sub-assembly inserted into an assembly is called a component.

Create the AirCylinder Linkage assembly consisting of the following components:

- Axle

- Shaft-collar

- Flatbar

- AirCylinder sub-assembly

Establishing the correct component relationship in an assembly requires forethought on component interaction. Mates are geometric relationships that align and fit components in an assembly.

Mates remove degrees of freedom from a component. Mates reflect the physical behavior of a component in an assembly. The components in the AirCylinder Linkage assembly utilize Standard Mate types only.

To activate the Quick Mates functionality, click Tools, Customize. On the toolbars tab, under Context toolbar settings, select Show quick mates.

Tutorial: AirCylinder Linkage Assembly 1-3

Redeem the code on the inside cover of the book. View the provided videos and models. Create the AirCylinder Linkage assembly using existing components and sub-assemblies from the book.

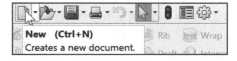

1) **Copy** the provided models to your SOLIDWORKS 2018 folder on your hard drive. Work from this folder.

2) Click **New** from the Menu bar.

3) Double-click the **Assembly** icon from the New SOLIDWORKS Document dialog box. The Begin Assembly PropertyManager is displayed.

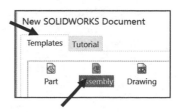

The Begin Assembly PropertyManager is displayed when the Start command when creating new assembly option box is checked.

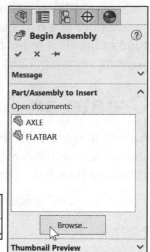

Insert a Component.

4) Click **Browse** from the Part/Assembly to Insert box.

5) Double-click the **AirCylinder** assembly from the SOLIDWORKS 2018 folder that you created on your computer.

Quick Filter buttons in the Open dialog box allow quicker access to commonly used SOLIDWORKS file types.

Press the **Tab** key to rotate 90° or **Shift+Tab** to rotate 90° in the most recently selected direction.

Fix the AirCylinder assembly to the origin.

6) Click **OK** ✔ from the Begin Assembly PropertyManager. Assem1 is displayed in the FeatureManager design tree and the AirCylinder sub-assembly is fixed to the origin.

7) Click **Save As** from the Menu bar.

8) Enter **AirCylinder Linkage** for File name.

9) Click **Save**. The AirCylinder Linkage FeatureManager design tree is displayed. The AirCylinder is the first sub-assembly in the AirCylinder Linkage assembly and is fixed (f). The (f) symbol is placed in front of the AirCylinder name in the FeatureManager.

Insert the Axle part into the assembly document.

10) Click the **Insert Components** tool from the Assembly toolbar. The Insert Component PropertyManager is displayed.

11) Click **Axle** from the Open documents box in the Insert Component PropertyManager.

12) Click a **position** to the front of the AirCylinder assembly as illustrated.

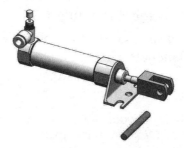

Insert a Concentric mate between the RodClevis and the Axle. A Concentric mate forces two cylindrical faces to become concentric. The faces can move along the common axis, but can't be moved away from this axis.

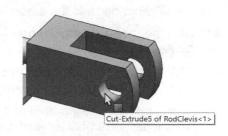

Cut-Extrude5 of RodClevis<1>

13) Click the **inside hole face** of the RodClevis as illustrated.

14) Hold the **Ctrl** key down.

15) Click the **long cylindrical face** of the Axle. The cursor displays the Face feedback symbol.

16) Release the **Ctrl** key. The Mate Pop-up Context box is displayed.

17) Click the **Concentric mate** icon. The Axle is positioned concentric to the RodClevis hole.

18) Click and drag the **Axle**. The Axle translates in and out of the RodClevis holes.

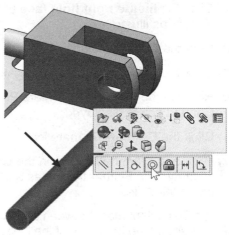

Position the mouse pointer in the middle of the face to select the entire face. Do not position the mouse pointer near the edge of the face.

Insert a Coincident mate. A Coincident mate forces two planar faces to become coplanar. The faces can move along one another but cannot be pulled apart. Use a different Mate technique.

19) Click the **Mate** icon from the Assembly toolbar.

20) **Expand** the fly-out AirCylinder Linkage FeatureManager.

21) Click the **Front Plane** of the AirCylinder assembly from the AirCylinder Linkage fly-out FeatureManager.

22) Click the **Front Plane** of the Axle part from the fly-out FeatureManager. Coincident mate is selected by default.

23) Click the **green check mark** in the Mate pop-up box. The AirCylinder Front Plane and the Axle Front Plane are Coincident. The Axle is centered in the RodClevis.

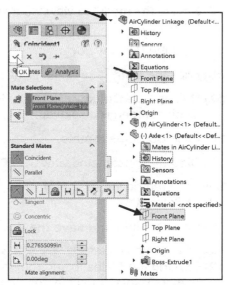

Display the Mates in the FeatureManager to check that the components and the mate types correspond to the original design intent. Mate icons are displayed in the Mates folder.

If you delete a mate and then recreate it, the mate numbers will be in a different order.

Insert the Flatbar part into the assembly document.

24) Click the **Insert Components** 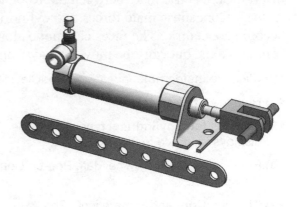 tool from the Assembly toolbar.

25) Click **Flatbar** from the Open documents box in the Insert Component PropertyManager.

26) Click a **position** to the front of the AirCylinder assembly as illustrated.

Mate the Flatbar component to the assembly.
27) Click the **inside right hole face** of the Flatbar as illustrated.

28) Hold the **Ctrl** key down.

29) Click the **cylindrical face** of the Axle.

30) Release the **Ctrl** key. The Mate Pop-up Context box is displayed.

31) Click the **Concentric** mate icon.

32) Click and drag the **Flatbar** in the Graphics window. The Flatbar translates and rotates along the Axle.

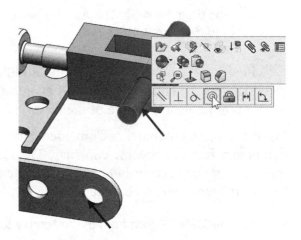

Insert a Coincident mate between the back face of the Flatbar and the front face of the RodClevis.
33) Click the **back face** of the Flatbar.

34) **Rotate** the model to view the front face of the RodClevis.

35) Hold the **Ctrl** key down.

36) Click the **front face** of the RodClevis.

37) Release the **Ctrl** key. The Mate Pop-up Content menu is displayed.

38) Click **Coincident** mate.

39) **Display** an Isometric view.

Save and rebuild the Assembly.
40) Click **Save**.

41) Click **Rebuild and save the document**.

View the created mates.
42) **Expand** the Mates folder from the FeatureManager design tree. View the inserted mates.

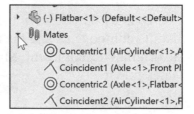

43) Perform the **same procedure** above to insert the second Flatbar component on the back side of the RodClevis.

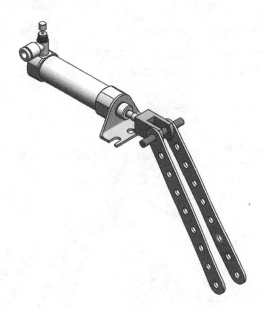

44) Insert a Parallel mate between the **top narrow face** of the first Flatbar and the **top narrow face** of the second Flatbar. A Parallel mate places the selected items so they lie in the same direction and remain a constant distance apart from each other.

45) Click and drag the **second Flatbar**. Both parts move together.

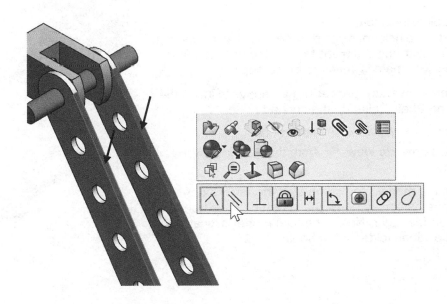

Insert the Shaft-collar part into the assembly document.

46) Click the **Insert Components** tool from the Assembly toolbar.

47) Click **Browse** from the Part/Assembly to Insert box.

48) Double-click the **Shaft-collar** part from the SOLIDWORKS 2018 folder. The Shaft-collar is displayed in the Graphics window.

49) Click a **position** to the front of the Axle.

Insert a Concentric mate.

50) Insert a Concentric mate between the inside **hole face** of the Shaft-Collar and the **long cylindrical face** of the Axle. Concentric mate is selected by default.

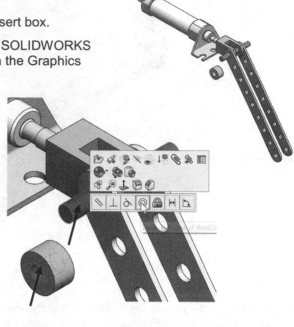

Press the **Shift-z** keys to Zoom in on the model.

Press the **f** key to fit the model to the Graphics window.

Press the **g** key to activate the Magnifying glass tool. Use the Magnifying glass tool to inspect a model and make selections without changing the overall view.

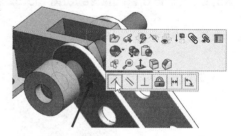

Insert a Coincident mate.

51) Insert a Coincident mate between the **back face** of the Shaft-collar and the **front face** of the first Flatbar. Coincident mate is selected by default.

52) Perform the **same procedure** as above to insert the second Shaft-Collar on the second Flatbar.

Display an Isometric view. Save the model.

53) Click **Isometric view** from the Heads-up View toolbar.

54) **Save** the model. You completed the AirCylinder Linkage assembly. View the inserted mates from the FeatureManager. Note: The final model is located in the Solutions folder in the book.

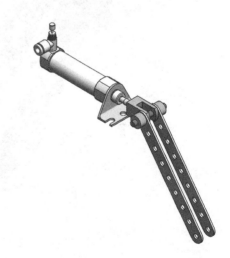

Create a New Assembly Drawing

A SOLIDWORKS drawing displays 2D and 3D views of a part or assembly. The foundation of a SOLIDWORKS drawing is the drawing template. Drawing size, drawing standards, company information, manufacturing, and or assembly requirements, units and other properties are defined in the drawing template. In this section, use the default drawing template.

The sheet format is incorporated into the drawing template. The sheet format contains the border, title block information, revision block information, company name and or logo information, Custom Properties and SOLIDWORKS Properties. Because this section of the book is a Quick Start section, you will not address these items at this time.

Custom Properties and SOLIDWORKS Properties are shared values between documents. Utilize an A (ANSI) Landscape size Drawing Template with a custom Sheet Format for the Air Cylinder Linkage assembly drawing.

A drawing contains views, geometric dimensioning and tolerances, notes and other related design information. When a part or assembly is modified, the drawing automatically updates. When a dimension in the drawing is modified, the part or the assembly is automatically updated.

Tutorial: AirCylinder Linkage Drawing 1-4

Create the ANSI AirCylinder Linkage assembly (Third Angle Projection) drawing. Display the Front, Top, Right and Isometric drawing views. Utilize the Model View tool from the View Layout tab in the CommandManager.

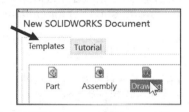

1) Click **New** ⬜ from the Menu bar.

2) Double-click **Drawing** from the Templates tab.

3) Click **OK** from the default Drawing Template dialog box.

4) If needed **uncheck** the Only show standard formats box.

5) Select **A (ANSI) Landscape** from the Sheet Format/Size dialog box.

6) Click **OK**. The Model View PropertyManager is displayed.

7) If needed, click **Cancel** ✖ from the Model View PropertyManager. Draw1 is displayed.

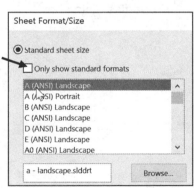

The book is written using the ANSI overall drafting standard and Third Angle projection.

🔅 Define drawing sheet zones on a sheet format for the purpose of providing locations where drawing views and annotations reside on the drawing.

Set the Sheet1 Properties.

8) Right-click **Properties** 🗏 in Sheet1. Expand the menu if needed. The Sheet Properties is displayed. Draw1 is the default drawing name. Sheet1 is the default first sheet name.

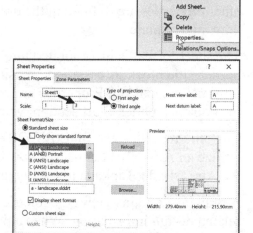

9) Enter Sheet Scale **1:3**.

10) Check **Third angle** for Type of projection.

11) Click **Apply Changes** from the Sheet Properties box. The A-Landscape paper is displayed in a new Graphics window. The sheet border defines the drawing size, 11" × 8.5" (or 279.4mm × 215.9mm). The View Layout toolbar is displayed in the CommandManager.

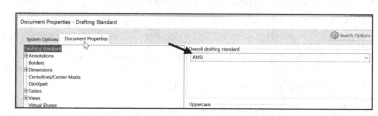

Set Document Properties. Set drafting standard, units and precision. Save the document.

12) Click **Options** ⚙ from the Main menu.

13) Click the **Document Properties** tab.

14) Select **ANSI** for Overall drafting standard.

15) Click **Units**.

16) Select **MMGS** (millimeters, gram, second) for Unit system.

17) Select **.12** for Length unit decimal places.

18) Select **None** for Angular unit decimal places.

19) Click **OK** from the Document Properties dialog box.

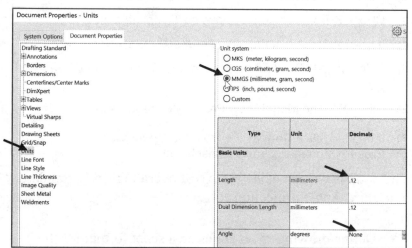

20) Save the drawing.

21) Enter **AirCylinder Linkage** for file name.

22) Click **Save**. The AirCylinder Linkage is displayed in the Drawing FeatureManager design tree.

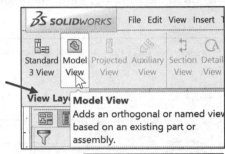

Insert the Drawing views. Use the Model View tool.

23) Click **Model View** 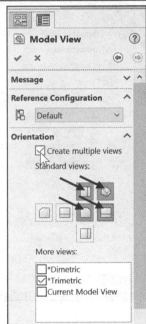 from the View Layout tab. The Model View PropertyManager is displayed.

24) Double-click **AirCylinder Linkage** from the Model View PropertyManager.

Insert a Front, Top, Right and Isometric view.

25) Check the **Create multiple views** box.

26) Select ***Front**, ***Top**, ***Right** and ***Isometric** view from the Orientation box.

27) Click **OK** ✔ from the Model View PropertyManager. The four views are displayed in Sheet1.

The Title block is located in the lower right hand corner of Sheet1.

A drawing contains two main modes:

- Edit Sheet

- Edit Sheet Format

Insert views and dimensions in the Edit Sheet mode. Modify the Sheet Format text, lines or title block information in the Edit Sheet Format mode. The CompanyName Custom Property is located in the title block above the TITLE box. There is no value defined for CompanyName. A small text box indicates an empty field.

Activate the Edit Sheet Format Mode.
28) Right-click in **Sheet1**. Do not select a view boundary.

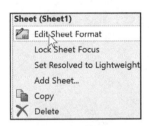

29) Click **Edit Sheet Format**. The Title block lines turn blue. View the right side of the Title block.

30) Double-click the **AirCylinder Linkage** text in the DWG NO. box.

31) Click the **drop-down arrows** to set the Text Font to **12** from the Formatting dialog box.

32) Click **OK** ✔ from the Note PropertyManager.

Return to the Edit Sheet mode. Save the model.
33) Right-click **Edit Sheet** in Sheet1.

34) Save the drawing. View the results.

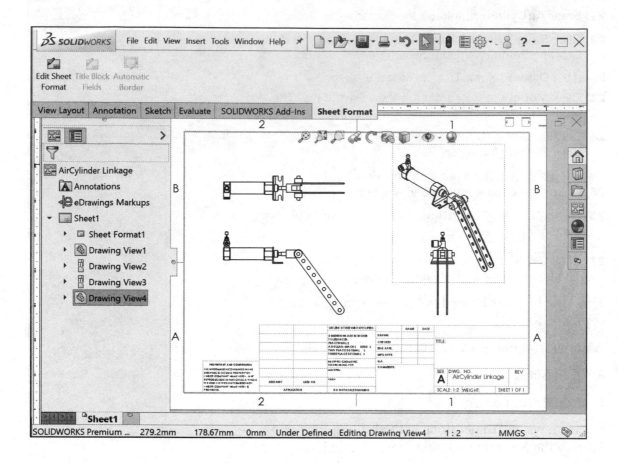

Fit and scale drawing text into the title box, Bill of Materials or any tight area on the drawing.
Select Fit text from the Formatting dialog box. Size the selected text.

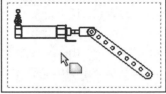

Check use specified color for drawings paper color in the System Options, Color section to display a different drawing sheet color.

Insert a Bill of Material into the AirCylinder Linkage assembly drawing.

35) Click inside the **Front view**, Drawing View1. Note the icon feedback symbol.

36) Click **Bill of Materials** from the Annotation tab. The Bill of Materials PropertyManager is displayed.

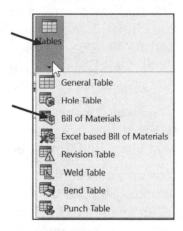

Edit the Bill of Materials table items in the Graphics window from the fly-out toolbar. Reorder rows and columns in the BOM with a drag and drop interface. View the Drawing chapter for detailed information.

37) Click **Top level only** from the BOM Type box. Accept all other defaults.

38) Click **OK** ✔ from the Bill of Materials PropertyManager. The AirCylinder Linkage assembly FeatureManager design tree is displayed.

Position the BOM. Save and close the document.
39) Click a position in the **top left corner** of Sheet1.

40) The Bill of Materials is incomplete. Click **inside Cell A2**. View the fly-out menu. Explore the toolbar as an exercise.

41) Click inside **Sheet1**.

	A	B	C	D	E
1	ITEM NO.	PART NUMBER	DESCRIPTION	MATERIAL	Default/ QTY.
2	1	GIDS-PC-10001	LINEAR ACTUATOR		1
3	2	Axle			1
4	3	Flatbar			2
5	4	GIDS-SC-10012-3-16	SHAFT-COLLAR	2014 Alloy	2

42) Save the drawing. As an exercise, fill in the rest of the BOM cells with Custom Properties. View SOLIDWORKS Help on Custom Properties and the Drawing chapter.

43) Close the model.

☼ Insert a Balloon directly into a Note annotation. Select the target Note annotation. Select the Balloon; the Balloon is inserted directly into the active note.

Summary

The SOLIDWORKS (UI) is designed to make maximum use of the Graphics window for your model. Displayed toolbars and commands are kept to a minimum.

SOLIDWORKS provides a new icon style. It also allows vector-based scaling for superior support of high resolution high pixel density displays. The new icon style standardized the perspective of icons. It also removes non-essential details and emphasizes primary elements. Consistent visual styling applies to all icons.

In this chapter, you started a SOLIDWORKS session and applied the SOLIDWORKS User Interface and CommandManager. You learned about the basic concepts in SOLIDWORKS: *Terminology, 2D Base sketch, Sketch tools, 3D Feature tools, Assemblies, Standard Mates, refining the design, associativity, 2D drawings, dimensions and geometric constraints.*

You created a simple Axle part. You selected the correct Sketch plane and applied the Circle sketch tool. You then applied the Extruded Boss/Base 3D feature with the Mid Plane End Condition option. You applied color (appearance) and material to the part.

You also created the Flatbar part. You selected the correct Sketch plane and applied various (Slot and Circle) Sketch tools. You then applied the Extruded Boss/Base 3D feature with the Blind End Condition option, Extruded Cut feature with the Through All option, and the Linear Pattern feature. You applied material to the Flatbar using the Material dialog box.

You copied key components and models from the book. You created the AirCylinder Linkage assembly by inserting six components and three Standard mates: Concentric, Coincident and Parallel.

You created an ANSI, Third Angle Projection AirCylinder Linkage drawing with a Front, Top, Right and Isometric view. In the drawing, you edited the Title block and inserted a simple Bill of Materials.

For commercial users, SOLIDWORKS Model Based Definition (MBD) is a separate application. For education users, SOLIDWORKS MBD is included in the SOLIDWORKS Education Edition as an Add In.

Quick Mate is a procedure to mate components together. No command (click Mate from the Assembly CommandManager) is required. Hold the Ctrl key down, and make your selections. Release the Ctrl key; a Quick Mate pop-up is displayed below the context toolbar. Select your mate and you are finished.

Utilize the Quick Mate procedure for Standard mates, Cam mate, Profile Center mate, Slot mate, Symmetric mate and Width mate. To activate the Quick Mate functionality, click Tools, Customize. On the toolbars tab, under Context toolbar settings, select Show Quick Mates. Quick Mate is selected by default.

Use the Quick Mate procedure for reference geometry (such as planes, axes, and points) along with model geometry (such as faces, edges and vertices).

Opening a SOLIDWORKS document from an earlier release can take extra time. After you open and save a file, subsequent opening time returns to normal. Use the SOLIDWORKS Task Scheduler (SOLIDWORKS Professional) to convert multiple files from an earlier version to the SOLIDWORKS 2018 format. Click Windows Start ➤ All Apps ➤ SOLIDWORKS 2018 ➤ SOLIDWORKS Tools 2018 ➤ SOLIDWORKS Task Scheduler.

Templates are part, drawing and assembly documents which include user-defined parameters. Open a new part, drawing or assembly. Select a template for the new document.

- *Parts*. The Parts default template is located in the C:\ProgramData\SolidWorks\\SOLIDWORKS 2018\templates\Part.prtdot folder.

- *Assemblies*. The Assemblies default template is located in the C:\ProgramData\SolidWorks\\SOLIDWORKS 2018\templates\Assembly.asmdot folder.

- *Drawings*. The Drawings default template is located in the C:\ProgramData\SolidWorks\\SOLIDWORKS 2018\templates\Drawing.drwdot folder.

In Chapter 2, explore the System Options in SOLIDWORKS. System Options provide the ability to customize SOLIDWORKS functionality for your needs.

Notes:

CHAPTER 2: SYSTEM OPTIONS

Chapter Objective

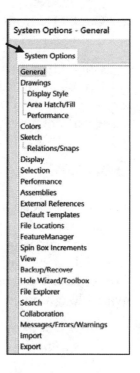

Chapter 2 provides a comprehensive understanding and the ability to use and modify the System Options in SOLIDWORKS.

On the completion of this chapter, you will be able to:

*   Setup and modify the available tools from the System Options section: *General, Drawings, Colors, Sketch, Display, Section, Performance, Assemblies, External References, Default Templates, File Locations, FeatureManager, Spin Box Increments, View, Backup/Recover, Touch, Hole Wizard/Toolbox, File Explorer, Search, Collaboration and Message/Errors/Warnings, Import, Export.*

*   Save custom interface settings:

    *   Registry file.

Systems Options

System Options are stored in the registry of your computer. System Options are not part of your document. Changes to the System Options affect all current and future documents. In Chapter 3, explore and address Document Properties Options.

 A part, assembly or drawing is referred to as a document in SOLIDWORKS.

The selections grouped under the System Options tab are displayed in a tree format on the left side of the System Options - General dialog box. Click an item in the tree; the options for the selected item are displayed on the right side of the dialog box. The title bar displays the title of the tab and the title of the options page.

System Options provides the ability to customize SOLIDWORKS functionality for your needs. Review the System Options - General dialog box structure.

Tutorial: Close all open models 2-1.

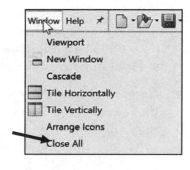

Close all parts, assemblies, and drawings. Access the System Options dialog box.

1. Click **Window** ➢ **Close All** from the Menu bar.

Access the System Options dialog box.

1. Click **New** 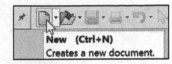 from the Menu bar. The Templates tab is the default tab. Part is the default template from the New SOLIDWORKS Document dialog box.

2. Click **OK** from the New SOLIDWORKS Document dialog box. The Part FeatureManager is displayed.

3. Click **Options** ⚙ from the Menu bar. The System Options - General dialog box is displayed. The General tab is selected by default. View the available selections for a Part document.

General

The General section provides specific system options such as enabling the performance feedback, the Confirmation Corner, etc. These are the available options:

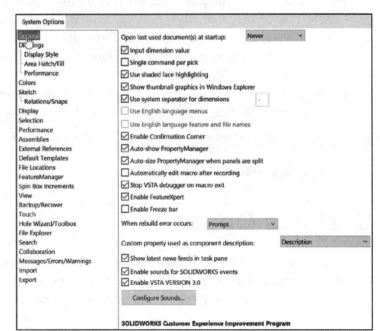

- *Open last used documents(s) at startup*. This option provides two selections:

  - **Always**. Provides the ability to open your last document automatically when a SOLIDWORKS session starts.

  - **Never**. Default setting. Provides the ability for SOLIDWORKS not to open any documents automatically.

- *Input dimension values*. Default setting. Provides the ability to specify that the Dimension Modify dialog box is displayed automatically for the input of a new dimension value.

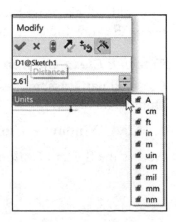

💡 You must double-click the dimension to modify the value if the Input dimension values option is not checked.

- *Single command per pick*. Provides the ability to specify that Sketch and Dimension tools are cleared after each time they are used.

☼ Double-clicking a tool will cause the tool to remain selected for additional use.

- *Use shaded face highlighting*. Default setting. Provides the ability to display the selected faces in a solid color. Modify the default color, solid green, by using the following commands: **Options** ➤ **Systems Options** ➤ **Colors** from the Menu bar. Some third-party applications may require that you clear this option.

- *Show thumbnail graphics in Windows Explorer*. Default setting. Provides the ability to display a thumbnail graphic instead of an icon in Windows Explorer for each SOLIDWORKS part or assembly document. The displayed thumbnail graphic is based on the view orientation of the model when you saved the document.

- *Use system separator for dimensions*. Default setting. Provides the ability to specify that the default system decimal separator is used when displaying decimal numbers. Use the Windows Control Panel to set your system default. To set a decimal separator different from the system default, **clear** this option and type a **symbol**, usually a period or comma.

- *Use English language menu*. Not active by default if you installed SOLIDWORKS 2018 or 2016 with English as your installed language. You must exit and re-start SOLIDWORKS for this change to take place.

- *Use English language feature and file names*. Not active by default if you installed SOLIDWORKS 2018 with English as your language. Provides the ability to display feature names in the FeatureManager design tree and to automatically create file names in English.

☼ Existing feature and file names in a foreign language do not update when you select this option.

- *Enable Confirmation Corner*. Default setting. Display the Confirmation Corner controls in the upper right corner of the Graphics window.

- *Auto-show PropertyManager*. Default setting. Display the PropertyManager when you select existing sketch entities, dimensions, and or annotations from the SOLIDWORKS Graphics window.

- *Auto-size PropertyManager when panels are split*. Default setting. Resizes the PropertyManager when you split the Graphics window into panels.

- *Automatically edit macro after recording*. Opens the macro editor after you recorded and saved your macro in SOLIDWORKS.

- *Stop VSTA debugger on macro exit*. Default setting. Stops the VSTA macro debugger as soon as a macro exits. When cleared, leaves the debugger running so that you can continue to debug the UI or event created by the macro.

- ***Enable FeatureXpert***. Default setting. FeatureXpert is based on the SOLIDWORKS Intelligent Feature Technology SWIFT™. This option provides the ability to automatically fix parts, so you can issue a successful rebuild. The Enable FeatureXpert option enables MateXpert, FilletXpert, DraftXpert, etc.

- ***Enable Freeze bar***. Provides the ability to freeze features to exclude them from rebuilds of the model. Uses a Freeze feature bar in the FeatureManager.

- ***When rebuild error occurs***. Prompt is the default option. There are three options from the drop-down menu. They are **Stop**, **Continue** and **Prompt**.

- ***Custom property used as component description***. Description is the default custom property. There are 39 custom property options from the drop-down menu. Provides the ability to set or type a name to define a custom description label. Example: Open dialog box has a Description label which displays the model description.

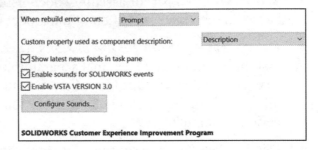

If you change the **Custom property used as component description** in a document that has already been saved, you must manually add the new description to the Summary Information dialog box.

- ***Show latest news feeds in task pane***. Default setting. Allows regular updates to be made to the items under **Latest News** on the **SOLIDWORKS Resources** tab of the **Task Pane**.

- ***Enable sounds for SOLIDWORKS events***. Enables sounds for events such as File, Open Complete, and Collision Detected.

- ***Enable VSTA VERSION 3.0***.

- ***Configure Sounds***. Opens the Windows Sound dialog box, where you can assign sounds to SOLIDWORKS events.

- ***Help make SOLIDWORKS products better by automatically sending your log files to DS SOLIDWORKS Corporation***. Silently sends your log files for review to the SOLIDWORKS Customer Experience Improvement Program.

Illustrations may vary depending on your SOLIDWORKS version and system setup.

Drawings

Customize setting for your drawings in this section. The Drawings section provides the following default options:

- **Eliminate duplicate model dimensions on insert**. Default setting. Provides the ability to duplicate dimensions which are not inserted into a drawing when you use the Model dimensions tool. Note: This option is overridden by Eliminate duplicates in the Model Items PropertyManager.

- **Eliminate duplicate model notes on inset**. Default setting. Duplicate notes are not inserted into drawings when model notes are inserted.

- **Mark all part/assembly dimension for import into drawings by default**. Default setting. Provides the ability to set any dimension inserted in a model as Mark For Drawing.

- **Automatically scale new drawing views**. Default setting. Provides the ability to automatically scale new drawing views to fit the drawing sheet, regardless of the selected paper size.

- **Enable symbol when adding new revision**. Display the revision symbol when adding a revision.

- **Display new detail circles as circle**. By default, the sketched profiles are displayed. When activated, new profiles, "loops" for detail views are displayed as circles. Note: A detail circle refers to any closed loop that is applied to create a Detail view. This option displays the loop as a circle, regardless of whether you drew it as a rectangle, hexagon, ellipse or any other closed sketch profile.

- **Select hidden entities**. When selected, you can select hidden tangent edges and edges that you have hidden manually. Move the mouse pointer over the hidden edge; the edge is displayed in phantom line font.

- **Disable not/dimension inference**. If cleared, when you place a note or dimension, a line appears to indicate horizontal or vertical alignment with other notes or dimensions.

- ***Disable note merging when dragging***. Default setting. Disables the merging of two notes or a note and a dimension when dragged to one another.

- ***Print out-of-sync water mark***. Default setting. Provides the ability to print an Out-of-Sync watermark on a detached drawing sheet that has not been updated since the last model change.

- ***Show reference geometry name in drawings***. Provides the ability to assign names to reference geometry, such as planes and axes and to control whether or not the selected entities are displayed on the drawing.

- ***Automatically hide components on view creation***. When selected, components of an assembly not visible in a new drawing view are hidden and are listed on the Hide/Show Components tab of the Drawing View Properties dialog box. The component names are transparent in the FeatureManager design tree.

- ***Display sketch arc centerpoints***. When selected, sketch arc center points are displayed in the drawing.

- ***Display sketch entity points***. When selected, the endpoints of the sketch entities are displayed as filled circles in drawing sheets and drawing sheet formats. They are not displayed in the drawing views.

- ***Display sketch hatch behind geometry***. When selected, the model's geometry displays over the hatch.

- ***Display sketch pictures on sheet behind geometry***. Displays sketch pictures as background images for drawing views as illustrated.

- ***Print breaklines in broken view***. Default setting. Prints the break lines in the Broken view.

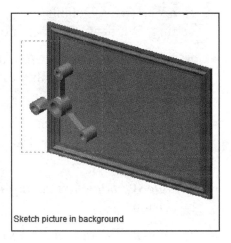

Sketch picture in background

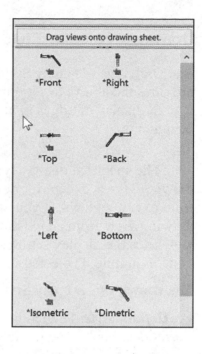

- ***Automatically populate View Palette with views***. Default setting. Provides the ability to display the available drawing views of an active document in the View Palette as illustrated.

- ***Show sheet format dialog when adding new sheet***. Displays the Sheet Format dialog when you add a new drawing sheet.

- ***Reduce spacing when dimensions are deleted or edited (add or change tolerance, text, etc.)***. Default setting. Automatically re-adjusts the space among the remaining dimensions if you delete a dimension or remove text from a dimension.

- ***Reuse view letters from deleted auxiliary, detail, and section views***. Reuses letters from deleted views (auxiliary, detail, section) in the drawing.

- ***Enable paragraph auto numbering***. Auto numbering for paragraph.

- ***Override quantity column name in Bill Of Materials***. When selected, uses the name you enter in Name to use for the quantity in a BOM.

- ***Detail view scaling***. 2X is the default setting. Specifies the scaling for your detail views. The scale is relative to the scale of the drawing views from where the detail view is generated from.

- ***Custom property used as Revision***. Revision is the default setting from the drop-down menu. This option provides the ability to specify the document custom property to be regarded as the revision data when checking a document into PDMWorks. There are numerous selections from the drop-down menu.

- ***Keyboard movement increment***. 0.39in (10mm) default setting. Provides the ability to specify the unit value of movement when you utilize the arrow keys to move or nudge drawing views, annotations or dimensions in small increments.

Drawings - Display Style

The Drawings - Display Style section provides the ability to set options for the default display of edges in all drawing documents.

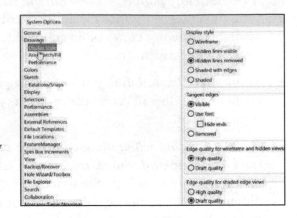

☼ The specified display types apply to new drawing views, except for new views created from existing views. If you create a new drawing view from an existing view, the new view uses the display settings of the source view. Example: Projected view.

The Drawings - Display Style section provides the following default options:

- *Default style*. Hidden lines removed selected by default. There are five display view modes. Each view mode specifies the way a part or assembly is displayed in a drawing. The display options are:

  - **Wireframe**. All edges are displayed in a drawing view.

  - **Hidden lines visible**. All visible edges as specified in the Line Font Options are displayed. Hidden edges are displayed in black.

  - **Hidden lines removed**. Default setting. Only edges that are visible at the chosen angle are displayed. Obscured lines are removed from all views.

  - **Shaded with edges**. Views in Shaded mode with Hidden lines Removed are displayed. Click **Options ➢ Colors** from the Menu bar to specify a color for the edges, or use the specified color or a color slightly darker than the model color.

  - **Shaded**. Parts are displayed shaded.

- *Tangent edges*. Visible selected by default. There are three options in this section. If you selected Hidden lines visible or Hidden lines removed, select one of the three options below for viewing the tangent edges, transition edges between rounded or filleted faces.

  - **Visible**. A solid line is displayed.

  - **Use font**. A line using the default font for tangent edges is displayed. A drawing document is required to activate this option.

    - **Hide ends**. Hides the start and end segments of tangent edges. You can also set the color for this type of tangent edge.

  - **Removed**. Tangent edges are not displayed.

- *Edge quality for wireframe and hidden views*. High quality selected by default. Provides the ability to display the quality for a new view. The available selections are:

- **High quality**. Model resolved.

- **Draft quality**. Model lightweight. Use for faster performance with large assemblies.

- *Edge quality for shaded edge views*. Draft quality selected by default. Provides the ability to display the quality for a new view.

Drawings - Area Hatch/Fill

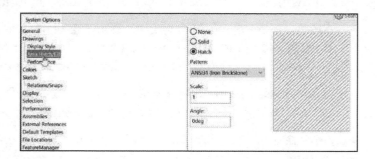

The Drawings - Area Hatch/Fill section provides the ability to set the hatch or fill options for an area hatch that you apply to a face, a closed loop, or to a sketch entity in a drawing.

The Drawings- Areas Hatch/Fill section provides the following options:

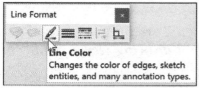

- *Type of hatch or fill*. Hatch selected by default. The default color of the solid fill is black. There are three pattern modes available. Modify the color of the fill by selecting the **area hatch** in the Graphics window, and click **Line Color** from the Line Format toolbar. This procedure does not work for a Section view. These are the three pattern modes:

  - **None**.

  - **Solid**.

  - **Hatch**.

- *Pattern*. Available only for the Hatch option. ANSI31 Iron BrickStone selected by default. Select a crosshatch pattern from the drop-down menu.

- *Scale*. Available only for the Hatch option. The default Scale value is 1.

- *Angle*. Available only for the Hatch option. The default Angle value is 0.

Material Hatch Pattern selected in a part propagates to a Section view in a drawing.

 The default color of the solid fill is black.

Drawings - Performance

The Drawings - Performance
section provides the ability to show
content while drawing.

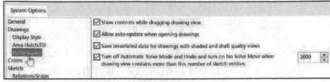

The Drawings - Performance section provides the following options:

- ***Show contents while dragging drawing view***. Default setting. Provides the ability to
  display the contents as you drag a view in the drawing. If this option is deselected,
  you will only see the view boundary as you drag the model in the drawing.

- ***Allow auto-update when opening drawings***. Default setting. Provides the ability for
  your drawing views to be updated automatically as the drawing is opened.

- ***Save tessellated data for drawings with shaded and draft quality views***. Default
  setting. When not selected, your file size is reduced by not saving the tessellated data
  in drawing documents with Shaded and draft quality views.

- ***Turn off Automatic Solve Mode and Undo and turn on No Solve Move when
  drawing view contains more than this number of sketch entities (default 2000)***.

Colors

The Colors section provides the ability to set colors
in the user interface: backgrounds, FeatureManager
design tree, drawing paper, sketch status, etc.

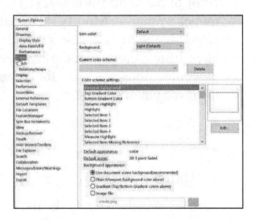

The Colors section provides the following default
options:

- ***Current color scheme***. Provides the ability to
  choose from various background color schemes.
  Corresponding image file names are displayed
  in the Image file under Background appearance.
  Note: Specifying an Image file takes
  precedence over schemes in the list. Schemes
  created with Save As Scheme are displayed in
  the list.

 A plain white Graphics window background is
displayed in this book.

- ***Color scheme settings***. Provides the ability to
  select an item in the list to display its color.
  Click **Edit** to modify the color of the selected
  item. Review the Color scheme settings by
  clicking on each item.

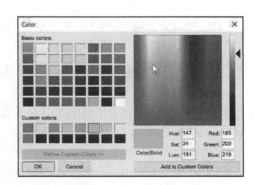

- *Background appearance*. Plain selected by default. There are four appearance options:

  - **Use document scene background (recommended)**. The scene background that is saved with the document is used when it is opened in SOLIDWORKS.

  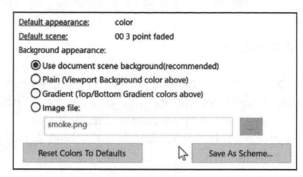

  - **Plain (Viewport Background color above)**. The color scheme selected for the Viewport Background is used as the background color.

  - **Gradient (Top/Bottom Gradient colors above)**. The color scheme selected for the Top and Bottom Gradient is used as the background color.

  - **Image file**. Various files are provided with the application, corresponding to the color schemes listed in Current color scheme. Browse to select a system file or any other image file.

- *Reset Colors To Defaults*. Resets the system's default colors to factory predefined conditions.

- *Save As Scheme*. Provides the ability to input a name to Save the various Color Scheme names.

- *Use specified color for drawing paper color*. Default setting. Provides the ability to set a specified background color on all drawing sheets.

  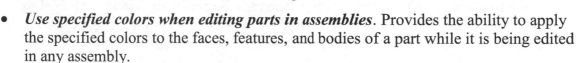

- *Use specified color for Shaded With Edges mode*. Default setting. Provides the ability to apply the specified color to model edges when the model is in the Shaded With Edges mode.

- *Use specified colors when editing parts in assemblies*. Provides the ability to apply the specified colors to the faces, features, and bodies of a part while it is being edited in any assembly.

- *Use specified color for changed drawing dimensions on open*. Applies the specified color to dimensions that have changed since the last time the drawing was saved. After the drawing with the changed dimensions is saved and closed, the changed dimension's highlighting is reset.

  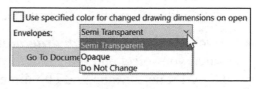

- *Envelopes*. Sets the transparency of envelope components. Select one of the following: Semi Transparent, Opaque, Do not Change.

- ***Go To Document Colors***. If a part or assembly is open when you set system color options, you can go to the Model Display options on the Document Properties tab to set the feature colors for models.

Sketch

The Sketch section provides the ability to set the default system options for a sketch. When you open a new part document, first you create a sketch. The sketch is the base for a 3D model.

The Sketch section provides the following default options:

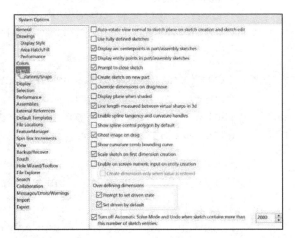

- ***Auto-rotate view normal to sketch plane on sketch creation***. Rotates views to be normal to the sketch plane whenever you open a sketch on a plane.

- ***Use fully defined sketches***. Requires sketches to be fully defined before they are used to create a feature in your model.

- ***Display arc centerpoints in part/assembly sketches***. Default setting. Displays arc center points in the sketch. Note: Arc center points can be useful to select for relations or dimensions.

- ***Display entity points in part/assembly sketches***. Default setting. Displays endpoints of sketch entities as filled circles. The color of the filled circles indicates the status of your sketch entity.

Black - Fully defined, Blue - Under defined, Red/Yellow- Over defined and Green - Selected.

- ***Prompt to close sketch***. Default setting. Provides the ability to display a dialog box with the question, Close Sketch With Model Edges? Use the model edges to close the sketch profile and to select the direction in which to close your sketch.

- ***Create sketch on new part***. Provides the ability to open a new part with an active sketch located on the Front plane. The Front plane is your active Sketch plane.

- ***Override dimensions on drag/move***. Provides the ability to override a dimension when you select and drag sketch entities or move your sketch entity in the Move or Copy PropertyManager. The dimension updates after the drag is finished.

- ***Display plane when shaded***. Provides the ability to display the Sketch plane when you edit your sketch in the Shaded With Edges or the Shaded mode.

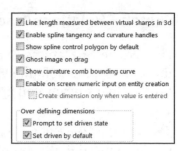

If the display is slow due to the shaded plane, it may be because of the Transparency options. With some graphics cards, the display speed improves if you use low transparency. Set a low transparency; click **Options ➢ System Options** from the Menu bar. Click **Performance**, clear **High quality for normal view mode** and **High quality for dynamic view mode**.

- *Line length measured between virtual sharps in 3d*. Default setting. Provides the ability to measure the line length from virtual sharps, as opposed to the end points in a 3D sketch.

- *Enable Spline Tangency and Curvature handles*. Default setting. Provides the ability to display spline handles for tangency and curvature.

- *Show spline control polygon by default*. Displays a control polygon to manipulate the shape of a spline.

A control polygon is a sequence of nodes (manipulators) in space used to manipulate an object's shape. The control polygon displays when you sketch and edit a 2D or 3D Spline sketch.

- *Ghost image on drag*. Default setting. Provides the ability to display a ghost image of a sketch entity's original position while you drag a sketch.

- *Show curvature comb bounding curve*. Provides the ability to display or hide the bounding curve used with curvature combs.

- *Enable on screen numeric input on entity creation*. Provides the ability to view dimensions during a sketch.

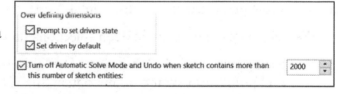

- *Over defining dimensions*. There are two options:

  - **Prompt to set driven state**. Provides the ability to display the Make Dimension Driven dialog box with the question, Make this dimension driven? This question is displayed when you add an over defining dimension to a sketch.

  - **Set driven by default**. Provides the ability to display the Make Dimension Driven dialog box with the question, Leave this dimension driving? when you add an over defining dimension to a sketch.

- *Turn off Automatic Solve Mode and Undo and turn on No Solve Move when drawing view contains more than this number of sketch entities (default 2000)*.

Sketch - Relations/Snaps

The Sketch - Relations/Snaps section provides the ability to set the default system options for various Sketch Snaps when you are sketching.

☀ Most of the Sketch Snaps functions can be accessed through the Quick Snaps toolbar.

The Sketch - Relations/Snaps section provides the following default options:

- **Enable snapping**. Default setting. Provides the ability to toggle all automatic relations, snapping and inferencing.

- **Snap to model geometry**. Default setting. Allows sketch entities to snap to model geometry.

- **Automatic relations**. Default setting. Provides the ability to create geometric relations as you add sketch elements.

- **Go To Document Grid Settings**. Sets options for the grid display and snap functionality.

☀ Relations to the global axes are called AlongX, AlongY and AlongZ. Relations that are local to a plane are called Horizontal, Vertical, and Normal.

- **Sketch Snaps**. There are sixteen selections under Sketch Snaps:

    - **End points and sketch points**. Default setting. Snaps to the center of arcs. The End points and sketch points option Snaps to the end of the following sketch entities: centerlines, chamfers, polygons, parabolas, rectangles, parallelograms, fillets, partial ellipses, splines, points and lines.

    - **Center Points**. Default setting. Snaps to the center of arcs, fillets, circles, parabolas, and partial ellipses sketch entities.

    - **Mid-points**. Default setting. Snaps to the midpoints of rectangles, chamfers, fillets, lines, polygons, parallelograms, splines, points, centerlines, arcs, parabolas and partial ellipses.

    - **Quadrant Points**. Default setting. Snaps to the quadrants of circles, arcs, partial ellipses, ellipses, fillets and parabolas.

- **Intersections**. Default setting. Snaps to the intersections of entities that meet or entities that intersect.

- **Nearest**. Default setting. Supports all entities. When selected, Nearest Snap and snaps are enabled only when the pointer is in the vicinity of the snap point.

- **Tangent**. Default setting. Snaps to tangents on fillets, circles, arcs, ellipses, partial ellipses, splines and parabolas.

- **Perpendicular**. Default setting. Snaps a line to another line.

- **Parallel**. Default setting. Creates a parallel entity to lines.

- **Horizontal/vertical lines**. Default setting. Snaps a line vertically to an existing horizontal sketch line, and horizontally to an existing vertical sketch line.

- **Horizontal/vertical to points**. Default setting. Snaps a line vertically or horizontally to an existing sketch point.

- **Length**. Default setting. Snaps lines to the increments that are set by the grid option.

- **Grid**. Snaps sketch entities to the grid's vertical and horizontal divisions.

  - **Snap only when grid is displayed**. Default setting.

- **Angle**. Default setting. Snaps to a selected angle.

- **Snap angle**: 45deg selected by default. Provides the ability to select your required angle from the spin box.

Display

The Display section provides the ability to set the default options for the display of edges planes, etc. The following options are available:

- *Hidden edges displayed as*. Dashed selected by default. Specifies how hidden edges are displayed in the Hidden Lines Visible (HLV) mode for a part and assembly document. There are two display modes:

  - **Solid**. Displays Hidden edges in solid lines.

  - **Dashed**. Default setting. Displays Hidden edges in dashed lines.

- *Part/Assembly tangent edge display*. As visible selected by default. Controls how tangent edges are displayed in various view modes. There are three display options:

  - **As visible**. Displays Tangent edges.

  - **As phantom**. Provides Tangent edges to be displayed using the Phantom style line font per ANSI standard.

- **Removed**. Default setting. Does not display Tangent edges.

- *Edge display in shaded with edges mode*. HLR is selected by default. There are two display options:

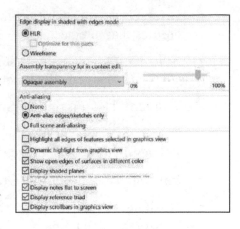

  - **HLR**. Default setting. Displays edges in the Hidden Lines Removed mode and in the Shaded With Edges mode.

  - **Wireframe**. Displays edges in the Shaded With Edges mode and in the Wireframe mode.

- *Assembly transparency for in context edit*. Force assembly transparency with a 90% transparency setting is selected by default. Provides the ability to control the transparency display when you edit your assembly components. There are three display options:

Move the transparency slider to the desired level; to the right increases transparency.

- *Anti-aliasing*. Determines the extent of anti-aliasing to apply to models in the Graphics window. Anti-aliasing smoothes jagged edges, making an image appear more realistic. The three options are **None, Anti-alias edges/sketches, and Full scene anti-aliasing**.

- *Highlight all edges of features selected in graphics view*. Provides the ability to specify that all edges on the selected feature are highlighted when selected.

- *Dynamic highlight from graphics view*. Default setting. Provides the ability to specify whether edges, model faces, and vertices are highlighted when you drag your mouse pointer over a model, sketch, or drawing.

The Dynamic highlight from graphics view option is not available when the Large Assembly Mode is activated.

- ***Show open edges of surfaces in different color***. Default setting. This option simplifies the process to differentiate between the open edges of a surface and any tangent edges or silhouette edges.

- ***Display shaded planes***. Default setting. Provides the ability to display transparent shaded planes with a Wireframe edge that have different front and back colors.

- ***Display scrollbars in graphics view***. This option is unavailable when a document is open. To change this setting, you must close all documents.

- ***Display notes flat to screen***. Select to display notes in the plane of your computer screen. Clear to display notes in the plane of the dimension's 3D annotation view. SOLIDWORKS part and assembly documents support 3D annotations according to the ASME Y14.41-2003 standard.

- ***Display reference triad***. Default setting. Provides the ability to display a reference triad. The triad is for reference only. You cannot select it or use it as an inference point.

Specify the triad colors; click **Options** ➢ **Systems Options** ➢ **Colors** from the Menu bar. Select any of the three axes to apply a color.

- ***Display scrollbars in graphics view***. This option is unavailable when a document is open. To change this setting, you must close all documents.

- ***Display notes flat to screen***. Select to display notes in the plane of your computer screen. Clear to display notes in the plane of the dimension's 3D annotation view. SOLIDWORKS part and assembly documents support 3D annotations according to the ASME Y14.41-2003 standard.

- ***Display draft quality ambient occlusion***. Draft quality renders faster but has less visual fidelity. Clear to use the default quality.

☐ Display draft quality ambient occlusion
☐ Display SpeedPak graphics circle
☑ Display pattern information tooltips
☑ Show breadcrumbs on selection
Projection type for four view viewport: First Angle

- ***Display Speedpak graphic circle***. See SOLIDWORKS Help for details.

- ***Display pattern information tooltips***. Selected by default.

- ***Show breadcrumbs on selection***. Selected by default.

- ***Projection type for four view viewport***. Default is dependent on your initial system install. This option controls which views are displayed in the viewports when you click the Four View command from the Heads-up View toolbar. There are two view options from the drop-down menu:

  - **First Angle**. Displays the Front, Left, Top and Isometric views. Note: ISO is typically a European drafting standard and uses First Angle Projection.

- **Third Angle**.
  Displays the Front,
  Right, Top and
  Isometric views.
  Note: ANSI is an
  American drafting
  standard and uses
  Third Angle
  Projection.

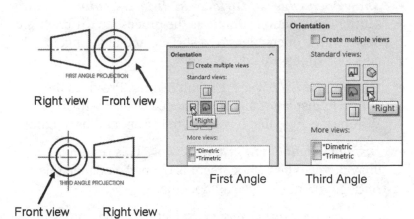

First Angle   Third Angle

Selection

The Selection section folder provides the ability to set
the default options for the selection of edges planes etc.
The following options are available:

- **Default bulk selection method**. Provides a method
  of bulk selection.

  - **Lasso**. Enables freehand selection.

  - **Box**. Enables box selection.

  - **Dashed**. Default setting. Displays Hidden edges in dashed lines.

- **Selection of hidden edges**. Allow selection in wireframe and HLV modes is selected
  by default. There are two display options:

  - **Allow selection in wireframe and HLV modes**. Default setting. Allows you to
    select hidden edges or vertices in Wireframe and Hidden Lines Visible modes.

  - **Allow selection in HLR and shaded modes**. Allows you to select hidden edges
    or vertices in Hidden Lines Removed (HLR), Shaded With Edges, and Shaded
    modes.

- **Enable selection through transparency**. Default setting. Provides the ability to select
  opaque objects behind transparent objects in the Graphics window. This includes
  opaque components through transparent components in an assembly as well as edges,
  interior faces, and vertices through transparent faces in a part. Move the pointer over
  opaque geometry which is behind transparent geometry; the opaque edges, faces, and
  vertices are highlighted.

- **Enhance small face selection precision**. Default setting See SOLIDWORKS Help
  for details.

- **Enhance selection on high resolution monitors**. Default setting See SOLIDWORKS
  Help for details.

Performance

The Performance section provides the ability to set system options as it relates to the performance of the software on your system.

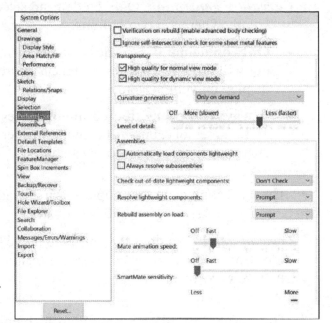

Changes to the performance settings will not affect documents which are open.

The Performance section provides the following default options:

- *Verification on rebuild (enable advanced body checking)*. Provides the ability to control the level of error checking when you create or modify features. In most applications, the default setting is adequate and will result in a faster model rebuild time.

- *Ignore self-intersection check for some sheet metal features*. Suppresses the warning messages for certain sheet metal parts. Example: A flange shares a common edge and the part flattens correctly but displays a warning message.

- *Transparency*. High quality for normal view mode and High quality for dynamic view mode selected by default. There are two display options:

  - **High quality for normal view mode**. Default setting. If the part or assembly is not moving, the transparency mode is of high quality. Low quality is applied when the part or assembly is moved. This is important if the part or assembly is complex.

  - **High quality for dynamic view mode**. Default setting. High quality transparency mode is applied when the part or assembly is moved. Depending on your graphics card, this option may result in slower performance.

The Transparency option is not available when the Large Assembly Mode is activated.

- *Curvature generation*. Only on demand selected by default. There are two options:

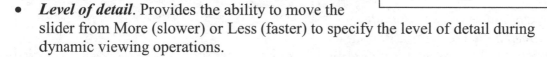

  - **Only on demand**. Uses less system memory, but will provide a slower initial curvature display on your system.

  - **Always (for every shaded model)**. Uses more system memory, but will provide a faster initial curvature display on your system.

☀ The Curvature generation option is not available when the Large Assembly Mode is activated.

- *Level of detail*. Provides the ability to move the slider from More (slower) or Less (faster) to specify the level of detail during dynamic viewing operations.

☀ The Level of detail option is not available when the Large Assembly Mode is activated.

- *Automatically load components lightweight*. Loads the individual components into assemblies which are opened as lightweight. Sub-assemblies are not lightweight, but the parts that they contain are.

- *Always resolve sub-assemblies*. Sub-assemblies are resolved when an assembly opens in a lightweight mode. The components in the sub-assemblies are lightweight.

- *Check out-of-date lightweight components*. Don't check selected by default. Provides the ability to specify how you want the system to load lightweight components which are out-of-date. There are three options:

  - **Don't check**. Loads the assemblies without checking for out-of-date components.

  - **Indicate**. Loads the assemblies and marks them with a lightweight icon, only if the assemblies contain an out-of-date component.

  - **Always Resolve**. Resolves the out-of-date assemblies during the loading process.

- *Resolve lightweight components*. Prompt selected by default. Some operations require certain model data that is not loaded in lightweight components. This option controls what happens when you request one of these operations in an assembly which has lightweight components. There are two selections:

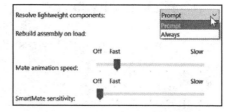

  - **Prompt**. Resolves lightweight components each time one of these operations is requested. In the dialog box that appears, click **Yes** to resolve the components and continue, or click **Cancel** to cancel the operation.

- **Always**. Automatically resolves lightweight components.

- *Rebuild assembly on load*. Prompt selected by default. Provides the ability to specify whether or not you want your assemblies to be rebuilt so the components are updated the next time you open them. There are three options:

  - **Prompt**. Asks if you want to rebuild each time an assembly is opened. Click **Yes** or **No** in the dialog box. If you check the Don't ask me again box, the option is updated to reflect your choice. Yes changes the option to Always. No will change the option to Never.

  - **Always**. Rebuilds an assembly when it is open.

  - **Never**. Opens your assembly without performing a rebuild.

- *Mate animation speed*. Fast selected by default. Enables animation of mates and controls the speed of your animation. When you add a mate, Click **Preview** or **OK** in the PropertyManager to view an animation of the mate.

- *SmartMate sensitivity.* Off selected by default. Sets the speed at which the software applies SmartMates.

- *Purge cashed configuration data*. Deletes the cashed configuration data.

- *Update mass properties while saving document*. Updates the Mass properties information when you save a document. If the document did not change, the next time you access the Mass properties, the system does not need to recalculate them.

☀ The Update Mass properties while saving document option is not available when the Large Assembly Mode is activated.

- *Use shaded preview*. Default setting. Displays shaded previews to help you visualize features that you create. Rotate, pan, zoom, and set standard views while maintaining the shaded preview.

- *Use Software OpenGL*. Disables the graphics adapter hardware acceleration and enables graphics rendering using only software. For many graphics cards, this will result in slower system performance.

☀ Only activate the Use Software OpenGL option when instructed by SW technical support.

- *No preview during open (faster)*. Select this option to disable the interactive preview. This will reduce the time to load your models. Clear to display the interactive preview while the model is loading.

Assemblies

The Assemblies section provides the ability to set the behavior options for dragging components in an assembly.

The Assemblies section provides the following default options:

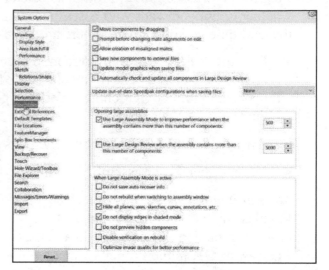

- **Move components by dragging**. Default setting. Provides the ability to move or rotate components within their degrees of freedom. When you deselect, you can still move or rotate a component with the Move with Triad function or the Move Component tool and Rotate Component tool located in the Assembly toolbar.

- **Prompt before changing mate alignments on edit**. When changes that you make to mates result in errors that the software can fix by flipping mate alignments, the software asks if you want it to make the changes. Otherwise, the software makes the changes automatically (without asking).

- **Allow creating of misaligned mates**. Default setting.

- **Save new component to external file**. If selected, prompts you to name and save new in-context components to external files. If cleared, saves new in-context components in the assembly file as *virtual components*.

- **Large assembles**. The selections you make under Large assemblies apply only when Large Assembly Mode is on. Set options for normal use (with Large Assembly Mode off) as indicated in the option descriptions below.

- **Use Large Assembly Mode to improve performance whenever working with an assembly containing more than this number of components**. Default setting. Set the number of resolved components above which Large Assembly Mode automatically activates when opening or working in an assembly. When the Large Assembly Mode is active, select the following options to improve performance:

  - **Do not save auto recover info**. Default setting. Disables automatic save of your model. (Set in Backup Options for normal use.)

  - **Hide all planes, axes, sketches, curves, annotations, etc**. Default setting. Selects Hide All Types on the View menu. When this option is selected, you can override it by clearing Hide All Types on the View menu, then selecting to show or hide individual types.

- ***Do not display edges in shaded mode***. Default setting. Turns off edges in shaded mode. If the display mode of the assembly is **Shaded With Edges**, it changes to **Shaded**. When this option is selected, you can override it by clicking **Shaded With Edges** from the Heads-up View toolbar.

- ***Do not rebuild when switching to assembly window.*** When you switch back to the assembly window after editing a component in a separate window, skips the message that asks if you want to rebuild. Skips the rebuild of the assembly even if you have previously selected Don't show again and clicked Yes (to rebuild).

- ***Use Large Design Review whenever working with an assembly containing more than this number of components***. Default setting 5000. Large Design Review is primarily intended as an environment for quick design reviews. If you want to ensure that all items are updated properly, you must open your assembly as lightweight or fully resolved.

- ***When Large Design Review is active:***

  - **Do not save auto recover info**.

  - **Do not rebuild when switching to assembly window.**

  - **Hide all planes, axes, sketches, curves, annotations, etc**. Selected by default.

  - **Do not display edges in shaded mode**. Selected by default.

  - **Do not preview hidden components.**

  - **Disable verification on rebuild.**

  - **Optimize image quality for better performance.**

  - **Suspend automatic rebuild.**

- ***Envelope Components***. Sets the mode in which envelope components are loaded when you open an assembly. Select one or both:

  - ***Automatically load lightweight***. Loads all envelopes as lightweight.

  - ***Load read-only***. Loads all envelopes as read-only.

External References

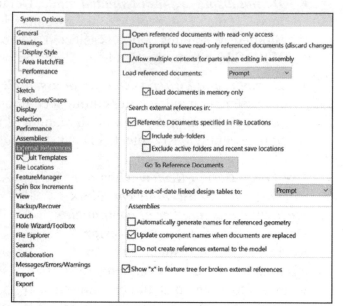

The External References section provides the ability to specify how a part, assembly or a drawing with external references is opened and managed.

An external reference is created when a document depends on another document for a solution.

In an assembly (Top-down) when a component references geometry from another component, an in-context feature and an external reference are created.

The External References section provides the following default options:

- ***Open referenced documents with read-only access***. Specifies that all referenced documents will be opened with read-only access.

- ***Don't prompt to save read-only referenced documents (discard changes)***. Specifies that when a parent document is saved or closed, no attempt will be made to save its read-only, referenced documents.

- ***Allow multiple contexts for parts when editing in assembly***. You can create external references to a single part from more than one assembly context. However, any individual feature or sketch within the assembly may only have one external reference.

🔆 Any individual feature or sketch within an assembly can only have one External reference.

- ***Load referenced documents***. Changed only selected by default. Specifies whether to load the referenced documents when you open a part that is derived from another document. There are four options:

  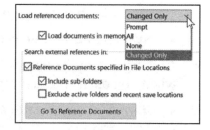

  - **Prompt**. Prompts you about loading externally referenced documents each time you open a document with External references.

  - **All**. Opens all of the externally referenced documents.

  - **None**. Does not open any External referenced documents. External references can be shown as out of context until you open the External referenced documents.

- **Changed Only**. Opens the Externally referenced documents which changed since the last time you opened the original document.

- *Reference Documents specification in File locations*. Default setting. Displays a message for modified external reference models. The message will ask if you want to save the referenced model. If the option is deselected, no message is displayed and the reference model is saved automatically. Include sub folders. Selected by default. Exclude active folder and resend save locations. See SOLIDWORKS Help for additional information.

- *Update out-of-date linked design tables to*. Prompt selected by default. The Update out-of-date linked design tables to option determines what happens to linked values and parameters if the model and the design table are out-of-sync. There are three options:

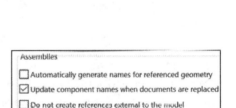

  - **Prompt**. Default setting. Prompts you when you open a document with a design table that is out-of-sync with the model.

  - **Model**. The design table updates with the model's values.

  - **Excel File**. The model updates with the design table's values.

- *Automatically generate names for referenced geometry*. When this option is not selected, you can mate to parts for which you have read-only access because you are only using the internal face IDs of the parts.

In a multi-user environment, leave this option deactivated.

- *Update component names when documents are replaced*. Default setting. Clear this option only if you use the Component Properties dialog box to assign a component name in the FeatureManager design tree that is different from the filename of the component.

- *Do not create references external to the model*. Select this option to NOT create External references when designing In-Context of an assembly. No In-Place mates are created when you create a new component in a Top-down assembly.

- *Show "x" in feature tree for broken external references*. Default setting. Flags items that have broken external references with an indicator (x) in the FeatureManager design tree. Clear this option if you want to hide the indicators (x).

Default Templates

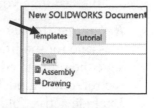

The Default Templates section specifies the folder and template file for parts, assemblies and drawings which are created automatically. Example: When you import a file from another application or create a derived part, the default template is used as the new document if you do not want to apply a template that you created.

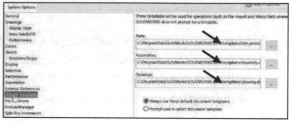

Templates are part, drawing and assembly documents which include user-defined parameters. Open a new part, drawing or assembly. Select a template for the new document.

- **_Parts_**. The Parts default template is located in the C:\ProgramData\SolidWorks\SOLIDWORKS 2018\templates\Part.prtdot folder.

- **_Assemblies_**. The Assemblies default template is located in the C:\ProgramData\SolidWorks\SOLIDWORKS 2018\templates\Assembly.asmdot folder.

- **_Drawings_**. The Drawings default template is located in the C:\ProgramData\SolidWorks\SOLIDWORKS 2018\templates\Drawing.drwdot folder.

☼ The sheet format and sheet size are set in the default Drawing Template. The Drawing Sheets document property lets you specify a default sheet format for when you add new sheets to drawing documents. This property lets you automatically have one sheet format for the first sheet and a separate sheet format for all additional sheets. To specify a different sheet format for a new sheet, click Tools > Options > Document Properties > Drawing Sheets, select Use different sheet format, and browse to select a sheet format file (file ending in .slddrt).

File Locations

The File locations section provides the ability to specify the location of the folders to be searched for using the specified document type. Folders are searched in the order in which they are listed. Each folder is listed under **Options ➢ System Options ➢ File Locations**. The tab is visible when the folder contains one or more SOLIDWORKS Part, Assembly or Drawing Templates.

When you open an assembly drawing, SOLIDWORKS searches for the referenced assembly document. If the assembly document cannot be located, SOLIDWORKS performs a search to locate the missing document. In the file open process, the search order is as follows: Documents loaded in memory, Optional user-defined search lists, Show folders for. The Show folders for option displays the various available search paths in SOLIDWORKS.

Tutorial: Document Templates Location 2-2

Add a new Document Templates tab.

1. **Start** a SOLIDWORKS session.

2. Click **New** ⬜ from the Menu bar. The Templates tab is the default tab. Part is the default template from the New SOLIDWORKS Document dialog box.

3. Click **OK** from the New SOLIDWORKS Document dialog box. The Part FeatureManager is displayed.

4. Click **Options** ⚙ ➢ **System Options** tab ➢ **File Locations**.

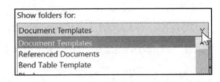

5. Select **Document Templates** from the show folders for box as illustrated.

6. Click the **Add** button.

7. Browse to the **SOLIDWORKS 2018\MY-TEMPLATES** folder. Create the folder if required. The path to the MY-TEMPLATES folder is added to the Folders list. If required, click the Move Down button to position the MY-TEMPLATES folder at the bottom of the Folders list. The MY-TEMPLATES folder will be the third tab in the New dialog box. The new tab is only displayed if there is a document in the folder. Note: Part and Drawing templates are provided in the MY-TEMPLATES folder in the book.

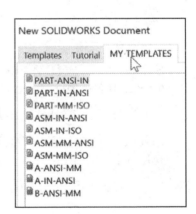

8. Click **OK** from the Browse For Folder dialog box.

9. Click **OK** from the System Options - File Locations dialog box.

10. Click **OK** to change search paths.

11. **Close** all models.

Tutorial: Referenced Document Location 2-3

Add a folder for the Referenced Documents.

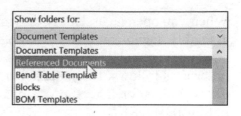

1. **Start** a SOLIDWORKS session. Click **Options** ⚙ ➤ **System Options** ➤ **File Locations**.

2. Select **Referenced Documents** in the Show folders for box.

3. Click the **Add** button.

4. Select the **SOLIDWORKS 2018\Delivery** folder. Create the folder if required.

5. Click **OK** from the Browse For Folder dialog box.

6. Click **OK** from the System Options - File Locations dialog box.

7. Click **OK** to change search paths.

8. **Close** all models.

Tutorial: Design Library Location 2-4

Add two folders to the Design Library.

1. **Start** a SOLIDWORKS session.

2. Click **Options** ⚙ ➤ **System Options** ➤ **File Locations**.

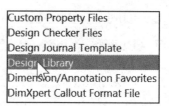

3. Select **Design Library** in the Show folders for box.

4. Click the **Add** button.

5. Select the **SOLIDWORKS 2018\MY-TOOLBOX** folder. Create the folder if required.

6. Click **OK** from the Browse For Folder dialog box.

7. Select the **SOLIDWORKS 2018\SMC** folder. Create the folder if required.

8. Click **OK** from the Browse For Folder dialog box.

9. Click **OK** from the System Options - File Locations dialog box.

10. Click **OK** to change search paths.

11. **Close** all models. **View** the two new folders in your Design Library.

FeatureManager

The FeatureManager section provides the ability to configure the FeatureManager design tree.

The FeatureManager design tree is located on the left side of the SOLIDWORKS Graphics window. The FeatureManager provides an outline view of the active part, assembly, or drawing. This provides insight on how the document was constructed and aids you to examine and edit the document.

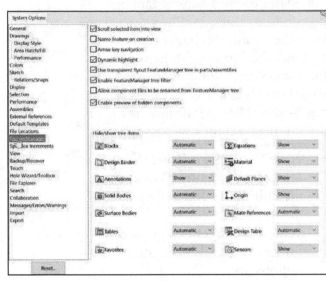

The FeatureManager design tree and the Graphics window are dynamically linked. Select sketches, features, drawing views, and construction geometry in either pane. The FeatureManager section provides the following default options:

- *Scroll selected item into view*. Default setting. Provides the ability for the FeatureManager design tree to scroll automatically to display features corresponding to items selected in the Graphics window.

- *Name feature on creation*. Provides the ability to name features and sketches for design intent.

- *Arrow key navigation*. When you create a new feature, the feature name in the FeatureManager design tree is automatically selected and is ready to enter a name.

- *Dynamic highlight*. Default setting. This option provides the ability for the geometry in the graphics area; edges, faces, planes, axes, etc. are highlighted when the pointer passes over the item in the FeatureManager design tree.

- *Use transparent fly-out FeatureManager in parts/assemblies*. Default setting. The fly-out design tree is transparent. When cleared, the fly-out design tree is not transparent.

- *Enable FeatureManager tree filter*. Default setting. Displays a box at the top of the FeatureManager design tree so you can type text to filter.

- *Allow component files to be renamed from FeatureManager tree*.

- *Enable preview of hidden components*. Default setting.

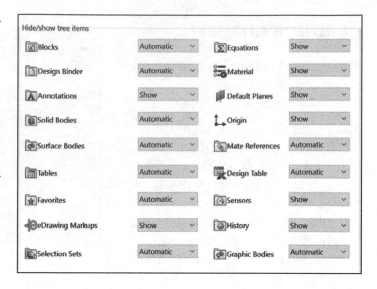

⟡ The fly-out FeatureManager design tree allows you to view both the FeatureManager design tree and the PropertyManager at the same time. Sometimes it is easier to select items in the fly-out FeatureManager design tree than in the Graphics window.

- *Hide/Show Tree Items*. View the default options. Provides the ability to control the display of the FeatureManager design tree folders and tools. The three options are:

  - **Automatic**. Displays the item if present. Otherwise, it is hidden.

  - **Hidden**. Always hides the item.

  - **Show**. Always shows the item.

Spin Box Increments

The Spin Box Increments section provides the ability to set the spin box increment value for both English and Metric units.

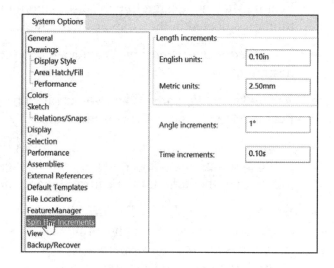

The Spin Box Increments section provides the following default options:

- *Length increments*. Provides the ability to specify the units added or subtracted when you click the spin box arrow to modify a linear dimension value. The selections are:

  - **English units**. Specifies English units in inches.

  - **Metric units**. Specifies Metric units in mm.

  - **Angle increments**. Specifies the angle increments, 1 degree added or subtracted when you click a spin box arrow to change an angular dimension value.

  - **Time increments**. Specifies the seconds added or subtracted when you click a spin box arrow to change a time value.

View

The View section provides the ability to set the default view rotation and transitions.

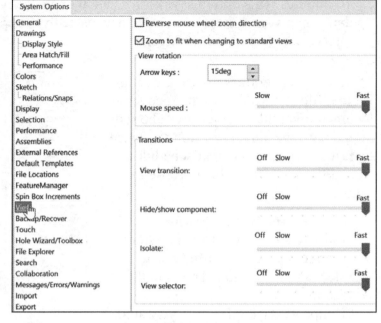

- ***Reverse mouse wheel zoom direction***. Changes the direction of the mouse wheel for Zooming in and out.

- ***Zoom to fit when changing to standard views***. Default setting.

- ***Arrow keys***. 15deg selected by default. Specifies the angle increment for view rotation when you use the arrow keys to rotate the model.

- ***Mouse speed***. Mouse speed by default is set to Fast. Provides the ability to specify the speed of rotation when you use the mouse to rotate the model or assembly component. Move the slider to the left to obtain finer control and slower rotation.

- ***View Transitions***. Changes from one view orientation to another. For example, from a front view to an isometric view.

- ***Hide/Show Component***. In assemblies, when you turn the visibility of selected components off or on.

- ***Isolate***. In assemblies and multi-body parts, when you isolate selected components.

- ***View selector***. Controls the animation when you activate the View Selector.

Backup/Recover

The Backup/Recover section provides the ability to set time frequency and folder locations for auto-recovery, backup, and save notification. Auto-recovery and save notification are controlled by a specified number of minutes or the number of changes. A change is defined by the following:

1. An action in a part or assembly document which requires a rebuild. Example: An addition of a feature.

2. An action in a drawing document which requires a rebuild. Example: A modification to a dimension.

- ***Save auto-recover info every***. Selected by default. The default time interval is every 10 minutes. This option saves information on the active document to prevent loss of data when your system terminates unexpectedly. There are two auto-recover intervals: *changes* and *minutes*.

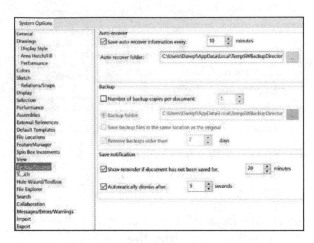

Auto-recover does not save over the original file. You can save the recovered document over the original file. The status bar displays the number of minutes since the last save and to the next scheduled save, if the save interval is specified in minutes.

- ***Auto recover folder***. Provides the ability to specify the folder to store auto recovered files or browse to select a new location. The default location: \TempSW\BackupDirectory\swauto.

- ***Backup***. Stores a backup of the original document before any changes are saved to the file. It is a version before the last saved version of the document. Backup files are named Backup of *<document_name>*. Multiple backups of a document are named with the most recent version in Backup (1), the next most recent in Backup (2), etc.

If changes to an active document are saved in error, opening the backup file brings the document back to the point before the changes were made.

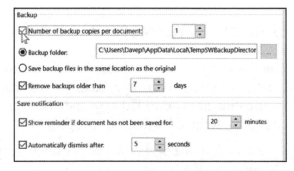

If you save a document without making any changes, the backup file is the same as the original.

- ***Number of backup copies per document***. Not selected by default. When this option is selected, you can specify from 1 to 10 copies to be saved for each document.

- **Save notification**. Provides a transparent message, Un-Saved Document Notification, displayed in the lower right corner of the Graphics window if the active document has not been saved within the specified interval, time or number of changes.

Click the commands in the message to save the active document or save all open documents. The message fades after a few seconds.

- **Show reminder if document has not been saved for**. Default setting is 20 minutes. When notification is enabled, you can specify the interval in minutes or number of changes.

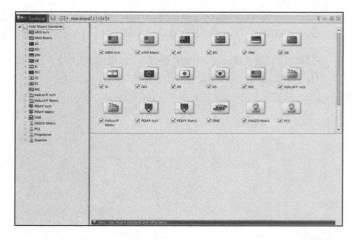

Hole Wizard/Toolbox

The Hole Wizard/Toolbox section provides the ability to create new standards or to edit existing standards used by the Hole Wizard holes and SOLIDWORKS Toolbox components. You can add administrative access to these standards and the options of the SOLIDWORKS Toolbox Add-in.

The Hole Wizard/Toolbox section provides the following default options:

- **HoleWizard\ToolBox folder**. The default HoleWizard\ToolBox folder location is C:\program files\SOLIDWORKS data\.

- **Make this folder the default search location for Toolbox components**.

- **Configure**. Provides the ability to access the Getting Started\Configure Data dialog box to configure contents, settings, properties, etc. for the SOLIDWORKS toolbox.

- **Display Toolbox Favorites folder**. Selected by default.

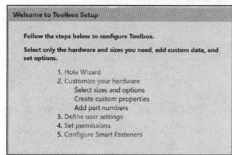

File Explorer

The File Explorer section duplicates Windows Explorer from your local computer and displays the following directories: Recent Documents and Open in SOLIDWORKS. The File Explorer section provides the following default options:

- ***Show in File Explorer view***. Provides the ability to show or hide the following folders and files on the File Explorer tab of the Task Pane. The following options are:

  - **My Documents**. Default setting.

  - **My Computer**. Default setting.

  - **My Network Places**. Default setting.

  - **Recent Documents**. Default setting.

  - **Hidden referenced documents**. Default setting.

  - **Samples**. Default setting. The Samples option is for the Online Tutorial and *What's New* examples.

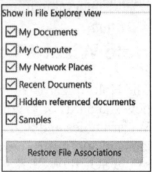

Search

The File and Model Search section locates key items. The section provides the following default options:

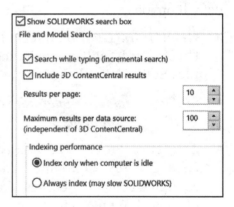

- ***Show SOLIDWORKS search box***. Default setting. Displays the SOLIDWORKS Search box.

- ***Search while typing (incremental search)***. Default setting. Starts the search as you type the search string.

- ***Include 3D ContentCentral results***. Default setting. Includes 3D ContentCentral results from the search.

- ***Results per page***. The default Results per page = 10. Specifies the number of documents to be displayed on each page of the Task Pane Search Results tab.

- ***Maximum results per data***. The default Maximum results per data = 100. Specifies the number of results for a search attempt.

- *Indexing performance*. Default setting. Updates the index to improve performance. All locations in the list on the Task Pane Search Results tab are indexed. There are two selections: **Index only when computer is idle** and **Always index (may slow SOLIDWORKS)**.

Collaboration

The Collaboration section provides the ability to specify options for a multi-user environment. The Collaboration section provides the following options:

- *Enable multi-user environment*. Enables the other options below. Note: The Enable multi-user environment option is activated in the illustration to improve visibility of the screen shot.

- *Add shortcut menu items for multi-user environment*. Provides menu items; Make Read-Only and Get Write Access are available on the File menu part documents and for assemblies.

- *Check if files opened read-only have been modified by other users*. Default time is 20 minutes. Checks files you have opened as read-only at the interval specified in Check files every X minutes to view if the files have been modified in one of the following ways:

  - Another user saves a file that you have open in SOLIDWORKS, making your file out of date.

  - Another user relinquishes write access to a file that you have open in SOLIDWORKS by making the file read-only, allowing you to take write access.

- *Comments and Markup*. The options are:

  - **Automatically add timestamp to comments**. Default setting.

  - **Show Comments in PropertyManager**. Default setting.

Messages/Errors/ Warnings

Restore messages that have been suppressed. You can suppress messages that you see frequently if you know that you will always choose the default response.

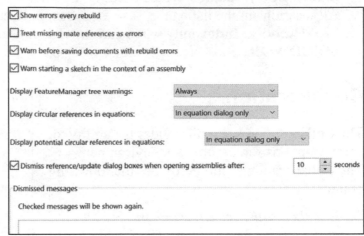

- **Show errors every rebuild**. Default setting.
- **Treat missing mate references as errors**.
- **Warn before saving documents with rebuild error**. Default setting.
- **Warn starting a sketch in the context of an assembly**. Default setting.
- **Display FeatureManager tree warnings**. Options: Always, Never and All but top level.
- **Display circular references in equations**. Options: In equation dialog only, Everywhere, Never.
- Display potential circular references in equations. Options are In equation dialog only, Everywhere, and Never.
- **Dismiss reference/update dialog boxes when opening assemblies after**. Default setting: 10 seconds.

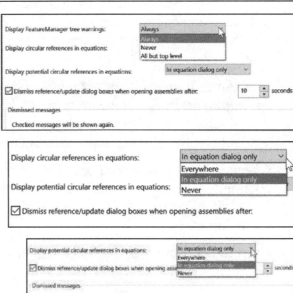

Import

This section provides various options for file import into SOLIDWORKS. The following options are available:

- **File Format**. Options:

  - **General**.

  - **STL/OBJ/OFF/PLY/PLY2**.

  - **VRML**.

  - **3MF**.

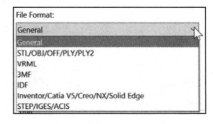

- **IDF**.

- **Inventor/Catia V5/Creo/NX/Solid Edge**.

- **STEP/IGES/ACIS**.

- *Enable 3D interconnect*.

- *Automatically run Import Diagnostics (Healing)*. Default setting. When importing a file, Import Diagnostics runs automatically. When cleared, a prompt appears for each import action asking if you want to run Import Diagnostics.

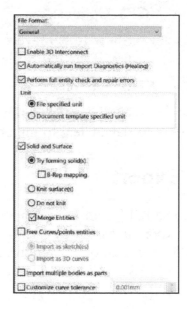

- *Perform full entity check and repair errors*. Default setting. Checks and repairs errors. Import performance is slower because the software spends time checking and (when possible) repairing the model entities.

- *Unit*. Options:

  - **File specified unit (default setting)**. Use the units of the imported file.

  - **Document template specified unit**. Use the units specified in the SOLIDWORKS template files under Tools, Options, System Options, Default Templates.

- *Solid and Surface*. Default setting. Options:

  - **Try forming solid(s)**, default setting.

  - **B-Rep mapping**.

  - **Knit surface(s)**.

  - **Do not knit**.

  - **Merge Entities**. Default setting.

- *Free Curves/points entities*. Options:

  - **Import as sketch(es)**.

  - **Import as 3D curves**.

- *Import multiple bodies as parts*. Imports a multi-body part as separate part documents in an assembly document. When cleared, the multi-body part imports as a part document with multiple bodies.

- *Customize curve tolerance*. Default setting .001mm. Customizes the tolerance when importing models with very small entities (smallest values on the order of 1.0e-6 to 1.0e-7 meters). When cleared, SOLIDWORKS uses internal tolerance settings, which are too large to properly import and display these small models. Enter a tolerance in the box.

- *IGES*. Options:

  - **Show IGES levels**. Displays the IGES-In Surfaces, Curves, and Levels dialog box if the IGES file contains curves or different levels (or layers).

- *STEP*. Options:

  - **Map configuration data**. Imports STEP file configuration data plus geometric data. Clear to import only geometric data.

Export

The IGES translator supports layers when you export an assembly document as an IGES Representation/System preference export option to MASTERCAM. The following options are available:

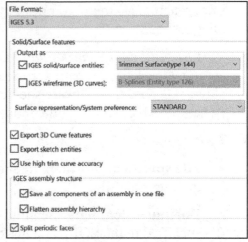

- *File Format*. Options:

  - **IGES 5.3, STEP, ACIS, Parasolid, VRML, IFC, EDRW/EPRT/EASM, PDF, 3DPDF**.

- *Surface/Solid features*. Exports the data as solid or surface entities. Select Trimmed Surface (type 144) to convert the faces of the part, assembly, or the selected surfaces to trimmed surfaces in the IGES file. Select Manifold Solid (type 186) to export boundary representation (BREP) data to the IGES file. You can also select Bounded Surface (type 143).

  - **IGES solid/surface entities**. Options are Trimmed Surface(type 144), Manifold Solid(type 186), Bounded Surface(type 143).

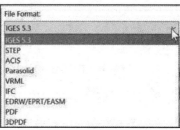

  - **IGES wireframe (3D curves)**. Converts the solid body to a 3D wireframe representation in the IGES file. Unlike surfaces or faces, you cannot export individually selected model edges to an IGES file.

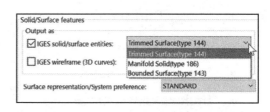

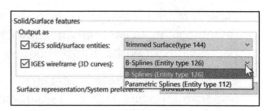

Select either B-Splines (Entity type 126) or Parametric Splines (Entity type 112) depending on the entity types required by the target system. For more information, see the IGES 3D Curves table.

- ***Surface representation/System preference***. The IGES entity types that compose the trimmed surfaces depend on the export format chosen. The Surface representation table shows the IGES entity types that compose the trimmed surfaces. View the options.

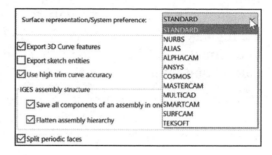

- ***Export 3D Curve features***. Default setting. Includes 3D curve features in the exported file.

- ***Export sketch entities***. Includes sketch entities in the exported file. All 2D and 3D sketch entities are included.

- ***Use high trim curve accuracy***. Default setting. Exports using high trim curve accuracy; the file size is larger than if the check box is cleared. This setting affects files exported both with Trimmed surfaces and with 3D curves. High trim curve accuracy can sometimes help if the target system has trouble importing the IGES file, or cannot knit the surfaces into a useful solid.

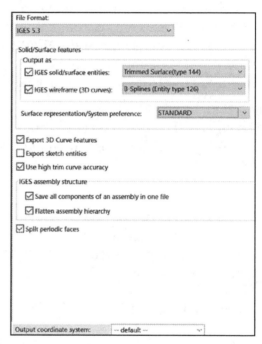

- ***IGES assembly structure***. Default setting. Saves all assembly components, subassemblies, and subassembly components in one file. Otherwise, the assembly components and the subassembly components are saved as individual IGES files in the same directory.

- ***Split periodic faces***. Default setting. Splits periodic faces, such as cylindrical faces, into two. Splitting a periodic face can improve the quality of the export but can affect performance.

- ***Output coordinate system***. Select a coordinate system to apply for export. If you select default, no transformation matrix is applied.

Summary

In this chapter, you learned about using and modifying the System Options in SOLIDWORKS. System Options are stored in the registry of your computer. System Options are not part of your document. Changes to the System Options influence all current and future documents.

You added a new Document Template in the File Locations section. You added a new folder to the Referenced Documents and two new folders to the Design Library.

You learned about creating a registry file (*sldreg) using the SOLIDWORKS Copy Settings Wizard to save your custom settings. Do not use the SOLIDWORKS Copy Settings Wizard unless you are familiar with the Windows register and you may require administrator access on a network system to apply the SOLIDWORKS registry file.

To change the default location, click Tools ➢ Options ➢ System Options ➢ File Locations. In Show folder for, select Sheet Formats.

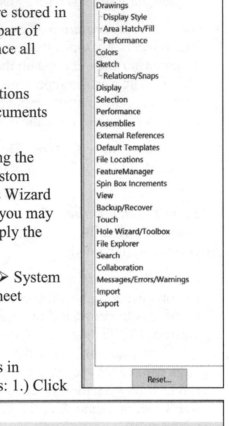

Enable or disable the creation of misaligned mates in System Options. To allow creation of misaligned mates: 1.) Click Tools ➢ Options ➢ System Options ➢ Assemblies. 2.) Select Allow creation of misaligned mates.

In Chapter 3, explore and address Templates and Document Properties for a Part, Assembly and Drawing document.

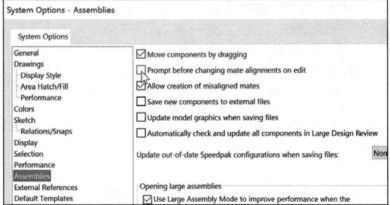

CHAPTER 3: DOCUMENT PROPERTIES

Chapter Objective

Chapter 3 provides a comprehensive understanding of the Document Properties section.

The available options under the Document Properties tab are document dependent as illustrated from left to right: Part, Assembly and Drawing.

On the completion of this chapter, you will be able to:

- Understand a document template.

- Set up and modify the available tools from the Document Properties section.

💡 Utilize the Search feature in the Systems Options dialog box for fast interactive searching for System Options or Document Properties.

💡 ANSI drafting standard using the IPS unit system is the default standard in this text.

The first half of the chapter addresses the Document Properties for a Part and Assembly document. The second half addresses the Document Properties for a Drawing document.

Document Properties/Templates

Templates are documents (part, drawing and assembly documents) which form the basis of a new document. They can include user-defined parameters, annotations, predefined views, geometry, etc. You can maintain numerous different document templates. Example:

- A document template using an **ANSI** Overall drafting standard.

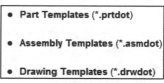

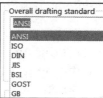

- A document template using an ISO Overall drafting standard.

- A document template using inches and another document template using millimeters with predefined views and notes.

Organize your document templates by placing them on different tabs in the New SOLIDWORKS Document dialog box. Create custom tabs to organize your templates.

Tutorial: Close all open models 3-1

Close all parts, assemblies, and drawings. Access the Document Properties dialog box.

1. Click **Window** ➢ **Close All** from the Menu bar. Access the Document Properties Drafting Standard dialog box. All models are located in the book. Copy all models to your hard drive.

2. Open the **Axle3** part from the SOLIDWORKS 2018 folder.

3. **Right-click** on Axle3. Click **Document Properties** or Click **Options** ⚙ ➢ **Document Properties** tab.

4. **View** the default Document Properties selection for a Part document. Note: The ANSI overall drafting standard and the IPS Unit System are selected for this part document with a precision of .12 decimal places.

The available options under the Document Properties tab are document dependent: Part, Assembly and Drawing.

Overall Drafting Standard

The Overall drafting standard feature provides the ability to select default standards for an active Part, Assembly, or Drawing document from the drop-down menu or to create a custom drafting standard.

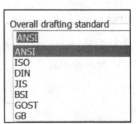

- *Overall drafting standard*. The default setting was chosen and set during the initial software installation. The default standard during this initial software installation was ANSI. The drop-down menu provides the ability to select the following default drafting standards:

- **ANSI**. American National Standards Institute.

- **ISO**. International Standards Organization.

- **DIN**. Deutsche Institute fur Normumg.

- **JIS**. Japanese Industry Standard.

- **BSI**. British Standards Institution.

- **GOST**. Gosndarstuennye State standard.

- **GB**. Guo Biao.

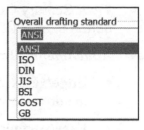

The Overall drafting standard provides the ability to create or modify a default standard. Modification to a standard drafting standard creates a modified standard.

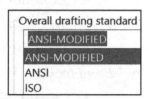

- *Rename*. Only available for a modified drafting standard. Provides the ability to rename the default modified drafting standard.

- *Copy*. Only available for a modified drafting standard. Provides the ability to copy the modified drafting standard.

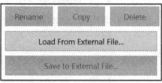

- *Delete*. Only available for a modified drafting standard. Provides the ability to delete the modified drafting standard from the drop-down menu.

- *Load From External File*. Load a drafting standard from an external file.

- *Save to External File*. Only available for a modified drafting standard. Saves a modified drafting standard to an external file.

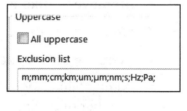

- *Uppercase*. Sets the default case used in notes and balloons to all uppercase.

- *Exclusion list*. Lists strings that should not be automatically capitalized.

Annotations

The Annotations section provides seven sub-sections: *Balloons, Datums, Geometric Tolerances, Notes, Revision Clouds, Surface Finishes* and *Weld Symbols*.

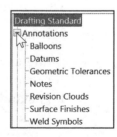

The Annotations section provides the ability to select the following:

- *Overall drafting standard*. Displays the selected drafting standard.

- *Text*. Century Gothic selected by default. Click the Font button to select a custom font style and size.

- *Attachments*. Provides the following arrow style options:

  - **Edge/vetex**. Select from various arrow styles from the drop-down menu.

  - **Face/surface**. Select from various arrow styles from the drop-down menu.

  - **Unattached**. Select from various arrow styles from the drop-down menu.

- *Bend leaders*. Provides the following selections:

  - **Use bent leaders**. Selected by default. Deselect to enter leader length. Note: Modify from a default standard creates a modfied standard as illustrated.

  - **Leader length**. Default leader length is 0.25in for the IPS system and 6.35mm for the MMGS system.

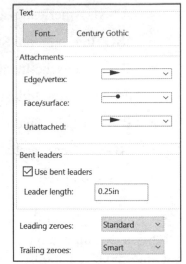

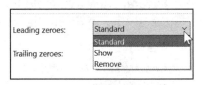

- *Leading zeroes*. Standard selected by default. Select one of three options from the drop-down menu:

  - **Standard**. Zeroes are displayed based on the selected dimensioning standard.

  - **Show**. Zeroes before the decimal points are displayed on the selected dimensioning standard.

  - **Remove**. Zeroes are not displayed.

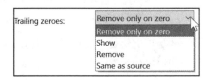

- *Trailing zeroes*. Provides the following selections:

  - **Remove only on zero**. Default setting.

  - **Show**. Dimensions have trailing zeroes up to the number of decimal places specified in **Options**, **Document Properties** and **Units**.

  - **Remove**. All trailing zeroes are removed.

  - **Same as Standard**. Trailing zeroes are trimmed for whole metric values. Conforms to ANSI and ISO standards.

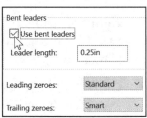

Annotations - Balloons

The Annotations - Balloons section provides the ability to set the default properties of balloons in a drawing document or in a note.

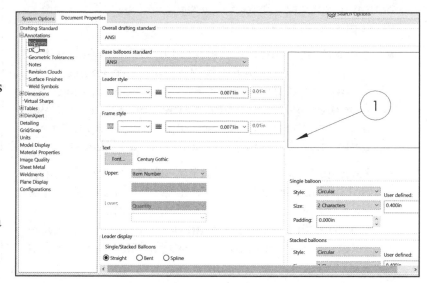

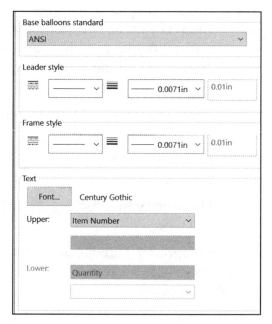

Create balloons in a drawing document or in a note. The balloons label the parts in the assembly and relate them to item numbers on the bill of materials (BOM).

You do not have to insert a BOM in order to add balloons. If the drawing has no BOM, the item numbers are the default values that the software would use if you did insert a BOM. If there is no BOM on the active sheet, but there is a BOM on another sheet, the numbers from that BOM are applied.

- **Overall drafting standard.** Displays the selected drafting standard.

- **Base balloons standard.** Provides the ability to select a default standard for an active Part, Assembly, or Drawing document from the drop-down menu or to create a custom Base balloons standard.

- **Leader style.** Provides the ability to use the default Leader style or to select a custom Leader style and thickness from the drop-down menus.

- **Frame style.** Provides the ability to use the default Frame style or to select a custom Frame style and thickness from the drop-down menus.

- ***Text***. Century Gothic selected by default. Click the Font button to select a custom font style and size. There are two options for Balloon text:

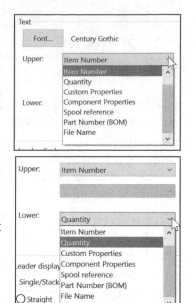

  - **Upper**. Item Number selected by default. Select **Text, Item Number, Quantity, Custom Properties, Component Properties, Spool reference, Part Number (BOM) or File Name** from the drop-down menu for the upper section of a Circular Split Line balloon or for the whole balloon of all other styles.

  - **Lower**. Quantity selected by default. Select Circular Split Line for balloon style to activate the Balloon text Lower option. Select **Text, Item Number, Quantity, Custom Properties, Component Properties, Spool reference, Part Number (BOB), File Name or Cut List Properties** from the drop-down menu.

  - **Leader display**. Provides the ability to select the following options: **Straight**: Selected by default. Straight Leader line, **Bent**: Bent Leader line. Provides the ability to access the Use document leader length option and **Spline**: Create a spline.

  - **Auto Ballons**. Provides the ability to select the following options: **Straight** and **Bent**.

  - **Use document leader length**. The default setting is 0.25in for the IPS system and 6.35mm for the MMGS system.

  - **Quantity gap**. Provides the ability to enter a distance.

- ***Single balloon***. Circular and 2 Characters selected by default. There are three selections for the Single balloon option:

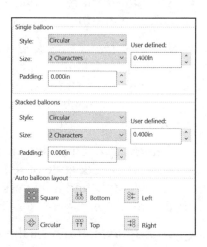

  - **Style**. Circular selected by default. There are 13 Style selections from the drop-down menu. Select a balloon style option.

  - **Size**. 2 Characters selected by default. There are six Size selections from the drop-down menu. Select a size to accommodate your requirements. Note: Select Tight Fit to adjust balloon size to the text or select the number of characters.

  - **Padding**. Padding is a tight-fit border option for annotation notes and balloons where you can specify a distance to offset the border from the selected text or annotation note.

- *Stacked balloons*. Circular and 2 Characters selected by default. There are three selections for the Stacked balloons option:

  - **Style**. There are 13 Style selections from the drop-down menu. Select a balloon style.

  - **Size**. There are six Size options from the drop-down menu. Select a balloon size. Note: Select Tight Fit to adjust balloon size to the text or select the number of characters.

  - **Padding**. Padding is a tight-fit border option for annotation notes and balloons where you can specify a distance to offset the border from the selected text or annotation note.

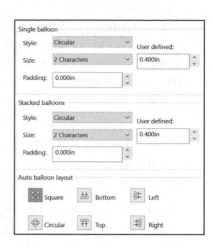

- *Auto Balloon Layout*. Square layout selected by default. There are six selections for the Auto Balloon Layout option. The option provides the ability to select a layout as the default for inserting Auto Balloons. The six selections are **Square**, **Bottom**, **Left**, **Circular**, **Top** and **Right**.

Annotations - Datums

You can attach a datum feature symbol to the following items:

- A part or assembly.

- On a planar model surface or on a reference planc.

- In a drawing view.

- On a surface that appears as an edge (not a silhouette) or on a section view surface.

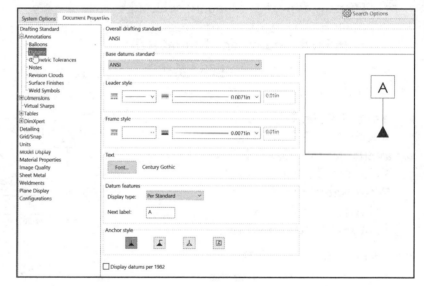

- A geometric tolerance symbol frame.

- A note.

- On a dimension, except for Ordinate dimensions, Chamfer dimensions, Angular dimensions and Arc dimensions.

The Annotations - Datums section provides the following selections:

- *Overall drafting standard*. Displays the selected drafting standard.

- ***Base datums standard***. Provides the ability to select a default standard for an active Part, Assembly or Drawing document from the drop-down menu or to create a custom Base datums standard.

- ***Leader style***. Provides the ability to use the default Leader style or to select a custom Leader style and thickness from the drop-down menus.

- ***Frame style***. Sets the style for the shape surrounding the text. Select a thickness or select Custom Size and enter a thickness.

- ***Text***. Century Gothic selected by default. Click the Font button to select a custom font style and size.

- ***Datum features***. The following selections are available:

  - **Display type**. Per Standard selected by default. The available options are **Per Standard, Square** and **Round (GB)**.

  - **Next Label**: A is the first label followed by B, C, etc.

- ***Anchor style***. Filled Triangle selected by default. The available options are **Filled Triangle, Filled Triangle With Shoulder, Empty Triangle** and **Empty Triangle With Shoulder**.

- ***Display datums per 1982***. Select this option to use the 1982 standard for the display of datums.

The Display datums per 1982 option is only available if you select the American National Standards Institute, ANSI dimensioning standard.

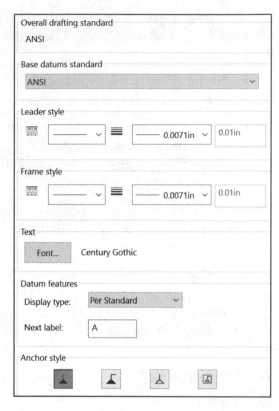

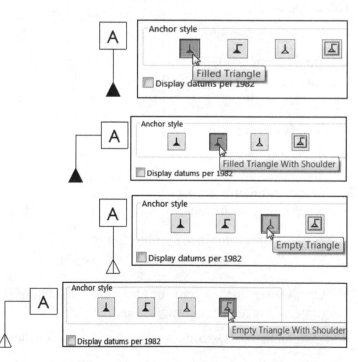

Annotations - Geometric Tolerance

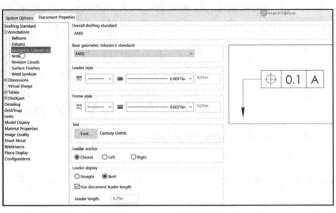

The Annotations - Geometric Tolerance section provides the ability to specify the properties when you create a geometric tolerance symbol. The SOLIDWORKS software supports the ANSI Y14.5 Geometric and True Position Tolerancing guidelines.

Place Geometric tolerancing symbols, with or without leaders, anywhere in a drawing, part, assembly, or sketch, and you can attach a symbol anywhere on a dimension line.

The Annotations - Geometric Tolerance section provides the following selections:

- *Overall drafting standard*. Displays the selected drafting standard.

- *Base geometric tolerance standard*. Provides the ability to select a default standard for an active Part, Assembly or Drawing document from the drop-down menu or to create a custom Base geometric tolerance standard.

- *Leader style*. Provides the ability to use the default Leader style or to select a custom Leader style and thickness from the drop-down menu.

- *Frame style*. Provides the ability to use the default Frame style or to select a custom Frame style and thickness from the drop-down menu.

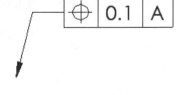

- *Text*. Century Gothic selected by default. Click the Font button to select a custom font style and size.

- *Leader anchor*. Anchor the leader to the specified location on the geometric tolerance. Closest is selected by default. The options are **Closest**, **Left** and **Right**.

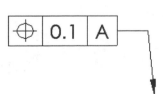

The pointer changes to 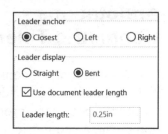 when it is on a geometric tolerancing symbol.

- ***Leader display***. Provides the ability to display either a Straight or Bent leader on the geometric tolerance. Bent is selected by default. The default Leader length is .25in for the IPS system and 6.35mm for the MMGS system.

  - **Use document leader length**. Selected by default.

Annotations - Notes

A note can be free floating or fixed, and it can be located with a leader pointing to an item, face, edge or vertex in the document. It can also contain simple text, symbols, parametric text and hyperlinks.

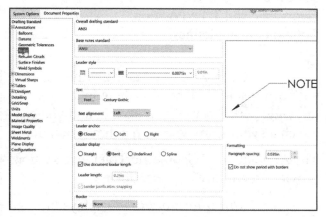

The Annotations - Notes section provides the following selections:

- ***Overall drafting standard***. Displays the selected drafting standard.

- ***Base notes standard***. Provides the ability to select a default standard for an active Part, Assembly or Drawing document from the drop-down menu or to create a custom Base notes standard.

- ***Leader style***. Provides the ability to use the default Leader style or to select a custom Leader style and thickness from the drop-down menus.

- ***Text***. Century Gothic selected by default. Click the Font button to select a custom font style and size.

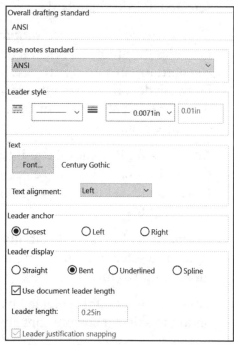

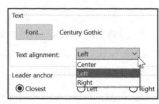

- **Text alignment**. Left is selected by default. The three options are **Center**, **Left** and **Right**.

- *Leader anchor*. Anchor the leader to the specified location on the note. Closest is selected by default. The three options are **Closest**, **Left** and **Right**.

- *Leader display*. Provides the ability to display either a Straight, Bent, Underlined or Spline leader on the note. Straight is selected by default. The default Leader length is .25in for the IPS system and 6.35mm for the MMGS system.

  - **Use document leader length**. Selected by default.

- *Border*. Provides the ability to insert a border around the note.

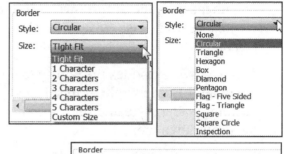

  - **Style**. Circular selected by default. Select a border style from the drop-down menu as illustrated.

  - **Size**. Select a border size from the drop-down menu as illustrated.

  - **Padding**. Padding is a tight-fit border option for annotation notes and balloons where you can specify a distance to offset the border from the selected text or annotation note.

Annotations - Surfaces Finishes

Specify the surface texture of a part face by using a surface finish symbol. Select the face in a part, assembly or drawing document.

Drag a surface finish symbol with a leader to any location. If you attach a leaderless symbol to an edge, then drag it off the model edge, an extension line is created.

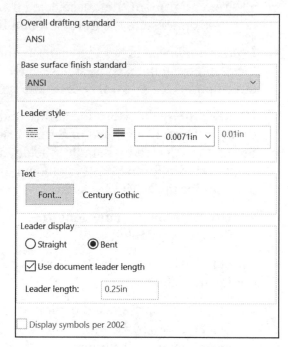

To keep the surface finish symbol locked on an edge, drag the symbol from anywhere except the bottommost handle.

The Annotations - Surface Finishes section provides the following options:

- *Overall drafting standard*. Displays the selected drafting standard.

- *Base surface finish standard*. Provides the ability to select a default standard for an active Part, Assembly or Drawing document from the drop-down menu or to create a custom Base surface finish standard.

- *Leader style*. Provides the ability to use the default Leader style or to select a custom Leader style and thickness from the drop-down menus.

- *Text*. Century Gothic selected by default. Click the Font button to select a custom font style and size.

- *Leader display*. Provides the ability to display either a Straight or Bent leader on the surface finish. Bent is selected by default. The default Leader length is .25in for the IPS system and 6.35mm for the MMGS system.

    - **Use document leader length**. Selected by default.

    - **Display symbols per 2002**. For ISO and related drafting standards, you can specify Display symbols per 2002.

Annotations - Weld Symbols

Create weld symbols in a Part, Assembly, or Drawing document.

Many properties are available both above and below the weld symbol line and are described in order from the line. A preview of the symbol is displayed in the Graphics window.

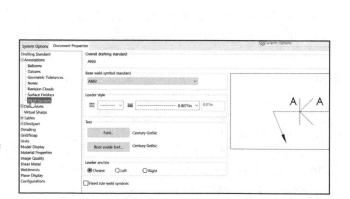

The Annotations - Weld Symbol section provides the following options:

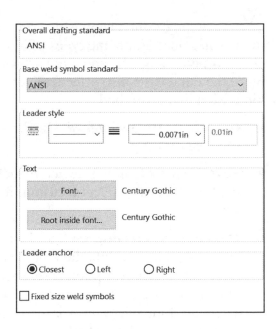

- **Overall drafting standard**. Displays the selected drafting standard.

- **Base weld symbol standard**. Provides the ability to select a default standard for an active Part, Assembly or Drawing document from the drop-down menu or to create a custom Base weld symbol standard.

- **Leader style**. Provides the ability to use the default Leader style or to select a custom Leader style and thickness from the drop-down menu.

- **Text**. Century Gothic selected by default. Click the Font button to select a custom font style and size.

- **Leader anchor**. Anchor the leader to the specified location on the note. Closest selected by default. The options are **Closest**, **Left** and **Right**.

- **Fixed size weld symbols**. Clear to scale the size of the symbol to the symbol font size.

Dimensions

The Dimensions section has the following sub-sections: *Angle, Angular Running, Arc Length, Chamfer, Diameter, Hole Callout, Linear, Ordinate, and Radius.*

See the Drawing document section for additional information.

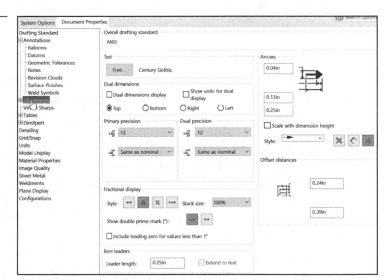

The Dimensions section provides the ability to set various options for dimensioning in an active document. The Dimensions/Relations toolbar and the Tools, Dimensions and Tools, Relations menus provide tools to dimension and to add and delete Geometric relations.

The type of dimension is determined by the items you select. For some types of dimensions, point-to-point, angular, and circular, where you place the dimension also affects the type of dimension that is applied.

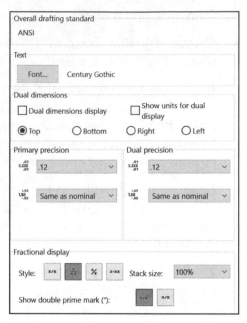

- *Overall drafting standard*. Displays the selected drafting standard.

- *Text*. Century Gothic selected by default. Click the Font button to select a custom font style and size.

- *Dual dimensions display*. The Top display option selected by default. When the Dual dimensions display option is selected, dimensions are displayed in two unit types. There are four selections for the Dual dimensions display option:

  - **Top**. Selected by default. Displays the second dimension on the top of the first dimension.

  - **Bottom**. Displays the second dimension on the bottom of the first dimension.

  - **Right**. Displays the second dimension on the right side of the first dimension.

  - **Left**. Displays the second dimension on the left side of the first dimension.

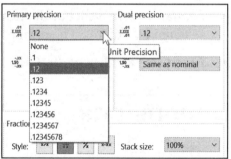

- *Primary precision*. Two selections are provided: **Primary Unit Precision** and **Tolerance Precision**. Select the number of digits after the decimal point from the drop-down menus for the Primary Unit Precision and select the number of digits after the decimal point for Tolerance Precision.

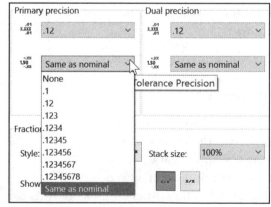

- *Dual precision*. Two selections are provided:

  - **Primary Unit Precision**. Select the number of digits after the decimal point from the drop-down menu.

  - **Tolerance Precision**. Select the number of digits after the decimal point from the drop-down menu.

- *Fractional display*. Fractional dimensions are available only for lengths in IPS (inch-pound-second) units. View the illustration for Style and Stack size.

- *Bent leaders*. Enter the length of the unbent portion of the leader. The default leader length is 0.25in for the IPS system and 6.35mm for the MMGS system.

- *Leading zeroes*. Standard selected by default. Select one of three options from the drop-down menu:

  - **Standard**. Zeroes are displayed based on the selected dimensioning standard.

  - **Show**. Zeroes before the decimal points are displayed on the selected dimensioning standard.

  - **Remove**. Zeroes are not displayed.

- *Trailing zeroes*. Provides the following selections:

  - **Dimension**: **Smart**. Selected by default. Trailing zeroes are trimmed for whole metric values. Conforms to ANSI and ISO standards. **Show**. Zeroes before the decimal points are displayed on the selected dimensioning standard. **Remove**. Zeroes are not displayed. **Standard**. Trims trailing zeroes to the ASME Y14.5M-1994 standard.

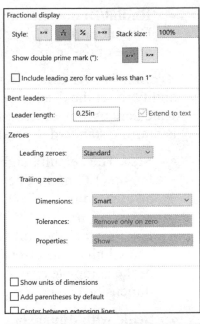

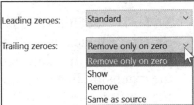

  - **Tolerances**: **Remove only on zero**. Selected by default. Options: Show, Remove, Remove only on zero, Same as Dimension.

  - **Properties**: **Show**. **Selected by default**. **Options**: Smart, Show, Standard, Remove.

  - **Show units of dimensions**. Select to show dimension units in drawings.

  - **Add parentheses by default**. Provides the ability to display reference dimensions in a drawing with parentheses.

  - **Center between extension lines**. Provides the ability to center dimensions between the extension lines.

  - **Included prefix inside basic tolerance box**. Not selected by default.

  - **Display dual basic dimensions in one box**. Select to include dual dimensions in one basic tolerance box.

  - **Show dimensions as broken in broken views**. Selected by default.

- **Radial/Diameter leader snap angle**.
  Default setting 15deg. Specifies the radial
  angle of the snap.

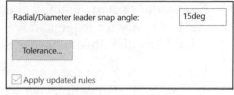

- *Arrows*. The Arrows selection provides the
  ability to set the display options for arrows
  which are used in drawings. The default option
  is drafting standard dependent.

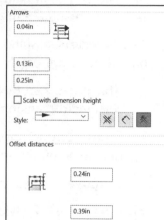

  - **Size**. The default height of the arrowhead is
    0.04in (1.02mm). The default width of the
    arrowhead is 0.13in (3.3mm). The default
    complete arrow is .25in (6.35mm). The size
    box provides the ability to specify the height
    and width of the arrowhead, and the length
    of the complete arrow, for leaders on notes,
    dimensions and other drawing annotations.

  - **Scale with dimension height**.

- *Style*. The default style is solid filled with
  Smart. Provides the ability to specify the
  placement of the dimension arrow in relation to
  the extension lines. There are three selections
  for the Style option:

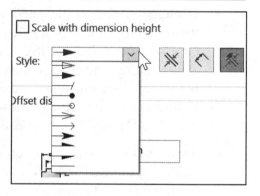

  - **Outside**. Specifies the outside placement of
    the dimension arrow in relation to the
    extension lines.

  - **Inside**. Specifies the inside placement of
    the dimension arrow in relation to the
    extension lines.

  - **Smart**. Selected by default. Specifies the
    placement of the dimension arrow in relation to the
    extension lines.

Smart arrows are displayed outside of extension lines
if the space is too small to accommodate the dimension
and the arrowheads.

- *Offset distances*. The default Offset distances are
  0.24in (6mm) and 0.39in (10mm), respectively,
  using the ANSI standard. Provides the ability to
  specify values for the distances between baseline
  dimensions. Baseline dimensions are reference
  dimensions used in a drawing. You cannot change
  their values or use the values to drive the model.

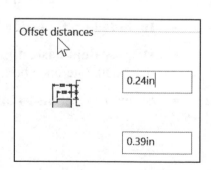

Dimensions - Angle

The Dimensions - Angle section provides the ability to place an angular dimension between two lines or a line and a model edge. Select the two entities and move the pointer to observe the dimension preview.

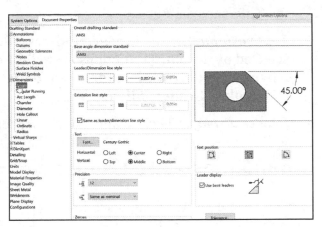

The angle to be dimensioned changes based on the pointer position.

The Dimensions - Angle section provides the following options:

- **Overall drafting standard**. Displays the selected drafting standard.

- **Base angle dimension standard**. Provides the ability to select a default standard for an active Part, Assembly or Drawing document from the drop-down menu or to create a custom Base angle dimension standard.

- **Leader/Dimension line style**. Provides the ability to use the default Leader style or to select a custom Leader style and thickness from the drop-down menu.

- **Extension line style**. Select a style. Select a thickness or select Custom Size and enter a thickness.

- **Text**. Century Gothic selected by default. Click the Font button to select a custom font style and size. The Text options provide the following locations relative to the angle.

  - **Horizontal: Left, Center, Right**.

  - **Vertical: Top, Middle, Bottom**.

- **Precision**. Two selections are provided: **Unit Precision** and **Tolerance Precision**. Select the number of digits after the decimal point from the drop-down menu for the Primary Unit Precision and select the number of digits after the decimal point for Tolerance Precision.

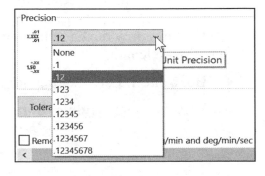

- *Tolerance*. Select the Tolerance button to access the Tolerance dialog box.

- *Text position*. Broken Leader, Horizontal Text selected by default. The options are **Solid Leader, Aligned Text**, **Broken Leader, Horizontal Text**, and **Broken Leader, Aligned Text**.

- *Leader display*. Check to apply a bent leader line.

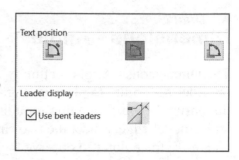

Dimensions - Arc Length

The Dimensions - Arc Length section provides the ability to dimension the true length of an arc. The Dimensions - Angle section provides the following options:

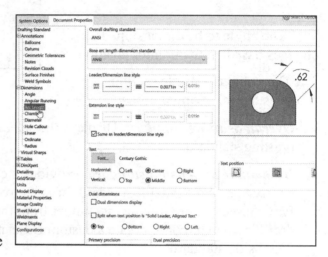

- *Overall drafting standard*. Displays the selected drafting standard.

- *Base arc length dimension standard*. Provides the ability to select a default standard for an active Part, Assembly or Drawing document from the drop-down menu or to create a custom Base arc length dimension standard.

- *Leader/Dimension line style*. Provides the ability to use the default Leader style or to select a custom Leader style and thickness from the drop-down menus.

- *Extension line style*. Select a style. Select a thickness or select Custom Size and enter a thickness.

- *Text*. Century Gothic selected by default. Click the Font button to select a custom font style and size. The Text options provide the following locations relative to the angle length:

  - **Horizontal: Left, Center, Right**.

  - **Vertical: Top, Middle, Bottom**.

- ***Dual dimensions display***. The Top display option selected by default. When the Dual dimensions display option is selected, dimensions are displayed in two unit types. There are four selections for the Dual dimensions display option:

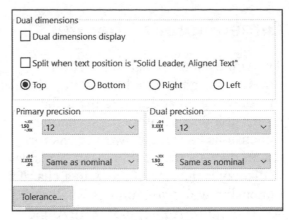

  - **Top**. Selected by default. Displays the second dimension on the top of the first dimension.

  - **Bottom**. Displays the second dimension on the bottom of the first dimension.

  - **Right**. Displays the second dimension on the right side of the first dimension.

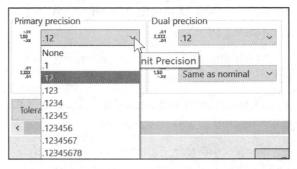

  - **Left**. Displays the second dimension on the left side of the first dimension.

- ***Primary precision***. Selections are **Primary Unit Precision** and **Tolerance Precision**. Select the number of digits after the decimal point from the drop-down menus for the Primary Unit Precision and select the number of digits after the decimal point for Tolerance Precision.

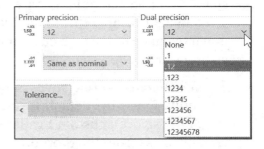

- ***Dual precision***. Selections are:

  - **Primary Unit Precision**. Select the number of digits after the decimal point from the drop-down menu.

  - **Tolerance Precision**. Select the number of digits after the decimal point from the drop-down menu.

- ***Tolerance***. Select the Tolerance button to access the Tolerance dialog box.

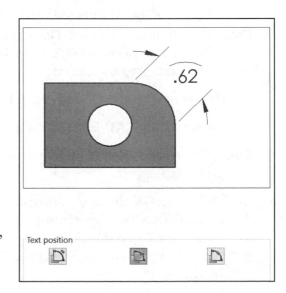

  - ***Text position***. Broken Leader, Horizontal Text selected by default. The options are **Solid Leader**, **Aligned Text**, **Broken Leader**, **Horizontal Text**, **Broken Leader**, **Aligned Text**.

Dimensions - Chamfer

The Dimensions - Chamfer section provides the ability to dimension a chamfer in a drawing. In addition to the usual dimension display properties, chamfer dimensions have their own options for leader display, text display, and X display. X display is the size of the X in a chamfer dimension with two numbers, such as 1 X 45° (Length X Angle), 45° X 1 (Angle X Length), 1 X 1 (Length X Length) or C1 (chamfers of 45°).

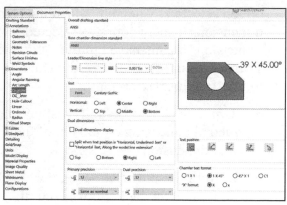

The Dimensions - Chamfer section provides the following options:

- **Overall drafting standard**. Displays the selected drafting standard.

- **Base chamfer dimension standard**. Provides the ability to select a default standard for an active Part, Assembly, or Drawing document from the drop-down menu or to create a custom Base chamfer dimension standard.

- **Leader/Dimension line style**. Provides the ability to use the default Leader style or to select a custom Leader style and thickness from the drop-down menus.

- **Text**. Century Gothic selected by default. Click the Font button to select a custom font style and size. The Text options provide the following locations relative to the chamfer:

  - **Horizontal**: **Left**, **Center**, **Right**.

  - **Vertical**: **Top**, **Middle**, **Bottom**.

- **Dual dimensions**. Select to display dimensions in dual units.

- **Split when text position is "Horizontal, Underlined Text" or "Horizontal Text, Along the model line extension."** The provided options are **Top**, **Bottom**, **Right** and **Left**. Top is selected by default.

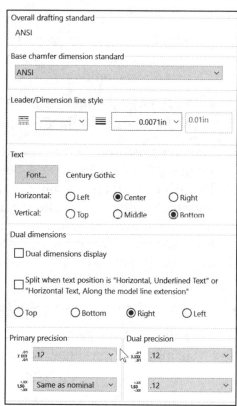

- *Primary precision*. Selections are **Primary Unit Precision** and **Tolerance Precision**. Select the number of digits after the decimal point from the drop-down menu for the Primary Unit Precision and select the number of digits after the decimal point for Tolerance Precision.

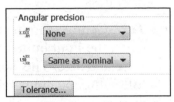

- *Angular precision*. Selections are **Primary Unit Precision** and **Tolerance Precision**. Select the number of digits after the decimal point from the drop-down menus for the Primary Unit Precision and select the number of digits after the decimal point for Tolerance Precision.

- *Tolerance*. Select the Tolerance button to access the Tolerance dialog box.

- *Text position*. Provides the following text position for the chamfer dimension. Horizontal Text, Underlined Text selected by default. Selections are **Horizontal Text**, **Horizontal**, **Underlined Text**, **Angled Text**, **Angled**, **Underlined Text**.

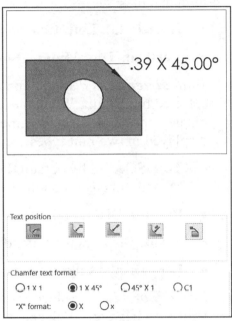

- *Chamfer text format*. Provides the ability to dimension the chamfer with a leader display, text display, and X display. X display is the size of the X in a chamfer dimension with two numbers, such as 1 X 45° (Length X Angle), 45° X 1 (Angle X Length), 1 X 1 (Length X Length) or C1 (chamfers of 45°).

Tolerance types for chamfer dimensions are limited to **None**, **Bilateral** and **Symmetric**.

Dimensions - Diameter

The Dimensions - Diameter section provides the following options:

- *Overall drafting standard*. Displays the selected drafting standard.

- *Base diameter dimension standard*. Provides the ability to select a default standard for an active Part, Assembly or Drawing document from the drop-down menu or to create a custom Base diameter dimension standard.

- *Leader/Dimension line style*. Provides the ability to use the default Leader style or to select a custom Leader style and thickness from the drop-down menus.

- *Extension line style*. Select a style. Select a thickness or select Custom Size and enter a thickness.

- *Text*. Century Gothic selected by default. Click the Font button to select a custom font style and size. The Text options provide the following locations relative to the diameter dimension:

  - **Horizontal**: **Left**, **Center**, **Right**.

  - **Vertical**: **Top**, **Middle**, **Bottom**.

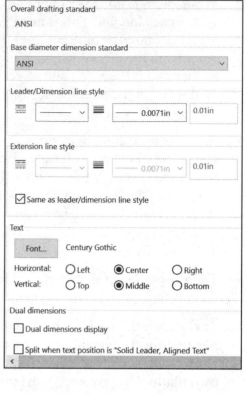

- *Dual dimensions display*. Top display option selected by default. When the Dual dimensions display option is selected, dimensions are displayed in two unit types. Selections are:

  - **Top**. Selected by default. Displays the second dimension on the top of the first dimension.

  - **Bottom**. Displays the second dimension on the bottom of the first dimension.

  - **Right**. Displays the second dimension on the right side of the first dimension.

  - **Left**. Displays the second dimension on the left side of the first dimension.

- *Primary precision*. Selections are **Primary Unit Precision** and **Tolerance Precision**. Select the number of digits after the decimal point from the drop-down menu for the Primary Unit Precision and select the number of digits after the decimal point for Tolerance Precision.

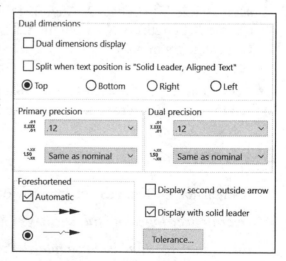

- *Dual precision*. Selections are:

  - **Primary Unit Precision**. Select the number of digits after the decimal point from the drop-down menus.

  - **Tolerance Precision**. Select the number of digits after the decimal point from the drop-down menu.

- *Foreshortened diameter*. The default option is Zigzag selected by the ANSI standard. The Foreshortened diameter option provides two selections. The selections are **Double arrow** and **Zigzag arrow**.

Create foreshortened diameters in drawing documents. When SOLIDWORKS detects that the diameter is too large for the drawing view, the dimension is automatically foreshortened.

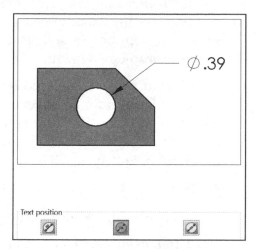

- *Tolerance*. Select the Tolerance button to access the Tolerance dialog box.

- *Text position*. Provides the following text position for the diameter dimension. Broken Leader, Horizontal Text selected by default. Selections are **Solid Leader, Aligned Text, Broken Leader, Horizontal Text, Broken Leader** and **Aligned Text**.

Dimensions - Hole Callout

If you change a hole dimension in the model, the callout updates automatically.

The callout contains a diameter symbol and the dimension of the hole diameter. If the depth of the hole is known, the callout also contains a depth symbol and the dimension of the depth. If the hole is created

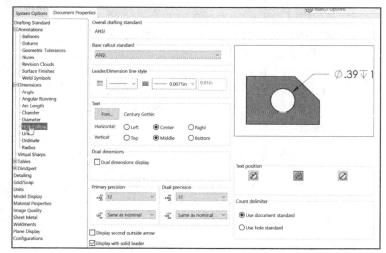

in the **Hole Wizard**, the callout contains additional information (the dimensions of a countersink or number of hole instances, for example).

The Dimensions - Hole Callout section provides the following options:

- *Overall drafting standard*. Displays the selected drafting standard.

- ***Base callout standard***. Provides the ability to select a default standard for an active Part, Assembly, or Drawing document from the drop-down menu or to create a custom Base callout standard.

- ***Leader/Dimension line style***. Provides the ability to use the default Leader style or to select a custom Leader style and thickness from the drop-down menu.

- ***Text***. Century Gothic selected by default. Click the Font button to select a custom font style and size. The Text options provide the following locations relative to the Hole Callout:

  - **Horizontal: Left, Center, Right**.

  - **Vertical: Top, Middle, Bottom**.

- ***Dual dimensions display***. Check to display dual dimensions.

- ***Primary precision***. Selections are **Primary Unit Precision** and **Tolerance Precision**. Select the number of digits after the decimal point from the drop-down menus for the Primary Unit Precision and select the number of digits after the decimal point for Tolerance Precision.

- ***Dual precision***. Selections are:

  - **Primary Unit Precision**. Select the number of digits after the decimal point from the drop-down menu.

  - **Tolerance Precision**. Select the number of digits after the decimal point from the drop-down menu.

- ***Display second outside arrow***. Select to display two outside arrows.

- ***Display with solid leader***. For standards other than ANSI, displays a leader through the center, rather than to the circumference.

- ***Tolerance***. Select the Tolerance button to access the Tolerance dialog box.

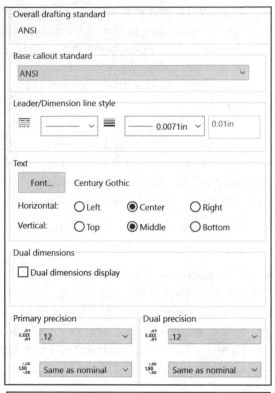

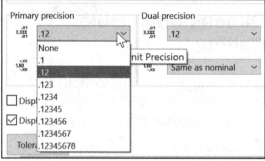

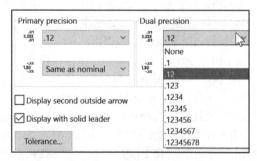

- **Text position**. Provides the following text position for the diameter dimension. Broken Leader, Horizontal Text selected by default. Selections are **Solid Leader**, **Aligned Text**, **Broken Leader**, **Horizontal Text**, **Broken Leader** and **Aligned Text**.

- **Use document standard**. Selected by default.

- **Use hole standard**.

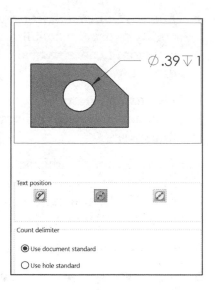

💡 If you attach a hole callout to a tapped hole in ANSI inch standard and the current drawing units are millimeters, the drill diameter and hole depths are reported in mm but the thread description retains the ANSI inch size designation. You might want to replace the thread description with other variables from the **Callout Variables** dialog box.

Dimensions - Linear

The Dimensions - Linear section provides the ability to measure the distance between two points. Since two points define a line, the units of distance are sometimes called "linear" units or dimensions.

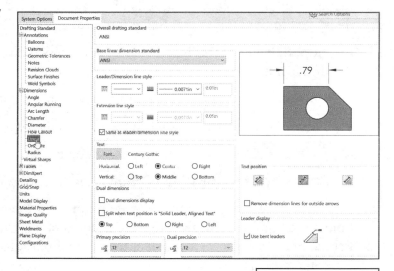

The Dimensions - Linear section provides the following options:

- **Overall drafting standard**. Displays the selected drafting standard.

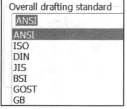

- **Base linear dimension standard**. Provides the ability to select a default standard for an active Part, Assembly or Drawing document from the drop-down menu or to create a custom Base linear dimension standard.

- *Leader/Dimension line style*. Provides the ability to use the default Leader style or to select a custom Leader style and thickness from the drop-down menu.

- *Extension line style*. Select a style. Select a thickness or select Custom Size and enter a thickness.

- *Text*. Century Gothic selected by default. Click the Font button to select a custom font style and size. The Text options provide the following locations relative to the Linear dimension:

  - **Horizontal**: **Left**, **Center**, **Right**.

  - **Vertical**: **Top**, **Middle**, **Bottom**.

- *Dual dimensions display*. Top display option is selected by default. When the Dual dimensions display option is selected, dimensions are displayed in two unit types. Selections are:

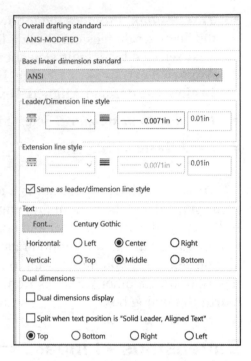

- **Top**. Selected by default. Displays the second dimension on the top of the first dimension.

- **Bottom**. Displays the second dimension on the bottom of the first dimension.

- **Right**. Displays the second dimension on the right side of the first dimension.

- **Left**. Displays the second dimension on the left side of the first dimension.

- **Split when text position is "Solid Leader, Aligned Test."**

- *Primary precision*. Selections are **Primary Unit Precision** and **Tolerance Precision**. Select the number of digits after the decimal point from the drop-down menus for the Primary Unit Precision and select the number of digits after the decimal point for Tolerance Precision.

- *Dual precision*. Selections are:

  - **Primary Unit Precision**. Select the number of digits after the decimal point from the drop-down menu.

  - **Tolerance Precision**. Select the number of digits after the decimal point from the drop-down menu.

- *Foreshortened*. Automatic is selected by default.

- *Tolerance*. Select the Tolerance button to access the Tolerance dialog box.

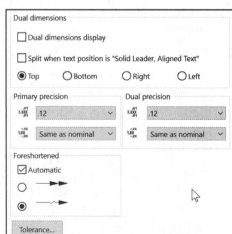

- ***Text position***. Provides the following text position for the Linear dimension placement. Broken Leader, Horizontal Text selected by default. Selections are **Solid Leader, Aligned Text, Broken Leader, Horizontal Text, Broken Leader** and **Aligned Text**.

- ***Remove dimension lines for outside arrows***. See illustration for results.

- ***Leader display***. Use bent leaders selected by default.

Dimensions - Ordinate

Ordinate dimensions are a set of dimensions measured from a zero ordinate in a drawing or sketch. In drawings, they are reference dimensions and you cannot change their values or use the values to drive the model.

Ordinate dimensions are measured from the axis you select first. The type of ordinate dimension (horizontal or vertical) is defined by the orientation of the points you select.

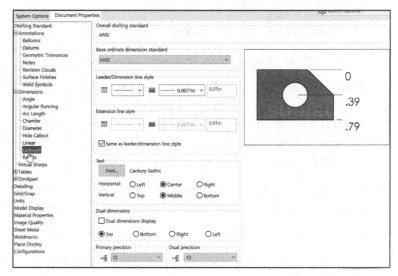

💡 Ordinate dimensions are automatically grouped to maintain alignment. When you drag any member of the group, all the members move together. To disconnect a dimension from the alignment group, right-click the dimension, and select Break Alignment.

The Dimensions - Ordinate section provides the following options:

- ***Overall drafting standard***. Displays the selected drafting standard.

- ***Base ordinate dimension standard***. Provides the ability to select a default standard for an active Part, Assembly, or Drawing document from the drop-down menu or to create a custom Base ordinate dimension standard.

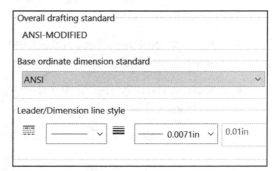

- *Leader/Dimension line style*. Provides the ability to use the default Leader style or to select a custom Leader style and thickness from the drop-down menus.

- *Extension line style*. Select a style. Select a thickness or select Custom Size and enter a thickness.

- *Text*. Century Gothic selected by default. Click the Font button to select a custom font style and size. The Text options provide the following locations relative to the Ordinate dimension:

  - **Horizontal**: **Left, Center, Right**.

  - **Vertical**: **Top, Middle, Bottom**.

- *Dual dimensions display*. Top display option selected by default. When the Dual dimensions display option is selected, dimensions are displayed in two unit types. Selections are:

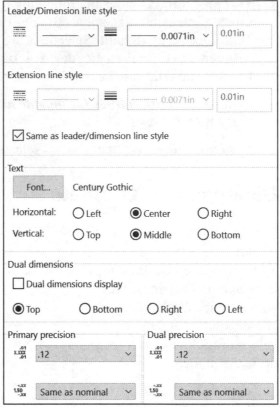

  - **Top**. Selected by default. Displays the second dimension on the top of the first dimension.

  - **Bottom**. Displays the second dimension on the bottom of the first dimension.

  - **Right**. Displays the second dimension on the right side of the first dimension.

  - **Left**. Displays the second dimension on the left side of the first dimension.

- *Primary precision*. Selections are **Primary Unit Precision** and **Tolerance Precision**. Select the number of digits after the decimal point from the drop-down menu for the Primary Unit Precision and select the number of digits after the decimal point for Tolerance Precision.

- *Dual precision*. Selections are:

  - **Primary Unit Precision**. Select the number of digits after the decimal point from the drop-down menu.

  - **Tolerance Precision**. Select the number of digits after the decimal point from the drop-down menu.

  - **Automatically jog ordinates**. Automatically bends the leader line of the ordinates.

- *Tolerance*. Select the Tolerance button to access the Tolerance dialog box.

⛯ You can drag the zero dimension to a new position, and all the ordinate dimensions update to match the new zero position.

Dimensions - Radius

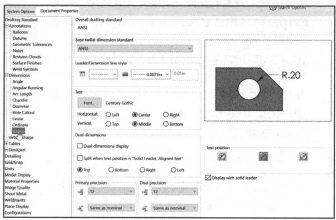

The Dimensions - Radius section provides the following options:

- **Overall drafting standard.** Displays the selected drafting standard.

- **Base radial dimension standard.** Provides the ability to select a default standard for an active Part, Assembly, or Drawing document from the drop-down menu or to create a custom Base radial dimension standard.

- **Leader/Dimension line style.** Provides the ability to use the default Leader style or to select a custom Leader style and thickness from the drop-down menus.

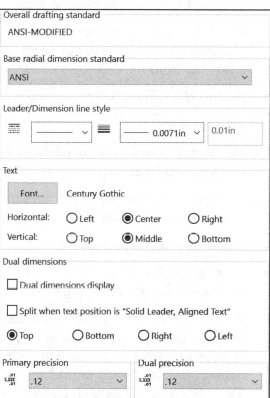

- **Text.** Century Gothic selected by default. Click the Font button to select a custom font style and size. The Text options provide the following locations relative to the radius dimension:

  - **Horizontal**: **Left**, **Center**, **Right**.

  - **Vertical**: **Top**, **Middle**, **Bottom**.

- **Dual dimensions display.** Top display option selected by default. When the Dual dimensions display option is selected, dimensions are displayed in two unit types. The four selections for the Dual dimensions display are:

  - **Top**. Selected by default. Displays the second dimension on the top of the first dimension.

- **Bottom**. Displays the second dimension on the bottom of the first dimension.

- **Right**. Displays the second dimension on the right side of the first dimension.

- **Left**. Displays the second dimension on the left side of the first dimension.

- *Primary precision*. Selections are **Primary Unit Precision** and **Tolerance Precision**. Select the number of digits after the decimal point from the drop-down menus for the Primary Unit Precision and select the number of digits after the decimal point for Tolerance Precision.

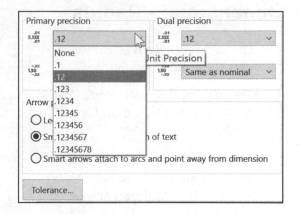

- *Dual precision*. Selections are:

  - **Primary Unit Precision**. Select the number of digits after the decimal point from the drop-down menu.

  - **Tolerance Precision**. Select the number of digits after the decimal point from the drop-down menu.

- *Arrow placement*. Selections are:

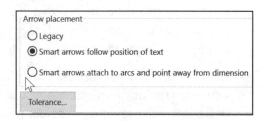

  - **Legacy**.

  - **Smart arrows follow position of text**. Selected by default.

  - **Smart arrows attach to arcs and point away from dimension**.

- *Display with solid leader*. Selected by default.

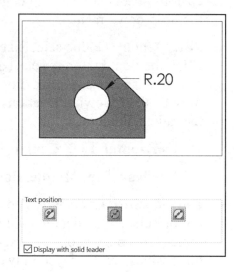

- *Tolerance*. Select the Tolerance button to access the Tolerance dialog box.

- *Text position*. Provides the following text position for the Linear dimension placement. Broken Leader, Horizontal Text selected by default. Selections are **Solid Leader**, **Aligned Text**, **Broken Leader**, **Horizontal Text**, **Broken Leader** and **Aligned Text**.

Virtual Sharps

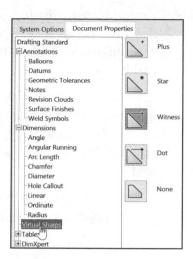

- *Virtual Sharps*. Witness style selected by default. There are five style selections: **Plus**, **Star**, **Witness**, **Dot** and **None**. This option provides the ability to set the display options for the virtual sharps. A virtual sharp creates a sketch point at the virtual intersection point of two sketch entities. Dimensions and relations to the virtual intersection point are retained, even if the actual intersection no longer exists. Example: When a corner is removed by using a Fillet or a Chamfer feature.

Tables

Bill of Materials

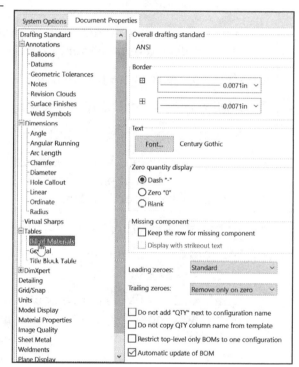

The Tables section for a Part or Assembly document provides the ability to select text font and options for a Bill of Materials, General or a Title Block Table. Note: The second half of this chapter addresses specific Document Properties for a Drawing document in SOLIDWORKS.

The Bill of Materials section for a Part or Assembly document provides the following options:

- *Overall drafting standard*. Displays the selected drafting standard.

- *Border*. Click the **Box Border** option or the **Grid Border** option and select a corresponding thickness. The default thickness is 0.007in for the IPS system and .18mm for the MMGS system.

- *Text*. Century Gothic selected by default. Click the Font button to select a custom font style and size.

- *Zero quantity display*. Display with dash '-' is selected by default. Provides the ability to select whether to display zero quantities with a dash (-) or a zero (0), or to leave the cell blank.

- *Missing component*. A Missing component created by having deleted or suppressed parts or sub-assemblies in the top-level assembly. There are two options: **Keep the row for missing component** and **Display with strikeout text**.

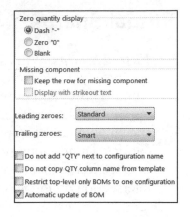

- *Leading zeroes*. Standard selected by default. Select one of three options from the drop-down menu:

  - **Standard**. Zeroes are displayed based on the selected dimensioning standard.

  - **Show**. Zeroes before the decimal points are displayed on the selected dimensioning standard.

  - **Remove**. Zeroes are not displayed.

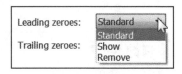

- *Trailing zeroes*. Provides the following selections:

  - **Remove only on zero**. Selected by default.

  - **Show**. Dimensions have trailing zeroes up to the number of decimal places specified in **Options**, **Document Properties**, **Units**.

  - **Remove**. All trailing zeroes are removed.

  - **Standard**. Trims trailing zeroes to the ASME Y14.5M-1994 standard.

  - **Don't add "QTY" next to configuration name**. Eliminates the word, **QTY**, that is displayed in the configuration column. Note: You must select this option prior to inserting a BOM. If you select it after a BOM exists, the option has no effect.

  - **Don't copy QTY column name from template**. Select to use the configuration name appended by the string /QTY for the quantity column header.

  - **Restrict top level only BOMs to one configuration**. Select to limit Top-level only BOMs to one configuration. When you change the configuration in the BOM, the quantity column label does not change.

- **Automatic update of BOM**. Select to update the BOM when you add or delete components in the associated assembly.

General Table

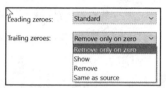

The General Table section provides the ability to specify document-level drafting settings for tables. Available for drawings only. Options:

- *Overall drafting standard*. Displays the selected drafting standard.

- *Border*. Click the **Box Border** option or the **Grid Border** option and select a corresponding thickness. The default thickness is 0.007in for the IPS system and .18mm for the MMGS system.

- *Text*. Century Gothic selected by default. Click the Font button to select a custom font style and size.

- *Leading zeroes*. Options: **Standard, Show, Remove**.

- *Trailing zeroes*. Options: **Remover only on zero, Show, Remove, Same as source**.

Title Block Table

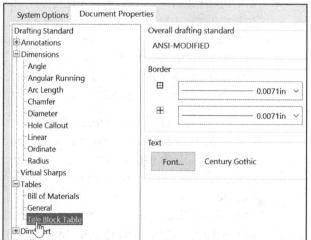

The Title Block Table section provides the ability to specify document-level drafting settings for title block tables. Available for all document types. Options:

- *Overall drafting standard*. Displays the selected drafting standard.

- *Border*. Click the **Box Border** option or the **Grid Border** option and select a corresponding thickness. The default thickness is 0.007in for the IPS system and .18mm for the MMGS system.

- *Text*. Century Gothic selected by default. Click the Font button to select a custom font style and size.

DimXpert

The DimXpert section provides the ability to define whether DimXpert uses Block Tolerances, General Tolerances or General Block Tolerances on dimensions that do not contain tolerances. Available for parts only.

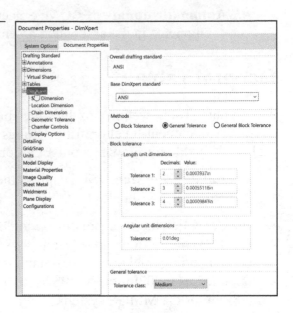

- *Overall drafting standard*. Displays the selected drafting standard. Provides the ability to select various drafting standards.

- *Base DimXpert Standard*. Displays the selected drafting standard. Provides the ability to select various drafting standards.

- *Methods*. The Methods section provides three options:

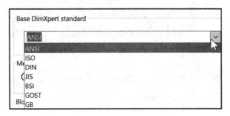

  - **Block Tolerance**. A common form of tolerancing used with inch units. The tolerance is based on the given precision of each dimension, so you must specify trailing zeroes. In Tools > Options > Document Properties > Dimensions, set Trailing zeroes to Standard.

  - **General Tolerance**. A common form of tolerancing used with metric units in conjunction with the ISO drawing standard. General Tolerancing is based on ISO 2768-1 Tolerances for linear and angular dimensions without individual tolerance indications.

  - **General Block Tolerance**. A form of tolerancing used with metric units, but may be used with inch units as well. The tolerance is typically shown in the title block or a note and applies to all untoleranced dimensions. The values for the general block tolerances are supported by custom properties.

- *Block tolerance*. Provides two options:

  - **Length unit dimensions**. Provides the ability to set up three block tolerances, each having a number of decimal places and tolerance value. DimXpert applies the Value as a symmetric plus and minus tolerance.

  - **Angular unit dimensions**. Default is 0.01deg. Provides the ability to set the tolerance value to use for all angular dimensions, including those applied cones and countersinks, and angle dimensions created between two features. DimXpert applies the Tolerance value as a symmetric plus and minus tolerance.

- **General tolerance**. Default is Medium. Provides the ability to set the part tolerance for the following: **Fine (f)**, **Medium (m)**, **Coarse (c)**, and **Very Coarse (v)**.

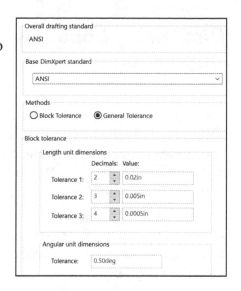

DimXpert - Size Dimension

The DimXpert - Size Dimension section defines the tolerance type and value to apply to newly created size dimensions, including dimensions created with the *Size Dimension* tool and the *AutoDimension Scheme* tool.

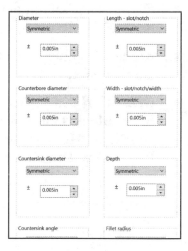

🔅 Click Location Dimension from the DimXpert toolbar or **Tools**, **DimXpert**, **Location Dimension**.

🔅 These options do not affect pre-existing features, dimensions, or tolerances.

The DimXpert - Size Dimension options are **Diameter** (Symmetric, Bilateral and General), **Counterbore diameter** (Symmetric, Bilateral and General), **Countersink diameter**, **Countersink angle**, **Length - slot/notch**, **Width - slot/notch/width**, **Depth** (Symmetric, Bilateral and General), and **Fillet radius** (Symmetric, Bilateral and General).

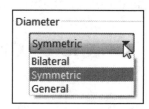

🔅 **Block** or **General** appears based on whether you select **Block Tolerance** or **General Tolerance** for the **DimXpert** method.

DimXpert - Location Dimension

The DimXpert - Location Dimension section provides the ability to apply the tolerance type and values to newly created linear and angular dimensions defined between two features.

The DimXpert - Location Dimension sections are **Distance** and **Angle**. Each section provides the following Tolerance Type:

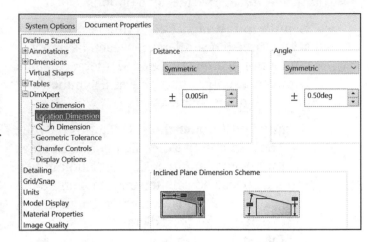

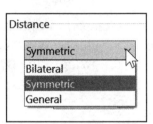

- **Symmetric**. Default option. Value is interpreted as plus and minus.

- **Bilateral**. Values are added or subtracted from the feature's nominal size.

- **General**. **Block** or **General** appears based on whether you select **Block Tolerance** or **General Tolerance** for the DimXpert **Method**. See **Tools**, **Options**, **Document Properties**, **DimXpert**.

- **Inclined Plane Dimension Scheme**. Applies angle dimensions to planes at an angle to a datum or reference plane.

Apply the Inclined Plane Dimension Scheme tool from the DimXpert toolbar to insert linear and angular dimensions between two DimXpert features, excluding surface, fillet, chamfer, and pocket features.

DimXpert - Chain Dimension

The DimXpert - Chain Dimension section provides the ability to define the type of dimension scheme to apply to pattern and pocket features and the tolerance type and values applied to chain dimension schemes.

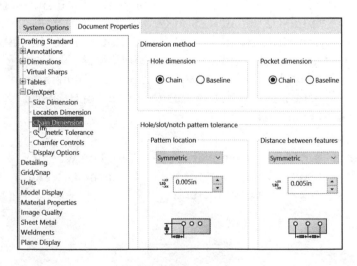

The options in this section only apply to dimensions that were created with the Auto Dimension Scheme tool when you set the Tolerance Type to Plus and Minus.

- *Dimension Method*. Provides the ability to define the dimension scheme used for pattern and pocket features.

  - **Hole dimension**. Chain selected by default. Provides the ability to define the type of dimension used for patterns of counterbores, countersinks, cylinders, holes, slots, and notches. Selections are **Chain** and **Baseline**.

  - **Pocket dimension**. Chain selected by default. Provides the ability to define the type of dimension used for patterns of counterbores, countersinks, cylinders, holes, slots, and notches. The available options are **Chain** and **Baseline**.

  - **Hole/slot/notch pattern tolerance**. Provides the ability to set the tolerance type and values used when creating chain dimension schemes. Selections are **Symmetric**, **Bilateral** and **Block**. Symmetric is selected by default.

  - **Pattern location**. Provides the ability to set the tolerance type and value used for the features locating the pattern from the origin features. Note: For two features, the feature nearest the origin is used to locate the pattern.

  - **Distance between features**. Provides the ability to set the tolerance type and value used for the dimensions applied between the features in the pattern.

DimXpert - Geometric Tolerance

The DimXpert - Geometric Tolerance section provides the ability to set the tolerance values and criteria for generating geometric tolerance schemes created by the Auto Dimension Scheme tool.

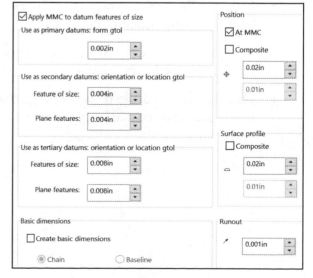

- *Apply MMC to datum features of size*. Selected by default. Defines whether an MMC symbol is placed in the datum fields when the datum feature is a feature of size.

  - **Use as primary datums: form gtol**. Sets the tolerance value for the tolerances that are applied to primary datum features. DimXpert uses this option when the primary datum feature is a plane, in which case a flatness tolerance is applied.

  - **Use as secondary datums: orientation or location gtol**. Sets the tolerance value for the orientation and location tolerances that are applied to secondary datum features.

  - **Use as tertiary datums: orientation or location gtol**. Sets the tolerance value for the orientation and location tolerances that are applied to tertiary datum features.

  - **Basic dimensions**. Provides the ability to enable or disable the creation of basic dimensions, and to select whether to use Chain or Baseline dimension schemes. This option only applies to position tolerances created by the **Auto Dimension Scheme**, **Geometric Tolerance** and **Recreate basic dim** commands.

    - **Chain**. Creates chain dimensions between parallel pattern features. When the features are not parallel, bascline dimensions are used.

    - **Baseline**. Creates baseline dimensions that can be applied to any pattern regardless of their orientation to one another. In the example, the features within thc pattern are not all parallel.

  - **Position**. Defines the tolerance values and criteria to use when creating position tolerances.

    - **At MMC**. Places an MMC (maximum material condition) symbol in the Tolerance 1 compartment of the feature control frame when applicable.

    - **Composite**. Creates composite position tolerances. Note: Clear Composite to create single scgmented position tolerances.

  - **Surface Profile**. Defines the tolerance values and criteria to use when creating surface profile tolerances.

    - **Composite**. Creates composite profile tolerances. Note: Clear Composite to create single segmented profile tolerances.

  - **Runout**. Defines the tolerance to use when creating runout tolerances. Runout tolerances are created only when the Part type is Turned and the Tolerance type is Geometric.

DimXpert - Chamfer Controls

The DimXpert - Chamfer Controls section provides the ability to influence how Chamfer features are recognized by DimXpert, and define the tolerance values used when size tolerances are created by the Auto Dimension Scheme tool or the Size Dimension tool.

- *Width settings*. Controls when faces can be considered as candidates for chamfer features.

    - **Chamfer width ratio**. Sets the chamfer width ratio, which is computed by dividing the width of a face adjacent to a candidate chamfer by the width of the candidate chamfer.

    - **Chamfer maximum width**. Sets the maximum chamfer width.

- *Tolerance settings*. Provides two options: **Distance** and **Angle**. Offers three tolerance types: *Symmetric*, *Bilateral* and *General*. Symmetric is selected by default.

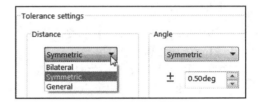

DimXpert - Display Options

The DimXpert - Display Options section provides the ability to define the dimensioning style used for slot dimensions, hole dimensions, Gtol linear dimension attachments and Datum gtol attachments and how duplicate dimensions and instance counts are managed.

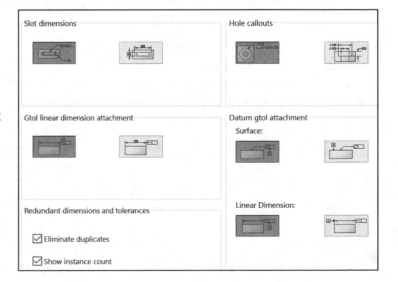

- *Slot dimensions*. Combined selected by default. These options define whether the length and width dimensions applied to slots are combined as a callout or are placed separately.

- *Hole callouts*. These options define whether hole callouts are displayed as combined or separate dimensions.

- *Gtol linear dimension attachments*. These options define whether the geometric tolerance feature control frames are combined with the size limits or placed separately.

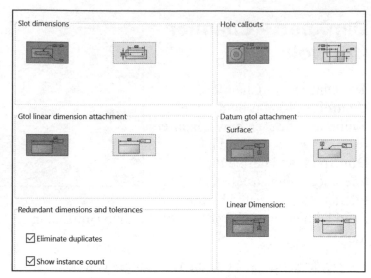

- *Datum gtol attachments*. These options define whether datums are attached to the surface of the feature, to the dimension, or to the feature control frame.

- *Redundant dimensions*. These options define how redundant dimensions and tolerances are displayed when you use the Auto Dimension Scheme tool. Note: You can also manually combine and break duplicates.

  - **Eliminate duplicates**. Selected by default. Specifies if dimensions are individually stated or combined into a group.

  - **Show instance count**. Selected by default. Defines whether instance counts are displayed with grouped dimensions.

The available options under the Document Properties tab is document dependent: part, assembly, drawing. In the next section, address the six additional options that are only available for a drawing document under the Document Properties tab: **DimXpert**, **Tables**, **View Labels**, **Line Font**, **Line Style** and **Sheet Metal**.

Detailing

Set options for detailing in the active document. You can also set the detailing options in Document Templates.

- ***Display filter***. The Display filter option provides the ability to specify the detailing display filters. The Display filter option provides twelve selections. To apply a filter, select **Display all types**, or clear **Display all types** and select one of the following filters:

  - **Cosmetic threads**. Selected by default.

  - **Datums**. Selected by default.

  - **Datum targets**. Selected by default.

  - **Feature dimensions**. Not selected by default.

  - **References dimensions**. Selected by default.

  - **DimXpert dimensions**. Not selected by default.

  - **Shaded cosmetic threads**. Not selected by default.

  - **Geometric tolerances**. Selected by default.

  - **Notes**. Selected by default.

  - **Surface finish**. Selected by default.

  - **Welds**. Selected by default.

  - **Display all types**. Not selected by default.

- ***Point, Axis and Coordinate System.*** Set font and display options for reference geometry names and labels for points, axes, and coordinate systems.

  - **Hide names**. Hides reference geometry names for points, axes, and coordinate systems.

  - **Name font**. Sets the font for the names of points, axes, and coordinate systems.

- **Label font**. Sets the font for the labels of coordinate system arrows.

- *Text scale*. Text scale 1:1. Selected by default. Use the text scale option only for part and assembly documents. Deselect the Always display text at the same size option to activate the Text scale and to specify a scale for the default size of the annotation text.

- *Always display text at the same size*. Selected by default. When selected, all annotations and dimensions are displayed at the same size regardless of zoom.

- *Display items only in the view in which they are created*. When selected, annotation is displayed only when the model is viewed in the same orientation as when the annotation was added. Rotating the part or selecting a different view orientation removes the annotation from the display.

- *Display annotations*. When selected, all annotation types which are selected in the Display filter are displayed. For assemblies, this includes the annotations that belong to the assembly, and the annotations that are displayed in the individual part documents.

- *Use assembly setting for all components*. When selected, the display of all annotations matches the setting for the assembly document. This is regardless of any setting in the individual part documents. Select this option along with the Display assembly annotations check box to display various combinations of annotations.

- *Hide dangling dimensions and annotations*. When selected, if you delete features in a part or an assembly, dangling dimensions and annotations are automatically hidden in the drawing. If a feature is suppressed, this option will automatically hide any dangling reference dimensions in the drawing.

- *Highlight associated elements on reference dimension selection*.

Grid/Snap

The Grid/Snap selection provides the ability to display a sketch grid in an active sketch or drawing and to set the options for the snap functionality and grid display. The options for grid spacing and minor grid lines per major lines apply to the rulers in drawings, drawing grid lines, and sketching.

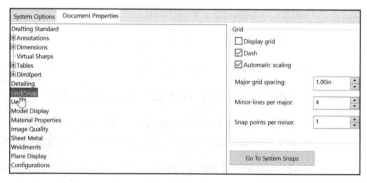

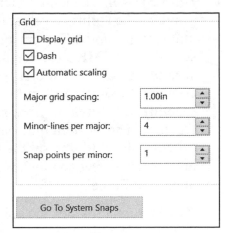

A drawing window displays rulers at the top and left side of the Graphics window. The rulers and the status bar display the position of your mouse pointer on the active sheet.

- *Grid*. The Grid option provides the following selections:

  - **Display grid**. Activates the sketch grid on or off.

  - **Dash**. Selected by default. Toggles between solid and dashed grid lines.

  - **Automatic scaling**. Selected by default. Automatically adjusts the display of the grid when you zoom in and out on a view.

  - **Major grid spacing**. Provides the user the ability to specify the space between major grid lines.

  - **Minor-lines per major**. The Minor-lines per major option default is 4. Provides the user the ability to specify the number of minor grid lines between major grid lines.

  - **Snap points per minor**. The Snap point per minor option default is 1. Provides the ability to set the number of snap points between the minor grid lines.

  - **Go To System Snaps**. This option locates you to the System Options, Relations/Snap section. See Chapter 2 for detailed information on System Options.

Units

The Units option provides the ability to specify units and precision for an active part, assembly, or drawing document.

If you use small units such as angstroms, nanometers, microns, mils, or microinches, it is helpful to create a custom template as the basis for your document.

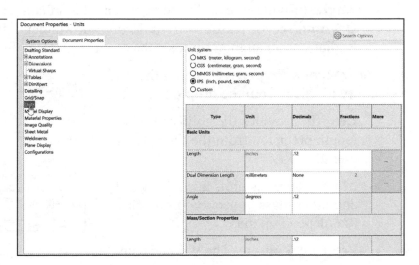

- *Units System*. The Unit system option provides the ability to select from four unit systems and a Custom system. The selections are **MKS (meter, kilogram, second), CGS (centimeter, gram, second), MMGS (millimeter, gram, second), IPS (inch, pound, second)** and **Custom**. The Custom unit system provides the ability to set the Length units, such as feet and inches, Density units, and Force.

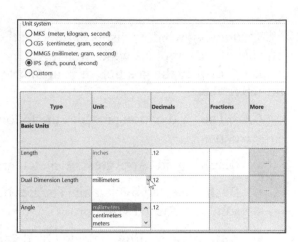

- *Basic Units*. Provides the ability to select from the drop-down list:

    - **Length**. Provides the ability to select length precision from the drop-down menu as illustrated.

    - **Dual Dimension Length**. Provides the ability to select a second type of unit just as you specify in the Length section

    - **Angle**. For degrees or radians only. Based on selection, displays from two to eight decimal places. None displays no decimal places.

 To display dual units in SOLIDWORKS, select **Dual dimensions display** in Detailing options under the Document Properties tab.

 Select up to eight (8) decimal places of precision.

    - **Angle**. Provides the ability to select angular units in **Degrees**, **Deg/min**, **Deg/min/sec**, or **Radians** from the drop-down menu. If you select Degrees or Radians as an angular unit, set the decimal place in the Decimal spin box.

- *Mass/Section Property*. Provides the ability to select from the drop-down list.

    - **Length**. Provides the ability to select length precision from the drop-down menu.

    - **Mass**. The mass units used for density and other mass or section properties that depend on mass.

    - **Per Unit Volume**. The volume units used for density.

- *Decimal rounding*. Your options are **Round half away from zero, Round half towards zero, Round half to even** and **Truncate without rounding**.

Model Display

The Model Display selection provides the ability to modify the color default options in an active part or assembly document. This selection provides the ability to modify the default color in a part or assembly document to be used as a template.

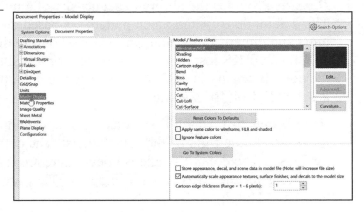

- *Model/feature colors*. Provides the ability to select a feature type or view mode. In an assembly document, the Shading, for Shaded view mode and the Hidden, Hidden Lines Visible view modes are available.

☼ If you edit the color for Shading, the Advanced button is active. Set values for Shininess, Transparency, etc.

- *Apply same color to wireframe, HLR and shaded*. Applies the same color.

- *Ignore feature colors*. Only available in a part document. Part colors take precedence over feature colors.

- *Go To System Colors*. Brings you directly to the System Options, Color box to set the colors for the system. See Chapter 2 for additional information.

- *Store Appearance, Decal and Scene data in model file*. Will increase file size.

- *Automatically scale appearance textures, surfaces finishes, and decals to the model size*. Scales surface finishes, decals, and textures to the model size. When cleared, uses a default scale that approximates the real-life appearance.

Material Properties

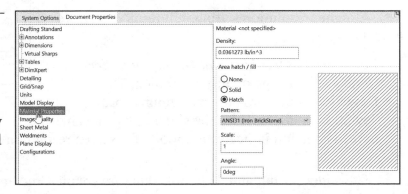

The Material Properties selection provides the ability to set crosshatch options and material density in an active part. These options are not available for drawing or assembly documents. The applied crosshatch pattern in the part document is displayed in the Section view of the part in the associative drawing.

☼ If you applied material to the part document, you must remove it before you set the Material Properties.

- *Density*. Provides the ability to set density of a material. The default density and hatch pattern are based on the units and dimensioning standard selected during the SOLIDWORKS installation. Example: For ANSI (IPS), the default density is 0.0361lb/in^3 and the default pattern is ANSI31 Iron Brickstone.

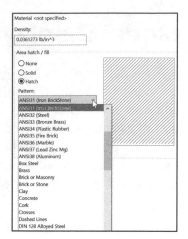

- *Area Hatch/Fill*. Hatch is selected by default and is illustrated to the right of the selection list. The default Scale is 1. The default Angle is 0deg. There are three selections for the Area Hatch/Fill option. They are **None**, **Solid** and **Hatch**.

- *Pattern*. Only available for the Hatch option. ISO (Steel) selected by default.

- *Scale*. Only available for the Hatch option. Default is 1.

- *Angle*. Only available for the Hatch option. Default is 0.

Image Quality

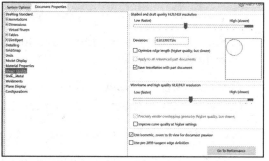

The Image Quality selection provides the ability to set the image display quality in the SOLIDWORKS Graphics window on your system.

- *Shaded and draft quality HLR/HLV resolution*. Provides the ability to control the tessellation of curved surfaces for shaded rendering output.

 A higher resolution setting will result in a slower model rebuild.

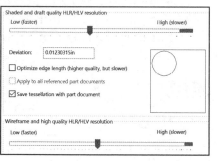

- *Low - to - High Slider and Deviation indicator*. Controls the image quality resolution. The Deviation indicator is the maximum chordal deviation in effect. Drag the slider or type a value in the Deviation indicator to modify the screen resolution.

 The slider setting and deviation value are coupled and are inversely proportional.

- *Optimize edge length (higher quality, but slower)*. Increases image quality if, after you move the slider to the highest setting, you still require a higher image quality.

- *Apply to all referenced part documents (assemblies only)*. Provides the ability to apply the settings to all of the part documents referenced by the active document.

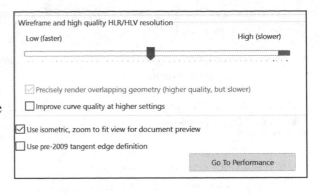

- ***Save tessellation with part document (parts only).*** Selected by default. Provides the ability to save the display information. When cleared, file size is reduced significantly. The model is not displayed when the file is opened in the view-only mode, in the SOLIDWORKS Viewer, or in SOLIDWORKS eDrawings. The display data is regenerated when the file is opened again in SOLIDWORKS.

- ***Precisely render overlapping geometry (higher quality, but slower).*** Applies to drawing documents only. When a drawing contains many tiny intersections, disable this option to enhance performance.

- ***Use isometric, zoom to fit view for document preview.*** Provides a standard view for preview images. When cleared, the preview displays the document using the last saved view.

- ***Use pre-2009 tangent edge definition.*** Leaves tangent edges visible as in the pre-2009 SOLIDWORKS implementation. When cleared, hides tangent edges when the angle between adjacent faces is less than one degree.

Sheet Metal

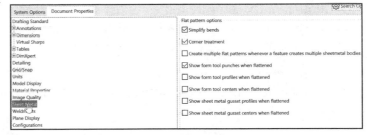

Sheet metal options vary depending on whether you are working with a part, assembly, or drawing.

The following options are available for a part:

- ***Simplify bends.*** Selected by default. Straightens curved edges in the flat pattern.

- ***Corner treatment.*** Selected by default. Applies smooth edges in the flat pattern.

- ***Create multiple flat patterns whenever a feature creates multiple sheet metal bodies.*** If you use a feature to create additional bodies in a sheet metal part, each new body gets a sheet metal and flat pattern feature.

- ***Show form tool punches when flattened.*** Selected by default. Displays the forming tool and its placement sketch in a flat pattern.

- ***Show form tool profiles when flattened.*** Displays the forming tool's placement sketch in a flat pattern.

- ***Show form tool centers when flattened.*** Displays the forming tool's center mark where the forming tool is located in a flat pattern.

- ***Show sheet metal gusset profiles when flattened***. Displays the sheet metal gusset profile when flattened.

- ***Show sheet metal gusset centers when flattened***. Displays the sheet metal gusset centers when flattened.

Weldments

On a per-document basis, use the Weldments Document Properties page to specify how the weldment software creates cut lists and configurations.

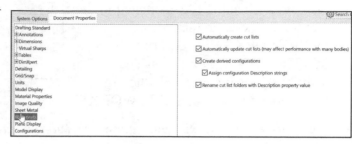

The following options are available for a part:

- ***Automatically create cut list***. Selected by default. Enables the Create Cut Lists Automatically setting on the cut list shortcut menu. This setting automatically groups similar bodies together within one cut list folder.

- ***Automatically update cut lists (may affect performance with many bodies)***. Selected by default. Enables the Update Automatically setting on the cut list shortcut menu. This setting updates the model's custom properties and internal supporting data when you make geometry changes or edits to weldments.

- ***Create derived configurations***. Selected by default. Creates the derived configuration Default[As Welded] when you create a structural member in a part.

- ***Assign configuration Description string***. Selected by default. Only available when Create derived configurations is enabled. Adds the As Welded and As Machined configuration descriptions when you insert a weldment feature into a new part.

- ***Rename cut list folders with Description property value***. Selected by default. Enables the option to rename cut list folders with the Description property value. The behavior of this option is as follows: 1.) When you create new parts in SOLIDWORKS 2016 or later using blank templates, the option is enabled. 2.) When you create new parts in SOLIDWORKS 2015 or later with existing/saved templates that were created using SOLIDWORKS 2015, the option is read from the templates you use. 3.) When you create new parts with existing/saved templates that were created in versions before SOLIDWORKS 2015, the option is disabled. 4.) For files created with a version of SOLIDWORKS that is older than 2015, the option is disabled. You must manually enable it for these files.

Plane Display

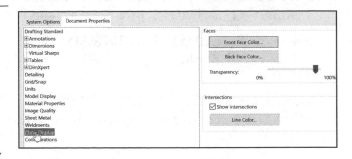

The Plane Display selection provides the ability to set the plane display options for a part or assembly document. The following options can be addressed: Face color, Transparency, and Intersection display and color.

- *Faces*. The Faces option provides the following selections:

  - **Front Face Color**. Sets the front face color of planes.

  - **Back Face Color**. Sets the back face color of planes using the Color dialog box.

- *Transparency*. The default setting is approximately 95%. Controls the planes' transparency. 0% displays a solid face color. 100% displays no face color.

- *Intersections*. Provides the following selections:

  - **Show intersections**. Selected by default. Displays the intersecting planes.

  - **Line Color**. Sets the plane intersection line color using the Color dialog box.

Tutorial: Assembly Template 3-2

Create a custom Assembly template. The custom Assembly template begins with the default assembly template. Templates require System Options, File Locations. System Options are addressed in Chapter 2. Note: The Assembly Template is located in the MY-TEMPLATES folder with many other templates.

Create a custom Assembly Template.

1. Click **New** ⬜ from the Menu bar.

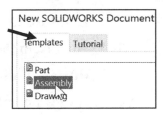

2. Double-click the **Assembly** icon from the default Templates tab. Click **Cancel** from the Open dialog box. The Begin Assembly PropertyManager is displayed. Click **Cancel** ✖. Set document properties for the assembly template.

☼ The Begin Assembly PropertyManager is displayed if the Start command when creating new assembly check box is active from the Options dialog box.

Set Document Properties.

3. Click **Options** ⚙ ➤ **Document Properties** tab.

4. Select **ANSI** from the Overall drafting standard drop-down menu.

5. Select **Units**. Select **MMGS (millimeter, gram, second)** for Unit system.

6. Select **.12** for Length Decimal places. Click **OK**.

7. Save the Assembly template. Select **Assembly Templates (*.ASMDOT)** for Save as type.

8. Select the **SOLIDWORKS 2018\MY-TEMPLATES** folder.

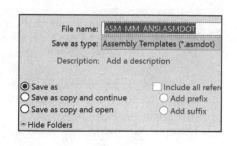

Create the SOLIDWORKS 2018\MY-TEMPLATES folder to save all custom templates for this book.

9. Enter **ASM-MM-ANSI** for File name.

10. Click **Save**. **Close** all files.

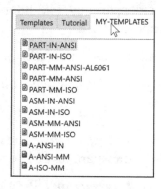

11. Click **New** 🗋 from the Menu bar. The MY-TEMPLATES tab is displayed. The MY-TEMPLATES folder was created in Chapter 2. The MY-TEMPLATES folder is only displayed when there is a document in the folder.

12. Click the **MY-TEMPLATE** tab. The ASM-MM-ANSI Assembly template is displayed. **Close** all models.

Tutorial: Part Template 3-3

Create a custom Part template. The custom Part template begins with the default part template.

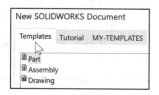

1. Click **New** 🗋 from the Menu bar.

2. Double-click **Part** from the default Templates tab.

Set document properties.

3. Click **Options** ⚙ ➤ **Document Properties** tab.

4. Select **ANSI** for Overall drafting standard.

5. Click **Units** from the left text box.

6. Click **MMGS (millimeter, gram, second)** for Unit system.

7. Select **.12** for Length Decimal places.

8. Click **OK**.

Apply material to the Part template.

9. Right-click **Material** from the FeatureManager design tree.

10. Click **Edit Material**.

11. Select **6061** from the Aluminum Alloys folder.

12. Click **Apply**.

13. Click **Close** from the Materials dialog box. 6061 Alloy is displayed in the Part FeatureManager design tree.

Save the Part template.

14. Click **Save**.

15. Select **Part Templates (*.PRTDOT)** from the Save As type list box.

16. Select the **SOLIDWORKS 2018\MY-TEMPLATES** folder from the Save in list box.

17. Enter **PART-MM-ANSI-AL6061** in the File name text box.

18. Click **Save**.

19. **Close** all files.

20. Click **New** 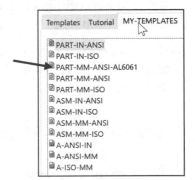 from the Menu bar. The MY-TEMPLATES tab is displayed. The MY-TEMPLATES folder was created in Chapter 2. The MY-TEMPLATES folder is only displayed when there is a document in the folder.

21. Click the **MY-TEMPLATE** tab. The PART-MM-ANSI-AL6061 Part template is displayed.

22. **Close** all models.

Configurations

Use the Configurations page to enable new configurations that are added to the current document to be marked for rebuild. The following option is available:

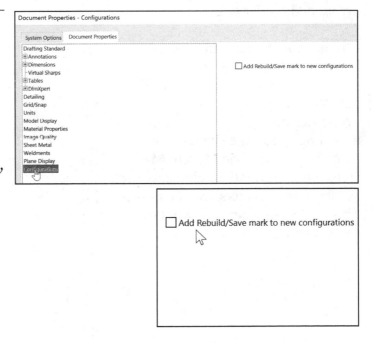

- *Add Rebuild/Save mark to new configurations*. *Faces*. When enabled, the software adds the Rebuild/Save mark to new configurations created in this document. This includes configurations with design tables. This option is disabled by default.

Drawing Document Properties Section

The available options under the Document Properties tab are document dependent as illustrated from left to right: *Part*, *Assembly* and *Drawing*.

In the first part of this chapter, Part Document Properties were addressed.

In this next section, Drawing document specific Document Properties are addressed.

Annotations - Borders

The Automatic Border tool lets you control every aspect of a sheet format's border, including zone layout and border size.

Using the Automatic Border tool, borders and zones automatically update to match changes in the Zone Parameters tab of the Sheet Properties dialog box without having to manually edit the sheet format. You can also include Margin Mask areas where formatting elements such as labels and dividers are not shown. This is helpful when you want to mask an area on a sheet for notes.

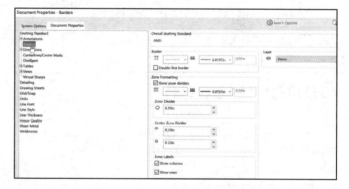

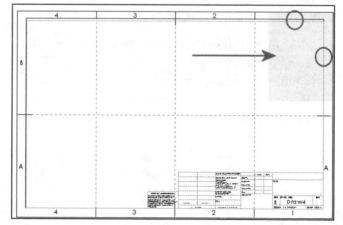

On the first page of the Automatic Border PropertyManager, select items to delete from the sheet's format. For example, you can delete existing format entities before creating a new smart border.

On the second page, define the margins, borders and zones.

On the third page, define Mask Out areas for zone labels and dividers that provide a convenient location for notes. In this example, the masked area masks the upper right zone labels.

💡 When you use the Automatic Border tool, the borders and zone dividers automatically align and update with existing zones.

The following options are available:

- *Overall drafting standard*. Uses settings from the Drafting Standard page.

- *Margins*. The following options are available:

  - **Line Style**. Sets the Style.

  - **Line Thickness**. Sets the thickness or select Custom size and enter a thickness.

  - **Double-Line Border**. Displays the border using two lines.

- *Zone Formatting*. The following options are available:

  - **Show Zone Dividers**. Displays zone dividers.

  - **Line Style**. Sets the style.

  - **Line Thickness**. Sets the thickness or select Custom size and enter a thickness.

  - **Zone Divider**. Sets the length for the zone divider.

  - **Center Zone Divider**. Provides the following options:

    o Sets the length for the outer center zone divider. In this example, the outer center zone divider is displayed in the blue box.

    o Sets the length for the inner center zone divider. In this example the inner center zone divider is displayed in the red box.

Displays zone dividers.

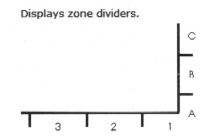

Sets the length for the zone divider.

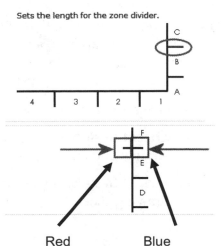

Red   Blue

Dimensions - Centerlines/Center Marks

The Dimensions - Centerlines/Center Marks section for a Drawing document provides the following options:

- **Overall drafting standard**. Displays the selected drafting standard.

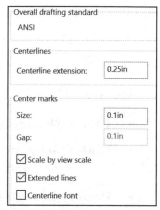

- **Centerlines extension**. The default is 0.25in for the IPS system and 6.35mm for the MMGS system. The default value is set according to the ANSI standards. The extension value controls the centerline's extension length beyond the section geometry in a drawing view.

- **Center marks**. Provides the ability to place center marks on circles or arcs in a drawing. The center mark lines can be used as references for dimensioning. Two options are provided:

  - **Extended lines**. Selected by default. Select to display center mark extension lines.

  - **Centerline font**. Select to make the center mark extension lines follow the font selected for centerlines.

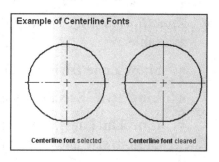

☼ Center marks are available as single marks, in linear patterns, or in circular patterns. Linear patterns can include connection lines. Circular patterns can include circular lines, radial lines, and base center marks. Display attributes include mark size, extended lines and specifying the centerline font for the center mark lines.

- **Slot center marks**. Selections are:

  - **Orient to slot**. Selected by default. Determines how straight and curved slots are created.

  - **Orient to sheet**. Determines the orientation to the drawing sheet.

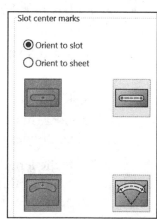

💡 Center marks are not automatically inserted into assembly drawings because of performance and screen clutter.

Dimensions - DimXpert

The Dimensions - DimXpert section provides the ability to insert dimensions in a drawing so that manufacturing features such as *slots*, *patterns*, *etc.* are fully-defined.

💡 The DimXpert tool is accessible using the Dimension PropertyManager or from the DimXpert tab in the CommandManager.

The Dimensions - DimXpert section provides the following options:

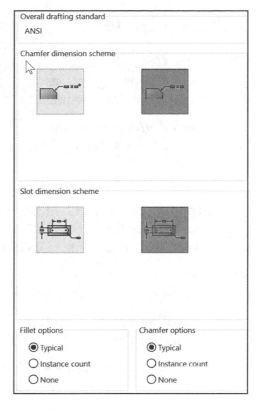

- *Overall drafting standard*. Displays the selected drafting standard.

- *Chamfer dimension scheme*. Distance X Distance is selected by default. Provides the ability to select between the Distance X Angle or the Distance X Distance dimension scheme for a chamfer.

- *Slot dimension scheme*. Overall Length is selected by default. Provides the ability to select between the Center to Center or the Overall Length dimension scheme for a slot.

- *Fillet options*. Selected by default. Provides the ability to select three fillet options:

  - **Typical**. Selected by default. Inserts single or multiple dimensions to fillets of the same size. The designation Typ is displayed after the dimension.

  - **Instance count**. Displays the number of instances of fillets of the same size.

  - **None**. Dimensions each fillet regardless if they are the same size.

- *Chamfer options*. Selected by default. Provides the ability to select three chamfer options:

  - **Typical**. Inserts single or multiple dimensions to chamfers of the same size. The designation Typ is displayed after the dimension.

- **Instance count**. Displays the number of instances of chamfers of the same size.

- **None**. Dimensions each chamfer regardless if they are the same size.

Tables - General

The Tables-General section provides the following options for an active Drawing document:

- *Overall drafting standard*. Displays the selected drafting standard.

- *Border*. Click the **Box Border** option or the **Grid Border** option and select a corresponding thickness. The default thickness is 0.007in for IPS and 0.18mm for MMGS unit system.

- *Text*. Century Gothic selected by default. Click the Font button to select a custom font style and size.

  - **Leading zeroes**. Standard selected by default. Selections are:

  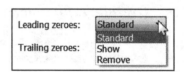

    - **Standard**. Zeroes are displayed based on the selected dimensioning standard.

    - **Show**. Zeroes before the decimal points are displayed on the selected dimensioning standard.

    - **Remove**. Zeroes are not displayed.

  - **Trailing zeroes**. Selections are:

  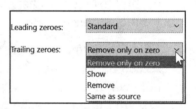

    - **Remove only on zero**. Selected by default. Trailing zeroes are trimmed for whole metric values. Conforms to ANSI and ISO standards.

    - **Show**. Dimensions have trailing zeroes up to the number of decimal places specified in **Options**, **Document Properties** and **Units**.

    - **Remove**. All trailing zeroes are removed.

    - **Same as source**. Trims trailing zeroes to the ASME Y14.5M-1994 standard.

- *Layer*. Select a Layer property to evaluate a drawing view.

Tables - Hole

The Tables - Hole section provides the ability to display the centers and diameters of holes on a face with respect to a selected table Origin in an active Drawing document.

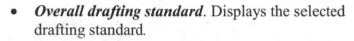 Hole Tables use a separate template.

The Tables - Hole section provides the following options:

- *Overall drafting standard*. Displays the selected drafting standard.

- *Border*. Click the **Box Border** option or the **Grid Border** option and select a corresponding thickness. The default thickness is 0.007in for the IPS system and .18mm for the MMGS system.

- *Text*. Century Gothic selected by default. Click the Font button to select a custom font style and size.

- *Location precision*. Specifies the number of significant digits for the hole location value.

- *Alpha/numerical control*. Specifies the tags identifying the holes to be letters or numbers.

- *Scheme*. The following options are available:

  - **Combine same tags**. Combines rows with holes in the same pattern, which also combines holes with the same sizes. The columns of location values are removed.

  - **Combine same size**. Available only if Combine same tags is cleared. Merges cells for holes of the same size. The columns of location values remain.

  - **Show ANSI inch letter and number drill sizes**.

- *Leading zeroes*. Standard selected by default. Selections are:

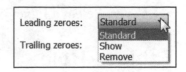

  - **Standard**. Zeroes are displayed based on the selected dimensioning standard.

  - **Show**. Zeroes before the decimal points are displayed on the selected dimensioning standard.

  - **Remove**. Zeroes are not displayed.

- *Trailing zeroes*. Selections are:

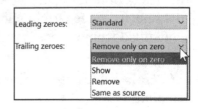

  - **Remove only on zero**. Selected by default. Trailing zeroes are trimmed for whole metric values. Conforms to ANSI and ISO standards.

  - **Show**. Dimensions have trailing zeroes up to the number of decimal places specified in **Options**, **Document Properties** and **Units**.

  - **Remove**. All trailing zeroes are removed.

  - **Same as source**.

  - **Show hole centers**. Selected by default.

  - **Automatic update of hole table**. Selected by default.

  - **Reuse deleted tags**.

  - **Add new row at the end of the table**.

- *Layer*. Select a Layer property to evaluate a drawing view.

- *Origin indicator*. Per Standard is selected by default. Provides the ability to specify the appearance of the indicator of the Origin from which the software calculates the hole position by selecting a standard. Selections are **Per Standard**, **ANSI, ISO, DIN, JIS, BSI, GOST** and **GB**.

- *Tag angle/offset from profile center*. Provides the ability to position the text at an angle and distance from the center of the hole. Selections are:

  - **Angle**. Angle from a vertical line through the center of the hole.

  - **Offset**. Distance from the hole profile.

- *Dual dimensions*. Provides the ability to display dual dimensions and location.

Tables - Punch

This option provides the ability to specify document-level drafting settings for punch tables.

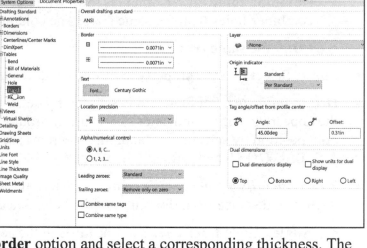

The Punch Table provides the following options:

- **Overall drafting standard**. Displays the selected drafting standard.

- **Border**. Click the **Box Border** option or the **Grid Border** option and select a corresponding thickness. The default thickness is 0.007in for IPS and 0.18mm for MMGS unit system.

- **Text**. Century Gothic selected by default. Click the Font button to select a custom font style and size.

- **Symbol shapes**. Circle is selected by default. Provides the ability to select a **circle**, **square**, **triangle** or **hexagon** for the revision symbol in a drawing.

- **Alpha/numerical control**. Alphabetic is selected by default. Provides the ability to specify whether the revision is alphabetic or numeric. If you switch from one to the other, you can change the revisions already in the table, or you can leave them unchanged and continue with future revisions in the new format. Any revision text you edit remains unchanged by automatic operations. The options are:

- **Leading zeroes**. Standard selected by default. These are the selections:

  - **Standard**. Zeroes are displayed based on the selected dimensioning standard.

  - **Show**. Zeroes before the decimal points are displayed on the selected dimensioning standard.

  - **Remove**. Zeroes are not displayed.

- **Trailing zeroes**. These are the selections:

  - **Remove only on zero**. Selected by default. Trailing zeroes are trimmed for whole metric values. Conforms to ANSI and ISO standards.

  - **Show**. Dimensions have trailing zeroes up to the number of decimal places specified in **Options**, **Document Properties** and **Units**.

  - **Remove**. All trailing zeroes are removed.

  - **Same as source**.

- *Scheme*. These are the selections:

  - **Combine same tags**. Combines rows with holes in the same pattern, which also combines holes with the same sizes. The columns of location values are removed.

  - **Combine same type**.

- *Layer*. Select a Layer property to evaluate a drawing view.

- *Origin indicator*. Per Standard is selected by default. Provides the ability to specify the appearance of the indicator of the Origin from which the software calculates the hole position by selecting a standard. Selections are **Per Standard, ANSI, ISO, DIN, JIS, BSI, GOST** and **GB**.

- *Tag angle/offset from profile center*. Provides the ability to position the text at an angle and distance from the center of the hole. Selections are:

  - **Angle**. Angle from a vertical line through the center of the hole.

  - **Offset**. Distance from the hole profile.

- *Dual Dimensions*. Provides the ability to display dual dimensions and location.

Tables - Revision

The Tables - Revision section provides the ability to track drawing revisions. The revision level is stored in a custom property in the file.

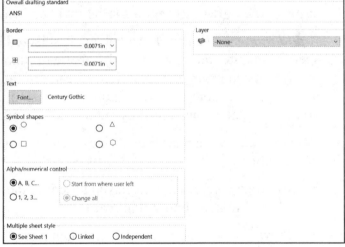

The Tables - Revision section provides the following options:

- ***Overall drafting standard***. Displays the selected drafting standard.

- *Border*. Click the **Box Border** option or the **Grid Border** option and select a corresponding thickness. The default thickness is 0.007in for IPS and .18mm for MMGS unit system.

- *Text*. Century Gothic selected by default. Click the Font button to select a custom font style and size.

- *Symbol shapes*. Circle is selected by default. Provides the ability to select a **circle, square, triangle** or **hexagon** for the revision symbol in a drawing.

- *Alpha/numerical control*. Alphabetic is selected by default. Provides the ability to specify whether the revision is alphabetic or numeric. If you switch from one to the other, you can change the revisions already in the table, or you can leave them unchanged and continue with future revisions in the new format. Any revision text you edit remains unchanged by automatic operations. Selections are:

  - **Start from where user left**. If you change control from alphabetic to numeric or vice versa, previous revisions remain as they are.

  - **Change all**. If you change control from alphabetic to numeric or vice versa, previous revisions, except any you have edited, change to the new format.

- *Multiple sheet style*. Provides the ability to control revision tables in drawings with multiple sheets. The three options are:

  - **See Sheet1**. Selected by default. The revision table on the first sheet is the active table. On all other drawing sheets, the revision table is labeled See Sheet 1.

  - **Linked**. A copy of the revision table from Sheet 1 is created on all sheets, and all revision tables update as one.

  - **Independent**. The revision table on each sheet is independent of any other revision table in the drawing. Updates to a revision table are not reflected in tables on other sheets.

- *Layer*. Select a Layer property to evaluate a drawing view.

Tables - Weld Tables

The Tables - Weld Tables section provides the ability to specify document-level drafting settings for title block tables. Selections are:

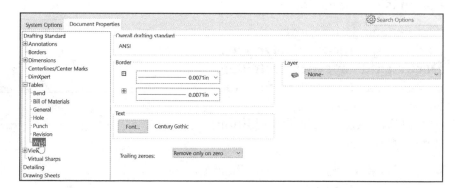

- *Overall drafting standard*. Displays the selected drafting standard.

- *Border*. Click the **Box Border** option or the **Grid Border** option and select a corresponding thickness. The default thickness is 0.007in for IPS and 0.18mm for MMGS unit system.

- ***Text***. Century Gothic selected by default. Click the Font button to select a custom font style and size.

- ***Trailing zeroes***. Selections are:

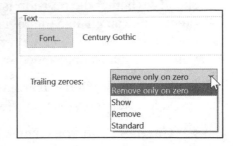

  - **Remove only on zero**. Selected by default. Trailing zeroes are trimmed for whole metric values. Conforms to ANSI and ISO standards.

  - **Show**. Dimensions have trailing zeroes up to the number of decimal places specified in **Options**, **Document Properties**, and **Units**.

  - **Remove**. All trailing zeroes are removed.

  - **Standard**.

- ***Layer***. Select a Layer property to evaluate a drawing view.

Views

The Views section provides the ability to set options for *Auxiliary, Detail, Section, Orthographic and Other* view labels in an active drawing document. The default option is Per standard.

The Per standard follows the standard specified in the **Options**, **Document Properties**, **Detailing**, **etc.** command procedure from the Menu bar toolbar. See SOLIDWORKS Help for additional detail information.

- ***Label options***. Select to adhere to the base standard. Clear to modify view label parameters:

  - **Name**. Select a title to appear in the view label, or type your own title.

  - **Label**. Select the way the label letter corresponding to the label on the parent view appears in the view label.

  - **Scale**. Select a label for the scale, or type your own label.

  - **Delimiter**. Select the delimiter between the two scale numbers and whether the scale is displayed in parentheses. If you select **#X**, the number in (#) can be an integer or a real number.

- **Preview**. Display the view label with name and label stacked on top of the scale (Stacked), or all data on the same line (In-line).

- **Display label above view**. Places view labels above the drawing view. This applies to new drawing views only.

Detailing

Additional tools are available for a Drawing document. See the Detailing section on Part/Assembly documents for additional information. Selections are:

- *Display filter*. View our options. **Cosmetic threads, Datums, Datum targets, Feature dimensions, Reference dimensions, DimXpert dimensions, Shaded cosmetic threads, Geometric tolerances, Notes, Surface finish, Welds, Display all types, Text scale, Always display text at the same size, Display items only in the view in which they are created, Display annotations, Use assembly setting for all components, Hide dangling dimensions and annotations, Use model color for HLR/HLV in drawings, Use Model color for HLR/HLV with SpeedPak configurations, Link child view to parent view configuration, Hatch density limit.**

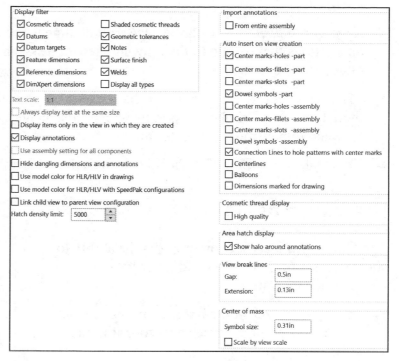

- *Import annotations*. Provides the ability to import dimensions, notes, etc. into a drawing.

- *Auto insert on view creation*. View your options. **Center marks-holes - part, Center marks-fillets-part, Center marks-slots - part, Dowel symbols - part, Center marks - holes - assembly, Center marks - slots - assembly, Dowel symbols - assembly, Connection Lines to hole patterns with center marks, Centerlines, Balloons Dimensions marked for drawing.**

- ***Cosmetic thread display***. Checks all cosmetic threads to determine if they should be visible or hidden. For example, if a hole (not a through hole) is on the back of a model and the model is in a front view, the cosmetic thread is hidden. You can set the display for each drawing view individually in the drawing view PropertyManager under **Cosmetic Thread Display**.

- ***Area hatch display***. Displays space around dimensions and annotations that belong to the drawing view or a sketch and are on top of an area hatch.

- ***View break lines***. Set a value for the distance between the model and dimension extension lines. This also controls the gap between dimensions and center marks.

- ***Center of mass***. Enter Symbol size to set a default symbol size.

Drawing Sheets

The Drawing Sheets section provides the ability to address sheet format, zones and letter layout in the zones. Selections are:

- ***Sheet format for new sheets.*** Provides the ability to use a different or same sheet format in the drawing.

- ***Zones***. Provides the ability to define drawing sheet zones on a sheet format for the purpose of providing locations where drawing views and annotations reside on the drawing. You can use annotation notes and balloons to identify which drawing zone they are in. As you move an annotation in the graphics area, the drawing zone updates to the current zone.

- ***Letter layout***. Provides the ability to select a column or row format for the letter layout.

- ***Multi-sheet custom properties source***. Provides the ability to use custom property values from this sheet on all sheets.

You can place location labels on detail, section, and auxiliary drawing views to provide the sheet and zone location for the parent view. Additionally, you can place location labels on detail circles, section cutting lines and auxiliary view arrows to provide the location of the corresponding child view.

Line Font

The Line Font section provides the ability to set the style and weight of a line for various edge types in a Drawing document. Selections are:

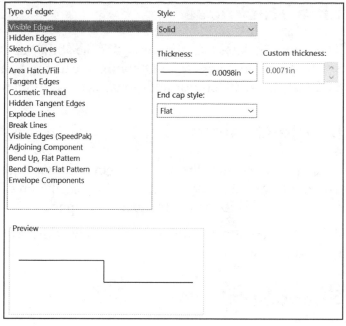

- *Type of edge*. Visible Edges selected by default. Select an edge type from the drop-down menu.

- *Style*. Solid selected by default. Select a style from the drop-down menu. Selections are **Solid**, **Dashed**, **Phantom**, **Chain**, **Center**, **Stitch** and **Thin/Thick Chain**.

- *Thickness*. Select a thickness from the drop-down menu or select custom size in input a dimension.

- *End Cap Style*. Flat selected by default. Select a style that defines the ends of edges: **Flat**, **Round** or **Square**.

Line Style

The Line Style section provides the ability to create or modify existing line styles for a Drawing document.

Apply a simple code which is illustrated to create a special line style. See SOLIDWORKS Help for additional detail information.

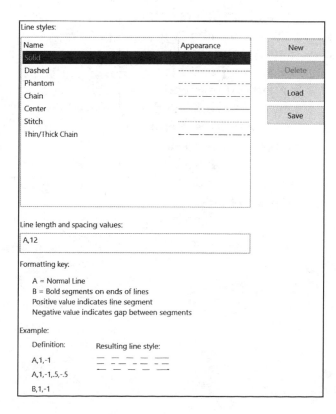

Line Thickness

The Line Thickness section provides the ability to modify existing line thickness for a Drawing document. See SOLIDWORKS Help for additional detail information.

Image Quality

This option provides the ability to control image quality. The image quality option controls the tessellation of curved surfaces for shaded rendering output.

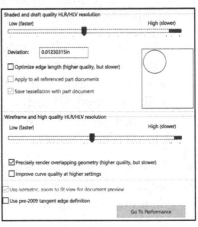

A higher resolution setting results in slower model rebuilding but more accurate curves.

Sheet Metal

The Sheet Metal section provides the ability to address Flat pattern colors and Bend notes to a Drawing document. These are the selections:

- **Flat pattern colors.** Bend Lines - Up Direction, color black is selected by default. The Flat pattern colors option provides the ability to modify colors for the following: **Bend Lines - Up Direction, Bend Lines - Down Direction, Form Features, Bend Lines - Hems, Model Edges, Flat Pattern Sketch Color and Bounding box.**

- ***Display sheet metal bend notes***. Above Bend Line selected by default. Provides the ability to display sheet metal notes above or below the bend line. Selections are **Above Bend Line**, **Below Bend Line** and **With Leader**.

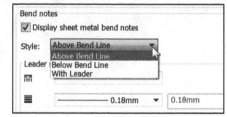

Sheet metal bend line notes provides the ability to 1.) Edit text outside the note parameters. 2.) Modify default format. Edit *install_dir>*\lang*<language>* \bendnoteformat.txt. 3.) Change the bend angle, bend direction, or bend radius, and the notes update in the drawing. 4.) Change the display position of the bend notes.

Summary

The chapter was intended to provide a general overview of the available Document Properties for a Part, Assembly and Drawing document.

The available options under the Document Properties tab are document dependent as illustrated from left to right: Part, Assembly and Drawing.

Templates are documents, (part, drawing, assembly) which include user-defined parameters and are the basis for a new document.

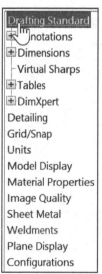

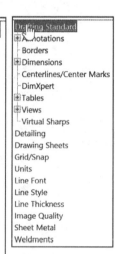

You created a custom Assembly template using the ANSI, MMGS Unit system and a Part template with an applied material.

The Document Properties applied to a template allows you to work efficiently. However, Document Properties in the current document can be modified at any time.

The first half of this chapter addressed the Document Properties for a Part or Assembly document. The second half of the chapter addressed Drawing document specific Document Properties.

In Chapter 4, explore and address Design Intent, 2D and 3D Sketching, Sketch states, Parent/Child relationships, Sketch Entities tools, and the Block and Spline toolbar. Design Intent is also explored in Chapter 18 under Intelligent Modeling Techniques.

Illustrations may vary depending on your SOLIDWORKS version and system setup.

Notes:

CHAPTER 4: DESIGN INTENT, SKETCHING AND SKETCH ENTITIES

Chapter Objective

Chapter 4 provides a comprehensive understanding of Design Intent, Sketching and the available Sketch Entities in SOLIDWORKS. On the completion of this chapter, you will be able to:

- Define and incorporate Design Intent in a:

  - Sketch, Feature, Part, Assembly and Drawing document.

- Utilize the available SOLIDWORKS Design Intent tools:

  - Comments, Design Binder, ConfigurationManager, Dimensions, Equations, Design Tables and Features.

- Identify the correct Reference planes:

  - 2D and 3D Sketching.

- Insert sketch Reference planes.

- Comprehend the Parent/Child relationship.

- Recognize and address Sketch states:

  - Fully Defined, Over Defined, Under Defined, No Solution Found and Invalid Solution Found.

- Identify and utilize the following Sketch Entities tools:

  - Line, Corner Rectangle, Center Rectangle, 3 Point Corner Rectangle, 3 Point Center Rectangle, Parallelogram, Straight Slot, Centerpoint Straight Slot, 3 Point Arc Slot, Centerpoint Arc Slot, Instant3D, Polygon, Circle, Perimeter Circle, Centerpoint Arc, Tangent Arc, 3 Point Arc, Ellipse, Partial Ellipse, Parabola, Conic, Spline, Style Spline, Spline on Surface, Equation Driven Curve, Point, Centerline, Text, Plane, Route line and Belt/Chain.

- Classify and utilize the following Block tools:

  - Make Block, Edit Block, Insert Block, Add/Remove, Rebuild, Saves Block, Explode Block and Belt/Chain.

- Recognize and utilize the following Spline tools:

- Add Tangency Control, Add Curvature Control, Insert Spline Point, Simplify Spline, Fit Spline, Show Spline Handle, Show Inflection points, Show Minimum Radius and Show Curvature Combs.

- Reuse an existing 2D Sketch.

Design Intent

What is design intent? All designs are created for a purpose. Design intent is the intellectual arrangements of features and dimensions of a design. Design intent governs the relationship between sketches in a feature, features in a part and parts in an assembly.

The SOLIDWORKS definition of design intent is the process in which the model is developed to accept future modifications. Models behave differently when design changes occur.

Design for change. Utilize geometry for symmetry, reuse common features, and reuse common parts. Build change into the following areas that you create: sketch, feature, part, assembly and drawing.

When editing or repairing geometric relations, it is considered best practice to edit the relation vs. deleting it.

Design Intent in a Sketch

Build design intent in a sketch as the profile is created. A profile is determined from the Sketch Entities. Example: Rectangle, Circle, Arc, Point, Slot, etc. Apply symmetry into a profile through a sketch centerline, mirror entity and position about the reference planes and Origin. Always know the location of the Origin in the sketch.

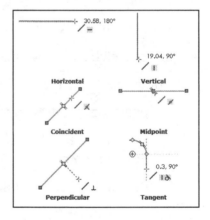

Build design intent as you sketch with automatic Geometric relations. Document the decisions made during the up-front design process. This is very valuable when you modify the design later.

A rectangle (Center Rectangle Sketch tool) contains Horizontal, Vertical and Perpendicular automatic Geometric relations.

Apply design intent using added Geometric relations if needed. Example: Horizontal, Vertical, Collinear, Perpendicular, Parallel, Equal, etc.

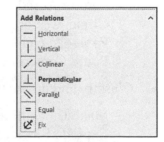

Example A: Apply design intent to create a square profile. Sketch a rectangle. Apply the Center Rectangle Sketch tool. Note: No construction reference centerline or Midpoint relation is required with the Center Rectangle tool. Insert dimensions to fully define the sketch.

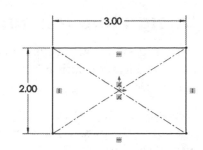

Example B: If you have a hole in a part that must always be 16.5mm ≤ from an edge, dimension to the edge rather than to another point on the sketch. As the part size is modified, the hole location remains 16.5mm ≤ from the edge as illustrated.

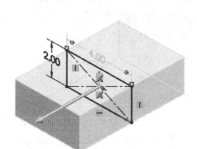

Design Intent in a Feature

Build design intent into a feature by addressing End Conditions (Blind, Through All, Up to Next, Up to Vertex, Up to Surface, Offset from Surface, Up to Body and Mid Plane), symmetry, feature selection, and the order of feature creation.

Example A: The Extruded Base feature remains symmetric about the Front Plane. Utilize the Mid Plane End Condition option in Direction 1. Modify the depth, and the feature remains symmetric about the Front Plane.

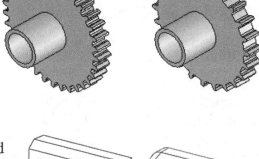

Example B: Create 34 teeth in the model. Do you create each tooth separately using the Extruded Cut feature? No.

Create a single tooth and then apply the Circular Pattern feature. Modify the Circular Pattern from 32 to 24 teeth.

Design Intent in a Part

Utilize symmetry, feature order and reusing common features to build design intent into a part. Example A: Feature order. Is the entire part symmetric? Feature order affects the part.

Apply the Shell feature before the Fillet feature and the inside corners remain perpendicular.

Design Intent in an Assembly

Utilizing symmetry, reusing common parts and using the Mate relation between parts builds the design intent into an assembly.

Example A: Reuse geometry in an assembly. The assembly contains a linear pattern of holes. Insert one screw into the first hole. Utilize the Component Pattern feature to copy the machine screw to the other holes.

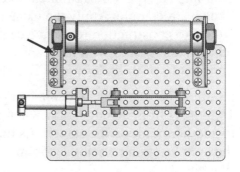

Design Intent in a Drawing

Utilize dimensions, tolerance and notes in parts and assemblies to build the design intent into a drawing.

Example A: Tolerance and material in the drawing. Insert an inside diameter tolerance +.000/-.002 into the Pipe part. The tolerance propagates to the drawing.

Define the Custom Property Material in the Part. The Material Custom Property propagates to your drawing.

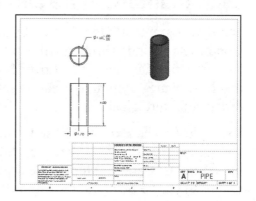

☼ Create a sketch on any of the default planes: Front, Top, Right or a created plane or face.

SOLIDWORKS Design Intent tools

Comments

Add comments, notes or additional information to features or sketches during the design period. This will aid you or your colleagues to recall and to better understand the fundamental design intent of the model and individual features later.

Right-click on the name, feature or sketch in the FeatureManager. Click Comment ➤ Add Comment. The Comment dialog box is displayed.

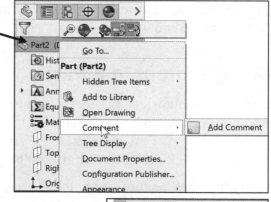

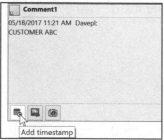

Enter the information. Click Save and Close. A Comments folder is displayed in the FeatureManager. Double-click the feature in the Comments folder to view the comment.

 You can also add a Date/Time stamp to your comment.

Design Binder

Activate the Design binder. Click **Options** ⚙ ➤ **System Options** tab ➤ **FeatureManager**. Select Show from the drop-down menu. The Design Binder is an embedded Microsoft Word document that provides the ability for the user to capture a screen image and to incorporate text into a document.

ConfigurationManager

When you create various configurations for assemblies, design tables, etc., use the comment area to incorporate a comment for these configurations.

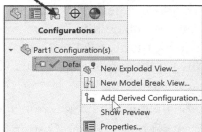

Dimensions

To be efficient, reuse existing geometry. Provide dimensions with descriptive names.

Equations

SOLIDWORKS provides a completely redesigned dialog box for creating and editing equations, global variables and dimensions.

Add a comment to the end of an equation to provide clarity for the future. Use descriptive names and organize your equations to improve clarity.

Create an Excel spread sheet to input your equations into SOLIDWORKS. Use the Link to external file option.

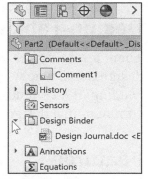

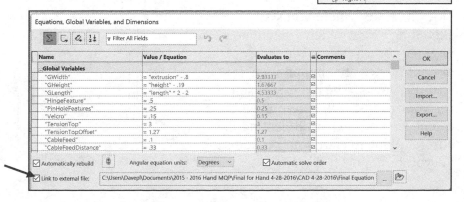

💡 You can specify the units of measurement for Global variables, and for the values and equations that define the global variables. You can define the units in the Equations and modify dialog boxes for dimensions, and in PropertyManagers that support equations.

Design Tables

There are a number of ways to incorporate a comment into a design table. One way is to add "$COMMENT" to the heading of a column. This provides the ability to add a comment in a design table.

Features

Always use descriptive names in the FeatureManager design tree, not the default feature names such as Boss-Extrude1, Cut-Extrude1, LPattern1, etc. Group important features toward the top of the FeatureManager design tree.

💡 Enable the FeatureManager Name feature on creation option: click **Options** ⚙ ➢ **System Options** ➢ **FeatureManager** ➢ **Name feature on creation box**. The Name feature on creation option highlights the feature name when created and allows the feature to be named.

Identify the Correct Reference Planes

Most SOLIDWORKS features start with a 2D sketch. Sketches are the foundation for creating features. SOLIDWORKS provides the ability to create either 2D or 3D sketches.

A 2D sketch is limited to a flat 2D Sketch Plane located on a Reference Plane, face or a created place. 3D sketches are very useful in creating sketch geometry that does not lie on an existing or easily defined Plane.

Does it matter where you start sketching? Yes. When you create a new part or assembly, the three default planes are aligned with specific views. The plane you select for your first sketch determines the orientation of the part. Selecting the correct plane to start your model is very important.

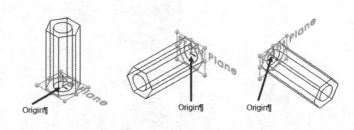

2D Sketching/Reference Planes

The three default ⊥ reference planes, displayed in the FeatureManager design tree, represent infinite 2D planes in 3D space:

- Front.

- Top.

- Right.

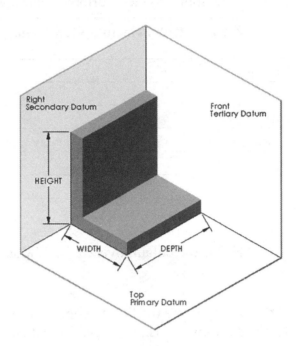

Planes have no thickness or mass. Orthographic projection is the process of projecting views onto parallel planes with ⊥ projectors. These are the default ⊥ datum planes:

- Primary.

- Secondary.

- Tertiary.

Use the following planes in a manufacturing environment:

- Primary datum plane: Contacts the part at a minimum of three points.

- Secondary datum plane: Contacts the part at a minimum of two points.

- Tertiary datum plane: Contacts the part at a minimum of one point.

The part view orientation depends on the sketch plane. Compare the Front, Top and Right Sketch planes for an L-shaped profile in the following illustration. Remember - the plane or face you select for the base sketch determines the orientation of the part.

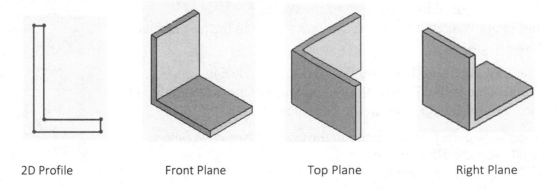

2D Profile   Front Plane   Top Plane   Right Plane

The six principle views of Orthographic projection listed in the ASME Y14.3 2009 standard are *Front, Top, Right, Bottom, Rear and Left*. SOLIDWORKS Standard view names correspond to these Orthographic projection view names.

ASME Y14.3 2009 Principle View Name:	SOLIDWORKS Standard View:
Front	Front
Top	Top
Right side	Right
Bottom	Bottom
Rear	Back
Left side	Left

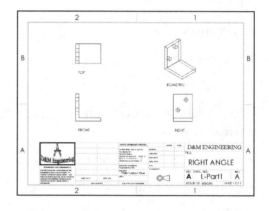

💡 ANSI is the default overall drafting standard used in the book.

💡 Displayed are the Front, Top, Right and Isometric views in Third Angle Projection.

Tutorial: Default Reference Planes 4-1

Create a New part. Display the default Reference planes in the Graphics window.

1. Click **New** 🗋 from the Menu bar. The Templates tab is the default tab. Part is the default template from the New SOLIDWORKS Document dialog box.

2. Click **OK**. The Part FeatureManager is displayed. Use the sketch Origin to understand the (x-,y-) coordinate location of the sketch and to apply sketch dimensions. Each sketch in the part has its own Origin.

3. Click **Front Plane** from the FeatureManager design tree. The Front Plane is the Sketch plane. The Front Plane is highlighted in the FeatureManager and in the Graphics window.

Show the default Reference planes in the Graphics window. Select the Front, Top and Right planes.

4. Hold the **Ctrl** key down.

5. Click **Top Plane** and **Right Plane** from the FeatureManager design tree.

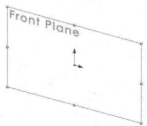

6. Release the **Ctrl** key.

7. Right-click **Show** 👁 from the Context toolbar.

8. Click **inside** the Graphics window. The three planes are displayed in the Graphics window. Note the Plane icon in the FeatureManager.

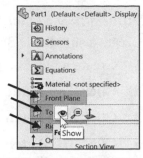

💡 Rename a feature or sketch for clarity. Slowly click the feature or sketch name twice and enter the new name when the old one is highlighted.

Rename the default Reference planes.

9. Rename the **Front Plane** to **Front-PlateA**.

10. Rename the **Top Plane** to **Top-PlateB**.

11. Rename the **Right Plane** to **Right-PlateC**.

12. **Rebuild** 🔘 the model.

13. **Close** the model.

3D Sketching/Reference Planes

Create a 3D Sketch in SOLIDWORKS. A 3D Sketch is typically used for advanced features such as Sweeps and Lofts or when the sketch does not lie on an existing or easily definable plane. Most basic features are created from a 2D Sketch.

There are two approaches to 3D Sketching. The first approach is called 2D Sketching with 3D Sketch planes. In this approach, you:

1. Activate a planar face by adding a 3D Sketch plane.

2. Sketch in 2D along the plane.

3. Add 3D Sketch planes each time you are required to move sketch entities to create a 3D sketch.

Using the 2D Sketching with 3D Sketch planes approach provides the ability to:

- Add relations:

    - Between planes.

    - Between sketch entities on different planes.

    - To planes.

    - Define planes.

    - Move and resize planes.

Tutorial: 3D Sketching 4-1

Create a simple 3D Sketch on plane using the line sketch tool.

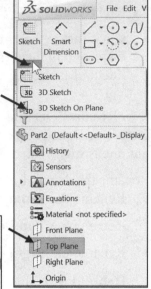

1. Create a **New** ⬜ part. Use the default ANSI Part template.

2. Click **Top Plane** from the FeatureManager.

3. Click the **Sketch** tab from the CommandManager.

4. Click **3D Sketch On Plane** from the Sketch Consolidated toolbar as illustrated. 3DSketch1 is active.

5. Click the **Line** ✏ Sketch tool.

6. Sketch a **vertical line** through the Origin.

7. Right-click **Exit Sketch** ⬑ from the Context toolbar. 3DSketch1 is displayed in the FeatureManager and is under defined.

8. **Close** the model.

Tutorial: 3D Sketching 4-2

Create the illustrated model. Insert two features: Boss-Extrude1 and Cut-Extrude1. Apply the 3D Sketch tool to create the Cut-Extrude1 feature. System units = MMGS.

1. Create a **New** ⬜ part in SOLIDWORKS. Use the default ANSI, MMGS Part template.

2. Right-click **Front Plane** from the FeatureManager. This is your Sketch plane.

Origin

3. Click **Sketch** ▣ from the Context toolbar. The Sketch toolbar is displayed.

4. Click the **Corner Rectangle** ▭ tool from the Consolidated Rectangle Sketch toolbar. The PropertyManager is displayed.

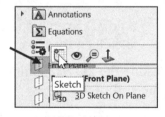

5. Sketch a **rectangle** as illustrated. The part Origin is located in the bottom left corner of the sketch.

Insert dimensions.

6. Right-click **Smart Dimension** ↖ from the Context toolbar.

7. **Dimension** the sketch as illustrated. The sketch is fully defined. The sketch is displayed in black.

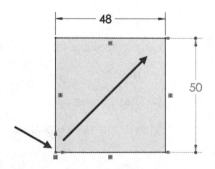

Create the first feature for the model. The Extruded Boss/Base feature adds material to a part. Extrude the sketch to create the first feature. Apply symmetry.

8. Click **Extruded Boss/Base**  from the Features toolbar.

9. Select the **Mid Plane** End Condition in Direction 1. Enter **100.00**mm for Depth in Direction 1. Accept the default conditions. Click **OK** ✔ from the Boss-Extrude PropertyManager.

The Instant3D tool provides the ability to drag geometry and dimension manipulator points to resize features or to create new features directly in the Graphics window. Click the face. Select a manipulator point. Click and drag. Click a location along the ruler for the required dimension. The rule increments are set in System Options.

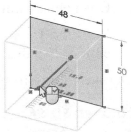

10. Create 3DSketch1. 3DSketch1 is a four point sketch. Click **3D Sketch On Plane** from the Consolidated Sketch toolbar as illustrated.

11. Click the **Line** ✐ Sketch tool. 3DSketch1 is a four point sketch as illustrated. 3DSketch1 is the profile for Cut-Extrude1.

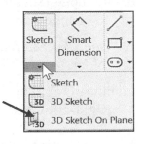

12. Locate the first point of the sketch. Click the **back right bottom point** as illustrated. Locate the second point of the sketch. Click the **top left front point** as illustrated. Locate the third point of the sketch.

13. Click the **bottom right front point** as illustrated.

Locate the fourth point of the sketch.

14. Click the **back right bottom** point as illustrated to close the sketch.

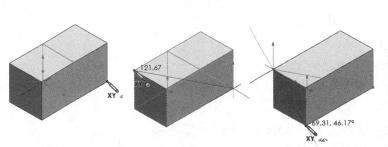

Create the Cut-Extrude1 feature.

15. Click **Extruded Cut** 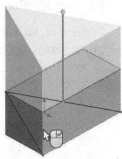 from the Features toolbar.

16. Click the **front right vertical edge** as illustrated to remove the material. Edge<1> is displayed in the Direction of Extrusions box.

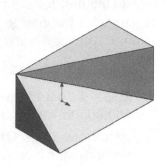

17. Click **OK** ✔ from the Cut-Extrude PropertyManager. Cut-Extrude1 is displayed in the FeatureManager.

18. **Close** the model.

💡 Select the front right vertical edge or the Top face to remove the required material in this tutorial.

The second approach to create a 3D Sketch in SOLIDWORKS is 3D Sketching. In this approach:

1. Open an existing 3D Sketch.

2. Press the Tab key each time you need to move your sketch entities to a different axis.

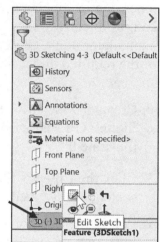

The sketch origin is placed wherever you start the sketch.

Tutorial: 3D Sketching 4-3

Apply the 3DSketch tool to create a 3DSketch using the Tab key to change Sketch planes.

1. Open **3D Sketching 4-3** from the SOLIDWORKS 2018 folder.

Edit the 3D Sketch.
2. Right-click **3DSketch** from the FeatureManager.

3. Click **Edit Sketch** 🖉 from the Context toolbar. 3DSketch1 is displayed in the Graphics window.

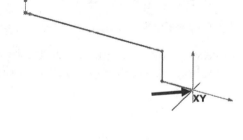

4. Click the **Line** ✏ Sketch tool. Note: The mouse feedback icon.

5. Click the **right endpoint** of the sketch as illustrated.

6. Press the **Tab** key to change the Sketch plane from XY to YZ. The YZ icon is displayed on the mouse pointer.

7. Sketch a **line** along the Z axis approximately 130mm as illustrated.

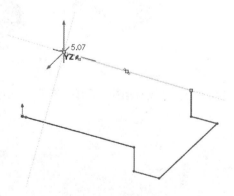

8. Press the **Tab** key twice to change the sketch plane from YZ to XY. The XY icon is displayed on the mouse pointer.

9. Sketch a **line** back along the X axis approximately 30mm.

10. Sketch a **line** on the Y axis approximately 30mm.

11. Sketch a **line** along the X axis of approximately 130mm.

Deselect the Line Sketch tool.
12. Right-click **Select**. View the 3D Sketch.

13. **Rebuild** the model.

14. **Close** the model.

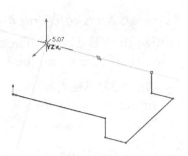

Tutorial: 3D Sketching 4-3A

Apply the circle sketch tool to create a Sweep along a 3D sketch.

1. Open **3D Sketching 4-3A** from the SOLIDWORKS 2018 folder.

Create the 2D sketch.
2. Right-click **Right Plane** from the FeatureManager.

3. Click **Sketch**.

4. Display a **Right** view.

5. Sketch a **circle** at the origin as illustrated.

Create a Swept Boss feature.
6. Click **Swept Boss/Base** from the Features CommandManager.

7. Click the **Circular Profile** box.

8. Enter **.10in** for diameter.

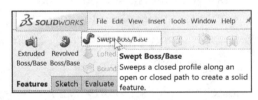

9. Click the **3D Sketch** (path) in the graphics window.

10. Click **OK**. View the results in the graphics window.

11. **Close** the model.

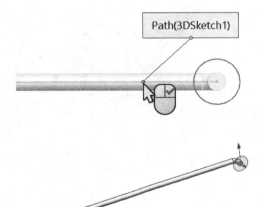

Tutorial: 3D Sketching 4-4

Utilize the 3D Sketch placement method. Insert a hole on a cylindrical face using the Hole Wizard.

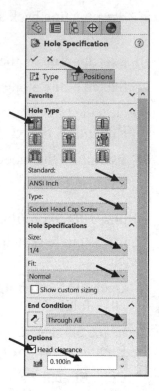

1. Open **3D Sketching 4-4** from the SOLIDWORKS 2018 folder. Note: With a 3D sketch, press the Tab key to move between planes.

2. Click the **Hole Wizard** 📷 feature tool.

3. Create a **Counterbore, ANSI Inch, Socket Head Cap Screw fastener Type**. Size - ¼. Fit - **Normal**. End Condition - **Through All** with a **.100in** Head clearance.

4. Click the **Positions** Tab.

5. Click the **3D Sketch** button. The Hole Position PropertyManager is displayed. SOLIDWORKS displays an active 3D interface with the Point  tool.

☀ When the Point tool is active, wherever you click, you will create a point.

Dimension the sketch.

6. Click the **cylindrical face** of the model as illustrated. The selected face is displayed in orange. This indicates that an OnSurface sketch relations will be created between the sketch point and the cylindrical face. The hole is displayed in the model.

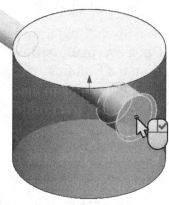

7. Insert a **.25in** dimension between the top face and the sketch point.

Locate the point angularly around the cylinder. Apply construction geometry.

8. Display the **Temporary Axes**.

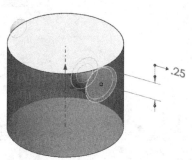

9. Click the **Line** ✏ Sketch tool. Note: 3D Sketch is still activated.

10. **Ctrl + Click** the top flat face of the model. This moves the red space handle origin to the selected face. This also constrains any new sketch entities to the top flat face.

    Note the mouse pointer 🖝⤢ icon.

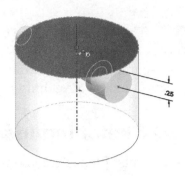

11. Move the **mouse pointer** near the center of the activated top flat face as illustrated. View the small black circle. The circle indicates that the end point of the line will pick up a Coincident relation.

12. Click the **center point** of the circle.

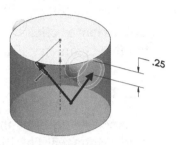

13. Sketch a line so it picks up the **AlongZ sketch relation**. The cursor displays the relation to be applied. **This is a very important step**.

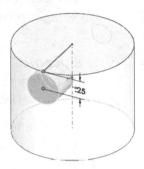

Deselect the Line Sketch tool.

14. Right-click **Select**.

15. Create an **AlongY** sketch relation between the center point of the hole on the cylindrical face and the endpoint of the sketched line. The sketch is fully defined.

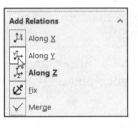

16. Click **OK** ✔ from the PropertyManager.

17. Click **OK** ✔ to return to the Part FeatureManager.

18. **Expand** the FeatureManager and view the results. The two sketches are fully defined. One sketch is the hole profile, the other sketch is to define the position of the feature.

19. **Close** the model.

💡 Create a second sketched line and insert an angle dimension between the two lines. This process is used to control the position of the center point of the hole on the cylindrical face as illustrated. Insert an AlongY sketch relation between the center point of the hole on the cylindrical face and the end point of the second control line. Control the hole's position with an angular dimension.

🔆 The main advantage of the 3D Sketch placement method is that it can place a set of holes on any set of solid faces, regardless of whether they are at different levels, or are non-parallel. A limitation: 3D sketches can be fairly cumbersome and difficult. See section on 3D Sketching in SOLIDWORKS Help for additional details.

2D Sketching/Inserting Reference Planes

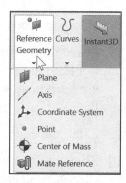

Reference Planes are geometry created from existing planes, faces, vertices, surfaces, axes, and or sketch geometry. To access the Plane PropertyManager, click Insert ➤ Reference Geometry ➤ Plane from the Menu bar or click the Plane 🔲 tool from the Consolidated Reference Geometry Features toolbar.

Plane Tool

The Plane 🔲 tool uses the Plane PropertyManager. The Plane PropertyManager provides the following selections:

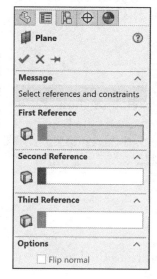

- *First Reference*. Select a face, plane or edge to create a reference plane. These options are available:

  - **Reference Entities**. Displays the selected planes either from the FeatureManager or from the Graphics window.

  - **Parallel**. Creates a plane parallel to the selected plane. For example, select a face for one reference and a point for another reference. The created plane is parallel to the face and coincident to the point.

  - **Perpendicular**. Creates a plane perpendicular to the selected reference. For example, select an edge or curve for one reference and a point or vertex for another reference.

- **Coincident**. Creates a plane through a point coincident to a plane or a face.

- **Project**. Projects a singular entity such as a point, vertex, origin, or coordinate system onto a non-planar surface. Select the sketch point and the model surface. Two options appear in the PropertyManager:

  - Nearest location on surface.

  - Along sketch normal.

- **Tangent**. Creates a plane tangent to cylindrical, conical, non-cylindrical, and non-planar faces. Select a surface and a sketch point on the surface. The plane is created tangent to the surface and coincident to the sketch point.

- **At Angle**. Select a plane or planar face and select an edge, axis, or sketch line. You can specify the number of planes to create. Enter the angle value in the angle box.

- **Offset Distance**. Creates a plane parallel to a plane or face, offset by a specified distance. You can specify the number of planes to create. Enter the Offset distance value in the distance box.

  - **Reverse direction**. Reverses the direction of the angle if required.

  - **Numbers of Planes to Create**. Displays the selected number of planes.

- **Mid Plane**. Creates a mid-plane between planar faces, reference planes, and 3D sketch planes. Select Mid Plane for both references.

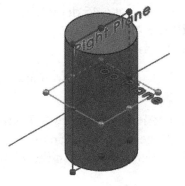

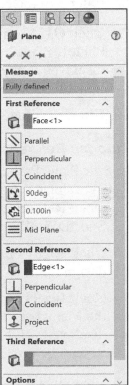

Tutorial: Reference Plane 4-1

Create three Reference planes using the Offset Distance option from the Plane PropertyManager.

1. Create a **New** 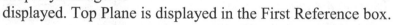 part. Use the default ANSI Part template. Set system units to MMGS.

2. Display an **Isometric view**.

3. Right-click **Top Plane** from the FeatureManager design tree.

4. Hold the **Ctrl** key down.

5. Click and drag the **boundary** of the Top Plane upwards in the Graphics window. The Plane PropertyManager is displayed. Top Plane is displayed in the First Reference box.

6. Release the **Ctrl** key and **mouse** pointer.

7. Enter **25**mm in the Distance. The Plane is offset by 25mm.

8. Enter **3** in the Number of planes to create box.

9. Click **OK** ✔ from the Plane PropertyManager. Plane1, Plane2 and Plane3 are created and displayed in the Graphics window. Each Plane is offset from the Top Plane a distance of 25mm.

10. **Close** the model.

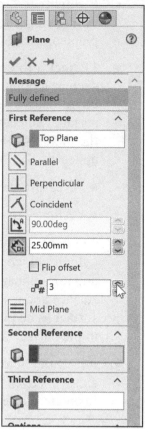

Tutorial: Reference Plane 4-2

Use the Plane PropertyManager to create two angled Reference planes from an edge at a 45 degree angle.

1. Open **Reference Plane 4-2** from the SOLIDWORKS 2018 folder.

2. Click **Right Plane** from the FeatureManager design tree.

3. Click **Insert ➤ Reference Geometry ➤ Plane** from the Menu bar. The Plane PropertyManager is displayed. Right Plane is displayed in the First Reference box.

4. Click the **At angle** button.

5. Enter **45**deg for Angle.

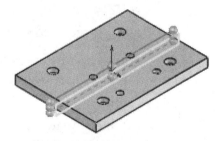
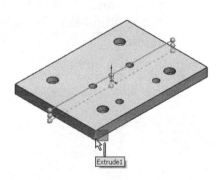

6. Click the **front vertical edge** of Extrude1 as illustrated. Do not select the midpoint. Edge<1> is displayed in the Selections box. The model is fully defined.

7. Enter **2** in the Number of planes to create box.

8. Click **OK** ✔ from the Plane PropertyManager. Plane1 and Plane2 are displayed in the FeatureManager and in the Graphics window.

9. **Close** the model.

Tutorial: Reference Plane 4-3

Use the Plane PropertyManager to create a Reference plane using three vertices (Coincident) to the Extrude1 feature.

1. Open **Reference Plane 4-3** from the SOLIDWORKS 2018 folder.

2. Click **Insert ➢ Reference Geometry ➢ Plane**. The Plane PropertyManager is displayed.

3. Click **three Vertices** as illustrated. Vertex<1>, Vertex<2> and Vertex<3> are displayed in the Reference Entities boxes. The model is fully defined.

4. Click **OK** ✔ from the Plane PropertyManager. The angled Plane1 is created and is displayed in the FeatureManager and the Graphics window.

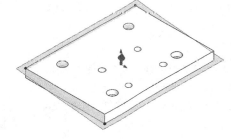

5. **Close** the model.

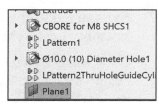

Fully define all sketches in the model. However, there are times when this is not practical, generally when using the Spline tool to create a complex freeform shape.

Tutorial: Reference Plane 4-4

Use the Plane PropertyManager to create a Reference plane parallel to an existing plane, through a selected point.

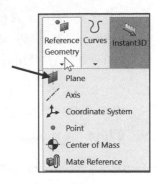

1. Open **Reference Planes 4-4** from the SOLIDWORKS 2018 folder.

2. Click the **Plane** tool from the Consolidated Reference Geometry Features toolbar. The Plane PropertyManager is displayed.

3. Click the **front top right vertex**.

4. Click the **Right Plane** from the fly-out FeatureManager. The model is fully defined.

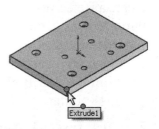

5. Click **OK** from the Plane PropertyManager. Plane1 is created and is displayed in the FeatureManager and Graphics window. **Close** the model.

Parent/Child Relationship

A Parent feature or sketch is an existing feature or sketch on which others depend. Example of a Parent feature: An Extruded feature is the Parent feature to a fillet which rounds the edges.

When you create a new feature or sketch which is based on other features or sketches, their existence depends on the previously built feature or sketch. The new feature or sketch is called a Child feature or a Child sketch. Example of a Child feature: A hole is the child of the Base-Extrude feature in which it is cut.

Tutorial: Parent-Child 4-1

View the Parent/Child relationship in an assembly for a feature and a sketch from a component.

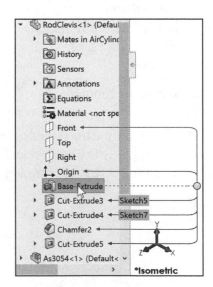

1. Open the **AirCylinder Linkage** assembly from the SOLIDWORKS 2018\Solutions folder.

2. **Expand** RodClevis<1> from the AirCylinder Linkage assembly FeatureManager.

3. Click **Base-Extrude**. View the information in the FeatureManager. Right-click **Base-Extrude**.

4. Click **Parent/Child**. The Parent/Child Relationships dialog box is displayed. View the Sketch2 relationships.

5. Click **Close** from the Parent/Child Relationships dialog box.

View the Parent/Child Relationships for a feature.

6. Right-click **Sketch2** from Base-Extrude in the FeatureManager design tree.

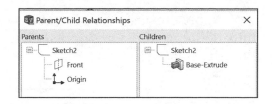

7. Click **Parent/Child**. View the Base-Extrude feature relationships.

8. Click **Close** from the Parent/Child Relationships dialog box.

9. **Close** the model.

Sketch States

Sketches can exist in any of five states. The state of the sketch is displayed in the status bar at the bottom of the SOLIDWORKS window. There are five sketch states in SOLIDWORKS:

 Color indicates the state of the individual Sketch entities.

1. *Under Defined.* Inadequate definition of the sketch, (blue). The FeatureManager displays a minus (-) symbol before the Sketch name.

2. *Fully Defined.* Complete information, (black). The FeatureManager displays no symbol before the Sketch name.

3. *Over Defined.* Duplicate dimensions and or relations, (orange -red). The FeatureManager displays a (+) symbol before the Sketch name. The What's Wrong dialog box is displayed.

4. *Invalid Solution Found.* Your sketch is solved but results in invalid geometry; for example, a zero length line, zero radius arc or a self-intersecting spline (yellow).

5. *No Solution Found.* Indicates sketch geometry that cannot be resolved, (Brown).

SketchXpert is designed to assist in an over defined state of a sketch. SketchXpert generates a list of causes for over defined sketches. The list is organized into Solution Sets. This tool enables you to delete a solution set of over defined dimensions or redundant relations without compromising your design intent.

Click **Options** ⚙ ➢ **System Options** ➢ **Sketch**, and click the **Use fully defined sketches** option to use fully defined sketches for created features.

☼ Apply dimensions and relations calculated by the SOLIDWORKS application to fully define sketches or selected sketch entities. Use the Fully Defined Sketch PropertyManager.

Sketch Entities

Sketch entities provide the ability to create lines, rectangles, parallelograms, circles, etc. during the sketching process. To access the available sketch entities for SOLIDWORKS, click **Tools ➤ Sketch Entities** from the Menu bar menu and select the required entity. This book does not go over each individual Sketch entity in full detail but will address the more advanced entities.

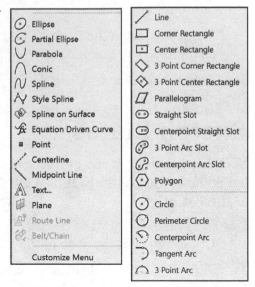

To obtain additional information of each sketch entity, activate the sketch entity and click the question mark ⑦ icon located at the top of the entity PropertyManager or click **Help ➤ SOLIDWORKS Help Topics** from the Menu bar and search by using one of the three available tabs: **Contents**, **Index**, or **Search**.

Line Sketch Entity

The Line Sketch entity ⟋ tool provides the ability to sketch multiple 2D lines in a sketch. The Line Sketch entity uses the Insert Line PropertyManager. The Insert Line PropertyManager provides the following selections:

- *Orientation*. The Orientation box provides the following options:

  - **As sketched**. Sketch a line in any direction using the click and drag method. Using the click-click method, the As sketched option provides the ability to sketch a line in any direction, and to continue sketching other lines in any direction, until you double-click to end your process.

  - **Horizontal**. Sketch a horizontal line until you release your mouse pointer.

  - **Vertical**. Sketch a vertical line until you release your mouse pointer.

  - **Angle**. Sketch a line at an angle until you release your mouse pointer.

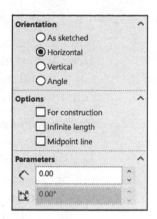

- *Options*. The Options box provides three line types:

  - **For construction**. Converts the selected sketch entity to construction geometry.

  - **Infinite length**. Creates a line of infinite length which you can later trim in the design process.

  - **Midpoint line**. Create a line that is symmetrical from the midpoint of the line.

- *Parameters*. You can specify any appropriate combination of the following parameters (or Additional Parameters) to define the line if the line is not constrained by relations. When you change one or more parameters, the other parameters update automatically.

💡 Use Construction geometry to assist in creating your sketch entities and geometry that are ultimately incorporated into the part. Construction geometry is ignored when the sketch is used to create a feature. Construction geometry uses the same line style as a centerline.

💡 The plane or face you select for the Base sketch determines the orientation of the part.

Rectangle and Parallelogram Sketch Entity

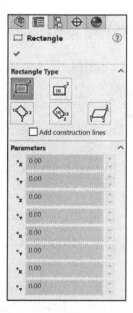

The Rectangle Sketch entity ✎ tool provides the ability to sketch a *Corner Rectangle, Center Rectangle, 3 Point Corner Rectangle* and a *3 Point Center Rectangle*. The Parallelogram Sketch entity provides the ability to sketch a *Parallelogram*.

The Rectangle and Parallelogram Sketch entity uses the Consolidated Rectangle PropertyManager. The Rectangle PropertyManager provides the following selections:

- *Rectangle Type*. The Rectangle Type box provides five selections:

  - **Corner Rectangle**. Sketch a standard rectangle at a corner point. Click to place the first corner of the rectangle, drag, and release when the rectangle is the correct size and shape.

  - **Center Rectangle**. Sketch a rectangle at a center point. Click to define the center. Drag to sketch the rectangle with centerlines. Release to set the four edges.

  - **3 Point Corner Rectangle**. Sketch a rectangle at a selected angle. Click to define the first corner. Drag, rotate, and then release to set the length and angle of the first edge. Click and drag to sketch the other three edges. Release to set the four edges.

- **3 Point Center Rectangle**. Sketch a rectangle with a center point at a selected angle. Click to define the first corner. Drag and rotate to set one half the length of the centerlines. Click and drag to sketch the other three edges and centerlines. Release to set the four edges.

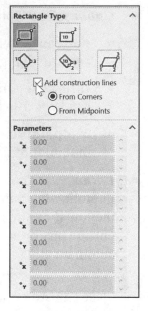

- **Parallelogram**. Sketch a standard parallelogram. A Parallelogram is a rectangle whose sides are not horizontal or vertical with respect to the sketch grid. Click to define the first corner. Drag, rotate, and then release to set the length and angle of the first edge. Click, rotate, and drag to set the angle and length of the other three edges. Release to set the four edges.

- *Add construction lines*. Converts the entities to construction geometry.

  - **From Corners**. Adds centerlines from corner to corner.

  - **From Midpoints**. Adds centerlines from midpoint of line segments.

- *Parameters*. Specify the appropriate combination of parameters to define the rectangle or parallelogram if they are not constrained by relations.

Slot Sketch Entity

The Slot Sketch entity 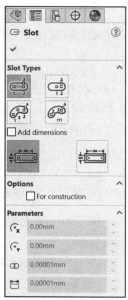 tool provides the ability to sketch a *Straight Slot*, *Centerpoint Straight Slot*, *3 Point Arc Slot* and a *Centerpoint Arc Slot* from the Consolidated Slot toolbar.

The Slot Sketch entity uses the Consolidated Slot PropertyManager. The Slot PropertyManager provides the following selections:

- *Slot Type*. The Slot Type box provides four selections:

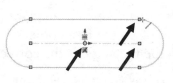

  - **Straight Slot**. Sketch a three point slot.

  - **Centerpoint Straight Slot**. Sketch a three point slot located at an Origin or centerpoint.

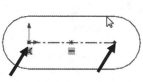

  - **3 Point Arc Slot**. Sketch a four point arc slot.

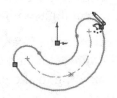

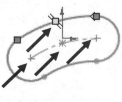

- **Centerpoint Arc Slot**. Sketch a four point arc slot located at an Origin or reference point.

- **Add dimensions**. Insert dimensions as you sketch the slot.

- **Center to Center**. Insert center to center dimension if the Add dimensions box is checked.

- **Overall Length**. Insert overall length if the Add dimensions box is checked.

- *Parameters*. Provides the ability to specify the appropriate combination of parameters to define slot if they are not constrained by relations. The parameters box provides the following options: **Center X Coordinate, Center Y Coordinate, Radius of Arc, Angle of Arc, Slot Width, Slot Length**.

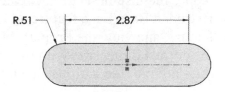

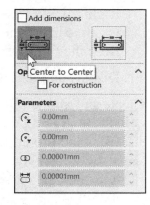

Tutorial: Slot Sketch - Instant3D 4-1

Utilize the Slot Sketch tool. Create an Extruded Boss/Base and Extruded Cut feature using the Instant3D tool.

1. Create a **New** part. Use the default ANSI, IPS Part template.

2. Right-click **Front Plane** from the FeatureManager.

3. Click **Sketch** from the Context toolbar. The Sketch toolbar is displayed.

4. Click **Centerpoint Straight Slot** from the Consolidated Slot toolbar as illustrated. The Slot PropertyManager is displayed. If needed, uncheck the Add dimension box.

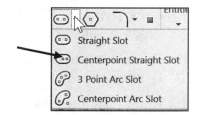

5. Click the **Origin**.

6. Click a **position** directly to the right of the Origin.

7. Click a **position** directly above the second point as illustrated.

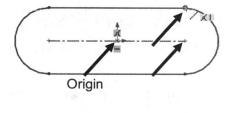

Dimension the slot sketch.

8. Right-click **Smart Dimension** from the Context toolbar in the Graphics window.

9. **Dimension** the model as illustrated. The sketch is fully defined. Note the location of the Origin.

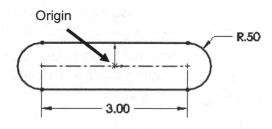

Apply the Instant3D tool. Create an Extruded Boss/Base feature. The Extruded Boss/Base feature adds material to a part. Note: Edit the feature, and view the PropertyManager to confirm that you picked the correct dimension with the on-screen ruler. The Instant3D tool does not provide various End Conditions for Design Intent.

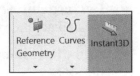

Exit Sketch1.

10. Right-click **Exit Sketch** . Note: Instant3D is active by default. If needed click Instant3D from the Features tab.

11. Display an **Isometric view**.

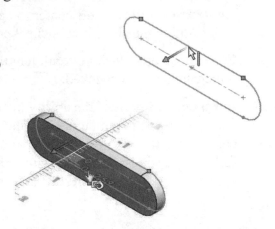

Create an Extruded Boss/Base feature with a default Blind End Condition.

12. Click the **top edge** as illustrated. A green arrow is displayed. Drag the **arrow** forward and click the on-screen ruler approximately **0.25in**.

13. **View** Boss-Extrude1 in the FeatureManager.

14. **Edit** Boss-Extrude1 in the FeatureManager.

15. Enter **Depth** = .30in. Click **OK** from the PropertyManager.

Create an Extruded Cut feature using Instant3D.

16. Display a **Front view**.

17. Right-click the **Front face**. This is your Sketch plane.

18. Click **Sketch** from the Context toolbar. The Sketch toolbar is displayed.

19. Click the **Circle** Sketch tool.

20. Sketch a **circle** centered at the Origin as illustrated.

Exit the Sketch.

21. Right-click **Exit Sketch**. Display an **Isometric view**.

Create an Extruded Cut feature.

22. Click the **circumference** of the sketch as illustrated.

23. Click and drag the arrow **backwards** to create a hole in the model. Note: The default End Condition is Blind.

24. **Release** the mouse icon behind the model. The Cut-Extrude1 feature is displayed. View the model.

25. **Close** the model.

Polygon Sketch Entity

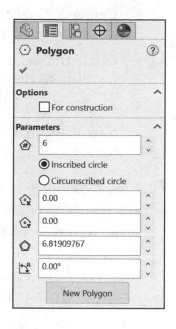

The Polygon Sketch entity ✎ tool provides the ability to create equilateral polygons with any number of sides between 3 and 40. The Polygon Sketch entity uses the Polygon PropertyManager. The Polygon PropertyManager provides the following selections:

- ***Existing Relations***. The Existing Relations box displays information on existing relations of the polygon sketch.

- ***Add Relation***. The Add Relation box displays the selected relations to the points and lines of your polygon sketch.

⛬ You cannot add a relation to a complete polygon. You can only add a relation to the points and lines of the polygon.

- ***Options***. The Options box provides the following selection:

  - **For construction**. Not selected by default. Converts the selected sketch entity to construction geometry.

- ***Parameters***. The Parameters box provides the ability to specify the appropriate combination of parameters to define the polygon. When you modify or change one or more parameters, the other parameters update automatically. The available selections are:

  - **Number of Sides**. Sets the number of sides in your polygon.

  - **Inscribed circle**. Selected by default. Displays an inscribed circle inside the polygon. This option defines the size of the polygon. The circle is construction geometry.

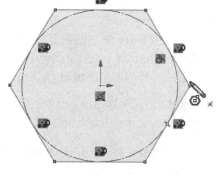

  - **Circumscribed circle**. Displays a circumscribed circle outside of the polygon. This option defines the size of the polygon. The circle is construction geometry.

  - **Center X Coordinate**. Displays the X coordinate for the center of your polygon.

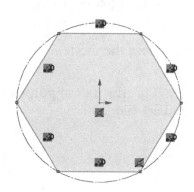

- **Center Y Coordinate**. Displays the Y coordinate for the center of your polygon.

- **Circle Diameter**. Displays the diameter of the inscribed or circumscribed circle.

- **Angle**. Displays the angle of rotation.

- **New Polygon**. Creates more than a single polygon without leaving the PropertyManager.

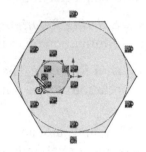

Tutorial: Polygon 4-1

Create a new part using the Polygon Sketch entity.

1. Create a **New** 🗋 part. Use the default ANSI Part template.

2. Right-click **Top Plane** from the FeatureManager design tree. This is your Sketch plane.

3. Click **Sketch** ✏ from the Context toolbar. The Top Plane boundary is displayed in the Top view.

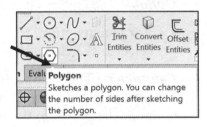

4. Click the **Circle** ⊙ Sketch tool. The Circle PropertyManager is displayed.

5. Sketch a **circle** centered at the origin.

Insert a polygon.

6. Click **Tools ➤ Sketch Entities ➤ Polygon** from the Menu bar or click **Polygon** from the Sketch toolbar. The Polygon PropertyManager is displayed.

7. Sketch a **Polygon** centered at the Origin larger than the circle as illustrated.

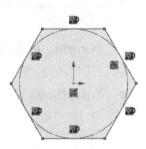

8. Click **OK** ✔ from the Polygon PropertyManager.

9. **Rebuild** 🗓 the model. View the created Polygon in the Graphics window.

10. **Close** the model.

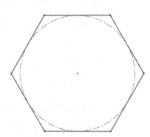

🔅 For illustration purposes, the Reference planes and grid are hidden.

Circle Sketch and Perimeter Circle Sketch Entity

The Circle Sketch and Perimeter Circle

Sketch entity 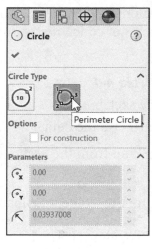 tool provides the ability to control the various properties of a circle.

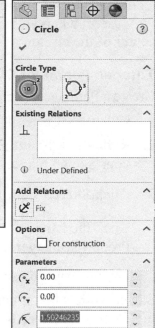

The Circle Sketch and Perimeter Circle Sketch entity uses the consolidated Circle PropertyManager. The consolidated Circle PropertyManager provides the following selections:

- *Circle Type*. Provides the ability to select either a center based circle or a perimeter-based circle sketch entity.

- *Existing Relations*. The Existing Relations box provides the following options:

  - **Relations**. Displays the automatic relations inferenced during sketching or created manually with the Add Relations tool. The callout in the Graphics window is highlighted when you select a relation in the dialog box. The Information icon in the Existing Relations box displays the sketch status of the selected sketch entity. Example: Under Defined, Fully Defined, etc.

- *Add Relations*. Provides the ability to add relations to the selected entity. Displays the relations which are possible for the selected entity.

- *Options*. **For construction**. Not selected by default. Converts the selected entity to construction geometry.

- *Parameters*. Specifies the appropriate combination of parameters to define the circle if the circle is not constrained by relations. The available selections are:

  - **Center X Coordinate**. Sets the Center X Coordinate value.

  - **Center Y Coordinate**. Sets the Center Y Coordinate value.

  - **Radius**. Sets the Radius value.

The Perimeter Circle Sketch entity provides the ability to define a circle using 3 points along its perimeter. This sketch entity is very useful when you are not concerned about the location of the center of the circle.

Tutorial: Perimeter Circle 4-1

Create a Perimeter Circle sketch.

1. Create a **New** ☐ part. Use the default ANSI Part template.

2. Right-click **Front Plane** from the FeatureManager. This is your Sketch plane.

3. Click **Sketch** ⌗ from the Context toolbar.

4. Click the **Line** ╱ Sketch tool. The Insert Line PropertyManager is displayed.

5. Sketch **two lines** as illustrated.

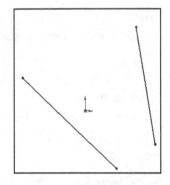

Sketch a circle that is tangent to both lines.

6. Click **Perimeter Circle** from the Consolidated Circle Sketch toolbar. The Circle PropertyManager is displayed.

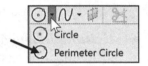

7. Click **each line** once.

8. Click a **position** as illustrated to determine the size of the circle. The X coordinate, Y coordinate, and Radius are displayed in the Parameters box. The two lines are tangent to the circle. You created a simple circle using the Perimeter Circle sketch entity.

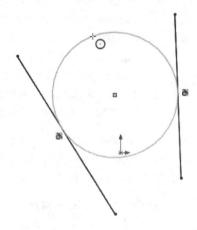

9. Click **OK** ✔ from the Circle PropertyManager. View the model.

10. **Rebuild** 🛈 the model.

11. **Close** the model.

🔅 If the two lines change, the circle remains tangent to the lines.

🔅 In System Options, under the Sketch folder, apply the Ghost image on drag option. The option displays a ghost image of a sketch entities' original position while you drag the sketch in the Graphics window.

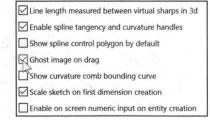

Centerpoint Arc Sketch Entity

The Centerpoint Arc Sketch entity 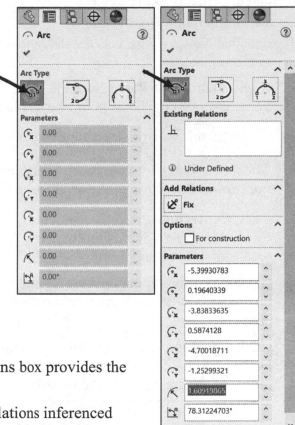 tool provides the ability to create an arc from a center point, a start point, and an end point. The Centerpoint Arc Sketch entity uses the consolidated Arc PropertyManager.

The Arc PropertyManager controls the properties of a sketched *Centerpoint Arc, Tangent Arc and 3 Point Arc*.

The consolidated Arc PropertyManager provides the following selections:

- *Arc Type*. Provides the ability to select a Centerpoint Arc, Tangent Arc or 3 Point Arc sketch entity.

- *Existing Relations*. The Existing Relations box provides the following options:

  - **Relations**. Displays the automatic relations inferenced during sketching or created manually with the Add Relations tool. The callout in the Graphics window is highlighted when you select a relation in the dialog box. The Information icon in the Existing Relations box displays the sketch status of the selected sketch entity. Example: Under Defined, Fully Defined, etc.

- *Add Relations*. Provides the ability to add relations to the selected entity. Displays the relations which are possible for the selected entity.

- *Options*. **For construction**. Converts the selected entity to construction geometry.

- *Parameters*. Provides the ability to specify any appropriate combination of parameters to define your arc if the arc is not constrained by relations. The available selections are **Center X Coordinate, Center Y Coordinate, Start X Coordinate, Start Y Coordinate, End X Coordinate, End Y Coordinate, Radius** and **Angle**.

Tutorial: Centerpoint Arc 4-1

Create a Centerpoint Arc sketch.

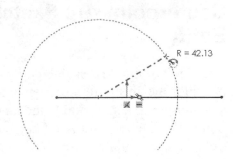

1. Create a **New** ☐ part. Use the default ANSI Part template.

2. Right-click **Front Plane** from the FeatureManager. This is your Sketch plane.

3. Click **Sketch** ⌐ from the Context toolbar.

4. Click the **Line** ✐ Sketch tool. Sketch a **horizontal line** through the origin as illustrated.

5. Click **Centerpoint Arc** ⌐ from the Consolidated Arc Sketch toolbar. The Arc icon is displayed on your mouse pointer.

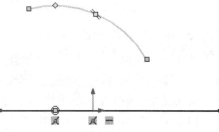

6. Click a location on the **horizontal line** left of the origin. Do not pick the origin or the endpoint.

7. Click a **position** to the right and up on the horizontal line. Drag to set the **angle** and the radius of the arc. Release the **mouse** button to place the arc. View the Arc PropertyManager and the coordinates.

8. Click **OK** ✔ from the Arc PropertyManager.

9. **Rebuild** 🔘 the model. **Close** the model.

Tangent Arc Sketch Entity

The Tangent Arc Sketch entity ⟋ tool provides the ability to create an arc, which is tangent to a sketch entity. The Tangent Arc Sketch entity uses the Arc PropertyManager. View the Centerpoint Arc Sketch entity section above for additional information on the Arc PropertyManager.

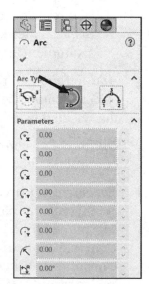

🔅 You can transition from sketching a line to sketching a Tangent Arc, and vice versa without selecting the Arc tool from the Sketch toolbar. Example: Create a line. Click the endpoint of the line. Move your mouse pointer away. The Preview displays another line. Move your mouse pointer back to the selected line endpoint. The Preview displays a tangent arc. Click to place the arc. Move the mouse pointer away from the arc endpoint. Perform the same procedure above to continue.

Tutorial: Tangent Arc 4-1

Create a Tangent Arc sketch.

1. Create a **New** ⬜ part. Use the default ANSI Part template.

2. Right-click **Front Plane** from the FeatureManager.

3. Click **Sketch** 🗔 from the Context toolbar.

4. Click the **Line** ✐ Sketch tool.

5. Sketch a **horizontal line** through the Origin.

6. Click **Tangent Arc** ⌓ from the Consolidated Arc Sketch toolbar. The Arc icon is displayed on the mouse pointer.

7. Click the **left end point** of the horizontal line.

8. Drag the **mouse pointer** upward and to the left for the desired shape.

9. Click an **end point**. You created a Tangent Arc. The Arc PropertyManager is displayed.

10. Right-click **Select** in the Graphics window to deselect the Sketch tool.

11. Click **OK** ✔ from the Arc PropertyManager.

12. **Exit** the sketch. **Close** the model.

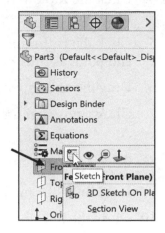

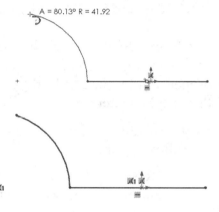

💡 You can select an end point of a line, ellipse, arc or a spline. The Tangent Arc tool creates both Normal and Tangent Arcs.

3 Point Arc Sketch Entity

The 3 Point Arc Sketch entity ⌒ tool provides the ability to create an arc by specifying three points: a starting point, an endpoint, and a midpoint. View the Centerpoint Arc Sketch entity section for additional information on the Arc PropertyManager.

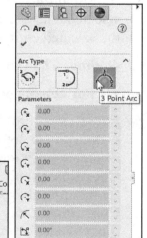

Tutorial: 3 Point Arc 4-1

Create a 3Point Arc sketch.

1. Open **3 Point Arc 4-1** from the SOLIDWORKS 2018 folder.

2. **Edit** Sketch1.

3. Click **3 Point Arc** ⌒ from the Consolidated Arc Sketch toolbar.

4. Click the **top point** of the vertical line for the start point as illustrated.

5. Click the **right point** of the open circle for the end point as illustrated.

6. Drag the **arc** downward to set the radius.

7. Click a **position** for the midpoint of the 3Point Arc.

8. Click **OK** ✔ from the Arc PropertyManager.

9. **Rebuild** the model.

10. **Close** the model.

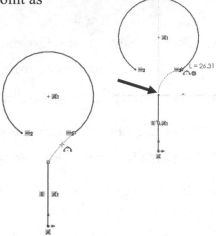

Ellipse Sketch Entity

The Ellipse Sketch entity ⌀ tool provides the ability to create a complete ellipse. The Ellipse PropertyManager controls the properties of a sketched Ellipse or a Partial Ellipse. The Ellipse PropertyManager provides the following selections:

- *Existing Relations*. The Existing Relations box provides the following options:

  - **Relations**. Displays the automatic relations inferenced during sketching or created manually with the Add Relations tool. The callout in the Graphics window is highlighted when you select a relation in the dialog box. The Information icon in the Existing Relations box displays the sketch status of the selected sketch entity. Example: Under Defined, Fully Defined, etc.

- *Add Relations*. Provides the ability to add relations to the selected entity. Displays the relations which are possible for the selected entity.

- *Options*. The Options box provides the following selection:

  - **For construction**. Converts the selected entity to construction geometry.

- *Parameters*. Provides the ability to specify any appropriate combination of parameters to define the ellipse if the ellipse is not constrained by relations. The available selections are **Center X Coordinate**, **Center Y Coordinate**, **Start X Coordinate** (only available for a Partial Ellipse), **Start Y Coordinate** (only available for a Partial Ellipse), **End X Coordinate** (only available for a Partial Ellipse), **End Y Coordinate** (only available for a Partial Ellipse), **Radius 1**, **Radius 2** and **Angle** (only available for a Partial Ellipse).

Tutorial: Ellipse 4-1

Create an Ellipse sketch.

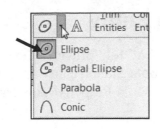

1. Create a **New** ▯ part. Use the default ANSI Part template.

2. Right-click **Front Plane** from the FeatureManager. This is your Sketch plane.

3. Click **Sketch** 🖉 from the Context toolbar.

4. Click the **Ellipse** ⊘ Sketch tool. The Ellipse icon is displayed on the mouse pointer.

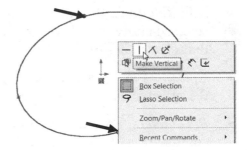

5. Click the **origin** in the Graphics window. This is the start location.

6. Drag and click to set the **major axis** of the ellipse. The ellipse PropertyManager is displayed.

7. Drag and click again to set the **minor axis** of the ellipse.

Deselect the sketch tool.

8. Right-click **Select** in the Graphics window.

Add a vertical geometric relation in the sketch.

9. Hold the **Ctrl** key down.

10. Click the **top vertical** point and the **bottom vertical** point of the Ellipse.

11. Right-click **Make Vertical** from the Pop-up menu.

12. **View** the sketch.

13. **Close** the model.

Partial Ellipse Sketch Entity

The Partial Ellipse Sketch entity 🖉 tool provides the ability to create a partial ellipse or an elliptical arc from a centerpoint, a start point, and an end point. You used a similar procedure when you created a Centerpoint Arc. The Ellipse PropertyManager controls the properties of a sketched Ellipse or a Partial Ellipse. View the Ellipse section for additional information on the Ellipse PropertyManager.

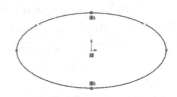

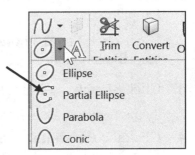

Parabola Sketch Entity

The Parabola Sketch entity $\cup$ tool provides the ability to create a parabolic curve. The Parabola Sketch entity uses the Parabola PropertyManager. The Parabola PropertyManager provides the following selections:

- **Existing Relations**. The Existing Relations box provides the following options:

  - **Relations**. Displays the automatic relations inferenced during sketching or created manually with the Add Relations tool. The callout in the Graphics window is highlighted when you select a relation in the dialog box. The Information icon in the Existing Relations box displays the sketch status of the selected sketch entity. Example: Under Defined, Fully Defined, etc.

- **Add Relations**. Provides the ability to add relations to the selected entity. Displays the relations which are possible for the selected entity.

- **Options**. **For construction**. Not selected by default. Converts the selected entity to construction geometry.

- **Parameters**. Provides the ability to specify the appropriate combination of parameters to define the parabola if the parabola is not constrained by relations. The available selections are **Start X Coordinate, Start Y Coordinate, End X Coordinate, End Y Coordinate, Center X Coordinate, Center Y Coordinate, Apex X Coordinate** and **Apex Y Coordinate**.

Tutorial: Parabola 4-1

Create a Parabola sketch.

1. Create a **New** part. Use the default ANSI, IPS Part template.

2. Right-click **Front Plane** from the FeatureManager. This is your Sketch plane.

3. Click **Sketch** from the Context toolbar.

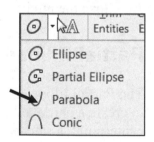

Apply the Parabola Sketch tool.

4. Click the **Parabola** $\cup$ Sketch tool from the Consolidated Sketch toolbar. The Parabola icon is displayed on the mouse pointer.

5. Click a **position** in the Graphics window to locate the focus of your parabola.

6. Drag to **enlarge** the parabola. The parabola is outlined.

7. Click on the **parabola** and drag to define the extent of the curve. The Parabola PropertyManager is displayed.

8. Click **OK** ✔ from the Parabola PropertyManager.

9. **Rebuild** 🔋 the model.

10. **Close** the model.

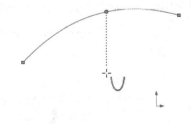

Conic Sketch Entity

The Conic Sketch entity ⋀ tool provides the ability to sketch conic curves driven by endpoints and Rho value. Depending on the Rho value, the curve can be elliptical, parabolic, or hyperbolic. In previous versions of the SOLIDWORKS software, you could sketch ellipses and parabolas. However, you could not sketch ellipses or parabolas by their endpoints, so it was difficult to make them tangent to existing geometry.

Conic curves can reference existing sketch or model geometry, or they can be standalone entities. You can dimension the curve with a driving dimension, and the resulting dimension displays the Rho value. The conic entity also includes a value for the radius of curvature.

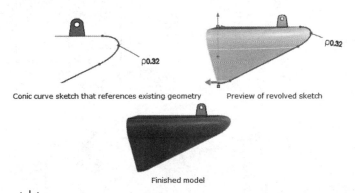

Conic curve sketch that references existing geometry   Preview of revolved sketch

Finished model

🔆 **Auto Tangency** accelerates the process of sketching a conic curve. When you select this option in the Conic PropertyManager, you set the first and second endpoints for the conic curve. The Conic tool applies tangent relations at each endpoint of the curve and selects the top vertex of the curve. This option is helpful when you reference existing geometry. If you clear this option, you must define the top vertex of the curve.

The Conic Sketch entity uses the Conic PropertyManager. The Conic PropertyManager provides the following selections:

- **Existing Relations**. The Existing Relations box provides the following options:

  - **Relations**. Displays the automatic relations inferenced during sketching or created manually with the Add Relations tool. The callout in the Graphics window is highlighted when you select a relation in the dialog box. The Information icon in the Existing Relations box displays the sketch status of the selected sketch entity. Example: Under Defined, Fully Defined, etc.

- **Add Relations**. Provides the ability to add relations to the selected entity. Displays the relations which are possible for the selected entity.

- **Options**. The Options box provides the following selection:

  - **For construction**. Not selected by default. Converts the selected entity to construction geometry.

- **Parameters**. Provides the ability to specify the appropriate combination of parameters to define the parabola if the parabola is not constrained by relations. The available selections are **Start X Coordinate, Start Y Coordinate, End X Coordinate, End Y Coordinate, Center X Coordinate, Center Y Coordinate, Apex X Coordinate** and **Apex Y Coordinate**.

Tutorial: Conic 4-1

Create a Conic sketch.

1. Open **Conic 4-1** from the SOLIDWORKS 2018 folder. The model contains an arc and spline.

Apply the Conic Sketch tool.

2. Click the **Conic** ⌒ Sketch tool from the Consolidated Sketch toolbar. The Conic icon is displayed on the mouse pointer.

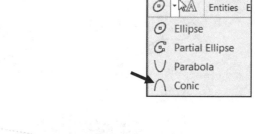

3. Click the **first endpoint** for the conic curve as illustrated.

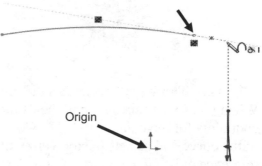

Origin

4. Drag the **pointer to the right** so it's tangent with the arc. Because the endpoint is attached to the existing arc sketch, a yellow inferencing line is displayed.

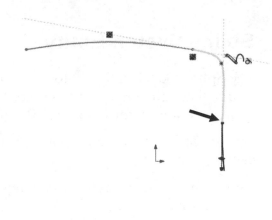

5. Click the **second endpoint** as illustrated.

6. Drag the **mouse pointer** upwards. A yellow inferencing line is displayed tangent to the spline sketch.

7. Drag the mouse pointer out to the **intersection** of the two yellow inferencing lines. You can snap to the intersection of both inference lines to choose the top vertex of the conic curve.

8. Click at the **intersection of both inferencing lines** to set the top vertex of the conic curve. By selecting the intersection of both inferencing lines, you ensure that the conic curve is tangent at both endpoints.

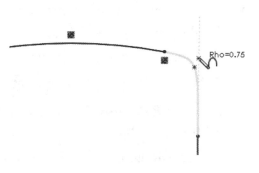

9. Drag the **pointer to the left** until the Rho value is 0.75. As you move the pointer, the conic curve's Rho value changes.

10. **Click** to set the Rho value to 0.75. Reference points are displayed for the curve's shoulder and top vertex. Tangent relations are created between the curve and the original sketches.

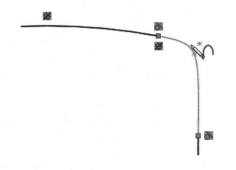

11. Click **OK** ✔ from the Conic PropertyManager.

12. **Rebuild** 🔘 the model.

13. **Close** the model.

Conic surfaces are very useful for precision flowing (aerodynamic) shapes. It is best to have no more than one singular point on a surface. The midline curve is usually shared by the top and bottom fuselage shape surfaces.

Rho is the squareness for conic, and translates into the squareness of the surface. In theory at 1.0 it is a dead sharp corner, 0.8 is a very modern squarish corner, 0.6 is most useful for airplane fuselages, 0.4 looks like an ellipse, and 0.1 is a slanted ellipse section that tends toward a line between the end points.

Spline Sketch Entity

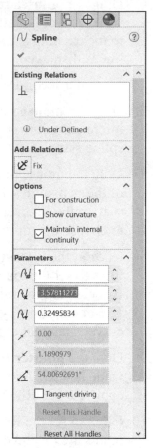

The Spline Sketch entity ∿ tool provides the ability to create a profile that utilizes a complex curve. This complex curve is called a Spline, (Non-uniform Rational B-Spline or NURB).

Create a spline with control points. With spline control points, you can:

- Use spline points as handles to pull the spline into the shape you want.

- Add dimensions between spline points or between spline points and other entities.

- Add relations to spline points.

The 2D Spline PropertyManager provides the following selections:

- **Existing Relations**. The Existing Relations box provides the following options:

  - **Relations**. Displays the automatic relations inferenced during sketching or created manually with the Add Relations tool. The callout in the Graphics window is highlighted when you select a relation in the dialog box. The Information icon in the Existing Relations box displays the sketch status of the selected sketch entity. Example: Under Defined, Fully Defined, etc.

- **Add Relations**. Provides the ability to add relations to the selected entity. Displays the relations which are possible for the selected entity.

- **Options**. The Options box provides the following selection:

  - **For construction**. Not selected by default. Converts the selected entity to construction geometry.

  - **Show Curvature**. Not selected by default. Provides the ability to control the Scale and Density of the curvature. Displays the Curvature of the spline in the Graphics window.

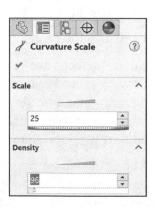

- **Maintain Internal Continuity**. Selected by default. Maintains the spline's internal curvature. When Maintain Internal Continuity is cleared, the curvature scales down abruptly.

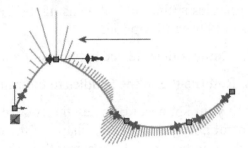

- **Raised degree**. Available only with splines that include curvature handles at each end. Raises or lowers the degree of the spline. You can also adjust the degrees by dragging the handles.

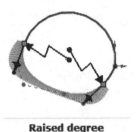

Raised degree

- *Parameters*. Provides the ability to specify the appropriate combination of parameters. The available parameters are:

  - **Spline Point Number**. Highlights your selected spline point in the Graphics window.

  - **X Coordinate**. Specifies the x coordinate of your spline point.

  - **Y Coordinate**. Specifies the y coordinate of your spline point.

  - **Radius of Curvature**. Controls the radius of curvature at any spline point. This option is only displayed if you select the Add Curvature Control option from the Spline toolbar, and add a curvature pointer to the spline.

  - **Curvature**. Displays the degree of curvature at the point where the curvature control was added. This option is only displayed if you add a curvature pointer to the spline.

  - **Tangent Weighting1**. Controls the left tangency vector by modifying the spline's curvature at the spline point.

  - **Tangent Weighting 2**. Controls the right tangency vector by modifying the spline's curvature at the spline point.

  - **Tangent Radial Direction**. Controls the tangency direction by modifying the spline's angle of inclination relative to the X, Y or Z axis.

- **Tangent Polar Direction**. Controls the elevation angle of the tangent vector with respect to a plane placed at a point perpendicular to a spline point. This is only for the 3D Spline PropertyManager.

- **Tangent Driving**. Enables spline control by using the Tangent Magnitude option and the Tangent Radial Direction option.

- **Reset This Handle**. Returns the selected spline handle to the initial state.

- **Reset All Handles**. Returns the spline handles to their initial state.

- **Relax Spline**. Sketch your spline and display the control polygon. You can drag any node on the control polygon to change shape. If the node you are dragging results in a spline which is not smooth, re-select the spline to display the PropertyManager. Click the Relax Spline selection under the Parameters option. This will re-parameterize or smooth the shape of your spline.

- **Proportional**. Retains the spline shape when you drag the spline.

Spline toolbar

Use the tools located on the Spline toolbar to control properties of a sketched spline. The Spline toolbar provides the following tools:

- *Add Tangency Control*. The Add Tangency Control tool adds a tangency control handle. Drag the control handle along the spline and position it to control the tangency at the selected point.

- *Add Curvature Control*. The Add Curvature Control tool adds a curvature control handle. Drag the control handle along the spline and position the control handle to control the spline shape at the selected point.

- *Insert Spline Point*. The Insert Spline Point tool adds a point to the spline. Drag spline points to reshape the spline.

You can insert dimensions between the spline points.

- *Simplify Spline*. The Simplify Spline tool reduces the number of points in the spline. This function improves the performance in the model for complex spline curves.

- *Fit Spline*. The Fit Spline tool L adds a spline based on the selected sketch entities and edges.

- *Show Spline Handles*. The Show Spline Handles tool 🕭 displays all handles of the selected spline in the Graphic window. Use the spline handles to reshape the spline.

- *Show Inflection Points*. The Show Inflection Points tool ⅄ displays all points where the concavity of a selected spline changes.

- *Show Minimum Radius*. The Show Minimum Radius tool ⋂ displays the measurement of the smallest radius in the selected spline.

- *Show Curvature Combs*. The Show Curvature Combs tool ⬦ displays the scalable curvature combs which visually enhance the curves of your selected spline.

Tutorial: 2D Spline 4-1

Create a 2D Spline sketch.

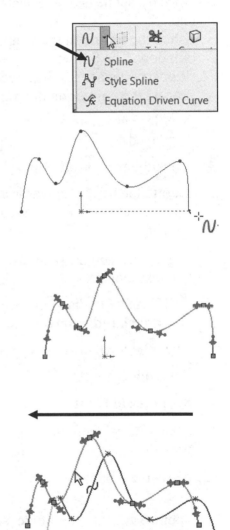

1. Create a **New** ▯ part. Use the default ANSI, IPS Part template.

2. Right-click **Right Plane** from the FeatureManager. This is your Sketch plane.

3. Click **Sketch** ⬒ from the Context toolbar.

Apply the Spline Sketch tool.
4. Click the **Spline** 𝒩 Sketch tool. The spline icon is displayed on the mouse pointer.

5. Create the Spline. This is an open profile. Create a **seven point control** spline as illustrated. Do not select the origin. The Spline PropertyManager is displayed.

6. Right-click **Select** on the last control point.

7. **Exit** the Spline 𝒩 tool.

8. Click and drag the **spline** from left to right. Do not select a control or handle. The spline moves without changing its shape.

9. Click **OK** ✔ from the Spline PropertyManager.

10. **Rebuild** 🛢 the model. View the results.

11. **Close** the model.

Spline handles are displayed by default. To hide or display spline handles, click the Show Spline Handles tool from the Spline toolbar or click Tools ➤ Spline Tools ➤ Show Spline Handles.

Tutorial: 2D Spline 4-2

Dimension a 2D Spline sketch handle.

1. Open **2D Spline 4-2** from the SOLIDWORKS 2018 folder.

2. **Edit** Sketch1.

Use the Smart Dimension tool to dimension a spline handle. Display the handles.

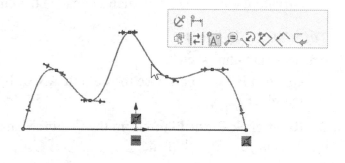

3. Click the **Spline** in the Graphics window. The Spline PropertyManager is displayed. Locate the pointer on the top handle. The rotate ↻ icon is displayed.

4. Click the **top handle tip**.

5. **Rotate** the handle upwards as illustrated to create a Tangent Driving.

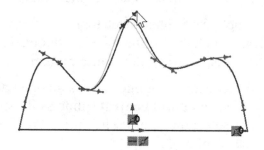

6. Right-click **Smart Dimension** from the Context toolbar.

Create a Tangent Radial dimension.

7. Click the **handle arrow tip**.

8. Click the **horizontal** line.

9. Enter **40**deg.

10. Set **precision** to none.

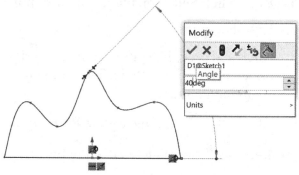

11. Click **OK** from the Dimension PropertyManager. View the results.

12. **Close** the model.

For a Tangent Magnitude, select the arrow tip on the handle to add the dimension.

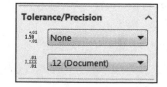

Illustrations may vary depending on your SOLIDWORKS version and system setup.

Tutorial: 3D Spline 4-1

Create a 3D Spline Sketch.

1. Create a **New** ⬚ part. Use the default ANSI Part template.

2. Click **Insert** ➤ **3D Sketch** from the Menu bar menu.

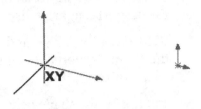

Apply the Spline Sketch tool.

3. Click the **Spline** ∿ Sketch tool.

4. Click a **location** in the Graphics window to place the first spline point. Each time you click a different location, the space handle is displayed to help you sketch on a different plane.

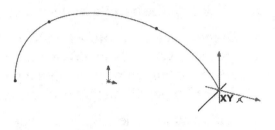

💡 Press the **Tab** key to change planes in a 3D Sketch.

5. Click **three** additional locations as illustrated. The Spline PropertyManager is displayed.

6. Right-click **Select** on the last control point.

7. **Exit** the 3D Sketch. 3DSketch1 is created and displayed in the FeatureManager.

8. **Close** the sketch.

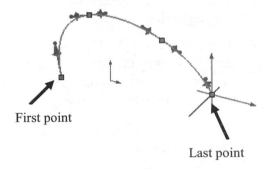

First point

Last point

Tutorial: 3D Spline 4-2

Edit a 3D spline sketch using the current elements.

1. Open **3D Spline 4-2** from the SOLIDWORKS 2018 folder.

2. **Edit** 3DSketch1.

3. Click the **Spline** in the Graphics window. The Spline PropertyManager is displayed.

Modify the Spline shape using the click and drag method on a control point.

4. Click a **spline control point** as illustrated and drag it upward. The Point PropertyManager is displayed. View the results.

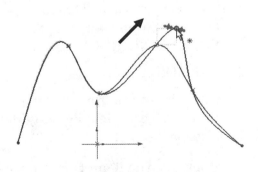

Modify the Spline shape using the x-, y- and
z- coordinates method on a control point.

5. Click a **spline point** as illustrated. The Point
    PropertyManager is displayed.

6. Set values for **x-, y-,** and **z-** coordinates from
    the Point PropertyManager. You modified the
    shape of the spline.

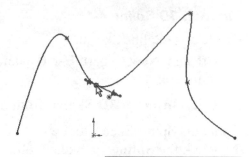

Modify the Spline shape using the click and drag
method on a handle.

7. Click a **spline handle** as illustrated.

8. Drag the selected **spline handle** to control the
    tangency vector by modifying the spline's
    degree of curvature at the spline point. The
    Spline PropertyManager is displayed.

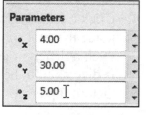

Reset the handle to the original location.

9. Click the **Reset This Handle** option. All values
    relative to the selected handle return to their
    original value.

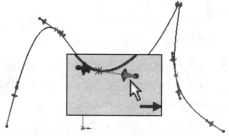

Move the spline.

10. Click and drag the **spline** from left to right. Do
    not select a control point. As the spline is
    moved, it retains its shape.

11. Click **OK** ✔ from the Spline
    PropertyManager.

12. **Exit** the 3D Sketch. View the results.

13. **Close** the model.

Tutorial: 3D Spline 4-3

Add new elements to a 3D Spline.

1. Open **3D Spline 4-3** from the
    SOLIDWORKS 2018 folder.

2. **Edit** 3DSketch1.

3. Click the **spline** from the Graphics
    window. The Spline PropertyManager is
    displayed.

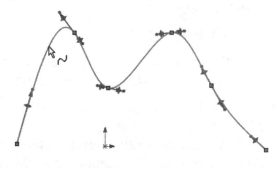

4. Click the **Add Tangency Control** 𝖸 tool
    from the Spline toolbar.

5. Drag to **place** the handle as illustrated.

6. Click a **position** for the handle.

7. Click **OK** ✔ from the Spline PropertyManager.

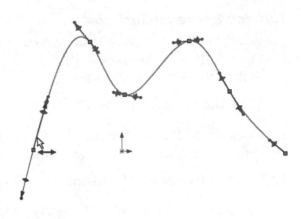

8. **Exit** the 3D Sketch. View the results.

9. **Close** the model.

Style Spline

The Style Spline 🖊 tool provides the ability to sketch single-span Bézier curves. These curves let you create smooth and robust surfaces, and are available in 2D and 3D sketches.

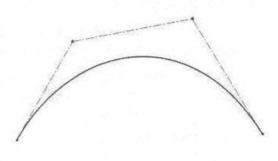

The Style Spline consists of only one span. You shape its curvature by selecting and dragging control vertex points. Sketch entities connect the control vertices which form the control polygon of the curve.

With the Style Spline tool, you can control the degree and continuity of the curves. You can inference style splines for tangency or equal curvature. You can constrain the points and dimension the curves' sides. These curves also support mirroring and self-symmetry.

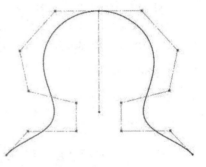

Spline on Surface Entity

The Spline on Surface entity 🖊 tool provides the ability to sketch splines on various surfaces. Splines sketched on surfaces include standard spline attributes, as well as the capabilities to add and drag points along the surface and to generate a preview that is automatically smoothed through the points.

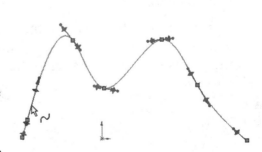

🔆 All spline points are bounded to the surface on which they are sketched.

Tutorial: Spline on Surface 4-1

Create a Closed Spline on Surface feature.

1. Open **Spline on Surface 4-1** from the SOLIDWORKS 2018 folder.

2. Click the **top left face** as illustrated.

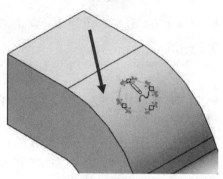

3. Click the **Spline on Surface** Sketch tool. The spline icon is displayed on the mouse pointer.

4. Create a **five point closed spline** as illustrated.

5. Click **OK** from Spline PropertyManager to close the sketch tool.

6. Click the **Extruded Boss/Base** Features tool. The Boss-Extrude PropertyManager is displayed.

7. Click the **left vertical line** as illustrated for Direction 1. Edge<1> is displayed in the Direction of Extrusion box.

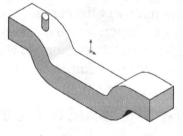

8. Enter **10**mm for Depth.

9. Click **OK** from the Boss-Extrude PropertyManager. Boss-Extrude1 is displayed. View the results in the Graphics window.

10. **Close** the model.

Splines on surfaces can:

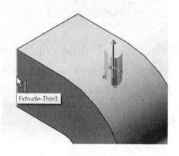

*   Span multiple faces.

*   Update if model geometry is updated (unless the splines are constrained).

*   You can move the points on the spline to different faces by dragging the points but not by dragging the curve itself.

Intelligence Modeling - Equation Driven Curve

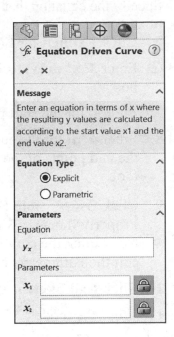

The Equation Driven Curve tool is a 2D Sketch spline driven by an equation. This Sketch entity provides the ability to create a curve by specifying the equations that define the curve. Equations defining a curve specify y- as a function of x-. You can use any function that is supported in the Equations dialog box. See Chapter 18 - Intelligence modeling techniques for additional information.

The Equation Driven Curve entity uses the Equation Driven Curve PropertyManager. The PropertyManager provides the following selections:

- **Equation Type**. There are equation types available: *Explicit* and *Parametric*.

- **Parameters**. The Parameters box provides the following options:

  - **Equation**. Provides the ability to enter an equation as a function of X-. If you enter an equation that can't be resolved, the text color is displayed in red.

  - **Parameters**. Provides the ability to enter the start and end X value of the equation.

  - **Transform**. Provides the ability to enter the translation of X-, Y- and angle for the curve.

Tutorial: Equation Driven Curve 4-1

Create an Equation Driven Curve on the Front Plane.

1. Create a **New** part. Use the default ANSI, IPS Part template.

Create a 2D Sketch on the Front Plane.

2. Right-click **Front Plane** from the FeatureManager.

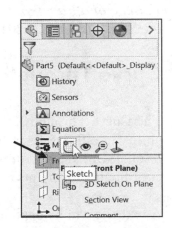

3. Click **Sketch** from the Context toolbar. The Sketch toolbar is displayed. Front Plane is your Sketch plane.

4. Click the **Equation Driven Curve** Sketch tool. The Equation Driven Curve PropertyManager is displayed. This tool provides the ability to build intelligences into a part.

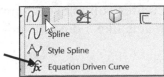

Specify the equation that defines the sketch. Note: y- is a function of x-.

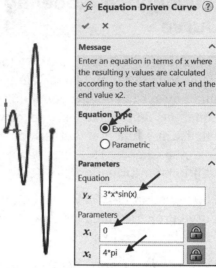

5. Click **Explicit** for Equation Type.

6. Enter **3*x*sin(x)** as illustrated in the Equation box.

7. Enter **0** for the start point as illustrated.

8. Enter **4*pi** for the end point. Note: You can fix the end points of the curve by applying the Fix tool.

9. Click **OK** ✔ from Equation Driven Curve PropertyManager. View the created curve. Note: click the curve in the Graphics window to edit the Parameters of the curve.

10. Click **Exit Sketch**. View the results.

11. **Close** the model.

🔅 Use regular mathematical notation and order of operations to write an equation. x_1 and x_2 are for the beginning and end of the curve. Use the transform options at the bottom of the PropertyManager to move the entire curve in x-, y- or rotation. To specify x=f(y) instead of y = f(x), use a 90 degree transform.

🔅 View the .mp4 provided movies in the book to better understand the potential of the Equation Driven Curve tool. The first one is *Calculating Area of a region bounded by two curves (secx)^2 and sin x* in SOLIDWORKS; the second one is *Determine the Volume of a Function Revolved Around the x Axis* in SOLIDWORKS. Both are located in the SOLIDWORKS 2018 folder.

Curve Through XYZ Points

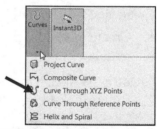

The Curve Through XYZ Points ⛎ feature provides the ability to either type in (using the Curve File dialog box) or click Browse and import a text file using x-, y-, z- coordinates for points on a curve.

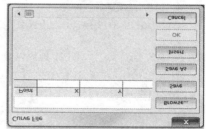

A text file can be generated by any program which creates columns of numbers. The Curve ⛎ feature reacts like a default spline that is fully defined.

Create a curve using the Curve Through XYZ Points tool. Import the x-, y-, z- data. Verify that the first and last points in the curve file are the same for a closed profile.

⚡ It is highly recommended that you insert a few extra points on either end of the curve to set end conditions or tangency in the Curve File dialog box.

Imported files can have an extension of either *.sldcrv or *.text.
The imported data: x-, y-, z-, must be separated by a space, a comma, or a tab.

Tutorial: Curve Through XYZ points 4-1

Create a curve using the Curve Through XYZ Points tool. Import x-, y-, z- data from a CAM program. Verify that the first and last points in the curve file are the same for a closed profile.

1. Open **Curve Through XYZ points 4-1** from the SOLIDWORKS 2018 folder.

2. Click the **Curve Through XYZ Points** ⟳ tool from the Features CommandManager. The Curve File dialog box is displayed.

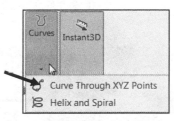

Import x-, y-, z- data. Verify that the first and last points in the curve file are the same for a closed profile.

3. Click **Browse** from the Curve File dialog box.

4. **Browse** to the SOLIDWORKS 2018 folder.

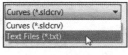

5. Set file type to **Text Files**.

6. Double-click **cam2.text**. View the data in the Curve File dialog box. View the sketch in the Graphics window. Review the data points in the dialog box.

7. Click **OK** from the Curve File dialog box. Curve1 is displayed in the FeatureManager. This curve can now be used in a sketch (fully defined profile). View the results of the fully defined sketch using the curve.

8. **Close** the model.

9. Open **Curve Through XYZ points 4-2** from the SOLIDWORKS 2018 folder to view the final model and FeatureManager.

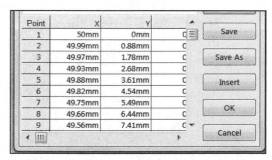

10. **Close** the model.

⚡ See Chapter 18 - Intelligence Modeling techniques for additional information.

Curve Through Reference Points

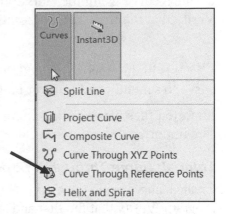

The Curve Through Reference Points tool provides
the ability to create a curve through points located on
one or more planes. See SOLIDWORKS Help for
additional information.

Point Sketch Entity

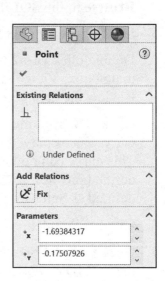

The Point Sketch entity tool provides the ability to insert points
in your sketches and drawings. To modify the properties of a
point, select the point in an active sketch, and edit the properties
in the Point PropertyManager. The Point PropertyManager
provides the following selections:

- ***Existing Relations***. The Existing Relations box provides the
  following options:

  - **Relations**. Displays the automatic relations inferenced
    during sketching or created manually with the Add
    Relations tool. The callout in the Graphics window is
    highlighted when you select a relation in the dialog box.
    The Information icon in the Existing Relations box
    displays the sketch status of the selected sketch entity.

- ***Add Relations***. Provides the ability to add relations to the
  selected entity. Displays the relations which are possible for
  the selected entity.

- *Parameters*. Provides the ability to specify the appropriate combination of parameters for your point only if the point is not constrained by relations. The available parameters are:

  - **X Coordinate**. Specifies the x coordinate of your point.

  - **Y Coordinate**. Specifies the y coordinate of your point.

Centerline Sketch Entity

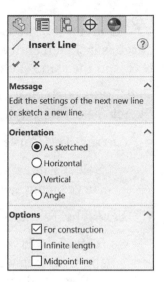

The Centerline Sketch entity tool provides the ability to use centerlines to create symmetrical sketch elements and revolved features, or as construction geometry. The Centerline sketch entity uses the Insert Line PropertyManager. The Insert Line PropertyManager provides the following selections:

- *Orientation*. The Orientation box provides the following options:

  - **As sketched**. Sketch a line in any direction using the click and drag method. Using the click-click method, the As sketched option provides the ability to sketch a line in any direction, and to continue sketching other lines in any direction, until you double-click to end your process.

  - **Horizontal**. Sketch a horizontal line until you release your mouse pointer.

  - **Vertical**. Sketch a vertical line until you release your mouse pointer.

  - **Angle**. Sketch a line at an angle until you release your mouse pointer.

The angle is created relative to the horizontal.

- *Options*. The Options box provides two line types:

  - **For construction**. Selected by default. Converts the selected sketch entity to construction geometry.

  - **Infinite length**. Not selected by default. Creates a line of infinite length which you can later trim in the design process.

  - **Midpoint line**. Creates a line that is symmetrical from the midpoint of the line.

Use Construction geometry to assist in creating your sketch entities and geometry that are ultimately incorporated into the part. Construction geometry is ignored when the sketch is used to create a feature. Construction geometry uses the same line style as a centerline.

Text Sketch Entity

The Text Sketch $\mathbb{A}$ entity tool provides the ability to sketch text on the face of a part and extrude or cut the text. Text can be inserted on any set of continuous curves or edges. This includes circles or profiles which consist of lines, arcs or splines.

Convert the sketch entities to construction geometry if the curve is a sketch entity, or a set of sketch entities, and your sketch text is in the same sketch as the curve.

The Text sketch entity uses the Sketch Text PropertyManager. The Sketch Text PropertyManager provides the following selections:

- *Curves*. The Curves box provides the following options:

  - **Select Edges, Curves Segment**. Displays the selected curves, edges, sketches, or sketch segments.

- *Text*. The Text box provides the following options:

  - **Text**. Displays the entered text along the selected entity in the Graphics window. If you do not select an entity, the text is displayed horizontally starting at the origin.

  - **Bold Style**. Bold the selected text.

  - **Italic Style**. Italicize the selected text.

  - **Rotate Style**. Rotate the selected text.

  - **Left Alignment**. Justify text to the left. Only available for text along an edge, curve or sketch segment.

  - **Center Alignment**. Justify text to the center. Only available for text along an edge, curve or sketch segment.

  - **Right Alignment**. Justify text to the right. Only available for text along an edge, curve or sketch segment.

  - **Full Justify Alignment**. Justify text. Only available for text along an edge, curve or sketch segment.

  - **Flip**. Provides the ability to flip the text.

- **Flip Vertical direction and back**. The Flip Vertical option is only available for text along an edge, curve or sketch segment.

- **Flip Horizontal direction and back**. The Flip Horizontal option is only available for text along an edge, curve or sketch segment.

- **Width Factor**. Provides the ability to widen each of your characters evenly by a specified percentage. Not available when you select the Use document's font option.

- **Spacing**. Provides the ability to modify the spacing between each character by a specified percentage. Not available when your text is fully justified or when you select the Use document's font option.

- **Use document's font**. Selected by default. Provides the ability to clear the initial font and to choose another font by using the Font option button.

- **Font**. Provides the ability to choose a font size and style from the Font dialog box.

Tutorial: Text 4-1

Apply text on a curved part. Use the Sketch Text tool.

1. Open **Text 4-1** from the SOLIDWORKS 2018 folder. The Lens Cap is displayed in the Graphics window.

2. Click **Front Plane** from the FeatureManager.

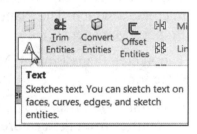

Apply the Text Sketch tool.

3. Click the **Text** 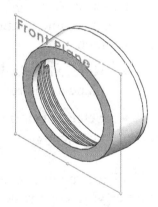 Sketch tool. The Sketch Text PropertyManager is displayed.

4. Click **Inside** the Text box.

5. Type **SOLIDWORKS**. SOLIDWORKS is displayed horizontally in the Graphics window starting at the origin because you do not select an entity.

6. Click **inside** the Curves option box. Click the **front edge** of Base Extrude. Edge<1> is displayed in the Curves option box. The SOLIDWORKS text is displayed on the selected edge.

7. Uncheck the **Use document font** box.

8. Enter **250**% in the Spacing box.

9. Click the **Font** Button. The Choose Font dialog box is displayed.

10. Select **Arial Black, Regular, Point 14**.

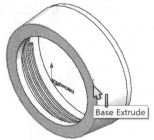

11. Click **OK** from the Choose Font dialog box.

12. Click **OK** ✔ from the Sketch Text PropertyManager. Sketch7 is created and is displayed in the FeatureManager. Zoom in on the created text.

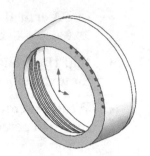

13. **Close** the model.

🔅 If the curve is a sketch entity, or a set of sketch entities, and the sketch text is in the same sketch as the curve, convert the sketch entities to construction geometry.

Plane Sketch Entity

The Plane Sketch entity tool provides the ability for the user to add Reference entities to a 3D sketch. Why? To facilitate sketching, and to add relations between sketch entities. Once you add the plane, you need to activate it to display its properties and to create a sketch. The Plane sketch entity uses the Sketch Plane PropertyManager.

The Sketch Plane PropertyManager provides the following selections:

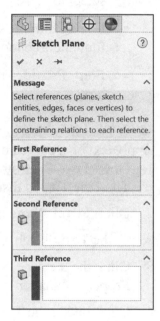

- *First Reference*. Provides the ability to select Reference entities from the FeatureManager or from the Graphics window. Select the constraining relations to each reference. The available relations selection is based on the selected geometry.

  - **Reference Entities**. Displays the selected sketch entity or plane as the reference to position the 3D Sketch plane.

  - **Select a relation**. Select a relation based on the selected geometry.

🔅 The available relations are displayed in the Sketch Plane PropertyManager under the First, Second and Third Reference. The relations are based on existing geometry.

- ***Second Reference***. Provides the ability to select Reference entities from the FeatureManager or from the Graphics window. Select the constraining relations to each reference. The available relations selection is based on the selected geometry.

    - **Reference Entities**. Displays the selected sketch entity or plane as the reference to position the 3D Sketch plane.

    - **Select a relation**. Selects a relation based on the selected geometry.

- ***Third Reference***. Provides the ability to select Reference entities from the FeatureManager or from the Graphics window. Select the constraining relations to each reference. The available relations selection is based on the selected geometry.

    - **Reference Entities**. Displays the selected sketch entity or plane as the reference to position the 3D Sketch plane.

    - **Select a relation**. Selects a relation based on the selected geometry.

With existing geometry, add a plane by referencing the entities that are present. Use any number of references required to achieve the desired results.

Tutorial: Sketch Plane 4-1

Add a 3D Sketch plane. Add a plane using references. Select two corner points of a 3D sketch. The corner points of the rectangle reference the position of the 3D sketch plane.

1. Open **Sketch Plane 4-1** from the SOLIDWORKS 2018 folder.

2. **Edit** 3DSketch1 from the FeatureManager.

3. Click **Tools ➢ Sketch Entities ➢ Plane**. The Sketch Plane PropertyManager is displayed.

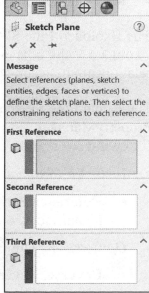

Rename a feature or sketch for clarity. Slowly click the feature or sketch name twice and enter the new name when the old one is highlighted.

4. Click the **top front corner point** of the rectangle as illustrated. Point3 is displayed in the First Reference box.

5. Click the **bottom back corner point** of the rectangle. Point4 is displayed in the Second Reference box.

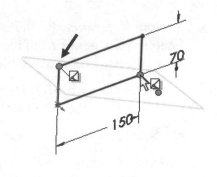

6. Click **OK** ✔ from the Sketch Plane PropertyManager.

7. **Rebuild** the model. Plane2@3DSketch1 is created and is displayed in the Graphics window.

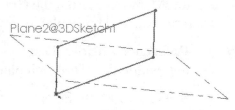

8. **Close** the model.

Route Line Sketch Entity

The Route Line Sketch entity tool provides the ability to insert a Route line between faces, circular edges, straight edges, or planar faces. The Route Line Sketch entity is active when you:

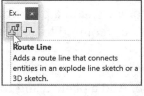

- Create an Explode Line Sketch.

- Edit an Explode Line Sketch.

- Select the Route Line tool in a 3D sketch.

The Exploded Line Sketch tool ⌖ is a 3D sketch added to an Exploded View in an assembly. The explode lines indicate the relationship between components in the assembly.

The Route Line Sketch entity ⌖ uses the Route Line PropertyManager. The Route Line PropertyManager provides the following selections:

- *Items To Connect*. The Items To Connect box provides the following option:

  - **Reference entities**. Displays the selected circular edges, faces, straight edges or planar faces to connect with your created route line.

- *Options*. The Options box provides the following selections:

  - **Reverse**. Reverses the direction of your route line. A preview arrow is displayed in the direction of the route line.

  - **Alternate Path**. Displays an alternate possible path for the route line.

  - **Along XYZ**. Creates a path parallel to the X, Y and Z axis directions.

Clear the Along XYZ option to use the shortest route.

Tutorial: Route Line 4-1

Create an Explode Line sketch using the Route Line Sketch entity.

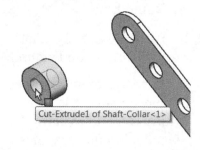

1. Open the **Route Line 4-1** assembly from the SOLIDWORKS 2018 folder. The assembly is displayed in an Exploded view.

2. Click the **Explode Line Sketch** ⚙ tool from the Assembly toolbar. The Route Line PropertyManager is displayed.

3. Click the **inside face** of Cut-Extrude1 of Shaft-Collar<1>. The direction arrow points towards the back. If required, check the **Reverse direction** box. Face<1> @Shaft-Collar-1 is displayed in the Items To Connect box.

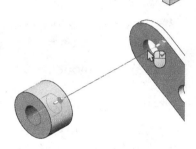

4. Click the **inside face** of the top hole of Cut-Extrude1 of the Flatbar<1> component. Face<2>@Flatbar-1 is displayed in the Items To Connect box. The arrow points towards the right.

5. Click the **inside face** of Cut-Extrude5 of RodClevis<1>. Face<3>@AirCylinder-1is displayed in the Items To Connect box. The arrow points towards the back.

6. Click the **inside face** of the second Cut-Extrude5 feature of RodClevis<1>. Face<4>@ AirCylinder-1 is displayed in the Items To Connect box.

7. Click the **inside face** of the top hole of Cut-Extrude1 of the Flatbar<2> component. Face<2>@Flatbar-2 is displayed in the Items To Connect box. The arrow points towards the back.

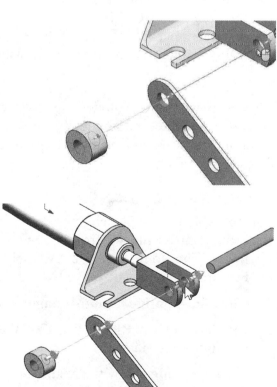

8. Click the **inside face** of Cut-Extrude1 of SHAFT-COLLAR<2>. The direction arrow points towards the back. Face<1> @Shaft-Collar<2> is displayed in the Items To Connect box.

9. Click **OK** ✔ from the Route Line PropertyManager. View the created Route Line.

10. Click **OK** ✔ from the PropertyManager to return to the FeatureManager.

Fit the model to the Graphics window.
Display an Isometric view.

11. Press the **f** key to fit the model to the Graphics window.

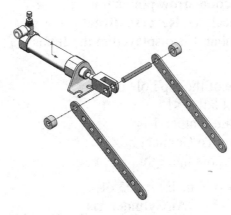

12. Click **Isometric view**. View the results.

13. **Close** the model.

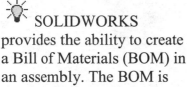 SOLIDWORKS provides the ability to create a Bill of Materials (BOM) in an assembly. The BOM is displayed in the Tables folder of the FeatureManager. See the Assembly section for additional information.

☀ In an assembly exploded view, you can create explode route lines automatically for selected components by using the Smart Explode Lines PropertyManager.

Tutorial: Route Line 4-2

Edit an Explode Line Sketch.

1. Open **Route Line 4-2** from the SOLIDWORKS 2018 folder. The Route Line 4-2 Assembly FeatureManager is displayed.

2. Click the **ConfigurationManager** tab.

3. **Expand** Default. Right-click **ExplView1** from the ConfigurationManager.

4. Click **Explode**. View the results in the Graphics window. **Expand** ExplView1.

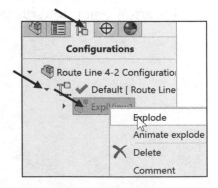

Edit the Explode Line Sketch.

5. Right-click **3DExplode1** from the
   ConfigurationManager.

6. Click **Edit Sketch**.

7. Display a **Top** view.

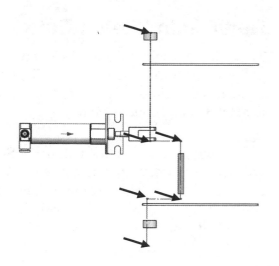

8. Click **Centerline** ✐ from the Sketch
   toolbar. The Insert Line PropertyManager is
   displayed.

9. Click the **endpoints** of the Route Line as
   illustrated.

Create a new path.

10. Click a **point** to the right of the Route Line
    endpoint and over the top of the axle.

11. Click a **point** below the axle as illustrated.

12. Click a **point** to the left of the axle over the Flatbar.

13. Click a **point** below the Flatbar.

14. Click a **point** above the SHAFT-COLLAR.

15. Click a **point** below the Shaft-Collar.

Deselect the active Sketch tool.

16. Right-click **Select** in the Graphics window.

17. **Rebuild** 🔘 the
    model. View the
    results.

18. **Return** to the
    FeatureManager
    design tree.

19. **Close** the model.

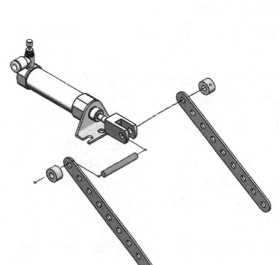

Belt/Chain Sketch Entity

This option provides the ability to create layout sketches for pulleys or sprocket mechanisms. The Belt/Chain feature simulates a cable-pulley mechanism. The Belt/Chain entity uses the Belt/Chain PropertyManager.

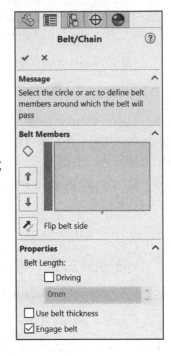

In the past, it was difficult using SOLIDWORKS to effectively model components such as cables, chains, and timing belts due to their flexible motion. The shapes of these items could change over time. This made it difficult to examine the way a belt would interact with two or more pulleys, or a cable system shifts a lever, or crank with a solid assembly model.

The Belt/Chain PropertyManager provides the following selections:

- *Belt Members*. The Belt Members box provides the following options:

  - **Pulley Components**. Displays the selected arcs or circles for the pulley components.

  - **Up arrow**. Moves the selected pulley component upwards in the displayed order.

  - **Down arrow**. Moves the selected pulley component downward in the displayed order.

  - **Flip belt side**. Changes the side on which the belt is located in the pulley system.

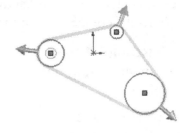

- *Properties*. The Properties box provides the ability to define the following belt conditions:

  - **Driving**. Cleared by default. Provides the ability to calculate the length of your belt. When checked, enter a value if you do not want the system to calculate your belt length. Based on the constraints, one or more of the components may move to adjust for the length of the belt.

  - **Use belt thickness**. Specifies the thickness value of the belt.

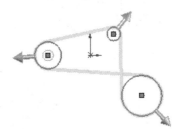

- **Engage belt**. Selected by default. Disengages the belt mechanism.

 Only circle entities in a block can be selected for a belt member.

Blocks

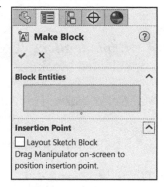

Blocks are a grouping of single or multiple sketch entities. In SOLIDWORKS, the origin of the block is aligned to the orientation of the sketch entity. Previously, blocks inherited their origin location from the Parent sketch. Additional enhancements to Blocks are:

- Modeling pulleys and chain sprockets.

- Modeling cam mechanisms.

Blocks are used in the Belt/Chain sketch entity. You can create blocks from a single entity or multiple sketch entities.

Why should you use Blocks? Blocks enable you to:

- Create layout sketches using the minimum of dimensions and sketch relations.

- Freeze a subset of sketch entities in a sketch to control as a single entity.

- Manage complicated sketches.

- Edit all instances of a block simultaneously.

 Use blocks in drawings to conserve drawing items such as standard label positions, notes, etc. See SOLIDWORKS Help for additional information.

Blocks toolbar

The Blocks toolbar controls blocks in sketching. The Blocks toolbar provides the following tools:

- **Make Block**. The Make Block 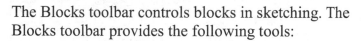 tool provides the ability to create a new block.

- **Edit Block**. The Edit Block tool provides the ability to add or remove sketch entities and to modify dimensions and relations in your sketch.

- **Insert Block**. The Insert Block tool provides the ability to add a new block to your sketch or drawing. This option provides the ability to either create multiple instances of existing blocks or to browse for existing blocks.

- **Add/Remove**. The Add/Remove tool provides the ability to add or remove sketch entities to or from your block.

- **Rebuild**. The Rebuild 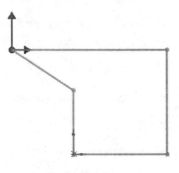 tool provides the ability to update the parent sketches affected by your block changes.

- **Saves Block**. The Saves Block tool provides the ability to save the block to a file. Adds an .sldblk extension.

- **Explode Block**. The Explode Block tool provides the ability to explode the selected block.

- **Belt/Chain**. The Belt/Chain tool provides the ability to insert a belt.

Exploding a single instance of the block only affects that instance of the block.

Tutorial: Block 4-1

Create a Block using the Blocks toolbar.

1. Create a **New** part. Use the default ANSI Part template.

2. Click **Top Plane** from the FeatureManager. This is your Sketch plane.

3. Sketch a **closed sketch** using sketch entities as illustrated. The sketch consists of five lines.

4. Click **Make Block** from the Block toolbar or click **Tools ➤ Blocks ➤ Make**. The Make Block PropertyManager is displayed.

5. Box-select the **sketch entities** from the Graphics window. The selected sketch entities are displayed in the Block Entities box.

6. **Expand** the Insertion Point box. Note the location of the insertion point.

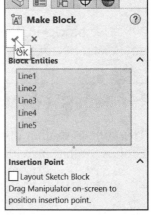

7. Click **OK** from the Make Block PropertyManager. Block1-1 is displayed in the FeatureManager.

8. **Rebuild** the model.

9. **Close** the model.

Create a block for a single or combination of multiple Sketch entities. Saving each block individually provides you with extensive design flexibility.

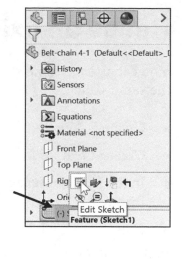

Tutorial: Belt-chain 4-1

Use the Belt/Chain Sketch entity function. Create a pulley sketch.

1. Open **Belt-chain 4-1** from the SOLIDWORKS 2018 folder. View the three created blocks in the Belt-chain FeatureManager.

2. Edit **Sketch1** from the FeatureManager.

3. Click **Tools ➤ Sketch Entities ➤ Belt/Chain**. The Belt/Chain PropertyManager is displayed.

4. Select the **three circles** from left to right. The select circles are displayed in the Belt Members dialog box.

5. Check the **Driving** box.

6. Enter **460**mm for Belt Length.

7. Check the **Use belt thickness** box.

8. Enter **20**mm for belt thickness.

9. Click **OK** ✔ from the Belt/Chain PropertyManager.

10. **Expand** Sketch1. Belt1 is created and is displayed in the FeatureManager. The calculated belt length was larger than the belt length entered in the PropertyManager. The system moved the entities to correct for the entered 460mm belt length.

11. **Rebuild** the model. View the results.

12. **Close** the model.

View SOLIDWORKS Help for additional information.

The Select Chain command has been enhanced to allow you to select construction entities. See SOLIDWORKS Help for additional information.

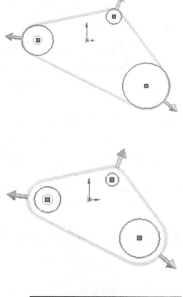

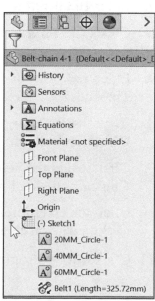

Reuse an existing 2D Sketch

You can use the same sketch to create different features. To create different features with the same sketch:

- Create a sketch.

- Create a feature from the sketch.

- In the FeatureManager design tree, select the same sketch used to create the first feature.

- Create a second feature from the sketch.

In the FeatureManager design tree, a sketch that is shared by multiple features is identified by the Shared 🔲 icon or by Shared 🔲 icon for a selected contour.

💡 Add additional sketches as required to create the features. For example, if one feature is an extrude and the other feature is a revolve use the same sketch for both features, but add a centerline to create the revolve.

Tutorial: Shared Sketch 4-1

Create an Extrude Thin1 feature with a shared sketch.

1. Open **Shared Sketch 4-1** from the SOLIDWORKS 2018 folder.

2. Click **Sketch1** from the FeatureManager. View the sketch in the Graphics window.

3. Click **Extruded Boss/Base** 📦 from the Feature toolbar.

4. Select **Mid Plane** for End Condition in Direction 1.

5. Enter **16.00mm** for Depth.

6. Check the **Thin Feature** dialog box.

7. Enter **4.50mm** for Thickness.

8. Click the **Reverse Direction** button.

9. Click inside the **Selected Contours** dialog box.

10. Click the **circumference of the large circle** in the Graphics window.

11. Click **OK** ✔ from the PropertyManager.

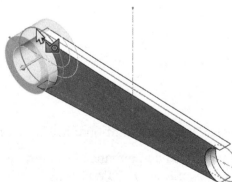

12. **Expand** Extrude-Thin1 in the
    FeatureManager. View the
    Shared sketch.

13. **Close** the model.

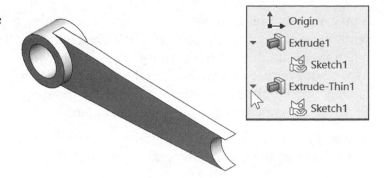

Summary

In this chapter, you learned about the available tools to incorporate Design Intent into a sketch, feature, part, assembly and drawing.

Sketching is the foundation of a model. You addressed 2D and 3D Sketching and how to recognize various Sketch states: *Fully Defined, Over Defined, Under Defined, No Solution Found and Invalid Solution Found* along with identifying the correct Reference plane.

You reviewed the following Sketch Entities: Line, Corner Rectangle, Center Rectangle, 3 Point Corner Rectangle, 3 Point Center Rectangle, Parallelogram, Straight Slot, Centerpoint Straight Slot, 3 Point Arc Slot, Centerpoint Arc Slot, Polygon, Circle, Perimeter Circle, Centerpoint Arc, Tangent Arc, 3 Point Arc, Ellipse, Partial Ellipse, Parabola, Conic, Spline, Style Spline, Spline on Surface, Equation Driven Curve, Point, Centerline, Text, Plane, Route line and Belt/Chain.

You utilized and addressed the following toolbars: Block toolbar and Spline toolbar and applied the Instant3D tool.

Fully define all sketches in the model. However, there are times when this is not practical, generally when using the spline tool to create a freeform shape.

Use Construction geometry to assist in creating your sketch entities and geometry that are ultimately incorporated into the part. Construction geometry is ignored when the sketch is used to create a feature. Construction geometry uses the same line style as a centerline.

 Press the Tab key to change planes in a 3D Sketch.

You can transition from sketching a line to sketching a Tangent Arc, and vice versa, without selecting the Arc tool from the Sketch toolbar. Example: Create a line. Click the endpoint of the line. Move your mouse pointer away. The Preview displays another line. Move your mouse pointer back to the selected line endpoint. The Preview displays a tangent arc. Click to place the arc. Move the mouse pointer away from the arc endpoint. Perform the same procedure above to continue.

Rename a feature or sketch for clarity. Slowly click the feature or sketch name twice and enter the new name when the old one is highlighted.

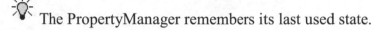 The plane or face you select for the Base sketch determines the orientation of the part.

The PropertyManager remembers its last used state.

Chapter 5 provides a comprehensive understanding of the available Sketch tools, Geometric relations, Dimensions/Relations tools and the DimXpertManager in SOLIDWORKS.

CHAPTER 5: SKETCH TOOLS, GEOMETRIC RELATIONS AND DIMENSIONS/RELATIONS TOOLS

Chapter Objective

Chapter 5 provides a comprehensive understanding of the available Sketch tools, Geometric relations, Dimensions/Relations tools and the DimXpertManager in SOLIDWORKS.

On the completion of this chapter, you will be able to:

- Understand and use the following Sketch tools:

  - Fillet, Chamfer, Offset Entities, Offset on Surface, Convert Entities, Intersection Curve, Face Curves, Segment, Trim, Extend, Split Entities, Jog Line, Construction Geometry, Make Path, Mirror Entities, Dynamic Mirror, Stretch Entities, Move Entities, Rotate Entities, Scale Entities, Copy Entities, Linear Pattern, Circular Pattern, Create Sketch From Selections, Repair Sketch, SketchXpert, Align, Modify, Close Sketch to Model, Check Sketch for Feature, 2D to 3D and Sketch Picture.

- Comprehend the 2D to 3D toolbar:

  - Front, Top, Left, Bottom, Back, Auxiliary, Sketch from Selections, Repair Sketch, Align Sketch, Extrude and Cut.

- Recognize and apply Geometric relations in 2D Sketches:

  - Automatic relations.

  - Manual relations.

- Know and apply Geometric relations in 3D Sketches.

- Understand and utilize the following Dimensions/Relations tools:

  - Smart Dimension, Horizontal Dimension, Vertical Dimension, Baseline Dimension, Ordinate Dimension, Horizontal Ordinate Dimension, Vertical Ordinate Dimension, Chamfer Dimension, Baseline Dimension, Add Relations, Display/Delete Relations and Fully Defined Sketch.

- Understand DimXpertManager and the following tools:

  - Auto Dimension Scheme, Basic Dimension, Show Tolerance Status, Copy Scheme and TolAnalyst.

Sketch tools

Sketch tools control the various aspects of creating and modifying a sketch. To access the available sketch tools for SOLIDWORKS, click **Tools ➢ Sketch Tools** from the Menu bar and select the required tool.

To obtain additional information on each sketch tool, activate the sketch tool and click the question mark 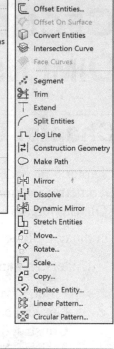 icon located at the top of the PropertyManager or click **Help ➢ SOLIDWORKS Help Topics** from the Menu bar and search by using one of the three available tabs: **Contents**, **Index** or **Search**.

Note: Un-check the Use SOLIDWORKS Web Help box to view the three default tabs in SOLIDWORKS Help.

Sketch Fillet Sketch tool

The Sketch Fillet ⌐ tool provides the ability to trim the corner at the intersection of two sketch entities to create a tangent arc. This tool is available for both 2D and 3D sketches.

The Sketch Fillet tool uses the Sketch Fillet PropertyManager. The Sketch Fillet PropertyManager provides the following selections:

- *Entities to Fillet*. Selected sketch entities are displayed in the list box.

- *Fillet Parameters*. The Fillet Parameters box provides the following options:

  - **Radius**. Displays the selected fillet sketch radius. Enter a radius value.

  - **Keep constrained corners**. Selected by default. Maintains the virtual intersection point, if the vertex has dimensions or relations.

  - **Dimension each fillet**. Adds dimensions to each fillet.

The Fillet feature tool from the Consolidated Features toolbar fillets entities such as edges in a part, not in a sketch. The Sketch Fillet tool is applied to a sketch, not a feature.

Tutorial: 2D Sketch Fillet 5-1

Create a Sketch Fillet 2D Sketch. Fillet the four corners of the rectangle with a 15mm Radius. Edit the Sketch Fillet and modify the corners.

1. Create a **New**  part. Use the default ANSI, MMGS Part template.

2. Right-click **Front Plane** from the FeatureManager. This is your Sketch plane.

3. Click **Sketch** from the Context toolbar.

Apply the Center Rectangle Sketch tool.

4. Click **Center Rectangle** from the Consolidated Rectangle Sketch toolbar.

5. Click the **Origin**.

6. Sketch a **rectangle** as illustrated.

Add dimensions.

7. Right-click **Smart Dimension** from the Context toolbar.

8. **Dimension** the sketch as illustrated.

Apply the Sketch Fillet Sketch tool.

9. Click **Sketch Fillet** on the Consolidated Sketch toolbar. The Sketch Fillet PropertyManager is displayed.

10. Enter **15**mm in the Radius box.

11. Click the **first** corner of the rectangle. The 15mm radius is applied to the first corner.

12. Click the other **three corners** of the rectangle. The four corners are displayed with a Sketch Fillet.

13. Click **OK** from the Sketch Fillet PropertyManager.

Modify the sketch fillet radius.
14. Double-click the **R15** dimension from the Graphics window. The modify dialog box is displayed.

15. Enter **10**mm.

16. Click **OK** from the Dimension PropertyManager.

17. **Rebuild** the model. View the results.

18. **Close** the model.

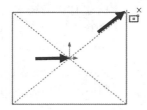

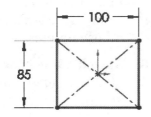

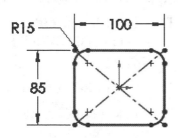

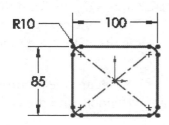

Tutorial: 3D Sketch Fillet 5-2

Create a 3D Sketch fillet.

1. Open **3D Sketch Fillet 5-2** from the SOLIDWORKS 2018 folder.

2. **Edit** ✎ 3DSketch1. 3DSketch1 is displayed in the Graphics window.

3. Click the **Line** Sketch tool.

4. Click the **right endpoint** of the sketch as illustrated.

5. Press the **Tab** key to change the Sketch plane from XY to YZ. The YZ icon is displayed on the mouse pointer.

6. Sketch a **line** along the Z axis approximately 130mm as illustrated.

7. Press the **Tab** key twice to change the sketch plane from YZ to XY. The XY icon is displayed on the mouse pointer.

8. Sketch a **line** back along the X axis approximately 30mm. Sketch a **line** on the Y axis approximately 30mm.

9. Sketch a **line** along the X axis of approximately 130mm.

Deselect the active Sketch tool.

10. Right-click **Select** in the Graphics window.

Apply the Sketch Fillet Sketch tool.

11. Click **Sketch Fillet** ⌐ from the Consolidated Sketch toolbar. The Sketch Fillet PropertyManager is displayed.

12. Enter **4**mm for Radius.

13. Click the **six corner points** of the sketch as illustrated to fillet.

14. Click **OK** ✔ from the Sketch Fillet PropertyManager. **Exit** the 3DSketch. View the results. **Close** the model.

💡 You can select non-intersecting entities. The entities are extended, and the corner is filleted.

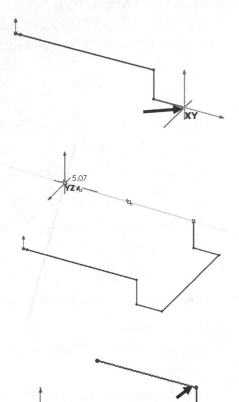

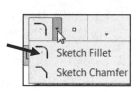

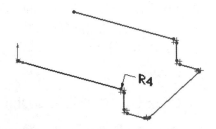

Sketch Chamfer Sketch tool

The Sketch Chamfer Sketch ⌐ tool provides the ability to apply a chamfer to an adjacent sketch entity either in a 2D or 3D sketch. The Sketch Chamfer tool uses the Sketch Chamfer PropertyManager. The Sketch Chamfer PropertyManager provides the following selections:

- **Chamfer Parameters**. Provides the ability to control the following options:

  - **Angle-distance**. The Angle-distance option provides two selections:

    - **Distance 1**. The Distance 1 box displays the selected value to apply to the first selected sketch entity.

    - **Direction 1 Angle**. The Direction 1 Angle box displays the selected value to apply from the first sketch entity toward the second sketch entity.

  - **Distance-distance**. Selected by default. Provides the following selections:

    - **Equal distance**. Selected by default. When selected, the Distance 1 value is applied to both sketch entities. When cleared, the Distance 1 value is applied to the first selected sketch entity. The Distance 2 value is applied to the second selected sketch entity.

    - **Distance 1**. Displays the selected value to apply to the first selected sketch entity.

    - **Distance 2**. Displays the selected value to apply to the second selected sketch entity.

☀ The Chamfer tool from the Features toolbar chamfers entities such as edges in a part, not in a sketch.

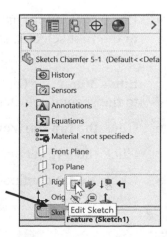

Tutorial: Sketch Chamfer 5-1

Create an Angle-distance Sketch chamfer for a 2D sketch.

1. Open **Sketch Chamfer 5-1** from the SOLIDWORKS 2018 folder. A rectangle is displayed in the Graphics window.

2. **Edit** ✎ Sketch1.

3. Click the **Sketch Chamfer** ⌐ tool from the Consolidated Sketch toolbar. The Sketch Chamfer PropertyManager is displayed.

4. Check the **Angle-distance** box.

5. Enter **40**mm for Distance 1. The value 40mm is applied to the first selected sketch entity.

6. Enter **45**deg in the Direction 1 Angle box. The 45deg angle is applied from the first selected sketch entity toward the second selected entity.

7. Click the **top horizontal line**. This is the first selected entity. Click the **left vertical line**.

8. Click **Yes** to confirm. This is the second selected entity. View the created chamfer.

9. Click **OK** ✔ from the Sketch Chamfer PropertyManager.

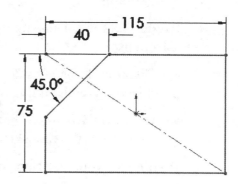

10. **Rebuild** 🛑 the model. View the results.

11. **Close** the model.

Tutorial: Sketch Chamfer 5-2

Create a Distance-distance Sketch chamfer for a 2D Sketch.

1. Open **Sketch Chamfer 5-2** from the SOLIDWORKS 2018 folder. The sketch is displayed in the Graphics window.

2. **Edit** 📝 Sketch1. Click the **Sketch Chamfer** ⌐ tool from the Sketch toolbar. The Sketch Chamfer PropertyManager is displayed.

3. Check the **Distance-distance** box.

4. Enter **20**mm for Distance 1. 20mm is applied to the first selected sketch entity. Note: if needed uncheck the Equal distance box.

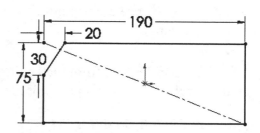

5. Enter **30**mm for Distance 2. 30mm is applied to the second selected sketch entity.

6. Click the **top horizontal line**.

7. Click the **left vertical line**.

8. Click **Yes** to continue. View the created chamfer.

9. Click **OK** ✅ from the Sketch Chamfer PropertyManager.

10. **Rebuild** 🎱 the model. View the results.

11. **Close** the model.

Tutorial: Sketch Chamfer 5-3

Create an Angle-distance Sketch chamfer for a 3D Sketch.

1. Open **Sketch Chamfer 5-3** from the SOLIDWORKS 2018 folder. The rectangle is displayed in the Graphics window.

2. **Edit** ✏️ 3DSketch1.

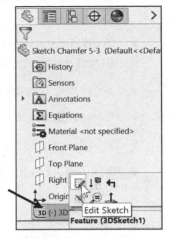

Apply the Sketch Chamfer Sketch tool.

3. Click the **Sketch Chamfer** ⌐ tool from the Sketch toolbar. The Sketch Chamfer PropertyManager is displayed.

4. Check the **Angle-distance** box.

5. Enter **10**mm for Distance 1. The value 10mm is applied to the first selected sketch entity.

6. Enter **25**deg in the Direction 1 Angle box. The 25deg angle is applied from the first selected sketch entity toward the second selected entity.

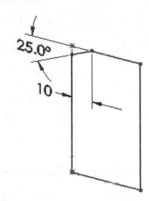

7. Click the **top horizontal line**. This is the first selected entity.

8. Click the **left vertical line**. View the created chamfer.

9. Click **OK** ✅ from the Sketch Chamfer PropertyManager.

10. **View** the model.

11. **Rebuild** 🎱 the model. View the results.

12. **Close** the model.

Offset Entities Sketch tool

The Offset Entities Sketch ⌐ tool provides the ability to offset one or more sketch entities, selected model edges, or model faces by a specified distance (using existing geometry). Example: You can offset sketch entities such as arcs, splines, loops, or sets of model edges etc.

The Offset Entities PropertyManager controls the following selections:

- *Parameters*. Provides the ability to control the following options:

  - **Offset Distance**. Displays the selected distance value to offset the sketch entity. To view a dynamic preview, hold the mouse button down and drag the pointer in the Graphics window. When you release the mouse button, the Offset Entity is complete.

  - **Add dimensions**. Includes the Offset Distance tool in the sketch. This does not affect any dimensions included with the original sketch entity.

  - **Reverse**. Reverses the direction of the offset if required.

  - **Select chain**. Creates an offset of all contiguous sketch entities.

  - **Bi-directional**. Creates offset entities in two directions.

  - **Make base construction**. Converts the original sketch entity to a construction line.

  - **Cap ends**. Extends the original non-intersecting sketch entities by selecting Bi-directional and adding a cap. You can create Arcs or Lines as extension cap types. The Cap ends check box provides two option types:

    - **Arcs**.

    - **Lines**.

    - **Construction geometry**. Converts the original sketch entity to a construction line using Base geometry, Offset geometry, or both options.

Tutorial: Offset Entity 5-1

Create an Offset Entity using the Offset Distance option using existing geometry.

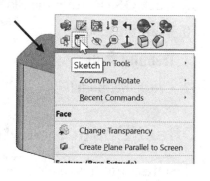

1. Open **Offset Entity 5-1** from the SOLIDWORKS 2018 folder.

Select the Sketch Plane.

2. Right-click the **Top face** of the model. Base Extrude is highlighted in the FeatureManager.

3. Click **Sketch** ⊞ from the Context toolbar.

4. Display a **Top view**.

Apply the Offset Entities Sketch tool.

5. Click the **Offset Entities** ⊑ tool from the Sketch toolbar. The Offset Entities PropertyManager is displayed.

6. Enter **.150**in for the Offset Distance.

7. The new Offset orange/yellow profile is displayed inside the original profile. Check the **Reverse** check box if needed.

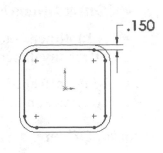

8. Click **OK** ✔ from the Offset Entities PropertyManager.

9. Drag the **dimension** off the model. **Rebuild** 🔋 the model. Sketch2 is created in the FeatureManager. View the results.

10. **Close** the model.

Tutorial: Offset Entity 5-2

Create an Offset Entity using the Cap ends option using existing geometry.

1. Open **Offset Entity 5-2** from the SOLIDWORKS 2018 folder. **Edit** 🖉 Sketch1.

2. Click the **spline construction** line. The Spline PropertyManager is displayed.

3. Click the **Offset Entities** ⊏ tool from the Sketch toolbar. The Offset Entities PropertyManager is displayed.

4. Enter **5mm** for the Offset Distance value.

5. Check the **Bi-directional** box.

6. Check the **CAP ends** box.

7. Check the **Arcs** box.

8. Click **OK** ✔ from the Offset Entities PropertyManager. View the results.

9. **Rebuild** 🔋 the model. View the results.

10. **Close** the model.

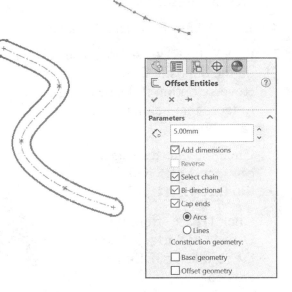

Offset On Surface

The Offset on Surface Sketch ⌀ tool provides the ability to offset 3D model edges and model faces in a 3D sketch. Previously, you had to create extra features for offsetting an edge.

The Offset on Surface PropertyManager controls the following selections:

- **Parameters**. Provides the ability to control the following options:

- **Offset Distance**. Displays the selected distance value to offset the sketch entity.

- **Add dimensions**. Includes the Offset Distance tool in the sketch. This does not affect any dimensions included with the original sketch entity.

- **Make offset base construction**. Converts the original sketch entity to a construction line.

Tutorial: *Offset on Surface 5-1*

Apply the Offset on Surface sketch tool. Create a sketch offset on 3D geometry surfaces.

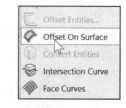

1. Open **Offset on Surface 5-1** from the SOLIDWORKS 2018 folder.

2. **Tools ➤ Sketch Tools ➤ Offset on Surface** from the Main menu. The Offset on Surface PropertyManager is displayed.

3. Click the **Surface-Loft4** edge in the Graphics area as illustrated.

4. Enter **10**mm for offset distance.

5. Check the **Reverse** box. The entity is projected on the opposite face. You can only use the Reverse option if the selected edge is connected to faces that belong to the same body.

6. Click the interior edges: **Surface-Loft5** and **Surface-Loft4**.

7. Click **OK** ✔ from the Offset On Surface. View the results. A 3DSketch is created.

8. **Close** the model.

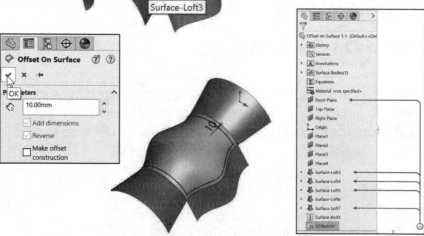

Convert Entities Sketch tool

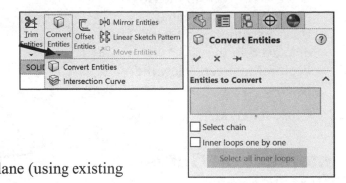

The Convert Entities Sketch tool provides the ability to create one or more curves in a sketch by projecting an edge, loop, face, curve, or external sketch contour, set of edges, or set of sketch curves onto a selected Sketch plane (using existing geometry).

You can convert an entity's internal loops or entities of a model face automatically with the Select all inner loops option of the Convert Entities tool.

The following relations are created when the Convert Entities tool is used:

- *Select Faces, Edges, or Sketch Entities to Convert.* Click a model edge, loop, face, curve, external sketch contour, set of edges, or set of curves.

- *Select Chain.* To convert all contiguous sketch entities.

- *Inner loops one by one.* Convert an entity's internal loops or entities of a model face automatically with the Select all inner loops option of the Convert Entities tool.

Tutorial: Convert Entity 5-1

Create a Convert Entities feature on a Flashlight lens using existing geometry.

1. Open **Convert Entities 5-1** from the SOLIDWORKS 2018 folder.

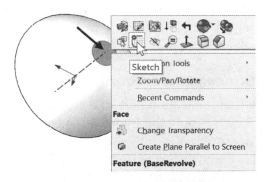

2. Right-click the **back face** of the model. This is the Sketch plane. The BaseRevolve feature is highlighted in the FeatureManager.

3. Click **Sketch** from the Context toolbar.

4. Click the **Convert Entities** tool from the Consolidated Sketch toolbar.

5. Click **Extruded Boss/Base** from the Features toolbar. The Boss-Extrude PropertyManager is displayed. The direction arrow points upwards.

6. Enter **.400**in for Depth. Accept all defaults.

7. Click **OK** ✔ from the Boss-Extrude PropertyManager. Boss-Extrude1 and Sketch2 are displayed in the FeatureManager. View the results.

8. **Close** the model.

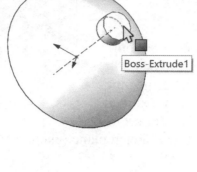

Boss-Extrude1

Intersection Curve Sketch tool

The Intersection ⊜ tool provides the ability to activate a sketch and creates a sketched curve at the following types of intersections:

- A plane and a surface or a model face, Two surfaces, A surface and a model face, A plane and the entire part, A surface and the entire part.

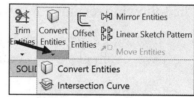

💡 In a 2D Sketch, use the Intersection Curve tool to extrude a feature. Select the plane, and click the Intersection Curve tool.

💡 In a 3D Sketch, use the Intersection Curve tool to extrude a feature. Click the Intersection Curve tool, and select the plane.

Tutorial: Intersection Curve 5-1

Measure the thickness of a cross section of a part for a 2D Sketch. Apply the Intersection Curve Sketch tool and Measure tool.

1. Open **Intersection Curve 5-1** from the SOLIDWORKS 2018 folder.

2. Click **Right Plane** from the FeatureManager.

Apply the Intersection Curve Sketch tool.

3. Click **Tools ➢ Sketch Tools ➢ Intersection Curve** or click the **Intersection Curve** ⊜ tool from the Consolidated Sketch toolbar.

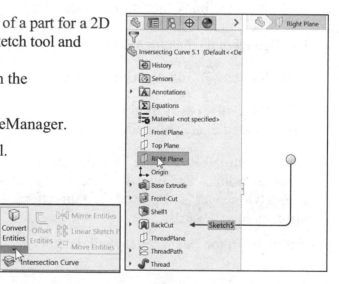

4. Click the **front outside face** of the part as illustrated.

5. Rotate the **part** to view the top inside face.

6. Click the **top inside** face of the part.

7. Click **OK** ✓ from the PropertyManager. Sketched splines are displayed on the top inside face and the outside face.

Measure the thickness of the part from face to face.

8. Click the **Evaluate** tab from the CommandManager.

9. Click the **Measure** 🔎 tool from the Evaluate toolbar. The Measure dialog box is displayed. Note: the Show XYZ Measurements button is active. Deactivate any other options at this time.

10. Click the **inside** and **outside** faces of the model to obtain the measured value. View the results.

11. **Close** the Measure dialog box.

12. **Rebuild** 🔘 the model. Sketch7 is created and is displayed in the FeatureManager.

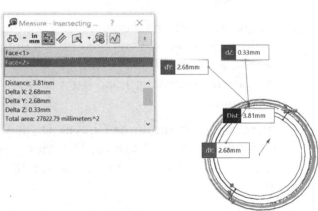

13. Display an **Isometric view**. View the results.

14. **Close** the model.

🔆 Configure the Measure tool to display two different units of measurement. Click the **Units/Precision** box in the Measure dialog box. Select **Use custom settings**. Select the **first** length unit. Click **Use Dual Units**. Select the **second** length unit. Click **OK**.

Face Curves Sketch tool

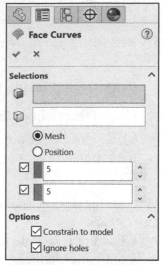

The Face Curves Sketch 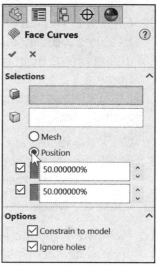 tool provides the ability to extract iso-parametric (UV) curves from a face or surface. Applications of this functionality include extracting the curves for imported surfaces and then performing localized cleaning using the face curves.

The Face Curves Sketch tool specifies a mesh of evenly spaced curves or a position which creates two orthogonal curves. The Face Curves Sketch tool uses the Face Curves PropertyManager. The Face Curves PropertyManager provides the following selections:

- *Selections*. Provides the ability to control the following options:

    - **Face**. The Face box displays the selected faces from the Graphics window.

    - **Position Vertex**. The Position Vertex box displays the selected vertex or point for the intersection of the two curves. You cannot drag the vertex.

    - **Mesh**. Selected by default. Evenly spaces the curves. Specify the number of curves for Direction 1 and Direction 2.

    - **Position**. Positions the intersection of the two orthogonal curves. Drag the position in the Graphics window or specify the percentage distance from the bottom for Direction 1 and from the right for Direction 2.

    - **Direction 1 Number of Curves**. Displays the selected number of curves. Clear the Direction 1 Number of Curves check box if a curve is not required in the first direction.

    - **Direction 2 Number of Curves**. Displays the selected number of curves. Clear the Direction 2 Number of Curves check box if a curve is not required in the second direction.

- *Options*. The Options box controls the following selections:

    - **Constrain to model**. Selected by default. Updates the curves if the model is modified.

    - **Ignore holes**. Selected by default. Generates curves across holes as though the surface was intact. When cleared, this option stops the curves at the edges of holes.

Each curve created by this process becomes a separate 3D sketch. However, if you are editing a 3D sketch when you invoke the Face Curves tool, all extracted curves are added to your active 3D sketch.

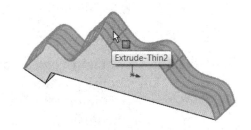

Tutorial: Face Curve 5-1

Extract iso-parametric curves using the Face Curve Sketch tool.

1. Open **Face Curve 5-1** from the SOLIDWORKS 2018 folder.

Apply the Face Curves Sketch tool.

2. Click **Tools** ➤ **Sketch Tools** ➤ **Face Curves**. The Face Curves PropertyManager is displayed.

3. Click the **top face** of the model. Face<1> is displayed in the Face box. A preview of the curves is displayed on the face in the Graphics window. The curves displayed, are one color in one direction and another color in the other direction.

4. Check the **Mesh** box.

5. Enter **7** for Direction 1 Number of Curves.

6. Enter **6** for Direction 2 Number of Curves.

7. Click **OK** ✔ from the Face Curves PropertyManager. The curves are displayed as 3D sketches in the FeatureManager. 3DSketch1 - 3DSketch13 are displayed.

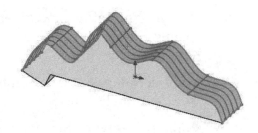

Tutorial: Face Curve 5-2

Extract iso-parametric curves using the Face Curve Sketch tool. Open **Face Curve 5-2** from the SOLIDWORKS 2018 folder.

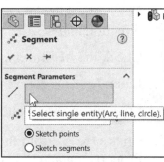

1. Click **Tools** ➢ **Sketch Tools** ➢ **Face Curves**. The Face Curves PropertyManager is displayed. Click the **top face** of the model. Face<1> is displayed in the Face box. A preview of the curves is displayed on the face in the Graphics window. The curves displayed are one color in one direction and another color in the other direction.

2. Check the **Position** box.

3. Click the **front right vertex** point of the model. Direction 1 position is 100%. Direction 2 position is 0%. Vertex<1> is displayed in the Position Vertex box.

4. Click **OK** ✔ from the Face Curve PropertyManager. The curves are displayed as 3D sketches in the FeatureManager. 3DSketch1 and 3DSketch2 are created and displayed.

Segment Sketch tool

The Segment Sketch ⬚ tool provides the ability to create equidistant points or segments in sketch entities.

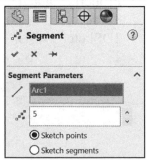

The Segment Sketch tool uses the Segment PropertyManager. The Segment PropertyManager provides the following selections:

*   ***Segment Parameters***. In an active sketch, select a single entity (Arc, line, circle.)

    *   **Number of points or segments**. Enter the number of points or segments to be created along the selected entity.

    *   **Sketch points**. Sets the number of sketch points on the selected single entity.

    *   **Sketch segments**. Sets the number of segments on the selected single entity.

Example 1: Five sketch points on the selected single entity (Arc).

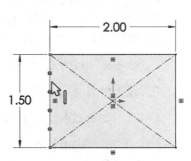

Example 1: Four sketch segments on the selected single entity (line).

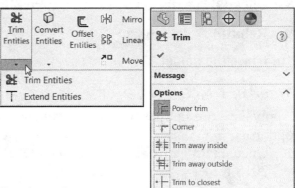

Trim Entities Sketch tool

The Trim Entities Sketch ⚒ tool provides the ability to select the trim option based on the entities you want to trim or extend. The Trim Entities tool uses the Trim PropertyManager. The Trim PropertyManager provides the following selections:

- *Options*. The Options box controls the following trim selections:

  - **Power trim**. Trims multiple, adjacent sketch entities by clicking and dragging the pointer across each sketch entity and to extend your sketch entities along their natural paths. Power trim is the default option.

  - **Corner**. Extends or trims two selected sketch entities until they intersect at a virtual corner.

  - **Trim away inside**. Trims open selected sketch entities that lie inside two bounding entities.

  - **Trim away outside**. Trims open selected sketch entities outside of two bounding entities.

  - **Trim to closest**. Trims each selected sketch entity or extended to the closest intersection.

Tutorial: Trim Entity 5-1

Use the Trim Entities Sketch tool with the Trim to closest option.

1. Open **Trim Entities 5-1** from the SOLIDWORKS 2018 folder.

2. **Edit** 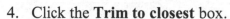 Sketch1.

Apply the Trim Entities Sketch tool.

3. Click the **Trim Entities** 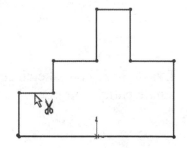 Sketch tool. The Trim Entities PropertyManager is displayed. The Trim icon is located on the mouse pointer.

4. Click the **Trim to closest** box.

5. Click the **first left horizontal line**.

6. Click the **second left horizontal line**.

7. Click **Yes**. Both horizontal lines are removed.

8. Click **OK** ✔ from the Trim PropertyManager.

9. **Rebuild** 🔘 the model.

10. **Close** the model.

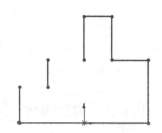

Tutorial: Trim Entity 5-2

Apply the Trim Entities Sketch tool with the Power Trim option.

1. Open **Trim Entities 5-2** from the SOLIDWORKS 2018 folder.

2. **Edit** 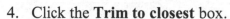 Sketch1.

Apply the Trim Entities Sketch tool.

3. Click the **Trim Entities** 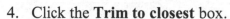 Sketch tool. The Trim PropertyManager is displayed.

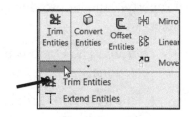

4. Click the **Power Trim** option box.

5. Click a **location** above the spline as illustrated.

6. Drag the **mouse point** to any location on the spline. The spline is removed.

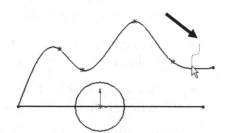

7. Click **OK** ✔ from the Trim PropertyManager.

8. **Rebuild** 🔘 the model. View the results.

9. **Close** the model.

Extend Entities Sketch tool

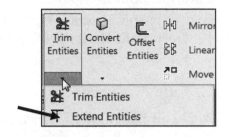

The Extend Entities Sketch $\top$ tool provides the ability to add to the length of your sketch entity. Example: line, centerline or arc. Use the Extend Entities sketch tool to extend a sketch entity to meet another sketch entity. The Extend Entities sketch tool does not use a PropertyManager.

Tutorial: Extend Entity 5-1

Create an Extend Entity sketch.
1. Open **Extend Entities 5-1** from the SOLIDWORKS 2018 folder.

2. **Edit** Sketch1.

Apply the Extend Entities Sketch tool.

3. Click the **Extend Entities** $\top$ tool from the Consolidated Sketch toolbar. The Extend $\top$ icon is displayed on the mouse pointer.

4. Drag the **mouse pointer** over the diagonal line. The selected entity is displayed in red. A preview is displayed in red in the direction to extend the entity. If the preview extends in the wrong direction, move the pointer to the other half of the line.

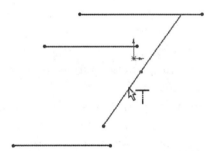

5. Click the **diagonal** line. The First selected entity is extended to the top horizontal line.

6. **Rebuild** the model. View the results.

7. **Close** the model.

Split Entities Sketch tool

The Split Entities Sketch Γ tool provides the ability to split a sketch entity to create two sketch entities. The Split Entities sketch tool does use a PropertyManager.

You can delete a split point to combine two sketch entities into a single sketch entity. Use two split points to split a circle, full ellipse or a closed spline.

Tutorial: Split Entity 5-1

Apply the Split Entities Sketch tool.

1. Open **Split Entity 5-1** from the SOLIDWORKS 2018 folder.

2. **Edit**  Sketch1.

Apply the Split Entities Sketch tool.

3. Click **Tools ➤ Sketch Tools ➤ Split Entities**. The Split Entities icon is displayed in the mouse pointer.

4. Click the **right bottom horizontal line**.

5. Right-click **OK**. The sketch entity splits into two entities at the selected location sketch. A split point is added between the two sketch entities.

6. **Rebuild** the model. View the results.

7. **Close** the model.

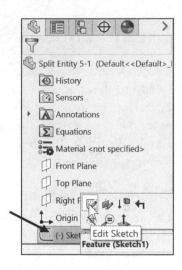

Construction Geometry Sketch tool

The Construction Geometry Sketch tool provides the ability to convert sketch entities in a sketch or drawing to construction geometry. Construction geometry is used to assist in creating the sketch entities and geometry that are ultimately incorporated into the part. Construction geometry is ignored when the sketch is used to create a feature. Construction geometry uses the same line style as centerlines. The Construction Geometry tool does not use a PropertyManager.

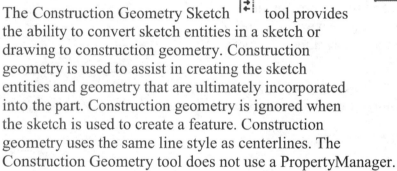

Any sketch entity can be specified for construction. Points and centerlines are always construction entities.

Tutorial: Construction Geometry 5-2

Convert existing geometry to Construction geometry.

1. Open **Construction Geometry 5-2** from the SOLIDWORKS 2018 folder.

2. **Expand** Cut-Extrude1 from the FeatureManager.

3. Right-click **Sketch 2** from the FeatureManager as illustrated.

4. Click **Edit Sketch** 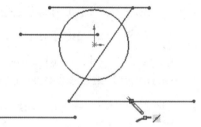 from the Context toolbar.

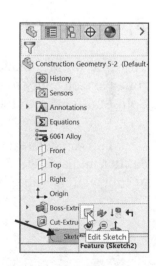

5. Click the **circular edge** of the circle as illustrated. The Circle PropertyManager is displayed.

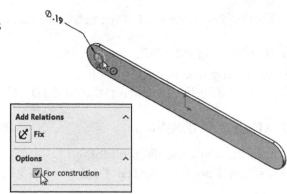

6. Check the **For construction** box in the Options section.

7. Click **Tools ➤ Sketch Tools ➤ Construction Geometry**.

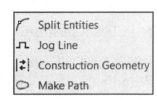

8. **Rebuild** the model. View the results.

9. **Close** the model.

Jog Line Sketch tool

The Jog Line Sketch ⊓ tool provides the ability to jog sketch lines in either a 2D or 3D sketch for a part, assembly, or drawing documents. Jog lines are automatically constrained to be parallel or perpendicular to the original sketch line. The Jog Line sketch tool does not use a PropertyManager.

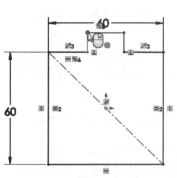

 You can drag and dimension Jog lines.

Tutorial: Jog line 5-1

Create a Jog line in a 2D sketch.

1. Open **Jog line 5-1** from the SOLIDWORKS 2018 folder.

2. **Edit** Sketch1.

Apply the Jog Line Sketch tool.

3. Click **Tools ➤ Sketch Tools ➤ Jog Line**.

4. Click a starting **point** midway on the top horizontal line.

5. **Drag** the **mouse pointer** directly above the starting point. This point determines the height of the Jog.

6. **Drag** the **mouse pointer** directly to the right of the second point.

7. Click the **third point** location. This point determines the width of the Jog. SOLIDWORKS created a Jog in the rectangle sketch.

8. Right-click **OK**.

9. **Rebuild** the model. View the results.

10. **Close** the model.

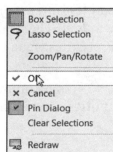

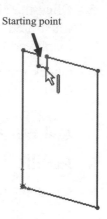

Starting point

💡 The Jog Line tool stays active so you can insert multiple jogs.

Tutorial: Jog line 5-2

Create a Jog line in a 3D sketch.
1. Open **Jog line 5-2** from the SOLIDWORKS 2018 folder.

2. **Edit** ✏️ 3DSketch1.

Apply the Jog Line Sketch tool.
3. Click **Tools ➢ Sketch Tools ➢ Jog Line**.

4. Click a starting **point** on the top horizontal, X line as illustrated. Drag the **mouse pointer** down and to the right to create the first Jog.

5. Click a **position** as illustrated.

Create a second Jog line sketch.
6. Click a starting **point** on the top horizontal, X line as illustrated.

7. Drag the **mouse pointer** down and to the right to create the second Jog.

8. Press the **Tab** key to change the plane of the Jog. The Jog is in the Z plane.

9. Click a **position**.

10. Create the **third** and **fourth** Jog as illustrated.

11. Right-click **OK** ✔️.

12. **Rebuild** 🎱 the model. View the results.

13. **Close** the model.

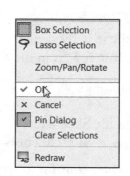

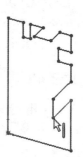

💡 Jog lines are automatically constrained to be perpendicular or parallel to the original sketch line.

Make Path Sketch tool

The Make Path Sketch ⬭ tool provides the ability to create a path with end-to-end coincident sketch entities. A path consists of sketch entities which are coincident end to end to form a single chain. The selected sketch entities are highlighted in the Graphics window. The Make Path Sketch tool uses the Path PropertyManager. The Path PropertiesManager provides the following selections:

💡 It is recommended to create the sketch entities into a block. A chain of sketch entities can belong only to a single path.

- *Existing Relations*. Provides the following option:

  - **Relations**. Displays the relations between the sketch entities that consist of the path and the sketch entities with which the path interacts.

- *Definition*. Provides the following option:

  - **Edit Path**. Adds sketch entities to create a path.

A Path consists of Sketch entities that are coincident end to end, forming a single chain.

Tutorial: Make Path 5-1

Create a machine design 2D layout sketch. Model a cam profile where the tangent relation between the cam and a follower automatically transitions as the cam rotates.

1. Open **Make Path 5-1** from the SOLIDWORKS 2018 folder.

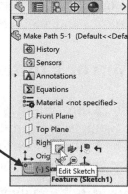

2. **Edit** Sketch1.

3. Click the **spline boundary** of the CAM block. The Block PropertyManager is displayed.

4. Right-click **Make Path**. Click **OK**.

5. Click **OK** from the Block Properties PropertyManager.

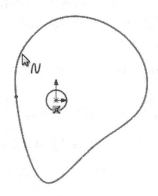

6. Click **Tools** ➤ **Blocks** ➤ **Insert**. The Insert Block PropertyManager is displayed.

7. Click the **Browse** button from the Blocks to Insert box.

8. Double-click **Follower.sldblk** from the SOLIDWORKS 2018 folder.

9. Click a **position** above the CAM. Follower and Block2 are displayed in the Open Blocks box.

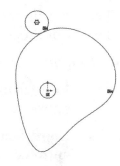

10. Click **OK** from the Insert Block PropertyManager.

11. Insert a Tangent relation between the **spline of the CAM** and the **circle of the Follower**.

12. Insert a Vertical relation between the **center point of the Follower** and the **center point of the CAM**.

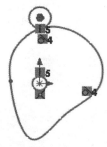

13. Slowly **rotate** the CAM block. View the model in the Graphics window.

14. **Rebuild** 🔘 the model. View Block 2-3 and Follower-1 in the FeatureManager and in the Graphics window.

15. **Close** the model.

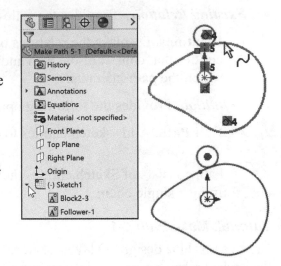

🔅 The Make Path tool enables you to create a tangent relation between a chain of Sketch entities and another Sketch entity.

Mirror Entities Sketch tool

The Mirror Entities Sketch ⊣⊢ tool provides the ability to mirror existing sketch entities. This tool is not available for 3D sketches. The tool provides the following capabilities which are the same as the Dynamic Mirror sketch tool:

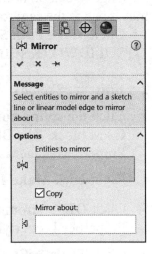

- Mirror to only include the new entity, or both the original and the mirrored entity.

- Mirror some or all of the sketch entities.

- Mirror about any type of line, not just a construction line.

- Mirror about edges in a drawing, part, or assembly.

🔅 When you create mirrored entities, SOLIDWORKS applies a symmetric relation between each corresponding pair of sketch points (the ends of mirrored lines, the centers of arcs, etc.). If you change a mirrored entity, its mirror image also changes.

The Mirror Entities Sketch tool uses the Mirror PropertyManager. The Mirror PropertyManager provides the following options:

- *Options*. Controls the following selections:

  - **Entities to mirror**. Displays the selected sketch entities to mirror.

  - **Copy**. Selected by default. Includes both the original and mirrored entities. Clear the Copy check box if you only want the mirrored entities and not the original entity.

- **Mirror about**. Displays the selected item to mirror about. You can Mirror about the following items: Centerlines, Lines, Linear model edges and Linear edges on drawings.

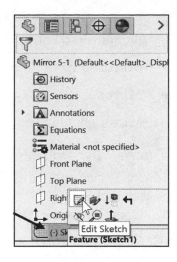

To mirror about a linear drawing edge, your sketch entities to mirror must lie within the boundary of the drawing.

Tutorial: Mirror Entities 5-1

Mirror existing sketch entities with the Mirror Sketch tool.

1. Open **Mirror 5-1** from the SOLIDWORKS 2018 folder.

2. **Edit** Sketch1.

Apply the Mirror Entities Sketch tool.

3. Click the **Mirror Entities** Sketch tool. The Mirror PropertyManager is displayed.

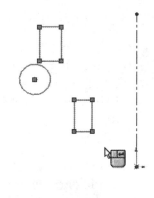

4. Box-Select the **two rectangles** and the **circle** on the left side of the centerline. The selected entities are displayed in the Entities to mirror box.

5. Uncheck the **Copy** box to add a mirror copy of the selected entities and to remove the original sketch entities. Note: if the Copy box is checked, you will include both the mirrored copy and the original sketch entities.

6. Click inside the **Mirror about** box.

7. Click the **centerline** to Mirror about. Line1 is displayed in the Mirror about box.

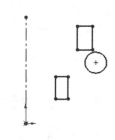

8. Click **OK** from the Mirror PropertyManager. The selected entities are mirrored about the centerline. No original entities are displayed due to the unchecked Copy box. **Rebuild** the model. View the results.

9. **Close** the model.

Dynamic Mirror Sketch tool

The Dynamic Mirror Sketch tool provides the ability to mirror sketch entities as you sketch them. Dynamics Mirror is not available for 3D sketches. The Dynamic Mirror sketch tool does not use a PropertyManager.

The Dynamic Mirror sketch tool provides the following capabilities which are the same as the Mirror sketch tool.

They are:

- Mirror to only include the new entity, or both the original and the mirrored entity.

- Mirror some or all of the sketch entities.

- Mirror about any type of line, not just a construction line.

- Mirror about edges in a drawing, part, or assembly.

Tutorial: Dynamic Mirror 5-1

Mirror Sketch entities as you sketch. Utilize the Dynamic Mirror Sketch tool.

1. Open **Dynamic Mirror 5-1** from the SOLIDWORKS 2018 folder.

2. **Edit**  Sketch1.

3. Click the **vertical line**. The Line Properties PropertyManager is displayed.

Apply the Dynamic Mirror Sketch tool.

4. Click **Tools ➢ Sketch Tools ➢ Dynamic Mirror**. The Mirror PropertyManager is displayed. A Symmetry symbol is displayed at both ends of the vertical line in the Graphics window.

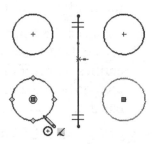

5. Create **two circles** on the left side of the vertical line. The circle entities are mirrored as you sketch them.

Deactivate the Dynamic Mirror Sketch tool.

6. Click **Tools ➢ Sketch Tools ➢ Dynamic Mirror**.

7. **Rebuild** the model. View the results.

8. **Close** the model.

Stretch Entities

The Stretch Entities tool provides the ability to stretch multiple sketch entities as a single group in a 2D sketch. The Stretch Entities tool uses the Stretch PropertyManager.

The Stretch PropertyManager provides the following options:

- **Entities To Stretch**. Displays the selected sketch entities.

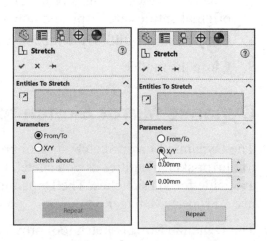

- *Parameters*. Provides the following options:

  - **From/To**. Selected by default. Adds a Base point to set the Start point. Drag the mouse pointer and click a location in the Graphics window to set the destination point.

  - **X/Y**. Sets the value for the Delta X and Delta Y coordinate.

  - **Base Point**. Displays the selected vertex to Stretch the sketch in the Graphics window.

  - **Repeat**. Moves your sketch entities again by the same distance when using the X/Y option.

Tutorial: Stretch Entities 5-1

Apply the Stretch Entities Sketch tool.

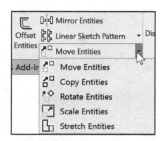

1. Open **Stretch Entities 5-1** from the SOLIDWORKS 2018 folder.

2. **Edit** Sketch1.

3. Click the **Stretch Entities** Sketch tool as illustrated from the drop-down menu. The Stretch PropertyManager is displayed.

Select the Sketch Entities to stretch.

4. Select the **5 sketch entities** as illustrated. The selected entities are displayed in the Entities to Stretch box.

5. Click the **From/To** box.

6. Click inside the **Base point** box.

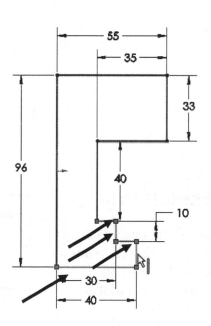

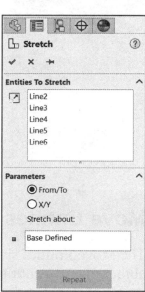

Select a Stretch about point.

7. Click the **left bottom vertex** in the Graphics window as illustrated. Base defined is displayed in the Base point box.

8. Drag the mouse pointer directly to the **right** for the stretch.

9. Click the stretch **end point**. View the results. The dimensions are updated.

10. **Close** the model.

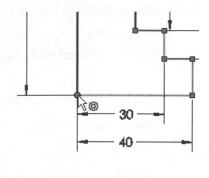

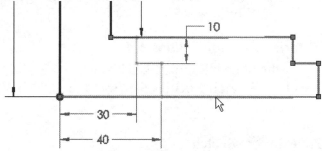

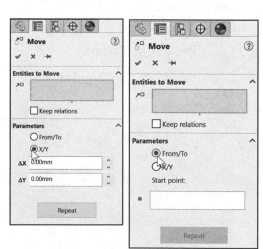

Move Entities Sketch tool

The Move Entities sketch 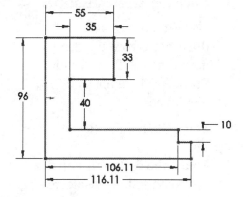 tool provides the ability to move entities by selecting from and to points or by using the X and Y destination coordinates. Using the Move sketch tool does not create relations.

The Move Entities sketch tool uses the Move PropertyManager. The Move PropertyManager provides the following selections:

- **Entities to Move**. Provides the following options:

  - **Sketch Items or annotations**. Displays the selected sketch entities to move.

- **Keep relations**. Maintains relations between sketch entities. If un-checked, sketch relations are broken only between selected entities and those that are not selected. The relations among the selected entities are still maintained.

- *Parameters*. The Parameters box controls the following options:

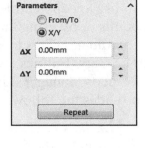

- **From/To**. Selected by default. Adds a Base point to set the Start point. Drag the mouse pointer and double-click a location in the Graphics window to set the destination point.

- **X/Y**. Sets the value for the Delta X and Delta Y coordinate.

- **Delta X coordinate**. Enter the Delta X coordinate value.

- **Delta Y coordinate**. Enter the Delta Y coordinate value.

- **Repeat**. Moves your sketch entities again by the same distance when using the X/Y option.

Tutorial: Move 5-1

Utilize the Move Entities Sketch tool using the From/To option.

1. Open **Move 5-1** from the SOLIDWORKS 2018 folder.

2. **Edit** Sketch1.

3. Box-Select the **two circles** on the left side of the vertical line. The Properties PropertyManager is displayed. Arc1 and Arc3 are displayed in the Selected Entities box.

4. Click **Tools ➢ Sketch Tools ➢ Move** from the Main menu. The two selected circles are displayed in the Entities to Move box. Note: The **From/To** box is checked.

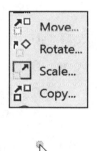

5. Click a position in the **upper right** section of the Graphics window. A base point is displayed in the Graphics window. From Point Defined is displayed in the Base point box.

6. Drag the **mouse pointer** in the Graphics window. The 4 circles move keeping their relations.

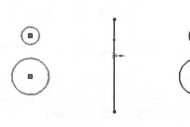

7. Click a **point** over the vertical line. The circles are moved and placed.

8. **Rebuild** the model. View the results.

9. **Close** the model.

Copy Entities Sketch tool

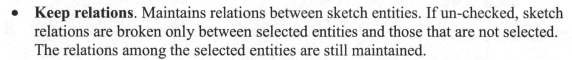

The Copy Entities sketch ⌐ tool provides the ability to copy entities by selecting from and to points or by using the X and Y destination coordinates. This tool does not create relations. The Copy Entities tool uses the Copy PropertyManager. The Copy PropertyManager provides the following selections:

- **_Entities to Copy_**. The Entities to Copy box provides the following options:

  - **Sketch Items or annotations**. Displays the selected sketch entities to copy.

  - **Keep relations**. Maintains relations between sketch entities. If un-checked, sketch relations are broken only between selected entities and those that are not selected. The relations among the selected entities are still maintained.

- **_Parameters_**. The Parameters box controls the following options:

  - **From/To**. Adds a Base point to set the Start point. Move the mouse pointer and double-click to set the destination point in the Graphics window.

  - **X/Y**. Sets the value for the Delta X and Delta Y coordinate.

  - **Delta X coordinate**. Enter the Delta X coordinate value.

  - **Delta Y coordinate**. Enter the Delta Y coordinate value.

  - **Repeat**. Moves your sketch entities again by the same distance when using the X/Y option.

Tutorial: Copy sketch 5-1

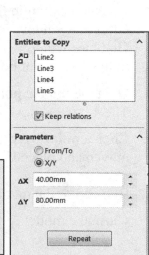

Utilize the Copy Entities Sketch tool using the X/Y option.
1. Open **Copy 5-1** from the SOLIDWORKS 2018 folder.

2. **Edit** ✎ Sketch1.

3. Box-Select the **rectangle** in your Graphics window. The Properties PropertyManager is displayed with the selected sketch entities.

4. Click **Tools** ➢ **Sketch Tools** ➢ **Copy** from the Main menu. The lines of the rectangle are displayed in the Entities to Copy box. Check the **Keep relations** box.

5. Enter **40**mm for Delta X.

6. Enter **80**mm for Delta Y. Zoom out if needed; view the copied rectangle displayed in yellow above and to the right of the original entity.

7. Click the **Repeat** button. The copied rectangle moves again by the same X and Y coordinate value. The new values are displayed in the Parameters box.

8. Click **OK** ✔ from the Copy PropertyManager.

9. Press the **f** key to fit the model to the Graphics window. View the copy sketch entity. **Rebuild** 🔴 the model. View the results.

10. **Close** the model.

Scale Entities Sketch tool

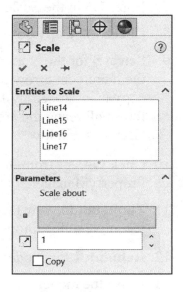

The Scale Entities sketch 🔲 tool provides the ability to scale and create copies of sketch entities by selecting a Base point, a Scale Factor, and the number of copies required. The Scale Entities tool uses the Scale PropertyManager. The Scale PropertyManager provides the following selections:

- **Entities to Scale**. The Entities to Scale box provides the following option:

  - **Sketch Items or Annotations**. Displays the selected sketch entities to scale from the Graphics window.

- **Parameters**. The Parameters box controls the following options:

  - **Scale about**. Display the selected Base point as the point to scale about.

  - **Scale Factor**. Display the selected scale required. Enter the Scale Factor.

  - **Number of Copies**. Display the selected value of the Number of Copies required. Enter the Number of Copies.

  - **Copy**. Selected by default. Creates one or more copies of your scaled entities.

Tutorial: Scale 5-1

Utilize the Scale Entities sketch tool using a 1.5 Scale Factor and multiple copies.

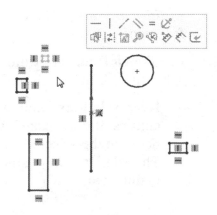

1. Open **Scale 5-1** from the SOLIDWORKS 2018 folder.

2. **Edit** 🖉 Sketch1.

3. Box-Select the **small rectangle** in the top left corner of the Graphics window. The Properties PropertyManager is displayed with the selected sketch entities.

4. Click **Tools** ➢ **Sketch Tools** ➢ **Scale** from the Main menu. The selected sketch entities are displayed in the Entities to Scale box.

5. Click inside the **Base point** box.

6. Click the **vertical** line as illustrated. Scale Point Defined is displayed in the Base point box. A base point is displayed in the Graphics window on the vertical line.

7. Enter **1.5** in the Scale Factor box.

8. Check the **Copy** box.

9. Select **5** for the Number of Copies using the increment arrow box. The scale of each copy is increased by 1.5. The 5 copies of the small rectangle are displayed in the Graphics window.

10. Click **OK** ✔ from the Scale PropertyManager.

11. **Zoom out** to view the created copies of the rectangle in the Graphics window.

12. **Rebuild** 🔴 the model. View the results.

13. **Close** the model.

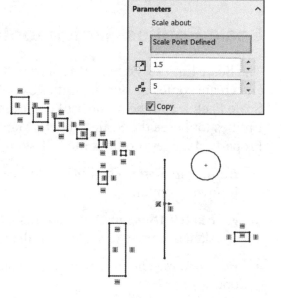

Rotate Entities Sketch tool

The Rotate Entities sketch ↻ tool provides the ability to rotate entities by selecting a center of rotation and the number of degrees to rotate. The Rotate Entities tool uses the Rotate PropertyManager. The Rotate PropertyManager provides the following selections:

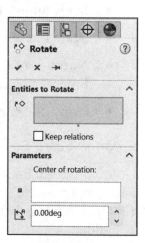

- *Entities to Rotate*. The Entities to Rotate box provides the following options:

  - **Sketch Items or Annotations**. Displays the selected sketch entities from the Graphics window.

  - **Keep relations**. Maintains relations between sketch entities. If un-checked, the sketch relations are broken only between the selected entities and those that are not selected. The relations among the selected sketch entities are still maintained.

- ***Parameters***. The Parameters box controls the following options:
  - **Center of rotation**. Displays the selected Base point as the Center of rotation.
  - **Angle**. Displays the selected value for the Angle.

Tutorial: Rotate 5-1

Utilize the Rotate Entities tool. Rotate a sketch 135 degrees.

1. Open **Rotate 5-1** from the SOLIDWORKS 2018 folder.

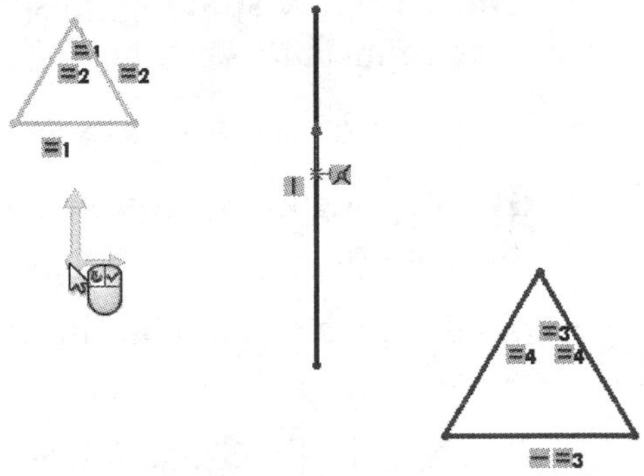

2. **Edit** Sketch1.
3. Box-Select the **small triangle** in the top left corner of the Graphics window. The Properties PropertyManager is displayed with the selected sketch entities.
4. Click **Tools ➢ Sketch Tools ➢ Rotate** from the Main menu. The selected entities are displayed in the Entities to Rotate box.
5. Click **inside** the Base point box.

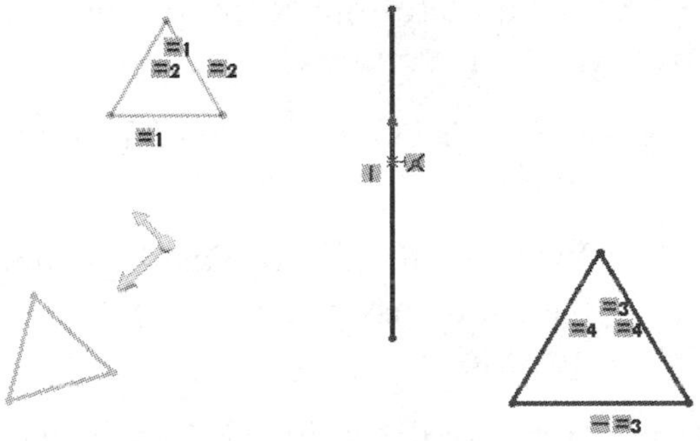

6. Click a **position** below the small triangle in the Graphics window. A point triad is displayed in the Graphics window.
7. Enter **135**deg for Angle. View the triangle rotated 135 degrees in the Graphics window.
8. Click **OK** from the Rotate PropertyManager.
9. **Rebuild** the model. View the results.
10. **Close** the model.

Linear Pattern Sketch tool

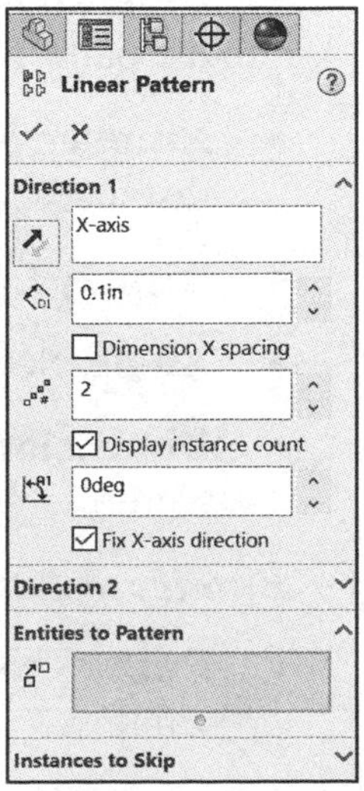

The Linear Pattern Sketch tool provides the ability to create a linear pattern along one or both axes. The Linear Pattern tool uses the Linear Pattern PropertyManager. The Linear Pattern PropertyManager provides the following selections:

- ***Direction 1***. The Direction 1 box provides the following options:

- **Pattern Direction**. Displays the selected edge from a part or an assembly to create a linear pattern in the X direction.

- **Reverse direction**. Reverses the pattern direction in the X direction if required.

- **Spacing**. Sets distance between the pattern instances for Direction 1.

- **Dimension X spacing**. Displays dimension between the pattern instances.

- **Number of Instances**. Sets the number of pattern instances.

- **Display instances count**. Shows the number of instances in the pattern.

- **Angle**. Sets an angular direction from the horizontal (X axis).

- **Fix X**-axis direction. Selected by default.

Set a value for the Number option to activate the Direction 2 box settings.

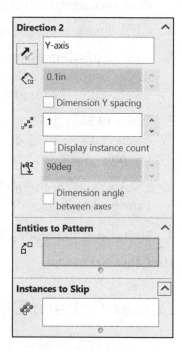

- *Direction 2*. The Direction 2 box provides the following options:

  - **Pattern Direction**. Displays the selected edge from a part or an assembly to create a linear pattern in the Y direction.

  - **Reverse direction**. Reverses the pattern direction in the X direction if required.

  - **Spacing**. Sets distance between the pattern instances for Direction 2.

  - **Dimension Y spacing**. Displays dimension between the pattern instances.

  - **Number of Instances**. Sets the number of pattern instances.

  - **Display instances count**. Shows the number of instances in the pattern.

  - **Angle**. Sets an angular direction from the horizontal (Y axis).

  - **Dimension angle between axes**. Displays dimension for the angle between the patterns.

- *Entities to Pattern*. The Entities to Pattern box provides the following option:

  - **Entities to Pattern**. Displays the selected sketch entities from the Graphics window.

- *Instances to Skip*. The Instances to Skip box provides the following option:

  - **Instances to Skip**. Displays the selected instances from the Graphics window which you want to skip. Click to select a pattern instance. The coordinates of the pattern instance are displayed in the Graphics window.

Tutorial: Linear Pattern sketch 5-1

Create a Linear Pattern Sketch using the Instances to Skip option.

1. Open **Linear Pattern 5-1** from the SOLIDWORKS 2018 folder. **Edit** 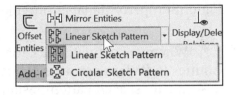 Sketch2.

Apply the Linear Sketch Pattern Sketch tool.

2. Click the **Linear Sketch Pattern** Sketch tool. The Linear Pattern PropertyManager is displayed. The Linear Sketch Pattern icon is displayed on the mouse pointer.

3. Click the **circumference** of the small circle. Arc 1 is displayed in the Entities to Pattern box. The pattern direction is to the right.

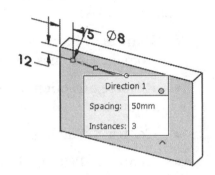

4. Click the **Reverse** direction check box option if required. Enter **50**mm for Spacing in Direction 1.

5. Enter **3** for Number of sketch entities in Direction 1.

6. Enter **355**deg for Angle at which to pattern the sketch entities in Direction 1.

Set Distance 2 between sketch entities along the Y-axis.

7. Click **inside** the Y-axis box.

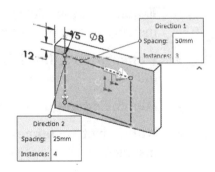

8. Enter **4** for Number of sketch entities.

9. Enter **25**mm for Spacing. The pattern direction arrow points downward. If required, click the **Reverse** direction check box.

10. **Expand** the Instances to Skip option box.

11. Click **inside** the Instances to Skip box.

12. Click the two bottom right corner pattern points, **(2,4)**, **(3,4)**. The selected sketch entities (2,4) and (3,4) are displayed in the Instances to Skip option box.

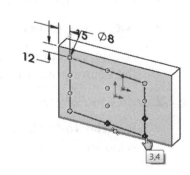

13. Click **OK** from the Linear Pattern PropertyManager.

14. **Rebuild** the model. View the results.

15. **Close** the model.

Circular Pattern Sketch tool

The Circular Pattern Sketch 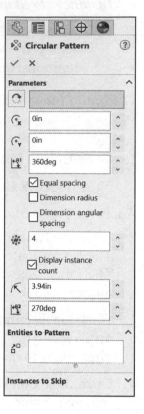 tool provides the ability to create a circular sketch pattern along one or both axes. The Circular Pattern tool uses the Circular Pattern PropertyManager. The Circular Pattern PropertyManager provides the following selections:

When you activate the Circular Pattern sketch tool, the Entities to Pattern option is selected. The circular pattern defaults to 4 equally spaced pattern instances.

- *Parameters*. The Parameters box controls the following options:

  - **Entities to Pattern**. Displays the selected entity to pattern from the Graphics window.

  - **Reverse direction**. Reverses the circular pattern direction if required.

  - **Center X**. Patterns the center along the X axis.

  - **Center Y**. Patterns the center along the Y axis.

  - **Spacing**. Displays the selected number of total degrees in the circular pattern.

  - **Equal Spacing**. Selected by default. Patterns the circular instances equidistant from each other.

  - **Dimension radius**. Displays the selected radius value of the pattern.

  - **Dimension angular spacing**. Displays the angular spacing.

  - **Number of Instances**. Sets the number of pattern instances.

  - **Display instances count**. Shows the number of instances in the pattern.

  - **Add dimensions**. Displays a dimension between the circular pattern instances.

  - **Angle**. Sets the selected angle measured from the center of the selected entities to the center point or vertex of the circular pattern.

- *Entities to Pattern*. The Entities to Pattern box provides the following option:

  - **Entities to Pattern**. Displays the selected sketch entities from the Graphics window.

- *Instances to Skip*. The Instances to Skip box provides the following option:

  - **Instances to Skip**. Displays the selected instances from the Graphics window which you want to skip. Click to select a pattern instance. The coordinates of the pattern instance are displayed in the Graphics window.

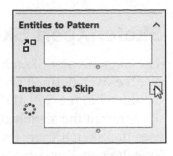

Tutorial: Circular Pattern sketch 5-1

Apply the Circular Pattern Sketch tool. Utilize the Seed sketch and check the Instances to Skip box.

1. Open **Circular Pattern 5-1** from the SOLIDWORKS 2018 folder.

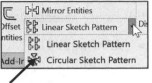

2. **Edit** Sketch1.

Apply the Circular Pattern sketch tool.

3. Click **Circular Sketch Pattern** from the Consolidated Sketch toolbar. The Sketch Circular Pattern PropertyManager is displayed.

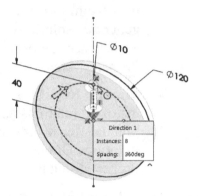

4. Click the **circumference** of the small circle in the Graphics window. Point-1 is displayed in the Parameters box. The direction arrow points downward. If required, click the Reverse direction box. Arc 2 is displayed in the Entities to Pattern box.

5. Enter **8** for spacing.

6. Check the **Equal spacing** box.

Center x-, Center y-, and Arc Angle are values generated by the location of your circular pattern's center. Each value can be edited independently.

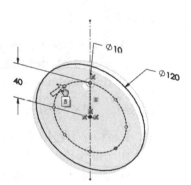

7. **Expand** the Instances to Skip box.

8. Click **inside** the Instances to Skip box.

9. Click **Point 8** and **Point 4** from the Graphics window. Point (8) and Point (4) are displayed in the Instances to Skip box.

10. Click **OK** from the Circular Pattern PropertyManager. The sketched Circular Pattern is displayed in the Graphics window.

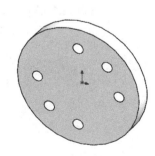

11. **Rebuild** the model. View the results.

12. **Close** the model.

SketchXpert Sketch tool

The SketchXpert Sketch tool provides the ability to resolve over-defined sketches. Color codes are displayed in the SOLIDWORKS Graphics window to represent the sketch states. The SketchXpert tool uses the SketchXpert PropertyManager. The SketchXpert PropertyManager provides the following selections:

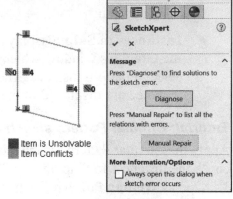

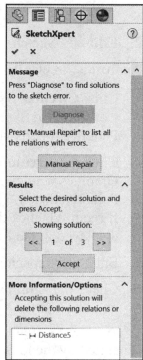

- *Message*. The Message box provides access to the following selections:

  - **Diagnose**. The Diagnose button generates a list of solutions for the sketch. The generated solutions are displayed in the Results section of the SketchXpert PropertyManager.

  - **Manual Repair**. The Manual Repair button generates a list of all relations and dimensions in the sketch. The Manual Repair information is displayed in the Conflicting Relations/Dimensions section of the SketchXpert PropertyManager.

- *More Information/Options*. Provides information on the relations or dimensions that would be deleted to solve the sketch.

  - **Always open this dialog when sketch error occurs**. Selected by default. Opens the dialog box when a sketch error is detected.

- *Results*. The Results box provides the following selections:

  - **Left or Right arrows**. Provides the ability to cycle through the solutions. As you select a solution, the solution is highlighted in the Graphics window.

  - **Accept**. Applies the selected solution. Your sketch is no longer over-defined.

- *More Information/Options*. The More Information/Options box provides the following selections:

  - **Diagnose**. The Diagnose box displays a list of the valid generated solutions.

  - **Always open this dialog when sketch error occurs**. Selected by default. Opens the dialog box when a sketch error is detected.

- *Conflicting Relations/Dimensions*. The Conflicting Relations/Dimensions box provides the ability to select a displayed conflicting relation or dimension. The selected item is highlighted in the Graphics window. The options include:

  - **Suppressed**. Suppresses the relation or dimension.

  - **Delete**. Removes the selected relation or dimension.

  - **Delete All**. Removes all relations and dimensions.

  - **Always open this dialog when sketch error occurs**. Selected by default. Opens the dialog box when a sketch error is detected.

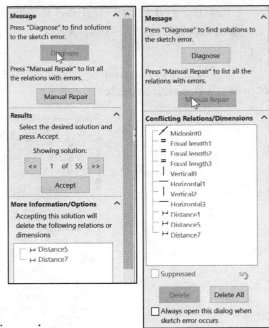

Tutorial: SketchXpert 5-1

Over-define an existing Sketch. Apply the SketchXpert tool to select a solution.

1. Open **SketchXpert 5-1** from the SOLIDWORKS 2018 folder.

2. **Edit** 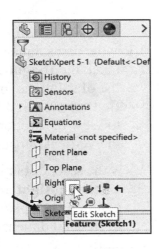 Sketch1. The Sketch is fully defined. The rectangle has a midpoint relation to the origin, and an equal relation with all four sides. The top horizontal line is dimensioned.

Insert a dimension to create an over defined sketch.

3. Click **Smart Dimension** ⟡.

4. Add a dimension to the **left vertical line**. This makes the sketch over-defined. The Make Dimension Driven dialog box is displayed.

5. Check the **Leave this dimension driving** box option.

6. Click **OK**. The Over Defined warning is displayed.

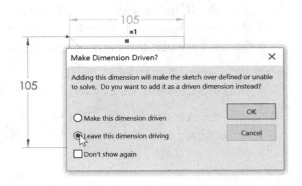

7. Click the **red Over Defined** message. The SketchXpert
   PropertyManager is displayed.

☼ Color codes are displayed in the Graphics window to
represent the sketch states.

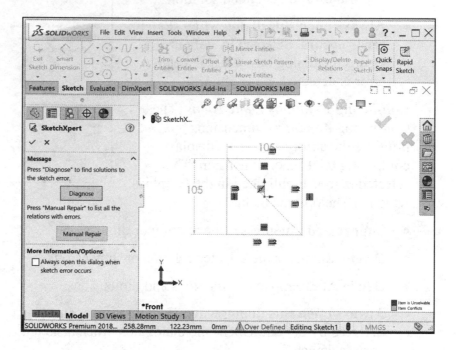

View the first solution.
8. Click the **Diagnose** button. The Diagnose button generates a
   list of solutions for your sketch. You can either accept the
   first solution or click the Right arrow key in the Results box
   to view the section solution. The first solution is to delete
   the vertical dimension of 105mm.

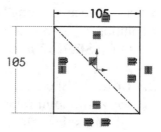

View the second solution.
9. Click the **Right arrow** key in the Results box. The second
   solution is displayed. The second solution is to delete the
   horizontal dimension of 105mm.

View the third solution.
10. Click the **Right arrow** key in the Results box. The third
    solution is displayed. The third solution is to delete the Equal
    relation between the vertical and horizontal lines.

Accept the second solution.
11. Click the **Left arrow** key to obtain the second solution.

12. Click the **Accept** button. The SketchXpert tool resolves the
    over-defined issue. A message is displayed.

13. Click **OK** ✔ from the SketchXpert PropertyManager.

14. **Rebuild** 🔵 the model. View the results. **Close** the model.

Align Sketch tool

The Align Sketch tool aligns your sketch grid with a selected model edge. There are three selections under the Align Sketch tool. They are **Sketch**, **Align Grid/Origin** and **Customize Menu**.

Sketch tool

The Sketch 🖵 tool option provides the ability to either select one sketch point, or two points or lines to align to. The Sketch option does not use a PropertyManager.

Align Grid/Origin tool

The Align Grid/Origin tool provides the ability to align the origin of a block to the orientation of the sketch entity. The Align Grid/Origin tool uses the Align Grid/Origin PropertyManager. The Align Grid/Origin PropertyManager provides the following selections:

- *Selections*. The Selections box provides the following options:

  - **Selections**. Displays the selected point or vertex from the Graphics window for the Sketch origin location.

  - **X-axis**. Displays the selected X-axis.

  - **Flip X-axis**. Flips the orientation of the origin by 180deg.

  - **Y-axis**. Displays the selected Y-axis.

  - **Flip Y-axis**. Flips the orientation of the origin by 180deg.

  - **Relocate origin only**. Moves the location of the origin.

  - **Relocate all sketch entities**. Selected by default. Moves the sketch's location based on relocating.

Custom Menu tool

The Custom Menu tool provides the ability to create a custom menu for the Align Sketch tool. The Custom Menu tool does not use a PropertyManager.

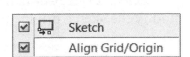

Tutorial: Align 5-1

Create a block. Display the origin location for the block.

1. Create a **New** ⬚ part in SOLIDWORKS. Accept the default Part template.

2. Right-click **Front Plane** from the FeatureManager. This is your Sketch plane.

3. Click **Sketch** ⬚ from the Context toolbar.

4. Sketch several **sketch entities** as illustrated with the Line Sketch tool.

5. Click **Insert** ➤ **Annotations** ➤ **Note**. The Note PropertyManager is displayed.

6. Click a **position** below the sketch entities.

7. Enter **BLOCK 1** for name. The Formatting toolbar is displayed to change font type and size.

8. Select **None** for Border style.

9. Click **OK** ✔ from the Note PropertyManager.

10. Box-Select the **sketch entities**. Line1 - Line5 are displayed in the Selected Entities box.

Create a Block.

11. Click **Tools** ➤ **Blocks** ➤ **Make**. The Make Block PropertyManager is displayed.

12. Click **OK** ✔ from the Make Block PropertyManager. Block1-1 is displayed in the FeatureManager.

13. **Rebuild** ❽ the model.

14. Right-click **Block1-1** from the FeatureManager.

15. Click **Edit Block**. You are in Edit mode. The Blocks toolbar is displayed.

16. Click **Tools** ➤ **Sketch Tools** ➤ **Align** ➤ **Align Grid/Origin**. The Align Grid/Origin PropertyManager is displayed.

💡 Select a vertex or point to modify the sketch origin.

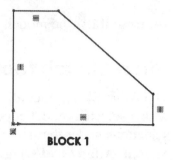

BLOCK 1

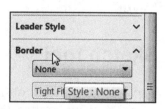

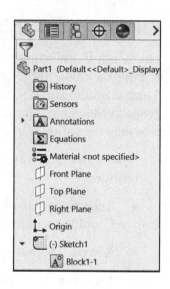

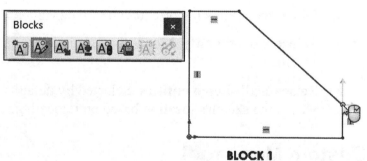

BLOCK 1

17. Click the **right middle vertex** as illustrated for the sketch origin location. The point is displayed in the Selections box of the Align Grid/Origin PropertyManager.

To modify both the sketch and orientation, click inside the X-axis or Y-axis box. Select a line to modify the orientation of the sketch origin.

18. Click **OK** ✔ from the Align Grid/Origin PropertyManager.

19. Select **Save Block** 🔲 from the Blocks toolbar.

20. Click **Save**. Block1 is the default name.

21. **Rebuild** 🔴 the model. View the results.

22. **Close** the model.

Modify Sketch tool

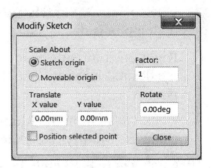

The Modify Sketch ⬦ tool provides the ability to move, rotate, or scale the entire sketch. To move, rotate, scale or copy an individual sketch entity, use the Move Entities, Rotate Entities, Scale Entities or Copy Entities sketch tools.

The available options are dependent on whether you are using the Modify sketch tool to move, rotate or scale an entire sketch. The Modify Sketch dialog box provides the following selections:

- *Scale About*. The Scale About section provides the following options:

  - **Sketch origin**. Selected by default. Applies a uniform scale about the origin of your sketch.

  - **Moveable origin**. Scales the sketch about the moveable origin.

  - **Factor**. The default Factor is 1. Enter the required factor scale.

- *Translate*. These are the available options:

  - **X value**. Moves the sketch geometry incrementally in the X direction.

  - **Y value**. Moves the sketch geometry incrementally in the Y direction.

  - **Position selected point**. The Position selected point option check box provides the ability to move a specified point of the sketch to a specific location.

- *Rotate*. These are the available options:
  - **Rotate**. Sets the value of the rotation.
- *Close*. Closes the dialog box.

Tutorial: Modify 5-1

Move and scale a sketch using the Modify Sketch tool.

1. Open **Modify 5-1** from the SOLIDWORKS 2018 folder.

2. **Edit**  Sketch1. Apply the modify Sketch tool.

3. Click **Tools ➢ Sketch Tools ➢ Modify**. The Modify Sketch dialog box is displayed. Note: The sketch can't contain a relation.

4. Click **outside** of the sketch in the Graphics window.

5. Drag the **mouse pointer** to the desired location. The Sketch moves with the mouse pointer.

6. Click **Close** from the Modify Sketch dialog box.

7. Click **Tools ➢ Sketch Tools ➢ Modify**. The Modify Sketch dialog box is displayed.

8. Right-click **outside** of the sketch. Rotate the **mouse pointer** to the desired location. The Sketch rotates with the mouse pointer.

9. Click **Close** from the Modify Sketch box. Scale the sketch using the Modify Sketch tool.

10. Click **Tools ➢ Sketch Tools ➢ Modify**. The Modify Sketch dialog box is displayed. Scale the sketch geometry.

11. Check the **Sketch origin** box.

12. Enter **1.3** for Factor. The model increases in size.

13. Click **Close** from the Modify Sketch box.

14. **Rebuild** the model. **Close** the model.

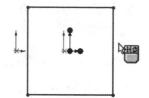

The Modify Sketch tool translates your entire sketch geometry in relation to the model. This includes the sketch origin. The sketch geometry does not move relative to the origin of the sketch.

2D to 3D Sketch tool

The 2D to 3D Sketch tool provides the ability to convert your 2D sketches into 3D models. Your 2D sketch can be an imported drawing, or it can be a simple 2D sketch.

You can copy and paste the drawing from a drawing document, or you can import the drawing directly into a 2D sketch in a part document.

The 2D to 3D Sketch tool does not use a PropertyManager. The 2D to 3D toolbar provides the following selections:

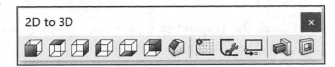

- **Add to Front Sketch**. The Front tool ⬚ displays the selected sketch entities in a front view when converting to a 3D part.

- **Add to Top Sketch**. The Top tool ⬚ displays the selected sketch entities in a top view when converting to a 3D part.

- **Add to Right Sketch**. The Right tool ⬚ displays the selected sketch entities in a right view when converting to a 3D part.

- **Add to Left Sketch**. The Left tool ⬚ displays the selected sketch entities in a left view when converting to a 3D part.

- **Add to Bottom Sketch**. The bottom tool ⬚ displays the selected sketch entities in a bottom view when converting to a 3D part.

- **Add to Back Sketch**. The Back tool ⬚ displays the selected sketch entities in a back view when converting to a 3D part.

- **Auxiliary Sketch**. The Auxiliary tool ⬚ displays the selected sketch entities in auxiliary view when converting to a 3D part. A line in another view must be selected to specify the angle of the auxiliary view.

- **Create Sketch from Selections**. The Create Sketch from Selections tool ⬚ creates a new sketch from the selected sketch entities. Example: You can extract a sketch and then modify it before creating a feature.

- **Repair Sketch**. The Repair Sketch tool ⬚ repairs the errors in your sketch. Example: Typical errors can be overlapping geometry, small segments which are collected into a single entity, or small gaps.

- **Align Sketch**. The Align Sketch tool ⬚ provides the ability to select an edge in one view and to align to the edge selected in a second view. The order of selection is very important.

- **Convert to Extrusion**. The Extrude tool ⬚ extrudes a feature from the selected sketch entities. You do not have to select a complete sketch.

- **Convert to Cut**. The Cut tool ⬚ cuts a feature from the selected sketch entities. You do not have to select a complete sketch.

Tutorial: 2D to 3D Sketch tool 5-1

Import a part and apply the 2D to 3D Sketch tool.

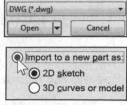

1. Open the **40-4325.DWG** drawing from the SOLIDWORKS 2018 folder.

2. Check the **Import to a new part** check box from the DXF/DWG Import dialog box.

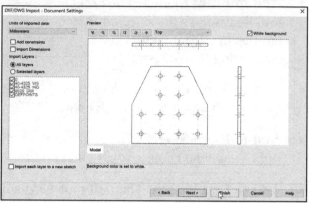

3. Check the **2D sketch** box.

4. Click **Next>**.

5. Check the **White background** check box.

6. Select **Millimeters** for Units for the Imported data.

7. Click **Next>**.

8. Click **Finish**.

9. Click **Yes**. Part1 is displayed in the Part FeatureManager. You are in the Edit Sketch mode.

The current sketch displays the Top, Front and Right views. SOLIDWORKS displays the part and the 2D to 3D toolbar in the Graphics window.

The Front view and Right view are required to extrude the sketch.

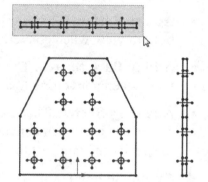

Delete the top view.

10. Box-Select the **Top view** of Sketch1. The Properties PropertyManager is displayed.

11. Press the **Delete** key.

Set the Front view.

12. Insert a Midpoint relation between the **origin** and the **bottom horizontal line**.

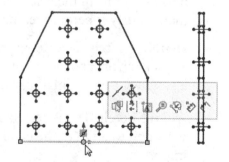

13. Click **OK** ✔ from the Properties PropertyManager.

14. Box-Select the **Front profile**.

15. Click **Fix** from the Add Relations box.

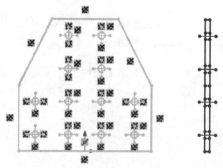

16. Click **OK** ✔ from the Properties PropertyManager.

17. Box-Select the **Front profile**.

18. Click **Add to Front Sketch** from the 2D to 3D toolbar. The view icons are highlighted in the 2D to 3D toolbar.

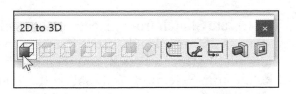

19. Box-Select the **Right profile**.

20. Click **Add to Right Sketch** from the 2D to 3D toolbar. The Right view moves and rotates.

21. Display an **Isometric** view.

Align the sketches.

22. Click the **Front edge** of the Right profile.

23. Hold the **Ctrl** key down.

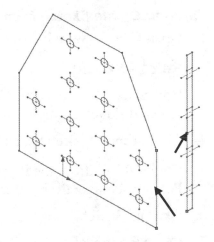

24. Click the **Right edge** of the Front profile.

25. Release the **Ctrl** key.

26. Click **Align Sketch** 🖳 from the 2D to 3D toolbar. The two profiles are aligned.

27. **Rebuild** 🛢 the model. Sketch2 and Sketch3 are created and are displayed in the FeatureManager.

28. Click **Sketch2** from the FeatureManager.

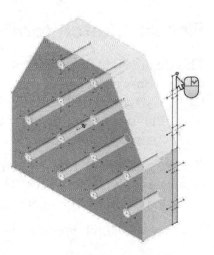

29. Click **Convert to Extrusion** 🗐 from the 2D to 3D toolbar. The PropertyManager is displayed. View the extrude direction arrow.

30. Click the **top back point** on the Right profile of Sketch3 as illustrated. The direction arrow points into the screen. Point4@Shetch3 is displayed in the Vertex box.

31. Click **OK** ✔ from the Boss-Extrude PropertyManager.

32. **Rebuild** 🛢 the model. View the results.

33. **Close** the model.

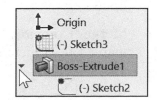

💡 The Save As option displays the new part name in the Graphics window. The Save as copy option copies the current document to a new file name. The current document remains open in the Graphics window.

Creates Sketch from Selections Sketch tool

The Create Sketch from Selections 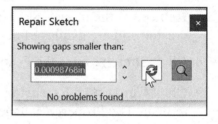 tool provides the ability to only extract the elements of your sketch, usually in an imported drawing that you need to create a feature. Example: Extract a sketch, and then modify it before creating a feature. The Create Sketch from Selections is a tool on the 2D to 3D toolbar. The Create Sketch from Selections does not use a PropertyManager.

Tutorial: Create Sketch from Selections 5-1

1. Open **Create Sketch from Selections 5-1** from the SOLIDWORKS 2018 folder.

2. **Edit** Sketch1.

3. Box-Select the **top view**. The Properties PropertyManager is displayed with the selected sketch entities.

4. Click **Tools ➢ Sketch Tools ➢ Create Sketch from Selections**.

5. Click **OK** from the Properties PropertyManager. A new sketch is displayed in the FeatureManager design tree, Sketch2.

6. **Exit** the Sketch.

7. **Close** the model.

Repair Sketch tool

The Repair Sketch tool provides the ability to find, and in some cases repair the sketch. Example: Some tools; Extrude and Cut repair sketches automatically when the tool detects an error which it can fix. You can also repair sketches manually.

The Repair Sketch tool is useful for sketches created by importing DXF/DWG files. The Repair Sketch tool does not use a PropertyManager.

The Repair Sketch proceeds automatically as follows:

- Deletes zero line length and arc segments. A zero line length is a segment that is less than ±1.0e-8 meters.

- Merges Coincident lines which are separated by less than ±1.0e-8 meters into a single line.

- Merges co-linear lines which overlap.

- Collects small segments in co-linear lines with no gaps that are greater than ±1.0e-8 meters.

- Eliminates gaps of less than ±1.0e-8 meters in co-linear lines.

Tutorial: Repair Sketch 5-1

Apply the Repair Sketch tool.
1. Open **Repair Sketch 5-1** from the SOLIDWORKS 2018 folder.

2. **Edit** 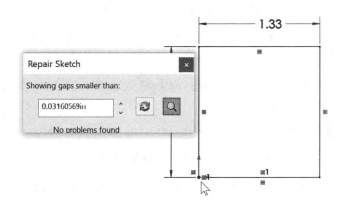 Sketch1 from the FeatureManager.

Apply the Repair Sketch tool.
3. Click **Tools** ➤ **Sketch Tools** ➤ **Repair Sketch**. The Repair Sketch dialog box is displayed.

4. Click the **up arrow** to increase the Showing gaps smaller than box.

5. Click the **magnifying glass** icon. View the results. Two point gap.

6. **Repair** the gap in the sketch.

7. **Close** the Repair Sketch dialog box.

8. **Exit** the sketch.

9. **Close** the model.

Sketch Picture Sketch tool

The Sketch Picture Sketch tool provides the ability to insert a picture in the following formats: (.bmp, .gif, .jpg, .jpeg, .tif and .wmf) on your Sketch plane. Use a picture as an underlay for creating 2D sketches. The Sketch Picture tool uses the Sketch Picture PropertyManager. The Sketch Picture PropertyManager provides the following selections:

- *Properties*. The Properties box provides the following options:

  - **Origin X Position**. Displays the selected X coordinate value for the origin of your picture.

  - **Origin Y Position**. Displays the selected Y coordinate value for the origin of your picture.

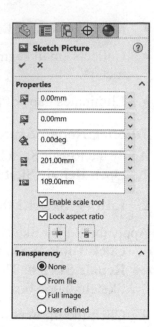

- **Angle**. Displays the selected angle value in degrees. A positive angle rotates your picture counterclockwise.

- **Width**. Displays the selected width value for the picture. If the Lock aspect ratio option check box is selected, the height adjusts automatically.

- **Height**. Displays the selected height value for the picture height. If the Lock Aspect Ratio option is selected, the width adjusts automatically.

- **Enable scale tool**. Selected by default.

- **Lock aspect ratio**. Selected by default. Keeps a fixed width and height aspect ratio.

- **Flip Horizontally**. Flips your picture horizontally within its borders.

- **Flip Vertically**. Flips your picture vertically within its borders.

- *Transparency*. The Transparency box provides the following options:

  - **None**. Selected by default. Does not use any transparency attributes.

  - **From file**. Keeps the transparency attributes which exist presently in the file.

  - **Full image**. Sets the whole image transparent.

Slide the slider to adjust the degree of transparency when you select the Full image option.

  - **User defined**. Provides the ability to select a color from the image, define a tolerance level for that color, and to apply a transparency level to the image.

The picture is inserted with its (0, 0) coordinate at the sketch origin, an initial size of 1 pixel per 1mm, and locked aspect ratio.

The picture is embedded in the document (not linked). If you change the original image, the sketch picture does not update. If you sketch on top of your picture, there is no snap to picture, inferencing, or autotracing capability. The sketch does not update if the image is moved, or deleted and replaced.

Tutorial: Sketch Picture 5-1

Insert a picture into an existing sketch.

1. Open **Sketch Picture 5-1** from the SOLIDWORKS 2018 folder.

2. **Edit** Sketch2 from the FeatureManager.

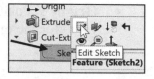

3. Click **Tools** ➤ **Sketch Tools** ➤ **Sketch Picture**.

4. **Browse** to locate a picture file in the SOLIDWORKS 2018 folder. Double-click the **logo** picture file. The SOLIDWORKS logo is displayed in the Graphics window.

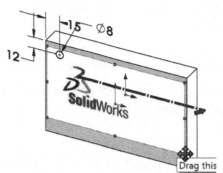

5. Set the **properties** from the Sketch Picture PropertyManager or use the picture handles in the Graphics window to move and size the logo on the model.

6. Click **OK** ✔ from the Sketch Picture PropertyManager.

7. **Rebuild** 🔧 the model. Right-click **Sketch2** from the FeatureManager.

8. Click **Show**. View the results.

9. **Close** the model.

Geometric Relations 2D Sketches

Relations between sketch entities and model geometry, in either 2D or 3D sketches, are an important means of building in design intent. Sketch relations are geometric constraints between sketch entities or between a sketch entity and a plane, axis, edge, or vertex. Relations can be added automatically or manually.

🔆 To select or clear the Automatic relations option, click **Tools** ➤ **Sketch Settings** ➤ **Automatic Relations** or click **Options** ⚙ ➤ **System Option** ➤ **Relations/Snaps**, and click **Automatic relations**.

🔆 Sketch relation pop-ups are now available upon termination of a sketch entity, allowing you to immediately add sketch relations.

Automatic Geometric Relations

As you sketch, allow SOLIDWORKS to automatically add geometric relations. Automatic relations rely on Inferencing, Point display, and Sketch Snaps and Quick Snaps.

When the Automatic Geometric relations tool is activated, the following relations are added:

- *Horizontal*

  - Entities to select: One or more lines or two or more points.

  - Resulting Relations: The lines become horizontal as defined by the current sketch. Points are aligned horizontally.

- *Coincident*

  - Entities to select: A point and a line, arc or ellipse.

  - Resulting Relations: The point lies on the line, arc or ellipse.

- *Perpendicular*

  - Entities to select: Two lines.

  - Resulting Relations: The two items are perpendicular to each other.

- *Vertical*

  - Entities to select: One or more lines or two or more points.

  - Resulting Relations: The lines become vertical as defined by the current sketch. Points are aligned vertically.

- *Midpoint*

  - Entities to select: Two lines or a point and a line.

  - Resulting Relation: The point remains at the midpoint of the line.

- *Tangent*

  - Entities to select: An arc, ellipse, or spline and a line or arc.

  - Resulting Relations: The two items remain tangent.

Manual Geometric Relations

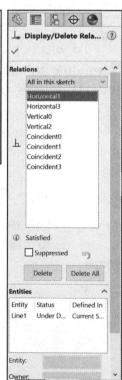

Manually add any required Geometric relations and then dimensions to fully define a sketch, or edit existing relations using the Display/Delete Relations tool.

The Properties PropertyManager is displayed when you select multiple Sketch entities in the Graphics window. Create geometric relations between sketch entities, or between sketch entities and planes, axes, edges or vertices.

For models with multiple configurations, you can apply the selected relations to This configuration, All configurations, or Specify configurations. If you select Specify configurations, select configurations in the Configurations list. Click All to select all the configurations in the list. Click Reset to reset the selections to the original settings.

The following are a few of the common 2D Sketch relations and how to create them:

- *Coincident*

    - Entities to select: A point and a line, arc or ellipse.

    - Resulting Relations: The point lies on the line, arc or ellipse.

- *Collinear*

    - Entities to select: Two or more lines.

    - Resulting Relations: The items lie on the same infinite line.

When you create a relation to a line, the relation is to the infinite line, not just the sketched line segment or the physical edge. As a result, some items may not touch when you expect them to.

- *Concentric*

    - Entities to select: Two or more arcs, or a point and an arc.

    - Resulting Relations: The arcs share the same centerpoint.

- *Coradial*

    - Entities to select: Two or more arcs.

    - Resulting Relations: The items share the same centerpoint and radius.

When you create a relation to an arc segment or elliptical segment, the relation is actually to the full circle or ellipse.

- *Equal*

    - Entities to select: Two or more lines or two or more arcs.

    - Resulting Relations: The line lengths or radii remain equal.

- *Fix*

    - Entities to select: An entity.

    - Resulting Relations: The entity's size and location are fixed. However, the end points of a fixed line are free to move along the infinite line that underlies it. The endpoints of an arc or elliptical segment are free to move along the underlying full circle or ellipse.

- *Horizontal or Vertical*
  - Entities to select: One or more lines or two or more points.
  - Resulting Relations: The lines become horizontal or vertical as defined by the current sketch. Points are aligned horizontally or vertically.

- *Intersection*
  - Entities to select: Two lines and one point.
  - Resulting Relations: The point remains at the intersection of the lines.

- *Merge Points*
  - Entities to select: Two sketch points or endpoints.
  - Resulting Relations: The two points are merged into a single point.

- *Midpoint*
  - Entities to select: Two lines or a point and a line.
  - Resulting Relations: The point remains at the midpoint of the line.

- *Parallel*
  - Entities to select: Two or more lines in a 2D sketch. A line and a plane in a 3D sketch. A line and a planar face in a 3D sketch.
  - Resulting relations: The items are parallel to each other in the 2D sketch. The line is parallel to the selected plane in the 3D sketch.

If you create a relation to an item that does not lie on the sketch plane, the resulting relation applies to the projection of that item as it is displayed on the Sketch plane.

- *Perpendicular*
  - Entities to select: Two lines.
  - Resulting Relations: The two items are perpendicular to each other.

- *Pierce*
  - Entities to select: A sketch point and an axis, edge, line or spline.
  - Resulting Relations: The sketch point is coincident to where the axis, edge, or curve pierces the sketch plane. The pierce relation is used in Sweeps with Guide Curves.

- *Symmetric*
  - Entities to select: A centerline and two points, lines, arcs or ellipses.
  - Resulting Relations: The items remain equidistant from the centerline, on a line perpendicular to the centerline.

- *Tangent*

  - Entities to select: An arc, ellipse, or spline, and a line or arc.

  - Resulting Relations: The two items remain tangent.

Geometric Relations in 3D Sketches

Many relations available in 2D sketches are available in 3D sketches. Additional sketch relations that are supported with 3D sketching include:

- Perpendicular relations between a line through a point on a surface.

- Symmetric relations about a line between 3D sketches created on the same plane.

- Midpoint relations.

- Equal relations.

- Relations between arcs such as concentric, tangent, or equal.

- Normal to applied between a line and a plane, or between two points and a plane.

- Relations between 3D sketch entities created on one sketch plane, and 3D entities created on other sketch planes.

3D Sketch Relations

- Example 1: Equal arcs created on perpendicular planes.

- Example 2: An Arc with a tangent line and perpendicular line to the midpoint between the sketch entities.

- Example 3: An Arc with a perpendicular line created on the perpendicular planes.

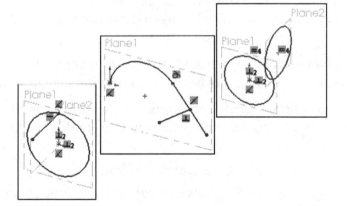

The following are a few of the common 3D Sketch relations and how to create them:

- *AlongZ*

  - Entities to select: A line and a plane in a 3D sketch. A line and a planar face in a 3D sketch.

  - Resulting Relations: The line is normal to the face of the selected plane.

 Relations to the global axes are called AlongX, AlongY and AlongZ. Relations that are local to a plane are called Horizontal, Vertical and Normal.

- *ParallelYZ*

  - Entities to select: A line and a plane in a 3D sketch. A line and a planar face in a 3D sketch.

  - Resulting relations: The line is parallel to the YZ plane with respect to the selected plane.

- *ParallelZX*

  - Entities to select: A line and a plane in a 3D Sketch. A line and a planar face in a 3D sketch.

  - Resulting relations: The line is parallel to the ZX plane with respect to the selected plane.

- *Parallel*

  - Entities to select: A line and a plane in a 3D sketch. A line and a planar face in a 3D sketch.

  - Resulting relations: The line is parallel to the selected plane in the 3D sketch.

Dimension/Relations toolbar

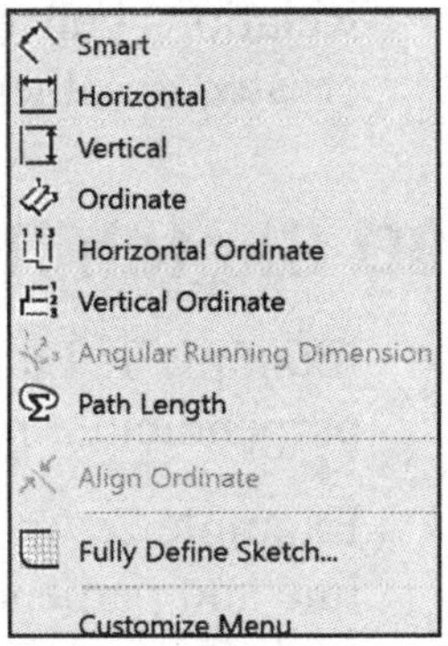

The Dimensions/Relations toolbar, **Tools ➤ Dimensions** menu, **Tools ➤ Relations** menu, and the Consolidated Sketch toolbar provide the required tools to dimension and to add and delete Geometric relations.

Not all toolbar buttons have corresponding menu items; conversely, not all menu items have corresponding toolbar buttons.

This book focuses on providing the most common tools and menu options. Not all dimension/relations tools are addressed.

⌖ To modify the dimension font size, click **Options** ⚙, **Document Properties**, **Detailing** and check the **Always display text at the same size** box.

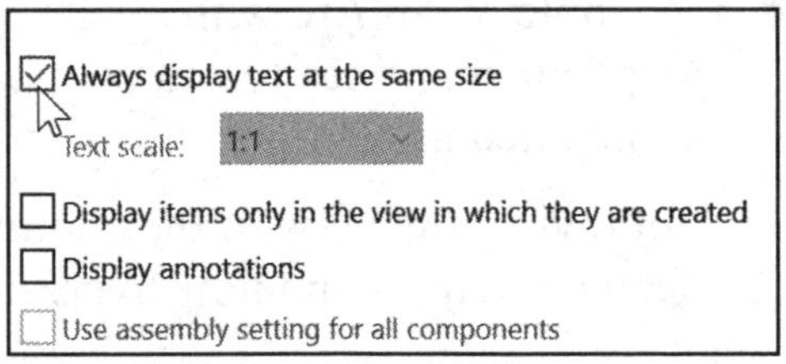

Smart Dimension tool

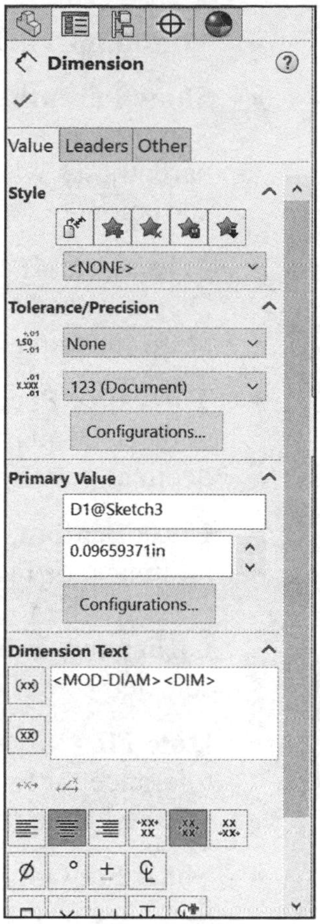

The Smart Dimension tool uses the Dimension PropertyManager. The Dimension PropertyManager provides the ability to either select the **Value, Leaders** or **Other** tab. Each tab provides a separate menu. Note: The Value tab is selected by default.

Smart Dimension tool: Value tab

The Value tab provides the following selections:

- ***Style***. The Style box provides the ability to define your favorite styles and various annotations (Notes, Geometric Tolerance Symbols, Surface Finish Symbols, and Weld Symbols). These are the available options:
  - **Apply the default attributes to selected dimensions**. Resets the selected dimension or dimensions to the document defaults.
  - **Add or Update a Style**. None is the default. Opens the Add or Updates dialog box. Review your styles using the drop-down menu.
  - **Delete a Style**. Delete the selected style from your document.
  - **Save a Style**. Opens the Save As dialog box with Style (.sldfvt) as the default file type. Save the existing style.
  - **Load Style**. Opens the Open dialog box with style (.sldfvt) as the active file type. Note: Use the Ctrl key or the Shift key to select multiple files.
  - **Set a current Style**. NONE is the default. Set a current style from the drop-down menu.

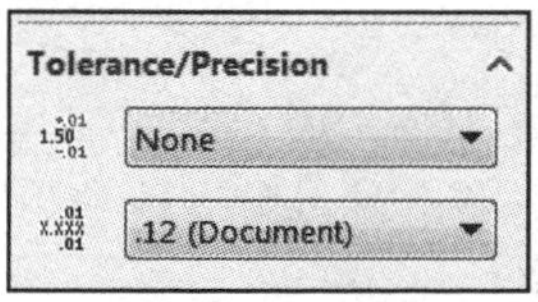

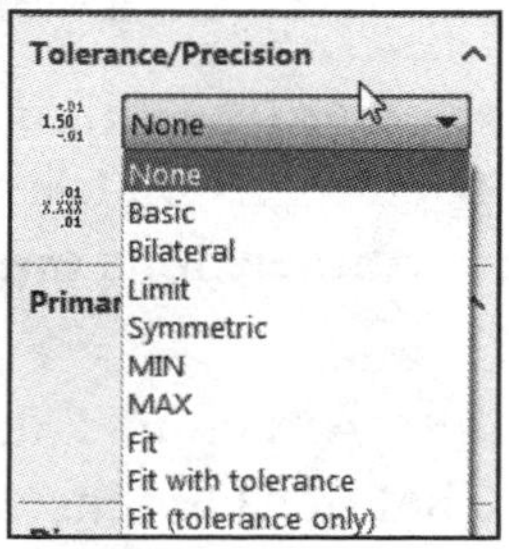

- ***Tolerance/Precision***. The Tolerance/Precision box provides the following selections:
  - **Tolerance Type**. Displays the selected tolerance Type. None is the default type. The available Types are dynamic in the list. Example: Types for chamfer dimensions are limited to **None, Bilateral** and **Symmetric**.

- **Maximum Variation**. Set the maximum variation value.

- **Minimum Variation**. Set the minimum variation value.

- **Show Parentheses**. Only available for Bilateral, Symmetric, and Fit with tolerance types. Displays parentheses when values are set for the Max and Min Variation.

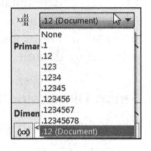

- **Primary Unit Precision**. .12 (Document) is the default. Displays the selected number of digits after the decimal point for primary unit precision.

- **Tolerance Precision**. Same as nominal (Document) is the default. Displays the selected number of digits after the decimal point for tolerance precision.

- **Classification**. Only available if you select Fit, Fit with tolerance, or Fit (tolerance only) types. This option provides the following selections: **User Defined**, **Clearance**, **Transitional** and **Press**. User Defined is selected by default.

- **Hole Fit**. Only available if you select Fit, or Fit with tolerance, or Fit (tolerance only) types. Displays the selected Hole fit type from the drop-down menu.

- **Shaft Fit**. Only available if you select Fit, or Fit with tolerance or Fit (tolerance only) types. Displays the selected Shaft Fit type from the drop-down menu.

- **Stacked with line display**. Only available if you select Fit, or Fit with tolerance or Fit (tolerance only) types.

- **Stacked without line display**. Only available if you select Fit, or Fit with tolerance or Fit (tolerance only) types.

- **Linear display**. Only available if you select Fit, or Fit with tolerance or Fit (tolerance only) types.

- *Configuration*. Parts and Assemblies only. Applies the dimension tolerance to specific configurations for driven dimensions only.

🔆 A second Tolerance/Precision section is available for Chamfer Dimensions.

- *Primary Value*. The Primary Value box provides the following two selections:

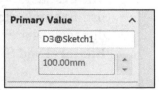

    - **Sketch name**. Displays the dimension name of the sketch.

- **Sketch entity dimension**. Displays the dimension of the selected entity in the Graphics window.

- *Dimension Text*. The Dimension Text box provides the following selections:

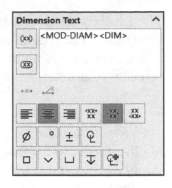

  - **Center Dimension**. Centers the displayed dimension.

  - **Offset Text**. Displays an offset text from the dimension.

  - **Add Parenthesis**. Displays the dimension with parentheses.

  - **Inspection Dimension**. Displays the dimension for inspection.

  - **Text Box**. Provides the ability for the dimension text to be automatically displayed in the center text box, represented by <DIM>. Place the mouse pointer anywhere in the text box to insert text. If you delete <DIM>, you can reinsert the value by clicking **Add Value**.

  - **Justify section**. The Justify section provides the ability to justify the text horizontally and for some standards, such as ANSI, you can justify the leader vertically. The available options are **Horizontal - Left Justify**, **Center Justify**, **Right Justify**, **Vertical - Top Justify**, **Middle Justify** and **Bottom Justify**.

  - **Dimension text symbols**. SOLIDWORKS displays eight commonly used symbols and a More button to access the Symbol Library. The eight displayed symbols from left to right are **Diameter**, **Degree**, **Plus/Minus**, **Centerline**, **Square**, **Countersink**, **Counterbore** and **Depth/Deep**.

- *Dual Dimension*. The Dual Dimension box provides the ability to display dual dimensions for the present model.

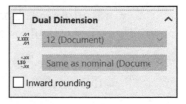

  - **Unit Precision**. Select the number of digits after the decimal point from the list for the dimension value.

  - **Tolerance Precision**. Select the number of digits after the decimal point for tolerance values.

  - **Inward rounding**. Controls the inward rounding for each individual dimension.

Smart Dimension tool: Leaders tab

The Leaders tab provides the following selections. The type of arrows and leaders available depends on the type of dimension selected (diameter, length, arc, etc.).

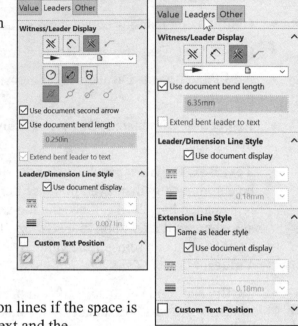

- ***Witness/Leader Display***. The Witness/Leader Display box provides the ability to display the selected type of arrows and leaders available, which depends on the type of dimension selected. There are three selections available for Witness/Leader placement. They are **Outside**, **Inside** and **Smart**. Smart is selected by default.

Smart specifies that arrows are displayed automatically outside of extension lines if the space is too small to accommodate the dimension text and the arrowheads.

- **Style**. Geometry type dependent. The Style drop-down menu provides the ability to select separate styles for each arrow when there are two arrows for a dimension. This feature supports the JIS dimensioning standard.

  Two lists are displayed in the Dimension PropertyManager only when separate styles are specified by the dimensioning standard.

  - **Radius**. Specifies that the dimension on an arc or circle displays the radius.

  - **Diameter**. Specifies that the dimension on an arc or circle displays the diameter.

  - **Linear**. Specifies the display of a diameter dimension as a linear dimension (not radial).

  - **Perpendicular to Axis**. Available if you select Linear for a radial dimension.

  - **Parallel to Axis**. Available if you select Linear for a radial dimension.

- **Foreshortened** 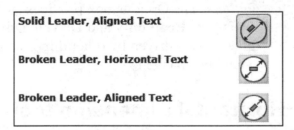. Specifies that the radius dimension line is foreshortened, broken. This is helpful when the centerpoint of a radius is outside of the drawing or interferes with another drawing view.

- **Solid Leader** . Specifies the display of a solid line across the circle for radial dimensions. Not available with ANSI standard.

- **Open Leader** .

- **Dimension to inside of arc**. Specifies that the dimension arrow is inside the arc. Use this option in combination with the **Arrows** setting, either **Inside** or **Outside** to meet your drawing standards.

- **Use document second arrow**. Checked by default. Specifies that a diameter dimension not displayed as linear with outside arrows follows the document default setting for a second arrow.

- **Use document bend length**. Checked by default. Uses the value for Bent leader length in the Options, Document Properties, Dimensions section.

- *Leader Style*. Provides the ability to use the default document style set in Options or to select a custom style for the drop-down menus.

- *Custom Text Position*. Provides the ability to locate and select leader type. The available selections are **Solid Leader**, **Aligned Text**, **Broken Leader**, **Horizontal Text**, and **Broken Leader Aligned Text**.

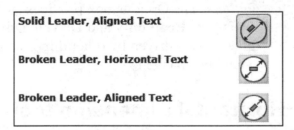

- *Arc Condition*. Provides the ability to set the dimension between an arc or circle. First Arc Condition specifies where on the arc or circle the distance is measured. Second Arc Condition specifies where on the second item the distance is measured, when both items are arcs or circles.

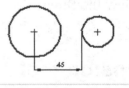 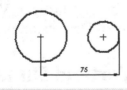

Center (default)   Min (closest point)   Max (farthest point)

Smart Dimension tool: Other tab

The Other tab provides the ability to specify the display of dimensions. If you select multiple dimensions, only the properties that apply to all the selected dimensions are available. The following options are available:

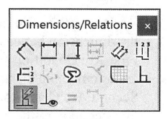

- *Override Units*. Provides the ability to override the units that were set in the Document Properties section of the active document. The available options depend on the type of dimensions that you select.

- *Text Fonts*. Provides the following options:

  - **Dimension font**. Specifies the font used for the dimension.

    - **Use document's font**. Selected by default. Provides the ability to use the current document's font.

    - **Font** button. Provides the ability to select a new font type, style and size for the selected items.

- *Options*. The Options box provides the following dimension selections: **Read only** and **Driven**. Driven specifies that the dimension is driven by other dimensions and conditions and cannot be changed.

Horizontal Dimension tool

The Horizontal dimension ⊟ tool creates a horizontal dimension between two selected entities. The horizontal direction is defined by the orientation of the current sketch. The Horizontal dimension tool uses the Dimension PropertyManager. View the Smart Dimension tool section for detail Dimension PropertyManager information.

Vertical Dimension tool

The Vertical dimension ⊟ tool creates a vertical dimension between two points. The vertical direction is defined by the orientation of the current sketch. The Vertical dimension tool uses the Dimension PropertyManager. View the Smart Dimension tool section for detailed Dimension PropertyManager information.

Baseline Dimension tool

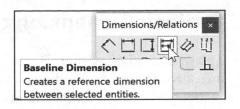

The Baseline Dimension tool creates reference dimensions used in a drawing. You cannot modify their values or use their values to drive the model. Baseline dimensions are grouped automatically. They are spaced at the distances specified by clicking **Tools ➤ Options ➤ Document Properties ➤ Dimensions**. Enter the Baseline dimensions under the Offset distances option.

The distance between dimension lines From last dimension (B) is used for the baseline dimension. This value also controls the snap distance when you drag a linear dimension.

The distance between the model and the first dimension From model (A) is used for the baseline dimension.

You can dimension to midpoints when you add baseline dimensions.

Tutorial: Baseline Dimension Drawing 5-1

Insert a Baseline Dimension into a drawing.
1. Open **Baseline 5-1.slddrw** from the SOLIDWORKS 2018 folder.

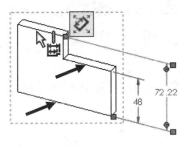

Activate the Dimensions/Relations toolbar.
2. Click **View ➤ Toolbars ➤ Dimensions/Relations**.

3. Click **Baseline Dimension** from the Dimensions/Relations toolbar.

4. Click the **bottom horizontal edge** as illustrated for the baseline reference.

5. Click the other two **horizontal edges** to dimension.

6. Select **Baseline Dimension** from the Dimensions/Relations toolbar to deactivate the tool.

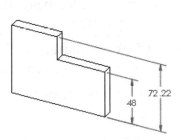

7. Click a **position** in Sheet1. View the drawing.

8. **Close** the drawing.

If you select a vertex, dimensions are measured point-to-point from the selected vertex. If you select an edge, dimensions are measured parallel to the selected edge.

Ordinate Dimension tool

The Ordinate Dimension ⟨⟩ tool provides the ability to create a set of dimensions measured from a zero ordinate in a drawing or sketch. You can dimension to edges, vertices, and arcs, centers and minimum and maximum points.

In drawings, the Ordinate dimension tool provides reference dimensions and you cannot change their values or use the values to drive your model. Ordinate dimensions are measured from the first selected axis. The type of ordinate dimension, vertical or horizontal, is defined by the orientation of the points you select. Ordinate dimensions are automatically grouped to maintain alignment. When you drag any member of the group, all the members move together. To disconnect a dimension from the alignment group, right-click the dimension, and select Break Alignment.

🔅 When you insert ordinate dimensions in a parent view, the dimensions continue in the detail view. The detail view uses the existing 0 point in the parent view. Conversely, when you insert ordinate dimensions in detail views and apply ordinate dimensions in parent views, the parent view uses the 0 point from the detail view.

Tutorial: Ordinate Dimension Drawing 5-1

Insert Ordinate dimensions into a drawing.

1. Open **Ordinate 5-1.slddrw** from the SOLIDWORKS 2018 folder.

2. Click **Ordinate Dimension** ⟨⟩ from the Dimensions/Relations toolbar.

🔅 You can also select the Horizontal Ordinate Dimension tool ⊏⊐ or the Vertical Ordinate Dimension tool ⊏⊐ to specify the direction of the dimensions.

3. Click the **centerline** from which all other sketch entities are measured from (the 0.0 dimension).

4. Click a **position** above the large circle to locate the 0.0 dimension.

5. Click the circumference of the **three small circles** as illustrated from left to right. As you click each circumference, the dimension is placed in the view, aligned to the zero ordinate.

6. Click **Ordinate Dimension** ⟨⟩ from the Dimensions/Relations toolbar to deactivate the tool.

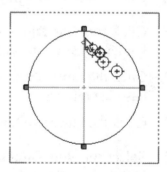

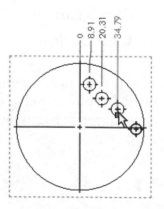

7. Click a **position** in Sheet1. View the dimensions. **Close** the drawing.

💡 To add additional dimensions along the same ordinate, Right-click an **ordinate dimension**, and select **Add To Ordinate**. Click the edges, or vertices, or arcs you want to dimension using the same ordinate.

Horizontal Ordinate Dimension tool

The Horizontal Ordinate dimension ⊔ tool provides the ability to create a horizontal dimension between two entities. The horizontal direction is defined by the orientation of the current sketch. View the above section on Ordinate Dimension tool for additional information.

Vertical Ordinate Dimension tool

The Vertical Ordinate dimension ⊟ tool provides the ability to create a vertical dimension between two points. The vertical direction is defined by the orientation of the current sketch. View the above section on Ordinate Dimension tool for additional information.

Chamfer Dimension tool

The Chamfer Dimension ⅄ tool provides the ability to create chamfer dimensions in drawings. The Chamfer Dimension tool uses the Dimension PropertyManager.

💡 Tolerance types for chamfer dimensions are limited to **None**, **Bilateral** and **Symmetric**. You must select the chamfered edge first. The dimension is not displayed until you select one of the lead-in edges.

Tutorial: Chamfer Dimension Drawing 5-1
Insert Chamfer dimensions into a drawing.
1. Open **Chamfer Dimension 5-1.slddrw** from the SOLIDWORKS 2018 folder.

2. Click **Chamfer Dimension** ⅄ from the Dimensions/Relations toolbar.

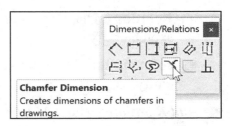

3. Click the **chamfered edge** as illustrated.

4. Click the **top horizontal edge**. The dimension is displayed.

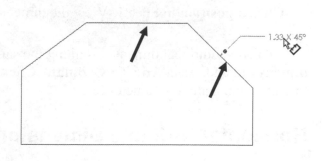

5. Click a **position** in the Graphics window to place the dimension. The Dimension PropertyManager is displayed.

6. Click **OK** ✔ from the Dimension PropertyManager. View the drawing.

7. **Close** the drawing.

Add Relation tool

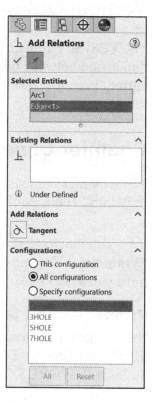

The Add Relation ⊥ tool provides the ability to create geometric relations between sketch entities, or between sketch entities and axes, planes, edges, or vertices. The Add Relation tool uses the Add Relations PropertyManager.

The Add Relations PropertyManager provides the following selections:

- **Selected Entities**. The Selected Entities box provides the following option.

  - **Selected Entities**. Displays the names of selected sketch entities from the Graphics window.

- **Existing Relations**. The Existing Relations box provides the following options:

  - **Relations**. Displays the existing relations for the selected sketch entity.

  - **Information**. Displays the status of the selected sketch entity.

- **Add Relations**. The Add Relations box provides the following option.

  - **Relations**. Provides the ability to add a relation to the selected entities from the list. The list includes only relations that are possible for the selected entities.

- **Configurations**. Provides that ability to address the following configurations.

  - **This configuration**.

  - **All configurations**.

  - **Specify configurations**.

Tutorial: Add Relation 5-1

Add a Horizontal relation to a 2D Sketch.

1. Open **Add Relation 5-1** from the SOLIDWORKS 2018 folder.

2. **Edit** 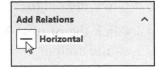 Sketch1.

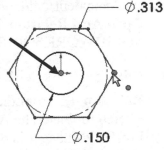

3. Click **Add Relation** ⊢ from the Consolidated Sketch toolbar. The Add Relations PropertyManager is displayed.

4. Click the **origin**. Click the **right most point** of the hexagon as illustrated.

5. Click **Horizontal** from the Add Relations box.

6. Click **OK** ✔ from the Add Relations PropertyManager. A Horizontal relation is applied.

7. **Rebuild** the model. View the results. **Close** the model.

☀ Insert Geometric relations without the Add Relations tool by using the Ctrl key and selecting two or more sketch entities. The Properties PropertyManager is displayed. The Add Relation tool is not required with this process.

Tutorial: Add Relation 5-2

Add a Midpoint, Equal, and Tangent geometric relation to a 2D Sketch.

1. Open **Add Relation 5-2** from the SOLIDWORKS 2018 folder.

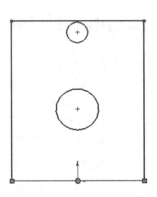

2. **Edit** Sketch1. Click **Add Relation** ⊢ from the Consolidated Sketch toolbar. The Add Relations PropertyManager is displayed.

3. Click the **origin**. Click the **bottom horizontal** line of the rectangle.

4. Click **Midpoint** from the Add Relations box. A Midpoint relation is added. Click the **circumference** of the two circles.

5. Click **Equal** from the Add Relations box.

6. Click **Tangent** from the Add Relations box. An Equal and Tangent relation is applied to the circles.

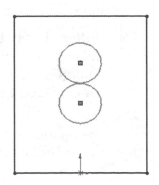

7. Click **OK** ✔ from the Add Relations PropertyManager.

8. **Rebuild** the model. View the results. **Close** the model.

Tutorial: Add Relation 5-3

Add a Coincident, Midpoint, and Equal relation in a 3DSketch.

1. Open **Add Relation 5-3** from the SOLIDWORKS 2018 folder.

2. **Edit** 3DSketch1. Click **Add Relation** from the Consolidated Sketch toolbar. The Add Relations PropertyManager is displayed.

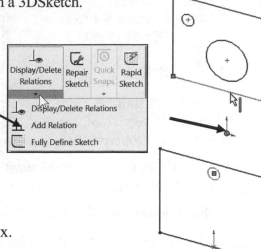

3. Click the **origin**.

4. Click the **bottom horizontal** X axis line as illustrated.

5. Select **Midpoint** from the Add Relations box.

6. Click the circumference of the **small circle**.

7. Click the circumference of the **large circle**.

8. Click **Equal** from the Add Relations box.

9. Click **OK** ✔ from the Add Relations PropertyManager.

10. **Rebuild** the model. View the results.

11. **Close** the model.

Display/Delete Relations tool

The Display/Delete Relations tool provides the ability to edit existing relations after you added the relations to your sketch. The Display/Delete Relations dimension tool uses the Display/Delete Relations PropertyManager. The Display/Delete Relations PropertyManager provides the following selections:

- *Relations*. The Relations box displays the selected relation based on the option selected from the filter box. The appropriate sketch entities are highlighted in your Graphics window. The Relations box provides the following options:

  - **Filter**. The Filter box provides the ability to specify which relation to display. The selections are **All in this sketch**, **Dangling**, **Over Defined/Not Solved**, **External**, **Defined In Context**, **Locked**, **Broken** and **Selected Entities**.

If the relation was created In Context of an assembly, the status can be Broken or Locked.

- **Information icon**. Displays the status of the selected sketch entity.

- **Suppressed**. Suppresses the selected relation for the current configuration. Example: The name of the suppressed relation is displayed in gray and the Information status changes from Satisfied to Driven.

- **Undo**. Provides the ability to change the last relation.

- **Delete**. Deletes or replaces your last action.

  - **Delete All**. Deletes your selected relations or deletes all of your relations.

- *Entities*. The Entities box provides the ability to view the selected relation on each selected entity. The Entities box provides the following information:

  - **Entity**. Lists each selected sketch entity.

  - **Status**. Displays the status of your selected sketch entity.

  - **Defined In**. Displays the location where the sketch entity is defined. Example: Same Model, Current Sketch, or External Model.

  - **Entity**. Information for external entities in an assembly. Displays the entity name for sketch entities in the Same Model or External Model.

  - **Owner**. Information for external entities in an assembly. Displays the part to which the sketch entity belongs.

  - **Assembly**. Information for external entities in an assembly. Displays the name of the top-level assembly where the relation was created for sketch entities in an External Model.

  - **Replace**. Replaces your selected entity with a different entity.

Tutorial: Display/Delete Relations 5-1

Delete and view the relations in a 3D Sketch.

1. Open **Display-Delete Relations 5-1** from the SOLIDWORKS 2018 folder.

2. Click **Close** to the Error message.

3. **Edit** 3DSketch1 from the FeatureManager.

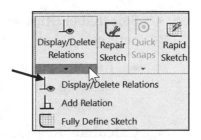

4. Click **Display/Delete Relations** ⊥⊚ from the Sketch toolbar. The Add Relations PropertyManager is displayed. View the relation in the Relations box.

5. **Delete** the Coincident6 relation.

6. Select **All in this sketch** from the Relations box. View the relations in the Relations box.

7. Click **OK** ✔ from the Display-Delete Relations PropertyManager.

8. **Rebuild** the model. View the results.

9. **Close** the model.

Fully Defined Sketch tool

The Fully Defined Sketch ⊞ tool provides the ability to calculate which dimensions and relations are required to fully define under defined sketches or selected sketch entities. You can access the Fully Define Sketch tool at any point and with any combination of dimensions and relations already added.

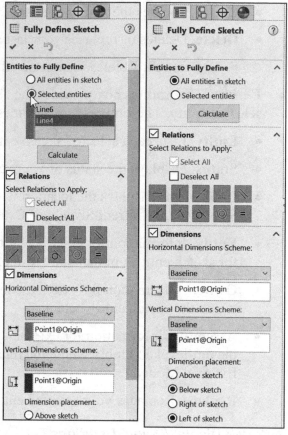

🔅 Your sketch should include some dimensions and relations before you use the Fully Define Sketch tool.

The Fully Defined Sketch tool uses the Fully Define Sketch PropertyManager. The Fully Define Sketch PropertyManager provides the following selections:

- ***Entities to Fully Define***. The Entities to Fully Define box provides the following options:

    - **All entities in sketch**. Fully defines the sketch by applying combinations of relations and dimensions.

    - **Selected entities**. Provides the ability to select sketch entities.

        - **Entities to Fully Define**. Only available when the Selected entities box is checked. Applies relations and dimensions to the specified sketch entities.

    - **Calculate**. Analyzes the sketch and generates the appropriate relations and dimensions.

- *Relations*. The Relations box provides the following selections:

  - **Select All**. Includes all relations in the results.

  - **Deselect All**. Omits all relations in the results.

  - **Individual relations**. Include or exclude needed relations. The available relations are **Horizontal, Vertical, Collinear, Perpendicular, Parallel, Midpoint, Coincident, Tangent, Concentric** and **Equal radius/Length**.

- *Dimensions*. The Dimensions box provides the following selections:

  - **Horizontal Dimensions**. Displays the selected Horizontal Dimensions Scheme and the entity used as the Datum - Vertical Model Edge, Model Vertex, Vertical Line or Point for the dimensions. The available options are **Baseline, Chain** and **Ordinate**.

  - **Vertical Dimensions**. Displays the selected Vertical Dimensions Scheme and the entity used as the Datum - Horizontal Model Edge, Model Vertex, Horizontal Line or Point for the dimensions. The available options are **Baseline, Chain** and **Ordinate**.

  - **Dimension**. Below sketch and Left of sketch is selected by default. Locates the dimension. There are four selections: **Above sketch, Below the sketch, Right of sketch** and **Left of sketch**.

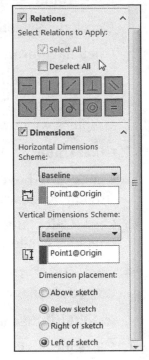

Tutorial: Fully Defined Sketch 5-1

Create a fully defined sketch. Fully defined sketches are needed for Manufacturing.

1. Open **Fully Defined Sketch 5-1** from the SOLIDWORKS 2018 folder.

2. **Edit** 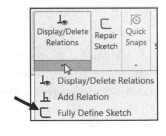 Sketch1.

3. Click **Fully Defined Sketch** from the Consolidated Sketch toolbar. The Fully Defined Sketch PropertyManager is displayed.

4. Check the **All entities in sketch** box.

5. Click **Calculate** from the Entities to Fully Define box. The dimensions are displayed in the Graphics window.

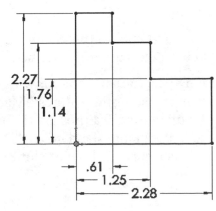

6. Click **Undo** 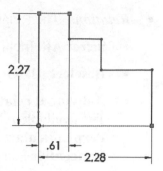 from the PropertyManager.

7. Check the **Selected entities** box.

8. Click the **bottom horizontal** line.

9. Click the **top horizontal** line.

10. Click **Calculate**. The dimensions are displayed in the Graphics window.

11. Click **Undo** from the PropertyManager.

12. Check the **Above sketch** box and the **Right of sketch** box from the Dimension section.

13. Click **Calculate**. The dimensions are displayed in the Graphics window.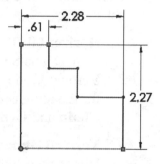

14. Click **OK** from the Fully Define Sketch PropertyManager.

15. **Rebuild** the model. View the results.

16. **Close** the model.

DimXpertManager

The DimXpertManager provides access to the DimXpert. DimXpert for parts is a set of tools you use to apply dimensions and tolerances to parts according to the requirements of the ASME Y14.41-2009 standard.

You can then use the tolerances with TolAnalyst to perform stack analysis on assemblies, or with downstream CAM, other tolerance analyses, or metrology applications.

Model dimensions are associative between parts, assemblies, and drawings. Reference dimensions and DimXpert dimensions are one-way associative from the part to the drawing or from the part to the assembly. DimXpert dimensions show up in a different color to help identify them from model dims and reference dims. DimXpert dims are the dimensions that are used when calculating tolerance stack-up using TolAnalyst.

DimXpert applies dimensions in drawings so that manufacturing features (patterns, slots, pockets, etc.) are fully-defined.

DimXpert for parts and drawings automatically recognize manufacturing features. What are manufacturing features? Manufacturing features are *not SOLIDWORKS features*. Manufacturing features are defined in 1.1.12 of the ASME Y14.5M-1994 Dimensioning and Tolerancing standard as "The general term applied to a physical portion of a part, such as a surface, hole or slot."

The DimXpertManager provides the following selections: **Auto Dimension Scheme** ⊕, **Basic Location Dimension** ⊢⊙⊣, **Basic Size Dimension** ⬡, **Show Tolerance Status** ⁺⊙, **Copy Scheme** ⊕, **TolAnalyst Study** ⏸.

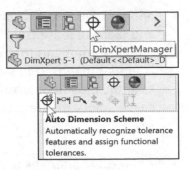

 TolAnalyst is available in SOLIDWORKS Premium.

DimXpert toolbar

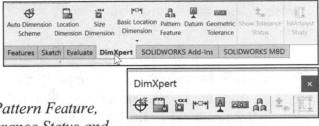

The DimXpert toolbar provides the following tools: *Auto Dimension Scheme, Location Dimension, Size Dimension, Basic Location Dimension, Pattern Feature, Datum, Geometric Tolerance, Show Tolerance Status and TolAnalyst Study.*

- *Auto Dimension Scheme* ⊕ provides the ability to automatically apply dimensions and tolerances to the manufacturing features of a part.

- *Location Dimension* ▤ provides the ability to apply linear and angular dimensions between two DimXpert features, (excluding surface, fillet, chamfer, and pocket features).

- *Size Dimension* ▢ provides the ability to place tolerance size dimensions on DimXpert features.

- *Basic Location Dimension* ⊢⊙⊣ adds DimXpert Basic Dimension location and size.

- *Pattern Feature* ⋒⋒ provides the ability to create or edit pattern features and collection features

- *Datum* ▣ provides the ability to define datum features. The tool supports these feature types: Boss, Cylinder, Notch, Plane, Simple Hole, Slot, Width.

- *Geometric Tolerance* ▣ provides the ability to apply geometric tolerances to DimXpert features. Note: When you apply geometric tolerances to features defined as datums or to features with pre-existing size tolerances, DimXpert automatically places the feature control frame and pre-existing annotation in an annotation group.

- *Show Tolerance Status* ⊥⊘ provides the ability to identify the manufacturing features and faces that are fully constrained, under constrained, and over constrained from a dimensioning and tolerancing perspective. Note: The DimXpert identification process is unlike that used in sketches, which utilizes dimensional and geometrical relationships to determine the constraint status of the sketch entities. DimXpert is based solely on dimension and tolerance constraints. Geometrical relationships, such as Concentric relationships, are not considered.

- *TolAnalyst Study* ⬚ provides the ability to create a new TolAnalyst study on an assembly.

DimXpert tools provide the ability to place dimensions and tolerances on parts. Access DimXpert from the DimXpertManager ⊕ tab or from the DimXpert tab in the CommandManager. The DimXpertManager provides the ability to list the tolerance features defined by DimXpert in chronological order and to display the available tools.

🔆 Care is required to apply DimXpert correctly on complex surfaces or with some existing models. See SOLIDWORKS help for detail information on DimXpert with complex surfaces.

Auto Dimension Scheme tool

The Auto Dimension Scheme tool/DimXpert automatically inserts dimensions and tolerances to the manufacturing features of a part. View SOLIDWORKS Help for additional detail information on "manufacturing features."

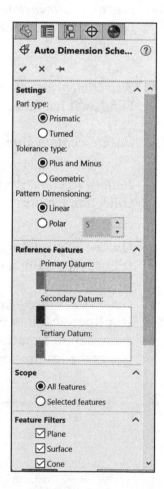

The Auto Dimension Scheme tool/DimXpert utilizes the Auto Dimensional PropertyManager. The Auto Dimensional PropertyManager provides the following selections:

- *Settings*. Set the PropertyManager options described below:

  - **Part type**. Defines the part based on how it is manufactured.

    - **Prismatic**. Selected by default. Used with Geometric Tolerance type. DimXpert applies position tolerances to locate holes and bosses. For annotation views, size dimensions are displayed as callouts with open leaders.

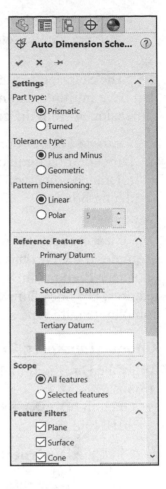

- **Turned**. Used with Geometric Tolerance type. Applies circular run-out tolerances to locate holes and bosses. For annotation views, size dimensions are displayed as linear dimensions, parallel to the axis.

- **Tolerance type**. Provides the ability to control how features of size (hole types, slots, notches, widths, and cones), pockets, and surfaces are located relative to the datum or reference features.

  - **Plus and Minus**. Selected by default. Locates the features with linear plus and minus dimensions. Does not apply dimensions to surface features in plus and minus mode.

  - **Geometric**. DimXpert locates axial features with position and circular runout tolerances. Pockets and surfaces are located with surface profiles.

- **Patterning Dimensioning**. Provides the ability to apply linear or polar plus and minus dimension schemes.

  - **Linear**. Locates all DimXpert features relative to the selected reference features as applicable, using linear plus-minus dimensions.

  - **Polar**. Apply to DimXpert pattern features that contain axial features that define a bolt circle. Set the **Minimum number of holes** to recognize as a pattern.

- *Reference Features/Datum Selection*. Defines the *Primary, Secondary and Tertiary* reference features for **Plus and Minus** schemes or datums for **Geometric** schemes. The selection provides the ability to select one to three reference features to use to generate linear plus and minus location dimensions. DimXpert validates reference features according to the rules defined for establishing datum reference frames, per ASME Y14.5.1M-1994.

- *Scope*. Controls which features DimXpert considers for dimensioning and tolerancing. The following selections are available:

  - **All features**. Selected by default. Provides the ability to apply dimensions and tolerances to the entire part. DimXpert considers all previously defined features, including those not listed under **Feature Filters**, as well as those features listed under **Feature Filters**.

The All features option is best suited for simple parts or parts for which all dimensions and tolerances are relative to a single datum reference frame or origin.

- **Selected features**. The objective is to dimension and tolerance the selected features.

The Selected features option is useful for more complex parts that have multiple datum schemes or require features to be dimensioned from different origins.

- *Feature Filters*. Provides the ability to control which features DimXpert recognizes and considers for dimensioning and tolerance when you select the All Features option in the Scope dialog box.

Clearing features omits further recognition of those types. For example, if the part has no DimXpert dimensions or tolerances and you clear **Plane**, DimXpert does not dimension or tolerance plane features, nor does DimXpert use the planar faces to derive a complex feature type such as a slot or notch.

Tutorial: DimXpert 5-1

Apply the DimXpert tool. Apply Prismatic and Geometric options.

1. Open the **DimXpert 5-1** part from the SOLIDWORKS 2018 folder.

2. Click the **DimXpertManager** ⊕ tab as illustrated.

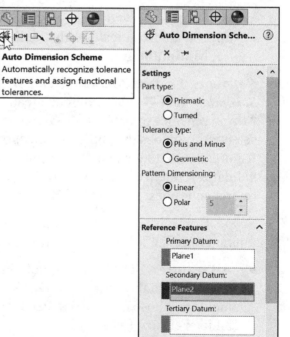

3. Click the **Auto Dimension Scheme** ⊕ tab from the DimXpertManager. The Auto Dimension Scheme PropertyManager is displayed.

4. Click the **Prismatic** box.

5. Click the **Plus and Minus** box.

6. Click the **Linear** box.

Select the Primary Datum.
7. Click the **front face** of the model. Plane1 is displayed in the Primary Datum box.

Select the Secondary Datum.
8. Click **inside** the Secondary Datum box.

9. Click the **left face** of the model. Plane2 is displayed in the Secondary Datum box.

Select the Tertiary Datum.

10. Click **inside** the Tertiary Datum box.

11. Click the **bottom face** of the model. Plane3 is displayed in the Tertiary Datum box. Accept the default options.

12. Display a **Front view**.

13. Click **OK** ✔ from the Auto Dimension Scheme PropertyManager. View the results in the Graphics window and in the DimXpertManager.

14. **Close** all models. Do not save the updates.

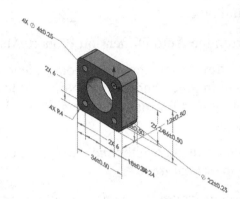

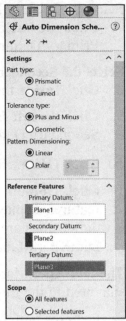

🔆 Right-click Delete to delete the Auto Dimension Scheme in the DimXpert PropertyManager.

Tutorial: DimXpert 5-2

Apply DimXpert: Plus and Minus option. Modify Tolerance and dimensions in the part. Remove Instance Count from the part.

1. Open the **DimXpert 5-2** part from the SOLIDWORKS 2018 folder. View the three Extrude features, Linear Pattern feature, and the Fillet feature.

2. Click the **DimXpertManager** ⊕ tab.

3. Click the **Auto Dimension Scheme** ⊕ tool from the DimXpertManager. The Auto Dimension PropertyManager is displayed.

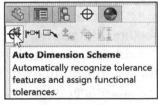

4. Click the **Prismatic** box.

5. Click the **Plus and Minus** box.

6. Click the **Linear** box.

A key difference between the *Plus and Minus* option versus the *Geometric* option is how DimXpert controls the four-hole pattern, and how it applies tolerances to interrelate the datum features when in *Geometric* mode.

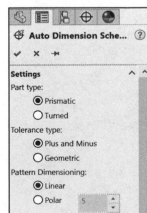

Select the Primary Datum.

7. Click the **back face** of the model. Plane1 is displayed in the Primary Datum box.

Select the Secondary Datum.

8. Click **inside** the Secondary Datum box.

9. Click the **left face** of the model. Plane2 is displayed in the Secondary Datum box.

10. Click **OK** ✔ from the Auto Dimension PropertyManager.

11. Click **Isometric view**. View the dimensions. View the features displayed in green and yellow. Note: Additional dimensions are required for manufacturing. View the DimXpertManager.

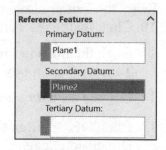

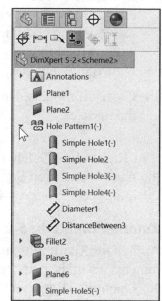

The DimXpertManager displays either *no mark*, *(+)*, or a *(-)* sign next to the Plane or Feature.

- Features with *no mark* after the name are fully constrained, as illustrated in the DimXpertManager, and are displayed in green.

- Features with the *(+)* sign following the name are over constrained and are displayed in red in the Graphics window.

- Features with the *(-)* sign following the name are under constrained and are displayed in yellow in the Graphics window.

🔅 For DimXpert, features mean manufacturing features. For example, in the CAD world, you create a "shell" feature, which is a type of "pocket" feature in the manufacturing world. See SOLIDWORKS Help for additional information.

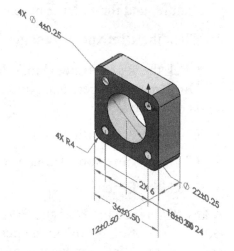

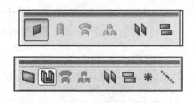

☀ When you apply DimXpert dimensions to manufacturing features, DimXpert uses the following two methods in this order to recognize features: Model feature recognition and then Topology recognition.

☀ The feature selector is a floating, context sensitive toolbar that you can use to distinguish between different DimXpert feature types. The available feature selector choices depend on the selected face and the active command.

The order of features in the feature selector is based on their complexity:

- Basic features like planes, cylinders, and cones are on the left.

- Composite features like counterbore holes, notches, slots, and patterns are in the middle.

- Compound features like compound holes and intersect points are on the right. Compound features require additional selections.

☀ Within DimXpert, a single face can typically define multiple manufacturing feature types that require different dimensions and tolerances.

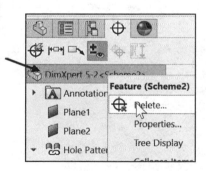

12. Click each **Plane** and **feature** in the Show Tolerance Status FeatureManager. The selected item is displayed in blue.

Delete the DimXpert Scheme.

13. Right-click **DimXpert 5-2** from the DimXpertManager.

14. Click **Delete**.

15. Click **Yes**.

Create a New Scheme which is fully defined.

16. Click the **Auto Dimension Scheme** ⊕ tool from the DimXpertManager. The Auto Dimension PropertyManager is displayed. Prismatic, Plus and Minus and Linear should be selected by default.

17. Click the **back face** of the model. The selected plane is displayed in the Primary Datum box.

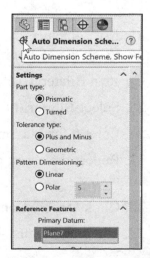

Select the Secondary Datum.
18. Click **inside** the Secondary Datum box.

19. Click the **left face** of the model. The selected plane is
    displayed in the Secondary Datum box.

Select the Tertiary Datum.
20. Click **inside** the Tertiary Datum box.

21. Click the **top face** of the model. The selected plane is
    displayed in the Tertiary Datum box.

22. Click **OK** ✔ from the Auto Dimension PropertyManager.

23. Click **Isometric view**. View the dimensions. All features are
    displayed in green.

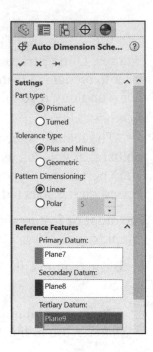

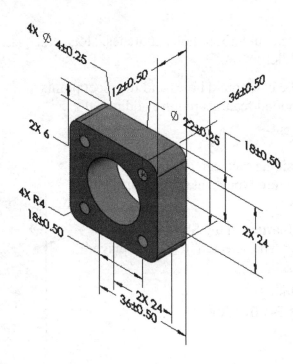

Modify Tolerance and dimensions in the part.
24. Click the **36 horizontal** dimension text.

Create a Bilateral Tolerance.

25. Select **Bilateral** from the Tolerance Type drop-down menu.

26. Enter **0** for Maximum Variation.

27. Repeat the above procedure for the **36 vertical** dimension
    text.

28. Click **OK** ✔ from the DimXpert PropertyManager.

29. **View** the results.

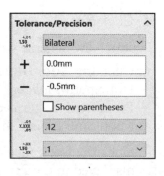

Remove Instance Count from the part.

30. Click the **2X 6 vertical** dimension text.

31. Hold the **Ctrl** key down.

32. Click the **2X 6 horizontal** dimension text.

33. Click the **2X 24 vertical** dimension text.

34. Click the **2X 24 horizontal** dimension text.

35. Release the **Ctrl** key.

36. **Uncheck** the Instance Count box from the Dimension Text dialog box. View the results. Remove the tolerance from the part.

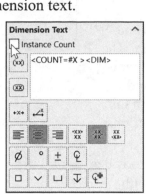

37. Click the **18 vertical** dimension text.

38. Hold the **Ctrl** key down.

39. Click the **18 horizontal** dimension text.

40. Release the **Ctrl** key.

41. Select **None** for Tolerance Type. View the results.

42. Click **OK** from the DimXpert PropertyManager.

43. **Close** the model. Do not save.

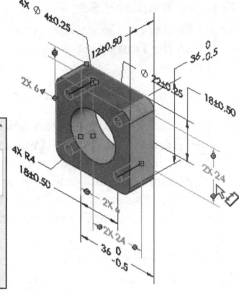

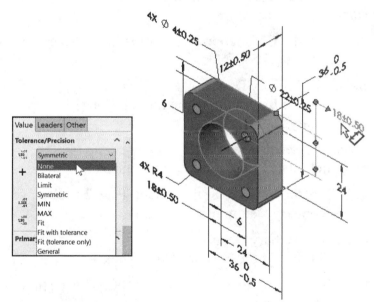

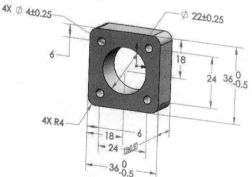

Tutorial: DimXpert 5-3

Apply DimXpert: Geometric option. Edit a Feature Control Frame.

1. Open the **DimXpert 5-3** part from the SOLIDWORKS 2018 folder.

2. Click the **DimXpertManager** ⊕ tab.

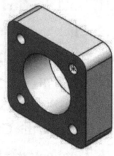

3. Click the **Auto Dimension Scheme** ⊕ tool from the DimXpertManager. The Auto Dimension PropertyManager is displayed. Prismatic and Plus and Minus are selected by default. In this section, select the Geometric option.

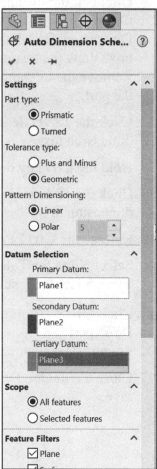

☀ DimXpert: Geometric option provides the ability to locate axial features with position and circular runout tolerances. Pockets and surfaces are located with surface profiles.

4. Click the **Geometric** box as illustrated. Note: the Prismatic box is checked.

Select the Primary Datum.

5. Click the **back face** of the model. Plane1 is displayed in the Primary Datum box.

Select the Secondary Datum.

6. Click **inside** the Secondary Datum box.

7. Click the **left face** of the model. Plane2 is displayed in the Secondary Datum box.

Select the Tertiary Datum.

8. Click **inside** the Tertiary Datum box.

9. Click the **top face** of the model. Plane3 is displayed in the Tertiary Datum box.

10. Click **OK** ✔ from the Auto Dimension PropertyManager.

11. Display an **Isometric view**. View the Datums, Feature Control Frames, and Geometric tolerances. All features are displayed in green.

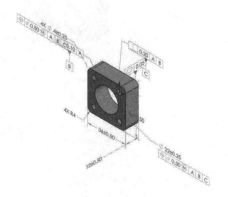

Edit a Feature Control Frame.

12. **Double-click** the illustrated Position Feature Control Frame. The Properties dialog box is displayed.

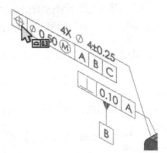

Modify the 0.50 tolerance.

13. Click **inside** the Tolerance 1 box.

14. **Delete** the existing text.

15. Enter **0.25**.

16. Click **OK** from the Properties dialog box.

17. **Repeat** the above procedure for the second Position Feature Control Frame. View the results.

18. **Close** the model. Do not save.

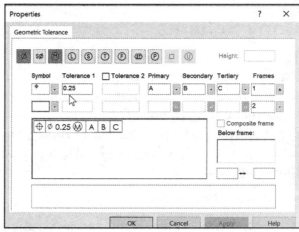

Show Tolerance Status

The Show Tolerance Status ⁺☉ option identifies the manufacturing features and faces, (different colors) which are fully constrained, under constrained, and over constrained from a dimensioning and tolerancing perspective. Example: All green means the model is fully constrained, red means over constrained and yellow means under constrained.

🔆 Set tolerance status color, click **Options** ⚙ ➢ **Colors, Color Scheme settings**. Click the required **setting** as illustrated.

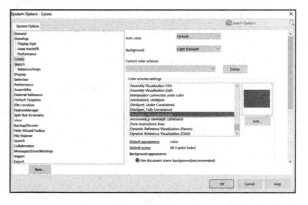

To exit tolerance status mode, click **Show Tolerance Status** or another SOLIDWORKS command.

In the DimXpertManager:

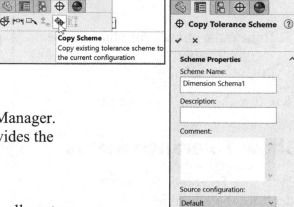

- Features with no mark after the name are fully constrained.

- Features with **(+)** following the name are over constrained.

- Features with **(-)** following the name are under constrained.

Copy Scheme

The Copy Scheme option provides the ability to copy a DimXpert tolerance scheme from one configuration of a part to another configuration. The Copy Scheme uses the Copy Tolerance PropertyManager. The Copy Tolerance PropertyManager provides the following selections:

- ***Scheme Properties***.

  - **Scheme Name**. DimXpert automatically enters a name (**Copy of** <*source configuration name*>), which you can modify.

  - **Description**. Add a description.

  - **Comment**. Add a comment.

  - **Source configuration**. Provides the ability to select the configuration with the tolerance scheme to copy.

TolAnalyst Study

TolAnalyst is available in SOLIDWORKS Premium. TolAnalyst is a tolerance analysis application that determines the effects that dimensions and tolerances have on parts and assemblies.

Use the DimXpert tool to apply dimensions and tolerances to a part or component in an assembly, then use the TolAnalyst tool to leverage that data for stack-up analysis.

TolAnalyst performs a tolerance analysis called a study, which you create using a four-step procedure:

- **Measurement**. Establish the measurement, which is a linear distance between two DimXpert features. The first step in creating a TolAnalyst study is to specify the measurement as a linear dimension between two DimXpert features.

- **Assembly Sequence**. Select the ordered set of parts to establish a tolerance chain between the two measurement features. The selected parts form the "simplified assembly." A simplified assembly includes, at a minimum, the parts necessary to establish a tolerance chain between the two measurement features. This step also establishes the sequence or order in which you place parts into the simplified assembly, which TolAnalyst replicates when computing the worst case conditions.

- **Assembly Constrains**. Define how each part is placed or constrained into the simplified assembly. Assembly constraints are analogous to mates. Constraints are derived by the relationships between DimXpert features, whereas mates are derived by the relationships between geometric entities. Additionally, constraints are applied in sequence, which can play an important role and have significant impact on the results.

- **Analysis Results**. Evaluate and review the minimum and maximum worst case tolerance stacks. When the Results PropertyManager is activated, the results are automatically computed with the default or saved settings.

TolAnalyst is an Add-In under the SOLIDWORKS Add-Ins tab.

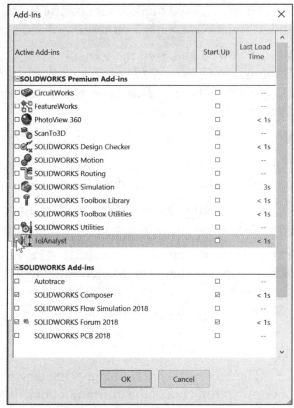

Summary

In this chapter, you learned about the available Sketch tools, Geometric Relations and Dimensions/Relations tools in SOLIDWORKS. Sketch tools control the various aspects of creating and modifying a sketch. You reviewed and addressed the following tools from the Sketch toolbar: *Fillet, Chamfer, Offset Entities, Offset on Surface, Convert Entities, Intersection Curve, Face Curves, Segment, Trim, Extend, Split Entities, Jog Line, Construction Geometry, Make Path, Mirror Entities, Dynamic Mirror, Stretch Entities, Move Entities, Rotate Entities, Scale Entities, Copy Entities, Linear Pattern, Circular Pattern, Create Sketch From Selections, Repair Sketch, SketchXpert, Align, Modify, Close Sketch to Model, Check Sketch for Feature, 2D to 3D and Sketch Picture*

You utilized the 2D to 3D toolbar and reviewed the following tools: *Front, Top, Left, Bottom, Back, Auxiliary, Sketch from Selections, Repair Sketch, Align Sketch, Extrude and Cut*.

You addressed and applied Geometric Relations for 2D sketches using both Automatic and Manual Relations. You reviewed and applied Geometric Relations to 3D sketches.

You reviewed and used the tools from the Dimensions/Relations toolbar: *Smart Dimension, Horizontal Dimension, Vertical Dimension, Baseline Dimension, Ordinate Dimension, Chamfer Dimension, Autodimension, Baseline Dimension, Add Relations and Display/Delete Relations*.

You also addressed the DimXpertManager tools and learned about TolAnalyst.

Think design intent. When do you use Geometric sketch relations? What are you trying to do with the design? How does the component fit into an assembly?

Fully define all sketches in the model. However, there are times when this is not practical, generally when using the Spline tool to create a freeform shape.

The Fillet tool from the Consolidated Features toolbar fillets entities such as edges in a part, not in a sketch. The Sketch Fillet tool is applied to a sketch, not a feature.

The plane or face you select for the Base sketch determines the orientation of the part.

The Save As option displays the new part name in the Graphics window. The Save as copy option copies the current document to a new file name. The current document remains open in the Graphics window.

Relations to the global axes are called AlongX, AlongY and AlongZ. Relations that are local to a plane are called Horizontal, Vertical and Normal.

If you select a vertex, dimensions are measured point-to-point from the selected vertex. If you select an edge, dimensions are measured parallel to the selected edge.

In Chapter 6 explore and address the Extruded Boss/Base feature, Extruded Cut feature, Extruded Solid/Thin feature and the Extruded Surface feature. Also, address the Fillet feature using the Fillet and FilletXpert PropertyManager and Cosmetic feature.

Notes:

CHAPTER 6: EXTRUDED BOSS/BASE, EXTRUDED CUT, FILLET AND COSMETIC FEATURE

Chapter Objective

Chapter 6 provides a comprehensive understanding of Extruded and Fillet features in SOLIDWORKS. On the completion of this chapter, you will be able to:

- Understand and utilize the following Extruded features:

  - Extruded Boss/Basc.

  - Extruded Cut.

  - Extruded Solid/Thin.

  - Extruded Surface:

    - Cut With Surface.

- Comprehend and utilize the following information on Fillets:

  - Fillets in General.

  - Fillet PropertyManager:

    - Manual tab.

      - Constant Radius Fillet.

      - Variable Radius Fillet.

      - Face Fillet.

      - Full Round Fillet.

    - Control Points.

    - FilletXpert PropertyManager:

      - Add tab.

      - Change tab.

      - Corner tab.

    - Fillet to Chamfer tool.

- Apply the Cosmetic Thread feature.

- Understand the difference between a Cosmetic Feature and a Cosmetic Pattern.

Extruded Boss/Base Features

Once a sketch is complete, you can extrude the sketch to create a feature. An Extruded Boss/Base feature is created by projecting a 2D sketch perpendicular to the Sketch plane to create a 3D shape. An Extrude Base feature (Boss-Extrude1) is the first feature that adds material in a part.

A Boss feature (Boss-Extrude2) adds material onto an existing feature. An Extruded Cut feature removes material from the part. The Extrude-Thin feature parameter option provides the ability to specify thickness of the extrusion. Extrude features can be solids or surfaces. Solids are utilized primary for machined parts and contain mass properties. Surfaces are utilized surfaces for organic shapes which contain no mass properties.

For an Extruded Boss/Base feature, you need to identify a Start Condition from the Sketch plane and an End Condition. The Boss-Extrude PropertyManager provides the ability to select four Start Conditions options. Sheet metal features are addressed later in the Sheet metal chapter of the book.

Extruded Boss/Base Feature

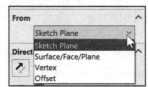

The Extruded Boss/Base 🗐 feature uses the Boss-Extrude PropertyManager. The Boss-Extrude PropertyManager provides the following selections:

- *From*. Displays the location of the initial extrude feature or the Start condition. The available options are:

  - **Sketch Plane**. Selected by default. Initiates that the extrude feature is from the selected Sketch plane.

  - **Surface/Face/Plane**. Initiates that the extrude feature is from a valid selected entity. A valid entity is a surface, face or plane.

💡 The entity can be planar or non-planar. Planar entities are not required to be parallel to the Sketch plane. The sketch must be entirely contained within the boundaries of the non-planar surface or face. The sketch follows the shape of the non-planar entity at the starting surface or face.

  - **Vertex**. Initiates that the extrude feature is from the selected vertex. Think design intent. When do you use the various options?

- **Offset**. Initiates that the extrude feature is from a plane that is offset from the current Sketch plane. Set the Offset distance in the Enter Offset Value box from the Offset option.

- *Direction 1*. Provides the ability to select an End Condition Type. Extrudes in a single direction from the selected option. The available End Condition Types for Direction 1 are feature dependent:

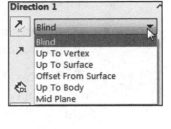

  - **Blind**. Selected by default. Extends the feature for a specified depth.

  - **Through All**. Extends the feature from the sketch plane through all existing geometry.

  - **Up To Next**. Only available for a Boss feature. Extrudes the feature from the Sketch plane to the next surface that intercepts the entire profile. The intercepting surface must be on the same part to use the Up To Next option.

  - **Up To Vertex**. Extends the feature from the Sketch plane to a plane which is parallel to the Sketch plane, and passes through the specified vertex. Sketch vertices are valid selections for the Up To Vertex option extrusions.

  - **Up To Surface**. Extends the feature from the Sketch plane to a selected surface. The extrusion is shaped by the selected surface. Only one surface selection is allowed.

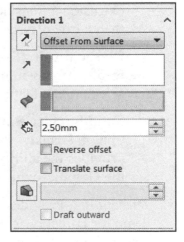

  - **Offset From Surface**. Extends the feature from the Sketch plane to the specified distance from the selected surface. An Offset From Surface value is required with this option as illustrated.

  - **Up To Body**. Extends the feature from the Sketch plane to a specified body. Use the Up To Body type option with assemblies, mold parts, or multi-body parts.

  - **Mid Plane**. Extends the feature from the Sketch plane equally in both directions. Direction 2 is not available with this option.

  - **Reverse Direction**. Only available for the following options: Blind, Up To Surface, Offset From Surface, and Up To Body. Reverses the direction of the extrude feature if required.

  - **Direction of Extrusion**. The Direction of Extrusion box provides the ability to display the selected direction vector from the Graphics Window to extrude the sketch in a direction other than normal to the sketch profile.

  - **Depth**. Displays the selected depth of the extruded feature.

- **Draft On/Off**. Adds a draft to the extruded feature. The Draft On/Off option provides two selections:

  - **Required draft angle**. Input the required draft angle value.

  - **Draft outward**. Modifies the direction of the draft if required.

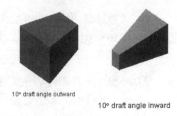

10° draft angle outward   10° draft angle inward

- **Up To Vertex**. Only available for the Up To Vertex option. Displays the selected vertex for the extruded feature.

- **Face/Plane**. Only available for the Up To Surface and Offset From Surface option. Displays the selected face or plane and provides the ability to select either a face or plane from the Graphics window or FeatureManager.

- **Reverse offset**. Only available for the Offset From Surface option. Reverses the offset direction if required.

- **Translate surface**. Only available for the Offset From Surface option. Provides the ability to make the end of the extrusion a translation of the reference surface, rather than a true offset.

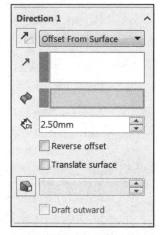

- **Merge result**. (Boss/Base extrudes only.) Merges resultant body into an existing body if possible. If not selected, the feature creates a distinct solid body.

- **Solid/Surface Body**. Only available for the Up To Body option. Displays the selected solid or surface body and provides the ability to select either a solid or surface from the Graphics window or FeatureManager.

- *Direction 2*. Provides the ability to select an End Condition Type for Direction 2. Extrudes in a single direction from the selected option. The available End Condition Types for Direction 2 are feature dependent:

  - **Blind**. Selected by default. Extends the feature for a specified depth.

  - **Through All**. Extends the feature from the Sketch plane through all existing geometry.

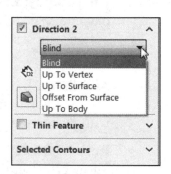

- **Up To Next**. Only available for a Boss feature. Extrudes the feature from the Sketch plane to the next surface that intercepts the entire profile. The intercepting surface must be on the same part to use the Up To Next option.

- **Up To Vertex**. Extends the feature from the Sketch plane to a plane which is parallel to the Sketch plane and passes through the specified vertex. Sketch vertices are valid selections for the Up To Vertex option extrusions.

- **Up To Surface**. Extends the feature from the Sketch plane to a selected surface. The extrusion is shaped by the selected surface. Only one surface selection is allowed.

- **Offset From Surface**. Extends the feature from the Sketch plane to the specified distance from the selected surface. An Offset From Surface value is required with this option.

- **Up To Body**. Extends the feature from the Sketch plane to a specified body. Use the Up To Body type option with assemblies, mold parts, or multi-body parts.

- **Reverse Direction**. Only available for the following options: Blind, Up To Surface, Offset From Surface, and Up To Body. Reverses the direction of the extrude feature if required.

- **Depth**. Displays the selected depth of the extruded feature.

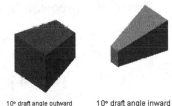

10° draft angle outward   10° draft angle inward

- **Draft On/Off**. Adds a draft to the extruded feature. The Draft On/Off option provides two selections:

  - **Required draft angle**. Input the required draft angle value.

  - **Draft outward**. Modifies the direction of the draft if required.

- **Up To Vertex**. Only available for the Up To Vertex option. Displays the selected vertex for the extruded feature.

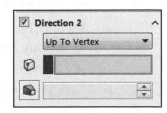

- **Face/Plane**. Only available for the Up To Surface and Offset From Surface option. Displays the selected face or plane and provides the ability to select either a face or plane from the Graphics window or FeatureManager.

- **Reverse offset**. Only available for the Offset From Surface option. Reverses the offset direction if required.

- **Translate surface**. Only available for the Offset From Surface option. Provides the ability to make the end of the extrusion a translation of the reference surface, rather than a true offset.

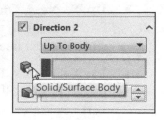

- **Solid/Surface Body**. Only available for the Up To Body option. Displays the selected solid or surface body and provides the ability to select either a solid or surface from the Graphics window or FeatureManager.

Set an End Condition Type for Direction 1 and Direction 2 to extrude in two directions with unequal depths. Direction 1 and Direction 2 are always separated by 180 degrees. Direction 2 becomes inactive if you select the Mid Plane End Condition in Direction 1.

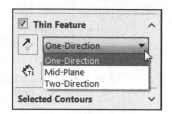

- *Thin Feature*. Controls the extrude thickness of the profile. This is not the depth. Create a Thin feature by using an active sketch profile and applying a wall thickness. Apply thickness to a Thin feature either to the inside or outside of the sketch, evenly on both sides of the sketch, or unevenly on either side of the sketch. Thin feature creation is automatically invoked for active contours that are extruded or revolved. Create a Thin feature from a closed contour. There are three conditions to select from:

  - **Type**. Displays the selected type of a thin extrude feature extrude. There are three options:

    - **One-Direction**. Sets the extrude thickness in one direction, outwards from the sketch. The Direction 1 Thickness option is available.

    - **Mid-Plane**. Sets the extrude thickness equally in both directions from the sketch. The Thickness option is available.

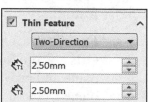

    - **Two-Direction**. Sets the extrude thickness in two directions, outward from the sketch. The Direction 1 and Direction 2 Thickness options are available.

  - **Direction 1 Thickness**. Displays the selected extrude thickness in one direction, outward from the sketch.

  - **Direction 2 Thickness**. Displays the selected extrude thickness in a second direction, outward from the sketch.

- **Cap ends**. Covers the end of the thin feature extrude, creating a hollow part. Provides the ability to set the Cap Thickness option. This option is available only for the first extruded body in a model.

- **Cap Thickness**. Displays the selected extrude thickness of the cap.

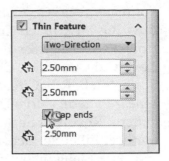

The Cap ends option is only available for the first extruded body in your model. This option specifies the end of the Thin feature extrude, which creates a hollow part. A thickness of the Cap is required.

- **Auto-fillet corners**. Selected by default. Only available for an open sketch. Rounds each edge where the lines meet at an angle.

- **Fillet Radius**. Only available if the Auto-fillet corners option is selected. Sets the inside radius of the round.

- *Selected Contours*. Displays the selected contours and provides the ability to use a partial sketch to create the extrude feature. Select sketch contours and model edges from the Graphics window or FeatureManager.

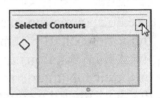

- *Feature Scope*. Apply features to one or more multi-body parts. The Feature Scope box provides the following selections:

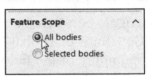

- **All bodies**. Applies the feature to all bodies every time the feature regenerates. If you add new bodies to the model that are intersected by the feature, these new bodies are also regenerated to include the feature.

- **Selected bodies**. Selected by default. Applies the feature to the selected bodies.

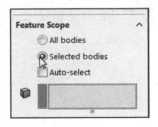

- **Auto-select**. Only available if you select the Selected bodies option. When you first create a model with multi-body parts, the feature automatically processes all the relevant intersecting parts. Auto-select is faster than All bodies because it processes only the bodies on the initial list and does not regenerate the entire model.

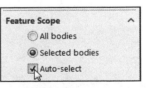

- **Solid Bodies to Affect**. Only available if Auto-select is not active. Displays the selected bodies to affect from the Graphics window.

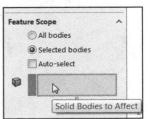

For many features (*Extruded Boss/Base*, Extruded Cut, Simple Hole, Revolved Boss/Base, Revolved Cut, Fillet, Chamfer, Scale, Shell, Rib, Circular Pattern, Linear Pattern, Curve Driven Pattern, Revolved Surface, Extruded Surface, Fillet Surface, Edge Flange and Base Flange) you can enter and modify equations directly in the PropertyManager fields that allow numerical inputs.

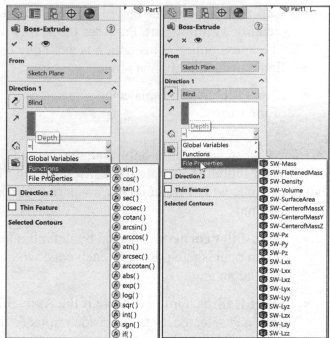

You can create equations with Global Variables, functions and file properties without accessing the Equations, Global Variables and Dimensions dialog box.

Create an equation in a numeric input field. Start by entering an = (equal sign). A drop-down list displays options for Global Variables, functions, and file properties. Numeric input fields that contain equations can display either the equation itself or its evaluated value.

Toggle between the equation and the value by clicking the Equations or Global Variable button that appears at the beginning of the field.

 Link to an external file (Excel spreadsheet).

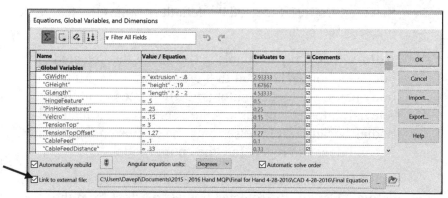

Tutorial: Extrude 6-1

Create a simple Rod part on the Front Plane using the Extruded Boss/Base feature.

1. Create a **New** □ part. Use the default ANSI, MMGS Part template.

2. Right-click **Front Plane** from the FeatureManager. This is your Sketch plane.

3. Click **Sketch** □ from the Context toolbar.

4. Click **Circle** ⊙ from the Sketch toolbar. The Circle PropertyManager is displayed.

5. Sketch **two circles**, one inside the other centered at the origin as illustrated. Center creation is checked in the Parameters box.

Add dimensions.

6. Click **Smart Dimension** ✎. Enter a **20**mm dimension for the small circle.

7. Enter a **50**mm dimension for the large circle as illustrated. The Sketch is fully defined.

8. Click **OK** ✔ from the Dimension PropertyManager.

9. **Rebuild** ▌ the model.

10. Rename **Sketch1** to **Sketch1-Profile**.

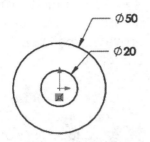

11. Right-click **Right Plane** from the FeatureManager.

12. Click **Sketch** □ from the Context toolbar.

13. Display a **Right view**. Right Plane is displayed in the Graphics window.

14. Click **Centerline** ⟋ from the Consolidated Line Sketch tool bar.

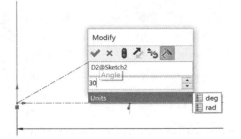

15. Sketch a **horizontal centerline** 50mm from the origin to the right.

16. Sketch a **30deg centerline** horizontal from the origin.

17. **Rebuild** ▌ the model. Rename **Sketch2** to **Sketch2-Vectors**.

18. Click **Sketch1-Profile** from the FeatureManager.

19. Display an **Isometric view** ▣. Click the **circular edge** of the 20mm circle. View the Instant3D arrow.

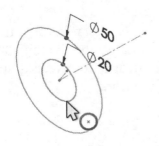

20. Click **Extruded Boss/Base** from the Features toolbar. The Boss-Extrude PropertyManager is displayed.

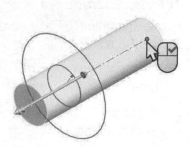

21. Select **Offset** for the Start Condition in the From dialog box.

22. Enter **20**mm for Offset value.

23. Select **Up To Vertex** for the End Condition Type in Direction 1.

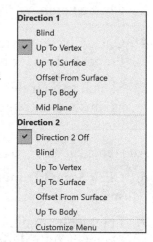

In Edit mode, right-click anywhere on the extrude feature to modify the end condition from the pop-up menu. The pop-up menu provides document dependent options for Direction 1 and Direction 2.

24. Click the **horizontal centerline** endpoint as illustrated. Point2@Sketch2-Vector is displayed in the vertex box. The extrude arrow points to the right.

25. Click inside the **Direction of Extrusion** box for Direction 1.

26. Click the **30deg centerline** from the Graphics window. Line2@Sketch2-Vector is displayed in the Direction of Extrusion box. The Extrusion is angled 30 degrees from horizontal.

27. **Expand** the Selected Contours box.

28. If needed, click the **circular edge** of the 20mm circle. The Sketch1-Profile Contour is displayed in the Selected Contours box.

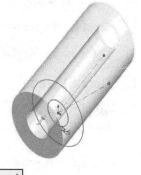

29. Click **OK** from the Boss-Extrude PropertyManager. The Extrusion is perpendicular to the Sketch plane. Sketch1-Profile is displayed under the Boss-Extrude1 feature in the FeatureManager design tree.

30. **Save** the part.

31. Enter the name **Extrude 6-1**.

32. **Close** the model.

By default, the direction of Extrusion is normal to the Sketch plane, but you can also select a linear entity such as an edge or axis as the direction.

Think design intent. When do you use the various End Conditions? What are you trying to do with the design? How does the component fit into an assembly?

Tutorial: Extrude 6-1A

Use an equation in the Chamfer PropertyManager. Bevel the edges of the model.

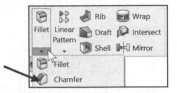

1. Open **Extrude 6-1A** from the SOLIDWORKS 2018 folder. The Extrude 6-1A FeatureManager is displayed.

2. Right-click the **Equations** folder in the FeatureManager.

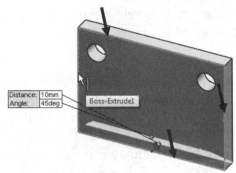

3. Click **Manage Equations**. View the Equations dialog box. Click **OK** to close the Equations dialog box.

4. Click **Chamfer** from the Features toolbar. Click **Angle Distance** for Chamfer type.

5. Click inside the **Edges and Faces or Vertex** box.

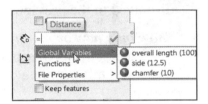

6. Select the **four edges** of the front face in the Graphics window as illustrated.

7. Create a new Global Variable for Distance. Click **inside** the Distance box.

8. Enter =.

9. Select **Chamfer** as illustrated. Click **OK** in the input field. The field displays an Equations button.

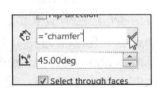

10. Click the **Equations** button to toggle the display between the equation and the value.

11. Modify the Distance value from **10mm** to **6mm**.

Create a Global Variable for Angle.
12. Click **inside** the Angle box.

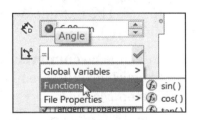

13. Enter =. Select **Functions > sin()** from the fly-out menu.

14. Enter sin(**90**)*10. Units: degrees.

15. Click in the input field. The field displays an Equations Σ button.

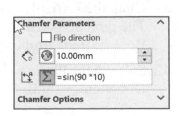

16. Right-click the **Equations** folder in the FeatureManager. Click **Manage Equations**. View the Equations dialog box. The Global Variable "chamfer" and the angle equation are listed in the Equations dialog box.

17. **Return** to the FeatureManager. View the results.

18. **Close** the model.

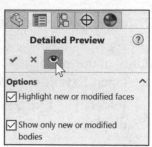

Detailed Preview PropertyManager

The Detailed Preview PropertyManager provides the ability to display detailed previews and control what is displayed in the Graphics window for the following features: Extrudes, Ribs and Drafts.

There are two selections in the Detailed Preview PropertyManager:

- *Highlight new or modified faces.* Displays new extrude, rib, or draft features, or the faces that were changed by your last edit. New or modified faces are highlighted in a different color, as opposed to being displayed shaded as they do in the standard PropertyManager if the Highlight new or modified faces check box is active. All individual bodies are displayed in the preview.

- *Show only new or modified bodies.* Displays new or modified bodies. All separate bodies are hidden in the preview. Only separate bodies are affected by this option.

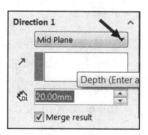

Tutorial: Extrude 6-2

Apply a few advanced Extruded Boss/Base feature options. Think about Design Intent. When do you insert the various End Conditions?

1. Open **Extrude 6-2** from the SOLIDWORKS 2018 folder. The Extrude 6-2 FeatureManager is displayed.

2. Click the **circumference** of the large circle. Sketch1-Profiles are highlighted in the FeatureManager.

Apply the Extruded Boss/Base feature.

3. Click the **Extruded Boss/Base** 🔲 Features tool. The direction arrow points upward. The Boss-Extrude PropertyManager is displayed.

4. Select **Mid Plane** for the End Condition in Direction 1. Think Design Intent.

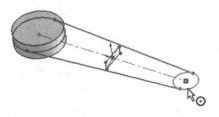

5. Enter **20**mm for Depth. This provides 10mm of extrusion on each side of the Sketch plane.

6. Click **OK** ✔ from the Boss-Extrude PropertyManager. Boss-Extrude1 is created and is displayed in the FeatureManager.

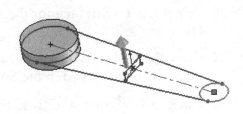

7. Click the **circumference** of the small circle. Sketch1-Profiles are highlighted in the FeatureManager under Extrude1.

8. Click the **Extruded Boss/Base** 🗔 Features tool. The Boss-Extrude PropertyManager is displayed.

9. Select **Mid Plane** for the End Condition in Direction 1. The extrude arrow points in an upward direction as illustrated.

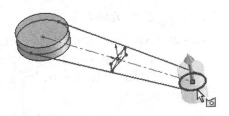

10. Enter **40**mm for Depth.

11. Click inside the **Selected Contours** box.

12. Click the **circumference** of the small circle. Sketch1-Profies-Contours<1> is displayed in the Selected Contours box.

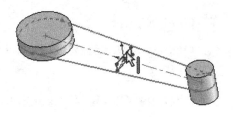

13. Click **OK** ✔ from the Boss-Extrude PropertyManager. Boss-Extrude2 is created and is displayed from the FeatureManager.

Extrude the sketch handle.

14. Click the **Top profile line** on the rectangular handle. Sketch2-Handle is highlighted in the FeatureManager.

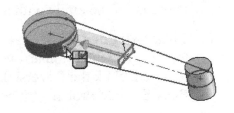

15. Right-click **Select Chain**. You selected all sketch entities of the handle.

16. Click **Extruded Boss/Base** 🗔 from the Features toolbar. The direction arrow points to the left. The Boss-Extrude PropertyManager is displayed.

17. Select **Up To Surface** for the End Condition in Direction 1.

18. Click the **cylindrical face** of the large cylinder. Face<1> is displayed in the Face/Plane box.

19. **Expand** the Direction 2 box from the Extrude PropertyManager.

20. Select **Up To Surface** for the End Condition in Direction 2.

21. Click the **inside cylindrical face** of the small cylinder. Face<2> is displayed in the Solid/Surface Body box.

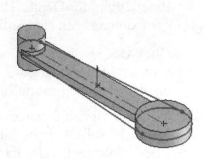

22. Click **OK** ✔ from the Boss-Extrude PropertyManager. Boss-Extrude3 is created.

23. Display an **Isometric view**. You created three extrusions using advance features of the Boss-Extrude PropertyManager. **Close** the model.

Tutorial: Extrude 6-3

Insert an Extruded Boss/Base feature in a 3D model.

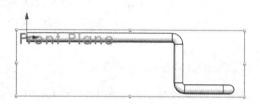

1. Open **Extrude 6-3** from the SOLIDWORKS 2018 folder. The Extruded 6-3 FeatureManager is displayed.

2. Right-click **Front Plane** from the FeatureManager. This is your Sketch plane.

3. Click **Sketch** from the Context toolbar.

4. Display a **Normal To view** from the Heads-up View toolbar.

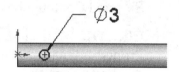

5. Click the **Circle** Sketch tool. The Circle PropertyManager is displayed.

6. Sketch a **circle** on the face of the tube as illustrated.

7. **Dimension** the circle with a 3mm diameter.

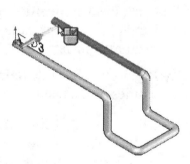

8. Click **Extruded Boss/Base** from the Features toolbar. The direction arrow points towards the back. If required, click the Reverse Direction box. The Boss-Extrude PropertyManager is displayed.

9. Display an **Isometric view**.

10. Select the **Up to Surface** End Condition for Direction1.

11. Click the **opposite side** of the frame as illustrated. Face<1> is displayed in the Face/Plane box.

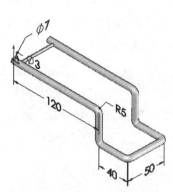

12. Check the **Merge result** box.

13. Click **OK** ✔ from the Boss-Extrude PropertyManager. Boss-Extrude1 is created and is displayed in the FeatureManager. View the model in the Graphics window.

14. **Close** the model.

Extruded Cut Feature

The Extruded Cut 🔲 feature procedure is very similar to the Extruded Boss/Base feature. First create a sketch, and second, select the required options from the Extrude PropertyManager. You need a Start Condition from the Sketch plane and an End Condition.

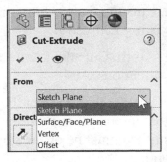

The Cut-Extrude PropertyManager provides four Start Condition options from the drop-down menu. Note: The four Start Condition options are the same as for the Extruded Boss/Base feature.

💡 In multi-body parts, use the Extruded Cut feature to create disjoint parts. Organize which parts to keep and which parts will be affected by the cut.

The Extruded Cut feature 🔲 uses the Cut-Extrude PropertyManager. The Cut-Extrude PropertyManager provides the following selections:

- *From*. Displays the start location of the initial extrude feature. The From dialog box provides the following available Start conditions:

  - **Sketch Plane**. Selected by default. Initiates the extrude feature from the selected Sketch plane.

  - **Surface/Face/Plane**. Initiates the extrude feature from a valid selected entity. A valid entity is a surface, face or plane.

  - **Vertex**. Initiates the extrude feature from the selected vertex.

  - **Offset**. Initiates the extrude feature on a plane that is offset from the current Sketch plane. Set the Offset distance in the Enter Offset Value box.

- *Direction 1*. Provides the ability to select an End Condition Type in Direction 1. This option provides the ability to cut in a single direction from the Sketch plane. The available End Condition Types for Direction 1 are feature dependent:

  - **Blind**. Selected by default. Cuts the feature for a specified depth.

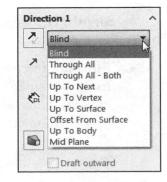

- **Through All**. Cuts the feature from the Sketch plane through all existing geometry.

- **Through All - Both**. Cuts the feature from the Sketch plane in both directions.

- **Up To Next**. Cuts the feature from the Sketch plane to the next surface that intercepts the entire profile. The intercepting surface must be on the same part to use the Up To Next option.

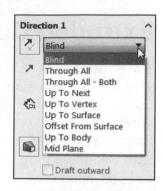

- **Up To Vertex**. Cuts the feature from the Sketch plane to a plane which is parallel to the Sketch plane and passing through the specified vertex. Sketch vertices are valid selections for Up To Vertex extrusions.

- **Up To Surface**. Cuts the feature from the Sketch plane to the selected surface. The extrusion is shaped by the selected surface. Only one surface selection is allowed.

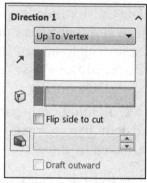

- **Offset From Surface**. Cuts the feature from the Sketch plane to a specified distance from the selected surface. You are required to enter a value with this option selected. Select the Reverse offset check box to offset the specified distance in the opposite direction if required.

- **Up To Body**. Cuts the feature from the sketch plane to a specified body. Use the Up To Body type with assemblies, mold parts, or multi-body parts.

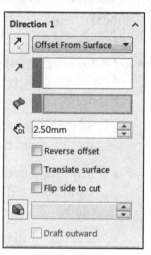

- **Mid Plane**. Cuts the feature from the Sketch plane equally in both directions.

- **Reverse Direction**. Only available for the following options: Blind, Through All, Up To Next, Up To Surface, Offset From Surface and Up To Body.

- **Direction of Extrusion**. The Direction of Extrusion box provides the ability to display the selected direction vector from the Graphics Window to extrude the sketch in a direction other than normal to the sketch profile.

- **Depth**. Displays the selected depth of the extruded cut feature.

- **Flip side to cut**. Removes all material from the outside of the profile. By default, material is removed from the inside of the profile.

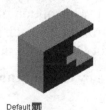

Default cut

Flip side cut

- **Draft On/Off**. Adds a draft to the extruded cut feature. The Draft On/Off option provides two selections:

  - **Required draft angle**. Input the required draft angle value.

  - **Draft outward**. Modifies the direction of the draft if required.

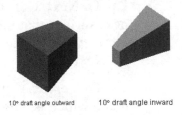

10° draft angle outward   10° draft angle inward

- **Vertex**. Only available for the Up To Vertex option. Displays the selected vertex for the extruded cut feature.

- **Face/Plane**. Only available for the Up To Surface and Offset From Surface option. Displays the selected face or plane and provides the ability to select either a face or plane from the Graphics window or FeatureManager.

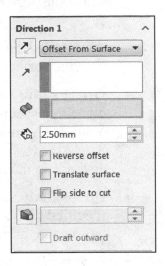

- **Reverse offset**. Only available for the Offset From Surface option. Reverses the offset direction if required.

- **Translate surface**. Only available for the Offset From Surface option. Provides the ability to make the end of the extrusion a translation of the reference surface, rather than a true offset.

- **Solid/Surface Body**. Only available for the Up To Body option. Displays the selected solid or surface body and provides the ability to select either a solid or surface from the Graphics window or FeatureManager.

- **Normal cut**. Only available for Sheet metal cut extrudes. Provides the ability to ensure that the cut is created normal to the Sheet metal thickness for folded Sheet metal parts.

Set an End Condition Type for Direction 1 and Direction 2 to cut in two directions with different depths.

- *Direction 2*. Provides the ability to select an End Condition Type in Direction 2. Extrudes in a second direction from the Sketch plane. The available End Condition Types for Direction 2 are feature dependent:

  - **Blind**. Selected by default. Extends the feature from the Sketch plane for a specified distance.

  - **Through All**. Cuts the feature from the Sketch plane through all existing geometry.

- **Up To Next**. Cuts the feature from the Sketch plane to the next surface that intercepts the entire profile. The intercepting surface must be on the same part to use the Up To Next option.

- **Up To Vertex**. Cuts the feature from the Sketch plane to a plane which is parallel to the Sketch plane, and passing through the specified vertex. Sketch vertices are valid selections for Up To Vertex extrusions.

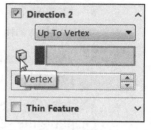

- **Up To Surface**. Cuts the feature from the Sketch plane to the selected surface. The extrusion is shaped by the selected surface. Only one surface selection is allowed.

- **Offset From Surface**. Cuts the feature from the Sketch plane to a specified distance from the selected surface. You are required to enter a value with this option selected. Select the Reverse offset check box to offset the specified distance in the opposite direction if required.

- **Up To Body**. Cuts the feature from the sketch plane to a specified body. Use the Up To Body type with assemblies, mold parts, or multi-body parts.

- **Reverse Direction**. Only available for the following options: Blind, Through All, Up To Next, Up To Surface, Offset From Surface and Up To Body.

- **Depth**. Displays the selected depth of the extruded feature.

- **Draft On/Off**. Adds a draft to the extruded feature. The Draft On/Off option provides two selections:

  - **Required draft angle**. Input the required draft angle value.

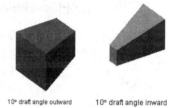

10° draft angle outward   10° draft angle inward

  - **Draft outward**. Modifies the direction of the draft if required.

- **Vertex**. Only available for the Up To Vertex option. Displays the selected vertex for the extruded feature.

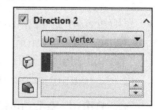

- **Face/Plane**. Only available for the Up To Surface and Offset From Surface option. Displays the selected face or plane and provides the ability to select either a face or plane from the Graphics window or FeatureManager.

- **Reverse offset**. Only available for the Offset From Surface option. Reverses the offset direction if required.

- **Translate surface**. Only available for the Offset From Surface option. Provides the ability to make the end of the extrusion a translation of the reference surface, rather than a true offset.

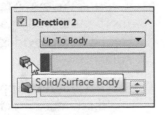

- **Solid/Surface Body**. Only available for the Up To Body option. Displays the selected solid or surface body and provides the ability to select either a solid or surface from the Graphics window or FeatureManager.

- *Thin Feature*. Controls the extrude thickness of the profile. This is not the depth. Create a Thin feature by using an active sketch profile and applying a wall thickness. Apply thickness to a Thin feature either to the inside or outside of the sketch, evenly on both sides of the sketch, or unevenly on either side of the sketch. Thin feature creation is automatically invoked for active contours that are extruded or revolved. Create a Thin feature from a closed contour. The following options are available:

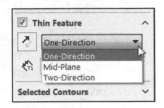

  - **Type**. Displays the selected type of thin feature cut. There are three conditions:

    - **One-Direction**. Displays the selected extrude thickness feature in one direction, outwards from the sketch. The Direction 1 Thickness option is available.

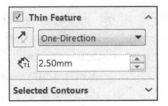

    - **Mid-Plane**. Displays the selected extrude thickness feature equally in both directions from the sketch. The Thickness option is available.

    - **Two-Direction**. Displays the selected extrude thickness in two directions, outward from the sketch. The Direction 1 and Direction 2 Thickness option is available.

  - **Reverse Direction**. Reverses the direction of the Thin Feature if required.

  - **Direction 1 Thickness**. Displays the selected cut extrude thickness in one direction, outward from the sketch.

  - **Auto-fillet corners**. Only available for an open sketch. Rounds each edge where the lines meet at an angle.

  - **Fillet Radius**. Only available if the Auto-fillet corners option is selected. Sets the inside radius of the round.

- *Selected Contours*. Displays the selected contours and to use a partial sketch to create extruded features. Select sketch contours and model edges from the Graphics window.

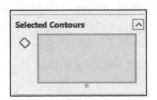

- *Feature Scope*. Apply features to one or more multi-body parts. The Feature Scope box provides the following selections:

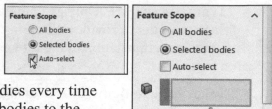

  - **All bodies**. Applies the feature to all bodies every time the feature regenerates. If you add new bodies to the model that are intersected by the feature, these new bodies are also regenerated to include the feature.

  - **Selected bodies**. Selected by default. Applies the feature to the selected bodies.

  - **Auto-select**. Only available if you select the Selected bodies option. When you first create a model with multi-body parts, the feature automatically processes all the relevant intersecting parts. Auto-select is faster than All bodies because it processes only the bodies on the initial list and does not regenerate the entire model.

  - **Solid Bodies to Affect**. Only available if Auto-select is not active. Displays the selected bodies to affect from the Graphics window.

For many features (Extruded Boss/Base, *Extruded Cut*, Simple Hole, Revolved Boss/Base, Revolved Cut, Fillet, Chamfer, Scale, Shell, Rib, Circular Pattern, Linear Pattern, Curve Driven Pattern, Revolved Surface, Extruded Surface, Fillet Surface, Edge Flange and Base Flange) you can enter and modify equations directly in the PropertyManager fields that allow numerical inputs.

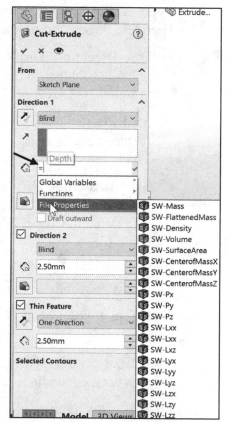

Create equations with Global Variables, functions, and file properties without accessing the Equations, Global Variables and Dimensions dialog box.

For example, in the Extruded Cut PropertyManager you can enter equations in:

- Depth fields for Direction 1 and Direction 2.

- Draft fields for Direction 1 and Direction 2.

- Thickness fields for a Thin Feature with two direction types.

- Offset Distance field.

Create an equation in a numeric input field. Start by entering an = (equal sign). A drop-down list displays options for Global Variables, functions and file properties. Numeric input fields that contain equations can display either the equation itself or its evaluated value.

Create extruded cuts in both directions with one click. In the Cut-Extrude PropertyManager, select Through All - Both Directions. Direction 2 automatically updates with the same value.

Tutorial: Extruded Cut 6-1

Create Extruded Cut features using various End Condition options. Think about design intent of the model. Why would you use these End Conditions in your design?

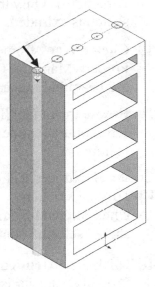

1. Open **Extruded Cut 6-1** from the SOLIDWORKS 2018 folder. The Extruded Cut 6-1 FeatureManager displays two Extrude features, a Shell feature and a Linear Pattern feature.

2. Click the circumference of the **front most circle**. Sketch3 is highlighted in the FeatureManager.

Create an Extruded Cut feature using the Selected Contours and Through All option.

3. Click **Extruded Cut** from the Features toolbar. The Cut-Extrude PropertyManager is displayed.

4. **Expand** the Selected Contours box if needed. Sketch3-Contour<1> is displayed in the Selected Contours box. The direction arrow points downward, if not click the Reverse Direction button.

5. Select **Through All** for End Condition in Direction 1. Only the first circle of your sketch is extruded.

6. Click **OK** from the Cut-Extruded PropertyManager. Cut-Extrude1 is displayed in the FeatureManager.

Create an Extruded Cut feature using the Selected Contours and the Up To Next option. Think about design intent.

7. Click **Sketch3** from the FeatureManager. Note the icon type for a contour sketch and its relationship to Boss-Extrude1.

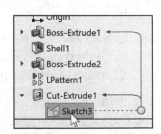

8. Click **Extruded Cut** from the Features toolbar. The Cut-Extrude PropertyManager is displayed.

9. **Expand** the Selected Contours box.

10. Delete **Sketch3** from the Selected Contours box.

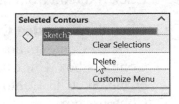

11. Click the circumference of the **second circle** as illustrated. Sketch3-Contour<1> is displayed in the Selected Contours box.

12. Select **Up To Next** for End Condition in Direction 1. Only the second circle of your sketch is extruded.

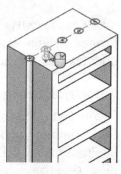

13. Click **OK** ✔ from the Cut-Extrude PropertyManager. Cut-Extrude2 is displayed in the FeatureManager.

14. **Rotate** the model (holding down the middle scroll button on the mouse) and view the created feature.

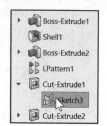

Create an Extruded Cut feature using the Selected Contours and the Up To Vertex option.

15. Click **Sketch3** from the FeatureManager. Note the icon. It's a shared selected contour sketch icon.

16. Click the **Extruded Cut** 🔲 Features tool. The Cut-Extrude PropertyManager is displayed.

17. **Expand** the Selected Contours box. Delete **Sketch3** from the Selected Contours box.

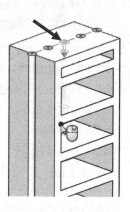

18. Click the circumference of the **third circle** as illustrated. Sketch3-Contour<1> is displayed in the Selected Contours box. Select **Up To Vertex** for End Condition in Direction 1. Only the third circle of your sketch is extruded.

19. Select a **vertex point** below the second shelf as illustrated. Vertex<1> is displayed in the Vertex box in Direction 1.

20. Click **OK** ✔ from the PropertyManager. Cut-Extrude3 is displayed in the FeatureManager. The third circle has an Extruded Cut feature through the top two shelves.

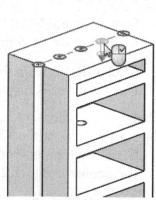

Create an Extruded Cut feature using the Selected Contours and the Offset From Surface option.

21. Click **Sketch3** from the FeatureManager.

22. Click the **Extruded Cut** 🔲 Features tool. **Expand** the Selected Contours box. Delete **Sketch3** from the Selected Contours box. Click the circumference of the **fourth circle**. Sketch3-Contour<1> is displayed in the Selected Contours box.

23. Select **Offset From Surface** for End
    Condition in Direction 1.

24. Click the **face** of the third shelf. Face<1> is
    displayed in the Face/Plane box in
    Direction1.

25. Enter **60**mm for Offset Distance.

26. Click the **Reverse offset** box.

27. Click **OK** ✔ from the PropertyManager.
    Cut-Extrude4 is displayed in the
    FeatureManager.

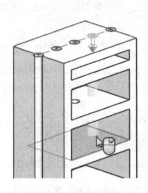

28. Display an **Isometric** view. View the
    created features. **Close** the model.

Tutorial: Extruded Cut 6-2

Create various Extruded Cut features using the
Through All option.

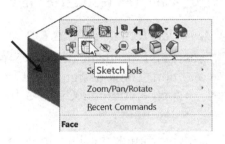

1. Open **Extruded Cut 6-2** from the
   SOLIDWORKS 2018 folder. View the
   Extruded Cut 6-2 FeatureManager.

2. Right-click the **Front face**. This is your Sketch plane.
   Click **Sketch** 🖿 from the Context toolbar. Extrude1
   is highlighted in the FeatureManager.

3. Click the **Line** ✏ Sketch tool. The Insert Line
   PropertyManager is displayed.

4. Sketch a **line** from the midpoint of the top edge to the
   midpoint of the right edge as illustrated.

Apply the Extruded Cut feature.

5. Click the **Extruded Cut** 🔲 Features tool. The
   Cut-Extrude PropertyManager is displayed.

6. Select **Through All** for End Condition. Note: You
   can right-click anywhere on an extruded feature and
   change the end condition from a shortcut menu. You
   can click in empty space, on geometry or on the
   handle.

7. Click **OK** ✔ from the Cut-Extrude
   PropertyManager. Cut-Extrude1 is displayed in the
   FeatureManager.

Create a second Cut-Extrude feature using the Right Plane.

8. Right-click **Right Plane** from FeatureManager. This is your
   Sketch plane.

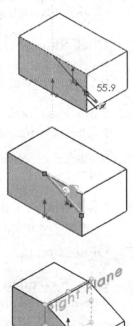

9. Click **Sketch** from the Context toolbar.

10. Click the **Line** Sketch tool. Select **Right view**.

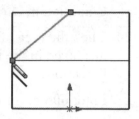

11. Sketch a **line** from the top midpoint of the model to the midpoint of the left edge as illustrated.

12. Click the **Extruded Cut** Features tool. The Cut-Extrude PropertyManager is displayed.

13. Select **Through All** for End Condition.

14. Check the **Flip side to cut** box in Direction 1. The cut is upwards and to the left.

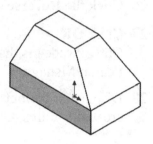

15. Click **OK** from the Cut-Extrude PropertyManager. Cut-Extrude2 is displayed in the FeatureManager.

16. Display an **Isometric view**. **Close** the model.

Tutorial: Extruded Cut 6-3

Create an Extruded Cut feature using the Offset from the start condition. Think design intent.

1. Open **Extruded Cut 6-3** from the SOLIDWORKS 2018 folder.

2. Right-click the **Top face** of the small cylinder in the Graphics window. This is your Sketch plane.

3. Click **Sketch** from the Context toolbar. The Top face is your Sketch plane. Extrude2 is highlighted in the FeatureManager.

4. Click the **Circle** Sketch tool. The Circle PropertyManager is displayed.

5. Sketch a **circle** concentric with the circular edge of the small cylinder.

6. Click **Smart Dimension**. The Dimension PropertyManager is displayed.

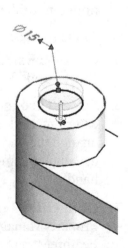

7. Dimension the circle to **15**mm.

8. Click **Extruded Cut** from the Features toolbar. The Cut-Extrude PropertyManager is displayed.

9. Select **Offset** for start condition as illustrated.

10. Enter **5**mm for Offset value.

11. Click the **Reverse direction** box. The Offset direction is downwards. Blind is the default End Condition in Direction 1.

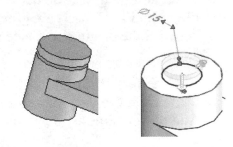

12. Enter **2**mm for Depth in Direction 1.

13. Check the **Flip side to cut** option in the Direction 1 box. The material on the outside of the 15mm circle is removed.

14. Click **OK** ✔ from the Cut-Extrude PropertyManager. Cut-Extrude1 is displayed in the FeatureManager.

15. **Close** the model.

Extruded Solid Thin Feature

An Extruded Solid Thin feature is a feature with an open profile and a specified thickness.

Tutorial: Solid Thin 6-1

Create an Extruded Solid Thin feature using the Mid Plane option.

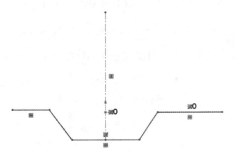

1. Create a **New** 🗋 part. Use the default ANSI, MMGS Part template.

2. Right-click **Front Plane** from the FeatureManager. This is your Sketch plane.

3. Click **Sketch** 🖉 from the Context toolbar.

4. Click the **Line** ✐ Sketch tool.

5. Sketch the **profile** on the Front plane as illustrated.

Apply the Centerline Sketch tool.

6. Click the **Centerline** ✐ Sketch tool. Sketch a **vertical centerline** through the origin and to the bottom of the profile.

7. Insert a Midpoint relation between the **bottom endpoint of the centerline** and the **bottom of the sketch profile**.

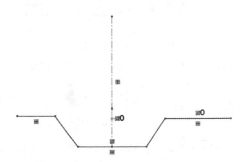

8. Insert an Equal relation between the **two horizontal lines**.

9. Insert a Horizontal relation between the **endpoint** of the left horizontal line and the **endpoint** of the right horizontal line.

10. Insert an Equal relation between the **two diagonal lines**. Insert a **Coincident** relation between the origin and the centerline.

Apply dimensions.

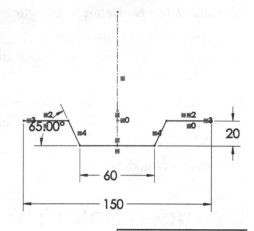

11. Click **Smart Dimensions** . The Smart Dimension PropertyManager is displayed. Add **dimensions** as illustrated.

12. Click **Extruded Boss/Base** from the Features toolbar. The Boss-Extrude PropertyManager is displayed.

13. Select **Mid Plane** for End Condition in Direction 1. Enter **25**mm for Depth. **Expand** the Thin Feature box. One-Direction is the default.

14. Enter **20**mm for Thickness.

15. Check the **Auto-fillet corners** box.

16. Enter **5**mm for Fillet Radius.

17. Click **OK** from the Boss-Extrude PropertyManager. The Extrude-Thin1 feature is created and is displayed in the FeatureManager.

18. **Close** the model.

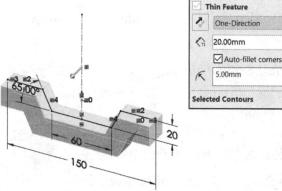

Extruded Surface Feature

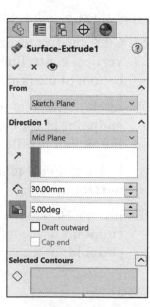

An Extruded Surface feature is created by projecting a 2D sketch perpendicular to the Sketch plane to create a 3D surface. The entity can be planar or non-planar. Planar entities do not have to be parallel to your sketch plane. Your sketch must be fully contained within the boundaries of the non-planar surface or face.

The Extruded Surface feature uses the Surface-Extrude PropertyManager. The Surface-Extrude PropertyManager provides the following selections:

- *From*. Displays the location of the initial extruded feature. The available options are:

  - **Sketch Plane**. Selected by default. Initiates the extrude feature from the selected Sketch plane.

- **Surface/Face/Plane**. Initiates the extrude feature from a valid selected entity. A valid entity is a surface, face, or plane.

- **Vertex**. Initiates the extrude feature from the selected vertex.

- **Offset**. Initiates the extrude feature on a plane that is offset from the current Sketch plane. Set the Offset distance in the Enter Offset Value box.

- *Direction 1*. Provides the ability to select an End Condition Type in Direction 1. The available End Condition Types in Direction 1 are feature dependent:

  - **Blind**. Selected by default. Extrudes the feature from the Sketch plane for a specified distance.

  - **Up To Vertex**. Extrudes the feature from the Sketch plane to a plane which is parallel to the Sketch plane, and passing through the specified vertex. Sketch vertices are valid selections for Up To Vertex extrusions.

  - **Up To Surface**. Extrudes the feature from the Sketch plane to the selected surface. The extrusion is shaped by the selected surface. Only one surface selection is allowed.

  - **Offset From Surface**. Extrudes the feature from the Sketch plane to a specified distance from the selected surface. You are required to enter a value with this option selected. Select the Reverse offset check box to offset the specified distance in the opposite direction if required.

  - **Up To Body**. Extrudes the feature from the sketch plane to a specified body. Use the Up To Body type with assemblies, mold parts or multi-body parts.

  - **Mid Plane**. Extrudes the feature from the Sketch plane equally in both directions.

  - **Reverse Direction**. Only available for the following options: Blind, Up To Surface, Offset From Surface and Up To Body.

  - **Direction of Extrusion**. The Direction of Extrusion box provides the ability to display the selected direction vector from the Graphics Window to extrude the sketch in a direction other than normal to the sketch profile.

  - **Depth**. Displays the selected depth of the extruded cut feature.

- **Draft On/Off**. Adds a draft to the extruded cut feature. The Draft On/Off option provides two selections:

  - **Required draft angle**. Input the required draft angle value.

  - **Draft outward**. Modifies the direction of the draft if required.

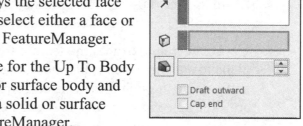

10° draft angle outward   10° draft angle inward

- **Vertex**. Only available for the Up To Vertex option. Displays the selected vertex for the extruded cut feature.

- **Face/Plane**. Only available for the Up To Surface and Offset From Surface option. Displays the selected face or plane and provides the ability to select either a face or plane from the Graphics window or FeatureManager.

- **Solid/Surface Body**. Only available for the Up To Body option. Displays the selected solid or surface body and provides the ability to select either a solid or surface from the Graphics window or FeatureManager.

☼ Set an End Condition Type for Direction 1 and Direction 2 to extrude in two directions with different depths.

- *Direction 2*. Provides the ability to select an End Condition Type in Direction 2. Extrude in a single direction from the Sketch plane. The available End Condition Types in Direction 2 are feature dependent:

  - **Blind**. Selected by default. Extends the feature from the Sketch plane for a specified distance.

  - **Up To Vertex**. Extends the feature from the Sketch plane to a plane which is parallel to the Sketch plane, and passing through the specified vertex. Sketch vertices are valid selections for Up To Vertex extrusions.

  - **Up To Surface**. Extends the feature from the Sketch plane to the selected surface. The extrusion is shaped by the selected surface. Only one surface selection is allowed.

  - **Offset From Surface**. Extends the feature from the Sketch plane to a specified distance from the selected surface. You are required to enter a value with this option selected. Select the Reverse offset check box to offset the specified distance in the opposite direction if required.

- **Up To Body**. Extends the feature from the sketch plane to a specified body. Use the Up To Body type with assemblies, mold parts, or multi-body parts.

- **Depth**. Displays the selected depth of the extruded feature.

- **Draft On/Off**. Adds a draft to the extruded feature. The Draft On/Off option provides two selections:

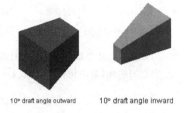

10° draft angle outward   10° draft angle inward

  - **Required draft angle**. Input the required draft angle value.

  - **Draft outward**. Modifies the direction of the draft if required.

- **Vertex**. Only available for the Up To Vertex option. Displays the selected vertex for the extruded feature.

- **Solid/Surface Body**. Only available for the Up To Surface and Up To Body option. Displays the selected solid or surface body from the Graphics window or FeatureManager.

- **Face/Plane**. Only available for the Offset From Surface option. Displays the selected face or plane from the Graphics window or FeatureManager.

- **Reverse offset**. Only available for the Offset From Surface option. Reverses the offset direction if required.

- **Translate surface**. Only available for the Offset From Surface option. Provides the ability to make the end of the extrusion a translation of the reference surface, rather than a true offset.

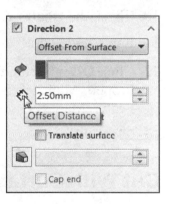

- *Selected Contours*. Displays the selected contours and to use a partial sketch to create extrude features. Select sketch contours and model edges from the Graphics window.

Tutorial: Surface 6-1

Create an Extruded Surface feature using the Spline Sketch tool.

1. Create a **New** □ part. Use the default ANSI, MMGS Part template.

2. Right-click **Front Plane** from the FeatureManager. This is your Sketch plane.

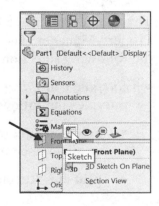

3. Click **Sketch** ⊞ from the Context toolbar. The Sketch toolbar is displayed.

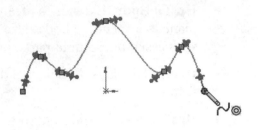

Apply the Spline Sketch tool.

4. Click the **Spline** $\mathcal{N}$ Sketch tool. The Spline icon is displayed on the mouse pointer.

5. Sketch a **Spline** as illustrated. The sketch remains under defined.

6. Deselect the active Sketch tool. Right-click **End Spline**.

Apply the Extruded Surface Feature tool.

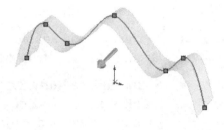

7. Click **Extruded Surface** ⬘ from the Surfaces toolbar. The Surface-Extrude PropertyManager is displayed. Blind is the Default End Condition.

8. Select **Mid Plane** for End Condition in Direction 1. Think design intent.

9. Enter **30**mm for Depth in Direction 1.

10. Check the **Draft on/off** box in Direction 1.

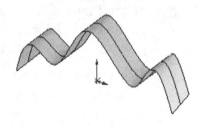

11. Enter a **5**deg draft angle.

12. Click **OK** ✔ from the Surface-Extrude PropertyManager. Surface-Extrude1 is displayed in the FeatureManager.

13. **Close** the model.

Cut With Surface Feature

A Cut With Surface feature ⬗ cuts a solid model by removing material with a plane or a surface. With multi-body parts, you can select which bodies to keep with the Feature Scope option.

The Cut With Surface feature uses the SurfaceCut PropertyManager. The SurfaceCut PropertyManager provides the following selections:

- *Surface Cut Parameters*. The Surface Cut Parameters box provides the following options:

  - **Selected surface for cut**. Displays the selected surface or plane to be used to cut the solid bodies.

  - **Flip cut**. Reverses the direction of the cut if required.

- *Feature Scope*. Provides the ability to select one of the following options:

  - **All bodies**. The surface cuts all bodies each time the feature is rebuilt.

  - **Selected bodies**. Selected by default. The surface cuts only the bodies which are selected.

  - **Auto-select**. Selected by default. The system automatically selects all relevant intersecting bodies. This option is faster than the All bodies option because the system only processes the bodies on the initial list. The system does not rebuild your entire model.

The Feature Scope option is only available with a multi-body part model.

The Auto-select option is only available with the Selected bodies option activated.

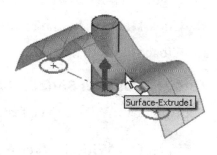

Tutorial: Cut With Surface 6-1

Apply the Cut With Surface feature.

1. Open **Cut With Surface 6-1** from the SOLIDWORKS 2018 folder.

2. Hold the **Ctrl** key down.

3. Click the circumference of the **left** and **right** circles as illustrated. Release the **Ctrl** key.

Activate the Extruded Boss/Base feature.

4. Click **Extruded Boss/Base** from the Features toolbar. The Boss-Extrude PropertyManager is displayed. Sketch2-Contour<1> and Sketch2-Contour<2> are displayed in the Selected Contours box.

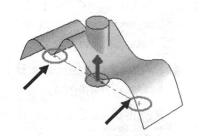

5. Select the **Up To Surface** End Condition for Direction 1. The Extrude2 feature points in an upward direction. Click the **top surface** as illustrated. Surface-Extrude1 is displayed in the Face/Plane box.

6. Click **OK** from the Boss-Extrude PropertyManager. Boss-Extrude1 is created.

7. Click the top face of the **center cylinder** from the Graphics window. Extrude2 is highlighted in the FeatureManager.

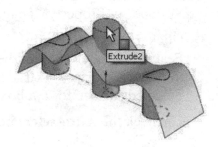

8. Click **Insert ➤ Cut ➤ With Surface** 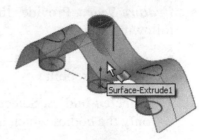. The
    SurfaceCut PropertyManager is displayed.

9. Click **Surface-Extrude1** from the Graphics window.
    Surface-Extrude1 is displayed in the Selected Surface
    for cut box. You are cutting the cylindrical surface
    with Surface-Extrude1.

10. Un-check **Auto-Select** from the Feature Scope option
    box.

Select the side of the cylinder that you want to keep.

11. Click the **Top face (Extrude2)** of the cylinder.

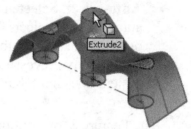

12. Click the **Flip Cut** option box. The direction arrow
    points to the front.

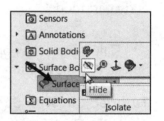

13. Click **OK** ✔ from the SurfaceCut PropertyManager.
    SurfaceCut1 is created and is displayed in the
    FeatureManager.

14. **Expand** the Surface Bodies folder from the
    FeatureManager.

15. Right-click **Surface-Extrude1**.

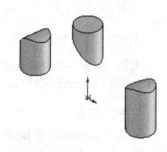

16. Click **Hide**.

17. Right-click **Sketch2** from the FeatureManager.

18. Click **Hide**. View the results.

19. **Close** the model.

Tutorial: Cut With Surface 6-2

Apply the Cut With Surface feature. Cut the body using
a plane.

1. Open **Cut With Surface 6-2** from the
    SOLIDWORKS 2018 folder.

2. Click **Insert ➤ Cut ➤ With Surface** 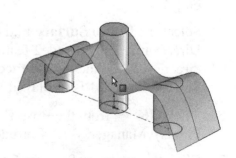. The
    SurfaceCut PropertyManager is displayed.

3. Click **Surface-Extrude1** from the fly-out
    FeatureManager or from the Graphics window as
    illustrated. Surface-Extrude1 is displayed in the
    Selected surface for cut box.

4. Uncheck the **Auto-select** box.

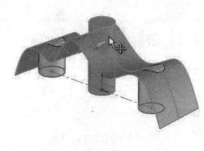

5. Click the **center cylinder** above the Surface-Extrude1. Extrude2 is displayed in the Feature Scope box. If required, click the **Flip cut** box. The arrow points towards the back.

6. Click **OK** ✔ from the SurfaceCut PropertyManager. SurfaceCut1 is created and is displayed in the FeatureManager. View the Graphics window.

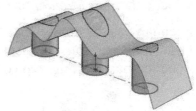

7. **Close** the model.

Fillets in General

Filleting refers to both fillets and rounds. The distinction between fillets and rounds is made by the geometric conditions and not by the command. Fillets add volume, and rounds remove volume.

Create fillets on a selected edge, selected edges, selected sets of faces, and edge loops. The selected edges can be selected in various ways. Options exist for constant or variable fillet radius and tangent edge propagation in the Fillet PropertyManager. There are a few general guidelines which you should know before creating fillets in a model:

- Create larger fillets before you create smaller ones.

- Insert drafts before you create fillets. Example: There are numerous filleted edges and drafted surfaces in a cast part or a molded part. Insert the draft features before you create the fillets in your model.

- Leave cosmetic fillets for the end. Insert cosmetic fillets after most of your other geometry is completed. If you add the cosmetic fillets too early in your modeling process, the system will take longer to rebuild your model.

- Use a single fillet operation to address more than one edge which requires an equal radius fillet. Why? A single operation enables the model to rebuild quicker. However, if you modify the radius of that fillet, all other fillets created in the same operation will also be modified.

- Group Fillets together into a Fillet folder. Positioning Fillets into a file folder reduces the time spent for your mold designer or toolmaker to locate each fillet in the FeatureManager design tree. This saves time and cost.

- Select Features to fillet from the FeatureManager.

Fillet Feature

The Fillet 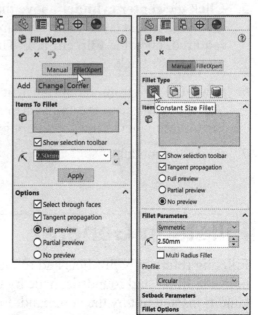 tool uses the Fillet PropertyManager. The Fillet PropertyManager provides the ability to create various fillet types.

The Fillet PropertyManager provides the ability to select either the *Manual* or *FilletXpert* tab.

Each tab has different menus and options. The Fillet PropertyManager displays the appropriate selection based on the type of fillet you create.

Fillet PropertyManager: Manual tab

Select the Manual tab to maintain control at the feature level. The following selections and options are available: *Constant Size Fillet*, *Variable Size Fillet*, *Face Fillet* and *Full Round Fillet* each with a unique menu.

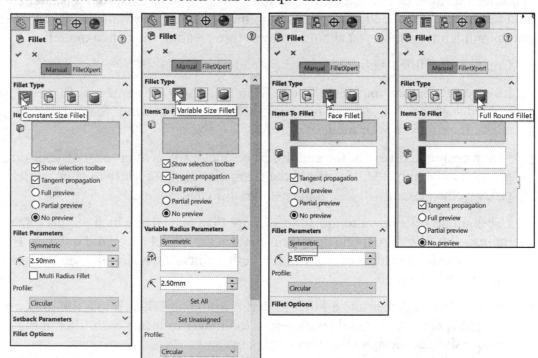

- *Fillet Type*. Provides the ability to select four fillet types:

  - **Constant Size Fillet**. Selected by default. Provides the ability to create a fillet with a constant radius for the entire length of the fillet. After you select the Constant radius option, set the PropertyManager options to create the following:

    - **Multiple Radius Fillets**. Creates fillets that have different radius values within a single fillet feature.

    - **Round Corner Fillets**. Creates a smooth transition where fillet edges meet.

    - **Setback Fillets**. Defines a setback distance from a vertex at which the fillets start to blend.

  - **Variable Size Fillet**. Provides the ability to create fillets with changeable radii values. Use control points to help define the fillet. When checked, the Variable Radius Parameters box is displayed in the PropertyManager. Use control points to help define the fillet.

  - **Face Fillet**. Provides the ability to blend non-adjacent, non-continuous faces.

  - **Full Round Fillet**. Provides the ability to create fillets that are tangent to three adjacent faces sets, with one or more faces tangent.

- *Items To Fillet*. Provides the ability to select the following options. The available options are dependent on the fillet type:

  - **Edges, Faces, Features, and Loops**. Displays the selected entities to fillet from the Graphics window.

  - **Show selection toolbar**. Displays the selection toolbar.

  - **Tangent Propagation**. Selected by default. Extends the fillet to all faces which are tangent to the selected face.

  - **Full preview**. Selected by default. Displays the fillet preview of all edges.

  - **Partial preview**. Displays a fillet preview of the first selected edge.

  - **No preview**. No preview is displayed.

  - **Face Set 1**. Only available with the face fillet type. Displays the selected face for set 1.

  - **Face Set 2**. Only available with the face fillet type. Displays the selected face for set 2.

  - **Side Face Set 1**. Only available with the Full round fillet type.

  - **Center Face Set**. Only available with the Full round fillet type.

- **Side Face Set 2**. Only available with the Full round fillet type.

- *Fillet Parameters*. Sets the Parameters for the fillet. The available options are for Constant size:

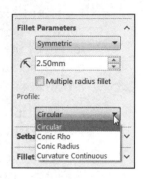

  - **Symmetric**, **Asymmetric**

  - **Radius**. Sets the fillet radius.

  - **Multiple radius fillet**. Creates fillets with different radius values for edges. You can create corners using three edges with different radii.

    - **Profile**. Sets the profile type to fillet. The profile defines the cross sectional shape of the fillet. The available options are **Circular, Conic Rho, Conic Radius and Curvature Continuous**.

- *Fillet Parameters*. Sets the Parameters for the fillet. The below options are available for a Variable Radius:

  - **Symmetric**, **Asymmetric**

  - **Radius**. Sets the fillet radius.

    - **Profile**. Sets the profile type to fillet. The profile defines the cross sectional shape of the fillet. The available options are **Circular, Conic Rho, Conic Radius and Curvature Continuous**.

  - **Attached Radii**. Lists the vertices of the edges selected under Items to Fillet for Edges, Faces, Features and Loops, and lists control points you selected in the graphics area.

  - **Set Unassigned**. The current Radius to all items that do not have assigned radii under Attached Radii.

  - **Set All**. Applies the current Radius to all items under Attached Radii.

  - **Number of Instances**. Sets the number of control points on the edges.

  - **Smooth transition. Creates a fillet that changes smoothly from one radius to another when** matching a fillet edge to the adjacent face.

  - **Straight transition**. Creates a fillet that changes from one radius to another linearly without matching edge tangency with an adjacent fillet.

- *Setback Parameters*. Creates an even transition connecting the blended surface, along the edge of the part, into the fillet corner. Select a vertex and radius, then assign a different or the same setback distances for each edge. The setback distance is the distance along each edge at which the fillet starts to merge into the three faces which combine into the common vertex. The Setback Parameters box is only available for the Constant radius type and Variable radius type fillet.

- The available options are dependent on the fillet type:

  - **Distance**. Sets the fillet setback distance from the vertex.

  - **Setback Vertices**. Displays the selected vertices from the Graphics window. The setback fillet edges intersect at the selected vertices.

  - **Setback Distances**. Displays the edge numbers with the corresponding setback distance value.

  - **Set Unassigned**. Applies the current Distance option value to all edges that do not have an assigned distance under the Setback Distances option.

  - **Set All**. Applies the current Distance option value to all edges that are under the Setback Distances option.

- *Fillet Options*. The Fillet Options box is only available for the Constant Radius type, Variable Radius type, and Face Fillet type. The available options are dependent on the fillet type:

  - **Select through faces**. Selected by default.

  - **Keep features**. Selected by default. Provides the ability to keep boss or cut features viewable from the Graphics window if you apply a large fillet radius to cover them.

Clear the Keep features option to cover an Extruded Boss or Extruded Cut feature with a fillet.

  - **Round corners**. Creates constant radius fillets with rounded corners. You need to select at two or more adjacent edges to fillet.

  - **Overflow type**. Controls the way fillets on single closed edges, splines, circles, ellipses, etc. behave when they meet edges. There are three options under this section:

  - **Default**. Provides the ability to select the Keep edge or Keep surface option.

  - **Keep edge**. Provides the ability for the edge of the model to remain unchanged, while the fillet is adjusted.

  - **Keep surface**. Provides the ability for the fillet edge to adjust to be uninterrupted and even, while the model edge alters to match the fillet edge.

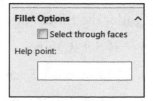

  - **Help point**. Provides the ability to control attachment of edges between the intersection features.

Select a feature in the FeatureManager design tree to display the Feature attachment option in the PropertyManager. View SOLIDWORKS Help for additional detail information.

Control Points

Apply control points to define a Variable Size Fillet type. Control and assign radius values to the control points between the vertices of the fillet. How do you use control point? Use the following as guidelines:

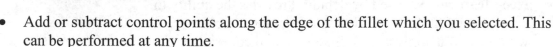

- Assign a radius value to each control point, or assign a value to one or both enclosing vertices.

- Modify the relative position of the control point by using the following two methods:

  - Modify the percentage of the control point in the callout section.

  - Select the control point and drag the control point to a new location.

- Add or subtract control points along the edge of the fillet which you selected. This can be performed at any time.

- Select more than one sketch entity to fillet. Complete each entity before selecting additional entities. Perform the following three step procedure to select sketch entities to fillet in the Items to Fillet section: 1.) *Apply a radius to each selected vertex*, 2.) *Apply a radius value to one or all control points*; *modify the position of a control point at any time, either before or after you apply a value for the radius.*

The system defaults to three control points located at equidistant increments of 25%, 50% and 75% along the edge between the two variable radii.

Adding or subtracting control points along the edge positions the control points in equidistant increments along that edge.

Use the Fillet tool to create symmetrical conic shaped fillets for parts, assemblies, and surfaces. You can apply conic shapes to *Constant Size*, *Variable Size*, and *Face* fillets.

Under the Fillet Parameters option:

- a) Set Radius.

- b) For Profile, select Conic Rho.

- c) Set Rho Radius.

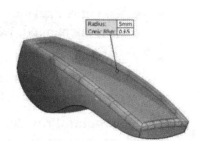

Rotate the model to view the curvature of the fillets. Use the callouts in the graphics area to change values for Radius and Conic Rho. The shape of the fillets adjust dynamically.

Tutorial: Fillet 6-1

Create a fillet feature using the Tangent propagation option with the manual tab in the PropertyManager.

1. Open **Fillet 6-1** from the SOLIDWORKS 2018 folder. The Fillet 6-1 FeatureManager is displayed.

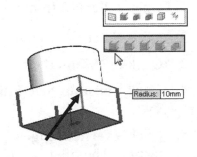

2. Click **Fillet** 🗇 from the Features toolbar. The Fillet PropertyManager is displayed.

3. Click the **Manual** tab.

4. Click **Constant Size Fillet** for Fillet Type.

5. Click a **vertical edge** of the Extrude1 feature. The fillet option pop-up toolbar is displayed. Options are model dependent.

6. Select the **Connected to start loop, 3 Edges icon**. The four selected edges are displayed in the Edges, Faces, Features, and Loop box.

7. Enter **5**mm for Radius.

8. Click **OK** ✔ from the Fillet PropertyManager. Fillet1 is displayed in the FeatureManager.

9. Display an **Isometric view**.

10. Click the **front face** of Extrude1.

11. Click **Fillet** 🗇 from the Features toolbar.

12. Check the **Tangent propagation** box and the **Full preview** box. This will pick all tangent faces and propagate around your model.

13. Click **OK** ✔ from the Fillet PropertyManager. The Fillet2 feature is displayed in the FeatureManager.

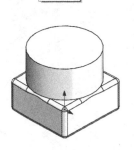

14. **Close** the model.

Tutorial: Fillet 6-2

Create a fillet feature using the Setback Parameters option.

1. Open **Fillet 6-2** from the SOLIDWORKS 2018 folder.

2. Click **Fillet** 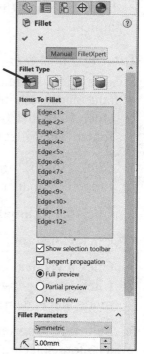 from the Features toolbar.

3. Click the **Manual** tab.

4. Click **Constant Size Fillet** for Fillet Type.

5. Click a **vertical edge** of the Extrude1 feature. The fillet pop-up toolbar is displayed. Options are model dependent.

6. Select the **Internal to feature, 11 Edges icon**. The 11 selected edges are displayed in the Edges, Faces, Features, and Loop box.

7. Enter **5**mm for Radius.

8. Enter **10**mm for Distance in the Parameters box.

9. Click inside the **Setback Vertices** box.

10. Click the **Right top Vertex** of Extrude1. Vertex<1> is displayed in the Setback Vertices box.

11. Enter **8**mm in the first Setback distance box in the Graphics window.

12. Enter **10**mm in the second Setback distance box in the Graphics window.

13. Enter **8**mm in the third Setback distance box in the Graphics window. View the updated PropertyManager.

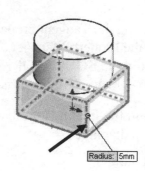

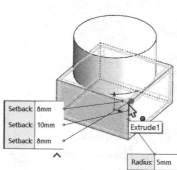

14. Click **OK** ✔ from the Fillet PropertyManager. Fillet1 is created and is displayed in the FeatureManager.

15. **View** the model in the Graphics window. Display Tangent Edges to view the Fillet features.

16. **Close** the model.

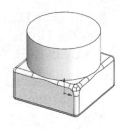

Tutorial Fillet 6-3

Create a full round fillet and a multiple radius fillet feature.

1. Open **Fillet 6-3** from the SOLIDWORKS 2018 folder. The Rollback bar is positioned below the Top Cut Feature in the FeatureManager as illustrated.

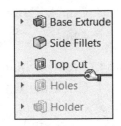

2. Click **Hidden Lines Visible** from the Heads-up View toolbar.

3. Click **Fillet** from the Features toolbar. The Fillet PropertyManager is displayed.

4. Click the **Manual** tab.

5. Click **Full Round Fillet** for Fillet Type.

6. Click the **inside Top Cut face** for Side Face Set 1.

7. Click **inside** the Center Face Set box.

8. Click the **top face** of Base Extrude for Center Face Set.

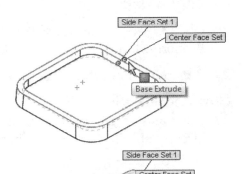

9. Click **inside** the Side Face Set 2 box.

10. **Rotate** the part and click the **outside Base Extrude face** for Side Face Set 2 as illustrated.

11. Click **OK** ✔ from the Fillet PropertyManager.

12. Rename **Fillet1** to **TopFillet**.

13. Drag the **Rollback** bar down to the bottom of the FeatureManager.

14. Display an **Isometric view**.

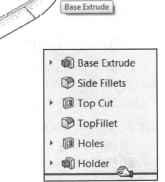

15. Display a **Shaded With Edges** view.

16. Click **Fillet** from the Features toolbar. The Fillet PropertyManager is displayed.

17. Click **Constant Size Fillet** for Fillet Type.

18. Click the **bottom outside circular edge** of the Holder as illustrated.

19. Enter **.050**in for Radius.

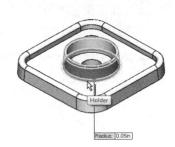

20. Click the **bottom inside circular edge**, "Top Cut" inside of the Holder as illustrated. A fillet option pop-up toolbar is displayed. Options are model dependent.

21. Select the **All concave, 8 Edges icon**. The selected edges are displayed in the Edges, Faces, Features, and Loop box.

22. Check **Tangent propagation**.

23. Check **Multiple radius fillet**.

Modify the Fillet values.

24. Click the **Radius** box for the Holder outside edge in the Graphics window as illustrated.

25. Enter **.04**in.

26. Click **OK** ✔ from the Fillet PropertyManager.

27. Rename **Fillet2** to **HolderFillet**.

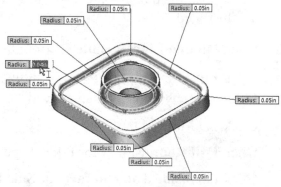

Group the Fillets into a Folder.

28. Click and drag **TopFillet** from the FeatureManager directly above HolderFillet in the FeatureManager.

29. Hold the **Ctrl** key down.

30. Click **TopFillet** from the FeatureManager.

31. Right-click **Add to New Folder**.

32. Release the **Ctrl** key.

33. Rename **Folder1** to **Fillet Folder**.

34. **Close** the model.

Tutorial: Fillet 6-4

Create a variable radius Fillet feature.

1. Open **Fillet 6-4** from the SOLIDWORKS 2018 folder.

2. Click the **Fillet** ⬡ Features tool. The Fillet PropertyManager is displayed.

3. Click the **Manual** tab.

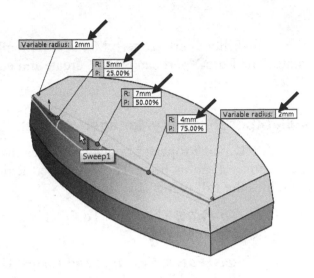

4. Click the **front top edge** as illustrated.

5. Click **Variable Size Fillet** for Fillet Type.

6. Enter **2**mm in the top left Variable radius box in the Graphics window.

7. Click the **second control** point.

8. Enter **5**mm for R.

9. Click the **third control** point.

10. Enter **7**mm for R.

11. Enter **4**mm in the remaining control point.

12. Enter **2**mm in the bottom right Variable radius box as illustrated.

13. Click **OK** ✅ from the Fillet PropertyManager. The VarFillet1 feature is displayed. View the results.

14. **Close** the model.

FilletXpert PropertyManager

Select the FilletXpert PropertyManager when you want the SOLIDWORKS software to manage the structure of the underlying features. The FilletXpert manages, organizes, and reorders only constant radius fillets. The FilletXpert provides the ability to

1. Create multiple fillets.

2. Automatically invoke the FeatureXpert.

3. Automatically reorder fillets when required.

4. Manage the desired type of fillet corner.

🔅 When manually adding or editing constant and variable radius fillets, a pop-up toolbar is available to help you select multiple edges or fillets. The same toolbar is used with FilletXpert.

The FilletXpert PropertyManager remembers the last used state. The FilletXpert can ONLY create and edit *Constant Radius* fillets.

FilletXpert PropertyManager: Add tab

The FilletXpert PropertyManager provides the ability to create a new constant radius fillet. The FilletXpert PropertyManager provides the following selections:

- *Items To Fillet*. The Items To Fillet box provides the following selections:

  - **Edges**, **Faces**, **Features and Loops**. Displays the selected items to fillet for Edges, Faces, Features and Loops.

  - **Show selection toolbar**. Displays the selection toolbar.

  - **Radius**. Enter the radius value.

  - **Apply**. Calculates and creates the fillets. The FilletXpert uses the FeatureXpert to create the fillet.

- *Options*. The Options box provides the following selections: **Select through faces**, **Tangent propagation**, **Full preview**, **Partial preview** and **No preview**.

FilletXpert PropertyManager: Change tab

Provides the ability to remove or resize a Constant Radius fillet. The FilletXpert PropertyManager provides the following selections:

- *Fillets to change*. The Fillets to change box provides the following selections:

  - **Filleted Faces**. Displays the selected fillets to resize or to remove.

  - **Radius**. Enter the radius value.

  - **Resize**. Modifies the selected fillets to the new radius value.

  - **Remove**. Removes the selected fillets from the model.

- *Existing Fillets*. Provides the following options:

  - **Sort by Size**. Filters all fillets by size. Select a fillet size from the list to select all fillets with that value in the model and display them under Filleted faces.

FilletXpert PropertyManager: Corner tab

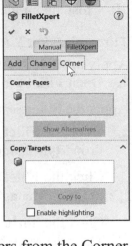

Provides the ability to create and manage the fillet corner feature where three filleted edges meet at a single vertex. The FilletXpert PropertyManager provides the following selections:

- *Corner Faces*. Displays the selected fillet corners from the Graphics window.

  - **Show Alternatives**. Displays alternative fillet corner previews in a pop-up.

- *Copy Targets*. Displays the selected target fillet corners. Click in the Copy Targets box, compatible target fillet corners are highlighted in the Graphics window.

  - **Copy to**. Provides the ability to copy selected target fillet corners from the Corner Faces box to a selected location.

  - **Enable highlighting**. Provides the ability to highlight compatible target fillet corners in the Graphics window.

The mouse pointer displays the fillet corner icon when you hover over a target fillet corner.

Tutorial: Fillet 6-5

Use the FeatureXpert tool to insert a Fillet feature.
1. Open **Fillet 6-5** from the SOLIDWORKS 2018 folder.

Apply the Fillet Feature tool.

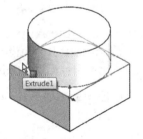

2. Click the **Fillet** Features tool. The Fillet PropertyManager is displayed.

3. Click the **Manual** tab.

4. Click **Constant Size** for Fillet Type.

5. Click the **top face** of Extrude1 as illustrated. Extrude1 is highlighted in the FeatureManager.

6. Enter **5**mm for Radius in the Item To Fillet box.

7. Click **OK** from the Fillet PropertyManager. The What's Wrong box is displayed. You failed to create a Fillet. Use the FeatureXpert.

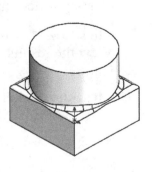

8. Click the **FeatureXpert** button from the What's Wrong box. The FeatureXpert solved your fillet problem and created two fillet features: Fillet1 and Fillet2. The 5mm Radius fillet was not possible with the selected Extrude1 face. The FeatureXpert solved the problem with edges for a smoother transition.

9. **Edit** both Fillet1 and Fillet2 from the FeatureManager to review the created fillets.

10. **Close** the model.

Tutorial: Fillet 6-6

Apply the FilletXpert tab option to create a Fillet feature.

1. Open **Fillet 6-6** from the SOLIDWORKS 2018 folder.

Apply the Fillet Feature tool.

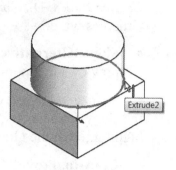

2. Click the **Fillet** ⬡ Features tool. The Fillet PropertyManager is displayed.

3. Click the **FilletXpert** tab. The FilletXpert PropertyManager is displayed. Click the **Add** tab.

4. Click the **bottom circular edge** of Extrude2 as illustrated. The selected edge is displayed in the Edges, Faces, Features and Loop box.

5. Enter **10**mm for Radius.

6. Click the **Apply** button.

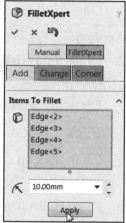

7. Click a **vertical corner** of Extrude1. The fillet option pop-up toolbar is displayed. Options are model dependent.

8. Select the **Connected to end face, 3 Edges icon**. The four selected edges are displayed in the Edges, Faces, Features, and Loop box.

9. Click the **Apply** button. FilletXpert adds two fillet features without leaving the FilletXpert PropertyManager. Modify the first fillet using the FilletXpert.

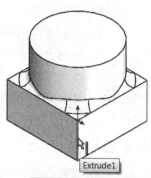

10. **Expand** the fly-out FeatureManager to view Fillet1 and Fillet2.

11. Click the **Change tab**.

12. Click **Fillet2** from the fly-out FeatureManager.

13. Enter **5.00**mm for Radius.

14. Click the **Resize** button. The model is modified and the existing Fillets are sorted by size.

15. Click **OK** ✔ from the FilletXpert PropertyManager. View the results. Display Tangent Edges to view the Fillet feature.

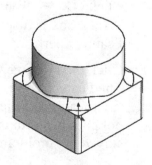

16. **Close** the model.

Tutorial: Fillet 6-7 fillet corner tab

Apply the FilletXpert Corner Tab option. Remember, the Corner tab option requires three constant radius filleted edges of mixed convexity meeting at a single vertex.

1. Open **Fillet 6-7** from the SOLIDWORKS 2018 folder.

2. Right-click **Fillet-Corner1** from the FeatureManager. Click **Edit Feature** from the Context toolbar. The FilletXpert Property Manager is displayed.

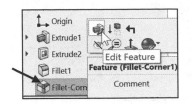

3. Click the **FilletXpert** tab.

4. Click the **Corner** tab.

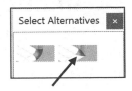

5. Click the **Show Alternatives** button. Two options are displayed in the Graphics window. Note if needed, expand the Select Alternatives dialog box.

6. Click the **right option**.

7. Click **OK** ✔ from the FilletXpert PropertyManager. View the results.

8. **Close** the model.

Fillet to Chamfer Tool

After you create a Fillet feature, you can edit the fillet in the FeatureManager and directly convert the fillets to chamfers and chamfers to fillets.

Tutorial: Fillet 6-8

Create chamfers from existing fillet features.

1. Open **Fillet 6-8** from the SOLIDWORKS 2018 folder. The Fillet 6-8 FeatureManager is displayed.

2. Right-click **Fillet1** from the FeatureManager.

3. Click **Edit Feature**. The Fillet1 PropertyManager is displayed. There are two Feature Types: Fillet Type and Chamfer Type.

4. Click **Chamfer Type** for Feature Type.

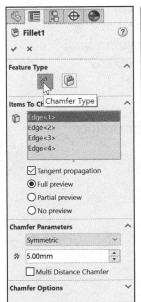

5. Click **OK** 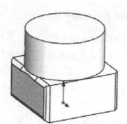 from the Fillet1
    PropertyManager. View the results.
    Fillet1 in the FeatureManager now
    displays the Chamfer ⬡ icon.

6. **Close** the model.

Cosmetic Thread Feature

A cosmetic thread represents the inner diameter of a thread on a
boss or the outer diameter of a thread. The properties of cosmetic
threads include:

*   You can represent threads on a part, assembly or drawing, and
    you can attach a thread callout note in drawings. You can add
    cosmetic threads to conical holes. If the conical thread does not
    end at a flat face, it is trimmed by the curved face.

*   A cosmetic thread differs from other annotations in that it is an
    absorbed feature of the item to which it is attached. For
    example, the cosmetic thread on a hole is in the
    FeatureManager design tree
    under the Hole feature, along
    with the sketches used to
    create the hole.

*   When the pointer is over a
    cosmetic thread, the pointer
    changes to  .

- Cosmetic threads in part documents are inserted automatically into drawing views. A thread callout is also inserted if the drawing document is in the ANSI standard. (You insert thread callouts in the Cosmetic Thread PropertyManager, but they appear only in drawing documents.) Note: When creating a cosmetic thread, you can now define cosmetic threads using international standard sizes. To insert cosmetic threads from assembly documents into drawings, click Insert, Model Items and click Cosmetic thread.

- You have the ability to turn off the annotation of the cosmetic thread in a complex assembly and still display the shaded ones. This is a great way of cleaning up the model while still understanding which holes are tapped.

- In drawings, Insert Callout appears in the Context toolbar. If a cosmetic thread callout is defined in the part or assembly but is not displayed in the drawing, you can display the callout by selecting this menu item. A leader attaches to the thread by default. The callout is a note. You can edit the callout as you would edit any note.

- If you add a cosmetic thread while working in a drawing view, the part or assembly is updated to include a Cosmetic Thread feature.

- You have the ability to turn off the annotation of the cosmetic thread in an assembly or drawing and display the shaded ones.

- You can dimension both the circular cosmetic thread and the linear dimension of the sides in drawings. You cannot dimension cosmetic threads in part or assembly documents.

- The visibility of cosmetic threads follows the visibility of the parent feature. When you change display mode, add features to the Show Hidden Edges list, or hide a component, the visibility of cosmetic threads changes automatically.

- You can set High quality cosmetic threads to check all cosmetic threads to determine if they should be visible or hidden.

- You can reference patterned cosmetic threads.

- For tap and pipe tap holes, you can add cosmetic threads in the Hole Wizard. See the Hole Wizard section in the book and SOLIDWORKS Help for additional information.

The Cosmetic Thread ⑂ feature uses the Cosmetic Thread PropertyManager. The Cosmetic PropertyManager provides the following default selections:

- *Thread Settings*. Provides the following options:

  - **Circular Edge**. Select a circular edge of the model in the Graphics window.

  - **Standard**. Provides the ability to set the dimensioning standard for the cosmetic thread.

  - **Type**. Provides the ability to select the type of threads.

  - **Size**. Select the size of the cosmetic thread based on the dimensioning standard you selected.

  - **Minor Diameter**, **Major Diameter**, or **Conical Offset**. Sets the diameter for the dimension corresponding to the entity type with the cosmetic thread.

  - **End Condition**. The cosmetic thread extends from the edge selected above to the end condition. The following End Conditions are available:

    - **Blind**. A specified depth. Specify the depth below.

    - **Through**. Completely through the existing geometry.

    - **Up To Next**. To the next entity that intercepts the thread.

  - **Depth**. Enter a value when the End Condition is Blind.

- *Thread Callout*. Type text to appear in the thread callout. (The thread callout appears in drawing documents only.)

💡 Cosmetic thread callouts are editable, and therefore configurable, only when Standard is set to None in the Cosmetic Thread PropertyManager.

Tutorial: Cosmetic Thread 6-1

Apply a Cosmetic thread to simulate an ANSI Inch, Machine Thread, ¼-20.

1. Open **Cosmetic thread 6-1** from the SOLIDWORKS 2018 folder.

Create a Cosmetic thread. You can represent cosmetic threads on a part, assembly or drawing. You can attach a thread callout note in a drawing. You can add cosmetic threads to conical holes. If the conical thread does not end at a flat face, it is trimmed by the curved face.

Start the Cosmetic thread feature.

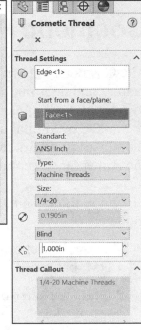

2. Click the **bottom edge** of the part as illustrated.

3. Click **Insert**, **Annotations**, **Cosmetic Thread** from the Menu bar menu. View the Cosmetic Thread PropertyManager. Edge<1> is displayed.

4. Click the **bottom** face.

5. Select **ANSI Inch** for Standard.

6. Select **Machine Threads** for Type.

7. Select **¼-20** for Size.

8. Select **Blind** for End Condition.

9. Enter **1.00** for depth.

10. Click **OK** ✔ from the Cosmetic Thread PropertyManager.

11. **Expand** the FeatureManager. View the Cosmetic Thread feature and icon. If needed, right-click the **Annotations** folder, click **Details**. Check the **Cosmetic threads** and **Shaded cosmetic threads** box to view the cosmetic thread in the Graphics window on the model.

12. Click **OK**. View the cosmetic thread on the model.

13. **Close** the model.

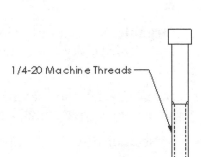

☀ The thread callout ¼-20 machine thread is automatically inserted into a drawing document, if the drawing document is in the ANSI drafting standard.

☀ Cosmetic thread callouts are editable, and therefore configurable, only when Standard is set to None in the Cosmetic Thread PropertyManager.

1/4-20 Machine Threads

¼-20-2 UNC 2A: - ¼ inch drill diameter - 20 threads/inch, 2 inches, Unified National Coarse thread series, Class 2, A -External thread. Pitch - 1/20 threads per inch.

You can dimension both the circular cosmetic thread and the linear dimension of the sides in drawings. You cannot dimension cosmetic threads in part or assembly documents.

For tapped holes with cosmetic threads created in the Hole Wizard, the hole diameter is the diameter of the tap drill. For tapped holes without cosmetic threads, the hole diameter is the outer diameter of the thread.

Cosmetic Pattern

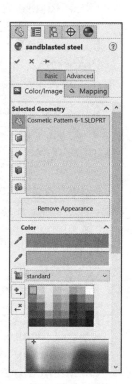

The Cosmetic Pattern differs from the Cosmetic feature as it is not a feature and does not produce a thread. The Cosmetic Pattern tool is located in the **Appearance library** - which includes multiple folders organized by similar appearances. Each appearance includes thumbnail previews in the Task Pane and in the Color/Image tab of the Appearances(color) PropertyManager.

The main folders and sub-folders of the Appearance(color) Library include:

- **Plastic**. Varieties of high gloss, textured, clear plastic, satin finish, EDM, patterned, composite and mesh.

- **Metal**. Varieties of steel, chrome, aluminum, bronze, brass, copper, nickel, zinc, magnesium, iron, titanium, tungsten, gold, silver, platinum, lead and galvanized.

- **Painted**. Car, sprayed, powder coat.

- **Rubber**. Matte, gloss, texture.

- **Glass**. Gloss and textured.

- **Lights**. Emissive appearances.

- **Fabric**. Varieties of cloth (cottons, canvas, burlap) and carpet.

- **Organic**. Varieties of wood (ash, cherry, etc.), water (still and rippled), sky (clear and cloudy), and miscellaneous (leather, sand, etc.).

- **Stone**. Varieties of paving (asphalt, concrete, etc.), stoneware (Bone china, ceramic, etc.), and architectural (granite, marble, etc.).

- **Miscellaneous**. Studio appearances and patterns.

Tutorial: Cosmetic Pattern 6-1

Apply a Cosmetic Pattern to the entire part using the Appearance (color) Library. In a part, you can add appearances to faces, features, bodies, and the part itself. In an assembly, you can add appearances to components. A hierarchy applies to appearances, based on where they are assigned on the model.

1. Open **Cosmetic Pattern 6-1** from the SOLIDWORKS 2018 folder.

Create a Cosmetic Pattern using the Task Pane.

2. Click the **Appearances, Scenes, and Deals** tab from the Task Pane.

3. **Expand** the Metal folder.

4. Click **Steel**.

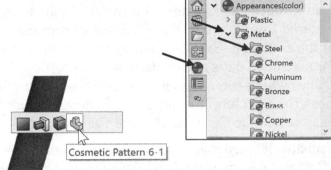

5. Drag and drop the **Sandblasted steel** icon from the lower left panel on the part in the Graphics window. View the pop-up toolbar. The Pop-up toolbar provides options to apply the appearance to the model.

6. Click the **Part** icon as illustrated.

Apply a Pattern.

7. **Expand** the Miscellaneous folder from the Task Pane.

8. Click the **Pattern** folder. View your options.

9. Drag and drop the **checker pattern** option on the head as illustrated.

10. Click the **Face** icon as illustrated.

11. Click the **DisplayManager** tab to view the summary. View SOLIDWORKS Help for additional information.

12. **Close** the model.

Every material is assigned a default appearance that emulates the look of the material.

In a part, you can add appearances to faces, features, bodies, and the part itself. In an assembly, you can add appearances to components. A hierarchy applies to appearances, based on where they are assigned on the model.

Summary

In this chapter, you learned about the Extruded Boss/Base, Extruded Cut, Fillet and the Cosmetic Thread features in SOLIDWORKS. An extrusion is the Base feature which adds material and is the first feature created in a part.

An Extruded Boss feature adds material onto an existing feature. A cut feature removes material from the part. The Thin feature parameter option allows you to specify thickness of the extrusion. Extrude features can be solids or surfaces. Solids are utilized primarily for machined parts and contain mass properties. Surfaces are utilized surfaces for organic shapes which contain no mass properties. You reviewed and applied the Extruded Boss/Base Feature, Extruded Cut Feature, Extruded Solid/Thin Feature, Extruded Surface Feature and the Cut With Surface Feature.

You also learned that Filleting refers to both fillets and rounds. The distinction between fillets and rounds is made by the geometric conditions and not by the command. Fillets add volume, and rounds remove volume.

Create fillets on a selected edge, selected edges, selected sets of faces, and edge loops. The selected edges can be selected in various ways. Options exist for constant or variable fillet radius and tangent edge propagation in the Fillet PropertyManager.

You reviewed the Manual tab in the Fillet PropertyManager, using control points, surfaces, and the FilletXpert PropertyManager with the Add Change and Corner tab.

The Fillet to Chamfer tool and the Chamfer to Fillet tool provides the option to edit directly in the FeatureManager.

You addressed adding a Cosmetic Thread and a Cosmetic Pattern to a simple part. View SOLIDWORKS online help for additional information on Cosmetic threads and the Hole Wizard section in the next chapter.

In Chapter 7, explore and address the Revolved Boss/Base feature, Revolved Cut feature, Hole Wizard feature, Advance Hole feature, Dome feature, Curve and Thread feature.

CHAPTER 7: REVOLVED, HOLE WIZARD, ADVANCED HOLE, DOME, CURVE AND THREAD FEATURES

Chapter Objective

Chapter 7 provides a comprehensive understanding of Revolved, Hole Wizard, Advanced Hole, Dome, Curve and Thread features.

On the completion of this chapter, you will be able to:

- Comprehend and utilize the following Revolved features:
  - Revolved Boss/Base.
  - Revolved Cut.
  - Revolve Boss Thin.
  - Revolved Surface.
- Understand and utilize the following Hole features:
  - Hole Wizard.
  - Advanced Hole.
- Know and utilize the Dome feature.
- Knowledge of various Curve features and their options: Split Line, Projected, Composite, Curve Through XYZ Points, Curve Through References Points and Helix/Spiral.
- Understand and utilize the Thread Wizard Feature.

Revolved Boss/Base Feature

The Revolved Boss/Base  feature adds material by revolving one or more profiles about a centerline. SOLIDWORKS provides the ability to create Revolved Boss/Base, Revolved Cut or Revolved Surface features.

The Revolved feature can be a solid, thin or surface. There are a few general guidelines that you should be aware of when creating a Revolved feature:

- The sketch for a solid revolved feature can contain multiple intersecting profiles. Activate the Selected Contours box from the Revolve PropertyManager.

- The Selected Contours icon is displayed on the mouse pointer. Click one or more intersecting or non-intersecting sketches to create the Revolved feature.

- The sketch for a thin or surface revolved feature can contain multiple open or closed intersecting profiles.

- Profiles cannot cross the centerline.

The Revolved Boss/Base feature 🍥 uses the Revolve PropertyManager. The Revolve PropertyManager provides the following selections:

- *Axis of Revolution*. Displays the selected axis around which the feature revolves. The Axis of Revolution can be a line, centerline, or even an edge. Your selection is dependent on the type of revolve feature you create.

- *Direction 1*. Creates the revolve feature in one direction from your Sketch plane. The available End Condition Types for Direction 1 and options are feature dependent: **Blind, Up To Vertex, Up To Surface, Offset from Surface, Mid Plane**.

  - **Angle**. Defines the angle value covered by the revolve type feature. The default angle is 360 degrees. The angle is measured clockwise from the selected Sketch plane.

  - **Reverse Direction**. Reverses the revolve feature direction if required.

  - **Reverse offset**. Reverses the offset of the feature direction.

  - **Offset Distance**. Distance of the offset.

- *Direction 2*. Creates the revolve feature in a clockwise and counter-clockwise direction from your Sketch plane. Set the Direction 1 Angle and the Direction 2 Angle. The total of the two angles can't exceed 360 degrees.

  - **Reverse Direction**. Reverses the revolve feature direction if required.

  - **Angle**. Defines the angle value covered by the revolve type feature. The default angle is 360 degrees. The angle is measured clockwise from the selected Sketch plane.

  - **Reverse Direction**. Reverses the revolve feature direction if required.

- **Reverse offset**. Reverses the offset of the feature direction.

- **Offset Distance**. Distance of the offset.

- *Thin Feature*. The Thin Feature controls the revolve thickness of your profile. This is not the depth. Create Thin Features by using an active sketch profile and then applying a wall thickness. Apply the wall thickness to the inside or outside of your sketch, evenly on both sides of the sketch or unevenly on either side of your sketch. Thin Feature creation is automatically invoked for active contours which are revolved or extruded. You can also create a Thin Feature from a closed contour. The Thin Feature box provides the following selections:

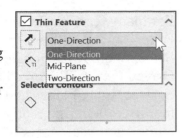

  - **Type**. Defines the direction of thickness. There are three selections:

  - **One-Direction**. Adds the thin-walled volume in one direction from your Sketch plane.

  - **Mid-Plane**. Adds the thin-walled volume using your sketch as the midpoint. This option applies the thin-walled volume equally on both sides of your sketch.

  - **Two-Direction**. Adds the thin-walled volume to both sides of your sketch. The Direction 1 Thickness adds thin-walled volume outward from the sketch. Direction 2 Thickness adds thin-walled volume inward from your sketch.

    - **Reverse Direction**. Reverses the direction of the thin-walled volume if required. View the thin-walled volume from the Graphics window.

    - **Direction 1 Thickness**. Scts thc thin-wallcd volumc thickness for the One Direction option and the Mid-Plane option for the Thin Feature option.

- *Selected Contours*. The Selected Contours box provides the following option:

  - **Select Contours**. Displays the selected contours. Select the contours from the Graphics window.

- *Feature Scope*. Apply features to one or more multi-body parts. The Feature Scope box provides the following selections:

  - **All bodies**. Applies the feature to all bodies every time the feature regenerates. If you add new bodies to the model that are intersected by the feature, these new bodies are also regenerated to include the feature.

  - **Selected bodies**. Selected by default. Applies the feature to the selected bodies.

  - **Auto-select**. Only available if you select the Selected bodies option. When you first create a model with multi-body parts, the feature automatically processes all the relevant intersecting parts. Auto-select is faster than All bodies because it processes only the bodies on the initial list and does not regenerate the entire model.

- **Solid Bodies to Affect**. Only available if Auto-select is not active. Displays the selected bodies to affect from the Graphics window.

For many features (Extruded Boss/Base, Extruded Cut, Simple Hole, *Revolved Boss/Base*, Revolved Cut, Fillet, Chamfer, Scale, Shell, Rib, Circular Pattern, Linear Pattern, Curve Driven Pattern, Revolved Surface, Extruded Surface, Fillet Surface, Edge Flange and Base Flange) you can enter and modify equations directly in the PropertyManager fields that allow numerical inputs.

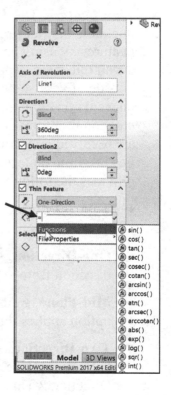

Create equations with Global Variables, functions, and file properties without accessing the Equations, Global Variables and Dimensions dialog box.

In the Revolved Boss/Base PropertyManager you can enter equations in:

- Depth fields for Direction 1 and Direction 2.

- Draft fields for Direction 1 and Direction 2.

- Thickness fields for a Thin Feature with two direction types.

- Offset Distance field.

Create an equation in a numeric input field. Start by entering an = (equal sign). A drop-down list displays options for Global Variables, functions and file properties. Numeric input fields that contain equations can display either the equation itself or its evaluated value.

Toggle between the equation and the value by clicking the Equations or Global Variable button that appears at the beginning of the field.

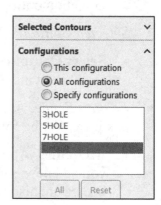

Address Configurations in the Configurations dialog box as illustrated.

Tutorial: Revolved Boss/Base 7-1

Create a Revolve Base feature and a Revolve Boss feature for a light bulb.

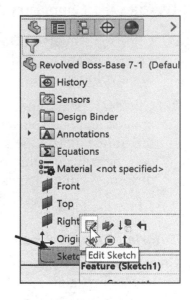

1. Open **Revolve Boss-Base 7-1** from the SOLIDWORKS 2018 folder. The Revolve Boss-Base 7-1 FeatureManager is displayed.

2. **Edit** ✏️ Sketch1. Display an **Isometric** view.

Insert a Revolved Boss/Base feature.

3. Click the **Revolved Boss/Base** 🐦 Features tool. The Revolve PropertyManager is displayed.

4. Click the **centerline** from the Graphics window. Line1 is displayed in the Axis of Revolution box. Accept the defaults.

5. Click **OK** ✔ from the Revolve PropertyManager. Revolve1 is created and is displayed in the FeatureManager.

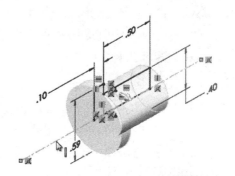

6. Display the **Temporary Axes**.

7. Display the **Right** view.

8. Click **Right Plane** from the FeatureManager.

9. Click the **Spline** 〜 Sketch tool. The Spline icon is displayed on the mouse pointer.

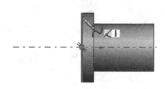

10. Click the **left vertical edge** of the Base feature for the start point.

11. **Drag** the mouse pointer to the left.

12. Click a **position** above the Temporary Axis. This is your control point.

13. Double-click the **Temporary Axis** to end the Spline. This is the end point.

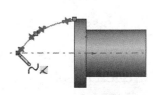

14. Click **OK** ✔ from the Spline PropertyManager.

15. Click the **Line** ✏ Sketch tool.

16. Sketch a **horizontal line** from the Spline endpoint to the left edge of the Base Revolved feature.

17. Sketch a **vertical line** to the start point of the Spline.

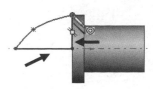

Deselect the active sketch tool.

18. Right-click **Select**.

19. Click the **Temporary Axis** in the Graphics window.

20. Click the **Revolved Boss/Base** 🍲 Features tool. The Revolve PropertyManager is displayed. Axis1 is displayed in the Axis of Revolution box.

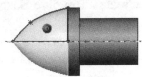

21. Click **OK** ✔ from the Revolve PropertyManager. Revolve2 is created and is displayed in the FeatureManager. The points of the Spline dictate the shape of the Spline. You can edit the control point in the sketch later to produce different results/shapes for the Revolved feature.

22. **Close** the model.

Tutorial: Revolved Boss/Base 7-2

Create a pulley wheel using the Revolved Boss/Base Feature tool.

1. Open **Revolved Boss-Base 7-2** from the SOLIDWORKS 2018 folder. The Revolve Boss-Base 7-2 FeatureManager is displayed.

2. **Edit** Sketch1.

Insert a Revolved Boss/Base feature.

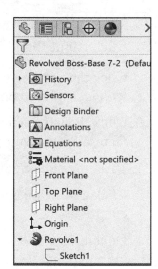

3. Click the **Revolved Boss/Base** 🍲 Feature tool. The Revolve PropertyManager is displayed.

4. Click the **horizontal centerline** from the Graphics window. Line13 is displayed in the Axis of Revolution box.

5. Enter **360**deg in the Angle box.

6. Click **OK** ✔ from the Revolve PropertyManager. Revolve1 is created and is displayed in the FeatureManager. View the results in the Graphics window.

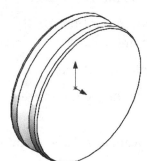

7. **Close** the model.

Tutorial: Revolved Boss/Base 7-3

Create a Revolve Thin Feature using the Revolve Boss/Base Feature tool.

1. Open **Revolve Boss-Base 7-3** from the SOLIDWORKS 2018 folder. The FeatureManager is displayed.

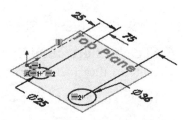

2. **Edit** 🖉 Sketch1.

Insert a Revolved Boss/Base feature.

3. Click the **Revolved Boss/Base** Features tool. The Revolve PropertyManager is displayed.

4. **Expand** the Thin Feature box.

5. Check the **Thin Feature** box. The Thin Feature box displays Type and Direction thickness.

6. Enter **5**mm for Direction 1 Thickness.

7. **Expand** the Selected Contours box.

8. Click the circumference of the **36mm circle**. Sketch1-Contour<1> is displayed in the Selected Contours box.

9. Click the circumference of the **25mm circle**. Sketch1-Contour<2> is displayed in the Selected Contours box.

10. Enter **180**deg in the Angle box.

11. Click **OK** ✔ from the Revolve PropertyManager. Revolve-Thin1 is created and is displayed in the FeatureManager.

12. **Close** the model. Note: You can click on the first contour and rotate using the on-screen ruler to modify the Angle.

Revolved Cut Feature

The Revolved Cut 🪣 feature removes material by rotating an open sketch profile around an axis. Sketch a centerline to create a diameter dimension for a revolved profile. The Temporary axis does not produce a diameter dimension.

The Revolved Cut 🪣 feature uses the Cut-Revolve PropertyManager. The Cut-Revolve PropertyManager provides the following selections:

•   *Axis of Revolution*. Displays the selected axis around which the feature revolves. The Axis of Revolution can be a line, centerline, or even an edge. Your selection is dependent on the type of revolve feature you create.

- ***Direction 1***. Creates the revolve feature in one direction from your Sketch plane. The available End Condition Types for Direction 1 and options are feature dependent: **Blind, Up To Vertex, Up To Surface, Offset From Surface, Mid Plane**.

  - **Angle**. Defines the angle value covered by the revolve type feature. The default angle is 360 degrees. The angle is measured clockwise from the selected Sketch plane.

  - **Reverse Direction**. Reverses the revolve feature direction if required.

  - **Reverse offset**. Reverses the offset of the feature direction.

  - **Offset Distance**. Distance of the offset.

- ***Direction 2***. Creates the revolve feature in a clockwise and counter-clockwise direction from your Sketch plane. Set the Direction 1 Angle and the Direction 2 Angle. The total of the two angles can't exceed 360 degrees.

  - **Reverse Direction**. Reverses the revolve feature direction if required.

  - **Angle**. Defines the angle value covered by the revolve type feature. The default angle is 360 degrees. The angle is measured clockwise from the selected Sketch plane.

  - **Reverse Direction**. Reverses the revolve feature direction if required.

  - **Reverse offset**. Reverses the offset of the feature direction.

  - **Offset Distance**. Distance of the offset.

- ***Thin Feature***. The Thin Feature controls the revolve thickness of your profile. This is not the depth. Create Thin Features by using an active sketch profile and then applying a wall thickness. The Thin Feature box provides the following selections:

  - **Type**. Defines the direction of thickness. There are three selections:

  - **One-Direction**. Adds the thin-walled volume in one direction from your Sketch plane.

  - **Mid-Plane**. Adds the thin-walled volume using your sketch as the midpoint. This option applies the thin-walled volume equally on both sides of your sketch.

- **Two-Direction**. Adds the thin-walled volume to both sides of your sketch. The Direction 1 Thickness adds thin-walled volume outward from the sketch. Direction 2 Thickness adds thin-walled volume inward from your sketch.

  - **Reverse Direction**. Reverses the direction of the thin-walled volume if required. View the thin-walled volume from the Graphics window.

  - **Direction 1 Thickness**. Sets the thin-walled volume thickness for the One-Direction option and the Mid-Plane option for the Thin Feature option.

- *Selected Contours*. The Selected Contours box provides the following option:

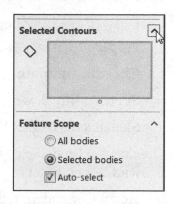

  - **Select Contours**. Displays the selected contours. Select the contours from the Graphics window.

- *Feature Scope*. Apply features to one or more multi-body parts. The Feature Scope box provides the following selections:

  - **All bodies**. Applies the feature to all bodies every time the feature regenerates. If you add new bodies to the model that are intersected by the feature, these new bodies are also regenerated to include the feature.

  - **Selected bodies**. Selected by default. Applies the feature to the selected bodies.

  - **Auto-select**. Only available if you select the Selected bodies option. When you first create a model with multi-body parts, the feature automatically processes all the relevant intersecting parts. Auto-select is faster than All bodies because it processes only the bodies on the initial list and does not regenerate the entire model.

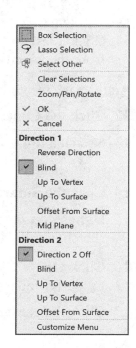

In Edit mode, right-click anywhere on the revolved feature to modify the end condition from a pop-up menu. The pop-up menu provides the document dependent options for Direction 1 and Direction 2.

Tutorial: Revolved Cut 7-1

Insert a Revolved Cut Thin feature in a bulb part.

1. Open **Revolved Cut 7-1** from the SOLIDWORKS 2018 folder.

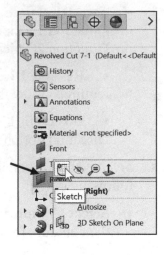

2. Right-click **Right Plane** from the FeatureManager. This is your Sketch plane.

3. Click **Sketch** from the Context toolbar.

4. Click the **Line** Sketch tool. The Insert Line PropertyManager is displayed.

5. Click the **midpoint** of the top silhouette edge.

6. Sketch a **line** downward and to the right.

7. Sketch a horizontal **line** to the right vertical edge.

Deselect the Line Sketch tool.

8. Right-click **Select**.

9. Add a Coincident relation between the **end point of the line** and the **right vertical edge**.

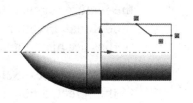

10. Click **OK** from the Properties PropertyManager.

Deactivate the Temporary Axes.

11. Click **View** ➤ **Hide/Show** ➤ uncheck **Temporary Axes** from the Menu bar.

Activate the Centerline Sketch tool.

12. Click the **Centerline** Sketch tool.

13. Sketch a **horizontal** centerline through the origin.

Insert dimensions.

14. Click the **Smart Dimension** Sketch tool.

15. **Dimension** the model as illustrated. The sketch is fully defined.

16. Click **OK** from the Dimension PropertyManager.

17. Click the **centerline** in the Graphics window.

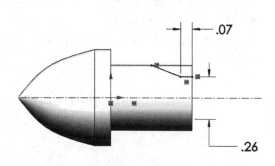

18. Click the **Revolved Cut**  Features tool. The Cut-Revolve PropertyManager is displayed.

19. Click **No** to the Warning message in the Graphics window. The Cut-Revolve PropertyManager is displayed.

20. Click the **Thin Feature** check box. Enter **.150**in for Direction 1 Thickness. Click **Reverse Direction** from the Thin Feature box.

21. Click **Reverse Direction** from the Direction 1 Parameters box.

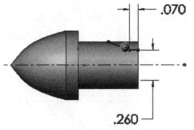

22. Click **OK** ✔ from the Cut-Revolve PropertyManager. View the results. Cut-Revolve-Thin1 is displayed in the Graphics window.

23. **Close** the model.

Tutorial: Revolved Cut 7-2

Insert a Revolved Cut and Mirror feature on the Right plane.

1. Open **Revolved Cut 7-2** from the SOLIDWORKS 2018 folder.

2. Click **Sketch2** from the FeatureManager.

3. Click the **Revolved Cut**  Features tool. The Cut-Revolve PropertyManager is displayed. Accept the defaults.

4. Click **OK** ✔ from the Cut-Revolve PropertyManager. The Cut-Revolve1 feature is displayed in the FeatureManager.

5. Click **Right Plane** from the FeatureManager. This is your Sketch plane.

6. Click the **Mirror** Features tool. The Mirror PropertyManager is displayed.

7. Click **Cut-Revolve1** from the fly-out FeatureManager. Cut-Revolve1 is displayed in the Features to Mirror box.

8. Click **OK** ✔ from the Mirror PropertyManager. The Mirror1 feature is displayed.

9. **Close** the model.

Revolved Boss Thin Feature

The Revolved Boss feature  adds material by rotating an open profile around an axis. A Revolved Boss Thin feature requires a Sketch plane, a Sketch profile, Axis of Revolution, Angle of Rotation and thickness.

The Revolved Boss Thin feature uses the Revolve PropertyManager.

Tutorial: Revolve Boss Thin 7-1

Insert a Revolve Boss Thin feature. Use the Temporary Axis.

1. Open **Revolve Boss Thin 7-1** from the SOLIDWORKS 2018 folder.

2. **Edit** Sketch5.

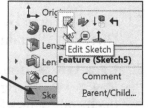

💡 In an active sketch, click **Tools ➢ Sketch Tools ➢ Check Sketch for Feature Usage** to determine if a sketch is valid for a specific feature and to understand what is wrong with a sketch.

3. Click the **Revolved Boss/Base** 🍥 Features tool. The Revolve PropertyManager is displayed.

4. Select **Mid-Plane** from the Thin Feature Type box.

5. Enter **.050**in for Direction1 Thickness.

6. **Display** the Temporary Axis.

7. Click inside the **Temporary Axis** box.

8. Click the **Temporary Axis** for the Graphics window as illustrated. Axis<1> is displayed in the Axis of Revolution box.

9. Click **OK** ✅ from the Revolve PropertyManager. Revolve-Thin1 is created and is displayed in the FeatureManager.

10. **Close** the model.

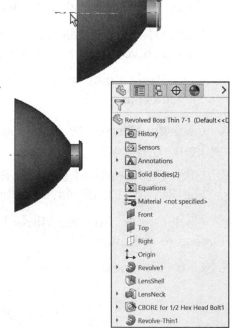

R.100

A Revolved sketch that remains open results in a Thin-Revolve feature ———⌐ . A Revolved sketch that is automatically closed results in a line drawn from the start point to the end point of the sketch. The sketch is closed and results in a non-thin Revolve feature ———⌐ .

Revolved Surface Feature

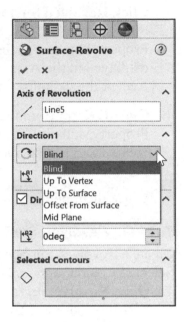

The Revolved Surface feature 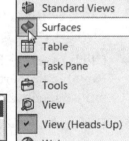 adds material by rotating a profile about an axis. The Revolved Surface feature uses the Surface-Revolve PropertyManager. The Surface-Revolve PropertyManager provides the same selections as the Revolve PropertyManager. View the section on Revolved Boss/Base for detail PropertyManager information. Revolve Types:

- Blind.

- Up To Vertex.

- Up To Surface.

- Offset From Surface.

- Mid Plane.

Click **Views** ➤ **Toolbars** ➤ **Surfaces** to display the Surfaces toolbar. View the illustrated Surfaces toolbar.

Tutorial: Revolved Surface 7-1

Create a Revolved Surface feature.

1. Open **Revolve Surface 7-1** from the SOLIDWORKS 2018 folder.

2. **Edit** Sketch1.

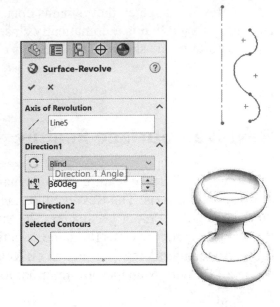

3. Click **Revolved Surface** from the Surfaces toolbar. The Surface-Revolve PropertyManager is displayed. Line5 is displayed in the Axis of Revolution box. Accept the defaults.

4. Click **OK** ✔ from the Surface-Revolve PropertyManager. Surface-Revolve1 is created and is displayed in the FeatureManager.

5. **Close** the model.

Tutorial: Revolved Surface 7-2

Create a Revolved Surface feature using the Selected Contour option.

1. Open **Revolve Surface 7-2** from the SOLIDWORKS 2018 folder.

2. Right-click **Sketch1**.

3. Click **Edit Sketch** from the Context toolbar. Sketch1 is the active sketch.

4. Click **Insert ➤ Surface ➤ Revolve**. The Surface-Revolve PropertyManager is displayed. Note the mouse feedback icon.

5. **Expand** the Selected Contours box.

6. Click inside the **Selected Contours** box.

7. Click the **C-shape** contour in the Graphics window as illustrated. Sketch1-Region<1> is displayed in the Selected Contours box.

8. Click the **Reverse Direction** button. The arrow points upwards as illustrated.

9. Enter **360**deg for Angle.

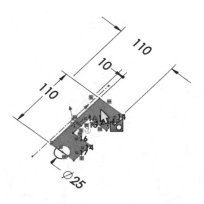

10. Click **inside** the Axis of Revolution box.

11. Select the **centerline** from the Graphics window. Line10 is displayed in the Axis of Revolution box.

12. Click **OK** ✔ from the Surface-Revolve PropertyManager. Surface-Revolve1 is created and is displayed in the FeatureManager.

13. **Close** the model.

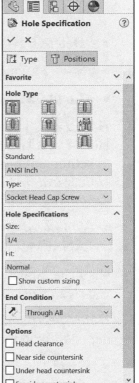

Hole Wizard Feature

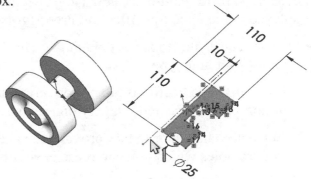

The Hole Wizard 🕮 feature provides the ability to create cuts in the form of holes for fasteners. This provides the ability to create holes on a plane as well as holes on planar and non-planar faces for various fasteners.

The hole type you select determines the capabilities, available selections, and graphic previews. After you select a hole type, determine the appropriate fastener. The fastener dynamically updates the appropriate parameters.

The Hole Wizard 🕮 feature creates either a 2D or 3D sketch for the placement of the hole in the FeatureManager. You can consecutively place multiple holes of the same type. The Hole Wizard creates 2D sketches for holes unless you select a non-planar face or click the 3D Sketch button in the Hole Position PropertyManager.

Hole Wizard creates two sketches. The first sketch is the cut profile of the selected hole type and the second sketch - center placement of the profile. Both sketches should be fully defined.

The placement sketch is listed first under the Hole Wizard feature. The placement sketch contains one or more sketch points marking the center of the hole. It may also contain construction geometry with geometry relations and dimensions to parametrically locate the hole centers.

The hole sketch or the revolved cut profile sketch is not in an identifiable sketch plane. You can modify (Edit Sketch) the hole sketch dimensions outside of the hole wizard.

 You can import and export data from Microsoft Excel files for Hole Wizard holes. This is helpful when you edit an entire standard or create several new sizes of holes.

The Hole Wizard feature uses the Hole Specification PropertyManager. Two tabs are displayed in the Hole Specification PropertyManager:

- *Type*. The Type tab is the default tab. The Type box provides the ability to select various set hole types.

- *Positions*. The Positions tab options provides the ability to locate the hole on a planar or non-planar face and to apply dimensions. You can switch between these two tabs.

- *Favorite*. The Favorite box provides the ability to manage a list of favorite Hole Wizard holes which you can reuse in your models. The available selections are:

    - **Apply Defaults/No Favorites**. Resets to No Favorite selected and the system default settings.

    - **Add or Update Favorite**. Adds the selected Hole Wizard hole to your Favorites list.

    - **Delete Favorite**. Deletes the selected favorite for your Favorites list.

    - **Save Favorite**. Saves the selected favorite to your Favorites list.

    - **Load Favorite**. Loads the selected favorite.

- *Hole Type*. The Hole Specification options vary depending on the Hole Type. Use the PropertyManager images and descriptive text to set the options. Your options are *Counterbore, Countersunk, Hole, Straight Tap, Tapered Tap, Legacy Hole, Counterbore Slot, Countersunk Slot or Slot*.

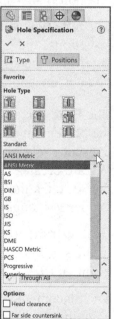

    - *Standard*. Specifies the hole standard. For example, select ANSI Metric or JIS.

    - *Type*. Specifies drill sizes, tap drills, dowel holes, or screw clearance.

- *Hole Specifications*. The Hole Specification selections vary depending on the selected Hole Type. The available options are:

    - **Size**. The Size drop-down menu provides the ability to select the size for the selected fastener. The Size options are dependent on Hole type.

    - **Fit**. Only available for the Counterbore and Countersink Hole type. The Fit drop down-menu provides the following options: **Close**, **Normal** and **Loose**.

    - **Show custom sizing**. Not selected by default. The Custom Sizing box selection varies depending on your selected hole type. Use the PropertyManager images and descriptive text to set options such as **diameter**, **depth** and **angle at bottom**.

- *End Condition*. The End Condition box provides the following feature dependent options:

    - **Blind**. Extends the feature from your Sketch plane for a specified distance.

- **Through All**. Extends the feature from your Sketch plane through all existing geometry.

- **Up To Next**. Extends the feature from your Sketch plane to the next surface that intercepts the entire profile. The intercepting surface must be on the same part.

- **Up To Vertex**. Extends the feature from your Sketch plane to a plane which is parallel to your Sketch plane and passes through a specified vertex.

- **Up To Surface**. Extends the feature from your Sketch plane to the selected surface.

- **Offset From Surface**. Extends the feature from your Sketch plane to a specified distance from the selected surface. Select a face to specify the surface. Specify an offset distance.

- **Reverse Direction**. Reverses the End Condition direction if required.

- **Blind Hole Depth**. Only available for the Blind End Condition. Sets the hole depth. For tap holes, set the thread depth and type. For pipe tap holes, set the thread depth.

- **Vertex**. Only available for the Up to Vertex Condition. Select a vertex.

- **Face/Surface/Plane**. Only available for the Up to Surface and Offset From Surface End Condition. Displays the selected face, surface, or plane from the Graphics window.

- **Offset Distance**. Only available for the Offset From Surface End Condition. Displays the offset distance from the selected face, surface, or plane in the Graphics window.

- *Options*. Options vary depending on the selected Hole type. The option selections are *Head clearance, Near side countersink, Under head countersink, Far side countersink, Cosmetic thread and Thread class*. If you select any of the choices from the Options section, the revolved sketch profile is altered to accommodate the modification.

- *Feature Scope*. The Feature Scope applies Hole Wizard holes to one or more multi-body parts. Select which bodies should include the holes.

  - **All bodies**. Applies the feature to all bodies each time the feature is regenerated. If you add a new body or bodies to your model that are intersected by a feature, the new body will regenerate to include the feature.

  - **Selected bodies**. Displays the selected bodies. If you added new bodies to the model that are intersected by the feature, use the Edit Feature function to edit the extrude feature. Displays the selected bodies.

- **Auto-select**. Only available if you select the Selected bodies option. When you first create a model with multi-body parts, the feature automatically processes all the relevant intersecting parts.

- **Solid Bodies to Affect**. Only available if you clear the Auto-select option. Displays the selected solid bodies to affect from the Graphics window.

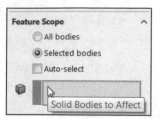

SOLIDWORKS Toolbox is integrated with the Hole Wizard feature. You can import and export data from Microsoft Excel files for Hole Wizard holes. This is helpful when you edit an entire standard or create several new sizes of holes. See SOLIDWORKS Help for additional information.

Tutorial: Hole Wizard 7-1

Apply the Hole Wizard tool to create a custom Counterbore hole.

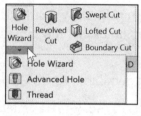

1. Add-in the **SOLIDWORKS Toolbox Library** and **SOLIDWORKS Toolbox Utilities**.

2. Open **Hole Wizard 7-1** from the SOLIDWORKS 2018 folder. The FeatureManager is displayed.

3. Click the **Hole Wizard** 🕳 Features tool. The Hole Specification PropertyManager is displayed.

4. Select **Counterbore** for Hole Type. Select **ANSI Inch** for Standard. Select **Hex Bolt** for fastener Type.

5. Select **1/2** for Size. Select **Normal** for Fit.

6. Check the **show custom sizing** box. Enter **.600**in in the Counter bore Diameter box.

7. Enter .**200**in in the Counter bore Depth box.

8. Select **Through All** for End Condition.

9. Click the **Positions tab**. The Hole Position PropertyManager is displayed.

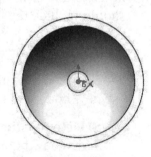

10. Click the small **inside circular** back face of the LensShell to create the hole placement (do not click the origin).

11. Click the **origin** to locate the placement of the hole and to fully define the sketch.

Deselect the point ▱ tool.

12. Right-click **Select**. Click **OK** ✔ from the Properties PropertyManager.

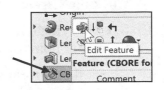

Add a new hole type to your Favorites list.

13. **Edit** the CBORE feature.

14. Select the **Add or Update Favorite** button from the Favorites box. Enter **CBORE for bulb**.

15. Click **OK**. Click **OK** ✔ from the Hole Specification PropertyManager.

16. **Expand** CBORE for 1/2 Hex Head Bolt1 from the FeatureManager. Sketch3 and Sketch4 create the CBORE feature. View the results.

17. **Close** the model.

Tutorial: Hole Wizard 7-2

Create a Countersunk hole located on a bolt circle using the 2D Sketch method.

1. Add-in the **SOLIDWORKS Toolbox**.

2. Open **Hole Wizard 7-2** from the SOLIDWORKS 2018 folder. Click the **Hole Wizard** 🖰 Features tool. The Hole Specification PropertyManager is displayed.

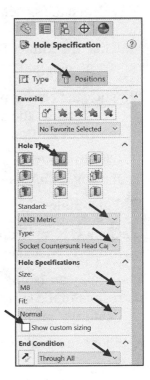

3. Select the **Countersink** Hole Type.

4. Select **ANSI Metric** for Standard.

5. Select **Socket Countersunk Head** for fastener Type.

6. Select **M8** for Size.

7. Select **Normal** for Fit.

8. Uncheck the **Show custom sizing** box.

9. Select **Through All** for End Condition.

10. Click the **Positions** Tab. The Hole Position PropertyManager is displayed.

11. Display a **Right view**.

12. Click a **position** on the part as illustrated. The Point sketch tool is activated.

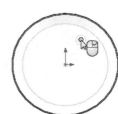

13. Click **again** to place the hole.

14. Click the **Circle** ⊙ Sketch tool.

15. Sketch a **circle** from the center point of the Revolve1 feature to the center point of the Countersink as illustrated.

16. Check the **For construction** box from the Circle PropertyManager.

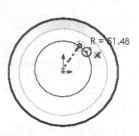

17. Click the **Centerline** ✎ Sketch tool.

18. Sketch a **horizontal centerline** from the center point of the Revolve1 feature to the quadrant point of the construction circle as illustrated.

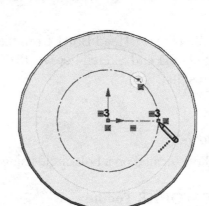

Fully define the sketch for the location of the hole.

19. Click the **Smart Dimension** ↖ Sketch tool.

20. Click the **center point** of the Countersink, the **origin** and the **right end point** of the centerline.

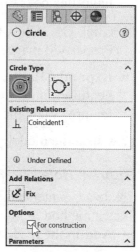

21. Click a **position** as illustrated to position the angle dimension.

22. Enter **30**deg.

23. Click **OK** ✔ from the Hole Position Property Manager.

24. Display an **Isometric view**. View the results in the Graphics window and FeatureManager.

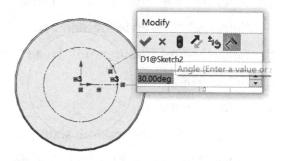

25. **Close** the model.

The main advantage of the 3D Sketch placement method is that it can place a set of holes on any set of solid faces, regardless of whether they are at different levels, or are non-parallel. A limitation, 3D Sketches can be fairly cumbersome and difficult. See section on 3D Sketching for additional details.

Note: With a 3D sketch, press the Tab key to move between planes.

Tutorial: Hole Wizard 7-3

Create a Counterbore using the 3D Sketch placement method.

1. Add-in the **SOLIDWORKS Toolbox**.

2. Open **Hole Wizard 7-3** from the SOLIDWORKS 2018 folder.

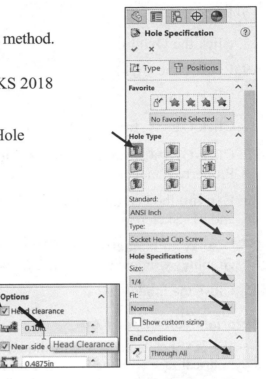

3. Click the **Hole Wizard** 🖉 Features tool. The Hole Specification PropertyManager is displayed.

4. Select the **Counterbore** Hole Type.

5. Select **ANSI Inch** for Standard.

6. Select **Socket Head Cap Screw** for fastener Type.

7. Select **1/4** for Size.

8. Select **Normal** for Fit.

9. Select **Through All** for End Condition.

10. Enter **.100** for Head clearance in the Options box.

11. Click the **Positions** Tab. The Hole Position PropertyManager is displayed.

12. Click the **3D Sketch** button. SOLIDWORKS displays a 3D interface

    with the Point ✏ XY ⤨ tool active. Note: When the Point tool is active, wherever you click you will create a point.

13. Click the **top face** of the model as illustrated. The selected face is displayed in orange/blue. This indicates that an OnSurface sketch relations will be created between the sketch point and the top face. The hole is displayed in the model.

14. Display a **Top view**.

15. Right-click **Select** to deselect the Point tool.

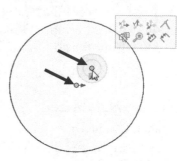

Add a Geometric Relation.

16. Click the **center point** of the hole.

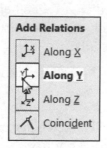

17. Hold the **Ctrl** key down.

18. Click the **Origin**.

19. Release the **Ctrl** key.

20. Click **Along Y**.

21. Click **OK** ✔ from the Properties PropertyManager.

22. Click **OK** ✔ from the Hole Position PropertyManager.

23. **Expand** the FeatureManager. View the created hole and 3D Sketch.

24. Display an **Isometric view**. View the results in the Graphics window and FeatureManager.

25. **Close** the model.

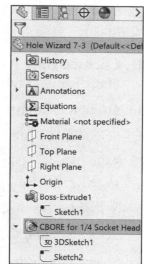

Tutorial: Hole Wizard 7-4

Use the 3D Sketch placement method to apply dimensions on a cylindrical face with the Hole Wizard feature.

1. Open **Hole Wizard 7-4** from the SOLIDWORKS 2018 folder.

2. Click the **Hole Wizard** 🔧 feature tool.

Note: With a 3D sketch, press the Tab key to move between planes.

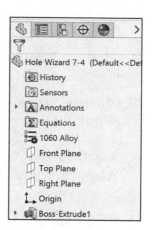

3. Create a **Counterbore, ANSI Inch, Socket Head Cap Screw fastener Type**. Size - ¼. Fit - **Normal**.

4. End Condition - **Through All** with a **.100in** Head clearance.

5. Click the **Positions** Tab.

6. Click the **3D Sketch** button. The Hole Position PropertyManager is displayed. SOLIDWORKS displays an active 3D interface with the Point ✎ XY tool.

When the Point tool is active, wherever you click, you will create a point.

Dimension the sketch.

7. Click the **cylindrical face** of the model as illustrated. The selected face is displayed in blue. This indicates that an OnSurface sketch relations will be created between the sketch point and the cylindrical face. The hole is displayed in the model.

8. Insert a **.25in** dimension between the top face and the sketch point.

Locate the point angularly around the cylinder. Apply construction geometry.

9. Display the **Temporary Axes**.

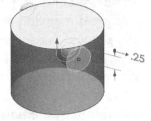

10. Click the **Line** ⟋ Sketch tool.
    Note: 3D Sketch is still activated.

11. **Ctrl + Click** the top flat face of the model. This moves the red space handle origin to the selected face. This also constrains any new sketch entities to the top flat face. Note the mouse pointer icon.

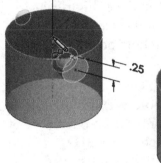

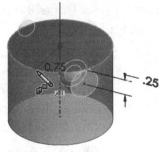

12. Move the **mouse pointer** near the center of the activated top flat face as illustrated. View the small black circle. The circle indicates that the end point of the line will pick up a Coincident relation.

13. Click the **center point** of the circle.

14. Sketch a line so it picks up the **Along Z sketch relation**. The cursor displays the relation to be applied. **This is a very important step**. If needed, add an Along Z relation.

Add Relations	
⤴	**Along X**
⤸	Along Y
⤵	Along Z
人	Coincident
✦	Fix

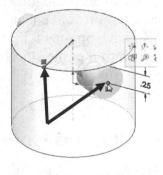

Deselect the Line Sketch tool.

15. Right-click **Select**.

16. Create an **Along Y** sketch relation between the center point of the hole on the cylindrical face and the endpoint of the sketched line. The sketch is fully defined.

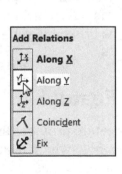

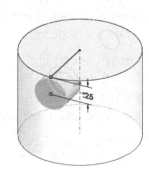

17. Click **OK** from the PropertyManager.

18. Click **OK** to return to the Part FeatureManager.

19. **Expand** the FeatureManager and view the results. The two sketches are fully defined. One sketch is the hole profile; the other sketch is to define the position of the feature.

20. **Close** the model.

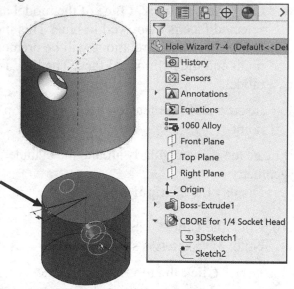

Create a second sketched line and insert an angle dimension between the two lines. This process is used to control the position of the centerpoint of the hole on the cylindrical face as illustrated. Insert an Along Y sketch relation between the centerpoint of the hole on the cylindrical face and the end point of the second control line. Control the hole position with the angular dimension.

🔅 When creating a Hole Wizard hole, you no longer have to preselect a planar face to create a 2D sketch. For a 3D sketch, request a 3D sketch as illustrated.

Slots have been added to the Hole Wizard. You can create regular slots as well and counterbore and countersink slots. You also have options for position and orientation of the slot. If you have hardware already mated in place, the mates will not be broken if you switch from a hole to a slot.

Advanced Hole Feature

The Advanced Hole 🔲 Feature provides the ability to create a complex hole which requires two or more manufacturing processes.

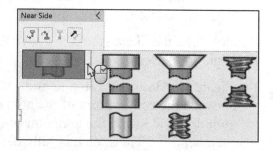

The Advanced Hole feature uses a similar Hole Wizard PropertyManager, but with Hole element fly-outs to guide the user through the process of Near Side and Far Side Elements.

Define and modify complex holes from the near and far side faces. The default hole type is Counterbore.

Tutorial: Advanced Hole 7-1

Apply the Advanced Hole tool to create and modify a Counterbore hole with three elements positioned at the origin.

1. Open **Advanced Hole 7-1** from the SOLIDWORKS 2018 folder.

2. Click the **Advanced Hole** 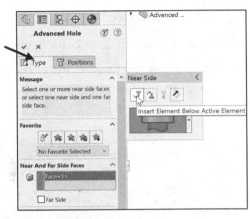 feature tool. The Advanced Hole PropertyManager is displayed. View your options. The default hole type is Counterbore.

3. Click the **Type** tab.

4. Click inside the **Near And Far Side Faces** dialog box.

5. Click the **top face** of the model in the Graphic window. Face<1> is displayed in the dialog box. Add a Near Side element with Element specification.

6. Click the **Insert Element Below Active Element** 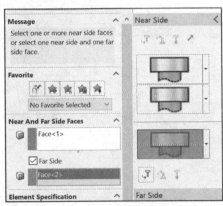 icon.

7. Select the Far side for the Advanced hole. Click the **Far Side** box.

8. Click the **bottom face** of the model. Face<2> is displayed in the dialog box.

9. Set the first Element Specification. Click the **Top Element Specification** for Near Side as illustrated.

10. Select **ANSI Inch** for Standard.

11. Select **Socket Head Cap Screw** for Element Type.

12. Select ¼ for Size.

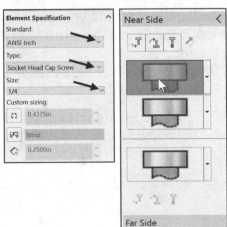

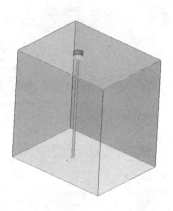

Create and modify the second Near Side Element.

13. **Expand** the second Near Side Element drop-down menu as illustrated.

14. Click the **Straight** icon as illustrated.

Set the second Element Specification.

15. Select **ANSI Inch** for Standard.

16. Select **Screw Clearances** for Type.

17. Select ¼ for Size.

18. Select **Blind** for End Condition.

19. Enter **1.0in** for Depth. View the updated model in the Graphics area.

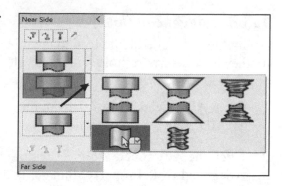

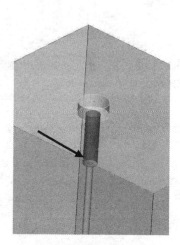

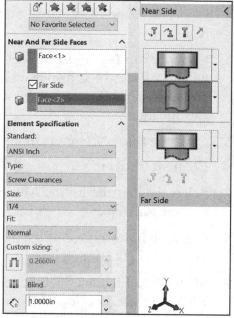

The Far Side element is currently set to Counterbore. Modify the Counterbore to a Straight tapped thread element.

20. **Expand** the Far Side Element drop-down menu as illustrated.

21. Click the **Straight Tapped Thread** icon as illustrated.

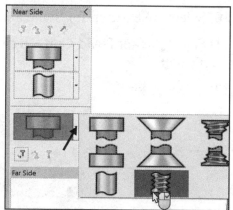

Set the Far Side Element Specification.

22. Select **ANSI Inch** for Standard.

23. Select **Tapped hole** for Type.

24. Select **¼-20** for Size.

25. Select **Up To Next Element** for End Condition.
The Far Side Element updates to the middle Near
Side Element due to the selected End Condition.

Position the hole in the center of the block.

26. Click the **Positions** tab.

27. Click the **top face** at the origin.

28. Click **OK** from the Hole Position
PropertyManager.

29. **View** the results in the Graphics area.

30. **Close** the model.

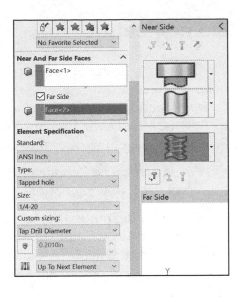

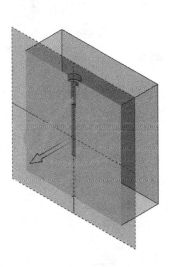

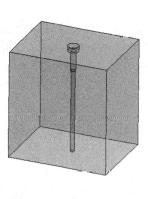

Dome Feature

The Dome 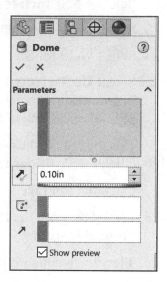 Feature provides the ability to create one or more dome features concurrently on the same model. A Dome feature creates spherical or elliptical shaped geometry. The Dome feature uses the Dome PropertyManager. The Dome PropertyManager provides the following selections:

- *Parameters*. The Parameters box provides the ability to select the following options:

  - **Faces to Dome**. Displays the selected planar or non-planar faces. Apply domes to faces whose centroid lies outside the face. This allows you to apply domes to irregularly shaped contours.

  - **Distance**. Set a value for the distance in which the dome expands. The Distance option is disabled when you use a sketch as a constraint.

Set the Distance option to 0 on your cylindrical and conical models. SOLIDWORKS calculates the distance using the radius of the arc as a basis for the dome. SOLIDWORKS creates a dome which is tangent to the adjacent cylindrical or conical face.

  - **Reverse Direction**. Reverses the direction of the Dome feature if required. Provides the ability to create a concave dome. The default is convex.

  - **Constraint Point or Sketch**. Provides the ability to control the dome feature. Select a sketch to constrain the shape of the sketch.

  - **Direction**. Provides the ability to choose a direction vector from the Graphics window. The direction is to extrude the dome other than normal to the face. You can choose a linear edge or the vector created by two sketch points as your direction vector.

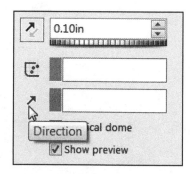

  - **Elliptical dome**. Specifies an elliptical dome for a cylindrical or conical model. An elliptical dome's shape is a half an ellipsoid with a height that is equal to one of the ellipsoid radii.

☀ The Elliptical dome option is not available when you use the Constraint Point or Sketch option or when you use the Direction option.

- **Continuous dome**. Specifies a continuous dome for a polygonal model. A continuous dome's shape slopes upwards, evenly on all sides. If you clear Continuous dome, the shape raises normal to the edges of the polygon.

☀ The Continuous dome option is not available when you use a Constraint Point or Sketch or a Direction vector.

- **Show preview**. Displays the total preview of the model.

Tutorial: Dome 7-1

Create an Elliptical dome using the Dome feature.

1. Open **Dome 7-1** from the SOLIDWORKS 2018 folder.

2. Click the **back face** of Revolve1.

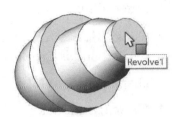

Activate the Dome feature.

3. Click **Insert ➤ Features ➤ Dome** from the Main menu. The Dome PropertyManager is displayed. Face<1> is displayed in the Faces to Dome box.

4. Click the **Elliptical dome** box.

5. Enter **.100**in in the Distance box.

6. Click **OK** ✔ from the Dome PropertyManager. Dome1 is created and is displayed in the FeatureManager.

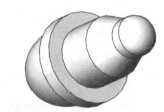

7. **Close** the model.

Tutorial: Dome 7-2

Create a dome on a non-planar face using the Dome Feature to create irregular shaped contours.

1. Open **Dome 7-2** from the SOLIDWORKS 2018 folder. The Dome 7-2 FeatureManager is displayed.

2. Click the **top curved face** of Extrude1.

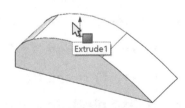

3. Click **Insert ➤ Features ➤ Dome** from the Main menu. The Dome PropertyManager is displayed. Face<1> is displayed in the Faces to Dome box.

4. Enter **50**mm for Distance.

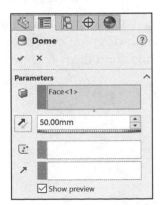

5. Click **inside** of the Direction box.

6. Select the **front left vertical edge**. Edge<1> is displayed. The Dome direction is downward.

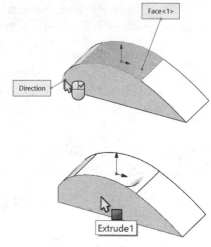

7. Click **OK** ✔ from the Dome Feature Manager. Dome1 is created.

8. Click the **front face** of Extrude1.

9. Click the **Dome** ⬗ Features tool. The Dome PropertyManager is displayed.

10. Click the **back face**. Face<1> and Face<2> are displayed in the Faces to Dome box.

11. Enter **40**mm for Distance.

12. Click **OK** ✔ from the Dome Feature Manager. Dome2 is created and is displayed.

13. Display an **Isometric** view. View the results.

14. **Close** the model.

Curve Overview

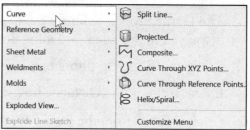

Create several types of 3D curves by using the following methods:

- **Split Line**. From a sketch projected to planar or curved (silhouette) faces.

- **Projected**. From a sketch projected to a model face or surface, or from sketched lines on intersecting planes.

- **Composite**. From a combination of curves, sketch geometry, and model edges.

- **Curve Through XYZ Points**. From a list of X,Y, Z coordinates for points.

- **Curve Through References Points**. From user-defined points or existing vertices.

- **Helix/Spiral**. Specifying a circular sketch, pitch, number of revolutions, and height.

Use the curves to create solid model features. For example, you can use a curve as the path or guide curve for a sweep feature, as the guide curve for a loft feature, as a parting line for a draft feature, and so on. You can also turn the display of curves on or off.

Project a sketched curve onto a model face to create a 3D curve. You can also create a 3D curve that represents the intersection of two extruded surfaces generated by creating sketches on two intersecting planes.

Split Line Curve tool

The Split Line ⬚ Curve tool projects an entity (sketch, solid, surface, face, plane, or surface spline) to surfaces, or curved or planar faces. It divides a selected face into multiple separate faces. You can split curves on multiple bodies with a single command. You can create the following split lines:

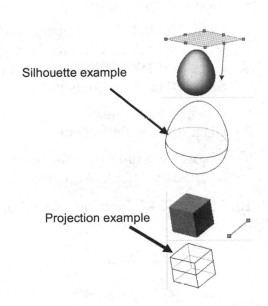

Silhouette example

- **Silhouette**. Creates a split line on a cylindrical part.

- **Projection**. Project a sketch on a surface. You can pattern split line features that were created using projected curves. You can create split lines using sketched text. This is useful for creating items such as decals or applying a force or pressure when using SOLIDWORKS Simulation.

Projection example

- **Intersection**. Splits faces with an intersecting solid, surface, face, plane, or surface spline.

☀ When you create a split line with an open profile sketch, the sketch must span at least two edges of the model.

Tutorial: Split Line 7-1

Create a Split Line feature to apply a distributed load to a selected face. The Split Line feature in this case does not add depth to the part.

1. Open **Split Line 7-1** from the SOLIDWORKS 2018 folder. The Split Line 7-1 FeatureManager is displayed.

2. Create a **rectangular sketch** as illustrated. Use the sketch for the Split Line feature.

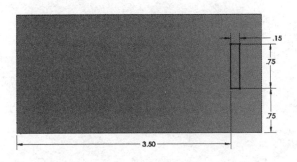

.15
.75
.75
3.50

Create the Split Line feature.

3. Click **Insert ➤ Curve ➤ Curve Split Line**. The Split Line PropertyManager is displayed. The current Sketch is displayed. Projection is selected by default.

4. Click the **top face** of Boss-Extrude1. Face 1 is displayed in the Faces to Split box.

5. Click **OK** ✔ from the Split Line PropertyManager. Split Line1 is created and is displayed.

6. **Save** the part.

Start a SOLIDWORKS Simulation study and utilize the Split Line feature as your selected face for the applied force on the model.

7. Click **SOLIDWORKS Add-Ins tab ➤ SOLIDWORKS Simulation**. The Simulation tab is displayed.

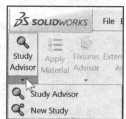

Create a new Static Study.

8. Click the **Simulation** tab.

9. Click **New Study** from the Study drop-down menu. Static is selected by default. Accept the default study name.

10. Click **OK** ✔ from the Study PropertyManager.

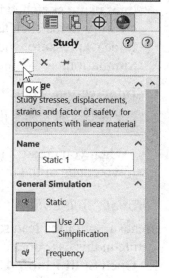

Fix the model at one end and apply a
distributed load using the Split Line
feature.

11. Right-click the **Fixtures** folder.

12. Click **Fixed Geometry**. The Fixture
PropertyManger is displayed.

13. Click the **left face of Boss-Extrude1**
as illustrated. The left face is fixed.

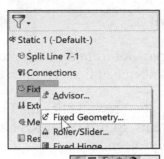

14. Click **OK** ✔ from the Fixture
PropertyManager. Fixed-1 is displayed.

Apply an External load (force) to the model using the
Split Line feature.

15. Right-click the **External Load** folder.

16. Click **Force**. The Force PropertyManager is
displayed. In this example, the area of the
applied force is the Split Line feature.

17. Click the **Split Line1** feature on the model in
the Graphics window.

18. Enter **100lbf** for force.

19. Click **OK** ✔ from the Force
PropertyManager. Force-1 is created and is
displayed.

20. Apply **Alloy Steel** to the part.

21. **Mesh and Run** the model.

22. **View** the results.

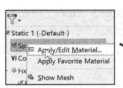

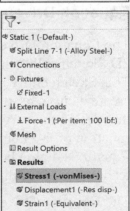

23. **Close** the part. The Split Line feature divides
a selected face into multiple separate faces.

💡 View the Split Line folder for additional
information. When you create a split line with an
open profile sketch, the sketch must span at least
two edges of the model.

Projected Curve tool

The Projected  Curve tool from a sketch projects to a model face or surface or from sketched lines on intersecting planes.

You can preselect items before you click the Projected Curve tool. If you preselect items, SOLIDWORKS attempts to select the appropriate type of projection.

- If you preselect two sketches, the **Sketch on sketch** option is activated with the two sketches shown under **Sketches to Project**.

- If you preselect a sketch and one or more faces, the **Sketch on faces** option is activated with the selected items shown in the correct boxes.

- If you preselect one or more faces, the **Sketch on faces** option is activated.

Composite Curve tool

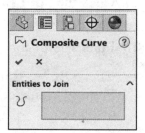

The Composite Curve tool provides the ability to create composite curves by combining curves, sketch geometry, and model edges into a single curve. Use a composite curve as a guide curve when creating a loft or a sweep. See chapter 18: Intelligent Modeling techniques for additional information.

Equation Driven Curve tool

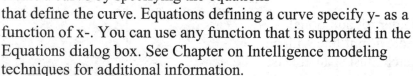

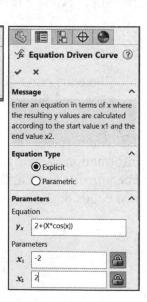

The Equation Driven Curve tool is a 2D Sketch spline driven by an equation. This Sketch entity provides the ability to create a curve by specifying the equations that define the curve. Equations defining a curve specify y- as a function of x-. You can use any function that is supported in the Equations dialog box. See Chapter on Intelligence modeling techniques for additional information.

The Equation Driven Curve entity uses the Equation Driven Curve PropertyManager. The PropertyManager provides the following selections:

- **Equation Type**. There are equation types available: *Explicit* and *Parametric*.

- ***Parameters***. The Parameters box provides the following options:

  - **Equation**. Provides the ability to enter an equation as a function of X-. If you enter an equation that can't be resolved, the text color is displayed in red.

  - **Parameters**. Provides the ability to enter the start and end X value of the equation.

  - **Transform**. Provides the ability to enter the translation of X-, Y- and angle for the curve.

Curve Through XYZ Points

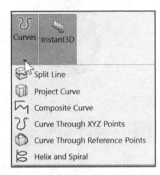

The Curve Through XYZ Points ⛎ feature provides the ability to either type in (using the Curve File dialog box) or click Browse and import a text file using x-, y-, z- coordinates for points on a curve.

A text file can be generated by any program which creates columns of numbers. The Curve ⛎ feature reacts like a default spline that is fully defined.

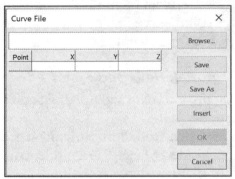

Create a curve using the Curve Through XYZ Points tool. Import the x-, y-, z- data. Verify that the first and last points in the curve file are the same for a closed profile.

It is highly recommend that you insert a few extra points on either end of the curve to set end conditions or tangency in the Curve File dialog box.

Imported files can have an extension of either *.sldcrv or *.text. The imported data, x-, y-, z-, must be separated by a space, a comma, or a tab.

⚡ See Chapter on Intelligence Modeling techniques for additional information.

Curve Through Reference Points

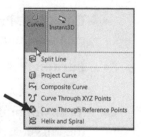

The Curve Through Reference Points 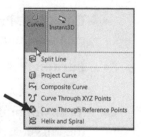 tool provides the ability to create a curve through points located on one or more planes. See SOLIDWORKS Help for additional information.

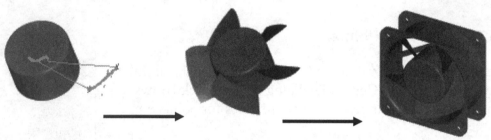

Helix and Spiral

The Helix and Spiral tool provides the ability to create a helix or spiral curve in a part. The curve can be used as a path or guide curve for a swept feature, or a guide curve for a lofted feature.

Tutorial: Helix and Spiral 7-1

Apply the Helix and Spiral feature to create a Variable Pitch helix spring.

1. Open **Helix and Spiral 7-1** from the SOLIDWORKS 2018 folder. Sketch1 is a circle, coincident to the Origin on the Top plane with a .235in dimension.

2. **Edit** Sketch1.

Create a Variable Pitch Helix/Spiral.

3. Click **Helix and Spiral** from the Consolidated Curves menu.

4. **Enter** the following information as illustrated in the Region Parameters table. The pitch needs to be slightly larger than the wire.

5. Enter **.021**in for the Pitch.

6. Enter Start angle of **0**deg.

7. Select **Clockwise**.

8. Create a **Sweep** feature. Select **Circular Profile**.

9. Click **Helix/Spiral1** in the Graphics window as illustrated.

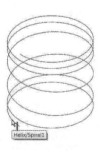

Region parameters table:

	P	Rev	H	Dia
1	0.021in	0	0in	0.235in
2	0.021in	1	0.021i	0.235in
3	0.08in	2	0.0715i	0.235in
4	0.08in	3	0.1515i	0.235in
5	0.021in	4	0.202i	0.235in
6	0.021in	5	0.223i	0.235in
7	0.021in	6	0.244i	0.235in
8				0.235in

10. Enter **.015in** for Depth as illustrated.

11. **View** the results.

12. **Close** the model.

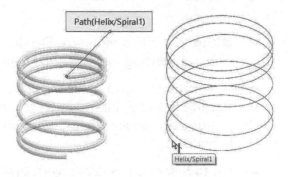

Thread Feature

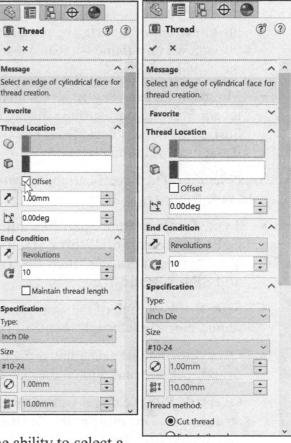

The Thread 🔩 Feature provides the ability to create threads for standard and non-standard threads for shafts and holes. For prototypes and digital simulation use the Thread feature. Select the start thread location, specify an offset, set end conditions, specify the type, size, diameter, pitch and rotation angle and choose options such as right-hand or left-hand thread.

The Thread feature uses the Thread PropertyManager. The Thread PropertyManager provides the following selections:

- *Thread Location*. The Thread Location box provides the ability to select the following options:

    - **Edge of Cylinder**. Select a circular edge.

    - **Optional Start Location**. Provides the ability to select a Vertex, Edge, Plane or Planar Surface.

    - **Offset**. The offset value must be positive. You can also enter an equation by starting with = (equal sign).

    - **Offset Distance**. Distance of the Offset.

    - **Start Angle**. Defines the starting location of the helix. The start angle must be positive. You can also enter an equation by starting with = (equal sign).

- ***End Condition***. The End Condition box provides the ability to select the following options:

  - **Blind**. Specify a value for Depth. Terminates the thread at the desired distance from the starting location, taking into account any input for offset.

  - **Revolutions**. Terminates the helix at the desired number of revolutions from the starting location, taking into account any input for offset. The value must be positive and greater than 0.00. You can also enter an equation by starting with = (equal sign).

  - **Up to Selection**. Select a vertex (sketch, model, or reference points), edge (sketch, model or reference axis), plane, or planar surface. A plane, planar face, or edge must be parallel to the circular edge (i.e., perpendicular to the thread axis). A vertex or point acts as the point through which a plane is created that is perpendicular to the cylindrical axis.

    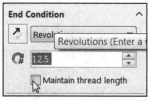

  - **Maintain thread length**. Keeps the thread at a constant length from the start surface. It only displays if the Start Surface Offset is selected and End Condition is set to Blind or Revolutions.

- ***Specification***. The Specification box provides the ability to select the following options:

  - **Type**. Displays library part files initially found in C:\ProgramData\SolidWorks\SOLIDWORKS YYYY\Thread Profiles. **Inch Die, Inch Tap, Metric Die and SP4xx Bottle**.

  - **Size**. Displays each of the configurations found in the library part file selected from the Type list.

  - **Override diameter**. Display diameter of the cylindrical face and will be the diameter of the helix.

    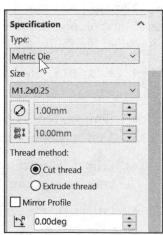

  - **Override Pitch**. Enter a value or start with = to create an equation. The value in the dimension input field is determined from the selected configuration's profile sketch and the length of a vertical construction line drawn from the origin of the model.

  - **Thread method**. Select one of the following:

    - **Cut thread**.

    - **Extruded thread**.

  - **Mirror Profile**. Select one of the following:

    - **Mirror horizontally**.

- **Mirror vertically**.

- **Locate Profile**. Select a new point on the thread profile sketch from the graphics area to attach to the helical path. You can select any vertex created by the entities and sketch points in the profile sketch. The button is disabled when no circular edge has been selected.

- *Thread Options*. The Thread Options box provides the ability to select the following options:

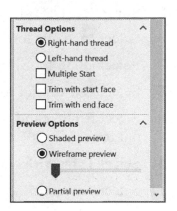

  - **Right-hand thread**. Creates a right hand thread.

  - **Left-hand thread**. Creates a left hand thread.

  - **Multiple Start**. Set the number of starts to define the number of times the thread is created in an evenly-spaced circular pattern around the hole or shaft. The image shows a four-start thread with a different color per thread.

  - **Trim with start face**. Provides the ability to align threads to start faces.

  - **Trim with end face**. Provides the ability to align threads to end face.

Align threads to end faces. A Cut thread is extended and cut to match the end face. An Extrude thread is cut to match the end face.

- *Preview Options*. The Preview Options box provides the ability to select the following options:

  - **Shaded preview**. Displays a fully tessellated preview of the thread.

  - **Wireframe preview**. Displays a wireframe preview of the thread.

  - **Partial preview**. Adjusts the number of wires displayed in the wireframe.

Tutorial: Thread Wizard 7-1

Apply the Thread Wizard to create a thread feature.
1. Open **Thread Wizard 7-1** from the SOLIDWORKS 2018 folder.

Apply the Thread Feature.
2. Click **Thread** from the Hole Wizard drop-down menu. Click **OK** if needed.

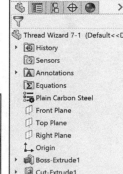

3. Click the **front circular edge** as illustrated. Edge1 is displayed in the Thread Location box.

4. Select **Up To Selection** for End Condition. Click inside the **End Condition** box.

5. Click the **front circular face** as illustrated. Face<1> is displayed.

6. Select **Inch Die** for Specification Type.

7. Select **#8-36** for Size.

8. Select **Cut thread** for Thread method.

9. Select **Right-hand thread** for Thread options.

10. Select **Trim with start face**.

11. Select **Trim with end face**.

12. Select **Shaded preview**.

13. Click **OK** ✔ from the Thread PropertyManager. Thread1 is created and is displayed.

14. **Zoom in** on the thread section. View the Trim with start face option.

15. **Close** the part.

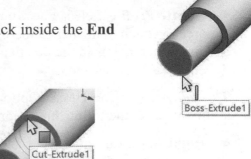

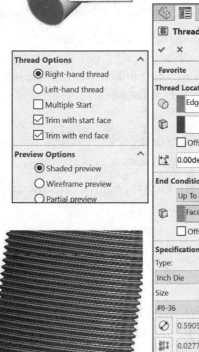

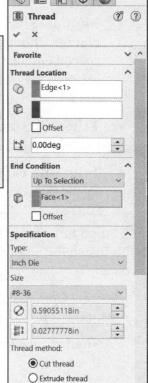

A Cut thread is extended and cut to match the end face. An Extrude thread is cut to match the end face. In the PropertyManager, under Thread Options, select Trim with start face and Trim with end face.

Summary

In this chapter, you learned about the Revolved Boss/Base feature, Revolved Cut feature, Hole Wizard feature, Advanced Hole feature, Dome feature, Thread feature and Curve feature. A Revolved feature adds or removes material by revolving one or more profiles about a centerline. SOLIDWORKS provides the ability to create Revolved Boss/Base, Revolved Cut, or Revolved Surface features. The Revolve feature can be a solid, thin, or a surface.

You learned about the Hole Wizard feature. The Hole Wizard feature provides the ability to create cuts in the form of standard holes. This feature provides the ability to create holes on a plane as well as holes on planar and non-planar faces. The Hole Wizard information from the model propagates to the drawing when the Hole Callout button is selected.

You learned about the Advanced Hole feature. The Advanced Hole feature provides the ability to create a complex hole which requires two or more manufacturing processes. The Advanced Hole feature uses a similar Hole Wizard PropertyManager but with Hole element fly-outs to guide the user through the process.

The Dome Feature provides the ability to create one or more dome features concurrently on the same model. A Dome feature creates spherical or elliptical shaped geometry.

You also covered various Curve features using a range of options: Split Line, Projected, Composite, Curve Through XYZ Points, Curve Through References Points and Helix/Spiral.

The Thread Feature provides the ability to create standard and non-standard threads for shafts and holes. For prototypes and digital simulation use the Thread feature. Select the start thread location, specify an offset, set end conditions, specify the type, size, diameter, pitch and rotation angle and choose options such as right-hand or left-hand thread.

Think design intent. When do you use the various End Conditions? What are you trying to do with the design? How does the component fit into an Assembly?

In Chapter 8, explore and address the Shell, Draft, Rib, Scale and Intersect feature. The Intersect feature provides the ability to intersect solids, surfaces, or planes to modify existing geometry, or to create new geometry with the Intersect tool.

Notes:

CHAPTER 8: SHELL, DRAFT, RIB, SCALE AND INTERSECT FEATURE

Chapter Objective

Chapter 8 provides a comprehensive understanding of Shell, Draft, Rib, Scale and Intersect features in SOLIDWORKS. On the completion of this chapter, you will be able to:

- Understand and utilize the following Shell features:

  - Shell uniform thickness.

  - Shell Error Diagnostics.

  - Shell with Multi-thickness.

- Understand and utilize the following Draft features:

  - Draft PropertyManager: Manual tab.

  - DraftXpert PropertyManager:

    - Add tab.

    - Change tab.

    - Draft Analysis.

- Understand and utilize the Rib feature.

- Knowledge of the Scale feature.

- Knowledge of the Intersect feature.

Shell Feature

The Shell ▧ feature provides the ability to "hollow out" a solid part. Select a face or faces on the model. The Shell feature leaves the selected faces open and creates a thin-walled feature on the remaining faces.

If a face is not selected on the model, the Shell feature provides the ability to shell a solid part, creating a closed hollow model.

You can also shell a model using the multiple thicknesses option for different faces. You can remove faces, set a default thickness for the remaining faces, and then set the different thicknesses for the faces that you selected from the remaining faces.

Most plastic parts have rounded corners. If you insert fillets to the edges prior to shelling, and the fillet radius is larger than the wall thickness, the inside corners of the part will automatically be rounded. The radius of the inside corners will be the same as the fillet radius minus the selected wall thickness. If the wall thickness is greater than the fillet radius, the inside corners will be sharp.

Insert fillets into your model before you apply the Shell feature.

The Shell feature uses the Shell PropertyManager. The Shell PropertyManager provides the following selections:

- *Parameters*. The Parameters box provides the ability to select the following options:

  - **Thickness**. Displays the selected thickness of the faces that you decide to keep.

  - **Faces to remove**. Displays the selected faces from the Graphics window to remove from the model.

To create a hollow part, do not select the Faces to remove option. Do not remove any faces from your model.

  - **Shell outward**. Increases the outside dimensions of the part.

  - **Show preview**. Displays a preview of the Shell feature. Clear the Show preview check box before selecting the faces. If you do not clear the Show preview check box, the preview will update with each face selection. This will slow your system operation down.

- *Multi-thickness Settings*. The Multi-thickness Settings box provides the following selections:

  - **Multi-thickness**. Displays the selected thickness for a multi-thickness model.

  - **Multi-thickness Faces**. Displays the selected faces of the multi-thickness model.

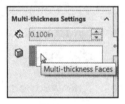

The Error Diagnostics section is displayed in the Shell PropertyManager if the Shell tool exhibits a modeling problem and a Rebuild error is displayed.

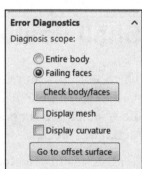

Tutorial: Shell 8-1

Create a Shell feature using the Check tool to remove the front face of the model.

1. Open **Shell 8-1** from the SOLIDWORKS 2018 folder. The Shell 8-1 FeatureManager is displayed.

2. Click the **Shell** 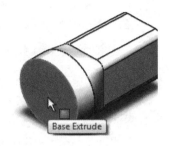 Features tool. The Shell PropertyManager is displayed.

3. Click the **front face** of Base-Extrude.

4. Enter **1.100**in for Thickness in the Parameters box.

5. Click **OK** ✅ from the Shell1 PropertyManager. A warning message is displayed. The Thickness value is greater than the Minimum radius of Curvature. Find the Minimum radius of Curvature.

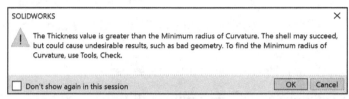

6. Click the **OK** button from the SOLIDWORKS dialog box. View the Rebuild Error message.

7. Click **Cancel** ✕ from the Shell PropertyManager.

8. Click **Check** 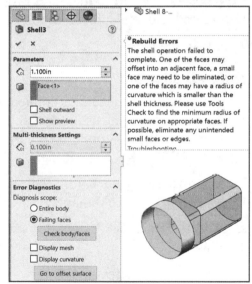 from the Evaluate tab in the CommandManager. The Check Entity dialog box is displayed.

9. Check the **Minimum radius of curvature** box.

10. Click the **Check** button. The Minimum radius of Curvature that the system calculated is .423in.

11. Click the **Close** button from the Check Entity dialog box.

12. Click the **Shell** Features tool. The Shell PropertyManager is displayed.

13. Click the **front face** of Base-Extrude.

14. Enter **.3**in for Thickness in the Parameters box.

15. Click **OK** ✅ from the Shell PropertyManager. Shell2 is created and is displayed in the FeatureManager.

16. **Close** the model.

Tutorial: Shell 8-2

Create a Shell feature using the Multi-thickness option.

1. Open **Shell 8-2** from the SOLIDWORKS 2018 folder. The Shell 8-2 FeatureManager is displayed.

2. Click the **front face** of Extrude1.

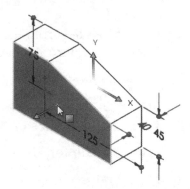

3. Click the **Shell** Features tool. The Shell PropertyManager is displayed. Face<1> is displayed in the Faces to Remove box.

4. Enter **2**mm for Thickness in the Parameters box.

5. Click inside of the **Multi-thickness Faces** box.

6. Click the **top two faces** of Extrude1. Face<2> and Face<3> are displayed in the Multi-thickness Faces box.

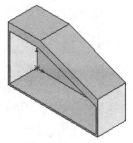

7. Enter **10**mm for Multi-thickness.

8. Click **OK** from the Shell PropertyManager. Shell1 is created and is displayed in the FeatureManager.

9. **Close** the model.

Tutorial: Shell 8-3

View how the Shell feature is applied on the rear support of the snowboard binding as illustrated.

1. Open **Shell 8-3 Snowboard** assembly from the SOLIDWORKS 2018 folder.

2. Open **RearSupport** from the assembly FeatureManager.

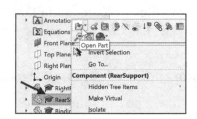

3. Roll the **Rollback bar** under Shell1.

4. Edit the **Shell1** feature. Three faces were selected with a depth of .125in. The shell tool hollows out a part, leaves open the faces you select and creates thin-walled features on the remaining faces. If you do not select any face on the model, you can shell a solid part creating a closed hollow model.

5. **Close** all models.

Draft Feature

A draft tapers a face using a specified angle to the selected face in a model. The Draft  feature tapers a selected model face with a specified angle by utilizing a:

- Neutral plane.
- Parting line.
- Step draft.

Apply the Draft feature to solid and surface models.

In order for a model to eject from a mold, all faces must draft away from the parting line which divides the core from the cavity.

Cavity side faces display a positive draft and core side faces display a negative draft. Design specifications include a minimum draft angle, usually less than 5deg.

For the model to eject successfully, all faces must display a draft angle greater than the minimum specified by the Draft Angle. The Draft PropertyManager and DraftXpert PropertyManager utilize the draft angle to determine what faces require additional draft based on the direction of pull.

Apply a draft angle when using the Extruded Boss/Base or Extruded Cut feature.

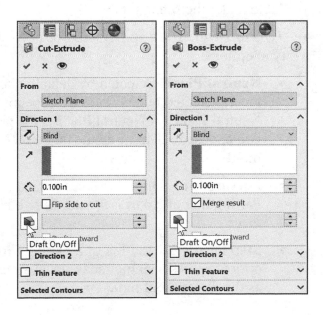

Draft PropertyManager

The Draft 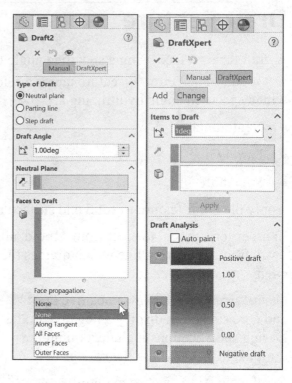 feature uses the Draft PropertyManager. The Draft PropertyManager provides the ability to create various draft types.

The Draft PropertyManager provides the ability to select either the *Manual* or *DraftXpert* tab.

Each tab has a separate menu and option selection. The PropertyManager displays the appropriate options based on the type of draft you create.

The PropertyManager remembers its last used state.

Draft PropertyManager: Manual tab

The Manual tab provides the ability to display and select various parameters for the Draft feature. Use this tab to maintain control at the feature level. The following selections and options are available under the Manual tab:

- ***Type of Draft***. The Type of Draft box provides the following type options:

  - **Neutral Plane**. Selected by default. Creates a feature that tapers selected model faces by a specified angle, using a Neutral Plane (A Neutral Plane is the plane or face that you choose to determine the pull direction when creating a mold).

  - **Parting Line**. Drafts surfaces around a Parting line. The Parting line of a part is where the two halves of the mold meet.

  - **Step Draft**. A variation of a Parting Line draft. This option creates a single face rotated around the plane used as the direction of pull.

- **Allow reduced angle**. Creates a draft when the sum of the angle produced by the maximum angle and the Draft Angle is 90° or greater.

- **Tapered steps**. Creates the surfaces in the same manner as the tapered surfaces.

- **Perpendicular steps**. Creates the surfaces perpendicular to the original main face.

- *Draft Angle*. Sets the draft angle, measured perpendicular to the neutral plane in 1 degree increments.

☀ The draft angle is measured perpendicular to the Neutral plane.

- *Neutral Plane*. (Neutral Plane only.) Select a planar face or a plane feature. If necessary, select Reverse direction to slant the draft in the opposite direction.

- *Direction of Pull*. (Parting Line and Step Draft option only.) Indicates the direction of pull. Displays the selected edge or face in the Graphics window. If necessary, select the Reverse direction option.

- *Faces to Draft*. Displays the selected faces to draft from the Graphics window.

  - **Face Propagation**. Only available for the Neutral Plane Draft option. The default option is None. If the Parting Line or Step Draft option is selected, only the None and Along Tangent selections are available. The available options are:

    - **None**. Selected by default. Drafts only the selected faces.

    - **Along Tangent**. Extends the draft to all faces which are tangent to the selected face.

    - **All Faces**. Drafts all faces extruded from the neutral plane.

    - **Inner Faces**. Drafts all inner faces extruded from the neutral plane.

    - **Outer Faces**. Drafts all outer faces next to the neutral plane.

- *Parting Lines*. Only available for the Parting Line and Step Draft options.

  - **Parting Lines**. Displays the selected parting lines from the Graphics window.

  - **Other Face**. Allows you to specify a different draft direction for each segment of the parting line. Click the name of the edge in the **Parting Lines** box, and click **Other Face**.

  - **Face Propagation**. Propagates the draft across additional faces. Select an item:

    - **None**. Drafts only the selected face.

    - **Along Tangent**. Extends the draft to all faces that are tangent to the selected face.

Tutorial: Draft 8-1

Create a Draft feature with a Neutral Plane.

1. Open **Draft 8-1** from the SOLIDWORKS 2018 folder. The Draft 8-1 FeatureManager is displayed.

2. Click the **Draft** 🔲 Features tool.

3. Click the **Manual** tab. The Draft PropertyManager is displayed. Neutral plane is displayed as the Type of Draft.

4. **Zoom in** and click the **thin front circular face** of the Base Extrude feature. This is your Neutral Plane.

5. Click the **outside circular face** of the Base Extrude feature. Face<2> is displayed in the Faces to Draft box.

6. Enter **8.5**deg for Draft Angle.

7. Click **OK** ✔ from the Draft PropertyManager. Draft1 is created and is displayed.

8. Display the **Right** view.

9. Click **Hidden Lines Removed**. View the created Draft Angle.

10. **Close** the model.

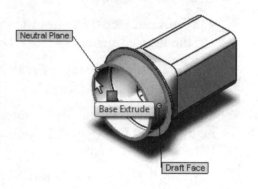

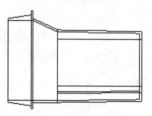

Tutorial: Draft 8-2

Create a Draft feature with a Parting Line.

1. Open **Draft 8-2** from the SOLIDWORKS 2018 folder. The Draft 8-2 FeatureManager is displayed.

2. Click the **Draft** 🔲 Features tool.

3. Click the **Manual** tab. The Draft PropertyManager is displayed.

4. Select **Parting line** for Type of Draft.

5. Enter **3.00**deg for Draft Angle.

6. Click inside the **Direction of Pull** box.

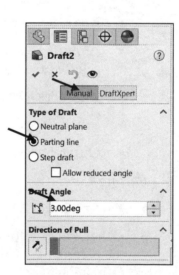

7. Click **Top Plane** from the fly-out FeatureManager in the Graphics window.

8. **Flip** the direction so that the Direction of Pull arrow points downward.

9. **Rotate** the model and right-click the **bottom front edge** Sweep1.

10. Click **Select Tangency** to display all bottom edges. Edge<1> - Edge<8> is displayed in the Parting Lines box.

11. Click **OK** ✅ from the Draft PropertyManager. Draft2 is created and displayed. With the Direction of Pull downward, the sides of the extruded cut feature are drafted inward by 3 degrees.

12. **Close** the model.

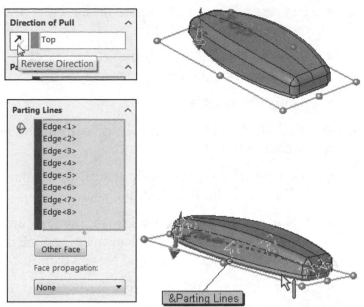

DraftXpert PropertyManager: Add/Change tabs

The DraftXpert PropertyManager provides the ability to manage the creation and modification of all Neutral Plane drafts. Select the draft angle and the references to the draft.

The DraftXpert provides the following capabilities: create multiple drafts, perform draft analysis, and edit drafts for (Neutral plane drafts only).

The DraftXpert PropertyManager provides the following selections:

DraftXpert PropertyManager/Add tab

- *Items to Draft*. The Items to Draft box provides the following selections:

  - **Draft Angle**. Sets the draft angle. The draft angle is measured perpendicular to the neutral plane.

  - **Neutral Plane**. Displays the selected planar face or plane feature from the fly-out FeatureManager.

- **Reverse Direction**. Reverses the direction of the draft if required.

- **Faces to Draft**. Displays the selected faces to draft from the Graphics window.

- **Apply**. Calculates the draft.

- *Draft Analysis*. The Draft Analysis box provides the following selections:

  - **Auto paint**. Enables the draft analysis of your model. Select a face for the Neutral plane.

  - **Color contour map**. Displays the range of draft in the model by color and numerical value, and the number of faces with Positive draft and Negative draft.

  - Click **Show** or **Hide** to toggle the display of faces that include or require draft.

DraftXpert PropertyManager/Change tab

Use the Change tab to modify or remove Neutral Plane drafts. The DraftXpert PropertyManager remembers the last used state.

- *Drafts to change*. The Drafts to change box provides the following selections:

  - **Draft Items**. Displays the selected faces with draft to change or remove from the Graphics window.

  - **Neutral Plane**. Displays the selected planar face or plane feature from the fly-out FeatureManager. Not required if you select the Draft Angle option.

  - **Reverse Direction**. Reverses the direction of the draft if required.

  - **Draft Angle**. Sets the draft angle. The draft angle is measured perpendicular to the Neutral plane.

  - **Change**. Changes the draft.

  - **Remove**. Removes the draft.

- *Existing Drafts*. The Existing Drafts box provides the following selections:

  - **Sort list by**. Filters all drafts by the following selections: **Angle**, **Neutral Plane** and **Direction of Pull**.

- *Draft Analysis*. The Draft Analysis box provides the following selections:

  - **Auto paint**. Enables the draft analysis of your model. Select a face for the Neutral plane.

  - **Color contour map**. Displays the range of draft in the model by color and numerical value, and the number of faces with Positive draft and Negative draft.

  - Click **Show** or **Hide** to toggle the display of faces that include or require draft.

Tutorial: DraftXpert 8-1

Apply DraftXpert to add and change a draft feature.

1. Open **DraftXpert 8-1** from the SOLIDWORKS 2018 folder.

2. Click the **Draft** 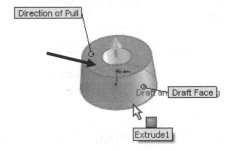 Features tool.

3. Click the **DraftXpert** tab. The DraftXpert PropertyManager is displayed.

4. Click the **Add** tab.

5. Enter **4**deg for Draft Angle.

6. Click the **top face** of Extrude1. Face<1> is displayed in the Neutral Plane box. The Direction of Pull is upwards.

7. Click the **circular face** of Extrude1. This is the Draft Face. Face<2> is displayed in the Items To Draft box.

8. Click the **Apply** button. The draft is applied and is displayed in the Graphics window.

9. Check the **Auto paint** box in the Draft Analysis box. The drafted face displays the Draft analysis color for 4 degrees of draft. Note: Yellow indicates no draft. View positive, required, and negative draft surfaces with the Show/Hide toggle tool.

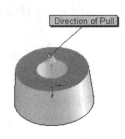

10. Drag the **mouse pointer** over the drafted face. View the feedback display.

11. Click the **Change** Tab from the DraftXpert PropertyManager.

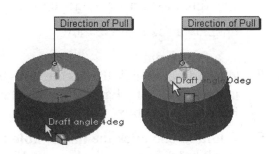

12. Click **4deg** inside the Existing Drafts box.

13. Enter **5deg** from the Draft Angle box.

14. Click the **Change** button in the Drafts to change box. View the changes in the draft in the Graphics window.

15. Click **OK** ✅ from the DraftXpert PropertyManager.

16. **Close** the model.

A Draft Analysis provides the designers of plastic parts and mold tooling manufacturers to check the correct application of draft to the faces of their part. With a draft analysis, you can verify draft angles, examine angle changes within a face, as well as locate parting lines, injection, and ejection surfaces in parts.

Tutorial: DraftXpert 8-2

Apply DraftXpert to analyze a draft.

1. Open **DraftXpert 8-2** from the SOLIDWORKS 2018 folder.

2. Click the **Draft** 🔲 Features tool.

3. Click the **DraftXpert** tab. The DraftXpert PropertyManager is displayed.

4. Click the **Add** tab.

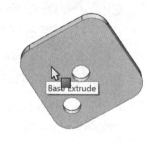

Create the Direction of pull.

5. **Rotate** the model and click the **back face**. Face<1> is displayed in the Direction of Pull box.

6. If needed, click the **Reverse direction** button. The direction arrow points upward.

7. Display an **Isometric** view.

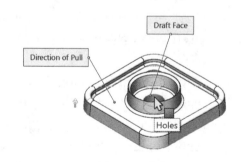

Select a face to draft.

8. **Zoom in** and click the **inside hole face** of the large center hole. Face<2> is displayed in the Faces to Draft box.

9. **Zoom in** and click the **inside hole face** of the small hole. Face<3> is displayed.

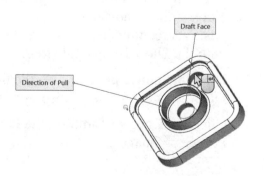

10. Enter **2**deg for Draft Angle.

11. Check the **Auto paint** button.

12. Click the **Apply** button. Rotate the part to display the faces that require a draft.

13. Click **OK** ✔ from the DraftXpert PropertyManager.

14. **Close** the model.

Rib Feature

The Rib 🔷 feature provides the ability to create Ribs in a SOLIDWORKS model. A Rib is a unique type of extruded feature. This extruded feature is created from the opening or closing of sketched contours.

A Rib adds material of a specified thickness in a specified direction between the contour and an existing part. A Rib can be created using single or multiple sketches. You can also create a rib feature with draft, or select a reference contour to a draft of sketches. A Rib requires a sketch, thickness, and an extrusion direction.

The Rib feature uses the Rib PropertyManager. The Rib PropertyManager provides the following selections:

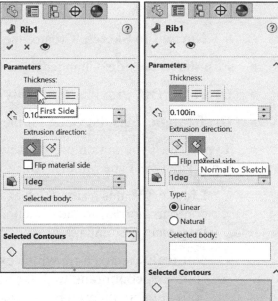

- *Parameters*. Provides the ability to select the following options:

    - **Thickness**. Adds thickness to the selected side of the sketch. Select one of the following three options:

        - **First Side**. Adds material only to one side of the sketch.

        - **Both Sides**. Selected by default. Adds material equally to both sides of the sketch.

        - **Second Side**. Adds material only to the other side of the sketch.

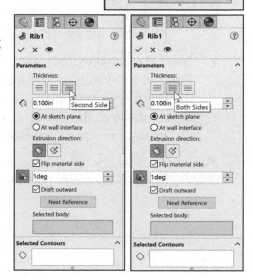

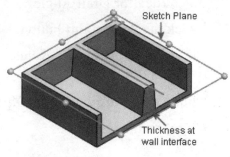

- **Rib Thickness**. Enter the rib thickness or select the thickness using the drop-down menu in .1inch increments using the ANSI, (IPS) unit standard.

🔆 Rule of thumb states that the Rib thickness should be ½ of the part wall thickness.

- **At sketch plane**. If you add draft, you can set the thickness at the sketch plane.

- **At wall interface**. If you add draft, you can set the thickness at the wall interface.

- **At sketch plane**
- **At wall interface**

- **Extrusion Direction**. Selects the Extrude direction. There are two options:

  - **Parallel to Sketch**. Selected by default. Creates the rib extrusion parallel to your sketch.

  - **Normal to Sketch**. Creates the rib extrusion normal to your sketch.

- **Flip material side**. Changes the direction of the extrusion.

- **Draft On/Off**. Provides the ability to add draft to the rib.

- **Draft Angle**. Sets the value for the Draft Angle for the rib.

- **Type**. The Type option provides two selections for Rib type. The two Type options are:

  - **Linear**. Creates a rib which extends the sketch contours normal to the direction of your sketch, until they meet a boundary.

  - **Natural**. Creates a rib which extends the sketch contours, continuing with the same contour equation until your rib meets a boundary. Example: The sketch is an arc of a circle. The Natural option extends your rib using the circle equation, until it meets a boundary.

- **Next Reference**. Only available when you select the Parallel to Sketch option for the Extrusion Direction and the Draft On/Off is also selected. Provides the ability to toggle through the sketch contours, to select the contour to use as your reference contour for the draft.

- *Selected Contours*. The Selected Contours box displays the selected sketch contours that you are using to create the Rib feature.

Tutorial: Rib 8-1

Create a Rib feature. Add a rib to increase structural integrity of the part. Apply the Parallel to Sketch option with a draft.

1. Open **Rib 8-1** from the SOLIDWORKS 2018 folder.

2. Right-click **Top Plane** from the FeatureManager. This is your Sketch plane.

3. Click **Sketch** from the Context toolbar.

4. Click the **Line** Sketch tool.

5. Sketch a **horizontal line** as illustrated on the model. The endpoints are located on either side of the Handle as illustrated.

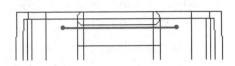

Add a dimension.

6. Click the **Smart Dimension** Sketch tool. The Dimension PropertyManager is displayed.

7. **Dimension** the model as illustrated.

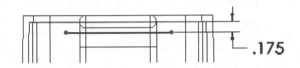
.175

Insert a Rib feature.

8. Click the **Rib** Features tool. The Rib PropertyManager is displayed.

9. Click the **Both Sides** button.

10. Enter **.100**in for Rib Thickness.

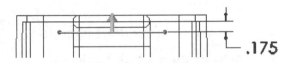
.175

11. Click the **Parallel to Sketch** button. The Rib direction arrow points to the back. If needed, click the **Flip material side** check box.

12. Click the **Draft On/Off** button.

13. Enter **1**deg for Draft Angle.

14. Display a **Front view**.

15. Click the **back inside face** of the HOUSING for the Body. The selected entity is displayed.

16. Click **OK** from the Rib PropertyManager.

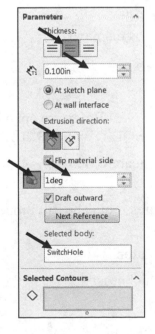

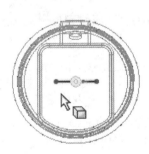

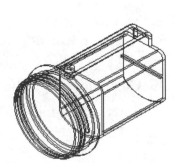

17. Display an **Isometric view**. Rib1 is created and is displayed in the FeatureManager. View the Rib in the Graphics window.

18. **Close** the model.

Tutorial: Rib 8-2

Create a Rib feature. Apply the Parallel to Sketch option with a draft.

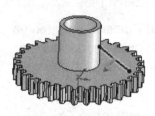

1. Open **Rib 8-2** from the SOLIDWORKS 2018 folder.

2. **Edit** Sketch9 from the FeatureManager.

Activate the Rib feature.

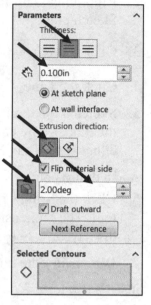

3. Click the **Rib** Features tool. The Rib PropertyManager is displayed.

4. Click the **Both Sides** button.

5. Enter **.100**in for Rib Thickness.

6. Click the **Parallel to Sketch** button. The Rib direction arrow points downwards.

7. Check the **Flip material side** box if required. The direction arrow points downwards.

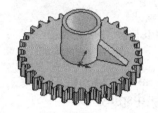

8. Click the **Draft On/Off** button.

9. Enter **2**deg for Draft Angle.

10. Click **OK** ✔ from the Rib PropertyManager. Rib1 is displayed.

11. **Close** the model.

Tutorial: Rib 8-3

Create a Rib feature with the Next Reference option.

1. Open **Rib 8-3** from the SOLIDWORKS 2018 folder.

2. **Edit** Sketch9 from the FeatureManager.

Activate the Rib feature.

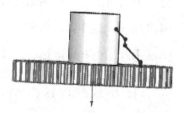

3. Click the **Rib** Features tool. The Rib PropertyManager is displayed.

4. Click the **Both Sides** button.

5. Enter **.150**in for Rib Thickness.

6. Click the **Parallel to Sketch** button. The Rib direction arrow points to the back.

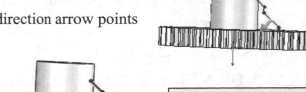

7. Click the **Flip material side** box. The direction arrow points downwards.

8. Click the **Draft On/Off** button.

9. Enter **3**deg for Draft Angle.

10. Click the **Next Reference** button until you select the contour as illustrated. You can only select one reference contour on which to base your draft.

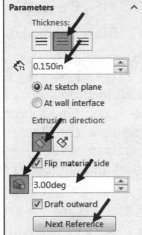

11. Click **OK** ✔ from the Rib PropertyManager. Rib1 is created and is displayed.

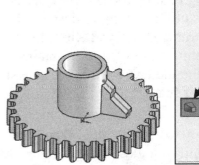

12. **Close** the model.

Tutorial: Rib 8-4

Create a Rib feature and apply the Normal to Sketch option with a Linear and Natural Rib Type.

1. Open **Rib 8-4** from the SOLIDWORKS 2018 folder.

2. Click **SketchRib** from the FeatureManager.

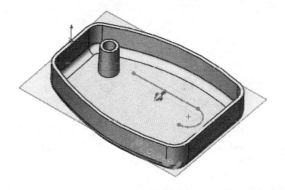

Activate the Rib feature.

3. Click the **Rib** ⬚ Features tool. The Rib PropertyManager is displayed. The direction arrow points toward the front as illustrated.

4. Click the **Both Sides** button.

5. Enter **2**mm for Rib Thickness.

6. Click the **Normal to Sketch** button. The direction arrow points downwards. Check the **Linear** box for Type.

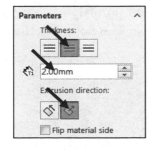

7. Click **OK** ✔ from the Rib PropertyManager.

8. Right-click **Rib1** from the FeatureManager.

9. Click **Edit Feature**.

10. Check the **Natural** box for Type.

11. Click **OK** ✔ from the Rib PropertyManager. View the results.

12. **Close** the model.

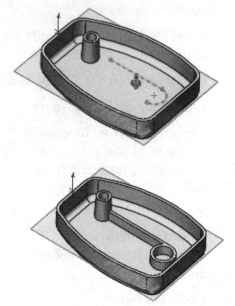

🔆 The Fastening Feature toolbar provides access to additional features utilized in mold design. These tools combine Extrude, Draft, and Rib features. For example, the Mounting Boss feature contains parameters to define the Extruded Boss, hole size, draft and number of fins.

🔆 Press the **g** key to activate the Magnifying glass tool. Use the Magnifying glass tool to inspect a model and make selections without changing the overall view.

Scale Feature

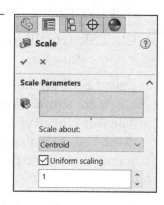

The Scale 🔲 feature provides the ability to scale a part or surface model about its centroid, the model origin, or a coordinate system.

A Scale feature is like any other feature in the FeatureManager design tree: it manipulates the geometry, but it does not change the definitions of features created before it was added. To temporarily restore the model to its unscaled size, you can roll back or suppress the Scale feature.

The Scale feature uses the Scale PropertyManager. The Scale PropertyManager provides the following selections:

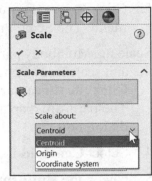

- *Scale Parameters*. Provides the ability to select the following options:

  - **(Multi-body parts only)**. Specifies which bodies to scale. Select bodies in the graphics area or in the FeatureManager design tree in the Solid Bodies and Surface Bodies folders.

  - **Scale About**. Specifies the entity about which to scale the model. Select one of the following:

    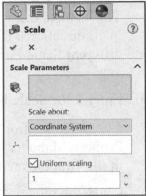

    - **Centroid**. Scales the model about its system-calculated centroid.

    - **Origin**. Scales the model about its origin.

    - **Coordinate System**. Scales the model about a user-defined coordinate system. For Coordinate System, select a coordinate system that you have previously defined.

    - **Uniform scaling**. Applies the same Scale Factor in all directions. Clear to specify a different Scale Factor for each axis.

    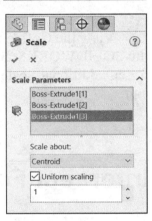

    - **Scale Factor**. Defines the factor to multiply by in each direction. If Uniform Scaling is selected, enter one Scale Factor to apply in all directions. Otherwise, enter a factor for each direction:

      o X Scale Factor.

      o Y Scale Factor.

      o Z Scale Factor.

Scaling occurs in both directions of axes when material is present on both sides of the entity selected for Scale About. When material is on one side only of an axis, the model scales in only one direction along that axis.

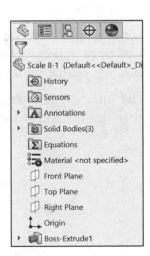

Tutorial: Scale 8-1

Scale the Multibody part about the Centroid. Modify the scale.

1. Open **Scale 8-1** from the SOLIDWORKS 2018 folder. The Scale 8-1 FeatureManager is displayed.

2. Click **Insert, Features, Scale** from the Main menu. The Scale PropertyManager is displayed.

Scale the Multibody part about the Centroid. Modify the unit scale from inches to mm.

3. Click the **three parts** from the Graphics window. The parts are displayed in the Scale Parameters dialog box.

4. Select **Centroid** for Scale about.

5. Check the **Uniform scaling** box. This applies the same Scale Factor in all directions. Clear to specify a different Scale Factor for each axis.

6. Enter **2** for Scale Factor.

7. Click **OK** ✔ from the Scale PropertyManager. View the results in the Graphics window.

8. **Close** the model.

☼ View the Coffee cup part in the SOLIDWORKS 2018 folder for an additional Scale example.

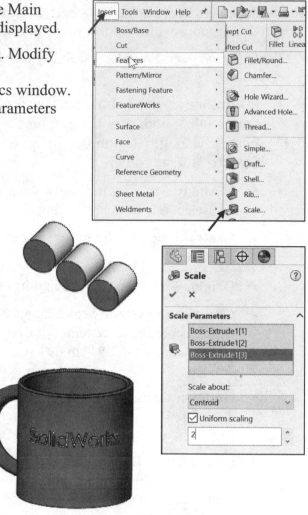

Intersect Feature

The Intersect feature provides the ability to intersect solids, surfaces, or planes to modify existing geometry, or to create new geometry with the Intersect tool.

For example, you can add open surface geometry to a solid, remove material from a model, or you can create geometry from an enclosed cavity. You can also merge solids that you define with the Intersect tool, or cap some surfaces to define closed volumes.

The Intersect feature uses the Intersect PropertyManager. The Intersect PropertyManager provides the following selections:

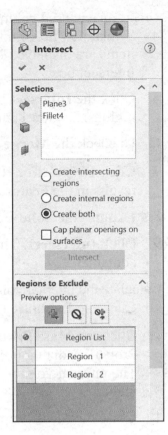

- *Selections box*. Provides the ability to select:

  - **Solids**.

  - **Surfaces**.

  - **Planes to intersect**.

    - **Cap planar openings on surfaces**. For surfaces with openings that are flat; closes the openings when you click Intersect.

    - **Intersect button**. Defines a distinct region from each closed volume created by the union or intersection of the selections. You can exclude one or more regions in Regions to Exclude.

  - **Regions to Exclude**. Provides the following options:

    - **Region List**. After clicking Intersect, displays the regions you can exclude from the final result.

    - **Select All**. Selects all regions for exclusion.

    - **Invert Selection**. Clears selected regions in the Region List and selects the cleared regions.

  - **Options**. Provides the following tools:

    - **Merge result**. After clicking OK, forms the union of the included regions. Touching regions are formed into one body, when possible. When cleared, creates a separate body for each included region.

    - **Consume surfaces**. After clicking OK, removes surfaces from the FeatureManager design tree for the part.

Tutorial: Intersection 8-1

Calculate the internal volume of a container (bottle) using the SOLIDWORKS Intersect feature.

1. Open **Intersection 8-1** from the SOLIDWORKS 2018 folder. The Intersection 8-1 FeatureManager is displayed.

Create a plane to cap off the top of the bottle.

2. Click **Plane** from the Feature toolbar. Create a plane coincident with the top face of the bottle with a zero offset distance.

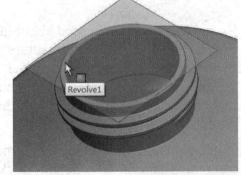

3. Click the **Intersect** Feature. The Intersect PropertyManager is displayed.

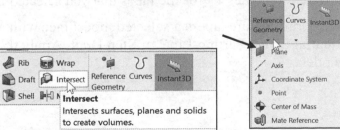

Create an enclosed volume.

4. Select the **Plane** and the **body** of the bottle. The two select items are displayed in the Solids, Surfaces or Planes to Intersect box.

5. Click the **Intersect** button. This creates two separate regions: Region 1 internal and Region 2 external.

6. Uncheck the **Merge result** options.

7. Click **OK** ✔ from the Intersect PropertyManager. View the results in the FeatureManager.

8. **Expand** the Solid Bodies folder.

9. Click **Intersect1**. This is the internal volume of the container.

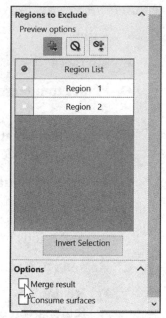

Calculate the internal volume of the container.

10. Click the **Evaluate** tab in the CommandManager.

11. Click the **Mass Properties** tool. The Mass Properties dialog box is displayed. View the results. The internal volume of the bottle is 80.2 cubic inches.

12. **Close** the Mass Properties dialog box.

13. **Close** the model.

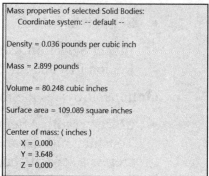

Summary

In this chapter, you learned about the Shell, Draft, Rib, Scale and Intersect feature in SOLIDWORKS. The Shell feature provides the ability to "hollow out" a solid part. Select a face or faces on the model. The Shell feature leaves the selected faces open and creates a thin-walled feature on the remaining faces. If a face is not selected on the model, the Shell feature provides the ability to shell a solid part, creating a closed hollow model.

You can also shell a model using the multiple thicknesses option for different faces. You can remove faces, set a default thickness for the remaining faces, and then set the different thicknesses for the faces that you selected from the remaining faces.

The Draft feature tapers a selected model face with a specified angle by utilizing a Neutral Plane, Parting Line, or Step Draft. Apply the Draft feature to solid and surface models.

In order for a model to eject from a mold, all faces must draft away from the parting line which divides the core from the cavity. Cavity side faces display a positive draft and core side faces display a negative draft. Design specifications include a minimum draft angle, usually less than 5 degrees.

The DraftXpert PropertyManager provided the ability to manage the creation and modification of all neutral plane drafts. Select the draft angle and the references to the draft. The DraftXpert manages the rest. It can verify draft angles, examine angle changes within a face, as well as locate parting lines, injection, and ejection surfaces in parts. This tool is ideal for mold manufacturing.

The Rib Feature provides the ability to create Ribs in a SOLIDWORKS model. A Rib is a unique type of extruded feature. This extruded feature is created from the opening or closing of sketched contours.

A Rib adds material of a specified thickness in a specified direction between the contour and an existing part. A Rib can be created using single or multiple sketches. You can also create a rib feature with draft, or select a reference contour to draft of sketches. A Rib requires a sketch, thickness and an extrusion direction.

The Scale feature provides the ability to scale a part or surface model about its centroid, the model origin, or a coordinate system.

A Scale feature is like any other feature in the FeatureManager design tree: it manipulates the geometry, but it does not change the definitions of features created before it was added.

The Intersect feature provides the ability to intersect solids, surfaces, or planes to modify existing geometry, or to create new geometry with the Intersect tool.

In Chapter 9, explore and address various Pattern Features: Linear Pattern, Circular Pattern, Curve Driven Pattern, Sketch Pattern, Table Driven Pattern, Fill Pattern along with the Mirror Feature and Coordinate System PropertyManager.

Notes:

CHAPTER 9: PATTERN FEATURES, MIRROR FEATURES AND COORDINATE SYSTEM

Chapter Objective

Chapter 9 provides a comprehensive understanding of Pattern and Mirror features in SOLIDWORKS along with the Coordinate System PropertyManager. On the completion of this chapter, you will be able to:

- Understand and utilize the following Pattern features:

  - Linear Pattern.

  - Circular Pattern.

  - Curve Driven Pattern.

  - Sketch Driven Pattern.

  - Table Driven Pattern.

  - Fill Pattern.

  - Variable Pattern.

- Recognize and utilize the Mirror feature.

- Understand and create a new Coordinate System.

Linear Pattern Feature

The Linear Pattern ⧉ feature provides the ability to create copies, or instances, along one or two linear paths in one or two different directions. The instances are dependent on the selected originals. Changes to the originals are passed onto the instanced features.

You can control linear patterns using reference geometry, such as vertices, edges, faces, or planes, to indicate how far you want to generate the pattern, and to designate the number of instances and spacing between instances.

Set an option in Direction 1 and Direction 2 to create a Linear Pattern feature in both directions from the Sketch plane.

The Linear Pattern feature uses the Linear PropertyManager. The Linear Pattern PropertyManager is feature dependent and provides the following selections:

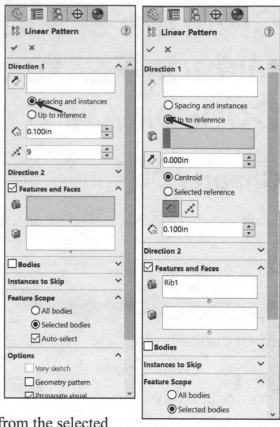

- *Direction 1*. Provides the following feature dependent options:

  - **Pattern Direction**. Displays the selected linear edge, line, axis, or dimension from the Graphics window.

  - **Reverse Direction**. Reverses the Direction 1 pattern if required.

  - **Spacing and instances**. Displays the set spacing value between the pattern instances for Direction 1, and displays the set value of the number of pattern instances for Direction 1. The Number of Instances includes the original features or selections.

- **Up to reference**. Set the offset distance of the patterned feature/body from the selected reference.

  - **Reference geometry (vertex, face, edge or plane)**.

- **Off set Distance**. Distance of offset.

- **Centroid**. Space from the Centroid.

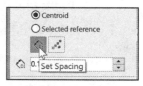

- **Set Spacing**. Space distance.

- **Set Number of Instances**. Number of Instances.

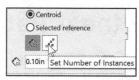

- **Selected reference**. Seed Reference. Select a vertex, edge, face or plane.

- **Configuration**. Only available with a model that has multiple configurations. The selections are **This configuration**, **All configurations** and **Specify configurations**.

The options in Direction 1 are the same as the options in Direction 2 except for the Pattern seed only check box.

- *Direction 2*. Provides the following feature dependent options:

  - **Pattern Direction**. Displays the selected linear edge, line, axis, or dimension from the Graphics window.

- **Reverse Direction**. Reverses the Direction 2 pattern if required.

- **Spacing and instances**. Displays the set spacing value between the pattern instances for Direction 1, and displays the set value of the number of pattern instances for Direction 1. The Number of Instances includes the original features or selections.

- **Up to reference**. Set the offset distance of the patterned feature/body from the selected reference.

- **Reference geometry (select a vertex, face, edge or plane).**

- **Off set Distance**. Distance of offset.

- **Set Spacing**. Space distance.

- **Set Number of Instances**. Number of Instances.

- **Centroid**. Space from the Centroid.

- **Selected reference**. Seed Reference. Select a vertex, edge, face or plane.

- **Configuration**. Only available with a model that has multiple configurations. The selections are **This Configuration**, **All Configurations** and **Specify Configurations**.

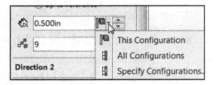

- **Pattern seed only**. Creates a linear pattern in Direction 2, only using the Seed feature, without duplicating the Direction 1 pattern instances.

- *Features and Faces*. Displays the selected seed to create the Linear Pattern and the selected faces of the seed feature to create the Linear Pattern utilizing the faces that compose the seed feature.

The Features and Faces option is very useful with models that import only the faces that compose the feature, and not the feature itself.

When using the Features and Faces option, your pattern must remain within the same boundary or face. The pattern cannot cross boundaries. Example: a cut across the complete face or different levels, such as a raised edge, would create separate faces and a boundary, stopping the pattern from spreading.

- *Bodies*. Displays the selected Solid/Surface Bodies and creates the pattern using the bodies that are selected in the multi-body part.

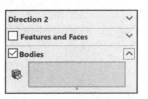

- ***Instances to Skip***. Displays the selected coordinates of the instances from the Graphics window which you do not want to include in the pattern. The mouse pointer changes when you float over each pattern instance in the Graphics window. Click to select a pattern instance.

💡 Use box or Lasso command to add or remove instances to skip.

- ***Feature Scope***. Applies various features to one or more multi-body parts. Check the Geometry pattern option box from the Options section to apply features to a multi-bound part. The available selections are:

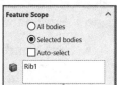

  - **All bodies**. Applies a feature to all bodies each time the feature is regenerated. If a new body is added to your model that is intersected by the feature, the new body is also regenerated to include that feature.

  - **Selected bodies**. Selected by default. Applies a feature to the body that is selected. If a new body is added to your model which is intersected by the feature, you are required to use the Edit Feature command to edit the pattern features. Select the bodies, and add them to the list of your selected bodies. If you do not add the new bodies to your selected list of selected bodies, they will remain intact.

  - **Auto-select**. Selected by default. Only available if you click the Selected bodies option. Provides the ability when you first create a model with multi-body parts, the features automatically process all the relevant intersecting parts.

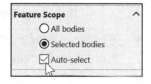

💡 The Auto-select option is faster than the All bodies option because the Auto-select option processes only the bodies on the initial list and does not regenerate your entire model.

  - **Bodies to Affect**. Only available if you un-check the Auto-select check box. Display the selected bodies from the Graphics window.

- ***Options***. The Options box provides the following selections:

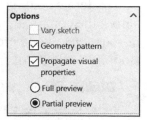

  - **Vary sketch**. Allows the pattern to change as it repeats.

  - **Geometry pattern**. Creates the pattern using only the geometry of faces and edges of the features, rather than patterning and resolving each instance of the features.

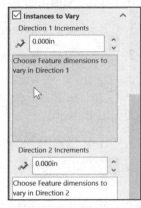

The Geometry Pattern option speeds up the creation and rebuilding of the pattern. You cannot create geometry patterns of features that have faces merged with the rest of the part.

- **Propagate visual properties**. Propagates SOLIDWORKS textures, colors, and cosmetic thread data to all pattern instances.

- *Instances to Vary*. Provides the ability to vary the dimensions and locations of instances in a feature pattern after it is created.

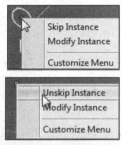

The Instance to Vary option in the Linear Pattern PropertyManager allows you to vary the dimensions and locations of instances in a feature pattern *after it is created*. You can vary the dimensions of a series of instances, so that each instance is larger or smaller than the previous one. You can also change the dimensions of a single instance in a pattern and change the position of that instance relative to the seed feature of the pattern. For linear patterns, you can change the spacing between the columns and rows in the pattern.

Tutorial: Linear Pattern 9-1

Create a Linear Pattern feature using the Pattern seed only option and the Geometry Pattern option.

1. Open **Linear Pattern 9-1** from the SOLIDWORKS 2018 folder.

2. Click the **Linear Pattern** ⬡ Features tool. Expand the **Features and Faces** option box.

3. Click inside the **Pattern Direction** box for Features and Faces. Click **Rib1** from the fly-out FeatureManager. Rib1 is displayed in the Features to Pattern box.

4. Click inside the **Pattern Direction** box for Direction 1.

5. Click the **hidden upper back vertical edge** of Shell1. Edge<1> is displayed in the Pattern Direction box. The direction arrow points upward. Click the Reverse direction button if required.

6. Click **Spacing and instances**.

7. Enter **.500**in for Spacing. Enter **3** for Number of Instances.

8. Expand the **Direction 2** option box.

9. Click the **hidden lower back vertical edge** of Shell1. The direction arrow points downward. Click the Reverse direction button if required.

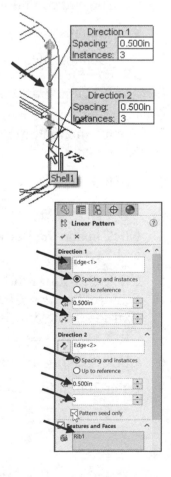

10. Enter **.500**in for Spacing. Enter **3** for Number of Instances.

11. Check the **Pattern seed only** box.

12. Check the **Geometry Pattern** box in the Options box.

Utilize the Geometry Pattern option to efficiently create and rebuild patterns.

13. Click **OK** ✅ from the Linear Pattern PropertyManager. LPattern1 is created and is displayed in the FeatureManager.

14. **Close** the model.

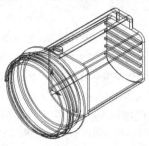

Know when to check the Geometry pattern box. Check the Geometry Pattern box when you require:

- An exact copy of the seed feature.

- Each instance is an exact copy of the faces and edges of the original feature.

- End conditions are not calculated.

- Saving rebuild time.

- Uncheck the Geometry Pattern box when you require:

  - The end condition to vary.

  - Each instance to have a different end condition.

  - Each instance is offset from the selected surface by the same amount.

Tutorial: Linear Pattern 9-2

Create a Linear Pattern feature using the Instances to skip option.

1. Open **Linear Pattern 9-2** from the SOLIDWORKS 2018 folder.

2. Click the **Linear Pattern** ⬚⬚ Features tool.

3. Expand the **Features and Faces** option box.

4. Click inside the **Pattern Direction** box for Features and Faces.

5. Click **Cut-Extrude1** from the fly-out FeatureManager. Click the **top horizontal edge** of Extrude1. The direction arrow points to the right. Click the Reverse direction button if required.

6. Click **Spacing and instances**.

7. Enter **20**mm for Spacing.

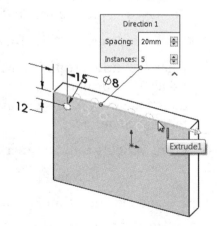

8. Enter **5** for Number of Instances.

9. Expand the **Direction 2** option box. Click inside the **Pattern Direction** box for Direction 2.

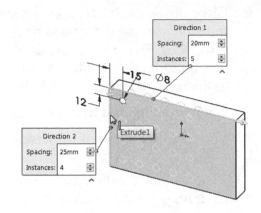

10. Click the **left vertical edge** of Extrude1. Edge<2> is displayed in the Pattern Direction box. The direction arrow points downward. Click the Reverse direction button if required.

11. Click **Spacing and instances**. Enter **25**mm for Spacing. Enter **4** for Number of Instances.

12. Un-check the **Pattern seed only** box.

13. Check the **Geometry pattern** box in the Options dialog box.

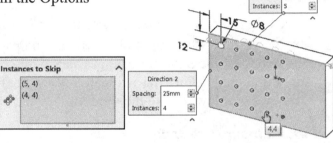

14. Expand the **Instances to Skip** box.

15. Select **(5,4)** & **(4,4)** from the Graphics window. (5,4) & (4,4) is displayed in the Instances to Skip box.

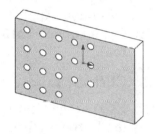

16. Click **OK** ✔ from the Linear Pattern PropertyManager. LPattern1 is created and is displayed in the FeatureManager.

17. **Close** the model.

Tutorial: Linear Pattern 9-3

Create a Linear Pattern feature using the Vary sketch option. Apply the seed feature to pattern three instances to the left.

1. Open **Linear Pattern 9-3** from the SOLIDWORKS 2018 folder.

2. Click the **Linear Pattern** ⊞ Features tool.

3. Expand the **Features and Faces** option box.

4. Click inside the **Pattern Direction** box for Features and Faces.

5. Click **Cut-Extrude2** from the fly-out FeatureManager.

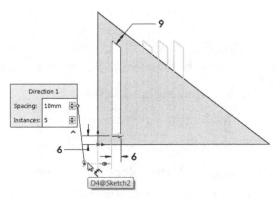

6. Click inside the **Pattern Direction** box for Direction 1. Click the **9**mm driving dimension from the Graphics window as illustrated. D4@Sketch2 is displayed in the Pattern Direction box for Direction 1. The pattern direction arrow is displayed to the right. If required, click the Reverse Direction box.

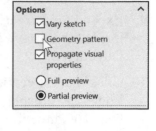

7. Click **Spacing and instances**. Enter **20**mm for Spacing.

8. Enter **4** for Number of Instances.

⚡ The feature sketch must be constrained to the boundary which defines the variation of the pattern instances. The angled top edge of the seed feature is parallel and is dimensioned to the angled edge on the base part. The feature sketch should be fully defined.

9. Check the **Vary sketch** box. Uncheck the **Geometry pattern** box. Click **OK** ✔ from the Linear Pattern PropertyManager. The height of the pattern instances vary while other relations are maintained. LPattern1 is created. View the results in the Graphics window. **Close** the model.

Tutorial: Linear Pattern 9-4

Create a Linear Pattern feature using the Feature Scope Selected body's option. Apply the skip instances option.

1. Open **Linear Pattern 9-4** from the SOLIDWORKS 2018 folder.

2. Click the **Linear Pattern** ⬚ Feature tool.

3. Expand the **Features and Faces** option box. Click inside the **Pattern Direction** box for Features and Faces.

4. Click **Extrude1** from the fly-out FeatureManager. Extrude1 is highlighted in the Graphics window.

5. Click inside the **Pattern Direction** box for Direction 1.

6. Click the **top edge** of Extrude-Thin1 as illustrated. Edge<1> is displayed in the Pattern Direction box for Direction 1. The direction arrow points to the right. Click the Reverse Direction button if required.

7. Click **Spacing and instances**.

8. Enter **20**mm for Spacing.

9. Enter **9** for Number of Instances.

10. Uncheck **Auto-select** in the Feature Scope box.

11. Check **Selected bodies** in the Feature Scope box.

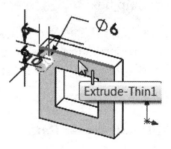

12. Click the **middle Extrude-Thin 1** feature in the Graphics window. Extrude-Thin1[2] is displayed in the Feature Scope box.

13. Click the **right Extrude-Thin1** feature in the Graphics window. Extrude-Thin1[3] is displayed in the Feature Scope box.

14. Check **Geometry pattern** in the Options box.

15. **Expand** the Instances to Skip box.

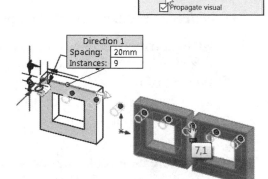

16. Select (**7,1**) from the Graphics window. (7,1) is displayed in the Instances to Skip box.

17. Click **OK** ✔ from the Linear Pattern PropertyManager. LPattern1 is displayed. View the results.

18. **Close** the model.

Circular Pattern Feature

The Circular Pattern ⬡ feature provides the ability to create copies or instances in a circular pattern. The Circular Pattern feature is controlled by a center of rotation, an angle, and the number of copies selected. The instances are dependent on the original. Changes to the originals are passed to the instanced features.

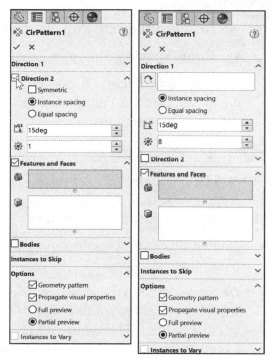

To create a Circular Pattern, select the features and an axis or edge as the center of rotation. Specify the total number of instances and the angular spacing between instances or the total number of instances and the total angle in which to create the circular pattern.

The Circular Pattern feature uses the Circular Pattern PropertyManager. The following selections are available:

- *Direction 1*. The Direction 1 box provides the following options:

    - **Pattern Axis**. Displays the selected axis, model edge, or an angular dimension from the Graphics window or from the fly-out FeatureManager. The circular pattern is created around the selected axis.

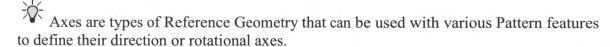 Axes are types of Reference Geometry that can be used with various Pattern features to define their direction or rotational axes.

    - **Reverse Direction**. Reverses the direction of the circular pattern if required.

- **Instance spacing**. Provides Instance spacing.

- **Equal spacing**. Provides equal spacing for instances.

- **Angle**. Displays the selected angle value in 1deg increments between each instance.

- **Number of Instances**. Displays the selected number of instances value for the seed feature.

- *Direction 2*. The Direction 2 box provides the following options:

  - **Symmetric**. Create circular patterns symmetrically or asymmetrically in both directions from the seed geometry. This is useful when the seed is not located at the end of the pattern arc. Adjust the angle, number of instances, and spacing settings for each direction independently. The Symmetric option applies the same settings to both directions.

⌁ You can create circular patterns symmetrically or asymmetrically in both directions from the seed geometry. This is useful when the seed is not located at the end of the pattern arc.

- **Instance spacing**. Provides Instance spacing.

- **Equal spacing**. Provides equal spacing for instances.

- *Features and Faces*. Displays the selected seed feature from the Graphics window to create the Circular Pattern and the selected faces from the Graphics window to create the circular pattern using the faces which compose the feature.

⌁ When using the Features and Faces, your pattern must remain within the same face or boundary. It cannot cross boundaries. Example: a cut across the complete face or different levels, such as a raised edge, would create a boundary and separate faces, stopping the pattern from spreading.

- *Bodies*. Displays the selected multi-body parts to create the pattern using bodies.

- *Instances to Skip*. Display the selected instances coordinates from the Graphics window which you want to skip in the circular pattern. The mouse pointer changes when you float over each pattern instance. To restore a pattern instance, click the instance again.

- *Feature Scope*. Applies features to one or more multi-body parts. Check the Geometry pattern box under the Options section of the Circular Pattern PropertyManager. Use this option to choose which bodies should include the feature. These are the Feature Scope options:

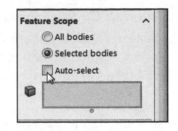

  - **All bodies**. Applies the feature to all bodies each time the feature is regenerated. If a new body is added to your model that is intersected by a feature, the new body is also regenerated to include that feature.

  - **Selected bodies**. Selected by default. Applies a feature to the body that is selected. If a new body is added to your model which is intersected by the feature, you are required to use the Edit Feature command to edit the pattern features. Select the bodies and add them to the list of your selected bodies.

  - **Auto-select**. Only available if you click the Selected bodies option. With the Auto-select option activated, when you first create a model with multi-body parts, the features automatically process all the relevant intersecting parts.

The Auto-select option is faster than the All bodies option because the Auto-select option processes only the bodies on the initial list and does not regenerate your entire model.

  - **Bodies to Affect**. Only available if you un-check the Auto-select option. Affects only the selected bodies from the Graphics window.

- *Options*. The Options box provides five selections:

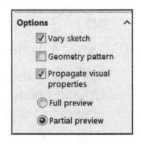

  - **Vary sketch**. Allows the pattern to change as it repeats.

  - **Geometry pattern**. Creates the pattern using only the geometry of face and edge of the feature, rather than patterning and resolving each instance of the feature.

  - **Propagate visual properties**. Propagates SOLIDWORKS textures, colors, and cosmetic thread data to all pattern instances.

  - **Full preview**. Displays a preview in the Graphics window.

  - **Partial preview**. Displays a partial preview in the Graphics window.

- *Instances to Vary*. Provides the ability to vary the dimensions and locations of instances in a feature pattern after it is created.

The Instance to Vary option in the Circular Pattern PropertyManager allows you to vary the dimensions and locations of instances in a feature pattern *after it is created.*

Tutorial: Circular Pattern 9-1

Create a Circular Pattern feature without using the Geometry pattern option. Apply the Skip instances option.

1. Open **Circular Pattern 9-1** from the SOLIDWORKS 2018 folder.

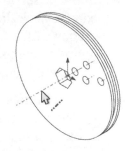

2. Click the **Circular Pattern**  Features tool. The Circular Pattern PropertyManager is displayed.

3. Click the **Temporary Axis** from the Graphics window at the center of the Hexagon as illustrated. Axis<1> is displayed in the Pattern Axis box.

4. Check the **Equal spacing** box.

5. Enter **360**deg for Angle.

6. Enter **8** for Number of Instances.

7. Click inside the **Features to Pattern** box.

8. Click **Cut-Extrude1** and **Cut-Extrude2** for Features to Pattern from the fly-out FeatureManager. Note: Cut-Extrude1 and Cut-Extrude2 are your seed features to pattern.

9. Un-check the **Geometry pattern** box.

10. **Expand** the Instances to Skip box.

11. Select hole section (**5**) from the Graphics window. (5) is displayed in the Instances to Skip box.

12. Click **OK** ✔ from the Circular Pattern PropertyManager. CirPattern1 is created and is displayed in the FeatureManager.

13. **Close** the model.

Tutorial: Circular Pattern Gear 9-1

Create a Circular Pattern feature using the Faces to Pattern option.

1. Open **Circular Pattern Gear 9-1** from the SOLIDWORKS 2018 folder.

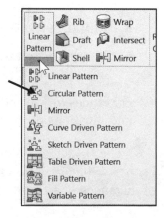

2. Click the **Circular Pattern**  Features tool. The Circular Pattern PropertyManager is displayed.

3. Click **Axis2** from the fly-out FeatureManager. Axis2 is displayed in the Pattern Axis box.

4. Check **Equal spacing**.

5. Enter **360**deg for Angle.

6. Enter **32** for Number of Instances.

7. Click inside the **Faces to Pattern** box.

8. Click the **9 faces** of the Gear tooth in the Graphics window as illustrated.

9. Check the **Geometry pattern** box and the **Propagate visual Properties** box in the Options section.

10. Click **OK** ✔ from the Circular Pattern PropertyManager. CirPattern1 is displayed. View the model in the Graphics window.

11. **Close** the model.

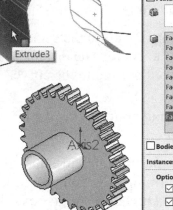

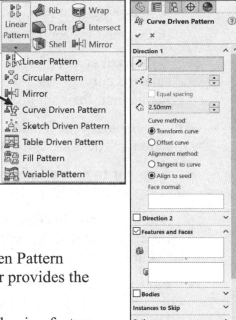

Curve Driven Pattern Feature

The Curve Driven Pattern ⟨icon⟩ feature provides the ability to create patterns along a planar or 3D curve. A pattern can be based on a closed curve or on an open curve, such as an ellipse or circle. Select a sketch segment, or the edge of a face, either a surface, (no mass) or solid, that lies along the plane to define the pattern.

To create a Curve Driven Pattern, select the features and a sketch segment or edge on which to pattern your feature. Then specify the curve type, curve method, and the alignment method.

The Curve Drive Pattern feature uses the Curve Driven Pattern PropertyManager. The Curve Drive PropertyManager provides the following selections:

- *Direction 1*. The Direction 1 box provides the following feature dependent options:

  - **Pattern Direction**. Displays the selected edge, sketch entity, or curve to use as the path for your pattern in Direction 1.

  - **Reverse Direction**. Reverses the direction of the pattern if required.

- **Number of Instances**. Displays the selected value for the number of instances of the seed feature in your pattern.

- **Equal spacing**. Sets spacing equal between each Curve Driven Pattern instance. The separation between instances is dependent on the curve that you select from the Pattern Direction box and on the Curve method option.

- **Spacing**. Displays the selected value for the distance between pattern instances along the curve. The distance between the curve and the Features to Pattern is measured normal to the curve.

The Spacing option is available only if you do not select the Equal spacing option.

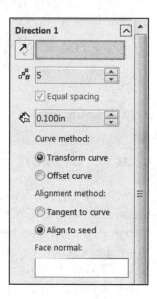

- **Curve method**. Defines the direction of the pattern by transforming how you use the curve selected for the Pattern Direction option box. There are two methods:

    - **Transform curve**. Selected by default. Uses the delta X and delta Y distances from the origin of the selected curve to the seed feature. These distances are maintained for each instance.

    - **Offset curve**. Normal distance from the origin of your selected curve to the seed feature. This distance is maintained for each instance.

- **Alignment method**. Defines the Alignment method for Direction 1. There are three methods:

    - **Tangent to curve**. Aligns each instance tangent to the curve selected for the Pattern Direction option box.

    - **Align to seed**. Selected by default. Aligns each instance to match the original alignment of the seed feature.

    - **Face normal**. Only available for 3D curves. Select the face on which the 3D curve lies to create the Curve Driven pattern.

- *Direction 2*. The Direction 2 box provides the following feature dependent options:

If you select the Direction 2 check box without selecting a sketch element or edge for the Pattern Direction option box from the Direction 2 section, an implicit pattern is created.

The created pattern is based on your selections from the Pattern Direction option under the Direction 1 section.

- **Pattern Direction**. Displays the selected edge, sketch entity, or curve to use as the path for your pattern in Direction 2.

- **Reverse Direction**. Reverses the direction of the pattern.

- **Number of Instances**. Displays the selected value for the number of instances of the seed feature in Direction 2.

- **Equal spacing**. Sets spacing equal between each pattern instance. The separation between instances is dependent on the curve that you selected for the Pattern Direction box and on the Curve method option in Direction 2.

- **Spacing**. Displays the selected value for the distance between pattern instances along your curve. The distance between instances depends on the curve selected for the Pattern Direction and on the Curve method option selected under Direction 1.

- **Pattern seed only**. Replicates only the seed pattern. This option creates a curve pattern under Direction 2, without replicating the curve pattern created under the Direction 1 section.

- *Features and Faces*. Displays the selected seed feature from the Graphics window and the selected faces from the Graphics window. This is a useful option with models that only import the faces that construct the feature, and not the feature itself.

- *Bodies*. Displays the selected bodies to pattern from the Graphics window for multi-body parts.

- *Instances to Skip*. Displays the selected pattern instance to skip from the Graphics window. Selected instances are removed from your pattern.

☀ Like other pattern types, such as Linear Pattern or Circular Pattern, you can select and skip pattern instances, and pattern either in one or two directions.

- *Options*. The Options box provides the following selections:

  - **Vary sketch**. Allows the pattern to change dimensions as it is repeated.

  - **Geometry pattern**. Creates a pattern using an exact copy of the seed feature. Individual instances of the seed feature are not solved. The End conditions and calculations are disregarded.

  - **Propagate visual Properties**. Propagates SOLIDWORKS colors, cosmetic thread data and textures to all pattern instances.

  - **Full preview**. Displays a preview in the Graphics window.

Tutorial: Curve Driven 9-1

Create a Curve driven pattern. Apply the equal spacing option.

1. Open **Curve Driven 9-1** from the SOLIDWORKS 2018 folder.

2. Click the **Curve Driven Pattern** Features tool.

3. Click the **sketched arc** on Extrude1. Arc1@Sketch3 is displayed in the Pattern Direction box.

4. Enter **5** for Number of Instances.

5. Check the **Equal spacing** box.

6. Click inside of the **Features to Pattern** box.

7. Click **Cut-Extrude1** for the fly-out FeatureManager. The Direction arrow points upward. If required, click the Reverse Direction button in the Direction 1 box.

8. Click **OK** ✔ from the Curve Driven Pattern PropertyManager. CrvPattern2 is created and is displayed in the FeatureManager.

9. **Close** the model.

Sketch Driven Pattern

The Sketch Driven Pattern feature provides the ability to use sketch points within a sketch for a pattern. The sketch seed feature propagates throughout the pattern to each point in the sketch pattern. You can use sketch driven patterns for holes or other feature instances.

The Sketch Driven Pattern feature uses the Sketch Driven Pattern PropertyManager. The following selections are available:

*   **Selections.** The Selections box provides the following options:

    *   **Reference Sketch.** Displays the selected sketch entity for the seed feature either from the Graphics window or from the fly-out FeatureManager.

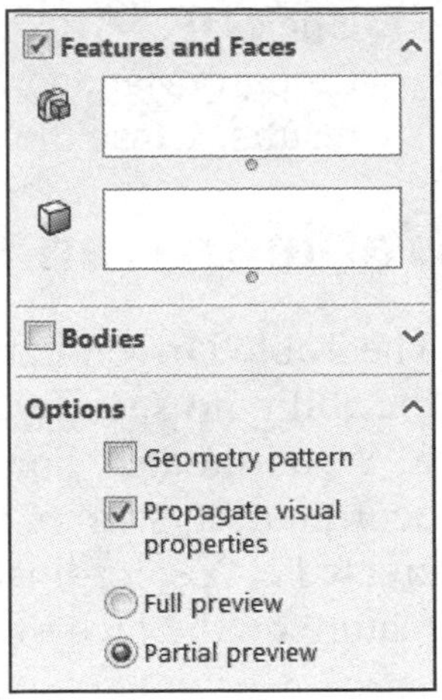

- **Centroid**. Provides the ability to use the centroid of the seed feature.
- **Selected point**. Provides the ability to use another point as the reference point.

- ***Features and Faces***. Displays the selected seed feature and the selected faces to create the Sketch Driven Pattern.
- ***Bodies***. Displays the selected bodies to pattern from the Graphics window for multi-body parts.
- ***Options***. The Options box provides the following selections:
  - **Geometry pattern**. Creates a pattern using an exact copy of the seed feature. Individual instances of the seed feature are not solved. The End conditions and calculations are disregarded.
  - **Propagate visual properties**. Propagates SOLIDWORKS colors, cosmetic thread data and textures to all pattern instances.
  - **Full preview**.

Tutorial: Sketch Driven Pattern 9-1

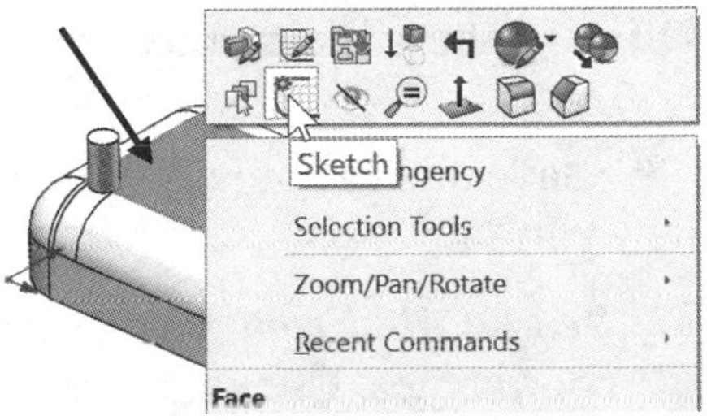

Create a Sketch Driven Pattern. Utilize a seed feature.

1. Open **Sketch Driven 9-1** from the SOLIDWORKS 2018 folder.
2. Right-click the **top face** of Extrude1.
3. Click **Sketch** from the Context toolbar.
4. Click the **Point** Sketch tool.
5. Sketch the illustrated **points** to represent a pattern, based on the seed feature.

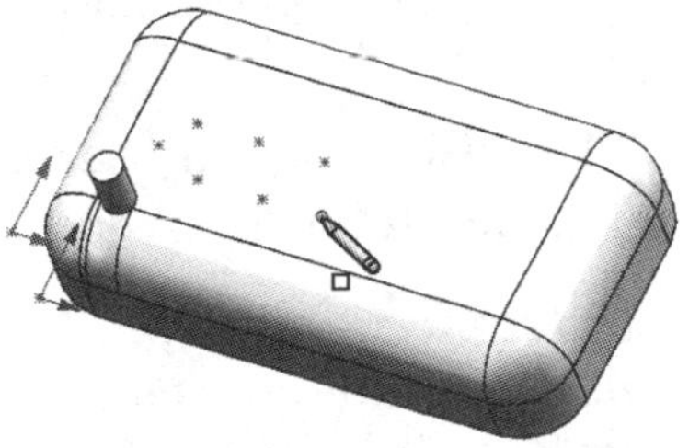

6. Click **OK** from the Point PropertyManager.
7. Right-click **Exit Sketch**.
8. Click the **Sketch Driven Pattern** Features tool. The Sketch Driven Pattern PropertyManager is displayed. Sketch3 is displayed in the Selections box.
9. Click **inside** the Features to Pattern box.
10. Click **Extrude2** from the fly-out FeatureManager. Extrude2 is your seed feature to pattern.

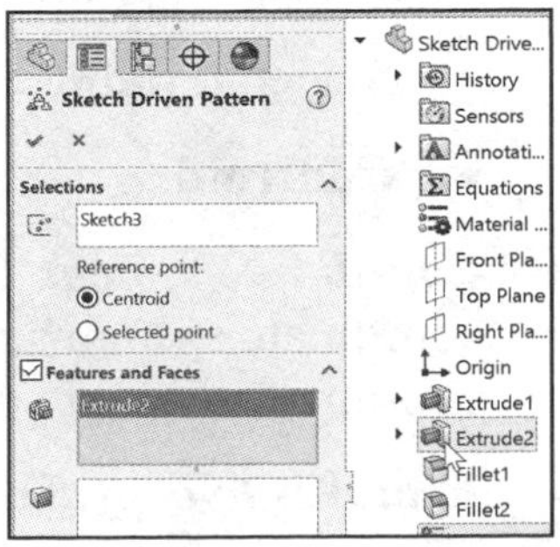

11. Click **OK** ✔ from the Sketch Driven Pattern PropertyManager. Sketch-Pattern1 is displayed. View the results. **Close** the model.

Table Driven Pattern Feature

The Table Driven Pattern ▦ feature provides the ability to specify a feature pattern using the X-Y coordinates. For a Table Pattern, create a coordinate system or retrieve previously created X-Y coordinates to populate a seed feature on the face of the model.

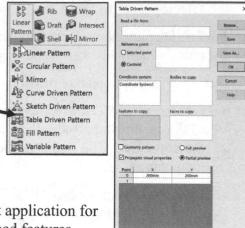

The origin of the created or retrieved coordinate system becomes the origin of the table pattern, and the X and Y axis define the plane in which the pattern occurs.

Hole patterns using X-Y coordinates are a frequent application for the Table pattern feature. You can also use other seed features, such as a boss, with your table driven patterns.

The Table Driven Pattern feature uses the Table Driven Pattern dialog box. The Table Driven Pattern dialog box provides the following selections:

🔅 Save and load the X-Y coordinates of a feature pattern, and then apply them to a new part later.

- **Read a file From**. Displays the selected imported pattern table or text file with X-Y coordinates. Click the Browse button on the right side of the Table Driven Pattern dialog box. Select a pattern table (*.sldptab) file or a text (*.txt) file to import into existing X-Y coordinates.

🔅 Text files can only have two columns. A single column for the X coordinate and a single column for the Y coordinate. The two columns must be separated by a delimiter.

- **Reference point**. Specifies the point to which the X-Y coordinates are applied when placing pattern instances. The X-Y coordinates of the reference point are displayed as Point 0 in the pattern table. There are two available selections:

  - **Selected point**. Sets the reference point to the selected vertex or sketch point.

  - **Centroid**. Sets the reference point to the centroid of the seed feature.

- **Coordinate system**. Displays the selected coordinate system, including the origin. The coordinate system and the origin are used to create the table pattern. Select the coordinate system you created from the FeatureManager design tree.

- **Features to copy**. Displays the created pattern based on the selected features. You can select multiple features from the Graphics window.

- *Faces to copy*. Creates the pattern based on the faces that make up the feature. Select the faces in the Graphics window. This is a useful option with models that only import the faces that make up the feature, and not the feature itself.

- *Bodies to copy*. Creates the pattern based on multi-body parts. Select the bodies to pattern from the Graphics window. Options:

  - **Geometry pattern**.

  - **Full preview**.

  - **Propagate visual properties.** Selected by default.

  - **Partial preview**. Selected by default.

☀ You cannot create geometry patterns of features that have faces merged with the rest of the part. The Geometry Pattern option speeds up the creation and rebuilding time of a pattern.

  - **Propagate visual properties**. Selected by default. Propagates SOLIDWORKS colors, cosmetic thread data, and textures to all pattern instances.

  - **X-Y coordinate table**. Creates the location points for the pattern instance using the X-Y coordinates. Double-click the area under Point 0 to enter the X-Y coordinates for each instance of the table pattern. The X-Y coordinates of the reference point are displayed for Point 0.

☀ You can use positive or negative coordinates. To enter a negative coordinate, precede the value with a minus (-).

Coordinate System

Before you create a Table Driven Pattern, review creating a Local or Relative Coordinate System and the Coordinate System PropertyManager. You can define a local or relative coordinate system for a part or assembly. Coordinate systems are useful:

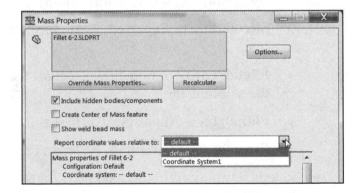

- With the Measure tool and Mass Properties tool.

- When exporting SOLIDWORKS documents to IGES, STL, ACIS, STEP, Parasolid, VRML and VDA.

- When applying assembly mates.

Coordinate PropertyManager

The Coordinate System PropertyManager is displayed when you add a new coordinate system to a part or assembly or when you edit an existing coordinate system. The Coordinate System PropertyManager provides the following selections:

- *Selections*. The Selections box provides the following options:

  - **Origin**. Displays the selected vertex, point, midpoint, or the default point of the origin on a part or assembly for the coordinate system origin.

  - **X axis**. Displays the selected X axis for the Axis Direction Reference. The available options are:

    - **Midpoint**, **vertex**, or **point**. Aligns the axis toward the selected point.

    - **Sketch line** or **linear edge**. Aligns the axis parallel to the selected edge or line.

    - **Sketch entity** or **Non-linear edge**. Aligns the axis toward the selected location on the selected entity.

    - **Planar face**. Aligns the axis in the normal direction of the selected face.

    - **Reverse Axis Direction**. Reverses the direction of an axis if required.

  - **Y axis**. Displays the selected Y axis for the Axis Direction Reference. The available options are:

    - **Midpoint**, **vertex**, or **point**. Aligns the axis toward the selected point.

    - **Sketch line** or **linear edge**. Aligns the axis parallel to the selected edge or line.

    - **Sketch entity** or **non-linear**. Aligns the axis toward the selected location on the selected entity.

    - **Planar face**. Aligns the axis in the normal direction of the selected face.

    - **Reverse Axis Direction**. Reverses the direction of an axis if required.

  - **Z axis**. Displays the selected Z axis for the Axis Direction Reference. The available options are:

    - **Midpoint**, **vertex**, or **point**. Aligns the axis toward the selected point.

    - **Sketch line** or **linear edge**. Aligns the axis parallel to the selected edge or line.

    - **Sketch entity** or **non-linear edge**. Aligns the axis toward the selected location on the selected entity.

- **Planar face**. Aligns the axis in the normal direction of the selected face.
- **Reverse Axis Direction**. Reverses the direction of an axis if required.

You can hide or show selected coordinate systems or all coordinate systems at once. To hide or show individual coordinate systems: Right-click the coordinate system and click Hide or Show.

Tutorial: Table Driven 9-1

Create a coordinate system. The origin of the coordinate system becomes the origin of the table pattern. The X and Y axes define the plane in which the pattern occurs. Create a Table Driven Pattern.

1. Open **Table Driven 9-1** from the SOLIDWORKS 2018 folder.

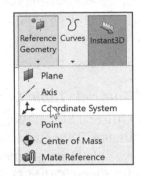

2. Click the **Coordinate System** icon from the Consolidated Reference Geometry Features drop-down toolbar. The Coordinate System PropertyManager is displayed.

Create the coordinate system.

3. Click the **bottom left vertex** as illustrated. Vertex<1> is displayed in the Origin box.

4. Click **OK** from the Coordinate System PropertyManager. Coordinate System1 is created and is displayed in the FeatureManager.

Create a Table Driven pattern.

5. Click the **Table Driven Pattern** Features tool. The Table Driven Pattern dialog box is displayed.

6. Click inside the **Coordinate system** box.

7. Click **Coordinate System1** from the FeatureManager.

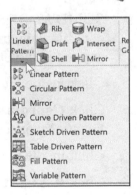

Use coordinate systems with the Measure and Mass Properties tools and for exporting SOLIDWORKS documents to STL, IGES, STEP, ACIS, Parasolid, VRML and VDA.

8. Click inside the **Features to copy** box.

9. Click **Boss-Extrude2** from the FeatureManager.

10. Double-click the **white text box** in the X column.

11. Enter **100**mm.

12. Double-click the **white text box** in the Y column.

13. Enter **50**mm.

14. Enter the remaining **X-Y** points as illustrated.

15. Click the **OK** from the Table Driven Pattern dialog box. TPattern1 is created and is displayed in the FeatureManager. View the results.

16. **Close** the model.

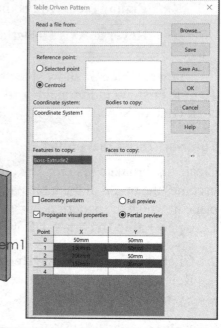

Tutorial: Table Driven 9-2

Utilize a SOLIDWORKS Pattern Table data file to create a Table Driven Pattern.

1. Open **Table Driven 9-2** from the SOLIDWORKS 2018 folder.

2. Click the **Table Driven Pattern** Feature tool. The Table Driven Pattern dialog box.

3. Click the **Browse** button from the Table Driven Pattern dialog box.

4. Double-click **PT1.sldptab** from the SOLIDWORKS 2018 folder. The Pattern Table data file contains X-, Y- coordinates and can be created in Notepad.

5. Click inside the **Coordinate system** box.

6. Click **Coordinate System1** from the FeatureManager. View the results.

7. Click inside the **Features to copy** box.

8. Click **Boss-Extrude2** from the FeatureManager.

9. **View** the results.

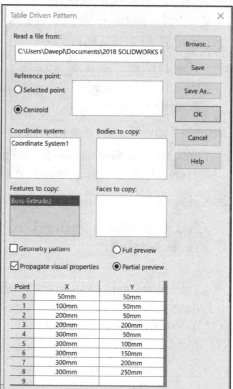

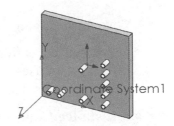

10. Click the **OK** from the Table Driven Pattern dialog box. TPattern1 is created and is displayed in the FeatureManager.

11. **Close** the model.

Fill Pattern Feature

The Fill Pattern 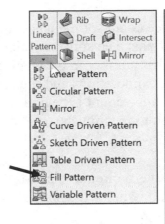 feature provides the ability to select a sketch which is on co-planar faces or an area defined by co-planar faces. The Fill Pattern feature fills the predefined cut shape or the defined region with a pattern of features. The Fill Pattern PropertyManager provides the ability to control the pattern layout. Typical uses for the Fill Pattern Feature are:

- Weight reduction.

- Grip surfaces.

- Ventilation holes.

The Fill Pattern feature uses the Fill Pattern PropertyManager. The Fill Pattern PropertyManager provides the following selections:

- *Fill Boundary*. Displays the selected sketches, planar curves on faces, a face, or co-planar faces. If you use a sketch for the boundary, you may need to select the pattern direction.

- *Pattern Layout*. Determines the layout pattern of the instances within the fill boundary. Select a customizable shape to pattern, or pattern a feature. Pattern instances are laid out concentrically from the seed feature. There are four Pattern layout selections:

  - **Perforation**. Creates a grid for a sheet metal perforation-style pattern. The perforation pattern is centered on the face if you do not select the vertex. If you select the vertex, the perforation pattern starts from the selected vertex. The following selections are available:

    - **Spacing**. Sets the distance value between the centers of your instances.

    - **Stagger Angle**. Sets the stagger angle value between rows of instances, starting at the vector used for your pattern direction.

    - **Margins**. Sets the margin value between the fill boundary and your outermost instance.

You can set a value of zero for margins.

- **Pattern Direction**. Sets the direction reference. If you do not specify a reference, the system uses the most appropriate reference (for example, the longest linear edge of the selected region).

- **Instance Count**. Fills the area using Number of Instances.

Obtain an instance count for the number of instances in a fill pattern for use in annotations and equations. The software calculates the number of instances in a fill pattern and displays that number in the graphics area. Use the instance count in a custom property and link it for annotations and equations. The number is not editable or configurable. The instance count is helpful for calculating open and closed volumes, conveying information to machinists, or determining a quantity based on a desired aesthetic.

- **Circular**. Creates a circular-shaped pattern. The following selections are available:

  - **Spacing**. Sets the distance between loops of instances using their centers.

  - **Target spacing**. Fills the area using the Spacing option value to set the distance between instances within each loop using their centers. Spacing can vary for each loop.

- **Instances per loop**. Fills the area using the Number of Instances option per loop.

- **Instances Spacing**. Sets the distance value between centers of instances within each loop. Check the Target spacing check box. Uncheck the Instances per loop check box.

- **Number of Instances**. Sets the number value of instances per loop. Check the Instances per loop check box. Uncheck the Target spacing check box.

- **Margins**. Sets the margin value between the fill boundary and the outermost instance. A zero value for margins can be set.

- **Pattern Direction**. Displays the selected direction reference. If you do not specify your reference, the system uses the most appropriate reference. Example: The longest linear edge of the selected region.

- **Instance Count**. Fills the area using Number of Instances.

- **Square**. Creates a square-shaped pattern. The following selections are available:

  - **Spacing**. Sets the distance between loops of instances using their centers.

  - **Target spacing**. Fills the area using the Spacing option value to set the distance between instances within each loop using their centers. Spacing can vary for each loop.

  - **Instances per side**. Fills the area using the Number of Instances option per side of each square.

  - **Instances Spacing**. Sets the distance value between centers of instances within each loop. Check the Target spacing check box. Uncheck the Instances per side check box.

  - **Number of Instances**. Sets the number value of instances per loop. Uncheck the Target spacing check box. Check the Instances per side check box.

  - **Margins**. Sets the margin value between the fill boundary and the outermost instance. A value of zero for the margins can be set.

  - **Pattern Direction**. Displays the selected direction reference. If you do not specify your reference, the system uses the most appropriate reference. Example: The longest linear edge of the selected region.

  - **Instance Count**. Fills the area using Number of Instances.

- **Polygon**. Creates a polygonal-shaped pattern. The following selections are available:

  - **Spacing**. Sets the distance between loops of instances using their centers.

  - **Sides**. Sets the number of sides in your pattern.

  - **Target spacing**. Fills the area using Spacing to set the distance between your instances within each loop using their centers. Actual spacing can vary for each loop.

  - **Instances per side**. Fills the area using the Number of Instances option, per side for each polygon.

  - **Instance Spacing**. Sets the distance value between centers of instances within each loop. Check the Target spacing check box. Uncheck the Instances per side check box.

  - **Number of Instances**. Sets the number value of instances per side of each polygon. Uncheck the Target spacing check box. Check the Instances per side check box.

- **Margins**. Sets the selected margin value between the fill boundary and the outermost instance. A value of zero for the margins can be set.

- **Pattern Direction**. Displays the selected direction reference. If you do not specify your reference, the system uses the most appropriate reference. Example: The longest linear edge of the selected region.

- **Instance Count**. Fills the area using Number of Instances.

- *Features and Faces*. Provides the ability to address selected features and predefined cut shapes and selected faces which you want to pattern. The selected faces must form a closed body which contacts the fill boundary face. The following selections are available:

  - **Selected features**. Provides the ability to select a feature to pattern in the Features to Pattern box.

  - **Create seed cut**. Provides the ability to customize a cut shape for the seed feature to pattern. The predefined cut shapes are **Circles**, **Squares**, **Diamonds**, **Polygons**.

  - **Diameter**. Sets the diameter.

  - **Vertex or Sketch point**. Locates the center of the seed feature at the selected vertex or sketch point and creates the pattern starting from this point. If you leave this box empty, the pattern is centered on the fill boundary face.

  - **Flip Shape Direction**. Reverses the direction of the seed feature about the face selected in the Fill Boundary option if required.

  - **Faces to Pattern**. Select faces to pattern. The faces must form a closed body that contacts the fill boundary face.

  If you select a vertex, the shape seed feature is located at the vertex. Otherwise, the seed feature is located at the center of the fill boundary.

- *Bodies*. Displays the selected bodies which you want to pattern.

- *Instances to Skip*. Displays the selected pattern instance to skip from the Graphics window. Selected instances are removed from your pattern.

- *Options*. The Options box provides the following selections:

  - **Vary sketch**. Allows the pattern to change dimensions as it is repeated.

- **Geometry pattern**. Creates a pattern using an exact copy of the seed feature. Individual instances of the seed feature are not solved. The End conditions and calculations are disregarded.

- **Propagate visual Properties**. Propagates SOLIDWORKS colors, textures, and cosmetic thread data to all pattern instances.

- **Show preview**. Displays a preview in the Graphics window.

- **Partial preview**. Displays a partial preview in the Graphics window.

Tutorial: Fill Pattern 9-1

Create a Polygon Layout Fill Pattern feature. Apply the seed cut option.

1. Open **Fill Pattern 9-1** from the SOLIDWORKS 2018 folder.

2. Click the **Fill Pattern** Features tool. The Fill Pattern PropertyManager is displayed.

3. Click the **Front face** of Extrude1. Face<1> is displayed in the Fill Boundary box. The direction arrow points to the right.

4. Click **Polygon** for Pattern Layout.

5. Click **Target spacing**.

6. Enter **15**mm for Loop Spacing.

7. Enter **6** for Polygon sides.

8. Enter **10**mm for Instances Spacing.

9. Enter **10**mm for Margins. View the direction arrow.

10. Click **Create seed cut**.

11. Click **Circle** for Features to Pattern.

12. Enter **4**mm for Diameter.

13. Click **OK** from the Fill Pattern PropertyManager. Fill Pattern1 is created and is displayed in the FeatureManager.

14. **Close** the model.

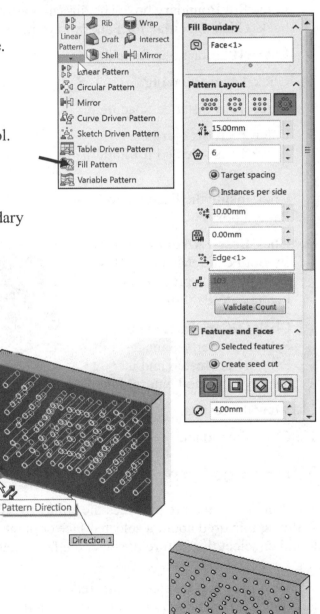

Tutorial: Fill Pattern 9-2

Create a Circular Layout Fill Pattern feature.

1. Open **Fill Pattern 9-2** from the SOLIDWORKS 2018 folder.

2. Click the **Fill Pattern** 🔲 Features tool. The Fill Pattern PropertyManager is displayed.

3. Click the **top face** of Boss-Extrude1. Face<1> is displayed in the Fill Boundary box. The direction arrow points to the back. Click **Circular** for Pattern Layout.

4. Enter **.10**in for Loop Spacing.

5. Click **Target spacing**.

6. Enter .**10**in Instance Spacing.

7. Enter .**05**in for Margins. Edge<1> is selected for Pattern Direction.

8. Click **inside** the Features to Pattern box.

9. Click **Boss-Extrude2** from the fly-out FeatureManager. Boss-Extrude2 is displayed in the Features to Pattern box.

10. Click **OK** ✔ from the Fill Pattern PropertyManager. Fill Pattern1 is created and is displayed in the FeatureManager.

11. **View** the results.

12. **Close** the model.

Mirror Feature

The Mirror 🗗 feature provides the ability to create a copy of a feature, or multiple features, mirrored about a selected face or plane. The Mirror feature provides the option to either select the feature or select the faces that compose the feature.

💡 A mirror part creates a mirrored version of an existing part. This is a good way to generate a left-hand version and a right-hand version of a part. The mirrored version is derived from the original version. The two parts always match.

The Mirror ⵌ feature uses the Mirror PropertyManager. The Mirror PropertyManager provides the following selections:

- **Mirror Face/Plane**. Displays the selected face or plane from the Graphics window.

- **Features to Mirror**. Displays the selected features, or faces to mirror. The Geometry Pattern check box is available when using the Features to Mirror option.

- **Faces to Mirror**. Displays the selected imported parts where the import process included the selected faces of the feature, but not the feature itself. In the Graphics window, click the faces that make up the feature you want to mirror. The Geometry Pattern check box is available when using the Faces to Mirror option.

- **Bodies to Mirror**. Displays the selected entire model. Select the model from the Graphics window or fly-out FeatureManager. The mirrored model attaches to the face you selected.

- **Options**. The Options box provides the following selections:

  - **Geometry Pattern**. Mirrors only the geometry, faces and edges of the features, rather than solving the whole feature.

  - **Propagate visual properties**. Mirrors the visual properties of the mirrored entities. Example: Select the Propagate Visual Properties option to propagate SOLIDWORKS colors, textures, and cosmetic thread data to all pattern instances and mirrored features.

  - **Full preview**. Displays the model in the Graphics window.

🔅 If you select the Bodies to Mirror option, the following options are displayed in the Option section:

  - **Merge solids**. Creates a mirrored box that is attached to the original body but is a separate entity. You need to select a face on a solid part, and then clear the Merge solids option check box. If you select the Merge solid option check box, the original part and the mirrored part become a single entity.

  - **Knit surfaces**. Provides the ability to knit two surfaces together. Select to mirror a surface by attaching the mirror face to the original face without intersections or gaps between the surfaces.

Example: Mirror 9-1

Create a Mirror feature about the Right plane. Mirror an existing rib to increase structural integrity of the model.

1. Open **Mirror 9-1** from the SOLIDWORKS 2018 folder.

2. Click **Rib2** from the FeatureManager.

3. Click the **Mirror** 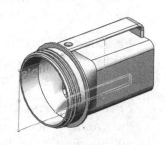 Features tool. The Mirror PropertyManager is displayed. Rib2 is displayed in the Features to Mirror box.

4. Click inside of the **Mirror Face/Plane** box.

5. Click **Right Plane** from the fly-out FeatureManager. Right is displayed in the Mirror Face/Plane box.

6. Address the **Options** box as illustrated.

7. Click **OK** ✔ from the Mirror PropertyManager. Mirror1 is created and is displayed in the FeatureManager.

8. **Close** the model.

Tutorial: Mirror 9-2

Create a Mirror feature about the Right plane using the Geometric Pattern option.

1. Open **Mirror 9-2** from the SOLIDWORKS 2018 folder. Mirror an existing extruded cut feature.

2. Click **Cut-Extrude2** from the FeatureManager.

3. Click the **Mirror** 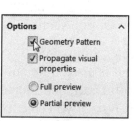 Features tool. The Mirror PropertyManager is displayed. Cut-Extrude2 is displayed in the Features to Mirror box.

4. Click inside of the **Mirror/Face/Plane** box.

5. Click **Right Plane** from the FeatureManager.

6. Address the **Options** box as illustrated. Click **OK** ✔ from the Mirror PropertyManager. Mirror1 is created and is displayed in the FeatureManager.

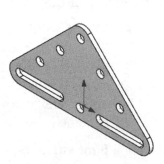

7. **Close** the model.

Coordinate System

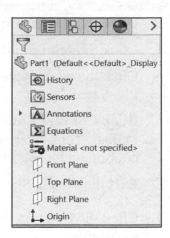

Global Coordinate System

Directional input refers by default to the global coordinate system (X, Y and Z), which is based on Plane1 with its origin located at the origin of the part or assembly. Plane1 (Front) is the first plane that appears in the FeatureManager design tree and can have a different name. The reference triad displays the global X-, Y- and Z-directions.

The figure below illustrates the relationship between the global coordinate system and Plane 1 (Front), Plane 2 (Top) and Plane 3 (Right).

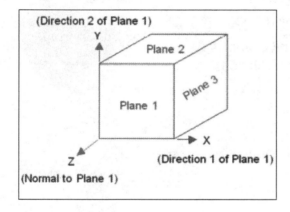

Where X is Direction 1 of Plane 1, Y is Direction 2 of Plane 1 and Z is the Normal to Plane 1.

Local Coordinate Systems

Local (Reference) coordinate systems are coordinate systems other than the global coordinate system. You can specify restraints and loads in any desired direction. For example, when defining a force on a cylindrical face, you can apply it in the radial, circumferential, or axial directions. Similarly if you choose a spherical face, you can choose the radial, longitude, or latitude directions. In addition, you can use reference planes and axes.

Tutorial: Coordinate System 9-1

Create a local coordinate system. Utilize the Mass Properties tool to calculate the Center of mass for a part located at a new coordinate location through a point.

1. Open **Coordinate System 9-1** from the SOLIDWORKS 2018 folder. View the default global coordinate system location.

Create a new coordinate system location. Locate the new coordinate system location at the center of the center hole as illustrated.

Base-Extrude

2. Right-click the **front face** of Base-Extrude.

3. Click **Sketch** from the Context toolbar.

4. Click the **edge** of the center hole as illustrated.

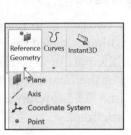

5. Click **Convert Entities** from the Sketch toolbar. The center point for the new coordinate location is displayed.

6. **Exit** the sketch. Sketch4 is displayed.

7. Click the **Coordinate System** tool from the Consolidated Reference Geometry toolbar. The Coordinate System PropertyManager is displayed.

8. Click the **center point** of the center hole in the Graphics window. Point2@Sketch4 is displayed in the Selections box as the Origin.

9. Click **OK** from the Coordinate System PropertyManager. Coordinate System1 is displayed.

10. **View** the new coordinate location of the center hole.

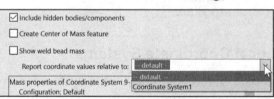
Coordinate System1

View the Mass Properties of the part with the new coordinate location.

11. Click the **Mass Properties** tool from the Evaluate tab.

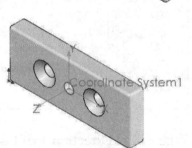

12. Select **Coordinate System1** from the Output box. The Center of mass relative to the new location is located at the following coordinates:

X = 0 millimeters

Y = 0 millimeters

Z = -3 millimeters

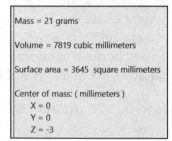

Mass = 21 grams

Volume = 7819 cubic millimeters

Surface area = 3645 square millimeters

Center of mass: (millimeters)
   X = 0
   Y = 0
   Z = -3

13. **Reverse** the direction of the axes as illustrated (+X, -Y, -Z).

14. **Close** the model.

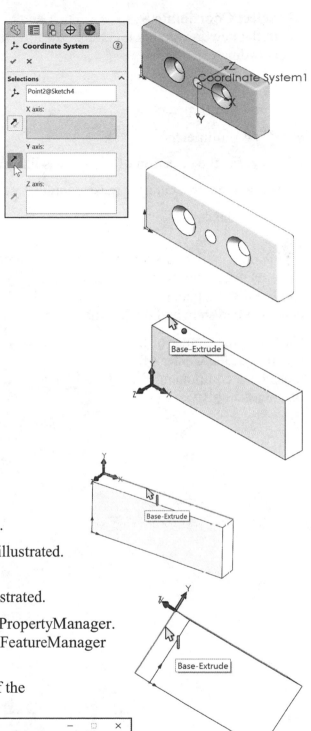

💡 To reverse the direction of an axis, click its **Reverse Axis Direction** button in the Coordinate System PropertyManager.

Tutorial: Coordinate System 9-2

Create a new coordinate system location. Locate the new coordinate system at the top back point as illustrated.

1. Open **Coordinate System 9-2** from the SOLIDWORKS 2018 folder.

2. **View** the location of the part Origin.

3. Drag the **Rollback bar** under the Base-Extrude feature in the FeatureManager.

4. Click the **Coordinate System** ⤴ tool from the Consolidated Reference Geometry toolbar. The Coordinate System PropertyManager is displayed.

5. Click the **back left vertex** as illustrated.

6. Click the **top back horizontal** edge as illustrated. Do not select the midpoint.

7. Click the **back left vertical** edge as illustrated.

8. Click **OK** from the Coordinate System PropertyManager. Coordinate System1 is displayed in the FeatureManager and in the Graphics window.

9. Drag the **Rollback bar** to the bottom of the FeatureManager.

10. **Calculate** the Center of mass relative to the new coordinate system.

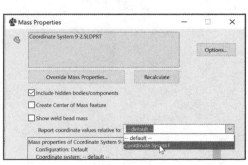

11. Select **Coordinate System1**. The Center of mass relative
to the new location is located at the following
coordinates:

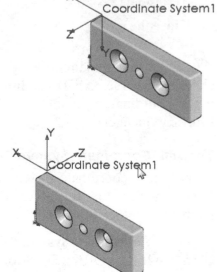

- X = 28 millimeters

- Y = 11 millimeters

- Z = 4 millimeters

12. **Reverse** the direction of the axes as illustrated.

13. **View** the new center of mass coordinates.

14. **Close** the model.

You can hide or show selected coordinate
systems or all coordinate systems at once. To hide or
show individual coordinate systems, Right-click the
coordinate system and click Hide or Show.

Calculate the Center of mass for an assembly or select the
location of the Center of
mass for a component
of the assembly.

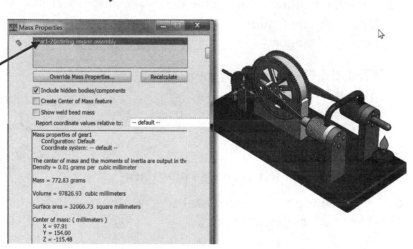

Summary

In this chapter, you learned about the Pattern Features (Linear Pattern, Circular Pattern, Curve Driven Pattern, Sketch Driven Pattern, Table Driven Pattern and Fill Pattern), Mirror Features and the Coordinate System PropertyManager.

The Linear Pattern feature provides the ability to create copies, or instances, along one or two linear paths in one or two different directions. The instances are dependent on the selected originals.

The Circular Pattern feature provides the ability to create copies or instances in a circular pattern. The Circular Pattern feature is controlled by a center of rotation, an angle, and the number of copies selected. The instances are dependent on the original.

The Instance to Vary option in the Linear and Circular Pattern PropertyManager allows you to vary the dimensions and locations of instances in a feature pattern. You can vary the dimensions of a series of instances so that each instance is larger or smaller than the previous one. You can also change the dimensions of a single instance in a pattern and change the position of that instance relative to the seed feature of the pattern.

For Linear patterns, you can change the spacing between the columns and rows in the pattern.

For circular patterns, you can arrange the instances to become closer together or farther apart. You can create circular patterns symmetrically or asymmetrically in both directions from the seed geometry. This is useful when the seed is not located at the end of the pattern arc.

The Curve Driven Pattern feature provided the ability to create patterns along a planar or 3D curve. A pattern can be based on a closed curve or on an open curve, such as an ellipse or circle.

The Sketch Driven Pattern feature provided the ability to use sketch points within a sketch for a pattern. The sketch seed feature propagates throughout the pattern to each point in the sketch pattern. You can use sketch driven patterns for holes or other feature instances.

The Table Pattern feature provided the ability to specify a feature pattern using the X-Y coordinates. For a Table Pattern, create a coordinate system or retrieve a previously created X-Y coordinates to populate a seed feature on the face of the model. The origin of the created or retrieved coordinate system becomes the origin of the table pattern, and the X and Y axes define the plane in which the pattern occurs.

The Fill Pattern feature provided the ability to select a sketch which is on co-planar faces or an area defined by co-planar faces. The Fill Pattern feature fills the predefined cut shape or the defined region with a pattern of features.

The Mirror feature provided the ability to create a copy of a feature, or multiple features, mirrored about a selected face or plane. The Mirror feature provides the option to either select the feature or select the faces that compose the feature.

A mirror part created a mirrored version of an existing part. This is a good way to generate a left-hand version and a right-hand version of a part. The mirrored version is derived from the original version. The two parts always match.

You addressed the Global and Local coordinate system. Global and local coordinate systems are used to locate geometry items. When you define a node or a key point, its coordinates are interpreted in the global Cartesian system by default. For some models, however, it may be more convenient to define the coordinates in a system other than global Cartesian. You can input the geometry in any of three predefined (global) coordinate systems, or in any number of user defined (local) coordinate systems.

A global coordinate system can be thought of as an absolute reference frame. Three predefined global systems are available: *Cartesian, cylindrical,* and *spherical*. All three of these systems are right-handed and, by definition, share the same origin.

In Chapter 10, explore and address the Swept Boss/Base, Swept Cut, Lofted Boss/Base, Lofted Cut, Wrap, Flex and Freeform Feature in SOLIDWORKS.

CHAPTER 10: SWEPT, LOFTED, WRAP, FLEX AND FREEFORM FEATURE

Chapter Objective

Chapter 10 provides a comprehensive understanding of the Swept, Lofted, Wrap, Flex and Freeform feature. On the completion of this chapter, you will be able to:

- Understand and utilize the following Swept features:
  - Swept Base with various options.
  - Swept Boss with Helix/Spiral.
  - Swept Thin with options.
  - Swept Cut with options.
- Recognize and utilize the following Lofted features:
  - Lofted Boss/Base with options.
  - Lofted Cut with options.
  - Adding a Lofted Section.
- Apply the Wrap feature: Analytical, Spline Surface
- Comprehend and apply the Flex PropertyManager.
- Understand and apply the Freeform feature.

Swept Feature

A Swept 🖋 feature creates a base, boss, cut, or surface by moving a profile along a path. There are a few rules that you should know when creating a Swept feature: Sweep creates a base, boss, cut, or surface by moving a profile (section) along a path, according to these rules: 1.) The profile must be closed for a base or boss sweep feature; 2.) The profile may be open or closed for a surface sweep feature; 3.) The path may be open or closed; 4.) The path may be a set of sketched curves contained in one sketch, a curve, or a set of model edges; 5.) The path must intersect the plane of the profile; 6.) Neither the section, the path, nor the resulting solid can be self-intersecting; 7.) The guide curve must be coincident with the profile or with a point in the profile sketch.

Swept Boss/Base Feature

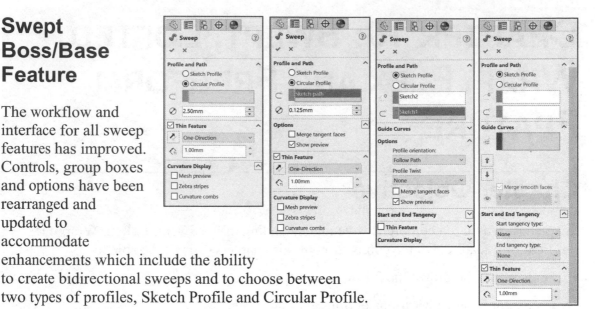

The workflow and interface for all sweep features has improved. Controls, group boxes and options have been rearranged and updated to accommodate enhancements which include the ability to create bidirectional sweeps and to choose between two types of profiles, Sketch Profile and Circular Profile.

Sketch Profile is similar to existing functionality where you create a sweep by moving a 2D sketch profile along a 2D or 3D sketch path, but it provides more options. Circular Profile lets you create a sweep directly on a model without having to work in a sketch.

The Swept Boss/Base 🐛 feature uses the Sweep PropertyManager. The Sweep PropertyManager provides the following selections:

- *Profile and Path*. The Profile and Path box provides two selections:

    - **Sketch Profile**. Select the profile sketch in the graphics area or FeatureManager design tree. The profile must be closed for a base or boss sweep feature. The profile may be open or closed for a surface sweep feature.

    - **Circular Path**. Displays the selected path along which the profile sweeps from the Graphics window or FeatureManager. It now lets you create a sweep directly on a model without having to work in a sketch.

- *Options*. The Options box provides the following feature dependent selections:

    - **Profile Orientation**. Provides the ability to select the type to control the orientation of the Profile as it sweeps along the Path. The Type options are:

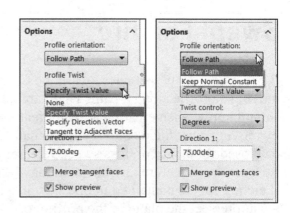

- **Follow Path**. Selected by default. Section remains at same angle with respect to path at all times.

- **Keep Normal Constant**. This option keeps the section parallel to the beginning section as it twists along the path.

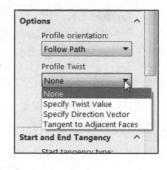

- **Profile Twist**. Provides the following options:

  - **None**. No twist.

  - **Specify Twist Value**. Define the value of the twist. Select Degrees, Radians or Turns.

  - **Specify Direction Vector**. Sets the direction of the vector.

  - **Tangent to Adjacent Faces**. If the sweep profile has tangent segments, causes the corresponding surfaces in the resulting sweep to be tangent. Faces that can be represented as a plane, cylinder, or cone are maintained. Other adjacent faces are merged, and the profiles are approximated. Sketch arcs may be converted to splines.

  - **Merge tangent faces**. Causes the corresponding surfaces in the resulting sweep to be tangent if the sweep profile has tangent segments. Faces that can be represented as a plane, cylinder, or cone are maintained. Other adjacent faces are merged, and the profiles are approximated.

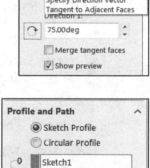

  - **Show preview**. Selected by default. Displays a shaded preview of the sweep. Clear the Show preview check box to only display the profile and path.

- **Profile and Path using a 2D and 3D sketch**.

  - **Merge tangent faces**. Causes the corresponding surfaces in the resulting sweep to be tangent if the sweep profile has tangent segments. Faces that can be represented as a plane, cylinder, or cone are maintained. Other adjacent faces are merged, and the profiles are approximated.

  - **Show preview**. Selected by default. Displays a shaded preview of the sweep. Clear the Show preview check box to only display the profile and path.

  - **Merge result**. Merges the solids into a single body.

  - **Align with end faces**. Continues the sweep profile up to the last face encountered by the path. The faces of the sweep are extended or truncated to match the faces at the ends of the sweep without requiring additional geometry. This option is commonly used with helices.

- *Guide Curves*. Displays the selected guide curves from the Graphics window. This option guides the profile as it sweeps along the path. Document dependent.

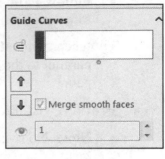

  - **Move Up Arrow**. Adjusts the profile order upwards.

  - **Move Down Arrow**. Adjusts the profile order downwards.

  - **Merge smooth faces**. Clear the Merge smooth faces check box option to improve your system performance of sweeps with guide curves and to segment the sweep at all points where the guide curve or path is not curvature continuous.

  - **Show Sections**. Displays the sections of the sweep. Select the arrows to view and troubleshoot the profile by the Section Number.

- *Start/End Tangency*. The Start/End Tangency box provides the following selections:

  - **Start tangency type**. The available options are:

    - **None**. No tangency is applied.

    - **Path Tangent**. Creates the sweep normal to the path at the start point.

  - **End tangency type**. The available options are:

    - **None**. No tangency is applied.

    - **Path Tangent**. Creates the sweep normal to the path at the end point.

- *Thin Feature*. Provides the ability to set the type of thin feature sweep. The available options are:

  - **One-Direction**. Creates the thin feature in a single direction from the profiles using the Thickness value.

  - **Reverse Direction**. Reverses the profile direction if required.

  - **Mid-Plane**. Creates the thin feature in two directions from the profiles, applying the same thickness value in both directions.

  - **Two-Direction**. Creates the thin feature in both directions from the profiles. Set individual values for both thicknesses.

- *Curvature Display*.

  - **Mesh preview**. Provides a preview mesh on the selected faces to better visualize the surfaces.

- **Zebra stripes**. Displays zebra stripes, to make it easier to see surface wrinkles or defects.

- **Curvature combs**. Activates the display of curvature combs.

Tutorial: Swept Base 10-1

Create an o-ring part. Create the part with the Swept Base feature. The Swept Base feature requires a sketch path and a circular profile diameter.

1. Open **Swept Base 10-1** from the SOLIDWORKS 2018 folder. The FeatureManager is displayed.

2. Click the **Swept Boss/Base**  Features tool. The Sweep PropertyManager is displayed.

3. Click the **Circular Profile** box.

4. Enter .**125**mm for Diameter.

5. Click the **Circular Profile** in the Graphics window. Sketch-path is displayed in the path (Sketch-path) box.

6. Click **OK** ✔ from the Sweep PropertyManager. Sweep1 is created and is displayed in the FeatureManager.

7. **Expand** Sweep1. View the results.

8. **Close** the model.

💡 For non-circular sketch profiles, create the sketch profile on a perpendicular plane to the path and use the pierce relation to locate the profile on the path.

Use the Circular Profile option to create a solid rod or hollow tube along a sketch line, edge, or curve directly on a model without having to sketch. This enhancement is available for Swept Boss/Base, Swept Cut, and Swept Surface features.

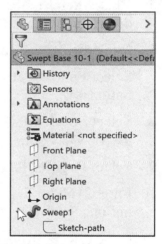

💡 The PropertyManager provides the ability to select a configuration after the Sweep feature is created.

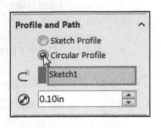

☀ Use the Circular Profile option to create a solid rod or hollow tube along a sketch line, edge, or curve directly on a model without having to sketch. This enhancement is available for Swept Boss/Base, Swept Cut and Swept Surface features.

Tutorial: Swept Base 10-2

Create a Swept Base feature. Create a simple spring (six turns).

1. Open **Swept Base 10-2** from the SOLIDWORKS 2018 folder. The FeatureManager is displayed. View two simple sketches on the Front Plane.

2. Click the **circumference** of the circle in the Graphics window, Sketch2.

3. Click the **Swept Boss/Base** 🖉 Features tool. The Sweep PropertyManager is displayed. Sketch2 is displayed in the Profile box.

4. Click inside the **Path** box.

5. Click **Sketch1** from the fly-out FeatureManager. Sketch1 is displayed in the Path box.

6. Click the **drop-down arrow** to expand the Options box.

7. Select **Follow Path** for Profile orientation. View your options.

8. Select **Specify Twist Value** for Profile Twist. View your options.

9. Select **Degrees** for Twist control.

10. Enter **2160** for Twist angle. View your options.

11. Click **OK** ✔ from the Sweep PropertyManager. Sweep1 is created and is displayed in the FeatureManager. View the results.

12. **Close** the model.

Tutorial: Swept Boss 10-1

Create a Swept Boss feature using the Helix/Spiral PropertyManager. Create a variable pitch helix for a spring.

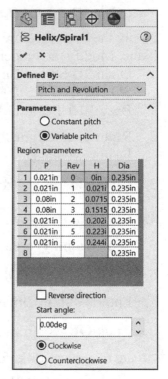

1. Open **Swept Boss 10-1** from the SOLIDWORKS 2018 folder. The FeatureManager is displayed.

2. Edit **Sketch1**.

3. Click **Insert ≻ Curve ≻ Helix/Spiral** from the Menu bar menu. The Helix/Spiral PropertyManager is displayed. The default direction of rotation is clockwise. Pitch and Revolution Type is selected by default.

4. Check the **Variable pitch** box.

5. Create a **Region parameters table**. Create a spring. Enter the following information as illustrated. Coils 1, 2, 5, 6 & 7 are the closed ends of the spring. The pitch needs to be slightly larger than the wire.

6. Enter **.021**in for the Pitch.

7. Enter Start angle of **0**deg.

8. Click **Clockwise**.

9. Click **OK** ✔ from the Helix/Spiral PropertyManager.

10. Display an **Isometric** view. View the results.

Create the Swept Boss feature.

11. Click the **Swept Boss/Base** ✍ Features tool. The Sweep PropertyManager is displayed.

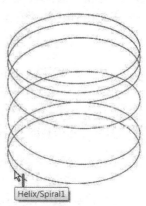

12. Click **Circular Profile**.

13. Click **Helix/Spiral1** from the Graphics window as illustrated.

14. Enter **.015in** for Depth.

15. Click **OK** ✔ from the PropertyManager. View the results.

16. **Close** the model.

Tutorial: 3D Swept Base 10-1

Apply the Swept feature in a 3D Sketch on the Right plane. Modify the model due to a failure.

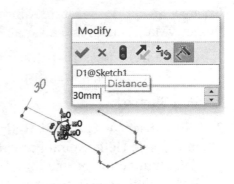

1. Open **3D Swept Base 10-1** from the SOLIDWORKS 2018 folder. The 3D Swept Base 10-1 FeatureManager is displayed.

2. Right-click **Right Plane** from the FeatureManager.

3. Click **Sketch** ⊞ from the Context toolbar.

4. Click the **Polygon** ⬡ Sketch tool.

5. Sketch a **polygon** at the origin as illustrated.

6. Click the **Smart Dimension** ⟡ Sketch tool.

7. Enter a **30**mm dimension for the construction circle.

8. Click **OK** ✔ from the PropertyManager.

Create the Swept Boss feature.

9. Click the **Swept Boss/Base** 🐛 Features tool. The Sweep PropertyManager is displayed.

10. Click **Sketch Profile**.

11. Click **Sketch1** from the fly-out FeatureManager. Sketch1 is displayed in the Path box.

12. Click **3D Sketch1** from the fly-out FeatureManager. 3DSketch1 is displayed in the Profile box.

13. Click **OK** ✔ from the Sweep PropertyManager. The feature failed due to the radius of the path. Modify the profile dimension.

14. Click **Cancel** ✗ from the Sweep PropertyManager.

Modify the profile dimension.

15. Double-click **Sketch1** from the FeatureManager. The Modify dialog box is displayed.

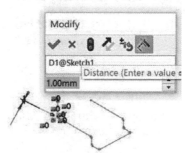

16. Enter **1mm** for dimension.

17. **Rebuild** 🔧 the sketch.

18. Click **OK** ✔ from the Modify dialog box.

19. Click **OK** ✔ from the dimension PropertyManager.

Create the Swept Boss feature.

20. Click the **Swept Boss/Boss** ∮ Features tool. The Sweep PropertyManager is displayed.

21. Click **Sketch1** from the fly-out FeatureManager. Sketch1 is displayed in the Path box.

22. Click **3D Sketch1** from the fly-out FeatureManager. 3DSketch1 is displayed in the Profile box.

23. Click **OK** ✔ from the Sweep PropertyManager. Sweep1 is displayed in the FeatureManager. View the results.

24. **Close** the model.

Sweep Thin Feature

A Sweep Thin feature adds material by moving an open profile with a specified thickness along a path. A Sweep Thin feature requires a sketched profile, thickness, and path. The Sweep Thin feature uses the Sweep PropertyManager.

Tutorial: Sweep Thin 10-1

Create a Sweep Thin profile. Remember, to create a sweep you need a sketch; a closed, non-intersecting profile on a face or a plane and you need to create the path for the profile.

1. Open **Sweep Thin 10-1** from the SOLIDWORKS 2018 folder.

2. Check **View ➤ Display ➤ Zebra Stripes** from the Menu bar. The Zebra Stripes PropertyManager is displayed. View your options.

3. Click **OK**.

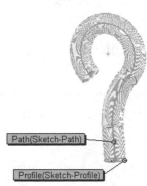

4. Click the **Swept Boss/Base** ∮ Features tool. The Sweep PropertyManager is displayed.

5. Click **Sketch Profile**.

6. Click inside the **Profile** box.

7. Click **Sketch-Profile** from the FeatureManager. Sketch-Profile is displayed in the Sweep-Profile text box.

8. Click inside the **Sweep-Path** box.

9. Click **Sketch-Path** from the fly-out FeatureManager. Sketch-Path is displayed in the Sweep-Path text box.

10. **Expand** the Options box.

11. Select **Follow Path** for Profile orientation.

12. Select **Specify Twist Value** for Profile Twist.

13. Select **Degrees** for Twist control.

14. Enter **0** for Angle. Review the surface.

15. Select **Two-Direction** for Thin Feature Type.

16. Enter **.010**in for Direction 1 and Direction 2 Thickness.

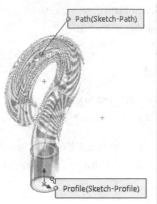

17. Click **OK** ✔ from the Sweep PropertyManager. Sweep-Thin1 is created and is displayed in the FeatureManager.

18. Display an **Isometric view**. View the results.

19. **Close** the model.

💡 The Zebra stripes function provides the ability to display small changes in a surface which may be difficult to view with a standard display. Zebra stripes simulate the reflection of long strips of light on a very shiny surface.

Tutorial: Sweep Guide Curves 10-1

Create a Swept feature utilizing the Guide Curves option.

1. Open **Sweep Guide Curves 10-1** from the SOLIDWORKS 2018 folder. The Sweep Guide Curves 10-1 FeatureManager is displayed.

2. Click the **Swept Boss/Base** 🗗 Features tool. The Sweep PropertyManager is displayed. Select **Sketch Profile**.

3. Click **Profile1** from the fly-out FeatureManager. Profile1 is displayed in the Profile box.

4. Click inside the **Path** box.

5. Click **Path1** from the fly-out FeatureManager. Path1 is displayed in the Path box. **Expand** the Guide Curves box.

6. Click the **Guide Curve1** Spline sketch from the Graphics window as illustrated. Guide Curve1 is displayed.

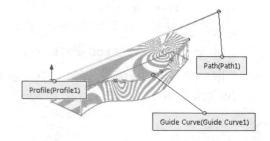

7. Click **OK** ✔ from the Sweep PropertyManager. Sweep1 is created and is displayed in the FeatureManager.

8. **Close** the model.

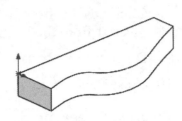

🔆 To insert an additional Guide Curve, insert a sketch before the Swept feature. Then edit the Swept feature to insert the new Guide Curve.

Tutorial: Sweep Guide Curves 10-2

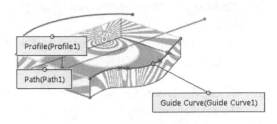

Create a Swept feature utilizing the Guide Curves option.

1. Open **Sweep Guide Curves 10-2** from the SOLIDWORKS 2018 folder.

2. Click **Sweep1** from the FeatureManager.

3. Right-click **Edit Feature**. The Sweep1 PropertyManager is displayed.

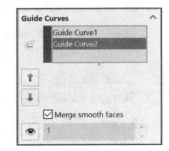

4. Click inside the **Guide Curves** box.

5. Click **Guide Curve2** from the fly-out FeatureManager. Guide Curve2 is displayed in the Guide Curves box.

6. Click **OK** ✔ from the Sweep PropertyManager. View the FeatureManager.

7. **Close** the model.

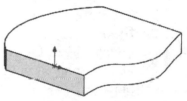

Tutorial: Sweep Twist 10-1

Create a Swept profile. Utilize the Sypecify Twist Value option to twist 75 degrees along the path.

1. Open **Sweep Twist 10-1** from the SOLIDWORKS 2018 folder.

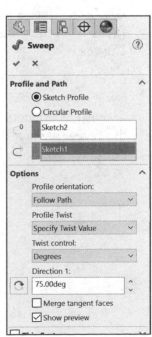

2. Click the **Swept Boss/Base** 𝒮 Features tool. The Sweep PropertyManager is displayed. Select **Sketch Profile**.

3. Click **Sketch2** from the fly-out FeatureManager. Sketch2 is displayed in the Profile box.

4. Click **Sketch1** from the fly-out FeatureManager. Sketch1 is displayed in the Path box.

5. **Expand** the Options box. Select **Follow Path** for Profile orientation. Select **Specify Twist Value**.

6. Select **Degrees** for Twist control.

7. Enter **75**deg for Angle.

8. Click **OK** ✔ from the Sweep PropertyManager. View Sweep1 from the FeatureManager.

9. Display an **Isometric view**. View the results.

10. **Close** the model.

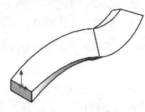

Tutorial: Sweep Merge Tangent Faces 10-1

Create a Sweep profile with Guide Curves. Utilize the Orientation Merge tangent faces option and Merge smooth faces option to create a single surface.

1. Open **Sweep Merge Tangent Faces 10-1** from the SOLIDWORKS 2018 folder.

2. Click the **Swept Boss/Base** 🐛 Features tool. The Sweep PropertyManager is displayed.

3. Select **Sketch Profile**.

4. Click **Sketch-Profile** from the fly-out FeatureManager. Sketch-Profile is displayed in the Profile box.

5. Click **Sketch-Path** from the fly-out FeatureManager. Sketch-Path is displayed in the Path box.

6. **Expand** the Guide Curves box.

7. Click **Guide Curve** from the fly-out FeatureManager. Guide Curve is displayed in the Graphics window.

8. Uncheck **Merge smooth faces** from the Guide Curves box.

9. Click **OK** ✔ from the Sweep PropertyManager. Sweep1 is created. View the faces displayed in the Graphics window.

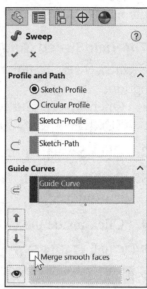

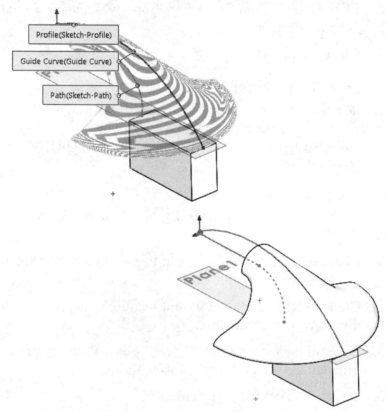

10. Right-click **Sweep1** from the FeatureManager.

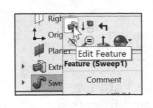

11. Click **Edit Feature**. The Sweep1 PropertyManager is displayed.

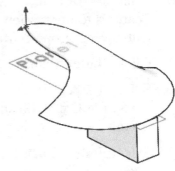

12. Check **Merge tangent faces** in the Options box.

13. Check **Merge smooth faces** in the Guide Curves box.

14. Click **Yes** to the warning message.

15. Click **OK** ✅ from the Sweep PropertyManager. A single face is displayed in the Graphics window.

16. **Close** the model.

Swept Cut Feature

A Swept Cut 🗔 feature removes material from a part. A Swept Cut requires a sketched profile and path. Utilize the Swept Cut feature to create threads. The Swept Cut feature uses the Cut-Sweep PropertyManager.

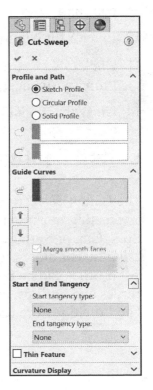

You can use the Circular Profile option to create a solid rod or hollow tube along a sketch line, edge, or curve directly on a model without having to sketch. This enhancement is available for Swept Boss/Base, Swept Cut, and Swept Surface features.

Tutorial: Sweep Cut Feature 10-1

Create a Simple Swept Cut. Create a Swept Cut feature along a handle.

1. Open **Sweep Cut 10-1** from the SOLIDWORKS 2018 folder.

2. Sketch a vertical **Line** from the Origin above the top of the model on the Front Plane as illustrated.

3. **Exit** the Sketch.

4. Sketch a **small circle below** the model on the Front Plane, near the left corner as illustrated. It is very important that you create the circle below the model. The diameter of the circle will impact the profile of the cut.

5. **Exit** the Sketch.

6. Click **Swept Cut** 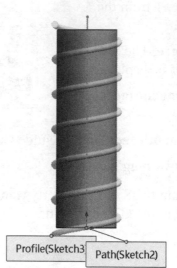 from the Features toolbar. View the Cut-Sweep PropertyManager.

7. Click **Sketch Profile**.

8. Select the **Profile** and the **Path** for the Swept Cut as illustrated from the fly-out FeatureManager.

9. Select **Follow Path** for Profile orientation.

10. Select **Specify Twist Value** for Profile Twist.

11. Select **Revolutions** for Twist control.

12. Enter **6** for Number of Revolutions.

13. Click **OK** from the Cut-Sweep PropertyManager. View the results.

14. **Close** the model.

Lofted Feature

The Lofted  feature provides the ability to create a feature by creating transitions between profiles. A loft can be a Base, Boss, Cut, or Surface feature. Create a Lofted feature by using two or more profiles. Only the first, last, or first and last profiles can be points. All sketch entities, including guide curves and profiles, can be enclosed in a single 3D Sketch. The Lofted feature uses the Loft PropertyManager.

The Loft PropertyManager provides the following selections:

- *Profiles*. Displays the selected faces, sketch profiles, or edges selected from the Graphics window or the fly-out FeatureManager.

🔆 Lofts are created based on the order of the profile selection. Select the point in which you want your path of the loft to travel for each profile.

- **Move Up Arrow**. Adjusts the profile order upwards.

- **Move Down Arrow**. Adjusts the profile order downwards.

- *Start/End Constraints*. Applies a constraint to control tangency for the start and end profiles. The selections for the Start constraints are:

  - **Default**. (Available with three profiles minimum). Approximates a parabola scribed between the first and last profiles. The tangency from this parabola drives the loft surface, which results in a more predictable and natural loft surface when matching conditions are not specified.

  - **None**. No tangency constraint. A zero curvature is applied.

  - **Direction Vector**. Applies a tangency constraint based on the selected entity which is used as a start direction vector. Select the Direction Vector option. Set the Draft angle option and the Start Tangent Length option.

  - **Start Tangent Length**. Not available when the None option is selected from the Start Constraint section. Controls the amount of influence on the loft feature. The effect of tangent length is limited up to the next section.

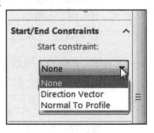

  - **Reverse Direction**. Reverses the tangent direction length if required.

  - **Normal to Profile**. Applies a tangency constraint normal to the selected start profile. Set the Draft angle option. Set the Start Tangent Length or End Tangent Length option.

- **Tangency to Face**. Only available when attaching a loft to existing geometry. This option makes the adjacent faces tangent at your selected start profile.

- **Curvature to Face**. Only available when attaching a loft feature to existing geometry. Applies a smooth curvature continuous loft at your selected start profile.

- **Next Face**. Only available with the Tangency to Face option selected or when the Curvature to Face option is selected from the Start section. Ability to toggle the loft feature between available faces.

- **Direction Vector**. Only available with the Direction Vector option selected for Start or End constraint. Applies a tangency constraint based on the selected entity used as a direction vector. The loft is tangent to the selected linear edge or axis, or to the normal of a selected face or plane.

- **Draft angle**. Only available with the Direction Vector or Normal to Profile option selected for the Start or End constraint. Applies a draft angle to the start or end of your profile.

- **Start and End Tangent Length**. Not available with the None option selected from the Start or End constraint. Controls the amount of influence on the loft.

- **Apply to all**. Displays one handle that controls all the constraints for the entire profile. Clear this option to display multiple handles that permit individual segment control. Drag the handles to modify the tangent length.

- ***End Constraints***. Applies a constraint to control tangency of the end profile. The options for the End constraints are:

  - **Default**. (Available with three profiles minimum.) Approximates a parabola scribed between the first profile and your last profile. The tangency from the parabola drives the loft surface. This results in a predictable loft surface when matching conditions are not specified.

  - **None**. Provides no tangency constraint. A zero curvature is applied.

  - **Direction Vector**. Only available with the Direction Vector option selected from the Start Constraint section. Applies a tangency constraint based on your selected entity used as a direction vector. The loft feature is tangent to your selected axis, linear edge or to the normal of a selected face or plane.

- **Draft angle**. Only available when the Direction Vector option or the Normal to Profile option is selected from the End Constraint section. Set a draft angle to the end of your profile.

- **Reverse Direction**. Reverses the direction of the applied draft angle if required.

- **End Tangent Length**. Not available when the None option is selected from the Start or End Constraint section. Controls the amount of influence on the loft feature. The effect of tangent length is limited up to the next section.

  - **Reverse Direction**. Reverses the tangent direction length if required.

  - **Apply to all**. Displays a handle which controls the constraints for the entire profile.

Clear the Apply to all check box option to display multiple handles. Clearing this option permits individual segment control. Drag the handles to modify the tangent length.

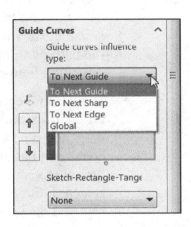

- *Guide Curves*. The Guide Curves influence box provides the ability to control the influence of the guide curves on the loft. The available options are:

  - **To Next Guide**. Extends your guide curve influence only to the next guide curve.

  - **To Next Sharp**. Extends your guide curve influence only to the next sharp.

  - **To Next Edge**. Extends your guide curve influence only to the next edge.

  - **Global**. Extends the guide curve influence to the entire loft.

A sharp is a hard corner of a profile. Example: any two contiguous sketch entities that do not have a tangent or an equal curvature relation with each other.

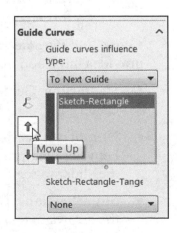

- **Global Curves**. The Global Curves box provides the ability to display the select guide curves to control the loft feature.

  - **Move Up Arrow**. Adjusts the profile order upwards.

  - **Move Down Arrow**. Adjusts the profile order downwards.

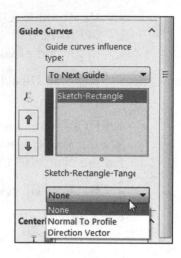

- **Guide tangency type**. Provides the ability to control the tangency where your loft feature meets the guide curves. The available options are document dependent:

  - **None**. No tangency constraint.

  - **Normal To Profile**. Applies a tangency constraint normal to your plane of the guide curve. Set the Draft angle option.

  - **Direction Vector**. Applies a tangency constraint based on your selected entity used as the direction vector. Select the Direction Vector option. Set the Draft angle option.

  - **Tangency to Face**. Only available when a guide curve lies on the edge of existing geometry. Adds side tangency between adjacent faces that lie along the path of a guide curve, creating a smoother transition between adjacent faces.

  - **Direction Vector**. Only available when the Direction Vector option is selected from the Guide tangency type section. Applies a tangency constraint based on your selected entity used as the direction vector. The loft feature is tangent to your selected linear axis or edge, or to the normal of a selected plane or face.

  - **Draft angle**. Only available when the Direction Vector option or the Normal to Profile option is selected from the Start or End Constraint section. Applies a draft angle to the loft feature along the guide curve.

  - **Reverse Direction**. Reverses the direction of the draft if required.

- *Centerline Parameters*. The Centerline Parameters box provides three options:

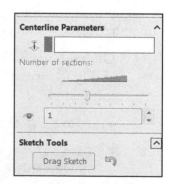

  - **Centerline**. Guides the loft feature shape using a centerline. Select a sketch from the Graphics window.

  - **Number of sections**. Provides the ability to add sections between your profile and the centerline. Click and drag the slider to regulate the number of sections.

  - **Show Sections**. Displays the loft feature sections. Click the arrows in this option to display the required sections.

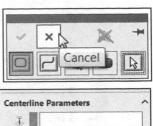

☀ Enter your **section number**, click the **Show Sections** option and directly go to your desired section.

- *Sketch Tools*. Use the Selection Manager to help select sketch entities.

  - **Drag Sketch**. Enables drag mode. When editing the loft feature, you can drag any 3D sketch segments, points, or planes from the 3D sketch from which contours have been defined for the loft. The 3D sketch updates as you drag. You can also edit the 3D sketch to dimension the contours using dimensioning tools. The loft preview updates when the drag ends or when you edit the 3D sketch dimensions. To exit drag mode, click Drag Sketch again or another selection list in the PropertyManager.

  - **Undo sketch drag**. Undoes the previous sketch drag and returns the preview to its previous state. You can undo multiple drags and dimension edits.

- *Options*. The Options box provides the following selections:

  - **Merge tangent faces**. Selected by default. Causes the corresponding surfaces in the resulting loft to be tangent if the corresponding lofting segments are tangent. Faces that can be represented as a plane, cylinder, or cone are maintained. Other adjacent faces are merged, and the sections are approximated. Sketch arcs may be converted to splines.

  - **Close loft**. Creates a closed body along the loft feature direction. The closed body connects your first and last sketch automatically.

  - **Show preview**. Selected by default. Displays the shaded preview of the loft feature. Uncheck the Show preview box to only view the path and the guide curves.

  - **Merge results**. Selected by default. Merges all the loft elements. Clear this option to not merge all the loft elements.

- *Thin Feature*. The Thin Feature box provides the ability to select the options required to create a thin loft feature. The options are:

  - **Thin feature type**. Sets the type of thin loft feature. The available options are:

    - **One-Direction**. Creates the thin feature in a single direction from the profiles using the Thickness value.

    - **Reverse Direction**. Reverses the direction of the profile if required.

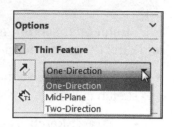

- **Mid-Plane**. Creates the thin feature in two directions from the profiles. This option applies the same Thickness value in both directions.

- **Two-Direction**. Creates the thin feature in two directions from the profiles. There are two selections:

- **Thickness 1**. Thickness value in the first direction.

- **Thickness 2**. Thickness value in the second direction.

- *Feature Scope*. Specifies which bodies you want the feature to affect. The options are:

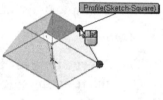

  - **All bodies**.

  - **Selected bodies**.

  - **Auto-select**.

Tutorial: Loft Thin 10-1

Create a Lofted Thin feature using the Normal To Profile End constraint with a 30deg draft angle.

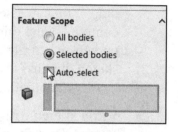

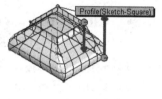

1. Open **Loft Thin 10-1** from the SOLIDWORKS 2018 folder. Sketch-Rectangle sketch plane is Top Plane. Sketch-Square sketch plane is Plane1.

2. Click **Lofted Boss/Base** 🛢 from the Features toolbar. The Loft PropertyManager is displayed.

3. Click inside the **Profiles** box. Click the **back right corner point** of Sketch-Rectangle.

4. Click the **back right corner point** of Sketch-Square. Sketch-Rectangle and Sketch-Square are displayed in the Profile box.

5. Select **Normal To Profile** for the Start constraint.

6. Enter **30**deg for Draft angle. Enter **1.5** for Start Tangent Length. Click **OK** ✔ from the Loft PropertyManager. Loft-Thin1 is created and is displayed in the FeatureManager.

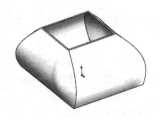

7. **Close** the model.

Tutorial: Loft Guide Curves 10-1

Create a Lofted base feature with profiles on different planes and two Guide Curves.

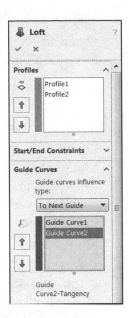

1. Open **Loft Guide Curves 10-1** from the SOLIDWORKS 2018 folder. The part contains four sketches. Profile1 is sketched on the Front plane. Profile2 is sketched on Plane1. Guide Curve1 and Guide Curve2 are sketched on the Right plane.

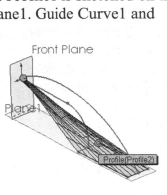

2. Click the **Lofted Boss/Base** Features tool. The Loft PropertyManager is displayed.

3. Click **Profile1** from the fly-out FeatureManager.

4. Click **Profile2** from the fly-out FeatureManager. Profile1 and Profilc2 are displayed in the Profiles box.

5. Click inside of the **Guide Curves** box.

6. Select **Guide Curve1** and **Guide Curve2** from the fly-out FeatureManager.

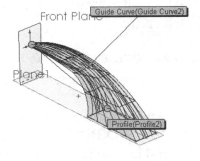

7. Click **OK** from the Loft PropertyManager. Loft1 is created and is displayed in the FeatureManager.

8. **Close** the model.

Tutorial: Loft Guide Curves 10-2

View the Lofted base feature with two profiles on different planes and three Guide Curves. The Guide Curves were created using the Curve Through XYZ points tool. The Loft feature provides the ability to create a feature by creating transitions between profiles.

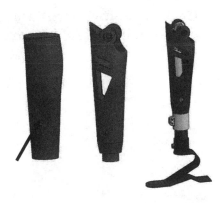

1. Open the **Loft Guide Curves 10-2** assembly from the SOLIDWORKS 2018 folder.

2. **Open** the Cylinder sub-assembly.

3. **Open** the NEW CASTING. View the NEW CASTING part in the Graphics window.

4. Roll the **rollback bar** under Loft2.

5. **Expand** Loft2. Sketch3 and Sketch4 are profiles. Curve2, Curve3 and Curve4 are Guide Curves. Sketch3 sketch plane is Plane1. Sketch4 sketch plane is Plane2. Curve2, Curve3 and Curve4 are used as Guide curves in the Loft feature. The Curves were created using the Curve through XYZ feature.

6. **Edit** Loft2. View the selected profiles and Guide Curves.

7. **Close** all models.

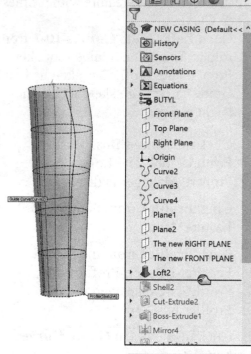

☀ Use the SelectionManager to help select sketch entities.

Tutorial: Loft to Point 10-1

Create a Boss Loft feature utilizing a point for a profile.

1. Open **Loft to Point 10-1** from the SOLIDWORKS 2018 folder. Sketch1 contains a single point on Plane2.

2. Click the **Rectangular** face of Loft1 as illustrated.

3. Click the **Lofted Boss/Base** ⬇ Features tool. Face<1> is displayed in the Profiles box.

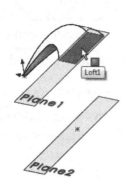

4. Click **Sketch1** from the fly-out FeatureManager. Sketch1 is displayed in the Profiles box.

5. Select **Tangency To Face** for Start constraint.

6. Enter **1.5** Start Tangent Length.

7. Display an **Isometric** view. Accept the default conditions.

8. Click **OK** ✔ from the Loft PropertyManager. Loft2 is created and is displayed in the FeatureManager.

9. **Close** the model.

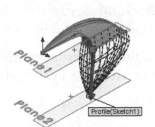

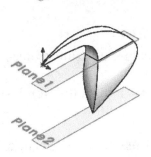

Tutorial: Loft Multi-body 10-1

Create a Lofted boss feature utilizing two individual bodies.

1. Open **Loft Multibody 10-1** from the SOLIDWORKS 2018 folder. The Loft Multibody 10-1 FeatureManager is displayed.

2. Click the **Lofted Boss/Base** ⬇ Features tool.

3. Click the **left circular face** of Extrude1 as illustrated. Face<1> is displayed in the Profiles box.

4. Click the **right bottom elliptical face** of Extrude2 as illustrated. Face<2> is displayed in the Profiles box.

5. Click inside the **Guide Curves** box.

6. Click the **bottom arc** for the first Guide Curve.

7. Click the **green check mark** to Select Open Loop. Open Group <1> is displayed in the Guide Curves box.

8. Click the **top arc** for the second Guide Curve.

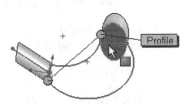

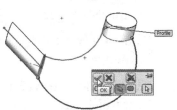

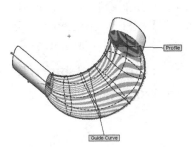

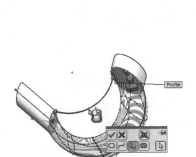

9. Click the **green check mark** to Select Open Loop. Open Group<2> is displayed in the Guide Curves box.

10. Click **OK** ✔ from the Loft PropertyManager. Loft1 is created and is displayed in the FeatureManager.

11. Display an **Isometric** view. View the results.

12. **Close** the model.

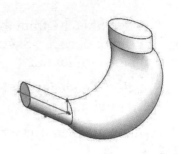

Tutorial: Loft Twist 10-1

Create a Lofted base feature utilizing profiles and the Guide Curve influence options.

1. Open **Loft Twist 10-1** from the SOLIDWORKS 2018 folder.

2. Click the **Lofted Boss/Base** Features tool. The Loft PropertyManager is displayed.

3. Click the **right bottom corner** point of Sketch1 in the Graphics window. Sketch1 is displayed in the Profiles box.

4. Click the **left top corner point** of Sketch2 in the Graphics window. Sketch2 is displayed in the Profiles box. The loft twists as it blends the two profiles.

5. **Deselect** Sketch2 in the Profiles box.

6. Click the **right top corner point** of Sketch1 to create a straight loft.

7. Click inside the **Guide Curves** box.

8. Click **GuideCurve1** from the fly-out FeatureManager.

9. Click **To Next Edge** for Guide curves influence.

10. Click **GuideCurve2** from the fly-out FeatureManager for the second Guide Curve.

11. Click **OK** ✔ from the Loft PropertyManager. Loft1 is created and is displayed in the FeatureManager.

12. Display an **Isometric** view. View the results.

13. **Close** the model.

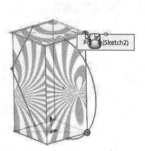

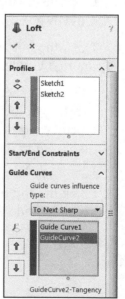

Lofted Cut Feature

The Lofted Cut  feature removes material by blending two or more sketches on different planes. A Lofted Cut feature requires a minimum of two sketched profiles. The Lofted Cut feature uses the Cut-Loft PropertyManager.

Tutorial: Loft Cut 10-1

Create a Lofted Cut feature utilizing two profiles.

1. Open **Loft Cut 10-1** from the SOLIDWORKS 2018 folder. The Cut Loft 10-1 FeatureManager is displayed.

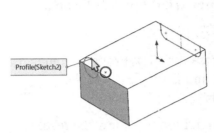

2. Click the **Lofted Cut** 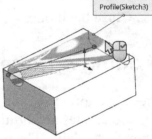 Features tool. The Cut-Loft PropertyManager is displayed.

3. Click the **circumference** of the circle. Sketch2 is displayed in the Profiles box.

4. Click the top edge of **Sketch3** as illustrated. Sketch3 is displayed in the Profiles box.

5. Click **OK** ✔ from the Cut-Loft PropertyManager. Cut-Loft1 is displayed in the FeatureManager.

6. Display an **Isometric** view. View the results.

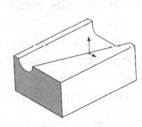

7. **Close** the model.

Tutorial: Loft with Flex 10-1

Create a Loft feature with Flex. A Flex feature deforms complex models in an intuitive manner.

1. Open **Loft Flex 10-1** from the SOLIDWORKS 2018 folder. The Loft Flex 10-1 FeatureManager is displayed.

2. Click the **Lofted Boss/Base** 🔱 Features tool. The Loft PropertyManager is displayed.

3. Click the **top right endpoint** for each profile in the Graphics window. Sketch1, Sketch2 and Sketch3 are displayed in the Profiles box.

4. Click **OK** ✔ from the Loft PropertyManager. Loft1 is created and is displayed in the PropertyManager.

5. Click **Insert > Features > Flex** ⑧. The Flex PropertyManager is displayed.

6. Click **Loft1** from the Graphics window. Solid Body<1> is displayed in the Bodies for Flex box.

7. Drag the **reference triad** towards Plane2.

8. Enter **15**deg for Angle.

9. Click **OK** ✔ from the Flex PropertyManager. Flex1 is created and is displayed in the FeatureManager.

10. Display an **Isometric** view. View the results.

11. **Close** the model.

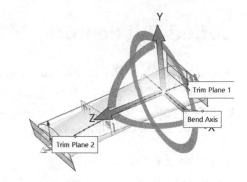

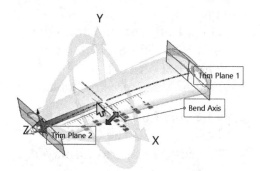

Adding A Lofted Section

Adding a lofted section to an existing Lofted feature creates an additional lofted feature and a temporary plane. The new loft section automatically creates pierce points at the end points. Drag to reposition the guide curve; the new loft section retains contact with the guide curve. Drag the plane to position the new loft section along the axis of the path. Use a pre-existing plane to position the new loft section.

Use the Context toolbar (Right-click) to edit the new loft section. Edit the loft section in the same way that you would edit any other sketch element: add relations, dimension, modify shape, etc.

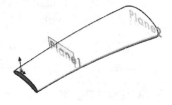

Tutorial: Add Loft Section 10-1

Create a Lofted Section and modify the sketch.

1. Open **Add Loft Section 10-1**
   from the SOLIDWORKS 2018
   folder. The Add Loft Section 10-
   1 FeatureManager is displayed.

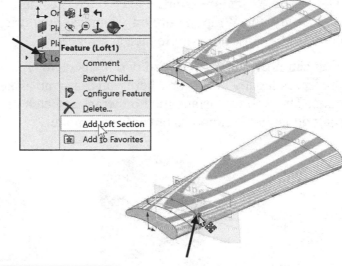

2. Right-click **Loft1** from the
   FeatureManager.

3. Click **Add Loft Section**. The
   Add Loft Section
   PropertyManager is displayed.

4. Click and drag the **plane** from
   the Graphics window by the
   arrow behind Planc1 as
   illustrated.

5. Click **OK** ✔ from the Add
   Loft Section
   PropertyManager.

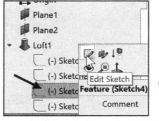

6. **Expand** Loft1 from the
   FeatureManager.

7. Click **Sketch4**.

8. Right-click **Edit Sketch**.

9. Drag the **top arc point** upward.

10. Click **OK** ✔ from the Point
    PropertyManager.

11. Click **Exit Sketch** from the Sketch toolbar to
    display the modified Loft1 feature.

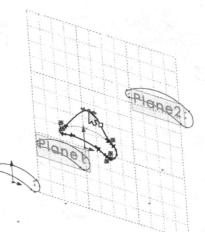

12. **Rebuild** the model.

13. Display an **Isometric** view. View the results.

14. **Close** the model.

💡 View the Lofted Feature
folder in the book for additional
information.

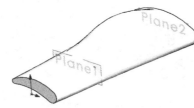

Wrap Feature

The Wrap 🗊 feature wraps a 2D Sketch (geometry) on any face type and multiple faces. Previously, the Wrap feature only handled cylindrical and cone shaped faces, and you could only work on a single face at a time.

You can now choose between two methods to create a wrap. The **Analytical** method preserves legacy behavior where you can wrap a sketch completely around a cylinder or cone. The Spline Surface method wraps a sketch on any face type. A Limitation with this method is that it cannot wrap around a model.

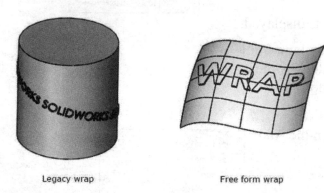

Legacy wrap   Free form wrap

The Wrap feature uses the Wrap PropertyManager. The Wrap PropertyManager provides the following selections:

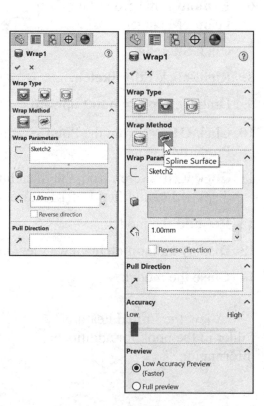

- *Wrap Type*. The Wrap Type box provides the following selections:

  - **Emboss**. Creates a raised feature on the face, and adds material inside the closed loop.

  - **Deboss**. Creates an indented feature on the face, and removes material inside the closed loop.

  - **Scribe**. Creates an imprint of the sketch contours on the face.

- *Wrap Method*. The Wrap Method box provides the following selections:

    - **Analytical**. Preserves legacy behavior where you wrap a sketch completely around a cylinder or cone.

    - **Spline Surface**. Provides the ability to wrap a sketch on any face type. A limitation with this method is that it cannot wrap around a model.

- *Wrap Parameters*. The Wrap Parameters provides the following selections:

    - **Source Sketch**. Displays the selected source.

    - **Face for Wrap Sketch**. Displays the selected face.

    - **Depth**. Depth of the selected feature.

- *Pull Direction*. Display the selected pull direction from the Graphics window. To wrap the sketch normal to the sketch plane, leave the Pull Direction box blank.

- *Accuracy*. Provides the ability to address display accuracy.

- *Preview*. Selected by default. Low Accuracy Preview (Faster).

- *Full preview*.

Tutorial: Wrap 10-1

Create a simple Wrap feature using the Deboss option.

1. Open **Wrap 10-1** from the SOLIDWORKS 2018 folder. The Wrap 10-1 FeatureManager is displayed.

2. Right-click the **Top Plane** from the FeatureManager.

3. Click **Sketch** from the Context toolbar.

4. Click the **Circle** ⊙ Sketch tool. Sketch a **circle (80mm)** as illustrated. Note the location of the origin.

5. Click the **Extruded Boss/Base** 🗊 Features tool. Blind is the default End Condition in Direction 1.

6. Enter **70**mm for Depth.

Create the sketch to Wrap.

7. Right-click the **Right Plane** from the FeatureManager.

8. Click **Sketch** from the Context toolbar.

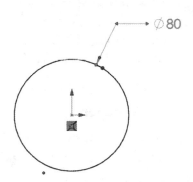
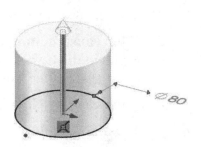

9. Click the **Center Rectangle** 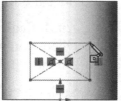 Sketch tool.

10. Display a **Right** view.

11. Sketch a **rectangle** as illustrated.

12. **Rebuild** 🛑 the model.

13. Click **Sketch2** from the FeatureManager.

14. Click the **Wrap**  Features tool. The Wrap PropertyManager is displayed. Sketch2 is displayed in the Source Sketch box.

15. Click the **Deboss** option.

16. Click the **Analytical** method option.

17. Click the **right cylindrical face** of Extrude1. Face<1> is displayed in the Face for Wrap Sketch box.

18. Enter **10**mm for Depth.

19. Click **OK** ✔ from the Wrap PropertyManager. View the results. Remember, The Sketch plane must be tangent to the face, allowing the face normal and the sketch normal to be parallel at the closest point and you cannot create a wrap feature from a sketch that contains any open contours.

Tutorial: Wrap 10-2

Create a simple Wrap feature using the Deboss option and the Sketch Text tool.

1. Open **Wrap 10-2** from the SOLIDWORKS 2018 folder. The Wrap 10-2 FeatureManager is displayed.

2. Right-click the **Right Plane** from the FeatureManager.

3. Click **Sketch**.

4. Display a **Right** view. Sketch a horizontal **construction line** midpoint on Boss-Extrude1 for the Wrap text.

Create text for the Wrap feature.

5. Click the **Tools**, **Sketch Entities**, **Text** 𝔸 from the Main menu. The Sketch Text PropertyManager is displayed.

6. Click the **construction line** from the Graphics window. Line1 is displayed in the Curves box. Click **inside** the Text box. Enter **Made in USA** in the Text box.

7. Click the **Center Align** button.

8. Uncheck the **Use document font** box.

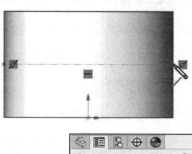

9. Enter **150%** in the Width Factor box.

10. Enter **120%** in the Spacing box.

11. Click **OK** ✔ from the Sketch Text PropertyManager.

12. **Rebuild** the model.

13. Click **Sketch2** from the FeatureManager.

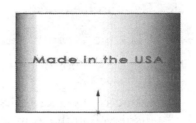

14. Click the **Wrap** 🗇 Features tool. The Wrap PropertyManager is displayed. Sketch2 is displayed in the Source Sketch box.

15. Select the **Deboss** option.

16. Click the **Analytical** method option.

17. Click the **cylindrical face** of Boss-Extrude1. Face<1> is displayed in the Face for Wrap Sketch box.

18. Enter **10**mm for Depth.

19. Click **OK** ✔ from the Wrap PropertyManager. View the results.

20. **Close** the model.

Tutorial: Wrap 10-3

Create a Wrap feature using the Emboss option and the Sketch Text tool.

1. Open **Wrap 10-3** from the SOLIDWORKS 2018 folder.

2. Right-click **Plane1** from the FeatureManager. This is your Sketch plane. Click **Sketch**.

Create text for the Wrap feature around the top section of the cone.

3. Click the **Tools**, **Sketch Entities**, **Text** 𝔸 from the Main menu. The Sketch Text PropertyManager is displayed.

4. Click **inside** the Text box.

5. Enter **Eat it all** in the Text box.

6. Click the **top section of the cone** (shell3). If needed, click the **Flip Horizontal** option.

7. Click **OK** ✔ from the Sketch Text PropertyManager.

8. **Rebuild** the model.

9. Click the **Wrap** 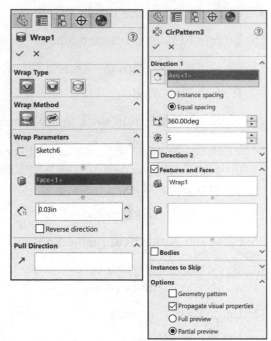 Features tool.

10. Click **Sketch6** from the fly-out FeatureManager. The Wrap PropertyManager is displayed. Sketch6 is displayed in the Source Sketch box.

11. Select the **Emboss** option. Click the **Analytical** method option.

12. Click the **top section of the cone** (shell3). Face<1> is displayed in the Face for Wrap Sketch box.

13. Enter **.03in** for thickness.

14. Click **OK** ✔ from the Wrap PropertyManager.

Create a Circular pattern of the Wrap feature.

15. Click **Circular Pattern** from the Features tab.

16. Check the **Equal spacing** box.

17. Enter **5** instances.

18. Click inside the **Features to Pattern** box.

19. Click **Wrap1** from the fly-out FeatureManager.

20. Click inside the **Pattern Access** box.

21. Click the **Temporary Axis**.

22. Uncheck the **Geometry pattern** option.

23. Click **OK** ✔ from the Circular Pattern PropertyManager. View the results.

24. **Close** the model.

Tutorial: Wrap 10-4

Create a Wrap feature using the Emboss option, and the Spline Surface method.

1. Open **Wrap 10-4** from the SOLIDWORKS 2018 folder.

2. Click **No** to feature recognition.

3. Click **Sketch 2** (SOLIDWORKS) from the FeatureManager.

4. Click the **Wrap** 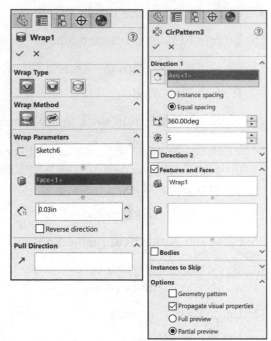 Features tool.

5. Select the **Emboss** option.

6. Click the **Spline Surface** method option.

7. Select the **six faces** along the handle as illustrated. The six faces are displayed in the Face for Wrap Sketch box.

8. Enter **4mm** for thickness.

9. Click **OK** ✔ from the Wrap PropertyManager. View the results.

Flex Feature

Flex 🖑 features are useful for modifying complex models with predictable, intuitive tools for many applications including concepts, mechanical design, industrial design, stamping dies, molds, and so on. Flex features can change single body or multibody parts.

The flex feature calculates the extents of the part using a bounding box. The trim planes are then initially located at the extents of the

bodies, perpendicular to the blue Z-axis of the triad 🔮.

- The flex feature affects the region between the trim planes only.

- The center of the flex feature occurs around the center of the triad location.

- To manipulate the extent and location of the flex feature, re-position the triad and trim planes. To reset all PropertyManager values to the state they were in upon opening the flex feature, right-click in the graphics area and select Reset flex.

The Flex feature uses the Flex PropertyManager. The Flex PropertyManager provides the following selections:

- *Flex Input*. The available selections are:

  - **Bodies for Flex** box. Displays the selected bodies to flex.

  - **Flex types**: Move the pointer over the edge of a trim plane to display pointers based on the flex type, then drag the pointer to modify the flex feature. The available selections are:

    - **Bending**. Bends one or more bodies about the triad's red X-axis (the bend axis). Selected by default. Position the triad and trim planes to control the degree, location, and extent of bending.

      - **Angle**. Displays the bend angle.

      - **Radius**. Displays the bend radius.

- **Twisting.** ^{Twisting} 🌀. Twists solid and surface bodies. Position the triad and trim planes to control the degree, location, and extent of twisting. Twisting occurs about the triad's blue Z-axis.

  - **Angle**. Displays the angle of twist.

- **Tapering.** ^{Tapering} ⇔. Tapers solid and surface bodies. Position the triad and trim planes to control the degree, location, and extent of tapering. Tapering follows the direction of the triad's blue Z-axis.

  - **Taper factor**. Sets the amount of taper. Note: The trim planes do not move when you adjust the Taper factor.

- **Stretching.** ^{Stretching} ↔. Stretches solid and surface bodies along the blue Z-axis of the triad. Specify a distance or drag the edge of a trim plane with the left mouse button.

  - **Stretch distance**. Displays the amount of stretch.

- **Hard edges.** Creates analytical surfaces (cones, cylinders, planes, and so on), which often result in split faces where the trim planes intersect the bodies. If cleared, results are spline-based, so surfaces and faces may appear smoother and original faces remain intact.

- *Trim Planes*. The Trim Plane section provides two options:

  - **Select a reference entity for Trim Plane**. Locates the origin of the trim plane to a selected point on the model.

  - **Trimming distance**. Moves the trim plane from the outer extents of the bodies along the trim plane axis (blue Z-axis) of the triad. Set a value.

- *Triad*. The Triad box provides the following selections:

  - **Select a coordinate system feature**. Locks the position and orientation of the triad to a coordinate system. Note: You must add a coordinate system feature to the model to use this option.

  - **Rotation Origin**. Moves the triad along the specified axis, relative to the triad's default location.

- **Rotation Angle**. Rotates the triad around the specified axis, relative to the triad itself. Set values. The angle represents rotations about the component coordinate system and are applied in this order: Z, Y, X.

- *Flex Options*. Provides the following selection:

  - **Flex accuracy**. Controls surface quality. Increasing the quality also increases the success rate of the flex feature. For example, if you get an error message, move the slider towards the right. Move the slider only as needed; increasing surface accuracy decreases performance.

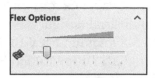

Tutorial: Flex 10-1

Create a Flex feature. Flex features are useful for modifying complex models with predictable, intuitive tools for many applications including concepts, mechanical design, industrial design, stamping dies, molds, and so on. Flex features can change single body or multibody parts.

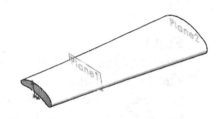

1. Open **Flex 10-1** from the SOLIDWORKS 2018 folder. The Flex 10-1 FeatureManager is displayed.

2. Click **Insert ➤ Features ➤ Flex** 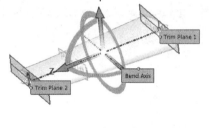. The Flex PropertyManager is displayed. The Bending option is selected by default.

3. Click **Loft1** from the Graphics window. Loft1 is displayed in the Bodies for Flex box.

4. Drag the **reference triad** towards Plane2.

5. Enter **15**deg for Angle.

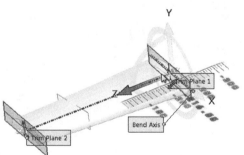

6. Click **OK** ✔ from the Flex PropertyManager. Flex1 is created and is displayed in the FeatureManager.

7. Display an **Isometric** view.

8. **View** the results.

9. **Close** the model.

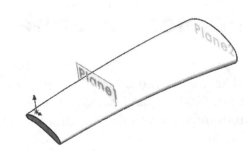

Tutorial: Flex 10-2

View how the Flex feature with the Twisting option was applied to the blades of the fan. Flex features are useful for modifying complex models with predictable, intuitive tools for many applications including concepts, mechanical design, industrial design, stamping dies, molds, and so on.

1. Open the **Flex 10-2** assembly from the SOLIDWORKS 2018\Flex folder. The Flex 10-2 Assembly FeatureManager is displayed.

2. Open the **Fan** part from the Fan assembly.

3. **Edit** Flex1. The Flex1 PropertyManager is displayed. The Twisting option is selected. You can Twist solid and surface bodies around the blue Z-axis of the triad. Position the triad and trim planes to control the degree, location, and extent of twisting. Angle of twist is 145degrees in this exercise.

4. **Close** the models.

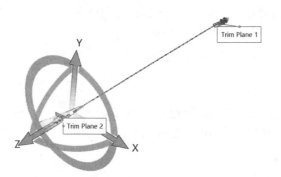

Freeform Feature

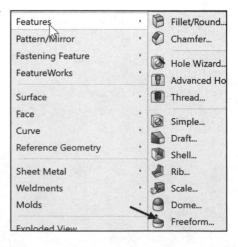

The Freeform is very useful in order to create desired shapes that would be next to impossible without this tool. Models that need comfortable grips or that require very complex curvatures can easily be created.

The Freeform feature allows points to be manipulated along control curves in order to obtain the desired shape.

The Freeform PropertyManager appears when you create a Freeform feature. You can modify only one face at a time. The face can have any number of sides.

Watch the below video on creating a Freeform feature:

http://blogs.SOLIDWORKS.com/teacher/2013/05/SOLIDWORKS-freeform-feature.html

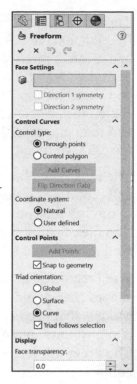

Summary

In this chapter, you learned about the Swept Boss/Base feature, Swept Cut feature, Swept Thin feature, Lofted Boss/Base feature, Wrap feature, Flex feature and Freeform feature.

The workflow and interface for all sweep features has improved. Controls, group boxes and options have been rearranged and updated to accommodate enhancements which include the ability to create bidirectional sweeps and to choose between two types of profiles, Sketch Profile and Circular Profile.

Sketch Profile is similar to existing functionality where you create a sweep by moving a 2D sketch profile along a 2D or 3D sketch path, but it provides more options. Circular Profile lets you create a sweep directly on a model without having to work in a sketch.

The Lofted feature provides the ability to create a feature by creating transitions between profiles. A loft can be a Base, Boss, Cut, or Surface feature. You created a Loft feature by using two or more profiles. Remember, only the first, last, or first and last profiles can be points. You addressed the Lofted Boss/Base feature, Lofted Cut feature and added a Loft Section.

Open the Wind Turbine assembly from the Loft Feature folder. View the lofted feature of the blades. The wind turbine assembly initially has the nacelle enclosure and generator enclosure transparent so as to allow viewing of the inner workings of the assembly. The generator is a simplified version so as to allow a less heavy assembly to run faster in SOLIDWORKS.

The Wrap feature provides the ability to wrap a 2D Sketch (Sketch text) or design onto a planar or non-planar face. You can create a planar face from either a cylindrical, conical, or extruded model. The wrap feature supports contour selection and sketch reuse. The Wrap feature provides the ability to flatten the selected face, relating the sketch to the flat pattern, and then mapping the face boundaries and sketch back onto the 3D face.

The Flex feature takes existing SOLIDWORKS geometry and modifies its shape. The Flex feature provides the ability to modify the entire model, or just a portion of the model. The Flex feature works on both solid and surface bodies as well as imported and native geometry.

The Freeform feature is very useful in order to create desired shapes that would be next to impossible without this tool. Models that need comfortable grips or that require very complex curvatures can easily be created.
The Freeform feature allows points to be manipulated along control curves in order to obtain the desired shape.

The Zebra stripes function provides the ability to display small changes in a surface which may be difficult to view with a standard display. Zebra stripes simulate the reflection of long strips of light on a very shiny surface.

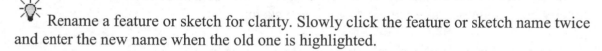 Rename a feature or sketch for clarity. Slowly click the feature or sketch name twice and enter the new name when the old one is highlighted.

Chapter 11 provides a comprehensive understanding of Bottom-up assembly modeling techniques, the ability to address Standard Mates, Advanced Mates, Mechanical Mates, InPlace Mates, SmartMates, Mate References, Large Assembly mode, Large Design Review, Assembly Visualization, Performance Evaluation, Center of Mass point, Speedpak, and Tree House.

CHAPTER 11: BOTTOM-UP ASSEMBLY MODELING AND MORE

Chapter Objective

Chapter 11 provides a comprehensive understanding of Bottom-up assembly modeling techniques, the ability to address Standard Mates, Advanced Mates, Mechanical Mates, InPlace Mates, SmartMates, Mate References, Large Assembly mode, Large Design Review, Assembly Visualization, Performance Evaluation, Center of Mass point, Speedpak, and Tree House. On the completion of this chapter, you will be able to:

- Apply Bottom-up assembly modeling technique.

- Understand various Assembly configuration methods:
  - Manual (Add Configuration command).
  - Configure Component/Configure Dimension tool.
  - Design Tables.

- Create an Assembly Task List - Before you begin.

- Comprehend an Assembly Template.

- Identify sections of the Assembly FeatureManager and various Component States.

- Recognize the various Mate types:
  - Standard, Advanced, Mechanical, SmartMates and InPlace.

- Apply the Mate PropertyManager:
  - Mates tab, Analysis tab, Mate Controller.

- Know and apply SmartMates.

- Understand and apply Mate References.

- Resolve Mating problems with the MateXpert tool.

- Employ the Large Assembly mode.

- Open an assembly using various options:
  - Resolved, Lightweight, Large Assembly Mode, Large Design Review, Speedpak.

- Utilize the Performance Evaluation tool.

- Apply the Assembly Visualization tool.

- Insert a Center of Mass point in an Assembly.

- Create a Tree House for an Assembly.

Bottom-up Assembly Modeling Technique

An assembly combines two or more parts. A part inserted into an assembly is called a component. A sub-assembly is a component contained within an assembly.

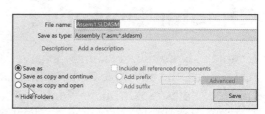

When you create your first assembly, Assem1.sldasm is the default document name. The Assembly document ends with the extension .SLDASM.

The foundation for a SOLIDWORKS assembly is the Assembly Template. Define drafting standards, units and other properties in the Assembly Template. The Assembly Template document ends with the .ASMDOT extension. See Chapter 2 for additional information on templates and file locations.

Terminology Review

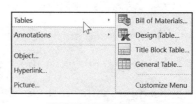

- *Design Table*: Design tables are used to control dimensions values for distance and angle mates, assembly features, the suppression state or visibility of components and the configurations of components within the assembly.

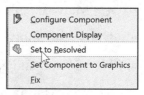

- *Hide Components*: The Hide Component tool provides the ability to remove a component's graphics without removing the component or its dependents. Mates associated with hidden components are still evaluated.

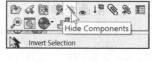

- *Lightweight Components*: Load an assembly with its active components *fully resolved* or *lightweight*. Both parts and subassemblies can be lightweight. When a component is *fully resolved*, all its model data is loaded in memory. When a component is lightweight only a subset of its model data is loaded in memory. The remaining model data is loaded on an as-needed basis. You can improve performance of large assemblies significantly by using lightweight components. Loading an assembly with lightweight components is faster than loading the same assembly with fully resolved components.

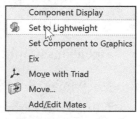

- *Suppress*: The Suppress tool provides the ability to temporarily remove a component. When a component is suppressed, the system treats the component as if it does not exist. This means other components and mates that are dependent will be suppressed.

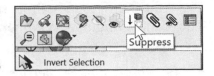

Assembly Configurations Methods

You can create assemblies using the bottom-up design approach, the top-down design approach, or a combination of both methods. This chapter focuses on the bottom-up design approach technique. The bottom-up design approach is the traditional method that combines individual components. Based on design criteria, the components are developed independently. There are several ways to create assembly configurations. Three ways are explored in this book.

- Manual (Add Configuration) command.

- Configure Component/Configure Dimension tool.

- Design Tables.

View Chapter 12 for additional information and short tutorials on these methods.

The three major steps in a bottom-up design approach are:

1. Create each component independent of any other component in the assembly.

2. Insert the components into the assembly.

3. Mate the components in the assembly as they relate to the physical constraints of the design.

The bottom-up design approach is the preferred technique for previously constructed, off-the-shelf parts, or standard components. Example: hardware, pulleys, motors, etc. These parts do not change their size and shape based on the design unless you choose a different component.

Manual (Add Configuration)

The Manual (Add Configuration) method uses the Add Configuration PropertyManager. Right-click the assembly name in the ConfigurationManager and click the Add Configuration command. The Add Configuration PropertyManager is displayed.

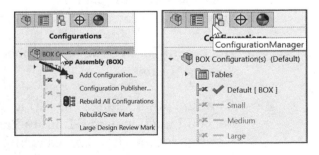

The Add Configuration PropertyManager is document dependent. The Add Configuration PropertyManager provides the following options:

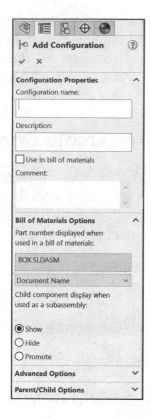

- ***Configuration Properties***. The Configuration Properties box provides the following dialog boxes:

  - **Configuration name**. Type a name for the configuration. The name must not include the forward slash (/) or "at" sign (@). A warning message is displayed when you close the dialog box if the name field contains either of these characters, if the field is blank, or if the name already exists. You can display component configuration names in the FeatureManager design tree.

  - **Description** (optional). Type a description that identifies the configuration. You can display component configuration descriptions in the FeatureManager and the ConfigurationManager.

  - **Use in bill of materials**. Specify how the assembly or part is listed in a Bill of Materials.

  - **Comment** (Optional). Type additional descriptive information about the configuration.

  - **Custom Properties** (available only when editing properties of an existing configuration). Click to access Configuration Specific properties in the Summary Information dialog box.

- ***Bill of Materials Options***. The Bill of Materials Options box provides the following dialog boxes:

  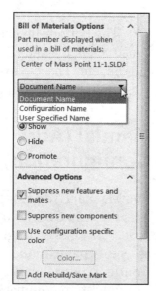

  - **Part number displayed when used in a bill of materials**. Provides the ability to specify how the assembly or part is listed in a Bill of Materials. Select one of the following:

    o **Document Name**. The part number is the same as the document name.

    o **Configuration Name**. The part number is the same as the configuration name, **Link to Parent Configuration**. (For derived configurations only.) The part number is the same as the parent configuration name.

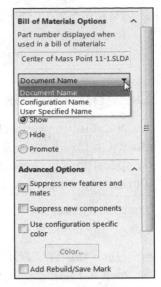

-    ○   **User Specified Name**. The part number is a name that you type.

- **Show**. Selected by default. Show child components in BOM when used as sub-assembly (assemblies only).

- **Hide**. When selected, the sub-assembly is always shown as a single item in the Bill of Materials.

- **Promote**. Dissolves the subassembly in the BOM and shows its child components, even if the BOM Type would not normally show them. For example, a Top-level only BOM usually lists just the subassembly and not its child components. If Promote is selected, the child components are listed and the subassembly is not.

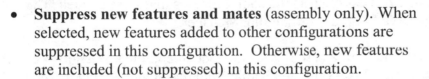

- *Advanced Options*. The Advanced Options box provides the following dialog boxes: Note: The following properties control what happens when you add new items to another configuration, and then activate this configuration again. Options are document dependent.

  - **Suppress new features and mates** (assembly only). When selected, new features added to other configurations are suppressed in this configuration. Otherwise, new features are included (not suppressed) in this configuration.

  - **Hide new components** (assemblies only). When selected, new components added to other configurations are hidden in this configuration. Otherwise, new components are displayed in this configuration.

  - **Suppress new components** (assemblies only). When selected, new components added to other configurations are suppressed in this configuration. Otherwise, new components are resolved (not suppressed) in this configuration.

  - **Use configuration specific color**. Provides the ability to specify a color for the configuration. Select the check box. Click **Color** to choose a color from the color palette.

- **Parent/Child Options** (assemblies only). Available when adding a new configuration to the assembly or one of its components. Select the components to which you want to add the new configuration.

Design Tables

A design table provides (adds intelligences to the model) the ability to build multiple configurations of parts or assemblies by specifying parameters in an embedded Microsoft Excel worksheet. Note: See Chapter 12 for additional information.

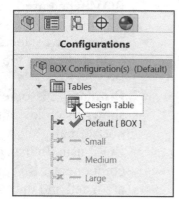

There are several different ways to insert a design table. You can:

- **Have the SOLIDWORKS Software automatically insert one for you**. In a part or assembly document, click **Design Table** from the Tools toolbar, or click **Insert**, **Design Table** from the Menu bar menu. In the PropertyManager, under Source, select **Auto-create**. Set any needed settings or Options. Click **OK**. Depending on the settings you selected, a dialog box may appear that asks which dimensions or parameters you want to add. An embedded worksheet is displayed in the Graphics window, and the SOLIDWORKS toolbars are replaced with Excel toolbars. Cell A1 identifies the worksheet as Design Table for <model_name>. Click anywhere **outside** of the worksheet in the Graphics area to close the design table. A message lists the configurations that were created. The Design Table is displayed in the ConfigurationManager.

- **Insert a blank design table**. In a part or assembly document, click **Design Table** from the Tools toolbar, or click **Insert**, **Design Table** from the Menu bar menu. The Design Table PropertyManager is displayed. In the PropertyManager, under Source, click **Blank**. Set any needed settings or Options. Click **OK**. Depending on the settings you selected, a dialog box may appear that asks which dimensions or parameters you want to add. An embedded worksheet is displayed in the Graphics window, and the SOLIDWORKS toolbars are replaced with Excel toolbars. Cell A1 identifies the worksheet as Design Table for <model_name>. Cell A3 contains the default name for the first new configuration, First Instance.

In row **2**, type the parameters that you want to control. Leave cell **A2** blank. Notice that cell B2 is active. In column **A** (cells **A3**, **A4**, etc.), type the names of the configurations that you want to create. The names can include numbers but must not include the forward slash (/) or at (@) characters. Type the parameter values in the worksheet cells. When you finish adding information to the worksheet, click **outside** the table to close it.

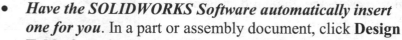 Enter parameters by double-clicking the feature or dimension in the graphics area or in the FeatureManager design tree. When you double-click a feature or dimension, its associated value is displayed in the **Default** row.

- ***Insert an external MS Excel table as a design table***. In a part or assembly document, click **Design Table** from the Tools toolbar, or click **Insert**, **Design Table** from the Menu bar menu. The Design Table PropertyManager is displayed. In the PropertyManager, under Source, click **From file**, click **Browse** to locate the Excel file. To link the design table to the model, select the **Link to file** check box. A linked design table reads all of its information from an external Excel file.

Note: If you update a linked design table in Microsoft Excel, then open the SOLIDWORKS model, you can choose to update either. Click **OK**. An embedded worksheet is displayed in the window, and the SOLIDWORKS toolbars are replaced with Excel toolbars. Click anywhere **outside** of the worksheet (but in the graphics window) to close the design table.

An External reference is created when one document is dependent on another document for its solution. If the referenced document changes, the dependent document also changes.

You can lock or break all external references in an entire assembly hierarchy (or in a selected sub-assembly within the hierarchy) at the same time. Previously, you had to find and address each component individually. Optionally, you can select to replace broken sketch relations with fixed relations when you break the references. Right-click a top-level assembly or one sub-assembly at a time and click List External.

Check **Options** ⚙, **System Options, External References**, **Do not create references external to the model**. New features or parts created in the assembly will not be created with External References.

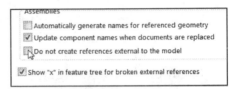

Know the default SW colors. A selected feature in the Graphics window is displayed in blue. The corresponding feature entry in the FeatureManager is displayed in blue. Fully defined sketched dimensions are displayed in black. Extruded depth dimensions are displayed in blue. Modify default colors under **Options** ⚙, **System Options, Colors, Color scheme setting**.

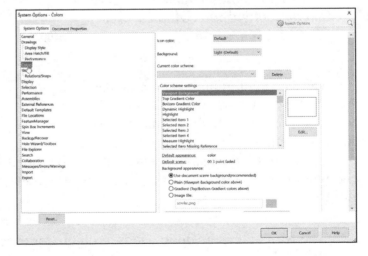

Configure Component tool/Configure Dimension tool

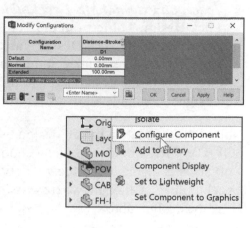

The Configure Component tool/Configure Dimension tool provides access to the Modify Configurations dialog box. The Modify Configurations dialog box facilitates creating and modifying configurations for commonly configured parameters in parts and assemblies. You can add, delete, and rename configurations and modify which configuration is active.

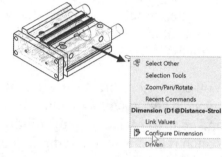

For features and sketches in parts, you can configure the following: Dimensions and Suppression states. In assemblies, you can configure the following: Which configurations of components to use, Suppression states of components, Assembly features, and mates, and Dimensions of assembly features and mates.

The Modify Configurations dialog box provides the following options:

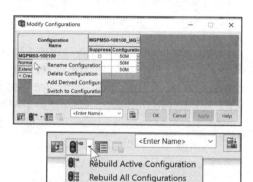

- **First Column**. List the configurations of the model and the configurable parameters of the selected item in the other columns. Note: Right-click any configuration and select the following option: *Rename Configuration, Delete Configuration, Add Derived Configuration, and Switch to Configuration.*

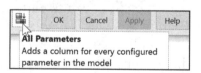

🔅 Derived configurations provide the ability to create a Parent-Child relationship within a configuration. By default, all parameters in the child configuration are linked to the parent configuration. If you change a parameter in the parent configuration, the change automatically propagates to the child.

🔅 The All Parameters option adds a column for every configured parameter in the model.

- **Parameter Columns**. Provides the ability to select one of the following: *Type to change the numeric values, Select from a list of component configurations or to Change the suppression state of features, sketches, components and mates.*

Assembly Task List - Before you begin

You are required to perform numerous tasks before you create a SOLIDWORKS assembly. Review the following tasks.

Review the Assembly Layout Diagram. Group components into sub-assemblies.

Comprehend the geometric and functional requirements of purchased components. Understand the interaction between components in the assembly. Examine the fit and function of each component. Obtain model files and data specifications from your vendors.

Place yourself in the position of the machinist, manufacturing technician, field service engineer, or customer. This will aid in the process to identify potential obstacles or design concerns before you start the assembly.

Prepare and create the part, assembly, drawing Templates ahead of time. Identify your units, dimensioning standards, and other document properties.

Organize your documents into file folders. Place the created templates, vendor components, library components, parts, assemblies, and drawings in a specific location with a descriptive name.

Acquire unique part numbers for your components. A unique part number will avoid duplication problems in the assembly later in the process.

Assembly Templates

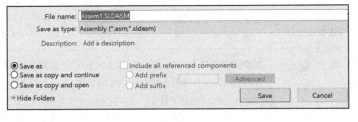

Templates are the foundation for assemblies, parts, and drawings.
Document Properties address: dimensioning standards, units, text style, center marks, witness lines, arrow styles, tolerance, precision, and other important parameters. Document Properties apply only to the current document.

The foundation of a SOLIDWORKS assembly is the Assembly Template. The custom Assembly Template begins with the default Assembly Template. You created a custom Assembly Template in Chapter 2 in the SOLIDWORKS 2018\MY-TEMPLATE folder.

 All templates are provided in the MY-TEMPLATES folder.

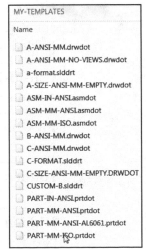

MY-TEMPLATES

Name

- A-ANSI-MM.drwdot
- A-ANSI-MM-NO-VIEWS.drwdot
- a-format.slddrt
- A-SIZE-ANSI-MM-EMPTY.drwdot
- ASM-IN-ANSI.asmdot
- ASM-MM-ANSI.asmdot
- ASM-MM-ISO.asmdot
- B-ANSI-MM.drwdot
- C-ANSI-MM.drwdot
- C-FORMAT.slddrt
- C-SIZE-ANSI-MM-EMPTY.DRWDOT
- CUSTOM-B.slddrt
- PART-IN-ANSI.prtdot
- PART-MM-ANSI.prtdot
- PART-MM-ANSI-AL6061.prtdot
- PART-MM-ISO.prtdot

💡 Conserve modeling time. Store Document Properties in the Assembly Template. Set the parameters for the Drafting standard and Units.

Assembly FeatureManager and Component States

How do you distinguish the difference between an assembly and a part in the FeatureManager? Answer: The assembly icon 🍱 contains a blue square block and an upside down yellow "T" extrusion. The part icon 🍱 contains an upside down yellow "T" extrusion. Entries in the FeatureManager design tree have specific definitions. Understanding syntax and states saves time when creating and modifying assemblies. Review the columns below.

First Column: A resolved component (not in a lightweight state 🍱) displays an Arrow ▸ icon. The Arrow ▸ icon ▸ 🍱 (f) HOUSING<1> indicates that additional feature information is available. A Down Arrow ▾ icon displays the fully expanded feature list.

Second Column: ▸ 🍱 (f) HOUSING<1> Identifies a component's (part or assembly) relationship with other components in the assembly.

Component States:	
Symbol:	**State:**
🍱	**Resolved part**. A part icon indicates a resolved state. A blue part icon indicates a selected, resolved part. The component is fully loaded into memory and all of its features and mates are editable.
🍱	**Lightweight part**. A blue feather on the part icon indicates a lightweight state. When a component is lightweight, only a subset of its model data is loaded in memory.
🍱	**Out-of-Date part**. An eye is displayed on the part icon. The part needs a rebuild in the assembly.
🍱	**Flexible state**. A blue box on the part icon indicates a flexible part in an assembly.
🍱	**Hidden**. A clear icon indicates the part is resolved but invisible.
🍱	**Hidden Lightweight**. A clear feather over a clear part icon indicates the part is hidden and lightweight.
🍱	**Smart Component**. A lightning bolt on the part icon indicates that the

	component is a Smart Component.
	Smart Fastener. A lightning bolt is displayed on the Fastener folder icon.
	Resolved assembly. Resolved (or unsuppressed) is the normal state for assembly components. A resolved assembly is fully loaded in memory, fully functional, and fully accessible.
	Inserted New Component. A new inserted component into an active assembly.

Third Column: (f) HOUSING<1>   The part is fixed (f). You can fix the position of a component so that it cannot move with respect to the assembly Origin. By default, the first part in an assembly is fixed; however, you can float it at any time.

It is recommended that at least one assembly component is either fixed or mated to the assembly planes or origin. This provides a frame of reference for all other mates and helps prevent unexpected movement of components when mates are added. The Component Properties are:

Component Properties in an Assembly	
Symbol:	**Relationship:**
(-)	A floating, under defined component has a minus sign (-) before its name in the FeatureManager and requires additional information.
(+)	An over-defined component has a plus sign (+) before its name in the FeatureManager.
(f)	A fixed component has an (f) before its name in the FeatureManager. The component does not move.
None	The Base component is mated to three assembly reference planes.
(?)	A question mark (?) indicates that additional information is required on the component.

Fourth Column: (f) HOUSING<1>   Name of the part.

Fifth Column: MOTHERBOARD<1> ->   The Resolved state displays the icon with an External reference symbol, "- >". The state of External references is displayed as follows:

If a part or feature has an external reference, its name is followed by –>. The name of any feature with external references is also followed by –>.

If an external reference is currently out of context, the feature name and the part name are followed by ->?.

The suffix ->* means that the reference is locked.

The suffix ->x means that the reference is broken.

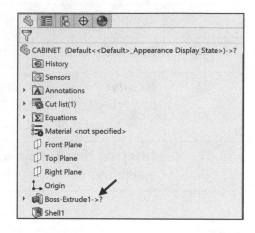

There are modeling situations in which unresolved components create rebuild errors. In these situations, issue the forced rebuild command, Ctrl+Q. The Ctrl+Q command rebuilds the model and all its features. If the mates still contain rebuild errors, resolve all the components below the entry in the FeatureManager that contains the first error.

General Mate Principles

Mates create geometric relationships between assembly components. As you add mates, you define the allowable directions of linear or rotational motion of the components. You can move a component within its degrees of freedom, visualizing the assembly's behavior. When creating mates, there are a few basic procedures to remember:

- Click and drag components in the Graphics window to assess their degrees of freedom.

- Remove display complexity. Hide components when visibility is not required.

- Utilize Section views to select internal geometry. Utilize Transparency to see through components required for mating.

- Apply the Move Component and Rotate Component tool from the Assembly toolbar before mating. Position the component in the correct orientation.

- Use a Coincident mate when the distance value between two entities is zero. Utilize a Distance mate when the distance value between two entities is not zero.

- Apply various colors to features and components to improve visibility for the mating process.

- Rename mates, key features, and reference geometry with descriptive names.

- Resolve a mate error as soon as it occurs. Adding additional mates will not resolve the earlier mate problem.

- Avoid redundant mates. Although SOLIDWORKS allows some redundant mates (all except distance and angle), these mates take longer to solve and make the mating scheme harder to understand and diagnose if problems occur.

If a component is causing problems, it is often easier to delete all its mates and re-create them instead of diagnosing each one. This is especially true with aligned/anti-aligned and dimension direction conflicts (you can flip the direction that a dimension is measuring).

Use the View Mates tool (Right-click a **component** and select **View Mates**) or expand the component in the FeatureManager design tree using **Tree Display**, **View Mates** and **Dependencies** to view the mates for each component.

☼ Use limit mates sparingly because they take longer to solve and whenever possible, mate all components to one or two fixed components or references. Long chains of components take longer to solve and are more prone to mate errors.

Mate PropertyManager

Mates provide the ability to create geometric relationships between assembly components. Mates define the allowable directions of *rotational* (3 degrees of freedom) and *linear motion* (3 degrees of freedom) of the components in the assembly.

Move a component within its degrees of freedom in the Graphics window to view the behavior of an assembly.

Mates are solved together as a system. The order in which you add mates does not matter. All mates are solved at the same time. You can suppress mates just as you can suppress features.

The Mate PropertyManager provides the ability to select either the *Mates* or *Analysis* tab. Each tab has a separate menu. The Mate PropertyManager displays the appropriate selections based on the type of geometry you select.

Mate PropertyManager: Mates tab

The Mates tab is the default tab. The Mates tab provides the ability to insert a *Standard, Advanced or Mechanical* mate.

- *Mate Selections*. The Mate Selections box provide the following selections:

  - **Entities to Mate**. Displays the selected faces, edges, planes, etc. that you want to mate.

  - **Multiple mate mode**. Mates multiple components to a common reference in a single operation. When activated, the following selections are available:

  - **Common references**. Displays the selected entity to which you want to mate several other components.

- **Component references**. Displays the selected entities on two or more other components to mate to the common reference. A mate is added for each component.

- **Create multi-mate folder**. Groups the resulting mates in a Multi-Mates folder.

- **Link dimensions**. Only available for Distance and Angle mates in a multi-mate folder. Provides the ability to link dimensions. The variable name in the Shared Values dialog box is the same as the multi-mate folder name.

- *Standard Mates*. The Standard Mates box provide the following selection:

  - **Coincident mate**. Locates the selected faces, edges, points or planes so they use the same infinite line. A Coincident mate positions two vertices for contact.

  - **Parallel mate**. Locates the selected items to lie in the same direction and to remain a constant distance apart. A parallel mate permits only translational motion of a single part with respect to another. No rotation is allowed.

  - **Perpendicular mate**. Locates the selected items at a 90 degree angle to each other. The perpendicular mate allows both translational and rotational motion of one part with respect to another.

🔆 You can apply a Perpendicular mate between a complex, non-analytic surface and a linear entity like a line, an edge, an axis, or an axial entity.

- **Tangent mate**. Places the selected items tangent to each other. At least one selected item must be either a conical, cylindrical, spherical face.

- **Concentric mate**. Locates the selected items so they can share the same center point. You can prevent the rotation of components that are mated with concentric mates by selecting the Lock Rotation option. Locked concentric mates are indicated by an icon in the FeatureManager design tree.

- **Lock mate**. Enables two components to be fully constrained to each other without needing to create a rigid sub-assembly. A real world example of a lock mate is a weld that holds two parts together.

- **Distance mate**. Locates the selected items with a specified distance between them. Use the drop-down arrow box or enter the distance value directly.

- **Angle mate**. Locates the selected items at the specified angle to each other. Use the drop-down arrow box or enter the angle value directly. You can add a reference direction to help prevent angle mates from flipping unexpectedly.

- **Mate alignment**. Provides the ability to toggle the mate alignment as necessary. There are two options:

- **Aligned**. Locates the components so the normal or axis vectors for the selected faces point in the same direction.

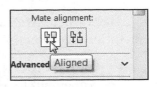

- **Anti-Aligned**. Locates the components so the normal or axis vectors for the selected faces point in the opposite direction.

- *Advanced Mates*. The Advanced Mates box provide the following selections:

- **Profile Center**. Mate to center automatically center-aligns common component types such as rectangular and circular profiles to each other.

- **Symmetric**. Forces two similar entities to be symmetric about a planar face or plane.

- **Width**. Centers a tab within the width of a groove. The Width mate is used to replace the Symmetric mate where components have tolerance and a gap rather than a tight fit.

- **Path Mate**. Enables any point on a component to be set to follow a defined path.

- **Linear/Linear Coupler**. Establishes a relationship between the translation of one component and the translation of another component.

- **Limit**. Provides the ability to allow components to move within a range of values for distance and angle. Select the angle and distance from the provided boxes. Specify a starting distance or angle as well as a maximum and minimum value.

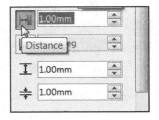

  o **Distance**. Locates the selected items with a specified distance between them. Use the drop-down arrow box or enter the distance value directly (Limit mate).

  o **Angle**. Locates the selected items at the specified angle to each other. Use the drop-down arrow box or enter the angle value directly.

- **Mate alignment**. Provides the ability to toggle the mate alignment as necessary. There are two options:

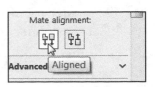

  o **Aligned**. Locates the components so the normal or axis vectors for the selected faces point in the same direction.

o **Anti-Aligned**: Locates the components so the normal or axis vectors for the selected faces point in the opposite direction.

- *Mechanical Mates*. The Mechanical Mates box provides the following selections:

- **Cam**. Forces a plane, cylinder, or point to be tangent or coincident to a series of tangent extruded faces. Four conditions can exist with the Cam mate: *Coincident, Tangent, CamMateCoincident* and *CamMateTangent*. Create the profile of the cam from lines, arcs and splines as long as they are tangent and form a closed loop. View SOLIDWORKS help for additional information.

- **Slot**. Constrains the movement of a bolt or a slot to within a slot hole. You can mate bolts to straight or arced slots and you can mate slots to slots. You can select an axis, cylindrical face, or a slot to create slot mates.

- **Hinge**. Limits the movement between two components to one rotational degree of freedom. It has the same effect as adding a Concentric mate plus a Coincident mate. You can also limit the angular movement between the two components.

- **Gear**. Forces two components to rotate relative to one another around selected axes. The Gear mate provides the ability to establish gear type relations between components without making the components physically mesh. Note: SOLIDWORKS provides the ability to modify the gear ratio without changing the size of the gears. Align the components before adding the Mechanical gear mate.

- **Rack and Pinion**. Linear translation of one component (the rack) causes circular rotation in another component (the pinion), and vice versa. You can mate any two components to have this type of movement relative to each other. The components do not need to have gear teeth.

- **Screw**. Constrains two components to be concentric, and adds a pitch relationship between the rotation of one component and the translation of the other. Translation of one component along the axis causes rotation of the other component according to the pitch relationship. Likewise, rotation of one component causes translation of the other component. Align the components before adding the Mechanical screw mate.

- **Universal Joint**. Permits the transfer of rotations from one rigid body to another. This mate is particularly useful to transfer rotational motion around corners, or to transfer rotational motion between two connected shafts that are permitted to bend at the connection (drive shaft on an automobile).

- *Mates*. The Mates box displays the activated selected mates.

- *Options*. The Options box provides the following selections:

  - **Add to new folder**. Provides the ability for new mates to be added and to be displayed in the Mates folder in the FeatureManager design tree.

  - **Show popup dialog**. Selected by default. Displays a standard mate when added in the Mate pop-up toolbar.

  - **Show preview**. Selected by default. Displays a preview of a mate.

  - **Use for positioning only**. When selected, components move to the position defined by the mate. A mate is not added to the FeatureManager design tree.

  - **Make first selection transparent**. Set an option in the Mate PropertyManager so that the first component you select from becomes transparent. Then selecting from the second component is easier, especially if the second component is behind the first. The option is supported for all mate types except those that might have more than one selection from the first component (width, symmetry, linear coupler, cam, and hinge).

A Cam-follower mate is a type of Tangent or Coincident mate. It allows you to mate a cylinder, plane or point to a series of tangent extruded faces, such as you would find on a cam. Create the cam profile from lines, arcs and splines as long as they are tangent and form a closed loop.

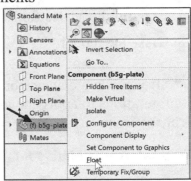

When starting a Gear or Screw mate, SOLIDWORKS creates the mate based on the current position of the models. Align the components before adding the Mechanical mate. To align the two components, move them manually or add a mate to get them where they need to be and then suppress the mate. The advantage in using this technique is if you need to make changes, you can suppress/unsuppress the mate to jog your component back to the starting location.

Tutorial: Coincident and Distance Mate 11-1

Modify the component state from fixed to float. Insert two Coincident mates, a Distance mate and apply the Align Standard mate option.

1. Open **Standard Mate 11-1** from the SOLIDWORKS 2018\BottomUpAssemblyModeling folder.

2. Right-click **bg5-plate** from the FeatureManager.

3. Click **Float**. The bg5-plate part is no longer fixed to the origin.

4. Click the **Mate** 🖉 tool from the Assembly tab. The Mate PropertyManager is displayed. **Pin** the Mate PropertyManager.

5. Click **Top Plane** of Standard Mate 11-1 from the fly-out FeatureManager.

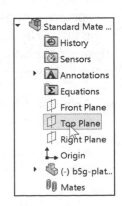

6. Click the **back face** of bg5-plate. Top Plane and Face<1>@b5g-plate-1 are displayed in the Mate Selections box. Coincident is selected by default.

7. Click **OK** ✔ from the PropertyManager.

8. Click **Front Plane** of the Standard Mate 11-1 from the fly-out FeatureManager.

9. Click **Top Plane** of bg5-plate from the fly-out FeatureManager.

10. Click **Aligned** from the Standard Mate box.

11. Click **Anti-Aligned** to review the position.

12. Display an **Isometric** view. The narrow cut faces front. Click **OK** ✔ from the Coincident PropertyManager.

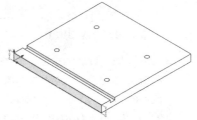

13. Click **Right Plane** of Standard Mate 11-1 from the fly-out FeatureManager. Click the **Right Plane** of bg5-plate from the fly-out FeatureManager.

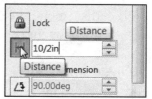

14. Select **Distance**. Enter **10/2**in. The part is in inches and the assembly is in millimeter units.

15. Click **OK** ✔ from the PropertyManager.

16. **Un-pin** the Mate PropertyManager. Click **OK** ✔ from the Mate PropertyManager.

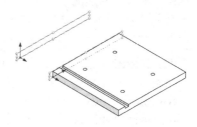

17. **Expand** the Mates folder. View the created mates.

18. **Close** the model.

💡 If you edit a distance or angle mate, you can select which configurations to apply the mate to directly from the Mate PropertyManager.

Tutorial: Angle Mate 11-1

Insert an Angle mate. The Angle ∠ mate places the selected items at the specified angle to each other. Toggle the mate alignment as necessary. The Mate alignment option provides the ability to set the alignment condition of standard and advanced mates in the Mate PropertyManager. The alignment conditions for a Standard mate are:

- **Aligned** ⊓⊓. Vectors normal to the selected faces point in the same direction.

- **Anti-Aligned** ⊓⊔. Vectors normal to the selected faces point in opposite directions.

☀ For cylindrical features, the axis vector (which you cannot see or determine) is aligned or anti-aligned. Click Aligned or Anti-Aligned to obtain the alignment you need.

1. Open **Angle Mate 11-1** from the SOLIDWORKS 2018\BottomUpAssemblyModeling folder. View the model.

2. Click the **Mate** ✎ tool from the Assembly tab. The Mate PropertyManager is displayed. Insert an Angle mate between the front face of the bracket and the back face of the crank.

3. Click the **Angle** ⌾ mate from the Standard Mate dialog box.

4. Click the **front face** of the bracket as illustrated. Click the **back face** of the (arm) crank as illustrated.

5. Enter **180** for Angle.

6. Click **Aligned**. Click **OK** ✔ from the Angle1 PropertyManager.

7. **Expand** the Mates folder. View the created Angle mate.

8. Suppress the **Angle mate** and turn the arm. View the results in the Graphics window. **Close** the model.

Tutorial: Angle Mate 11-2

Create an Angle mate between the Handle and the Side of the valve using Planes. The assembly was obtained from the SOLIDWORKS Tutorial.

1. Open **BALLVALVE 11-2** from the SOLIDWORKS 2018\BottomUpAssemblyModeling folder. The assembly is displayed. Turn the handle and view the internal ball moving to restrict flow. Create an Angle mate to restrict movement.

☀ An Angle mate places the selected items at the specified angle to each other.

2. Click the **Mate** ✎ tool from the Assembly tab.

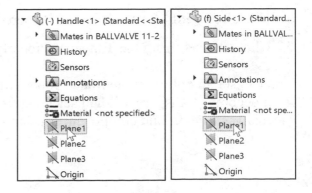

3. Click the **Angle** ⟁ mate from the Standard Mates dialog box. Select entities to Mate.

4. **Expand** the fly-out FeatureManager in the Graphics window.

5. Click **Plane1** from Side<1>.

6. Click **Plane1** from Handle<1>.

7. Enter **165** degrees for Angle.

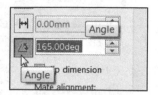

8. Click **OK** ✔ from the Angle PropertyManager.

9. Display an **Isometric** view.

10. **Zoom in** on the ball. The handle has a 165 degree Angle mate to restrict flow through the valve.

11. **Close** the model.

Tutorial: Tangent Mate 11-1

Create a Tangent mate between the roller-wheel and the cam on both sides to simulate movement per the avi file.

1. Open **TANGENT 11-1** from the SOLIDWORKS 2018\BottomUpAssemblyModeling\Tangent folder. The assembly is displayed.

Create a Tangent mate between the roller-wheel and the cam on both sides. The cam was created with imported geometry (curve).

💡 A Tangent mate places the selected items tangent to each other. At least one selected item must be either a conical, cylindrical, spherical face. An Angle mate places the selected items at the specified angle to each other.

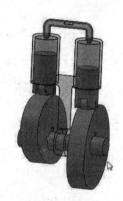

2. Click the **Mate** ✎ tool from the Assembly tab.

3. **Pin** the Mate PropertyManager.

Create the first Tangent mate.

4. Click the **contact face** of the left cam as illustrated.

5. Click the **contact face** of the left roller-wheel as illustrated.

6. Click **Tangent** ◌ mate from the PropertyManager.

7. Click **OK** ✔ from the Tangent1 PropertyManager.

You can set an option in the Mate PropertyManager so that the first component you select from becomes transparent. Then selecting from the second component is easier, especially if the second component is behind the first.

The option is supported for all mate types except those that might have more than one selection from the first component (width, symmetry, linear coupler, cam, and hinge).

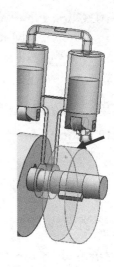

Create the second Tangent mate.

8. Click the **contact face** of the left cam. Click the **contact face** of the left roller-wheel. Click **Tangent** ⭕ mate from the PropertyManager.

9. Click **OK** ✔ from the Tangent2 PropertyManager.

10. **Un-Pin** the Mate PropertyManager.

11. Click **OK** ✔ from the Mate PropertyManager.

12. Display an **Isometric** view.

13. **Move** the cam.

14. **Close** the model.

Tutorial: Gear Mate 11-1

Insert a Mechanical Gear mate between two gears. Gear mates force two components to rotate relative to one another about selected axes. Valid selections for the axis of rotation for gear mates include cylindrical and conical faces, axes, and linear edges.

1. Open **Mechanical Gear Mate 11-1** from the SOLIDWORKS 2018\BottomUpAssemblyModeling folder.

2. Click and drag the **right gear**.

3. Click and drag the **left gear**. The gears move independently.

4. Click the **Mate** ✎ tool from the Assembly tab. The Mate PropertyManager is displayed.

5. **Clear** all selections.

6. **Align the components** before adding the Mechanical mate. To align the two components, select a Front view and move them manually.

7. Select the **inside cylindrical face** of the left hole.

8. Select the **inside cylindrical face** of the right hole.

9. **Expand** the Mechanical Mates box from the Mate PropertyManager.

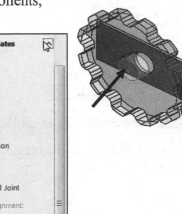

10. Click **Gear** mate from the Mechanical Mates box. The GearMate1 PropertyManager is displayed. Accept the default settings.

11. Click **OK** ✅ from the GearMate1 PropertyManager.

12. **Expand** the Mates folder. View the created GearMate1.

13. **Rotate** the Gears in the Graphics window. View the results.

14. **Close** the model.

🔆 View the Gear folder for additional information.

Tutorial: Cam Mate 11-1

Create a Tangent Cam follower mate and a Coincident Cam follower mate. A cam-follower mate is a type of Tangent or Coincident mate. It allows you to mate a cylinder, plane or point to a series of tangent extruded faces, such as you would find on a cam.

Create the profile of the cam from lines, arcs and splines as long as they are tangent and form a closed loop.

1. Open **Cam Mate 11-1** from the SOLIDWORKS 2018\BottomUpAssemblyModeling folder. View the model. The model contains two cams.

2. **Expand** the Mates folder to view the existing mates.

Insert a tangent Cam follower mate.

3. Click the **Mate** ✎ tool from the Assembly tab.

4. **Pin** the Mate PropertyManager.

5. **Expand** the Mechanical Mates box from the Mate PropertyManager.

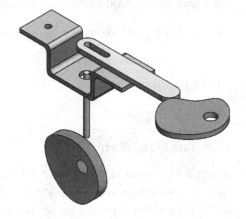

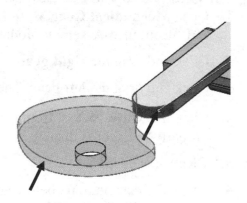

6. Click **Cam Follower** ⌀ mate.

7. Click inside of the **Cam Path** box.

8. Click the **outside face** of cam-s<1> as illustrated.

9. Click the **face of Link-c<1>** as illustrated. Face<2>@Link-c-1 is displayed.

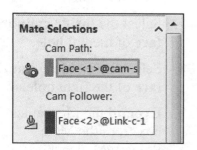

10. Click **OK** ✔ from the CamMateTangent1 PropertyManager.

Insert a Coincident Cam follower mate.

11. Click **Cam Follower** ⌀ mate from the Mechanical Mates box.

12. Click the **outside face** of cam-s2<1>.

13. Click the vertex of **riser<1>** as illustrated. Vertex<1>@riser-1 is displayed.

14. Click **OK** ✔ from the CamMateCoincident1 PropertyManager.

15. **Un-pin** the Mate PropertyManager.

16. Click **OK** ✔ from the Mate PropertyManager.

17. **Expand** the Mates folder. View the new created mates.

18. Display an **Isometric** view.

19. **Rotate** the cams. View the results in the Graphics window.

20. **Close** the model.

When creating a cam follower mate, make sure that your Spline or Extruded Boss/Base feature, which you use to form the cam contact face, does nothing but form the face.

In the next tutorial of the spur gear and rack, the gear's pitch circle must be tangent to the rack's pitch line, which is the construction line in the middle of the tooth cut.

☀ Use the Alt key to temporarily hide a face when you need to select an obscured face for mates.

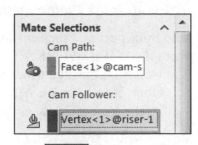

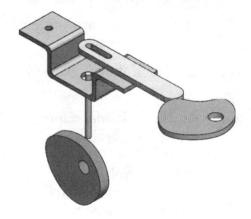

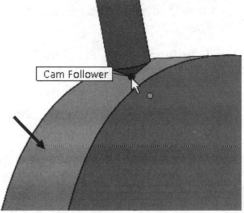

Tutorial: Rack Pinion Mate 11-1

The Spur gear and rack components for this tutorial were obtained from the SOLIDWORKS Toolbox folder marked Power Transmission. Insert a Mechanical Rack Pinion mate.

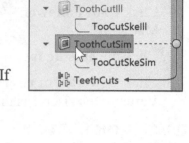

1. Open **Rack Pinion Mate 11-1** from the SOLIDWORKS 2018\BottomUpAssemblyModeling folder.

2. **View** the displayed: (TooCutSkeSim and ToothCutSim). If needed, display the illustrated sketches to mate the assembly.

The Diametral Pitch, Pressure Angle and Face Width values should be the same for the spur gear as the rack.

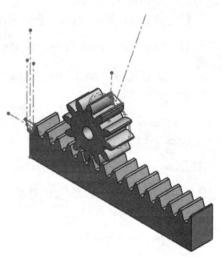

Tangency is required between the spur gear and rack.

3. **Expand** the Mates folder to view the existing mates. Four Coincident mates and a Distance mate were created in the assembly.

Insert a Rack Pinion mate.

4. Click the **Mate** ✎ tool from the Assembly tab.

5. **Expand** the Mechanical Mates box from the Mate PropertyManager.

6. Click **Rack Pinion** ⚖ mate from the Mechanical Mates box.

7. Mate the rack. Make sure the teeth of the spur gear and rack are **meshing properly**; they are not interfering with each other. It is important to create a starting point where the gear and rack have no interference.

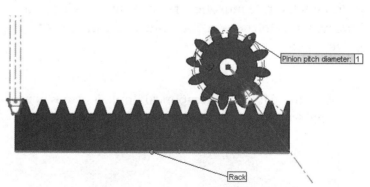

8. Click the **bottom linear edge of the rack** in the direction of travel. Any linear edge that runs in the direction of travel is acceptable.

9. Click the **spur gear pitch circle** as illustrated. The Pinion pitch diameter, 1in, is taken from the sketch geometry. Accept the default settings.

10. Click **OK** ✔ from the RackPinionMate1 PropertyManager.

11. **Rotate** the spur gear on the rack. View the results.

12. **Hide** the sketches in the model.

13. **Close** the model.

The Pitch Diameter is the diameter of an imaginary pitch circle on which a gear tooth is designed. Pitch circles of the two spur gears are required to be tangent for the gears to move and work properly.

The distance between the center point of the gear and the pitch line of the rack represents the theoretical distance required.

🔆 Diametral Pitch is the ratio equal to the number of teeth on a gear per inch of pitch diameter.

🔆 Before applying the Rack Pinion mate, make sure the teeth of the spur gear and rack are meshing properly.

Pressure Angle is the angle of direction of pressure between contacting teeth. Pressure angle determines the size of the base circle and the shape of the involute teeth. It is common for the pressure angle to be 20 or 14 1/2 degrees.

Tutorial: Hinge Mate 11-1

Create a Mechanical Hinge mate. The two components in this tutorial were obtained from the SOLIDWORKS What's New section.

1. Open **Hinge Mate 11-1** from the SOLIDWORKS 2018\BottomUpAssemblyModeling folder. Two components are displayed.

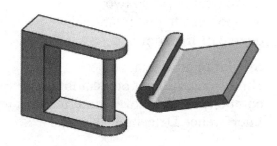

A Hinge mate has the same effect as adding a Concentric mate plus a Coincident mate. You can also limit the angular movement between the two components.

2. Click the **Mate** ✎ tool from the Assembly tab.

3. **Expand** the Mechanical Mates box from the Mate PropertyManager.

4. Click **Hinge** ▦ mate from the Mechanical Mates box.

Set two Concentric faces.

5. Click the **inside cylindrical face** of the first component as illustrated.

6. Click the **outside cylindrical face** of the second component as illustrated. The two selected faces are displayed.

Set two Coincident faces.

7. Click the **front flat face** of the first component as illustrated.

8. Click the **bottom flat face** of the second component as illustrated. View the results in the Graphics window.

You can specify a limit angle for rotation by checking the **Specify angle limits** box and selecting the required faces.

9. Click **OK** ✔ from the Hinge1 PropertyManager.

10. **Rotate** the flap component about the pin.

11. **Close** the model.

A Hinge mate limits the movement between two components to one rotational degree of freedom.

Similar to other mate types, Hinge mates do not prevent interference or collisions between components. To prevent interference, use Collision Detection or Interference Detection.

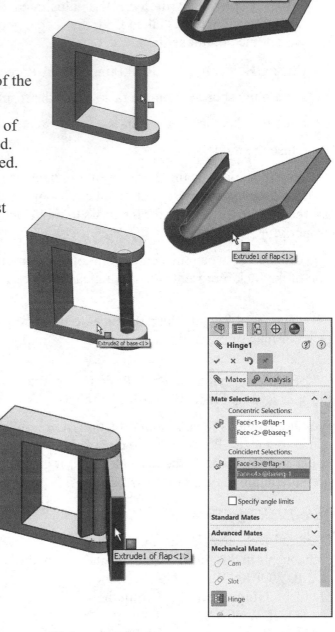

Tutorial: Slot Mate 11-1

Create a Mechanical Slot mate. Insert a slot mate between the Flange bolt and the right side slot on the Guide.

1. Open **Slot Mate 11-1** from the SOLIDWORKS 2018\BottomUpAssemblyModeling\Slot folder. The assembly is displayed.

You can select an axis, cylindrical face, or a slot to create a slot mate.

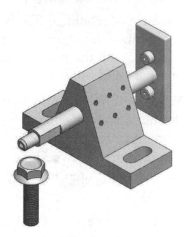

2. Click the **Mate** ✎ Assembly tool. The Mate PropertyManager is displayed.

3. Expand the **Mechanical Mates** section.

4. Select **Slot** ⬭ for mate type.

5. Select **Center in Slot** from the Constraint drop-down menu. This option centers the component in the slot. For Mate Selections, select a slot face and the feature to mate to it.

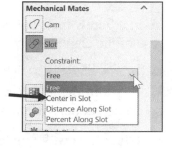

6. Click the **cylindrical face** of the Flange bolt as illustrated.

7. Click the **inside face** of the slot as illustrated. View the results. Slot1 is created.

8. Click **OK** ✔ from the Slot1 PropertyManager.

9. **Close** the model. A Coincident mate was created between the Flange bolt head and the right slot cut top face for proper fit.

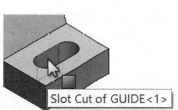

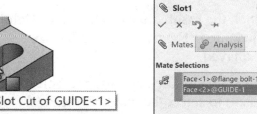

💡 Use the Mate Controller PropertyManager, to show and save the position of assembly components at various mate values and degrees of freedom. You can create simple animations between those positions. Supported mate types: Angle, Distance, LimitAngle, LimitDistance, Slot (Distance Along Slot, Percent Along Slot) and Width (Dimension, Percent).

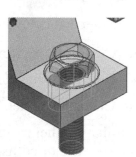

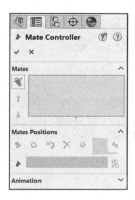

Tutorial: Screw Mate 11-1

Create a Mechanical Screw mate. A screw mate constrains one rigid body to rotate as it translates with respect to another rigid body.

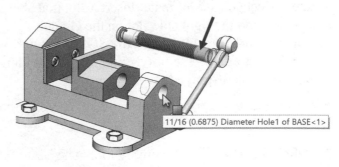

11/16 (0.6875) Diameter Hole1 of BASE<1>

When defining a screw mate, you can define the pitch. The pitch is the amount of relative translational displacement between the rigid bodies for each full rotation of the first rigid body.

1. Open **Screw Mate 11-1** from the SOLIDWORKS 2018\BottomUpAssemblyModeling\Screw folder.

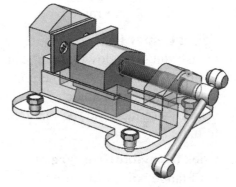

Insert a Screw mate between the screw bar and the base of the clamp.

2. Click the **Mate** tool from the Assembly tab.

3. **Expand** the Mechanical Mates box from the Mate PropertyManager.

4. Click **Screw** mate.

Select two Concentric faces.

5. Click the outside **cylindrical face** of the vice_screw bar as illustrated.

6. Click the **inside face of Diameter Hole1** as illustrated. The two selected faces are displayed.

If needed, Right-click and use the Select Other command to select the inside cylindrical face of the hole.

7. Enter **1** for Revolutions/mm.

8. Click **OK** from the Screw1 PropertyManager.

9. **Rotate** the handle of the screw-bar. View the results.

10. **Close** the model.

Similar to other mate types, Screw mates do not prevent interference or collisions between components. To prevent interference, use Collision Detection or Interference Detection.

Tutorial: Universal Joint Mate 11-1

Create a Mechanical Universal Joint Mate. The universal joint, also known as Hooke's coupling, is used to connect two intersecting shafts, and seems to have its widest use in the automotive industry.

The Universal Joint Mate is useful for cases when you need to transfer rotational motion around corners, in the case of the drive shaft of a car, or between two connected shafts that are allowed to bend at the connection point. The origin location of the universal mate represents the connection point of the two rigid bodies.

1. Open **Universal Joint Mate 11-1** from the SOLIDWORKS 2018\BottomUpAssemblyModeling folder.

Insert a Universal Joint Mate between the two bodies and the pin.

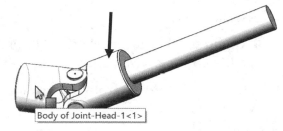

Body of Joint-Head-1<1>

2. Click the **Mate** 🖉 tool from the Assembly tab.

3. **Expand** the Mechanical Mates box from the Mate PropertyManager.

4. Click **Universal Joint** 🕭 mate. Select the two entities to mate.

5. Click the **face** of the first Joint Head component.

6. Click the **face** of the second Joint Head component.

7. Check the **Define Joint point** box. The joint point option represents the connection point between the two selected components to rotate. In this case, the point where the axis of one shaft intersects the axis of the other.

8. Click the **Origin** of the Connection pin from the fly-out FeatureManager. This is the joint point.

9. Click **OK** ✔ from the Universal Joint1 PropertyManager.

10. **Expand** the Mates folder and view the results.

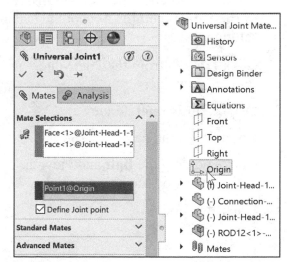

11. **Rotate** the ROD and view the movement of the Universal Joint mate.

12. **Close** the model.

🔆 Additional mates and an Axle reference are needed to simulate actual movement of a universal joint assembly of a drive shaft.

Tutorial: Path Mate 11-1

Create an Advanced Path mate. The Path mate constrains a selected point on a component to a path. In this example, the Path mate is the selected sketch point of the ball to the spiral trough.

1. Open **Path Mate 11-1** from the SOLIDWORKS 2018\BottomUpAssemblyModeling\Path folder.

2. Click the **Mate** ✎ tool from the Assembly tab.

3. **Expand** the Advanced Mates box from the Mate PropertyManager.

4. Click **Path Mate** ⌐ from the Advanced Mates box.

Select the component vertex of the ball (Bottom point of the Revolved Base sketch).

5. **Zoom in** and click the **Revolved Base sketch vertex** as illustrated.

Select the path for the Path Mate.

6. **Zoom in** and click the **bottom inside edge of the spiral trough**. Accept the default settings.

7. Click **OK** ✔ from the PathMate PropertyManager.

8. Display an **Isometric** view.

9. **Zoom in** on the ball.

10. Click and drag the **ball** and view the PathMate along the spiral trough.

11. **Close** the model.

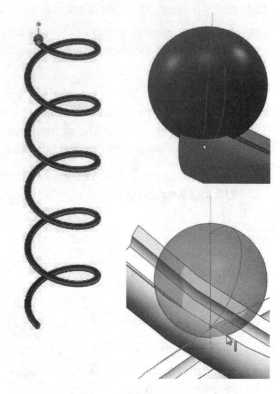

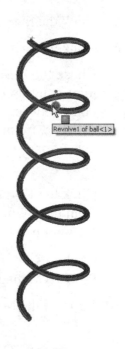

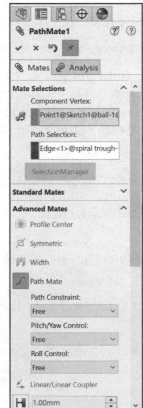

💡 Define pitch, yaw and roll of the component as it travels along the path. You can also add SOLIDWORKS Simulation Motion.

Tutorial: Limit Mate 11-1

Create an Advanced Distance Mate (Limit Mate) between the collar and the base of the drill. The Limit mate restricts overall height movement.

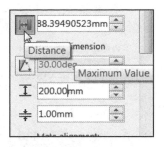

1. Open **Drill Press 11-1** from the SOLIDWORKS 2018\BottomUpAssemblyModeling\Limit folder. The assembly is displayed. Move the assembly up and down.

A Limit mate allows components to move within a range of values for distance and angle mates.

2. Click the **Mate** 📎 tool from the Assembly tab.

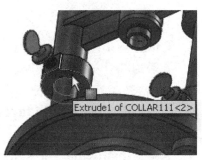

3. Click **Distance** 📐 mate from the Advanced Mates dialog box.

Select entities to Mate.

4. Click the **face of the bottom left COLLAR** as illustrated in the Graphics window.

5. Click the **left face of the base** as illustrated.

6. Enter **200**mm for MAX distance.

7. Enter **1**mm for MIN distance.

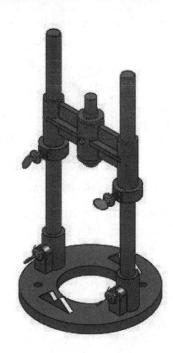

8. Click **OK** ✔ from the LimitDistance1 PropertyManager.

9. Display an **Isometric** view.

10. Click and drag the **Slider** up and down. View the limited movement on the model.

11. **Close** the model.

💡 The Defeature tool provides the ability to remove details from a part or assembly and save the results to a new file in which the details are replaced by dumb solids (that is, solids without feature definition or history). You can share the new file without revealing all the design details of the model.

Tutorial: Width Mate 11-1

Create an Advanced Width mate between the Base and the Handle in the assembly.

1. Open **Door Handle 11.1** from the SOLIDWORKS 2018\BottomUpAssemblyModeling folder. The assembly is displayed. Move the Handle. Create a Width mate to align the two components.

A Width mate centers a tab within the width of a groove.

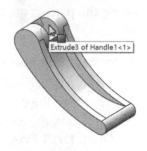

☀ You can select geometry to drive the limits of width mates, eliminating the need for numerical input.

2. Click the **Mate** ✎ tool from the Assembly tab.

3. Click **Width** 𝄃𝄃 mate from the Advanced Mates dialog box.

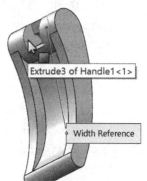

Select entities to Mate (Width Selection).

4. Click the **two inside faces** of the Handle as illustrated.

5. Click **inside** the Tab selections box.

6. Click the **two outside faces** of the base tab as illustrated.

7. Click **OK** ✔ from the Width PropertyManager.

8. Display an **Isometric** view.

9. Click and drag the **Handle up and down**. View the created Width mate.

10. **Close** the model.

The components align so that the tab is centered between the faces of the groove. The tab can translate along the center plane of the groove and rotate about an axis normal to the center plane. The width mate prevents the tab from translating or rotating side to side.

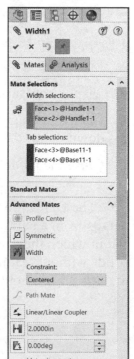

Tutorial: Symmetric Mate 11-1

Create an Advanced Symmetric mate between the two Guide
Rollers about the Front Plane.

1. Open **Guide Assembly 11-1** from the SOLIDWORKS
   2018\BottomUpAssemblyModeling folder. The assembly
   is displayed.

2. Click and drag the **Guide rollers**. The Guide Rollers
   move independently. Create a Symmetric mate between
   the two Guide Rollers about the Front Plane.

A Symmetric mate forces two similar entities to be
symmetric about a plane or planar face.

3. Click the **Mate** 🖉 tool from the Assembly tab.

4. Click **Symmetric** 🗗 mate from the Advanced Mates dialog
   box.

Select Symmetric Plane.

5. Click **Front Plane** from the fly-out FeatureManager. FRONT
   is displayed in the Symmetry plane dialog box.

Select entities to Mate.

6. Click the **face** of the first Guide Roller.

7. Click the **face** of the second Guide Roller.

8. Click **OK** ✔ from the
   Symmetric PropertyManager.

9. Display an **Isometric** view.

10. Click and drag the **Guide
    rollers**. They move together.
    Additional mates are needed for
    proper movement.

11. **Close** the model.

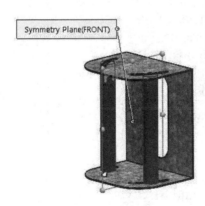

Symmetry Plane(FRONT)

Mate PropertyManager: Analysis tab

Assign mate properties for use in a SOLIDWORKS Motion analysis or SOLIDWORKS Simulation. Motion studies use mate location points to determine how parts move in relation to each other. Modify the mate location used for Motion studies. The Analysis tab option is *Mate dependent*.

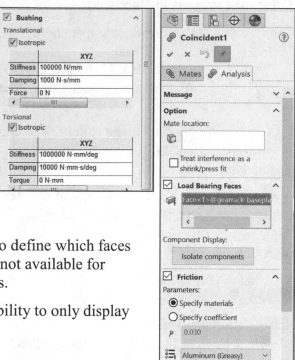

The Analysis tab provides the following options for a selected mate:

Load Bearing Faces. Associates additional faces with the selected mate to define which faces share in bearing the load. This option is not available for **Symmetric**, **Width**, **Path** or **Cam** mates.

- **Isolate components**. Provides the ability to only display the components referenced by the mate.

- ***Friction***. Provides the ability to associate friction properties with some types of mates. The following options are available:

  - **Parameters**. Select how to define the friction properties of the mate.

  - **Specify materials**. Select the materials of the components from the first and second lists.

  - **Specify coefficient**. Provides the ability to specify the following: **Dynamic Friction Coefficient** by typing a number or moving the slider between **Slippery** and **Sticky.**

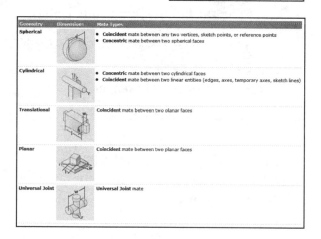

  - **Joint dimensions**. Available dimensions vary depending on geometry and mate type.

- ***Bushing***. Provides the ability to associate bushing properties with a mate. Bushing properties make a mate somewhat flexible by giving it spring and damper characteristics. Mates with bushing properties can produce a more realistic distribution of forces.

See SOLIDWORKS help for additional information on this topic.

Tutorial: Assign mate properties with the Analysis Tab 11-1

Apply the Analysis tab. Assign mate properties for use in a SOLIDWORKS Motion analysis.

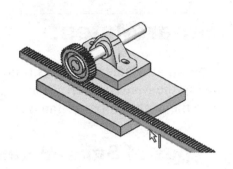

1. Open **Analysis 11-1** from the SOLIDWORKS 2018\BottomUpAssemblyModeling folder.

2. Click and drag the **Rack** in the Graphics window. The Rack moves linearly.

3. Click and drag the **Gear**. The Rack and Gear moves independently of each other.

Edit the Coincident1 Mate under the gearrack-baseplate component.

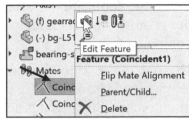

4. Right-click **Coincident1** as illustrated.

5. Click **Edit Feature**. The Coincident1 PropertyManager is displayed.

6. Click the **Analysis** tab.

7. Check the **Load Bearing Faces** dialog box as illustrated.

8. Click the **top face** of the gearrack-baseplate as illustrated. The selected face is displayed in the Load Bearing Faces dialog box.

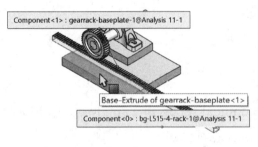

9. Check the **Friction** box. Specify material properties.

10. Select **Aluminum (Greasy)** for both material properties to define friction. Accept the default Bushing options.

11. Click **OK** ✔ from the Coincident1 PropertyManager. You assigned mate properties for a Motion analysis.

12. **Close** the model.

13. Do **not save**.

SmartMates:

The SmartMates tool saves time by allowing you to create commonly used mates without using the Mate PropertyManager. Standard Mates, Advance Mates and Mechanical Mates use the Mate PropertyManager.

Types of SmartMates

There are various SmartMates types that are available to you in SOLIDWORKS. The available SmartMates types are dependent on the application and your situation. In most cases, the application creates a single mate. The type of SmartMate created depends on the geometry that is selected, "to drag" and the type of geometry into which you drop the component.

Use one of the following entities to drag the component: a linear or circular edge, a temporary axis, a vertex, a planar face, or a cylindrical/conical face. The following types of automatic SmartMates are supported and are displayed on your mouse pointer:

Coincident SmartMate:

* Mate two linear edges .

* Mate two planar faces .

* Mate two vertices .

* Mate two axes, two conical faces, or a single conical face and a single axis .

Concentric & Coincident SmartMate:

* Mate two circular edges, (Peg-in-Hole SmartMate). The edges do not have to be complete circles . There are a few conditions that you need to know to apply the Peg-in-Hole SmartMate:

    * One feature must be a Base or Boss.

    * The other feature must be a Hole or a Cut.

    * The features must be Extruded or Revolved.

    * The faces that are selected in the mate must both be of the same type, either a cylinder or a cone. Both need to be the same. You cannot have one of each type.

- A planar face must be adjacent to the conical/cylindrical face of both features.

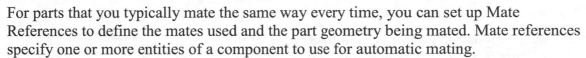

- Mates two circular patterns on flanges (Flange SmartMate) .

For parts that you typically mate the same way every time, you can set up Mate References to define the mates used and the part geometry being mated. Mate references specify one or more entities of a component to use for automatic mating.

Tutorial: SmartMate 11-1

Insert two Concentric SmartMates and a Coincident SmartMate into an assembly.

1. Open the **SmartMate 11-1** assembly from the SOLIDWORKS 2018\BottomUpAssemblyModeling folder.

2. Click the **left back cylindrical hole** face of the IM15-MOUNT part.

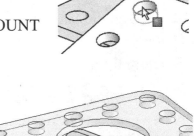

3. Hold the **Alt** key down.

4. Drag the mouse pointer to the **left cylindrical hole** face of the Angle Bracket part. The Concentric SmartMate icon is displayed.

5. Release the **Alt** key. Release the **mouse button**. Concentric is selected by default.

6. Click the **green check mark**.

Insert a second Concentric SmartMate.

7. Click the **right back cylinder hole** face of the IM15-MOUNT part.

8. Hold the **Alt** key down.

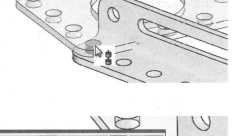

9. Drag the mouse pointer to the **right cylindrical hole** face of the Angle Bracket part. The Concentric SmartMate icon is displayed.

10. Release the **Alt** key.

11. Release the **mouse button**. Concentric is selected by default.

12. Click the **green check mark**.

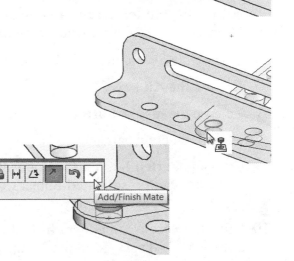

Insert a Coincident SmartMate.

13. Click the **right bottom edge** of the IM15-Mount part.

14. Hold the **Alt** key down.

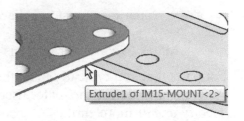

15. Drag the mouse pointer to the **top front edge** of the Angle Bracket part. The Coincident SmartMate icon is displayed.

16. Release the **Alt** key. Release the **mouse button**. Coincident is selected by default.

17. Click the **green check mark**.

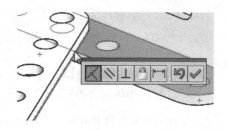

18. **Expand** the Mates folder. View the three inserted mates.

19. **Close** the model.

Tutorial: SmartMate 11-2

Insert a Concentric/Coincident SmartMate. Utilize the Pattern Driven Component Pattern tool with the Instances to Skip option.

1. Open the **SmartMate 11-2** assembly from the SOLIDWORKS 2018\BottomUpAssemblyModeling folder.

2. Open the **SCREW** part from the SOLIDWORKS 2018\BottomUpAssemblyModeling folder.

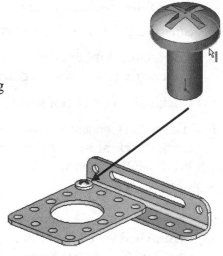

3. Click **Window ➢ Tile Horizontally**.

4. Hold the **Alt** key down.

5. Click and drag the **bottom edge** of the Screw head to the **top left circular edge** of the IM15-Mount part. The Concentric/Coincident SmartMate icon is displayed.

6. Release the **Alt** key.

7. Release the **mouse button**.

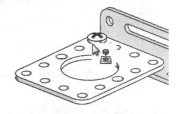

8. **Close** the Screw part.

9. Click the **Patten Driven Component Pattern** tool from the Consolidated Linear Component toolbar. The PropertyManager is displayed.

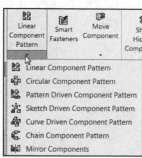

10. Click the **Screw** component from the Graphics window. Screw<1> is displayed in the Components to Pattern box.

11. **Expand** Angle-Bracket from the fly-out FeatureManager.

12. Click inside the **Driving Feature** box.

13. Click **LPattern1** from the fly-out FeatureManager. LPattern1 is displayed in the Driving Feature box.

14. **Expand** the Instances to Skip box.

15. Click inside the **Instances to Skip** box.

16. Select (**5, 1**), (**6, 1**), and (**7, 1**) from the Graphics window to skip. Note: Select the bottom control point.

17. Click **OK** ✔ from the Pattern Driven PropertyManager. The DrivedLPattern2 feature is created and is displayed.

18. **Close** all models.

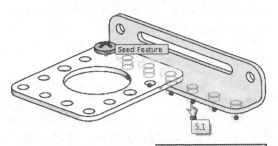

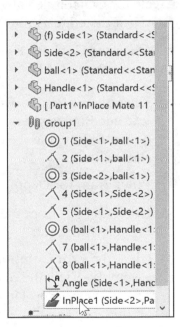

InPlace Mates

InPlace Mates are created automatically for In-context components. InPlace Mates are created to prevent movement of the component.

The In-context component is attached to geometry of components in the assembly through External references, references that cross between components at the assembly level.

An InPlace Mate is a Coincident Mate created between the Front Plane of a new component and the selected planar geometry of the assembly. The component is fully defined; no additional Mates are required to position the component.

By default, SOLIDWORKS uses the default templates for new parts and assemblies developed In-context of an existing assembly.

To select a custom Template, define the System Options, Document Templates option before you insert a new component into the assembly.

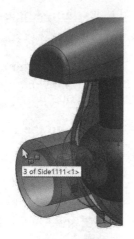

🔆 For the new user, utilize the Search feature in the Systems Options dialog box to quickly locate options for System Options or Document Properties.

Tutorial: InPlace Mate 11-1

Insert a New part (Cap) with an InPlace Mate using the Top-down Assembly method. View Chapter 12 for additional detail information on the Top-down assembly method.

1. Open the **InPlace Mate 11-1** assembly from the SOLIDWORKS 2018\BottomUpAssemblyModeling folder.

2. Click **Insert ➤ Component New Part** 📎 from the Main menu. The new part is displayed in the FeatureManager with a name in the form **[Part***n*^***assembly_name***]**. The square brackets indicate that the part is a virtual component. The mouse pointer displays the ✓ icon.

3. **Rotate** the assembly and click the **face** of the first side as illustrated.

Create the Cap for the valve. Select the Sketch Plane.

4. Click the **edge** as illustrated.

5. Click the **Convert Entities** 🗊 Sketch tool. Use existing geometry to create the CAP.

6. Click the **Extruded Boss/Base** 🗊 Features tool. The Boss-Extrude PropertyManager is displayed. The extrude direction is outward. **Reverse** the direction. Create a CAP that is inside the Ball valve.

7. Enter **2**mm for Depth.

8. Click **OK** ✓ from the Boss-Extrude PropertyManager.

9. Click the **Edit Component** tool to return to the assembly. View the results. You just created a part using the Top-Down Assembly method with an InPlace Mate.

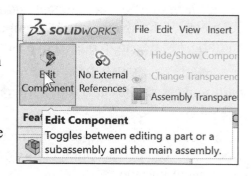

10. **Close** the model.

See Chapter 12 for additional information on InPlace Mates and Top-Down Assembly modeling.

Mate Reference

Mate references specify one or more entities of a component to use for automatic mating. When you click and drag a component with a mate reference into an assembly, the software tries to locate other combinations of the same mate reference name and type. If the name is the same, but the type does not match, the software does not add the mate.

The Mate Reference 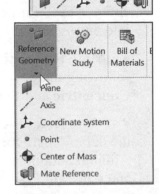 tool is located in the Reference Geometry toolbar and under the Assembly tab in the CommandManager. Below are a few items to be aware of when using the Mate Reference tool:

- *Components*. You can add mate references to parts and assemblies. Select assembly geometry, for example: a plane in the assembly or component geometry, for example: the face of a component.

- *Multiple mate references*. More than a single mate reference can be contained in a component. All mate references are located in the Mate References folder in the FeatureManager design tree. Example: You have a component in an assembly with two mate references: nut and bolt. When you click and drag a fastener with a mate reference named nut into your assembly, mates are inserted between the entities with the same mate reference name.

- *Multiple mated entities*. Each mate reference may contain one to three mated entities. The mated entities are a primary for the first, a secondary for the second, and tertiary for the three reference entity. Each of the entities can have an assigned mate type and alignment. For two components to mate automatically, their mate references must have the same *Number of entities*, *Name* and *Mate type for corresponding entities*.

- *SmartMates*. When the SmartMate PropertyManager is active, the software adds mates through the Mate References tool before it adds geometric SmartMates.

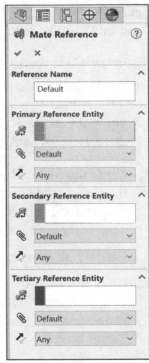

The Mate Reference 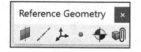 tool uses the Mate Reference PropertyManager. The Mate Reference PropertyManager provides the following selections:

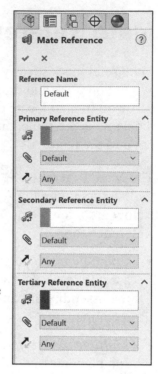

- **Reference Name**. Displays the name for the mate reference. Default is the default name reference. Accept Default or type a name in the mate reference box.

- **Primary Reference Entity**. Displays the selected face, edge, vertex, or plane for the Primary reference entity. The selected entity is used for potential mates when dragging a component into an assembly.

  - **Mate Reference Type**. Provides the ability to select the following mate types: **Default**, **Tangent**, **Coincident**, **Concentric** and **Parallel**.

  - **Mate Reference Alignment**. Provides the ability to define the default mate for the reference entity. The following Alignment options are available: **Any**, **Aligned**, **Anti-Aligned** and **Closest**.

Secondary and tertiary entity options are the same as the Primary Reference Entity box.

Tutorial: Mate Reference 11-1

Utilize the Reference Mate tool. Insert a Concentric and a Coincident Mate Reference Type.

1. Open the **Mate Reference 11-1** part from the SOLIDWORKS 2018\BottomUpAssemblyModeling folder.

2. Click the **Mate Reference** 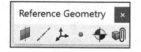 tool from the Reference Geometry toolbar. The Mate Reference PropertyManager is displayed. The Reference Name is Default.

3. Display an **Isometric** view.

4. Click the **cylindrical right face** in the Graphics window as illustrated. Face<1> is displayed in the Primary Reference Entity box.

5. Select **Concentric** for Mate Reference Type. Accept Any as the default option.

6. Click the **front face** in the Graphics window. Face<2> is displayed in the Secondary Reference Entity box.

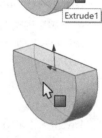

7. Select **Coincident** for Mate Reference Type. Accept Any as the default option.

8. Click the **back face** from the Graphics window. Face<3> is displayed in the Tertiary Reference Entity box.

9. Select **Coincident** for Mate Reference Type. Accept Any as the default option.

10. Click **OK** ✔ from the Mate Reference PropertyManager. The Mate References folder is displayed in the FeatureManager.

11. Display an **Isometric** view. Click **Save As** from the Menu bar.

12. Enter **Key** for file name. Open the **Mate Reference Shaft 11-1** assembly from the SOLIDWORKS 2018\BottomUpAssemblyModeling folder.

13. Click the **Insert Components** 🗗 tool from the Assembly toolbar. Click **Key** from the Part/Assembly to insert box.

14. Click the **cylindrical face** of Shaft-2 inside cut as illustrated. View the results. Three mates are created. **Close** the model.

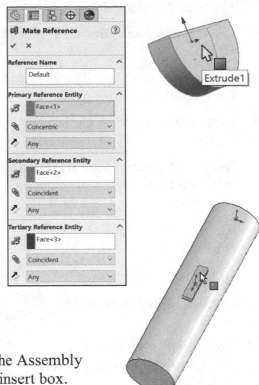

Quick Mate

Quick Mate is a procedure to mate components together. No command (click Mate from the Assembly CommandManager) is required. Hold the Ctrl key down, and make your selections. Release the Ctrl key, and a Quick Mate pop-up is displayed below the context toolbar. Select your mate and you are finished.

Utilize the Quick Mate procedure for Standard mates, Cam mate, Profile Center mate, Slot mate, Symmetric mate and Width mate. To activate the Quick Mate functionality, click Tools, Customize. On the toolbars tab, under Context toolbar settings, select Show Quick Mates. Quick Mate is selected by default.

Tutorial: Quick Mate 11-1

Utilize the Quick Mate procedure. Insert a Concentric mate between the Axle and the Collar. Insert a Distance mate between the back face of the Bushing and the face of the Axle.

1. Open the **Quick Mate 11-1** assembly from the SOLIDWORKS 2018\BottomUpAssemblyModeling\FlyWheel folder.

2. Click the **cylindrical face** of the Axle.

3. Hold the **Ctrl** key down.

4. Click the **inside cylindrical face** of the Collar.

5. Release the **Ctrl** key. The Pop-up Mate Context menu is displayed.

6. Click **Concentric**. A Concentric Mate is created.

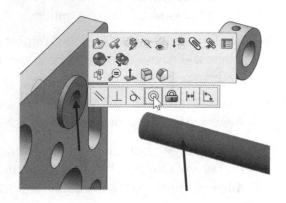

7. Insert a Distance Mate. **Rotate** the model to view the back face of the Bushing as illustrated.

8. Click the **back face** of the Bushing.

9. Hold the **Ctrl** key down.

10. Click the **back face** of the Axle.

11. Release the **Ctrl** key. The Pop-up Mate Context menu is displayed.

12. Click **Distance** Mate.

13. Enter **20mm**.

14. Click the **Flip Dimension** tool if needed.

15. Click the **Green Check mark** from the Pop-up Mate Context menu.

16. Display an **Isometric** view. You created a Concentric and Distance mate using the Quick mate procedure.

17. **Close** the model.

Mate Diagnostics/MateXpert

The Mate Diagnostics tool provides the ability to recognize mating problems in an assembly. The tool provides information to examine the details of mates that are not satisfied and identifies groups of mates that are over defined in your assembly.

The following icons are displayed to indicate errors or warnings in an assembly:

- *Error* ⬇. The Error icon indicates an error in your model. The Error icon is displayed on the document name at the top of the FeatureManager design tree and on the component that contains the error.

- *Warning* 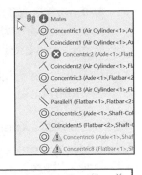. The Warning icon indicates a warning in your model. The Warning icon is displayed on the component that contains the feature that issued the warning. When it is displayed on the Mates folder, it indicates that all the mates are satisfied, but one or more mates are over defined.

View the What's Wrong dialog box for additional information on identifying errors and warnings in your parts or assemblies. The following icons indicate the status of the mates:

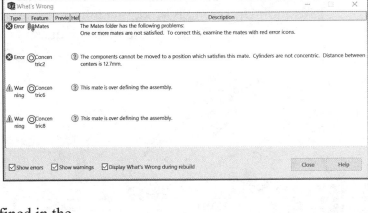

- *Not Satisfied* ⊗. The Not Satisfied icon indicates that all mates are not satisfied in your assembly. Use the Mate Diagnostics tool to address.

- *Satisfied but over defined* ⚠. The Satisfied icon indicates that the mates are satisfied, but they are over defined in the assembly. Use the Mate Diagnostic tool to address.

The MateXpert tool uses the MateXpert PropertyManager. To active the MateXpert PropertyManager, click **Tools ➤ MateXpert**, or right-click the **Mates** folder or any mate in the **Mates** folder, and click **MateXpert**.

The MateXpert PropertyManager provides the following selections:

- *Analyze Problem*. The Analyze Problem box provides the following options:

  - **Analyze**. Displays one or more subsets of mates with problems. The components that are not related to the current subset are displayed transparent in the Graphics window. A message is displayed with information on the mating problem.

  - **Diagnose**. Performs the diagnoses of the displayed mates in the Analyze box.

- *Not Satisfied Mates*. Displays the unsolved mates. The selected unsolved mate is highlighted in the Graphics window. A message informs you that the distance or angle by which the mated entities are currently misaligned.

Mates which are displayed in both the **Analyze** box and the **Not Satisfied Mates** box are displayed in bold.

Tutorial: MateXpert 11-1

View Mate errors. Utilize the Mate Diagnostics tool. Resolve a Mate conflict.

1. Open **MateXpert 11-1** assembly from the SOLIDWORKS 2018\BottomUpAssemblyModeling folder. Review the Mate errors on the What's Wrong dialog box. These are the Mates in question.

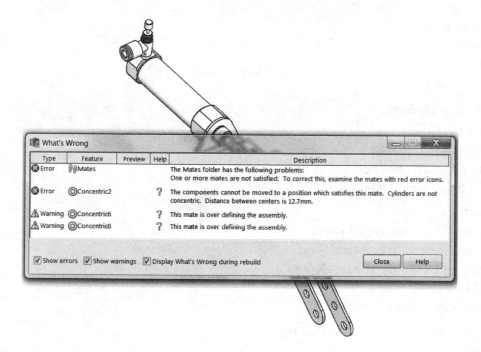

2. Click **Close** from the What's Wrong dialog box.

View Mate errors.

3. Right-click **Axle<1>** from the FeatureManager.

4. Click **View Mate Errors**. View the display in the Graphics window. Additional mates are displayed.

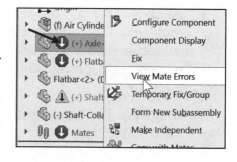

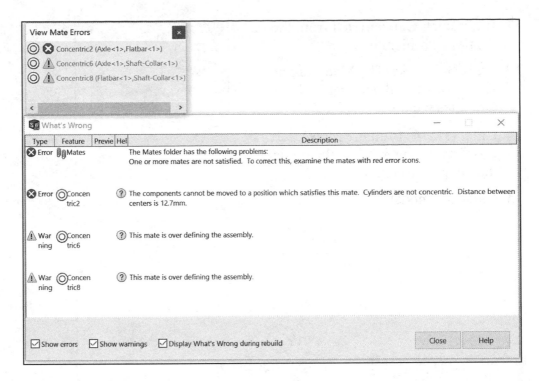

View the Mate PropertyManager.

5. Click **Close** from the What's Wrong dialog box.

6. **Close** the View Mate Errors dialog box to return to the standard FeatureManager and display in the Graphics window.

7. Right-click the **Mates** folder from the FeatureManager.

8. Click **MateXpert**. The MateXpert PropertyManager is displayed.

9. Click **Diagnose** from the Analyze Problem box.

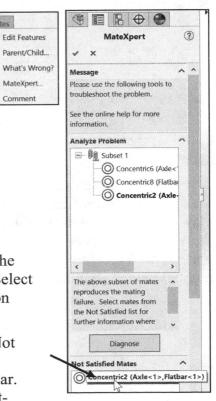

10. **Expand** Subset1 and review the mates. A message is displayed with information on the mating problem, "The above subset of mates reproduces the mating failure. Select mates from the Not Satisfied list for further information where applicable."

11. Click **Concentric2(Axle<1>, Flatbar<1>)** from the Not Satisfied Mates box. View the Axle in the Graphics window. Concentric6 mates the Axle to the Shaft-Collar. Concentric8 mates the Flatbar second hole to the Shaft-Collar. Concentric2 mates the Axle to the first hole of the Flatbar. The Shaft-Collar cannot be physically located concentrically at both the first hole and the second hole of the Flatbar.

12. Right-click **Concentric8** from the Analyze Problem box.

13. Click **Suppress**. The Mates are now satisfied. View the model in the Graphics window.

14. Click **OK** ✔ from the MateXpert PropertyManager. View the model in the Graphics window.

15. **Close** the model.

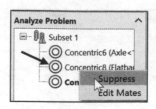

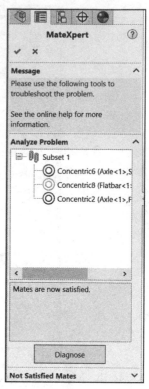

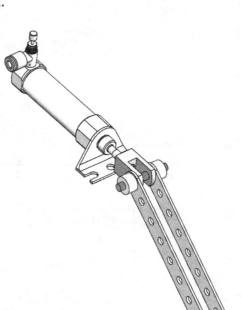

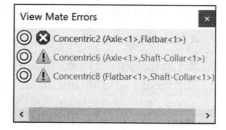 Click the **Over Defined** tab in the lower right corner of the Graphics window to view Mate Errors in the Graphics window and to obtain the View Mate Errors PropertyManager. Select **each** mate in the View Mate Errors PropertyManager to highlight in the Graphics window.

Performance Evaluation

Performance Evaluation analyzes performance of
assemblies and suggests possible actions you can take to
improve performance. This is useful when you work with
large and complex assemblies. In some cases, you can
select to have the software make changes to your assembly
to improve performance.

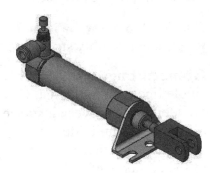

Although the conditions identified by the Performance
Evaluation can degrade assembly performance, they are
not errors. It is important that you understand and think
about the recommendations before you modify the model.
In some cases, implementing the system recommendations
could improve your assembly performance but could also
compromise the design intent of your model.

Tutorial: Performance Evaluation 11-1

Utilize the Performance Evaluation tool.

1. Open **Performance Evaluation 11-1**
   assembly from the SOLIDWORKS
   2018\BottomUpAssembly Modeling folder.

2. Click the **Performance Evaluation** tool
   from the Evaluate tab in the
   CommandManager. View the dialog box.

3. **Close** the model.

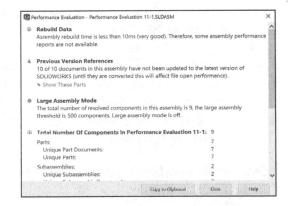

Assembly Visualization

In an assembly, the designer selects material based on cost,
performance, physical properties, manufacturing processes,
sustainability, etc. The SOLIDWORKS Assembly
Visualization tool includes a set of predefined columns to
help troubleshoot assembly performance. You can view the
open and rebuild times for the components, and the total
number of graphics triangles for all instances of components.
The Assembly Visualization tool provides the ability to rank
components based on the default values (**weight, mass,
density, volume, etc.**) or their custom properties (**cost,
sustainability, density, surface area, volume, etc.**) or an
equation and activate a spectrum of colors that reflects the
relative values of the properties for each component.

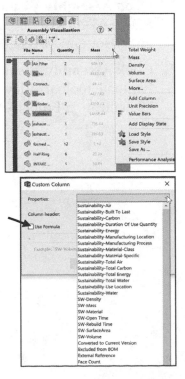

Hide/Show Value Bar ▤ **icon**. Available for numeric properties. Turns the value bars off and on. When the value bars are on, the component with the highest value displays the longest bar. You can set the length of the bars to be calculated relative to the highest-value component or relative to the entire assembly.

Flat Nested View ⬙ **icon**. Nested view, where subassemblies are indented. Flat view, where subassembly structures are ignored (similar to a parts-only BOM).

Grouped/Ungrouped View ⬙ **icon**. Groups multiple instances of a component into a single line item in the list. Grouped View is useful when listing values for properties that are identical for every instance of the component. Ungrouped views lists each instance of a component individually. Ungrouped View is useful when listing values for instance-specific properties, such as fully mated, which might be different for different instances of the component.

Performance Analysis ⬚ **icon**. Provides additional information on the open, display, and rebuild performance of models in an assembly.

Filter ▽ ˅ **icon**. Filters the list by text and by component show/hide state.

The Assembly Visualization ⬙ tab in the FeatureManager design tree panel contains a list of all components in the assembly, sorted initially by file name. There are three default columns: **File name**, **Quantity**, **Mass (Default)**.

Tutorial: Assembly Visualization 11-1

Utilize the Assembly Visualization tool.

1. Open **Assembly Visualization 11-1** from the SOLIDWORKS 2018\BottomUpAssemblyModeling folder.

2. Click **Assembly Visualization** ⬙ from the Evaluate toolbar. View the Assembly Visualization PropertyManager.

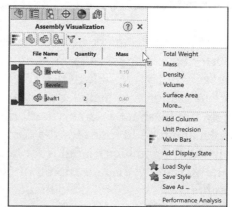

Click a column header to sort by its parameter.

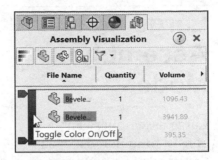

3. Click the **arrow** to the right of Mass. View your display options.

4. Click **Volume**. View the results.

5. Click the **arrow** under Quantity. This changes the order of the components in the assembly.

6. Click the **Slider** as illustrated. This provides the ability to add a color to the spectrum when you click in the blank area to the left of the vertical bar.

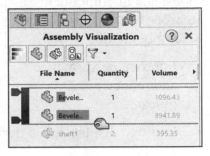

7. Position the **Rollback bar** below the first component. View the results in the Graphics window. This option provides the ability to hide or show items in the list and graphics area when you drag the bar up or down.

8. Position the **Rollback bar** to its original position.

9. **Close** the model.

Performance Evaluation Enhancements

The Assembly Open Progress indicator provides information on the status of operations while you open an assembly.

The indicator provides information on the following operations: Open Components, Update Assembly, Update Graphics, Elapsed time, Previous time to open.

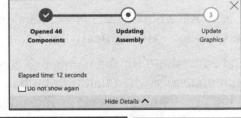

To view the performance information at a later time, click **Tools** > **Evaluate** > **Performance Evaluation**. The Performance Evaluation dialog box is displayed. Explore the various options.

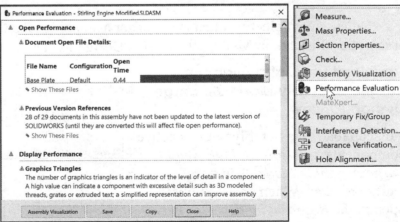

Large Assembly Mode

Large assemblies can consist of hundreds, even thousands of components. Reasons to simplify a large assembly include:

- Improve performance and reduce rebuild times.

- Improve display speed during dynamic view operations (zoom, pan, rotate, etc.).

- Focus your work on a subset of components.

Large Assembly Mode is a collection of system option settings that improves the performance of large assemblies. You can turn on Large Assembly Mode at any time, or you can set a threshold for the number of components and have Large Assembly Mode turn on automatically when that threshold is reached. See System Options chapter for additional details.

Large Assembly Mode is a collection of system settings that improves the performance of assemblies. Automatically load components lightweight is checked by default.

Set a threshold for the number of components and have Large Assembly Mode active automatically when that threshold is reached. The default setting for the Large Assembly Mode is 500 components as illustrated.

Quick view/Selective open is no longer available in the Open dialog box for assemblies. Instead, in Mode, select Large Design Review.

Resolved vs. Lightweight vs. Large Assembly Mode vs. Large Design Review

You can load an assembly with its active components in *Resolved, Lightweight, Large Assembly Mode or in Large Design Review*. When an assembly is open in the *Resolved mode*, all of its components and model data are loaded into memory.

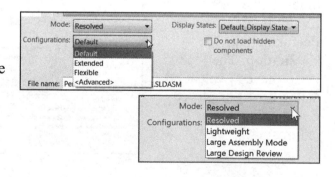

When an assembly or component is open in the *Lightweight mode*, only a subset of its model data is loaded in memory.

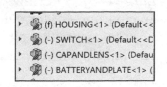

The remaining model data is loaded on an as-needed basis. A feather is displayed in the FeatureManager next to the lightweight components. This improves performance of large and complex assemblies.

Large Assembly Mode is a collection of system settings that improves the performance of assemblies.

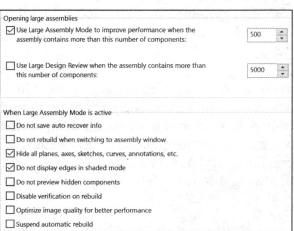

You can turn on Large Assembly Mode at any time, or you can set a threshold for the number of components and have Large Assembly Mode turn on automatically when that threshold is reached.

While using the Open dialog box to open an assembly whose number of components exceeds the specified threshold, mode is automatically set to Large Assembly Mode, but you can select another mode from the drop-down list.

To activate the Large Assembly mode, click Large Assembly Mode from the open menu.

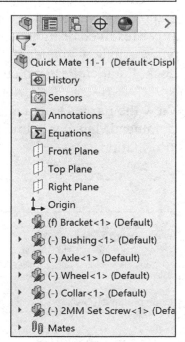

Large Design Review  is primarily intended as an environment for quick design reviews. If you want to ensure that all items are updated properly, you must open your assembly as lightweight or fully resolved.

Information about the following items is not shown in the FeatureManager design tree: **Assembly features**, **Component patterns**, and **Mates**.

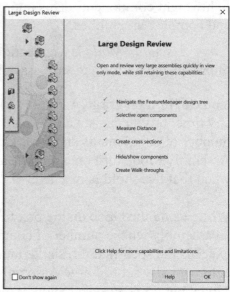

⌖ An assembly must be saved in SOLIDWORKS 2012 or later to open properly in Large Design Review.

In an assembly opened in Large Design Review mode, you can set a component to lightweight or resolved in order to work on the component. The other components in the assembly remain as graphics-only. You can change a component from lightweight or resolved to graphics-only.

⌖ Create a mate or mate reference between a graphics-only component and a component that is either resolved or lightweight. Supported mate types: Angle, Coincident, Concentric, Distance, Lock, Parallel, Perpendicular and Tangent.

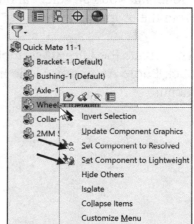

View the Large Design Review default CommandManager below for additional information.

Simplifying Large Assemblies

Large assemblies can consist of hundreds of components. Reasons to simplify a large assembly include:

- Improve performance and reduce rebuild times.

- Improve display speed during dynamic view operations (zoom, pan, rotate, and so on).

- Focus your work on a subset of components.

- When the computational resources do not permit for simulation on the exact model, you can consider the following simplification options when working with assemblies or multi-bodies.

To simplify assemblies:

- *Toggle the visibility of the components.* You can toggle the display of assembly components. You can remove the component completely from view or make it 75% transparent. Turning off the display of a component temporarily removes it from view, allowing you to work with underlying components. Hiding or showing a component affects only the visibility of the component. Hidden components have the same accessibility and behaviors as shown components in the same suppression state (see suppression states of components for additional help).

Toggling the visibility does not affect the rebuild or evaluation speed. Display performance improves, however.

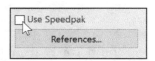

- *Create SpeedPak* 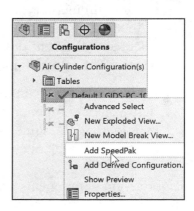 *configurations.* SpeedPak creates a simplified configuration of an assembly without losing references. If you work with very large and complex assemblies, using a SpeedPak configuration can significantly improve performance while working in the assembly and its drawing. A SpeedPak configuration is essentially a subset of the parts and faces of an assembly. Unlike regular configurations, where you can simplify an assembly only by suppressing components, SpeedPak simplifies without suppressing.

Therefore, you can substitute a SpeedPak configuration for the full assembly in higher level assemblies without losing references. Because only a subset of the parts and faces is used, memory usage is reduced, which can increase performance of many operations.

When you make changes to an assembly, such as adding, deleting, or moving components, the changes are not automatically incorporated into the SpeedPak configuration, even when you rebuild the assembly.

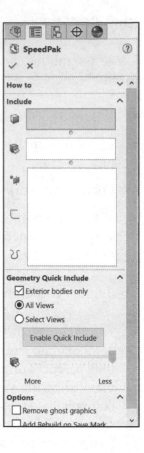

You must manually update the SpeedPak configuration to incorporate the changes. The Use Speedpak checkbox is shown only for modes that allow it.

- *Change the suppression state of components*. Depending on the scope of the work you plan to do at any given time, you can specify an appropriate suppression state for the components. This can reduce the amount of data that is loaded and evaluated as you work. The assembly displays and rebuilds faster, and you make more efficient use of your system resources. See Comparison of Components Suppression States for additional help.

There are three suppression states for assembly components:

- *Resolve*. Resolved (or unsuppressed) is the normal state for assembly components. A resolved component is fully loaded in memory, fully functional and fully accessible. All its model data is available, so its entities can be selected, referenced, edited, used in mates, etc.

- *Suppressed*. A Suppressed state removes a component temporarily from the assembly, without deleting it. It is not loaded into memory, and it is no longer a functional member of the assembly. You cannot see a suppressed component, or select any of its entities. A suppressed component is displayed in gray in the Assembly FeatureManager.

- *Lightweight*. You can improve performance of large assemblies significantly by using lightweight components. Loading an assembly with lightweight components is faster than loading the same assembly with fully resolved components. Assemblies with lightweight components rebuild faster because less data is evaluated.

Lightweight components are efficient because the full model data for the components is loaded only as it is needed. Only components that are affected by changes that you make in the current editing session become fully resolved.

To fully resolve a component, right-click the component name in the Assembly FeatureManager and click Set to Resolved.

You can perform the following assembly operations on lightweight components without resolving them: *Add/remove mates, Interference detection, Collision detection, Edge/Face/Component selection, Annotations, Measure, Dimensions, Section Properties, Assembly Reference Geometry, Mass Properties, Section View, Exploded View,* and *Physical Simulation.*

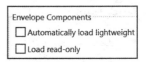

☀ The Level of detail option affects performance. Drag the slider to the right for a faster display and a reduced level of detail in the Performance section of System Options.

How large is a large assembly? Answer: There is no right or wrong answer. A goal in the chapter is to develop sound assembly modeling techniques and to understand the various assembly options in SOLIDWORKS.

☀ Set the mode in which envelope components are loaded when you open an assembly. Select one or both. Located under Options/Assemblies.

Open an Assembly Document

When opening a SOLIDWORKS Assembly document the following options are available:

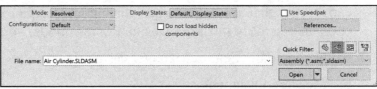

- *Mode, Configurations, Display State, Do not load hidden components, Use Speedpak, References, Quick Filter, (Assemblies, Top-Level Assemblies) and document type.*

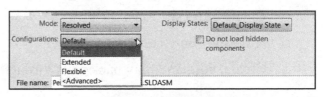

Mode

The Mode drop-down menu provides four options. Resolved is selected by default.

- *Resolved.* All individual components and sub-assemblies are loaded into the assembly.
- *Lightweight.* Only a subset of the data is loaded into memory. The remaining data of the model is loaded on an as-needed basis. You can perform the following assembly operations on lightweight components without resolving them: *Add/remove mates, Interference detection, Collision detection, Edge/Face/Component selection, Annotations, Measure, Dimensions, Section Properties, Assembly Reference Geometry, Mass Properties, Section View, Exploded View* and *Basic Motion.*

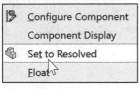

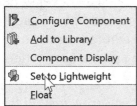

A 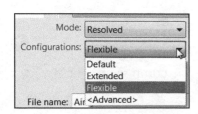 feather is displayed in the FeatureManager for a Lightweight component in an assembly. Right-click on the component and click **Set to Resolved** to resolve the component.

- *Large Assembly Mode.* Collection of system settings that improves the performance of assemblies. You can select Large Assembly Mode when you open an assembly, you can turn Large Assembly Mode on from System Options at any time, or you can set a threshold for the number of components and have Large Assembly Mode turn on automatically when that threshold is reached.

- *Large Design Review.* Opens very large assemblies quickly, while retaining capabilities that are useful when conducting design reviews of assemblies. When you open an assembly in Large Design Review mode, you can:

  o Navigate the FeatureManager design tree.

  o Measure Distances.

  o Create cross sections.

  o Hide and show components.

  o Create, edit, and play back walk-through.

Configurations

Provide the ability to open an assembly in the selected configuration.

Display State

Provides the ability to open an assembly in the selected display state.

- *Do not load hidden components.* Does not load selected hidden components in the assembly.

SpeedPak

Creates a simplified configuration of an assembly without losing references. If you work with very large and complex assemblies, using a SpeedPak configuration can significantly improve performance while working in the assembly and its drawing.

A SpeedPak configuration is essentially a subset of the parts and faces of an assembly. Unlike regular configurations, where you can simplify an assembly only by suppressing components, SpeedPak simplifies without suppressing. Therefore, you can substitute a SpeedPak configuration for the full assembly in higher level assemblies without losing references. Because only a subset of the parts and faces is used, memory usage is reduced, which can increase performance of many operations.

Center of Mass point

Add a Center of Mass (COM) ⊕ point to a part, assembly or drawing. In an assembly, COM points added in component files appear in the assembly.

The position of the COM point updates when the model's center of mass changes. For example, the position of the COM point updates as you add, modify, delete, or suppress components or add or delete assembly features.

Center of Mass icons

Various icons indicate a model's center of mass and related points in the Graphics window:

- ⊕ **Center of Mass**. Indicates the global center of mass of the entire model.

- ⊕ **Center of Mass for a part**. In an assembly, indicates the center of mass of a component that is a part. Defined in the part document.

- ⊕ **Center of Mass for a subassembly**. In an assembly, indicates the center of mass of a component that is a subassembly. Defined in the subassembly document.

- **Center of Mass - user defined**. Indicates the global center of mass of the entire model, as defined in the Override Mass Properties dialog box with user-specified coordinates.

- **Center of Mass - part user defined**. In an assembly, indicates a user-defined center of mass of a component that is a part. Defined in the part document, in the Override Mass Properties dialog box.

- **Subassembly - user defined**. In an assembly, indicates a user-defined center of mass of a component that is a subassembly. Defined in the subassembly document, in the Override Mass Properties dialog box.

- **Center of Mass Reference Point (COMRP)**. Indicates the center of mass of the features above the point in the FeatureManager design tree.

Various icons indicate a model's center of mass and related points in the FeatureManager tree:

- **Center of Mass**. Indicates the global center of mass of the entire model. Located just below Origin in the FeatureManager design tree.

- **Center of Mass Reference Point (COMRP)**. Indicates the center of mass of the features above the point in the FeatureManager design tree.

Create Center of Mass symbol

Utilize the following commands to create the Center of Mass symbol for SOLIDWORKS part, assembly or drawing document.

- Click **Insert ➤ Reference Geometry ➤ Center of Mass** ⊕ from the Main Toolbar menu.

- Click **Center of Mass** ⊕ from the Reference Geometry toolbar.

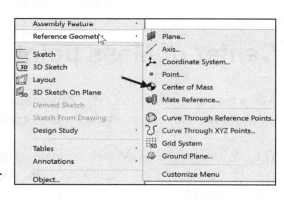

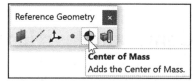

- Click **Create Center of Mass feature** from the Mass Properties box.

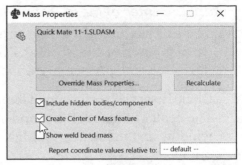

Display Center of Mass Symbol

If needed, utilize the following commands to display the Center of Mass symbol in a SOLIDWORKS part, assembly or drawing document.

You first need to create the Center of Mass symbol before you display it in a SOLIDWORKS document.

- Click **Center of Mass** ⊕ from the Heads-up View toolbar (part, assembly or drawing).

- Click **View ➤ Hide /Show ➤ View Center of Mass** ⊕ from the Main Toolbar menu (part, assembly or drawing).

- Click **Center of Mass** ⊕ from Model Items PropertyManager as illustrated (drawing only).

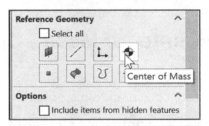

- Click **Insert ➤ Model Items** from the main Toolbar menu. The Model Items PropertyManager is displayed. Click **Center of Mass** ⊕ from the Reference Geometry box (drawing only).

The position of the COM point ⊕ updates when the model's center of mass changes. For example, the position of the COM point updates as you add, modify, delete, or suppress features or components. The COM point can be useful when you are designing assemblies requiring balanced mass (for example, to avoid excessive vibration).

You can mate to the COM points of an assembly's components in distance, coincident, and concentric mates. You cannot mate to the COM point of the assembly itself.

You can override moments of inertia properties for an assembly or its components. Adding custom inertia properties to a component can help you visualize the component's effect on the overall inertia and mass of the structure.

Tutorial: Center of Mass Point in an assembly 11-1

Apply the Center of Mass tool in an assembly to validate location.

1. Open **Center of Mass Point 11-1** from the SOLIDWORKS 2018\BottomUpAssemblyModeling\COM folder.

2. Click **Insert ➤ Reference Geometry ➤ ✛ Center of Mass**. The center of mass of the model is displayed in the Graphics window and in the FeatureManager design tree just below the origin.

3. Click the **Measure** 🔎 tool from the Evaluate tab. The Measure dialog box is displayed. The *Show XYZ Measurements* button should be the only button activated.

4. **Expand** the Measure dialog box if needed.

5. Click the **top edge** of the BATTERY assembly.

6. Click the **Center of Mass**. View the results.

7. **Close** the assembly.

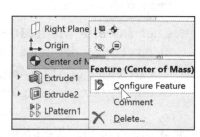

Center of Mass - Configuration

The Center of Mass (COM) point can be suppressed and unsuppressed for configurations.

- Right-click the **Center of Mass** icon in the FeatureManager tree.

- Click **Configure Feature**. The Modify Configurations dialog box is displayed.

- View your **Selections**.

💡 You can **measure distances** and **add reference dimensions** between the COM point and entities such as vertices, edges, and faces.

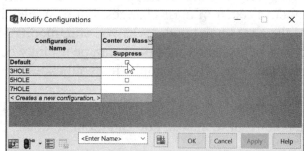

Center of Mass - Reference Point

If you want to display a reference point where the **CG** was located at some particular point in the FeatureManager, you can insert a Center of Mass Reference Point. A **COMRP** is a reference point created at the current center of mass of the part (only part documents). It remains at the coordinates where you create it even if the COM point moves due to changes in the geometry of the part.

For parts only, you can **measure distances** and **add reference dimensions**.

CM and CG coincides only if the acceleration due to gravity acting on all the particles of the body have same value.

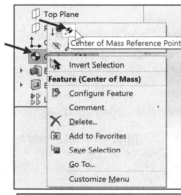

Create Reference Point

- Right-click the **Center of Mass** in the FeatureManager design tree or Graphics window.

- Click **Center of Mass Reference Point** ⊹ icon as illustrated. In the FeatureManager design tree, Center of Mass Reference Point is added as the next feature in the tree. In the Graphics window, Center of Mass Reference Point appears at the current center of mass. Initially, the icon is hidden by the Center of Mass.

- If you add additional features to the part, the Center of Mass ⊕ icon moves and the COMRP ⊹ icon becomes visible.

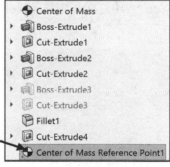

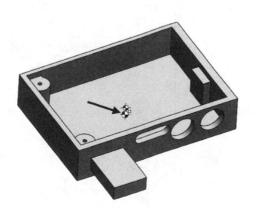

Tree House

With Treehouse, you can set up your file structure in a graphical user interface before you start building the models in SOLIDWORKS.

To run Treehouse in Windows, click **Start ➤ All apps ➤ SOLIDWORKS 2018 ➤ SOLIDWORKS Tools 2018 ➤ SOLIDWORKS Treehouse 2018**. The SOLIDWORKS Treehouse dialog box is displayed.

In Treehouse you can:

- Drag part, assembly, and drawing nodes into the graphics area to create a file structure.

- Edit file properties.

- Add configurations.

- Suppress components and component instances.

- Drag existing documents into Treehouse from Windows Explorer.

See SOLIDWORKS Help for additional information.

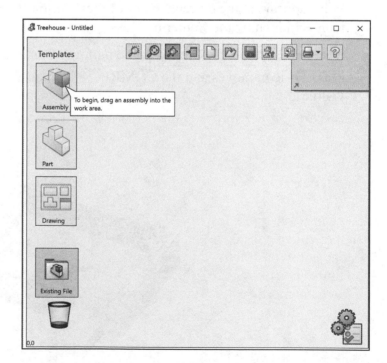

Summary

In this chapter, you learned about Assembly modeling and the Bottom-up Assembly Modeling techniques. An assembly combines two or more parts. A part inserted into an assembly is called a component. A sub-assembly is a component contained within an assembly. When you create your first assembly, Assem1 is the default document name. The Assembly document ends with the extension .sldasm.

You reviewed the various Assembly configuration methods: *Manual (Add Configuration command), Configure component/Configure dimension tool,* and *Design Tables* and addressed the before you begin an assembly task list. See Chapter 12 for additional tutorials on assembly configuration methods.

You reviewed the Assembly FeatureManager and the various Component States of an assembly. You also reviewed and applied the Mate PropertyManager, (Mates and Analysis tabs), reviewed Standard Mates, Mechanical Mates, Advanced Mates, SmartMates, InPlace Mates and Mate References.

Use the Mate Controller PropertyManager, to show and save the position of assembly components at various mate values and degrees of freedom. You can create simple animations between those positions. Supported mate types: Angle, Distance, LimitAngle, LimitDistance, Slot (Distance Along Slot, Percent Along Slot) and Width (Dimension, Percent).

The Mate Diagnostics tool provided the ability to recognize mating problems in an assembly. The tool provides information to examine the details of mates which are not satisfied and identifies groups of mates which are over defined in your assembly.

You explored the Large Assembly mode with its collection of system option settings that improves the performance of large assemblies.

By default, when you create a sub-assembly, it is in the rigid mode. Right-click the **component** in the assembly. Click **Component Properties** from the Context toolbar. Set the desired mode.

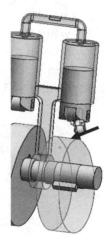

You can set an option in the Mate PropertyManager so that the first component you select from becomes transparent. Then selecting from the second component is easier, especially if the second component is behind the first. The option is supported for all mate types except those that might have more than one selection from the first component (width, symmetry, linear coupler, cam, and hinge).

Quick mate is a procedure to mate components together. No command (click Mate from the Assembly CommandManager) is required. Hold the Ctrl key down, and make your selections. Release the Ctrl key; a Quick Mate pop-up is displayed below the context toolbar. Select your mate and you are finished.

Quick mate is a procedure to mate Standard mates, Cam mate, Profile Center mate, Slot mate, Symmetric mate and a Width mate.

To activate the Quick Mate functionality, click Tools, Customize. On the toolbars tab, under Context toolbar settings, select Show Quick Mates. Quick Mate is selected by default.

You can measure distances and add reference dimensions between the COM point and entities such as vertices, edges, and faces.

You cannot create driving dimensions from the COM ✛ point. However, you can create a Center of Mass Reference Point (COMRP) and use that point to define driving dimensions. A COMRP is a reference point created at the current center of mass of the part. It remains at the coordinates where you create it even if the COM ✛ point moves due to changes in the geometry of the part.

Use the Alt key to temporarily hide a face when you need to select an obscured face for mates.

You can enable or disable the creation of misaligned mates in System Options. To allow creation of misaligned mates: 1.) Click Tools > Options > System Options > Assemblies. 2.) Select Allow creation of misaligned mates.

In Chapter 12, explore and learn how to create and apply Assembly modeling methods used in the Top-down Assembly modeling approach along with configurations using the following methods: *Manual(Add Configuration command), Design Tables* and *Configure Component tool/Configure Dimension tool*. You will also understand the difference between In-Context and Out-of-Context, create and apply equations with global variables and review SOLIDWORKS SpeedPak.

Chapter 12: Top-Down Assembly Modeling, Configurations and More

Chapter Objective

Chapter 12 provides a comprehensive understanding of the modeling methods used in the Top-down assembly approach, Motion Studies and Assembly configurations. On the completion of this chapter, you will be able to:

- Know and apply the three different Top-down assembly methods:

  - Individual features, Complete parts and Entire assembly.

- Recognize and utilize the following tools from the Assembly toolbar:

  - Insert Components, New Part, New Assembly, Copy with Mate, Mate, Linear Component Pattern, Smart Fasteners, Move Component, Rotate Component, Show Hidden Component, Assembly Features, Reference Geometry, Bill of Materials, Exploded View, Smart Exploded Line, and Exploded View Sketch.

- Create and address Motions Studies:

  - MotionManager, Animation, and Basic Motion.

- Understand and implement design Assembly configurations methods:

  - Manual (Add Configuration command).

  - Configure Component/Configure Dimension tool.

  - Design Tables.

- Comprehend the difference between In-Context and Out-of-Context.

- Create and apply equations with Global Variables.

- Understand SpeedPak.

Top-Down Assembly Design

In top-down assembly design, one or more features of a part are defined by something in an assembly, such as a layout sketch or the geometry of another part. The design intent (sizes of features, placement of components in the assembly, proximity to other parts, etc.) comes from the top (the assembly) and moves down (into the parts), hence the phrase "top-down."

For example, when creating a locating pin on a plastic part using the Extrude command, you might choose the Up to Surface option and select the bottom of a board (a different part). This selection would make the locating pin exactly long enough to touch the board, even if the board were moved in a future design change. Thus the length of the pin is defined in the assembly, not by a static dimension in the part.

Assembly Methods

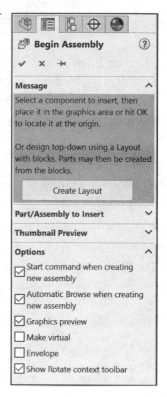

Designers usually use the Top-down assembly modeling approach to lay their assemblies out and to capture key design aspects of custom parts specific in the assemblies. There are three key methods to use for the Top-down assembly modeling approach:

- *Individual features*. Can be designed top-down by referencing other parts in the assembly, as in the case of the locating pin described above. In bottom-up design, a part is built in a separate window where only that part is visible. However, SOLIDWORKS also allows you to edit parts while working in the assembly window. This makes all of the other components' geometry available to reference (for example, copy or dimension to). This method is helpful for those parts that are mostly static but have certain features that interface with other assembly components.

- *Complete parts*. Can be built with top-down methods by creating new components within the context of the assembly. The component you build is actually attached (matcd) to another existing component in the assembly. The geometry for the component you build is based upon the existing component. This method is useful for parts like brackets and fixtures, which are mostly or completely dependent on other parts to define their shape and size.

- *An entire assembly*. Can be designed from the top-down as well, by first building a layout sketch that defines component locations, key dimensions, etc. Then build 3D parts using one of the methods above, so the 3D parts follow the sketch for their size and location. The speed and flexibility of the sketch allows you to quickly try several versions of the design before building any 3D geometry. Even after you build the 3D geometry, the sketch allows you to make a large number of changes in one central location.

The top-down assembly approach is also referred to as "In-Context."

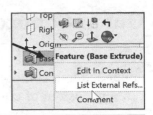

💡 You can lock, unlock, or break the external references of components or features created in the context of an assembly and the external references of various types of derived parts. When you lock the external references on a component, the existing references no longer update and you cannot add any new references to that component. Once you unlock the external references, you can add new references or edit the existing references.

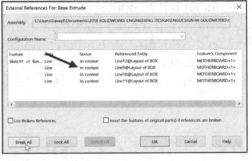

In-Context

You can create a new part In-Context of an assembly. That way you can use the geometry of other assembly components while designing the part. The new part is saved as a virtual component. Virtual components are saved internally in the assembly file in which they are created, instead of in a separate part file. When you create components in the context of an assembly, the software saves them inside the assembly file as virtual components. Later, you can save the components to external files or delete them.

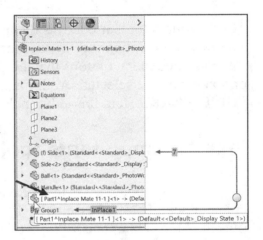

During the conceptual design process, when you frequently experiment with and make changes to the assembly structure and components, using virtual components has several advantages over the Bottom-up assembly method:

- You can rename these virtual components in the FeatureManager design tree, avoiding the need to open, save as a copy, and use the Replace Components command.

- You can make one instance of a virtual component independent of other instances in a single step.

- The folder where you store your assembly is not cluttered with unused part and assembly files resulting from iterations of component designs.

💡 Rename the virtual component in the assembly. Right-click the **virtual component** and click **Rename Part**.

Out of Context

Look at the External reference symbols. If you see the notation "->? appended to a component or feature, the ? indicates that the External reference is out of context.

To put an out of context component back into context, open the externally referenced document. The Edit In Context automatically opens the document that is referenced by an External reference. This is quite a time saver because you do not have to query the feature to identify the references file, browse to locate it, and then open it manually.

Assembly Toolbar

The Assembly toolbar controls the management, movement, and mating of components. The Assembly options are dependent on the views displayed in your Graphics window. The tools and menu options that are displayed in gray are called gray-out. The gray icon or text cannot be selected. Additional information is required for these options. View SOLIDWORKS Help for additional information on the Assembly toolbar and its options.

Insert Components tool

The Insert Components tool provides the ability to add a part or sub-assembly to the assembly. The Insert Components tool uses the Insert Component PropertyManager. The Insert Component PropertyManager provides the following selections:

- *Part/Assembly to Insert*. Provides the following selections:

  - **Open documents**. Displays the active parts and assembly documents.

  - **Browse**. Provides the ability to select a part or assembly to insert into your model. Click a location in the Graphics window to place the selected component in your model.

Active documents are displayed in the Part/Assembly to Insert box.

- *Options*. The Options box provides the following selections:

  - **Start command when creating new assembly**. Default setting. Opens the Insert Component PropertyManager when you create a new assembly.

- **Graphics preview**. Provides a preview of the selected document in the Graphics window under your mouse pointer.

- **Make Virtual**. Makes the component you are inserting virtual. Making a component virtual breaks the link to the external component file and stores the component definition inside the assembly file. Exception: If the external file is a derived part, the virtual component retains references to the parent of that part. You can manually break these external references in the virtual part.

- **Envelope**. Makes the component you are inserting an envelope component.

- **Show Rotate context dialog**. Displays the Rotate context toolbar when you insert a component. You can use the context toolbar to rotate the component around the X, Y or Z axis.

An assembly envelope is a special type of assembly component. Envelopes have two main functions. You can use envelopes as reference components and as selection tools. Envelopes are ignored in global assembly operations such as bills of materials and mass properties.

New Part

The New Part ![icon] tool provides the ability to create a new part In-Context of an assembly. The New Part tool provides the ability to apply the geometry of other assembly components while designing the part. The new part is saved internally in the assembly file as a virtual component. Later, you can save the part to its own part file.

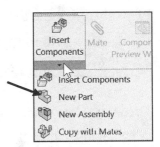

New Assembly

The New Assembly ![icon] tool provides the ability to insert a new, empty sub-assembly at any level of the assembly hierarchy.

Tutorial: Insert a feature In-Context of an assembly 12-1

Insert a feature In-Context of an assembly. Utilize the Individual features Top-Down assembly method.

1. Open **Individual Features 12-1** from the SOLIDWORKS 2018\TopDownAssemblyModeling folder.

2. Right-click the right face of the **b5g-plate** from the Graphics window as illustrated.

3. Click **Edit Part** ![icon] from the Context toolbar. The b5g-plate part is displayed in blue in the FeatureManager.

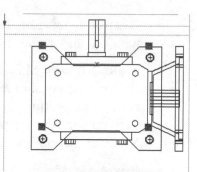

4. Right-click the **bottom face** of b5g-plate for the Sketch plane. Click **Sketch** ⬒ from the Context toolbar.

5. Display a **Bottom** view. Display **Wireframe**.

6. Ctrl-select the **4 circular edges** of the f718b-30-b5-g mounting holes.

7. Click the **Convert Entities** ⬚ Sketch tool. The 4 circles are projected on the Sketch plane of the b5g-plate part.

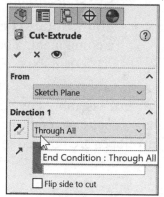

8. Click the **Extruded Cut** ⬓ Features tool. The PropertyManager is displayed. Select **Through All** for End Condition in Direction 1.

9. Click **OK** ✔ from the Cut-Extrude PropertyManager.

10. Click the **Edit Component** ⬓ tool to return to the assembly.

11. **Rebuild** ⬓ the assembly. The b5g-plate part is updated.

12. Display a **Shaded With Edges** view. Display an **Isometric** view. View the Through all mounting holes in the assembly. **Close** the model.

Tutorial: New Part In-Context of the assembly 12-1

Create a New Part In-Context of the assembly. Utilize the Complete Part Top-Down assembly method.

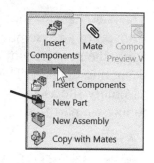

1. Open **Complete Parts 12-1** from the SOLIDWORKS 2018\TopDownAssemblyModeling folder.

2. Click **New Part** from the Consolidated Insert Components Assembly toolbar. Accept the default Part Template.

3. The new part is displayed in the FeatureManager with a name in the form **[Part*n*^*assembly_name*]**. The square brackets indicate that the part is a virtual component. The mouse pointer displays the ✔ icon.

4. Click the **top face** of Support Plate_PCS2B for your Sketch plane. Edit Component is selected. A new sketch opens.

5. Click the **top face** of Support Plate_PCS2B.

6. Click the **Convert Entities** ⬚ Sketch tool. The rectangular boundary of the Support Plate_PCS2B is projected onto the current Sketch plane of the Top Plate_PCS2B part.

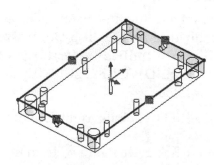

7. Click the **Extruded Boss/Base** 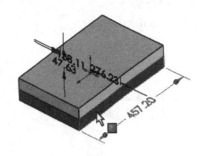 Features tool. The Boss-Extrude PropertyManager is displayed. The extrude direction is upward.

8. Enter **50**mm for Depth in Direction 1. Blind is the default End Condition.

9. Click **OK** ✔ from the Boss-Extrude PropertyManager.

10. Click the **Edit Component** 🖉 tool to return to the assembly.

11. Double-click the **right face** of Support Plate_PCS-2B as illustrated. Double-click **457.20**.

12. Enter **500**. **Rebuild** 🔴 the model to modify the Support Plate_PCS2B and the Top Plate. View the results. Note: To name the new virtual component, right-click the component, and click Open part. Click Save. Enter the desired name. View the created InPlace Mate.

13. **Close** the model.

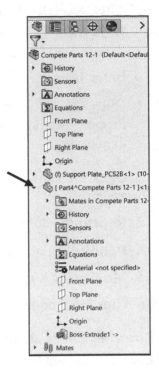

Tutorial: Layout Sketch Assembly 12-1

Create components from a Layout sketch utilizing blocks.

1. Open **Layout Sketch Assembly 12-1** from the SOLIDWORKS 2018\TopDownAssemblyModeling folder. Three blocks are displayed.

Activate the Make part from block tool.

2. Click the **Make Part from Block** 🔳 tool from the Layout tab. The Make Part from Block PropertyManager is displayed.

3. Ctrl-Select the **three blocks** from the Graphics window. The selected entities are displayed in the Selected Blocks box.

4. Click the **On Block** option.

5. Click **OK** ✔ from the Make Part From Block PropertyManager. Accept the default Part Template. The three blocks are displayed in the FeatureManager and the Layout icon is displayed in the assembly FeatureManager. Rebuild if needed.

6. Right-click the **20MM** part in the FeatureManager.

7. Click **Edit Part** from the Context toolbar. The 20MM-1 part is displayed in blue.

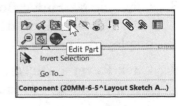

8. **Expand** the 20MM part from the FeatureManager.

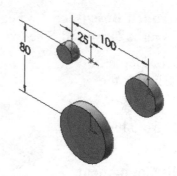

9. Click **Sketch1**. Click the **Extruded Boss/Base** 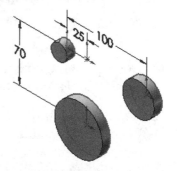 Features tool. Blind is the default End Condition in Direction 1.

10. Enter **10**mm.

11. Click **OK** ✔ from the Boss-Extrude PropertyManager. Boss-Extrude1 is displayed.

12. Click the **Edit Component** tool to return to the assembly.

13. **Repeat** the above steps for the **40MM** and **60MM** parts.

14. **Rebuild** the model. View the model.

15. **Modify** the 80mm dimension to 70mm.

16. **Close** the model.

Tutorial: Entire Assembly 12-2

Create an assembly from a Layout sketch. Utilize the Entire assembly method.

1. Open **Entire Assembly 12-2** from the SOLIDWORKS 2018\TopDownAssemblyModeling folder. Layout is the first sketch.

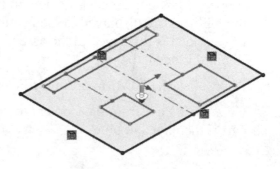

2. Double-click the **Layout** sketch from the FeatureManager to display the sketched dimensions.

3. Click **New Part** from the Consolidated Insert Components toolbar. Accept the default Part Template.

4. Click **Top Plane** from the Entire Assembly 12-2 FeatureManager. Click the **bottom horizontal line** from the Graphics window.

5. Click the **Convert Entities** Sketch tool.

6. Ctrl-Select the other **three edges**.

7. Click the **Convert Entities** Sketch tool.

8. Click the **Extruded Boss/Base** Features tool. The Boss-Extrude PropertyManager is displayed. Display an **Isometric** view.

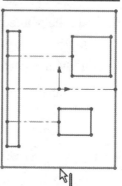

9. Click **Reverse Direction**. The direction arrow points downward into the screen.

10. Enter **20**mm for Depth in Direction 1. Blind is the default End Condition.

11. Click **OK** ✔ from the Boss-Extrude PropertyManager.

12. Click the **Edit Component** tool to return to the assembly.

13. **Rebuild** 🔘 the model. View the results.

14. Click **Save**.

15. Enter the desired **part name**.

16. **Close** the model.

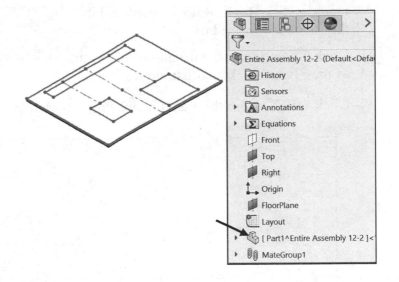

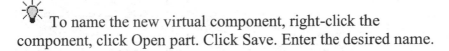

💡 To name the new virtual component, right-click the component, click Open part. Click Save. Enter the desired name.

Tutorial: Layout Tool with Block Assembly 12-3

Apply the Layout tool with blocks to address motion.

1. Open **Layout 12-3** from the SOLIDWORKS 2018\TopDownAssemblyModeling folder. Block B translates vertically and represents the motion of the Linear Transfer assembly.

2. Click the **Layout** 🔲 tool from the Layout tool bar in the CommandManager. Block B translates vertically and represents the motion of the Linear Transfer assembly.

3. Drag **Block B** vertically. Note: Block C translates horizontally and represents the motion of a 2AXIS-TRANSFER assembly.

4. Drag **Block C** horizontally a few mm.

5. **Return** Block B and Block C to their original positions.

6. Click the **Insert Block** 🔳 tool from the Layout toolbar. The Insert Block PropertyManager is displayed.

7. Click the **Browse** button.

8. **Browse** to the SOLIDWORKS 2018\TopDownAssemblyModeling folder.

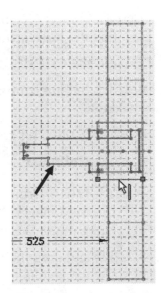

9. Double-click **Rack**. Rack is displayed in the Blocks to Insert box.

10. Click a **position** at the midpoint of the end of the Slider block as illustrated.

11. Click **OK** ✔ from the Insert Block PropertyManager.

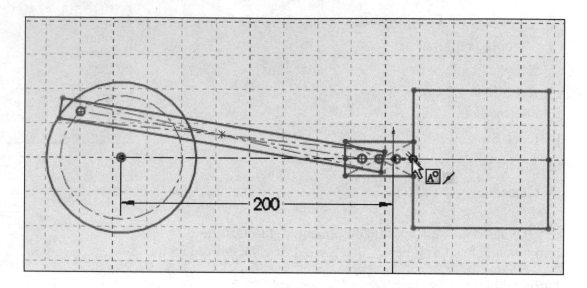

Insert a Collinear Geometric relation.

12. Click the **horizontal centerline** of the Rack.

13. Hold the **Ctrl** key down.

14. Click the **horizontal centerline** in the Layout sketch.

15. Release the **Ctrl** key.

16. Click **Collinear**.

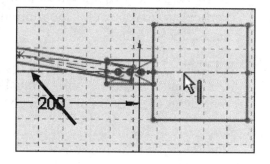

17. Click **OK** ✔ from the Properties PropertyManager. Verify the Rack motion relative to the Wheel.

18. Click and drag the **wheel** counterclockwise. View the results.

Exit the Layout Sketch.

19. Click the **Layout** ⬜ tool from the Layout toolbar in the CommandManager.

20. **Close** all models. Do not save.

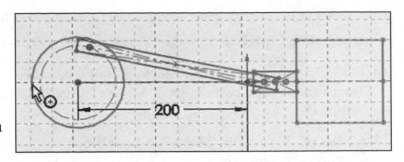

Tutorial: Layout Tool with Block Assembly 12-4

Apply the Layout tool with blocks to address motion. You can make a block from any single or combination of multiple sketch entities. Saving each block individually provides extensive design flexibility.

1. Open the **Wheel-Slider 12-4** assembly from the SOLIDWORKS 2018\TopDownAssemblyModeling folder. View the FeatureManager.

Create a Base feature for Wheel A-2.

2. Right-click the **Wheel A-2** block from the FeatureManager.

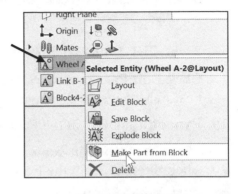

3. Click the **Make Part From Block** 🗔 tool. The Make Part from Block PropertyManager is displayed. On Block is selected. Accept the default settings.

4. Click **OK** ✔ from the Make Part from Block PropertyManager. The New SOLIDWORKS Document dialog box is displayed. Accept the default Part Template.

5. Right click **Wheel A** from the FeatureManager.

6. Click **Open Part** from the Context toolbar. The Part FeatureManager is displayed.

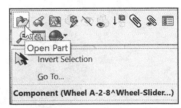

7. Click **Sketch1** from the FeatureManager.

8. Click **Extruded Boss/Base** from the Features toolbar. The Boss-Extrude PropertyManager is displayed.

9. Click the **Reverse Direction** arrow. The arrow points toward the back of the model.

10. Enter **10mm** for Depth. Blind is the default End Condition in Direction 1. Click **OK** ✔ from the Boss-Extrude PropertyManager. View the results.

11. **Close** the part.

12. Click **Yes** to rebuild. The Wheel-Slider assembly is displayed.

13. Right-click **Link B-1** from the FeatureManager.

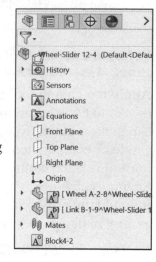

14. Click the **Make Part from Block** tool. The Make Part from Block PropertyManager is displayed. Accept the default settings.

15. Click **OK** ✔ from the Make Part from Block PropertyManager. The New SOLIDWORKS Document dialog box is displayed. Accept the default Part Template.

16. Click **OK**.

17. Right-click **Link B-1** from the FeatureManager.

18. Click **Edit Part** from the Context toolbar. The Link B-1 FeatureManager is displayed in blue.

19. **Expand** Link B-1 in the FeatureManager.

Create a Base feature in-context.

20. Click **Sketch1** from the FeatureManager.

21. Click **Extruded Boss/Base** from the Features toolbar. The Boss-Extrude PropertyManager is displayed. Blind is the default End Condition in Direction 1.

22. Enter **10mm** for Depth.

23. Click **OK** ✔ from the Boss-Extrude PropertyManager.

24. **Close** the part. Click **Yes**.

25. Display an **Isometric** view. View the results.

26. Drag **Wheel A** counterclockwise. View the results.

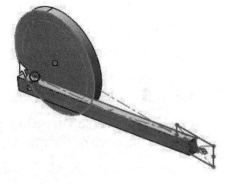

Insert a new part into the assembly. Extrude with a different Start Condition.

27. Right-click **Block 4-2** from the FeatureManager.

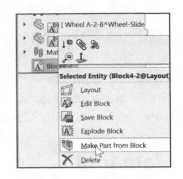

28. Click the **Make Part from Block** 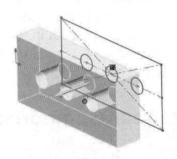 tool. The Make Part from Block PropertyManager is displayed. Accept the default settings.

29. Click **OK** ✔ from the Make Part from Block PropertyManager. The New SOLIDWORKS Document dialog box is displayed. Accept the default Part Template.

30. Click **OK**. Right-click **Block 4-2** from the FeatureManager.

31. Click **Edit Part**. The Block 4-2 FeatureManager is displayed in blue.

32. **Expand** Block4-2 in the FeatureManager.

33. Click **Sketch1**.

Create a Base feature in-context.

34. Click **Extruded Boss/Base** from the Features toolbar. The Boss-Extrude PropertyManager is displayed. Blind is the default End Condition in Direction 1.

35. Select **Vertex** for Start Condition.

36. Click the starting **vertex point**.

37. Enter **10mm** for Depth in Direction 1. Blind is the default End Condition,

38. Click **OK** ✔ from the Boss-Extrude PropertyManager. View the results. **Close** the part. Click **Yes**.

Deactivate displayed sketches.

39. Click **View**, **Hide/Show**, uncheck **Sketches** from the Menu bar menu.

40. Display an **Isometric** view. View the results.

41. **Close** the model. Do not save.

🔆 Utilize the Search feature in the Systems Options dialog box to quickly locate information for System Options or Document Properties.

Copy with Mates tool

The Copy with Mates 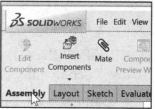 tool provides the ability to copy components and their associated mates. The Copy with Mates tool uses the Copy With Mates PropertyManager.

Mate tool

The Mate 🖇 tool provides the ability to create Geometric relations between assembly components. Mates define the allowable directions of rotational or linear motion of the components in the assembly. Move a component within its degrees of freedom in the Graphics window to view the behavior of an assembly. View Chapter 11 for additional information.

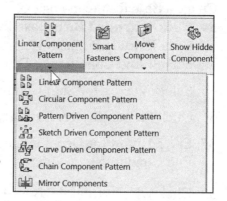

Linear Component Pattern tool

The Linear Component Pattern 🔡 tool provides the ability to create a linear pattern of components in an assembly in one or two directions. The Consolidated Linear Component Pattern toolbar provides access to the following tools: *Linear Component Pattern, Circular Component Pattern, Pattern Driven Component Pattern, Sketch Driven Component Pattern, Curve Driven Component Pattern, Chain Component Pattern* and *Mirror Components*. See SOLIDWORKS Help for additional information.

Smart Fasteners tool

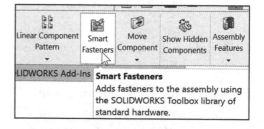

The Smart Fasteners 📝 tool automatically adds fasteners from the SOLIDWORKS Toolbox to your assembly if there is a hole, a series of holes, or a pattern of holes, which is sized to accept standard hardware.

🔆 Configure Smart Fasteners to add any type of bolt or screw as a default. The fasteners are automatically mated to the holes with Concentric and Coincident mates.

🔆 Install and Add-in the **SOLIDWORKS Toolbox Library** to access SOLIDWORKS Toolbox contents.

The Smart Fasteners feature uses the SOLIDWORKS Toolbox library. The SOLIDWORKS Toolbox provides a large variety of ANSI inch, Metric and other standard hardware.

The Smart Fasteners tool uses the Smart Fasteners PropertyManager. The Smart Fasteners PropertyManager provides the following selections:

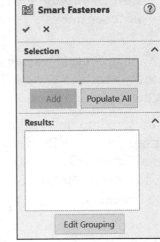

- *Selection*. The Selection box provides the following options:

  - **Selection**. Displays the selected hole, face or component to add a fastener.

  - **Add**. Adds fasteners to selected holes. You can select holes, faces, or components. If you select a face, the Smart Fasteners tool locates all available holes which pass through the surface. If you select a component, the Smart Fasteners tool locates all available holes in that component.

  - **Populate All**. Adds fasteners to all holes in the assembly.

☀ When you click the Add or Populate All option, new fasteners are added to the assembly. The length of a new fastener is the next smallest national standard length for a blind hole, and the next longest national standard length for a through hole. When holes are deeper than the longest fastener length, the longest one is used.

- *Results*. The Results box displays groups of fasteners you are adding or editing. Select a group to make changes to its fastener type and properties under the **Series Components** and **Properties** option box.

- *Series Components*. Displays the fastener type for the item that you select in the Results list.

  - **Fastener**. Right-click to change fastener type or to revert to the default fastener type.

  - **Top Stack/Bottom Stack**. Select items from **Add to Top Stack** and **Add to Bottom Stack** to add hardware to the top and bottom stacks.

☀ Bottom Stack (Available for through holes only). Washers and nuts at the end of the fastener.

  - **Auto size to hole diameter**. Selected by default. Automatically updates the hardware size whenever the hole diameter changes. In the FeatureManager design tree, fasteners and their associated mates are identified with a star if they have this option applied in System options.

- *Properties*. Displays the properties of the hardware selected under Series Components. Available properties vary depending on the hardware type. You can edit properties such as size, length, and display.

- *Edit Grouping*. Provides the ability to display and edit the fastener tree.

The Smart Fasteners tool does not automatically insert the needed washers or nuts. You must add the items by editing the fastener or the series.

Add standard nuts and washers to Smart Fasteners from the Smart Fasteners PropertyManager. Each Series of fasteners has an associated Top Stack (washers added under the head of the fastener) and Bottom Stack (washers and nuts added to the end of the fastener).

Expand each fastener series from the Fasteners box to view its Top Stack and Bottom Stack. If you add hardware to the Top Stack or Bottom Stack at the fastener level, that hardware is displayed in each Series.

If you add hardware to the Top Stack or Bottom Stack at the Series level, the hardware is only displayed in that Series.

Tutorial: Insert a Smart Fastener 12-1

Apply the Smart Fastener tool. Insert Smart Fasteners into an assembly.

1. Open **SmartFastener 12-1** from the SOLIDWORKS 2018\TopDownAssemblyModeling folder.

If required, install SOLIDWORKS Toolbox.

2. Check the **SOLIDWORKS Toolbox** from the SOLIDWORKS Add-Ins tab in the CommandManager.

3. Click the **Smart Fasteners** tool from the Assembly toolbar.

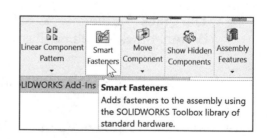

4. Click **OK** from the SOLIDWORKS dialog box. The Smart Fasteners PropertyManager is displayed.

5. Click **CBORE for M5 Hex Head Bolt1** of
   SF_plate1 from the fly-out FeatureManager.
   CBORE is the seed feature for the
   DerivedLPattern1.

6. Click the **Add** button from the Smart Fasteners
   PropertyManager. View the results.

🔅 Right-click in the Fastener box to change
fastener type or to revert to the default fastener
type.

7. Click **OK** ✔ from the Smart Fasteners
   PropertyManager. View the model with the
   inserted Smart Fasteners and the
   FeatureManager.

8. Click **OK** ✔ from the Smart Fasteners
   PropertyManager.

9. **Expand** the Smart Fastener1 folder in the
   FeatureManager. View the created
   DerivedLPattern feature.

10. **Close** the model.

🔅 Press the **g** key to activate the Magnifying
glass tool. Use the Magnifying glass to inspect a
model and make selections without changing the
overall view.

🔅 Hiding and Suppressing parts in
SOLIDWORKS assemblies can have similar
looking results, but both operations behave quite
differently from each other.

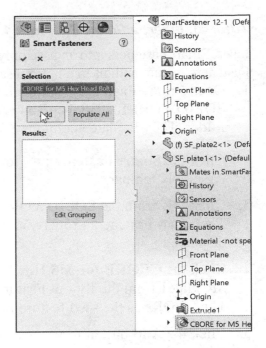

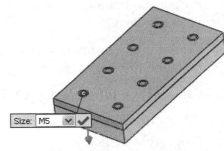

Tutorial: Insert a Smart Fastener 12-2

Insert Smart Fasteners into an assembly. Use the Smart Fasteners feature and insert a bottom stack.

1. Open **Smart Fastener 12-2** from the SOLIDWORKS 2018\TopDownAssemblyModeling folder. The SF_plate3 contains 8 Thru All holes.

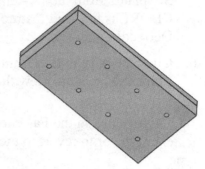

2. Click the **Smart Fasteners** 🖾 tool from the Assembly toolbar.

3. Click **OK** from the SOLIDWORKS dialog box. The Smart Fasteners PropertyManager is displayed.

4. Click the **CBORE for M5 Hex Head Bolt1** of SF_plate1 from the fly-out FeatureManager. The CBORE is the seed feature for LPattern1.

5. Click the **Add** button from the Smart Fasteners PropertyManager. View the results in the Fasteners box.

6. Click the **drop-down arrow** from the Bottom Stack box to select hardware.

7. Select **Hex Flange Nut - AMSI B18.2.4.4M**.

8. Click **OK** ✔ from the Smart Fasteners PropertyManager.

9. **Rotate** the model to view the washers and hex nuts inserted into the assembly.

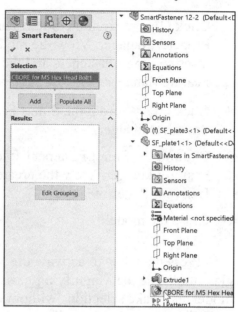

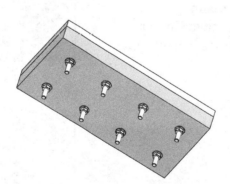

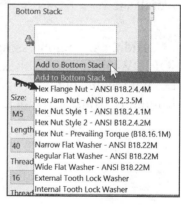

10. **Expand** SmartFastener1 from the FeatureManager. View the results.

11. **Close** the model.

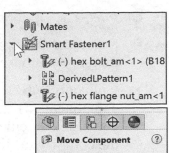

Move Component tool

The Move Component 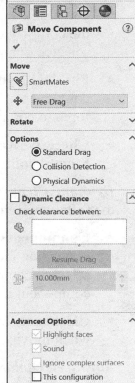 tool provides the ability to drag and move a component in the Graphics window. The component moves within its degrees of freedom.

The Move Component tool uses the Move Component PropertyManager. The Move Component PropertyManager provides the following capabilities: *Move a component*, *Add SmartMates while moving a component*, *Rotate a component*, *Detect collision with other components*, *Activate Physical Dynamics* and *Dynamically detect the clearance between selected components*.

The available selections are dependent on the selected options. The Move Component PropertyManager provides the following selections:

- *Move*. The Move box provides the ability to move the selected component with the following options:

  - **SmartMates**. Creates a SmartMate while moving a component. The SmartMates PropertyManager is displayed.

  - **Move**. The Move box provides the following options: *Free Drag*, *Along Assembly XYZ*, *Along Entity*, *By Delta XYZ* and *To XYZ Position*.

- *Rotate*. Provides the ability to rotate a component in the Graphics window. The Rotate box provides the following selections:

  - **Free Drag**. Provides the ability to drag a selected component in any direction.

  - **About Entity**. Select a line, an edge, or an axis. Drag a component from the Graphics window around the selected entity.

  - **By Delta XYZ**. Moves a component around an assembly axis by a specified angular value. Enter an X, Y, or Z value in the Move Component PropertyManager. Click Apply.

  - **To XYZ Position**. Moves a component to a specified XYZ position.

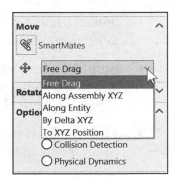

- *Options*. The Options box provides the following selections:

  - **Standard Drag**. Selected by default. Provides a standard drag to the mouse pointer.

  - **Collision Detection**. Detects collisions with other components when moving or rotating a component. Locate collisions for either the selected components or for all of the components that move as a result of mates to the selected components.

  - **Physical Dynamics**. View the motion of the assembly components. Drag a component. The component applies a force to components that it touches.

  - **Check between**: **All components**, **These components**, **Stop at collision** and **Dragged part only**.

- *Dynamic Clearance*. The Dynamic Clearance box provides the following selections:

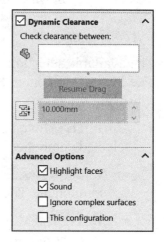

  - **Components for Collision Check**. Displays the dimension indicating the minimum distance between the selected components when moving or rotating a component in the Graphics window.

  - **Clearance**. Specify a distance between two components when moving or rotating.

- *Advanced Option*. The Advance Option box provides the following selections:

  - **Highlight faces**. Selected by default. Faces in the Graphics window are highlighted.

  - **Sound**. Selected by default. The computer beeps when the minimum distance in the Clearance box is reached.

  - **Ignore complex surfaces**. Clearances are only detected on the following surface types: planar, cylindrical, conical, spherical and toroidal.

  - **This configuration**. Apply the movement of the components to only the active configuration.

☼ The "This configuration" check box does not apply to Collision Detection, Physical Dynamics, or Dynamic Clearance. It applies only to Move Component or Rotate Component.

Rotate Component tool

The Consolidated Rotate 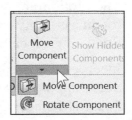 tool provides the ability to rotate a component within the degrees of freedom defined by its mates. The Rotate Component tool uses the Rotate Component PropertyManager. The Rotate Component PropertyManager provides the same selections as the Move PropertyManager. View the Move Component tool section for detail PropertyManager information.

Show Hidden Components tool

The Show Hidden Components 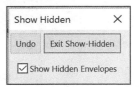 tool provides the ability to toggle the display of hidden and shown components in an assembly. The tool provides the ability to select which hidden component will be displayed in the Graphics window.

Consolidated Assembly Features toolbar

The Consolidated Assembly Features  toolbar provides the ability to access the following tools for an assembly: *Hole Series*, *Hole Wizard*, *Simple Hole*, *Extruded Cut*, *Revolved Cut*, *Swept Cut*, *Fillet*, *Chamfer*, *Weld Bead* and *Belt/Chain*.

Consolidated Reference Geometry toolbar

The Consolidated Reference Geometry  toolbar provides access to the following options: *Plane, Axis, Coordinate System, Point, Center of Mass* and *Mate Reference*.

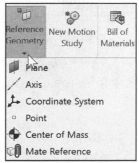

Bill of Materials tool

The Bill of Materials  tool provides the ability to create a Bill of Materials directly in an Assembly. The created BOM is displayed in the Tables folder in the FeatureManager design tree. The name of the configuration to which the BOM applies is displayed beside the BOM feature.

Motion Study tool

The Motion Study 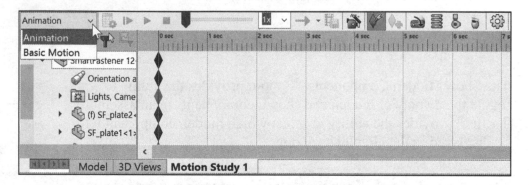 tool provides the ability to create graphical simulations of motion for an assembly. Access the MotionManager from the Motion Study tab. The Motion Study tab is located in the bottom left corner of the Graphics window.

Incorporate visual properties such as lighting and camera perspective. Click the Motion Study tab to view the MotionManager. Click the Model tab to return to the FeatureManager design tree.

The MotionManager displays a timeline-based interface and provides the following selections from the drop-down menu as illustrated:

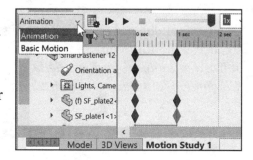

- ***Animation***. Default setting. Apply Animation to animate the motion of an assembly. Add a motor and insert positions of assembly components at various times using set key points. Use the Animation option to create animations for motion that do **not** require accounting for mass or gravity.

- ***Basic Motion***. Apply Basic Motion for approximating the effects of motors, springs, collisions, and gravity on assemblies. Basic Motion takes mass into account in calculating motion. Basic Motion computation is relatively fast, so you can use this for creating presentation animations using physics-based simulations. Use the Basic Motion option to create simulations of motion that account for mass, collisions or gravity.

- ***Motion Analysis***. Only available with SOLIDWORKS Motion as an Add-In. See SOLIDWORKS Help for additional information.

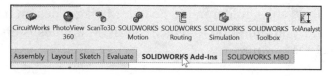

💡 If the Motion Study tab is not displayed in the Graphics window, click **View** ➢ **User Interface** ➢ **MotionManager** from the Menu bar.

Create a new Motion Study, click **Insert** ➢ **User Interface** ➢ **New Motion Study** from the Menu bar or click the **New Motion Study** tool from the Assembly tab in the CommandManager.

Animation Wizard

The Animation Wizard 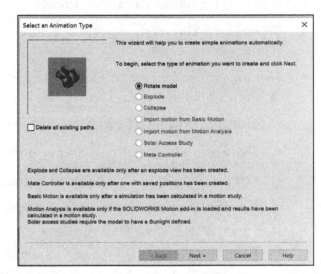 tool provides the ability to rotate parts or assemblies or explode or collapse assemblies using a simple wizard format. The following selections are available:

- *Rotate model*. Rotate parts or assemblies.

- *Explode model*. Explode assemblies.

- *Collapse model*. Collapse assemblies.

- *Import motion from Basic Motion*. Import calculated Basic Motion results into the animation.

- *Import motion from Motion Analysis*. Import calculated Motion Analysis results into the animation.

- *Solar Access Study*. Import a solar study. You can simulate the movement of the sun as it passes over models of buildings, solar panels, and outdoor equipment.

- *Mate Controller*. Controls selected mates during the motion study.

💡 To apply the Animation Wizard in the Explode and Collapse option, you must first create an Exploded view of your assembly.

Basic Motion

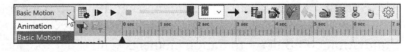

Basic Motion provides the ability to simulate the effects of motors, springs and gravity on your assemblies. When you record a simulation, the affected components move to a new location in the assembly. Physical Simulation uses the following options: Motor (Linear, Rotary), *Spring*, *Contacts* *and Gravity*.

Linear/Rotary Motor tool

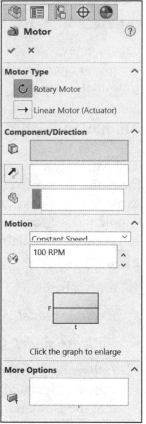

The Linear/Rotary Motor tool simulates elements that move components around an assembly. The Linear/Rotary Motor tool uses the Motor PropertyManager. The Motor PropertyManager provides the following selections:

- *Motor Type*. Select **Rotary** or **Linear**.

- *Component/Direction*. Select the component the motor will act on.

- *Motion*. Select the type of motion to apply with the motor and the corresponding value. The available options are:

    - **Constant speed**. The motor's value will be constant.

    - **Distance**. The motor will operate only for the set distance.

    - **Oscillating**. Set the amplitude and frequency.

    - **Interpolated**. Select the item to interpolate, *Displacement, Velocity, Acceleration*.

    - **Formula**. Select the type of formula to apply: *Displacement, Velocity or Acceleration* and enter the formula.

- *More Options*. Provides the ability to set motion relative to another part and to select components for Load-bearing Faces/Edges to transfer them to a Simulation analysis.

Motors move components in a selected direction, but they are not forces. Motor strength does not vary based on component size or mass. For example, a small cube moves at the same speed as a large cube if you set the velocity of the motor to a given value in both cases.

Spring

The Spring tool simulates elements that move components around an assembly using Physical Simulation. Physical Simulation combines simulation elements with other tools such as Mates and Physical Dynamics to move components within the components' degrees of freedom. The Spring PropertyManager provides the following selections:

- *Spring Type*. Select Linear Spring or Torsional Spring.

  - **Linear Spring**. Only available in Basic Motion and SOLIDWORKS Motion Analysis. Represents forces acting between two components over a distance and along a particular direction.

  - **Torsional Spring**. Only available in SOLIDWORKS Motion Analysis. Represents torsional forces acting between two components.

- *Spring Parameters*. Select the following options:

  - **Spring Endpoints**. Select two features for Spring Endpoints.

  - **Exponent of Spring Force Expression**. Based on the Functional Expressions for Springs.

  - **Spring Constant**. Based on the Functional Expressions for Springs.

  - **Free Length**. The initial distance is the distance between the parts as currently displayed in the Graphics window.

  - **Update to model changes**. Provides the ability to have the free length dynamically update to model changes while the PropertyManager is open.

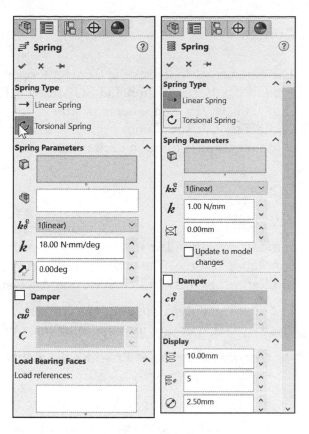

- *Damper*. Only available in SOLIDWORKS Motion Analysis. Provides two options:

  - **Exponent of Damper Force Expression**. Based on the Functional Expressions for Dampers.

  - **Damping Constant**. Based on the Functional Expressions for Dampers.

- *Display*. View the display values only when the **Spring** PropertyManager is open, or when you calculate the study. Provides three options:

  - **Coil Diameter**. Diameter of the coil.

  - **Number of Coils**. Number of twists.

  - **Wire Diameter**. Diameter of the wire.

- *Load Bearing Faces*. Select components for Load-bearing Faces/Edges to transfer them to a SOLIDWORKS Motion Analysis.

Contact

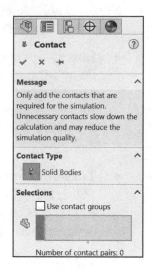

The Contact 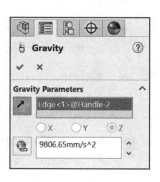 tool uses the Contact PropertyManager. Only available in Basic Motion and SOLIDWORKS Motion Analysis. Contact between components is ignored unless you apply the 3D Contact tool. If you do not use the Contact tool to specify contact, components pass through each other.

Gravity

The Gravity tool is a simulation element that moves components around an assembly by inserting a simulated gravitational force. The Gravity tool is only available in Basic Motion and SOLIDWORKS Motion Analysis.

The Gravity tool uses the Gravity PropertyManager. The Gravity PropertyManager provides the following options:

- *Gravity Parameters*. Provides the ability to set a Direction Reference for gravity.

  - A face to orient gravity parallel to the normal.

  - An edge to orient gravity parallel to the edge.

  - X, Y or Z to orient gravity in the chosen direction in the assembly reference frame.

  - **Numeric gravity value**. Set the Numeric gravity value. Default is standard gravity.

Tutorial: Motion Study 12-1

Perform a Motion Study with an assembly using the linear motor and gravity options.

1. Open **Motion Study 12-1** from the SOLIDWORKS 2018\TopDownAssemblyModeling folder.

2. Click the **Motion Study1** tab in the lower left corner of the Graphics window.

3. Select **Basic Motion** for Study type from the drop-down menu.

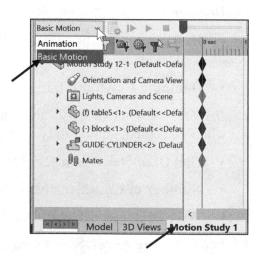

4. Click the **linear right front edge** of Plate of MGPRod<1> as illustrated.

5. Click the **Motor** 🔧 tool from the Basic Motion toolbar. The Motor PropertyManager is displayed. Edge<1>@GUIDE-CYLINDER is displayed in the Motor direction box.

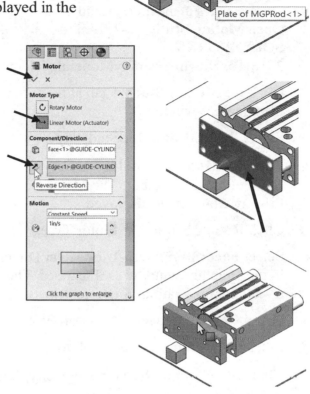

6. Click the **Linear Motor** button.

7. Click the **front face** of the plate. Accept the defaults.

8. Click the **Reverse Direction** button. The direction arrow points to the front.

9. Click **OK** ✔ from the Motor PropertyManager.

10. Click the **top face** of the plate part as illustrated.

11. Click the **Gravity** 🍎 tool from the Basic Motion toolbar. The Gravity PropertyManager is displayed. The direction arrow points downward. If required click the **Reverse Direction** button.

12. Click **OK** ✔ from the Gravity PropertyManager.

13. Click **Calculate** 📊 from the MotionManager. The piston moves and pushes the block of the table.

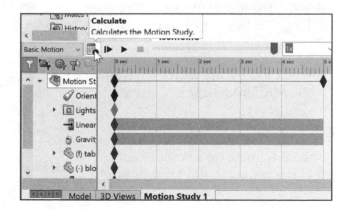

Return to the FeatureManager.

14. Click the **Model** tab to return to SOLIDWORKS.

15. **Close** the model.

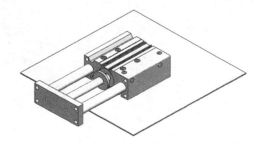

Tutorial: Motion Study 12-2

Apply a motor to move components in an assembly. Set up an animation motion study. Use the motion study time line and the MotionManager design tree to suppress mates in the motion study.

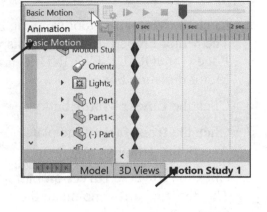

1. Open **Motion Study 12-2** from the SOLIDWORKS 2018\TopDownAssemblyModeling folder.

2. Click the **Motion Study 1** tab at the bottom of the Graphics window.

3. Select **Basic Motion** for Study type.

4. Click the **Motor** tool from the MotionManager toolbar. The Motor PropertyManager is displayed.

5. Click **Rotary Motor** for Motor Type.

6. Select **Face<1>@Part2-1** for Motor Direction in the Graphics window as illustrated. The direction arrow points counterclockwise.

7. Select **Constant speed** for Motor Type.

8. Enter **30** for Constant Speed Motor.

9. Click **OK** from the Motor PropertyManager.

Set the duration and run your animation on the 4bar-Linkage.

10. Drag the key for Motion Study 12-2 to **6** seconds as illustrated.

11. Click **Play from Start** from the MotionManager. View the results.

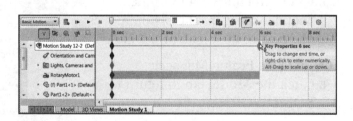

In the next section, suppress a Concentric mate on a component. View the results in the Motion Study.

12. **Expand** the Mate folder in the
    MotionManager Design tree.

13. Right-click at 2 seconds and select
    Place Key 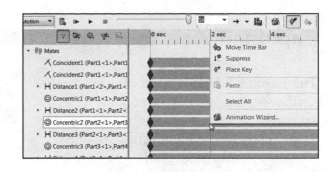 for the Concentric2
    mate.

14. Right-click at 4 seconds and select
    Place Key for the Concentric2
    mate.

15. Set the time bar to **2 seconds**.

16. Right-click **Concentric2** in the
    MotionManager design tree.

17. Click **Suppress**. The mate is suppressed
    between the 2 and 4 second marks.

18. Click the **Calculate** tool. After calculating
    a motion study, you can view it again.

19. Click **Play from Start** from the
    MotionManager toolbar. View the results.
    Explore your options and the
    available tools. View
    SOLIDWORKS Help for
    additional information on a
    Motion Study.

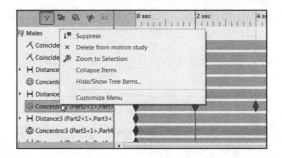

Return to the FeatureManager.

20. Click the **Model** tab to return
    to SOLIDWORKS.

21. **Close** the model.

Exploded View tool

The Exploded View 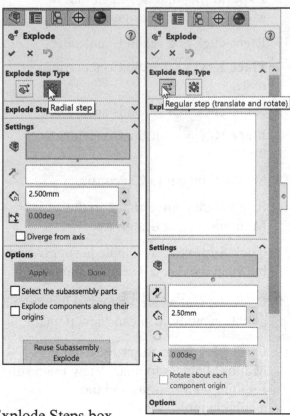 tool provides the ability to create an exploded view of an assembly by selecting and dragging parts in the Graphics window. The Exploded View tool uses the Explode PropertyManager. The Explode PropertyManager provides the following selections:

- *Explode Step Type*. Contains two options:

    - **Regular step (translate and rotate).**

    - **Radial step**. Explode components aligned radially/cylindrically about an axis.

- *Explode Steps*. The Explode Steps box displays the selected components exploded in the Graphics window. The Explode Steps box displays the following items:

    - **Explode Step<n>**. The Explode Step displays one or more selected components exploded to a single position.

    - **Chain<n>**. The Chain<n> displays a stack of two or more selected components exploded along an axis using the Auto-space components after drag option.

- *Setting*. The Settings box provides the following selections:

    - **Explode Step Components**. Displays the selected component for the current explode step.

    - **Explode direction**. Displays the selected direction (X, Y or Z) and component name for the current explode step.

    - **Reverse direction**. Reverses the explode direction if required.

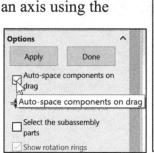

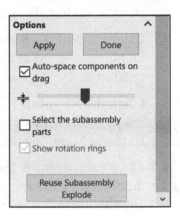

- **Explode distance**. Displays the selected distance to move the component for the current explode step.

- **Exploded Axis**. Specifies exploded axis.

- **Exploded Angle**. Specifies the rotation angle.

- **Rotate about each component origin**. If a subassembly has an exploded view, you can reuse that view in a higher-level assembly.

- *Option*. The Option box provides the following selections:

  - **Apply**. Previews the changes to the explode steps.

  - **Done**. Completes the new or changed explode steps.

  - **Auto-space components after drag**. Spaces a group of components equally along an axis.

  - **Adjust the spacing between chain components**. Adjusts the distance between components placed by the Auto-space components after drag option.

  - **Select the subassembly's parts**. Enables you to select individual components of a sub-assembly. When cleared, you can select an entire sub-assembly.

  - **Show rotation rings**. Displays the rotation rings on the Triad.

  - **Re-use Subassembly Explode**. Uses the explode steps that you defined previously in a selected sub-assembly.

Smart Explode Lines

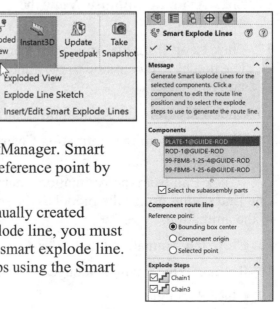

The Smart Explode line ![icon] tool provides the ability to create smart explode lines automatically for components in an exploded view. You can view the associated explode steps for a selected component in the Smart Explode Lines PropertyManager. Smart explode lines use the bounding box center as a reference point by default.

You can use smart explode lines along with manually created explode lines. To use a different path for an explode line, you must manually create the explode line or dissolve the smart explode line. You cannot add a component to the explode steps using the Smart Explode Lines PropertyManager.

Tutorial: Exploded View 12-1

Use the Exploded View tool. Insert an Exploded View into an assembly.

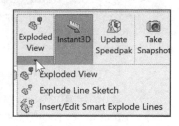

1. Activate **SOLIDWORKS Toolbox**.

2. Open **ExplodedView 12-1** from the SOLIDWORKS 2018\TopDownAssemblyModeling folder.

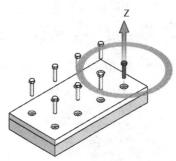

3. Click the **Exploded View** tool. The Explode PropertyManager is displayed. The rotation and translation handles appear.

4. Click each of the **8 Hex bolts** in the Graphics window and drag upward using the triad. Explode Step1 - 8 are created in the Explode Steps box.

5. Click the **SF_plate1** in the Graphics window and drag the plate backwards using the triad as illustrated. Explode Step9 is created.

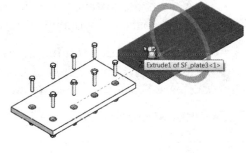

6. Click each of the **8 Hex Flange Nuts** and drag them downward using the triad from the Graphics window as illustrated.

7. Click **OK** ✔ from the Explode PropertyManager.

8. Click the **ConfigurationManager** tab.

9. **Expand** the Default configuration.

10. Right-click **ExplView1**.

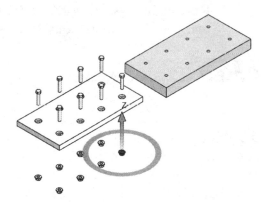

11. Click **Animate collapse**. View the Animation.

12. Click **Loop** from the Animation Controller. The model will continue to play until you stop it.

13. Click **Play** from the Animation Controller.

14. Click **Stop** from the Animation Controller.

🔆 Click End from the Animation Controller. The End option brings the model to the end point in the animation.

15. **Close** the Animation Controller.

16. **Return** to the FeatureManager.

17. **Close** the model.

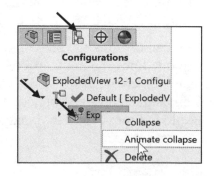

💡 For collapse and explode animation that resemble manufacturing procedures, create the exploded steps in the order of disassembling a physical assembly.

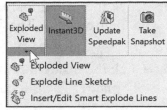

Explode Line Sketch tool

The Explode Line Sketch 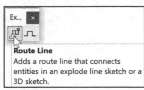 tool provides the ability to add explode lines, which is a type of 3D sketch that you add to an exploded view in an assembly. The explode lines indicate the relationship between components in the assembly.

The Explode Line Sketch tool uses the Route Line PropertyManager. The Route Line PropertyManager provides the following selections:

- *Items To Connect*. The Items To Connect box provides the following option:

    - **Reference entities**. Displays the selected circular edges, faces, straight edges, or planar faces to connect with your created route line.

- *Options*. The Options box provides the following selections:

    - **Reverse**. Reverses the direction of your route line. A preview arrow is displayed in the direction of the route line.

    - **Alternate Path**. Displays an alternate possible path for the route line.

    - **Along XYZ**. Selected by default. Creates a path parallel to the X, Y and Z axis directions.

Tutorial: Explode Line Sketch 12-1

Apply the Explode Line Sketch tool. Insert an Explode Line feature in an assembly.

1. Activate the **SOLIDWORKS Toolbox**.

2. Open **ExplodeLine Sketch 12-1** from the SOLIDWORKS 2018\TopDownAssemblyModeling folder.

💡 You require an Exploded view to create an Explode Line Sketch.

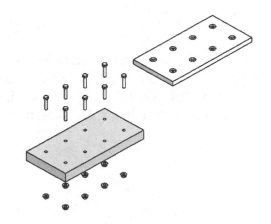

3. Click the **Explode Line Sketch** 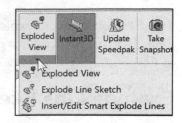 tool. The Route Line PropertyManager is displayed.

4. Click the cylindrical face of the **third hex bolt** in the front row. The direction arrow points upward. Click **Reverse** direction if needed.

5. Click the inside face of the **third front hole** of SF_plate1. The direction arrow points downward.

6. Click the inside face of the **third front hole** of SF_plate3. The direction arrow points downward.

7. Click the inside face of the **third hex flange nut** in the front row. The direction arrow points downward.

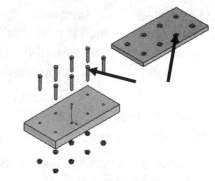

8. Click **OK** ✔ from the Route Line PropertyManager.

9. Repeat the above Explode Line Sketch procedure for the **fourth hex bolt** in the front row.

10. Click **OK** ✔ from the Route Line PropertyManager.

11. Click **OK** ✔ from the Route Line PropertyManager.

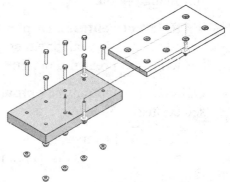

12. Click and drag the two **Explode Line Sketch** segments upward to create spacing between the lines.

13. Click **OK** ✔ from the Line Properties PropertyManager. View the results.

14. **Close** the model.

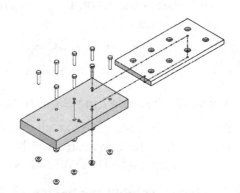

🔅 The Explode Line Sketch, 3DExplode<n>, is located in the ConfigurationManager under ExplView<n>. To edit the Explode Line, right-click **3DExplode<n>**. Click **Edit Sketch**.

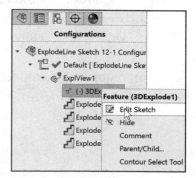

Interference Detection tool

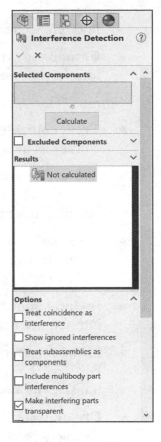

The Interference Detection 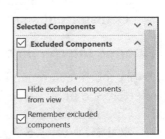 tool provides the ability to visually determine if there is an interference between components in an assembly. The Interference Detection tool can:

- Reveal the interference between components.

  - Display the <u>volume</u> of interference as a shaded volume. **Apply**. Previews the changes to the explode steps.

- Done. Completes the new or changed explode steps

- Alter the display settings of the interfering and non-interfering components to better display the interference.

- Select to ignore interferences that you want to exclude, such as interferences of threaded fasteners, press fits, etc.

- Include interferences between bodies within a multi-body part.

- Treat a sub-assembly as a single component so that interferences between the sub-assembly's components are not reported.

- Distinguish between coincidence interferences and standard interferences.

The Interference Detection tool uses the Interference Detection PropertyManager. The Interference Detection PropertyManager provides the following selections:

- *Selected Components*. The Selected Components box provides the following selections:

  - **Select**. Displays the selected components to perform the Interference Detection on. By default, the top-level assembly is displayed unless you pre-select another component.

  - **Calculate**. Performs the Interference detection process.

  - **Excluded Components**. Select Excluded Components to activate this group of commands:

    - Hide excluded components from view.

    - Remember excluded components.

- *Results*. The Results box provides the following selections:

  - **Results**. Displays the results of the Interference detection process. The volume of each interference is displayed to the right of each listing.

- **Ignore**. Switches between the ignored and un-ignored mode for the selected interference in the Results box. If an interference is set to the Ignore mode, the interference remains ignored during subsequent interference calculations.

  - **Component view**. Displays the interferences by component name instead of interference number.

- *Options*. The Options box provides the following selections:

  - **Treat coincidence as interference**. Reports coincident entities as interferences.

  - **Show ignored interferences**. Displays ignored interferences in the Results list, with a gray icon. When this option is cleared, ignored interferences are not listed.

  - **Treat subassemblies as components**. Treats sub-assemblies as single components. Interferences between a sub-assembly's components are not reported.

  - **Include multibody part interferences**. Reports interference between bodies within multi-body parts.

  - **Make interfering parts transparent**. Selected by default. Displays the components of the selected interference in transparent mode.

  - **Create fasteners folder**. Under Results, segregates interferences between fasteners (such as a nut and bolt) into a separate folder named Fasteners.

  - **Create matching cosmetic threads folder**. Under Results, segregates interferences between components with properly matched cosmetic threads into a separate folder named Matching cosmetic threads.

  - **Ignore hidden bodies**. Ignores interferences with hidden bodies.

- *Non-interfering Components*. The Non-interfering Components box provides the ability to display non-interfering components in a selected display mode. Use current is selected by default. The available display modes are **Wireframe**, **Hidden**, **Transparent** and **Use current**.

Tutorial: Interference Detection 12-1

Apply the Interference Detection tool. Calculate the interference in an assembly.

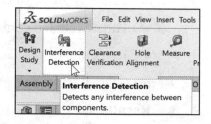

1. Open **Interference Detection 12-1** from the SOLIDWORKS 2018\TopDownAssemblyModeling folder.

2. Click the **Interference Detection** tool from the Evaluate tab in the CommandManager.

3. Click the **Calculate** button from the Selected Components box. The volume of Interference is displayed in red (X, Y, Z). The components that interfere are transparent.

4. Click **OK** ✔ from the Interference Detection PropertyManager. View the Interference. Resolve the interference issue.

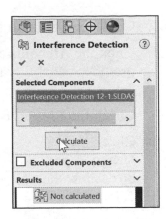

5. **Expand** the Mates folder from the FeatureManager.

6. Right-click **Distance1**.

7. Click **Edit Feature**. The Distance1 Mate PropertyManager is displayed.

8. Check the **Flip direction** box from the Standard Mates box.

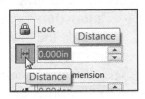

9. Enter **0** for Distance.

10. Click **OK** ✔ from the Distance1 PropertyManager.

Recalculate the Interference of the assembly with the new information.

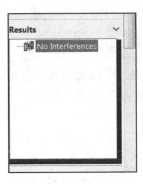

11. Click the **Interference Detection** 🔩 tool.

12. Click the **Calculate** button. There are no interferences.

13. Click **OK** ✔ from the Interference Detection PropertyManager.

14. **Close** the model.

Collision Detection

You can detect collisions with other components when moving or rotating a component. The software can detect collisions with the entire assembly or a selected group of components. You can find collisions for either the selected components or for all of the components that move as a result of mates to the selected components.

Tutorial: Collision Detection 12-1

Apply the Collision Detection tool to an assembly.

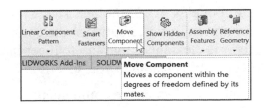

1. Open **Collision Detection 12-1** from the SOLIDWORKS 2018\TopDownAssemblyModeling folder.

2. Click the **Move Component** 🖫 tool from the Assembly tab in the CommandManager. View the Move Component PropertyManager.

3. Click the **Collision Detection** box.

4. Click the **All components** box. The **These components** option provides the ability to select individual components for collision detection.

5. Check the **Stop at collision** box.

6. Check the **Dragged part only** box. View the defaults.

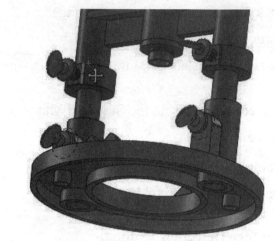

7. Click and drag the **Collar downward** towards the base as illustrated. The Collision Detection tool informs the user when there is a collision between the components.

8. Click **OK** ✔ from the Move Component PropertyManager.

9. **Close** the model.

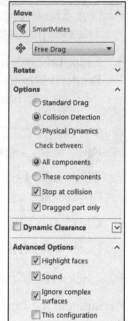

Physical Dynamics is an option in Collision Detection that allows you to see the motion of assembly components in a realistic way. With Physical Dynamics enabled, when you drag a component, the component applies a force to components that it touches and moves the components if they are free to move.

Clearance Verification

The Clearance Verification ⛉ tool provides the ability to check the clearance between selected components in assemblies. The software checks the minimum distance between the components and reports clearances that fail to meet the minimum acceptable clearance you specify.

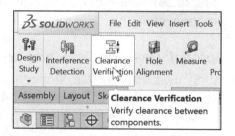

You can select entire components or particular faces of components. You can check between just the selected components, or between selected components and the rest of the assembly.

The Clearance Verification tool uses the Clearance Verification PropertyManager. The Clearance Verification PropertyManager provides the following selections:

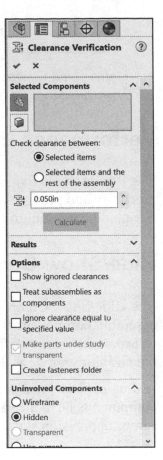

- *Selected Component*. The Selected Component box provides the following selections:

  - **Select**. Lists entities selected for the clearance check.

  - **Select Components**. Selected by default. To filter the type of entity you want to select.

  - **Select Faces**. To filter the type of entity you want to select.

  - **Selected items**. Selected by default. Specifies whether to check only between the entities you select, or between the entities you select and the rest of the assembly.

  - **Selected items and the rest of the assembly**. Specifies whether to check only between the entities you select, or between the entities you select and the rest of the assembly.

  - **Minimum Acceptable Clearance**. Clearances less than or equal to this value are reported in Results.

  - **Calculate**. Click to run the clearance check.

- *Results*. Lists clearances that fail to meet the minimum acceptable clearance. The value of each failed clearance appears in its listing. When you select a clearance under Results, it highlights in the graphics area.

- *Options*. Provides the following options:

  - **Show ignored clearances**. Select to show ignored clearances in the Results list with a gray icon. When this option is cleared, ignored clearances are not listed.

  - **Treat subassemblies as components**. Treats sub-assemblies as single components, so clearances between a sub-assembly's components are not checked.

  - **Ignore clearance equal to specified value**. Reports only clearances less than the specified value.

  - **Make parts under study transparent**. Selected by default. Displays in transparent mode the components whose clearances are being verified.

- **Create fasteners folder**. Segregates clearances between fasteners (such as a nut and bolt) into a separate folder under Results.

- *Uninvolved Components*. Provides the following options: *Wireframe, Hidden, Transparent and Use current display.*

Performance Evaluation

The Performance Evaluation 🔍 tool provides the ability to analyze the performance of an assembly and provides possible actions you can take to improve performance. View Chapter 11 for additional information.

Hide/Show Components tool from the Display Pane

The Hide/Show Components 🖉 tool provides the ability to hide or display the selected assembly component. You can hide the component completely from view or make it 75% transparent. Turning off the display of a component temporarily removes it from view in the Graphics window. This provides the ability to work with the underlying components.

Hiding or showing a component only affects the visibility of the component. Hidden components have the same accessibility and behaviors as displayed components in the same suppression state.

🔅 You can also right-click on a component in an assembly and apply the Hide / Show tool, the Change transparent tool, and obtain access to the color and texture options from the Context toolbar.

Tutorial: Component States 12-1

Utilize the Hide/Show Components tool. Use Show Display Pane for the Transparency, Color and Hide features.

1. Open **Component States 12-1** from the SOLIDWORKS 2018\TopDownAssemblyModeling folder.

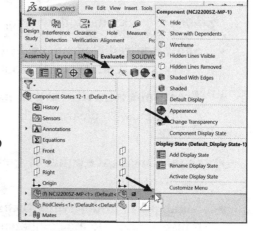

2. Click the **Show Display Pane** ˃ icon at the top of the Component States 12-1 FeatureManager.

3. Right-click the **Transparency** column to the right of the NCJ22005Z-MP assembly.

4. Click **Change Transparency**. The component is transparent in the model.

5. Right-click inside the **Appearances** column to the right of RodClevis<1>.

6. Click **Appearance**. The Appearance PropertyManager is displayed.

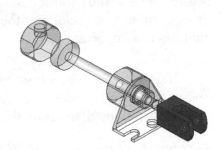

7. Select **blue** for a Color Swatch. The color set in the assembly overrides the part color.

8. Click **OK** ✔ from the Appearance PropertyManager.

9. **Expand** the NCJ22005Z-MP assembly.

10. Right-click the **Transparency** column next to NCJ2TUBES<1>.

11. Click **Hide**. View the assembly in the Graphics window.

12. **Close** the model.

🔆 Hiding and suppressing parts in SOLIDWORKS assemblies can have similar looking results, but both operations behave quite differently from each other. View SOLIDWORKS help for additional information.

Edit Component (Part) tool

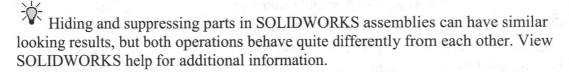

The Edit Component 🗇 tool provides the ability to move between editing a part or sub-assembly and the main assembly.

In the Top-Down assembly method, relationships are created while editing a part within an assembly. This method is referred to as In-Context editing. You create or edit your feature In-Context of the assembly, instead of in isolation, as you traditionally create parts.

In-Context editing provides the ability to view your part in its location in the assembly as you create new features in the assembly. Use geometry of the surrounding parts to define the size or shape of the new feature.

Relations that are defined In-Context are listed as External references. External references are created when one document is dependent on another document for its solution. If the referenced document is modified, the dependent document is also modified. In-Context relations and External references are powerful tools in the design phase.

In the Assembly FeatureManager, an item with an External reference has a suffix which displays the reference status. They are:

* -> The reference is In-Context. It is solved and up-to-date.

* ->? The reference is out-of-context. The feature is not solved or not up-to-date. To solve and update the feature, open the assembly that contains the update path.

* ->* The reference is locked.

* ->x The reference is broken.

Mastering assembly modeling techniques with In-Context relations requires practice and time. Planning and selecting the correct reference and understanding how to incorporate changes are important.

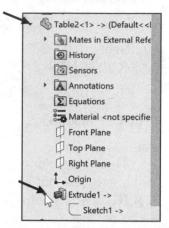

☼ Right-click a component, and click the Edit Part tool to obtain the ability to move between editing a part or sub-assembly and the main assembly.

Tutorial: Edit Component 12-1

Edit a Component In-Context to an assembly.

1. Open **External References 12-1** from the SOLIDWORKS 2018\TopDownAssemblyModeling folder.

2. **Expand** Table2<1> -> from the FeatureManager. The -> symbol indicates External references, In-Context. View the InPlace1(Top) mate from the Table2 component. The Table2 component references the Top Plane in the assembly. The InPlace1 mate is also displayed in the MateGroup1 folder.

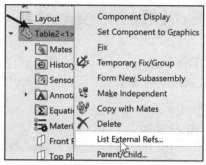

3. **Expand** Extrude1 -> from the FeatureManager. Extrude1 -> and Sketch1 -> in the FeatureManager contain External references. The references were created from four converted lines in an assembly Layout sketch.

List the External References in the Assembly.

4. Right-click **Table2<1> ->** from the FeatureManager.

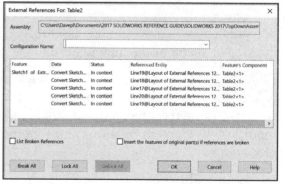

5. Click **List External Refs**. The External References For: Table2 dialog box lists the sketched lines of the Layout sketch. View your options. Click the HELP button to view additional information.

6. Click **OK** from the External References For: Table2 dialog box.

Delete the InPlace1 mate.

7. Click the **InPlace1(Top)** mate from the FeatureManager.

8. Right-click **Delete**. Click **Yes** to delete.

9. Click **Yes** to delete references. The Table2 component is free to move in the assembly.

10. **Rebuild** 🔘 the model.

11. **Expand** Extrude1 from the FeatureManager. Sketch1 is under defined and requires relations and dimensions. The External References are removed.

12. **Close** the model.

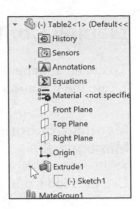

☀️ To list External references on a part or feature: Right-click the **component** or the **feature** with the External reference, ➢ **List External Refs**. The referenced components, features, and entities are listed.

Assembly ConfigurationManager

SOLIDWORKS provides the ability to create multiple variations of a part or assembly model within a single document. The ConfigurationManager provides the tools to develop and manage families of models with various dimensions, components, or other key design parameters.

The ConfigurationManager is located on the left side of the SOLIDWORKS window. You can split the ConfigurationManager to either display two ConfigurationManager instances, or combine the ConfigurationManager with the FeatureManager, PropertyManager, or a third party application that uses the panel. Use Configurations for part, assembly, and drawing documents.

In a part document, the part ConfigurationManager provides the ability to create a family of parts with unique dimensions, features, and properties, including custom properties.

In an assembly document, the assembly ConfigurationManager provides the ability to create simplified versions of the design by suppressing components and families of assemblies with unique configurations of the components, with various parameters for assembly features, dimensions, or custom properties.

In a drawing document, the drawing ConfigurationManager provides the ability to display views of the various configurations that were created in the part and assembly documents.

☀️ If a configuration has never been activated (such as one defined in a design table), only essential defining data about the configuration is carried in the model. When you activate a configuration for the first time, the full definition of the configuration's model data is generated. This full data set is updated and saved every time you save the document. By default, once the full data set is generated, it remains available every time you open the model. Having this full data set readily available for a configuration can save a significant amount of time when switching to that configuration from another. However, for each configuration that you activate, the file size of the model increases, as does the time required to rebuild and save the file.

Manual Configurations

To create a manual configuration, first specify the properties. Then modify the model to create the variation in the new configuration. You can add or edit manual configurations.

Manual Configuration/Add Configuration Property Manager

Use the Add Configuration PropertyManager to add a new configuration. To display the Add Configuration PropertyManager, click the **ConfigurationManager** tab, then right-click **Add Configuration**. The Add Configuration PropertyManager is document dependent and provides the following selections:

- *Configuration Properties*. The Configuration Properties box provides the following selections:

  - **Configuration name**. Displays the entered name for the configuration. The name cannot include the following: forward slash (/) or "at" sign (@). A warning message is displayed when you close the dialog box if the name field contains either of these characters, or if the field is blank.

  - **Description**. (Optional). Displays the entered description of the configuration.

  - **Use in bill of materials**. Specify how the assembly or part is listed in a Bill of Materials.

  - **Comment**. (Optional). Displays the entered additional descriptive information on the configuration.

  - **Custom Properties**. Only available only when editing properties of an existing configuration.

- *Bill of Materials Options*. Specifies how the part or assembly is listed in the Bill of Materials. The box provides the following selections:

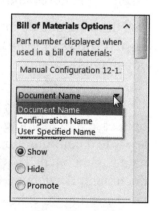

  - **Document Name**. Displays the part number. The Part number is the same as the document name.

- **Configuration Name**. Displays the part number. The Part number is the same as the configuration name.

- **User Specified Name**. The part number is a name that you type.

- **Show**. Selected by default. Show child components in BOM when used as sub-assembly (assemblies only).

- **Hide**. When selected, the sub-assembly is <u>always</u> shown as a single item in the Bill of Materials (assemblies only).

- **Promote**.

- *Advanced Options*. The following selections control what happens when you add new items to another configuration and then activate this configuration again:

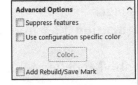

- **Suppress new features and mates**. Only for assemblies. When selected, new mates and features added to other configurations are suppressed in this configuration. Otherwise, new mates and features arc contained, not suppressed in this configuration.

- **Hide new components**. Only for assemblies. When selected, new components added to other configurations are hidden in this configuration. Otherwise, new components are displayed in this configuration.

- **Suppress new components**. Only for assemblies. When selected, new components added to other configurations are suppressed in this configuration. Otherwise, new components are resolved, not suppressed in this configuration.

- **Suppress features**. Only for parts. When selected, new features added to other configurations are suppressed in this configuration. Otherwise, new features are contained, not suppressed in this configuration.

- **Use configuration specific color**. Specifies a color for the configuration.

- **Color**. Choose a color from the color palette. If the color for wireframe and HLR modes is the same as the color for shaded mode, the configuration-specific color applies to all three modes. If the color is not the same for the three modes, the configuration specific color is applied only to the shaded mode.

  - **Add Rebuild/Save Mark**. See SOLIDWORKS Help for additional information.

- *Parent/Child Options*. Only available for assemblies and when adding a new configuration to the assembly or one of its components. Select the components to which you want to add the new configuration.

Tutorial: Manual Configuration 12-1

Create a manual assembly configuration with the Add Configuration PropertyManager.

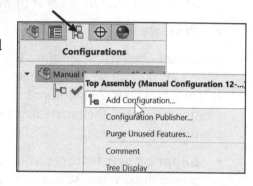

1. Open **Manual Configuration 12-1** from the SOLIDWORKS 2018\TopDownAssemblyModeling folder.

2. Click the **ConfigurationManager** tab.

3. Right-click **Manual Configuration 12-1** from the ConfigurationManager.

4. Click **Add Configuration**. The Add Configuration PropertyManager is displayed.

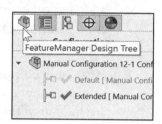

5. Enter **Extended** in the Configuration name box.

6. Enter **Extended Position 10mm** for Description.

7. Click **OK** ✔ from the Add Configuration PropertyManager. Extended Configuration is currently selected.

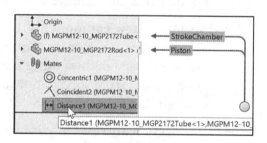

8. **Return** to the FeatureManager.

9. **Expand** the Mates folder in the FeatureManager.

10. Double-click the **Distance1** Mate. The 0 dimension is displayed in the Graphics window.

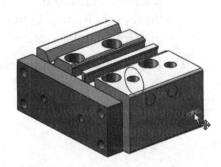

11. **Zoom-in** on the 0 dimension in the Graphics window.

12. Double-click the **0** dimension from the Graphics window.

13. Enter **10mm**.

14. Select **This Configuration** as illustrated in the Modify dialog box.

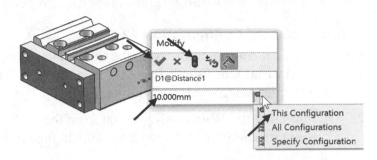

15. **Rebuild** the model from the Modify dialog box. The piston is extended 10mm in the Graphics window.

16. Click the **green check mark** in the Modify dialog box.

17. Click **OK** ✔ from the Dimension PropertyManager.

18. **Return** to the ConfigurationManager.

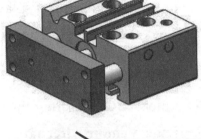

19. Double-click the **Default** Configuration in the ConfigurationManager. View the results in the Graphics window.

20. **Return** to the FeatureManager.

21. **Close** the model.

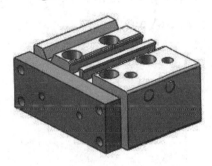

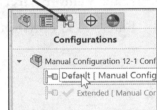

Manual Configuration/Edit Configuration

Use the Configuration Properties PropertyManager to edit an existing configuration. To display the Configuration Properties PropertyManager, click the **ConfigurationManager** tab ➤ right-click **Properties**.

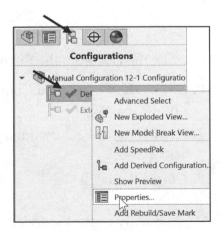

The Configuration Properties PropertyManager provides the same selections as the Add Configuration Properties PropertyManager.

Tutorial: Manual Configuration 12-2

Create a manual configuration using the Custom Properties option.

1. Open **Manual Configuration 12-2** from the SOLIDWORKS 2018\TopDownAssemblyModeling folder.

2. Double-click the **Blue** configuration from the ConfigurationManager. View the results in the Graphics window.

3. Right-click **Properties**. The Configuration Properties PropertyManager is displayed.

4. Click the **Custom Properties** button. Review the Cost and Finish Custom Properties.

5. Select **Vendor** for Property Name in Row 3.

6. Enter **ABC Finishing** for Evaluated Value.

7. Enter **Vendor Alternate** for Property Name in Row 4.

8. Enter **XYZ Finishing** for Evaluated Value.

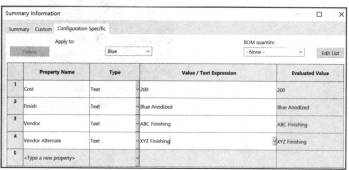

9. Click the **OK** from the Summary Information dialog box.

10. Click **OK** ✔ from the Configuration Properties PropertyManager.

11. **Return** to the FeatureManager. The Blue configuration is active.

12. **Close** all models.

Manually control individual color states for parts in an assembly. Select the color swatch in the FeatureManager. Select the configuration.

Automatic Configuration: Design Tables

To create a design table, define the names of the configurations, specify the parameters to control, and assign the value for each parameter.

There are several ways to create a design table:

- Insert a new, empty design table in the model. Enter the design table information directly in the worksheet. When you finish entering the design table information, the new configurations are automatically created in the model.

- Have SOLIDWORKS automatically create the design table. SOLIDWORKS loads all configured parameters and their associated values from a part or assembly.

- Create a design table worksheet as a separate operation in Microsoft Excel. Save the worksheet. Insert the worksheet in the model document to create the configurations.

- Insert a partially completed worksheet. Edit the partially completed worksheet later to add additional configurations, to control additional parameters, or to update values.

The Design Table PropertyManager provides the ability to create an excel file which automatically creates a design table. To Display the Design Table PropertyManager click **Insert ➢ Design Table**. The Design Table PropertyManager provides the following selections:

- *Source*. The Source box provides the following selections:

  - **Blank**. Inserts a blank design table where you fill in the parameters.

  - **Auto-create**. Selected by default. Automatically creates a new design table, and loads all configured parameters and their associated values from a part or assembly.

  - **From file**. References a Microsoft Excel table.

  - **Browse**. Browse to a file location to select a Microsoft Excel table.

  - **Link to file**. Links the Excel table to the model. When a design table is linked, any changes you make to the table outside of SOLIDWORKS are reflected in the table within the SOLIDWORKS model.

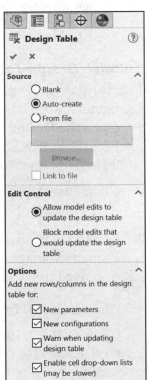

In a design table, equations must be preceded by a single apostrophe and an equal sign ('='). The single apostrophe ensures that if the equation is exported to Excel, it is interpreted as an equation and not as a text string.

You cannot use configurable equations and suppression/unsuppression in the same model. If you created a model with SOLIDWORKS 2013 or an earlier version, you must remove all suppressed and unsuppressed Equations and Global Variables, before adding configurable Equations and Global Variables. You must also remove all suppressed and unsuppressed Equations and Global Variables from design tables.

- *Edit Control*. The Edit Control box provides the following selections:

    - **Allow model edits to update the design table**. If you modify the model, the changes are updated in the design table.

    - **Block model edits that would update the design table**. You are not allowed to modify the model if the change updates the design table.

- *Options*. The Options box provides the following selections:

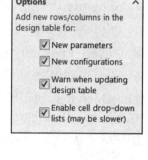

    - **New parameters**. Selected by default. Adds a new column to the design table if you add a new parameter to the model.

    - **New configurations**. Selected by default. Adds a new row to the design table if you add a new configuration to the model.

    - **Warn when updating design table**. Selected by default. Warns you that the design table will change based on the parameters you updated in the model.

    - **Enable cell drop-down lists (may be slower).**

Tutorial: Design Table 12-1

Insert an assembly design table with suppress and active configurations.

1. Open the **Design Table 12-1** assembly from the SOLIDWORKS 2018\TopDownAssemblyModeling folder.

2. Click **Insert** ➤ **Tables** ➤ **Design Table**. The Auto-create option is selected.

3. Click **OK** ✔ from the Design Table PropertyManager. A blank design table is displayed in the Graphics window.

4. Click **Cell A4**. Enter **No Shaft Collar**.

5. Click **Cell B2**. Enter **$State@SHAFT-COLLAR<2>**. Click **Cell B3**. Enter **R** for Resolved. Click **Cell B4**. Enter **S** for Suppressed.

6. Click a **position** outside the Design Table.

7. Click **OK** to display the No Shaft Collar configuration.

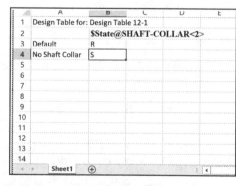

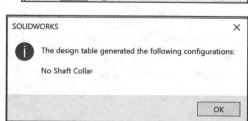

8. Double-click the **No Shaft Collar** configuration from the ConfigurationManager. View the model.

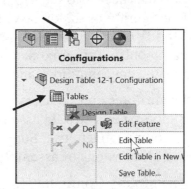

9. Double-click **Default** from the ConfigurationManager. View the model.

10. Display an **Isometric** view.

11. **Close** all models.

Tutorial: Design Table 12-2

Modify a design table. Apply Custom properties option.

1. Open **Design Table 12-2** from the SOLIDWORKS 2018\TopDownAssemblyModeling folder.

Edit the existing design table.

2. Right-click **Design Table** from the ConfigurationManager.

3. Click **Edit Table**.

4. Click **Cancel** to the message. The design table is displayed in the Graphics window.

5. Click **Cell C4**.

6. Delete **7HOLE**.

7. Enter **3HOLE**.

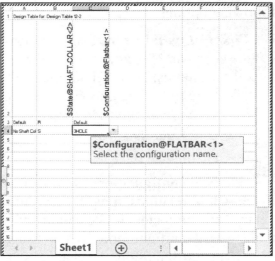

Close the design table.

8. Click a **position** outside the Design Table.

9. Double-click the **No Shaft Collar** configuration from the ConfigurationManager. The 3HOLE configuration is displayed in the Graphics window.

10. Double-click the **Default** configuration from the ConfigurationManager. The default configuration is displayed in the Graphics window.

11. Display an **Isometric** view.

12. **Return** to the FeatureManager.

13. **Close** all models.

Configure component tool/Configure dimension tool

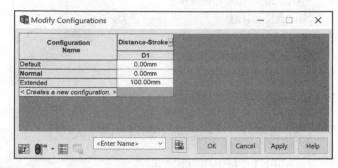

The Configure component tool/Configure dimension tool provides access to the Modify Configurations dialog box. The Modify Configurations dialog box facilitates creating and modifying configurations for commonly configured parameters in parts and assemblies. You can add, delete, and rename configurations and modify which configuration is active.

For features and sketches in parts, you can configure the following: Dimensions and Suppression states. In assemblies, you can configure the following: Which configurations of components to use, Suppression states of components, Assembly features, and mates, and Dimensions of assembly features and mates.

The Modify Configurations dialog box provides the following options:

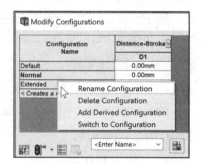

- **First Column**. List the configurations of the model and the configurable parameters of the selected item in the other columns. Note: Right-click any configuration and select the following option: *Rename Configuration, Delete Configuration, Add Derived Configuration* and *Switch to Configuration*.

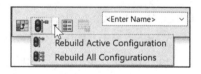

Derived configurations provide the ability to create a Parent-Child relationship within a configuration. By default, all parameters in the child configuration are linked to the parent configuration. If you change a parameter in the parent configuration, the change automatically propagates to the child.

The All Parameters option adds a column for every configured parameter in the model.

- **Parameter Columns**. Provides the ability to select one of the following: *Type to change the numeric values, select from a list of component configurations or to change the suppression state of features, sketches, components and mates.*

Tutorial: Configure Component/Dimension tool 12-1

Add a new configuration. Utilize the Configure dimension tool.

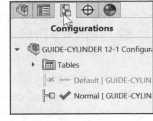

1. Open the **GUIDE-CYLINDER 12-1** assembly from the SOLIDWORKS 2018\TopDownAssemblyModeling folder.

2. Click the **ConfigurationManager** tab. View the existing configurations.

Create a new configuration.

3. **Return** to the FeatureManager.

4. Double-click the **Distance-Stroke** mate from the MateGroup1 folder. View the illustrated dimension in the Graphics window.

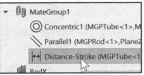

5. Right-click near the **0** dimension text in the Graphics window.

6. Click **Configure Dimension** from the Context toolbar. The Modify Configurations dialog box is displayed.

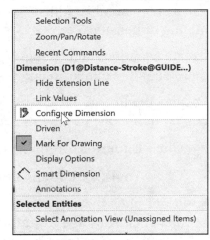

7. Click inside the **Creates a new configuration** box.

8. Enter **Extended**.

9. Press the **Tap** key.

10. Enter **100**mm as illustrated.

11. Click **OK** from the Modify Configuration dialog box.

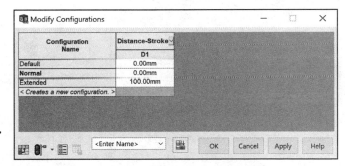

12. Click **OK** from the Dimension PropertyManager. View the results in the Graphics window.

13. Click the **ConfigurationManager** tab. View the new configuration.

14. Double-click the **Extended** configuration.

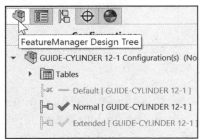

15. **View** the results in the Graphics window.

16. **Return** to the Normal configuration.

17. **Return** to the FeatureManager.

18. **Close** the model.

Equations

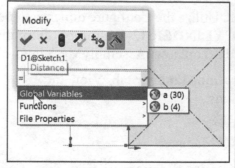

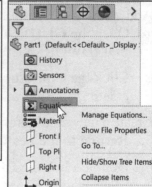

Use equations Σ to create mathematical relations between the dimensions of your model. Use dimension names as variables. Set equations in an assembly between multiple parts, a part and a sub-assembly with mating dimensions, etc.

Create equations and Global Variables directly in the Modify dialog box for dimensions.

Display the Equations, Global Variables and Dimensions dialog box by doing one of the following:

- Click **Equations** from the Equations toolbar.

- Click **Tools**, **Equations** from the Main menu.

- Right-click the **Equations** folder from the FeatureManager and click **Manage Equations**. Note: Show the Equations folder from System Options.

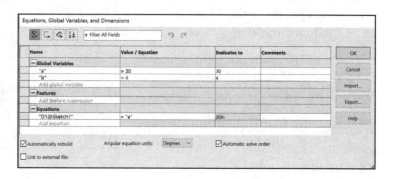

Create equation references between assembly configurations.

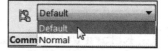

In a design table, equations should be preceded by a single apostrophe and an equal sign ('='). The single apostrophe ensures that if the equation is exported to Excel, it is interpreted as an equation and not as a text string.

Tutorial: Equations 12-1

Display the Equations, Global Variables and Dimensions dialog box.

1. Open **Equations 12-1** from the SOLIDWORKS 2018\TopDownAssemblyModeling folder.

2. Right-click **Equations**.

3. Click **Manage Equations**. View the Equations, Global Variables, and Dimensions dialog box. Click the Help button and explore the options.

4. **Close** the Help dialog box.

5. Click **OK** from the dialog box.

6. **Close** the model.

 Existing equations are listed in the dialog box.

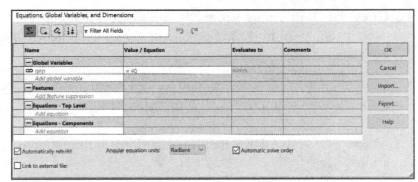

Equations tool

The Equations tool uses the Equations, Global Variables and Dimensions dialog box. The Equations, Global Variables and Dimensions dialog box provides the ability to define dimensions using global variables and mathematical functions, and create mathematical relationships between two or more dimensions in parts and assemblies.

You can use the following as variables in equations:

- Dimension names.

- Global Variables.

- Other equations.

- Mathematical functions.

- File properties.

- Dimension measurements.

The Equations, Global Variables, and Dimensions dialog box offers three views. Each view shows a different combination and sequence of equations, global variables and dimensions to help you perform tasks such as finding a specific equation, viewing all dimensions used in a part, and changing the order in which equations are solved.

- Click the Equations View Σ icon to see all global variables and equations for dimensions and features.

- Click the Sketch Equations View icon to see all Global Variables and sketch equations.

- Click the Dimensions View icon to see all Global Variables, equations, and dimensions used in the active part or assembly, whether they are associated with an equation or not.

- Click the Ordered View icon to see Global Variables and equations in the order they are solved and to see suppressed equations.

View SOLIDWORKS help for additional information on global variables and equations.

Tutorial: Equations 12-2

Create an Equation to define the Layout sketch in a Top-down assembly.

1. Open **Equations 12-2** from the SOLIDWORKS 2018\TopDownAssemblyModeling folder.

2. Right-click **Layout** from the FeatureManager.

3. Click **Edit Sketch** from the Context toolbar. View the Graphics window.

4. Right-click the **Equations** folder from the FeatureManager.

5. Click **Manager Equations**. The Equations, Global Variables and Dimensions dialog box is displayed.

6. Click inside the **Equations - Top Level box**.

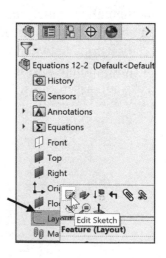

7. Click the mountplate_height dimension **1160** from the Graphics window. The variable "mountplate_height@Layout" is added to the cell. An equal sign is displayed in the Value/Equations cell as illustrated.

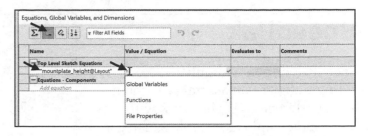

8. Click inside the = **cell** under Value/Equations. View your options.

9. Place the mouse pointer over **File Properties**. View your options.

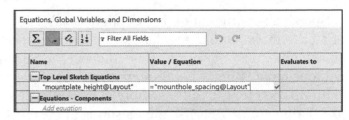

10. Click mounthole_spacing dimension, **880**. The variable "mounthole_spacing@Layout" is added to the Value/Equations cell. A green check mark is displayed.

11. Enter the following from the keyboard: **+2*(**

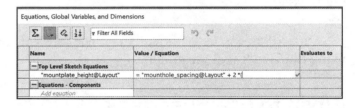

12. Click **40** from the Graphics window. The variable "gap@Layout" is added to the cell.

13. Enter the following from the keyboard: **+120)**. A green check mark is displayed.

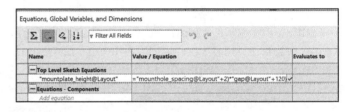

14. Click inside the **Evaluates to** box. The variable mountplate_height equals 1200.

15. Click **OK** from the dialog box.

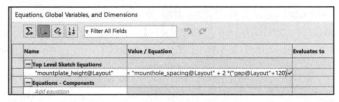

16. Click **Exit Sketch**.

17. **Close** the model.

SpeedPak

SpeedPak creates a simplified
configuration of an assembly without
losing references. If you work with very
large and complex assemblies, using a
SpeedPak configuration can significantly

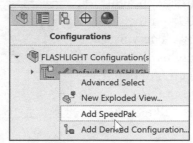

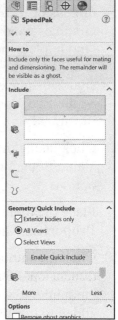

improve performance while working in the assembly and its drawing.

A SpeedPak configuration is essentially a subset of the parts and faces
of an assembly. Unlike regular configurations, where you can simplify
an assembly only by suppressing components, SpeedPak simplifies
without suppressing. Therefore, you can substitute a SpeedPak
configuration for the full assembly in higher level assemblies without
losing references. Because only a subset of the parts and faces are
used, memory usage is reduced, which can increase performance of
many operations.

SpeedPak can automatically exclude internal components when
using the Quick Include slider.

When to use SpeedPak

SpeedPak creates a simplified configuration of an assembly without
losing references. If you work with very large and complex
assemblies, using a SpeedPak configuration can significantly improve
performance while working in the assembly and its drawing.

You can also use SpeedPak to facilitate file sharing. The SpeedPak information is saved
entirely within the assembly file. Therefore, when sharing an assembly, you can send just
the assembly file. You do not need to include component files.

For example, suppose you design an engine assembly, and need to send it to your
customer's design team so they can insert it in their
vehicle design.

1. You create a SpeedPak configuration of the
   engine assembly, including all the faces and
   bodies your customer plans to reference in their
   vehicle model.
2. You send them just the engine assembly file. You
   do not need to send files for any of the engine
   component parts.

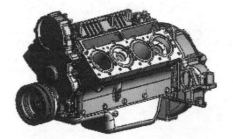

3. They insert your engine assembly file into their vehicle assembly file. They can add mates and dimensions to all faces and bodies you included in the SpeedPak definition.

Creating a SpeedPak for an Assembly

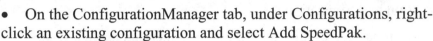

In an assembly file, you can derive a SpeedPak configuration from an existing configuration. To create a SpeedPak:

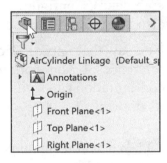

•   On the ConfigurationManager tab, under Configurations, right-click an existing configuration and select Add SpeedPak.

In the PropertyManager:

•   Select the faces and bodies that you want to be selectable in the SpeedPak configuration.

•   Optionally, select Remove ghost to hide all other faces, which improves performance even more.

•   Click OK.

A SpeedPak configuration is created as a child of the original configuration. It is identified with the SpeedPak 🖼 icon in the ConfigurationManager and speedpak is appended to its name.

No components appear in the FeatureManager design tree.

In the graphics area, when you move the pointer over the assembly, only the faces and bodies you selected for the SpeedPak are visible and selectable in the region surrounding the pointer.

SpeedPak in a Drawing

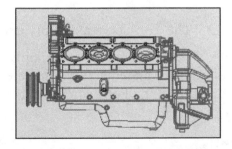

When you dimension SpeedPak configurations in drawings, you can only dimension to edges included in the SpeedPak, which are shown in black. Edges not included in the SpeedPak are shown in gray. When you print the drawing, all the lines print in black, unless you select Color/Gray scale in the Page Setup dialog box.

In a drawing document, when you create a bill of materials for an assembly whose active configuration is a SpeedPak configuration, you can include columns for other configurations in addition to the SpeedPak configuration.

💡 A SpeedPak configuration is derived from an existing configuration.

Summary

In this chapter, you learned about Assembly modeling with the Top-down assembly modeling method. You addressed three different Assembly methods for Top-down modeling: *Individual features method*, *An Entire assembly method* and the *Complete parts method*.

You reviewed and applied various tools from the Assembly toolbar: Insert Components, Hide/Show Components, Change Suppression State, Edit Component, No External References, Mate, Move Component, Smart Fasteners, Explode View, Explode Line Sketch, Interference Detection and Simulation.

Your addressed the New Motions Studies tool, the MotionManager and created Basic Motion studies.

You reviewed information on External references, In-Content components, and InPlace mates. Remember, an InPlace Mate is a Coincident Mate created between the Front Plane of a new component and the selected planar geometry of the assembly.

You addressed three types of assembly configurations: Manual (Add Configuration command), Configure component/Configure dimension tool, and Design Tables. You also created and applied an equation.

You saw that SpeedPak creates a simplified configuration of an assembly without losing references. If you work with very large and complex assemblies, using a SpeedPak configuration can significantly improve performance while working in the assembly and its drawing.

When you create your first assembly, Assem1.sldasm is the default document name. The Assembly document ends with the extension .sldasm.

Whenever you create a part or feature using the Top-Down approach, external references are created to the geometry you referenced.

The Top-down assembly approach is also referred to as "In-Context."

In Chapter 13, explore and apply various tools from the Drawing and Annotations toolbar along with understanding the View Palette, Inserting DimXpert Annotations into a drawing, Line Format tools and create an eDrawing.

CHAPTER 13: DRAWINGS AND DRAWING TOOLS

Chapter Objective

Chapter 13 provides a comprehensive understanding of document properties, settings and the ability to create dimensioned multi-view drawings from a part or assembly and more. On the completion of this chapter, you will be able to:

- Address Sheet format, size, borders, zones and Document Properties.

- Understand and apply the SOLIDWORKS View Palette.

- Address and utilize the following tools from the View Layout toolbar:

    - Model View, Projected View, Auxiliary View, Section View, Detail View, Standard 3 View, Broken-out Section View, Break View, Crop View and Alternate Position View.

- Know and utilize the following tools from the Annotation toolbar:

    - Smart Dimension, Model Items, Spell Checker, Format Painter, Note, Linear Note Pattern, Circular Note Pattern, Balloon, AutoBalloon, Magnetic Line, Surface Finish, Weld Symbol, Hole Callout, Geometric Tolerance, Datum Feature, Datum Target, Area Hatch/Fill, Blocks, Center Mark, Centerline, Revision Symbol, Revision cloud and Tables.

- Insert DimXpert Annotation using the SOLIDWORKS View Palette.

- Publish an eDrawing.

- Create a Detached drawing.

- Export drawings to another software package.

- Open a Drawing Document using various options.

- Insert a Center of Mass point.

Drawings

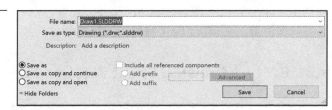

Create 2D drawings of a 3D part or assembly. Parts, assemblies and drawings are linked documents. This means that any changes that you incorporate into the part or assembly will modify the associate drawing document.

Drawings consist of one or more views produced from a part or assembly. The part or assembly connected with the drawing must be saved before you can create the drawing. Drawing files have the .SLDDRW extension. A new drawing takes the name of the first model inserted. The name is displayed in the title bar. When you save the drawing, the name of the model is displayed in the Save As dialog box as the default file name.

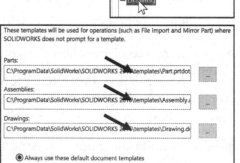

 Create a drawing within a part or assembly document.

The foundation of a SOLIDWORKS drawing is the Drawing Template. Drawing sheet size, drawing standards, company information, manufacturing and or assembly requirements; units, layers, line styles and other properties are defined in the Drawing Template.

The Sheet Format is incorporated into the Drawing Template. The Sheet Format contains the following: sheet border, title block and revision block information, company name, and or logo information, Custom Properties and SOLIDWORKS Properties.

SOLIDWORKS starts with a default Drawing Template, Drawing #.DRWDOT. The default Drawing Template is located in the\ProgramData\SOLIDWORKS\SOLIDWORKS 2018\templates folder. SOLIDWORKS is the name of the installation folder.

Sheet Format, Size and Properties

The Sheet Format\Size dialog box defines the Sheet Format and the paper size. The U.S. default Standard Sheet Format is A (ANSI)-Landscape.SLDDRT. The Display sheet format option toggles the sheet format display on/off. The Standard Sheet Formats are located in the \SOLIDWORKS\data folder.

ASME Y14.1 Drawing Sheet Size and Format

There are two ASME standards that define sheet size and format:

1. ASME Y14.1-1995 Decimal Inch Drawing Sheet Size and Format.

2. ASME Y14.1M-1995 Metric Drawing Sheet Size.

💡 The Hole Wizard information from the model propagates to the drawing when the Hole Callout button is selected.

Drawing size refers to the physical paper size used to create the drawing. The most common paper size in the U.S. is the A size: (8.5in. x 11in.). The most common paper size internationally is the A4 size: (210mm x 297mm). The ASME Y14.1-1995 and ASME Y14.1M-1995 standards contain both a horizontal and vertical format for A and A4 sizes respectively. The corresponding SOLIDWORKS format is Landscape for horizontal and Portrait for vertical.

SOLIDWORKS pre-defines U.S. drawing sizes A through E. Drawing sizes F, G, H, J & K utilize the Custom sheet size option. Enter values for Width and Height. SOLIDWORKS predefines metric drawing sizes A4 through A0. Metric roll paper sizes utilize the Custom sheet size option.

The ASME Y14.1-1995 Decimal Inch Drawing and ASME Y14.1M-1995 Metric Sheet Size standard are as follows:

Drawing Size: "Physical Paper"	Size in Inches: Vertical	Horizontal
A horizontal (landscape)	8.5	11.0
A vertical (portrait)	11.0	8.5
B	11.0	17.0
C	17.0	22.0
D	22.0	34.0
E	34.0	44.0
F	28.0	40.0
G, H, J and K apply to roll sizes, User Defined		

Drawing Size: "Physical Paper" Metric	Size in Millimeters: Vertical	Horizontal
A0	841	1189
A1	594	841
A2	420	594
A3	297	420
A4 horizontal (landscape)	210	297
A4 vertical (portrait)	297	210

Use caution when sending electronic drawings between U.S. and International colleagues. Drawing paper sizes will vary. Example: An A-size (11in. x 8.5in.) drawing (280mm x 216mm) does not fit an A4 metric drawing (297mm x 210mm). Use a larger paper size or scale the drawing using the printer setup options.

You can define drawing sheet zones on a sheet format for the purpose of providing locations where drawing views and annotations reside on the drawing.

Sheet Properties display properties of the selected sheet. Sheet Properties define the following: Name of the Sheet, Sheet Scale, Type of Projection (First angle or Third angle), Sheet Format, Sheet Size, View label and Datum label.

The sheet format and sheet size are set in the default Drawing Template. The Drawing Sheets document property lets you specify a default sheet format for when you add new sheets to drawing documents. This property lets you automatically have one sheet format for the first sheet and a separate sheet format for all additional sheets. To specify a different sheet format for a new sheet, click **Tools ➢ Options ➢ Document Properties ➢ Drawing Sheets**, select **Use different sheet format**, and browse to select a sheet format file (file ending in .SLDDRT).

Tutorial: Sheet Properties 13-1

Display the drawing sheet properties. Create a new drawing document.

1. Click **File ➢ New** from the Main menu. The New SOLIDWORKS Document dialog box is displayed.

2. Double-click **Drawing** from the New SOLIDWORKS Document dialog box. The Sheet Format/Size dialog box is displayed.

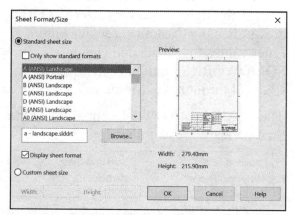

3. Uncheck the **Only show standards formats** box. View your options.

4. Click **A (ANSI) Landscape** for Standard sheet size.

5. Click **OK** from the Sheet Format/Size dialog box. View the results in the Graphics window.

6. If needed, click **Cancel ✖** from the Model View PropertyManager.

Define drawing sheet zones on a sheet format for the purpose of providing locations where drawing views and annotations reside on the drawing.

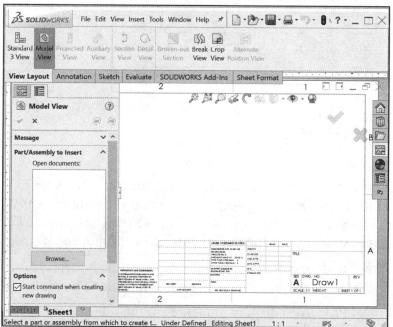

View the Sheet Properties.

7. **Right-click** Sheet1 in the Drawing PropertyManager.

8. Click **Properties**. The Sheet Properties dialog box is displayed. View the properties of the first sheet: Third angle projection, Scale: 1:1, Display sheet format, Sheet1 name, etc.

9. Click the **Zone Parameters** tab. View your options. The Go to Drawing Sheet Properties option provides the ability to go directly to the Document Properties - Drawing sheet section.

10. Click **Cancel** from the Sheet Properties dialog box.

11. **Close** the drawing.

Define drawing sheet zones (Automatic Border command) on a sheet format for the purpose of providing locations where drawing views and annotations reside on the drawing.

Use annotation notes and balloons to identify which drawing zone they are in. As you move an annotation in the Graphics area, the drawing zone updates to the current zone. You can add the current zone to an annotation by clicking an open space within the drawing view's bounding box while typing the annotation.

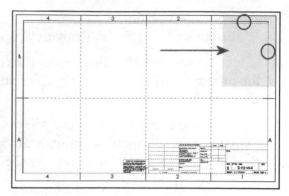

🔅 ANSI, Third Angle projection is illustrated and used in this book.

View Palette

The View Palette 🔲 is located in the Task Pane. Apply the View Palette tool to insert images of Standard views, Annotation views, Section views and flat patterns (sheet metal parts) of an active part or assembly, or click the Browse button to locate your desired model.

Click and drag the selected model view from the View Palette onto an active drawing sheet to create the drawing view. The View Palette provides the following options: *Import Annotations, Design Annotations, DimXpert Annotations, Include items from hidden features and Auto-start projected view*.

🔅 The **(A) Front** and **(A) Right** drawing views in the illustration are displayed with DimXpert Annotations which were applied at the part level.

Tutorial: View Palette/Import Annotations 13-1

Create a three (3) standard view drawing using the View Palette with DimXpert dimensions.

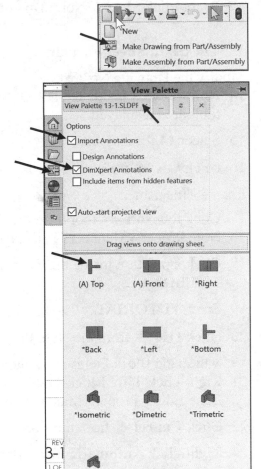

1. Open **View Palette 13-1.SLDPRT** from the SOLIDWORKS 2018\Drawings folder. View the model with the created DimXpert dimensions.

2. Click **Make Drawing from Part/Assembly** from the Menu bar. The New SOLIDWORKS Document dialog box is displayed. Drawing is selected by default.

3. Click **OK**. Accept the Standard Sheet size and format.

4. Click **OK** from the Sheet Format/Size dialog box.

5. Click the **View Palette** tab from the Task Pane. View Palette 13-1 is displayed with the available views.

6. Check the **Import Annotations** box.

7. Check the **DimXpert Annotations** box.

8. Click and drag the **(A) Front** drawing view into Sheet1.

9. Click a **position** directly above the Front view. The Top view is created.

10. Click a **position** directly to the right of the Front view. The Right view is created.

11. Click **OK** ✔ from the Projected View PropertyManager. You created a drawing with the default drawing template, using the View Palette and imported Annotations from DimXpert. Note: There are numerous ways to create a drawing. If needed modify the units to MMGS.

💡 SOLIDWORKS (SP0) does not let the user move DimXpert dimensions from one view to the other.

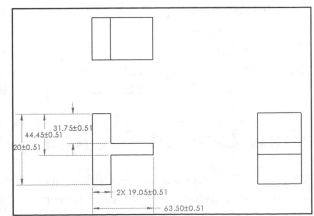

Modify the Title block.

12. Right-click **Sheet1**. Note: Do not click inside a drawing view.

13. Click **Edit Sheet Format**.

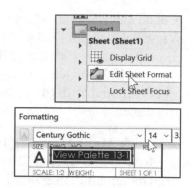

Modify the Drawing No. text.

14. **Double-click** inside the DWG. NO. box. The Formatting toolbar is displayed.

15. Select **14** for text size.

16. Click **OK** ✔ from the Note PropertyManager.

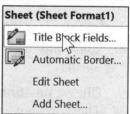

Enter a Title name.

17. Click the **Note** $\mathbf{A}$ tool from the Annotation tab. The Note PropertyManager is displayed.

18. Click a position inside the **Title box** below the TITLE: text.

19. Enter **TUTORIAL**.

20. Click **OK** ✔ from the Note PropertyManager.

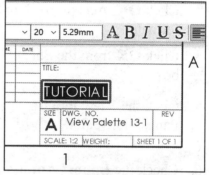

View the Title Block Fields and Automatic Border.

21. Right-click **Title Block Fields** in Sheet1. View the Zone Editor PropertyManager.

22. Click **Cancel** ✖ from the Zone PropertyManager.

23. Right-click **Automatic Border** in Sheet1. View the PropertyManager.

24. Click **Cancel** ✖ from the PropertyManager.

Return to the Edit Sheet mode.

25. Right-click **Sheet1**. Click **Edit Sheet**. View the drawing.

26. **Close** all models.

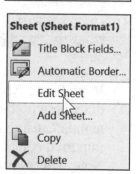

💡 To insert a picture into a drawing, click **Insert** ➤ **Picture** from the Menu bar menu. Select a **picture file**. The picture is inserted into the picture. The Sketch Picture PropertyManager is displayed.

View Layout Toolbar

The View Layout toolbar provides tools for aligning dimensions and creating drawing views. The drawing options are dependent on the views displayed in the drawing sheet. The tools and menu options that are displayed in gray cannot be selected. Additional information is required for these options.

Standard 3 Views tool

The Standard 3 Views  tool provides the ability to add three standard, orthogonal views. The type and orientation of the views can be 1st or 3rd Angle. The alignment of the top and side views is fixed in relation to the front view. The Standard 3 Views tool uses the Standard 3 View PropertyManager. The PropertyManager provides the following selections:

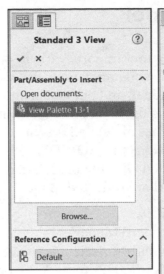

- ***Part/Assembly to Insert***. The Part/Assembly to Insert box provides the ability to select a document from the following selections:

  - **Open documents**. Displays the active part or assembly file. Select the part or assembly file.

  - **Browse**. Browse for the needed part or assembly file.

- ***Reference Configuration***. Provides the ability to select the model configuration.

Tutorial: Standard 3 Views 13-1

Apply the Standard 3 View drawing tool.

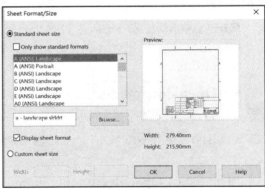

1. Create a **New** ⬚ drawing. Use the ANSI Drafting standard, Third Angle projection, A (ANSI) Landscape template.

2. Click the **Standard 3 View**  drawing tool from the View Layout toolbar.

3. Click the **Browse** button from the Part/Assembly to Insert box.

4. Double-click the **Shaft-Collar** part from the SOLIDWORKS 2018\Drawings folder. Three standard views are inserted on Sheet1.

5. If need, **modify** the Sheet scale to fit Sheet1. View the results. **Close** the drawing.

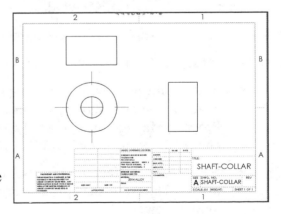

🔅 Modify the drawing view scale. Click inside the desired view. Click the **Use custom size** box. Enter the new scale.

Model View tool

The Model View 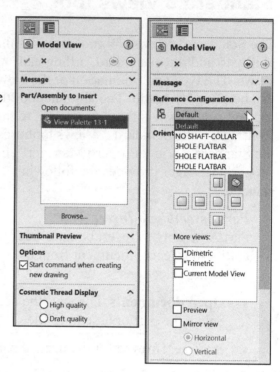 tool imports various dimensions, symbols and annotations that were used to create the model, and puts them into the drawing. Dimensions that come directly from the features or sketches are called driven dimensions. Dimensions in a SOLIDWORKS drawing are associated with the model, and changes in the model are reflected in the drawing.

The Model View PropertyManager is displayed when you insert or select a Model View, a Predefined View, an Empty View or when you drag a model with annotation views into a drawing. Note: Third Angle projection is illustrated and used in the book.

The available selections in the Model View are dependent on the type of view selected. The options are:

- *Part/Assembly to Insert*. The Part/Assembly to Insert box provides the ability to select a document from the following selections:

  - **Open documents**. Displays the open part or assembly files. Click to select an option document.

  - **Browse**. Browse to a part or assembly file.

The list of Open documents includes saved models, both parts and assemblies that are open in SOLIDWORKS windows, plus models that are displayed in drawing views.

- *Thumbnail Preview*. Displays a thumbnail view of the selected part or assembly.

- *Options*. The Options box provides the following selections:

  - **Start command when creating new drawing**. Selected by default. Only available when inserting a model into a new drawing. The Model View PropertyManager is displayed whenever you create a new drawing except if you check Make Drawing from Part/Assembly.

  - **Auto-start projected view**. Selected by default. Inserts projected views of the model. The Auto-start projected view option is displayed after you insert the model view.

- *Cosmetic Thread Display*. Sets either the High quality or Draft quality settings from the Model View PropertyManager for Assembly drawings only. The setting in the Cosmetic Thread Display, if different, will override the Cosmetic thread display option that is set in the Document Properties, Detailing section.

- *Reference Configuration*. Provides the ability to change drawing view configurations.

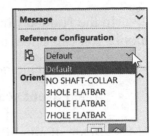

- *Select Bodies*. Provides the ability to select the bodies of a multi-body part for inclusion in the drawing view. For flat patterns of multi-body sheet metal parts, you can use one body per view.

- *Orientation*. The Orientation box provides the following selections:

  - **Create multiple views**. Ability to display multiple drawing views.

  - **Standard views**. Provides the following views: ***Front, *Back, *Top, *Bottom, *Right, *Left** and ***Isometric**.

  - **Annotation view**. Displays annotation views if created. The selections are ***Front, *Back, *Top, *Bottom, *Right, *Left** and ***Isometric**.

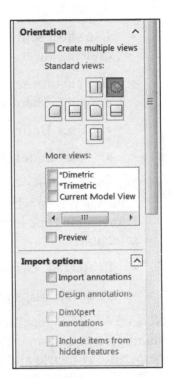

  - **More views**. This option is application dependent. Displays additional views such as **Current Model View, *Trimetric, *Dimetric, Annotation View 1** & **Annotation View 2**.

  - **Preview**. Only available with a single drawing view. Displays a preview of the model while inserting a view.

- *Import Options*. The available options are **Import annotations, Design annotations, DimXpert annotations** and **Include Items for hidden feature**.

- *Display State*. Only available for assemblies. Defines different combinations of settings for each component in an assembly. Saves the settings in Display States. The available settings are **Hide/Show state, Display Mode, Appearance** and **Transparency**.

- *Options*. Auto-start projected view is selected by default.

- *Display Style*. Provides the following view display styles: **Wireframe, Hidden Lines Visible, Hidden Lines Removed, Shaded With Edges** and **Shaded**.

The Use parent style check box option provides the ability to apply the same display style used from the parent.

In the Hidden Lines Visible or Hidden Lines Removed mode, you can select a style for Tangent Edge Display.

- *Scale*. The Scale box provides the following sections:

  - **Use Parent scale**. Applies the same scale used for the parent view. If you modify the scale of a parent view, the scale of all child views that use the parent scale are updated.

  - **Use sheet scale**. Applies the same scale used for the drawing sheet.

  - **Use custom scale**. Creates a custom scale for your drawing. There are two selections:

    - **User Defined**. Enter a scale value.

    - **Use model text scale**. Uses the scale of the model.

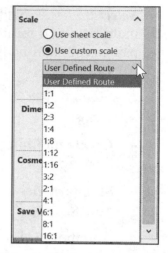

- *Dimension Type*. The Dimension Type box sets the dimension type when you insert a drawing view. The Dimension Type box provides two dimension type selections:

  - **Projected**. Displays 2D dimensions.

  - **True**. Displays accurate model values.

SOLIDWORKS specifies Projected type dimensions for standard and custom orthogonal views and True type for Isometric, Dimetric and Trimetric views.

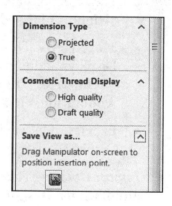

Move a dimension from one view to another view. Apply the Shift key to click and drag the required dimension into the new view.

Use the Model Break View tool to create configuration-based 3D break views (also known as interrupted views) of a model for individual drawing views.

Open a part from an active drawing view. Right-click inside the drawing view, click Open Part from the Context toolbar.

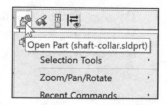

- **Cosmetic Thread Display**. Sets either the High quality or Draft quality settings from the Model View PropertyManager. The setting in the Cosmetic Thread Display, if different, will override the Cosmetic thread display option that is set in the Document Properties, Detailing section. The selections are:

  - **High quality**. Displays precise line fonts and trimming in cosmetic threads. If a cosmetic thread is only partially visible, the High quality option will display only the visible portion.

  - **Draft quality**. Displays cosmetic threads with less detail. If a cosmetic thread is only partially visible, the Draft quality option will display the entire feature.

- **Save View as**. Drag the Manipulator on-screen to position the insertion point.

  - **More Properties**. The More Properties button provides the ability to modify the bill or materials information, show hidden edges, etc. after a view is created.

Tutorial: Model View 13-1

Create a new drawing using the Model View tool.

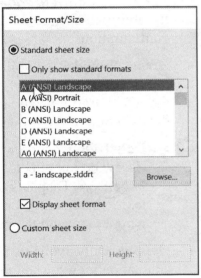

1. Create a **New** ⬜ drawing. Use the ANSI Drafting standard, Third Angle projection, A (ANSI) Landscape default drawing template.

2. Click **OK** from the Sheet Format/Size dialog box. The Model View 🔎 PropertyManager is displayed.

3. Click the **Browse** button.

4. Click **Shaft-Collar** from the SOLIDWORKS 2018\Drawings folder.

5. Check **Multiple views** from the Number of Views box.

6. Click ***Isometric**, ***Top** and ***Right** view. Front view should be selected by default.

7. Click **Shaded With Edges** from the Display Style box.

8. Check the **Use custom scale** box.

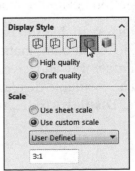

9. Select **User Defined**.

10. Enter **3:1**.

11. Click **OK** ✅ from the Model View PropertyManager.

12. **Close** the Drawing.

Projected View tool

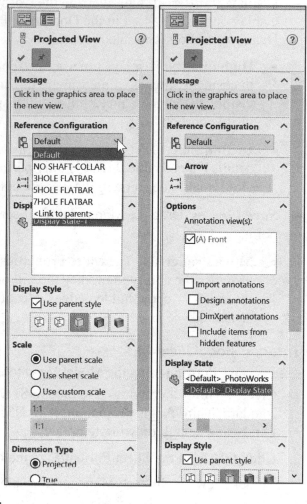

The Projected View ▦ tool adds a projected view by unfolding a new view from an existing view. The Projected View tool uses the Projected View PropertyManager. The Projected View PropertyManager is displayed when you create a Projected View in a drawing, or when you select an existing Projected View. The Projected View PropertyManager provides the following selections:

- **Reference Configuration**. Provides the ability to change drawing view configurations.

- **Arrow**. Provides the ability to display a view arrow, or a set of arrows in the ANSI drafting standard indicating the direction of the projection.

  - **Label**. Only active when the Arrow box is selected. The first Label is A. Displays the entered text to be displayed with both the parent view and the projected view.

- **Options**. Only available if the model was created with annotation views. The Options box provides the following selections.

  - **Annotation view(s)**. Select an annotation view, if the model was created with annotation views. The view will include annotations from the model.

- **Display State**. Only available for assemblies. Provides the ability to define different combinations of settings for each component in an assembly and saves these settings in Display States. The available display state settings are **Hide/Show state**, **Display Mode**, **Appearance** and **Transparency**.

- *Display Style*. Displays drawing views in the following modes. They are **Wireframe, Hidden Lines Visible, Hidden Lines Removed, Shaded With Edges** and **Shaded**.

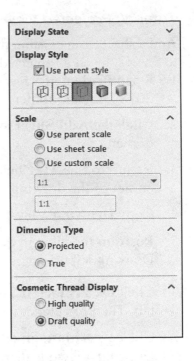

The Use parent style check box option provides the ability to apply the same display style used from the parent.

When in the Hidden Lines Visible or Hidden Lines Removed mode, you can select a style for Tangent Edge Display.

- *Scale*. The Scale box provides the following sections:

  - **Use Parent scale**. Applies the same scale used for the parent view. If you modify the scale of a parent view, the scale of all child views that use the parent scale are updated.

  - **Use sheet scale**. Applies the same scale used for the drawing sheet.

  - **Use custom scale**. Create a custom scale for your drawing. There are two selections:

    - **User Defined**. Enter a scale value.

    - **Use model text scale**. Uses model scale.

- *Dimension Type*. Sets the dimension type when you insert a drawing view. The Dimension Type box provides two dimension type selections:

  - **Projected**. Displays 2D dimensions.

  - **True**. Displays accurate model values.

SOLIDWORKS specifies Projected type dimensions for standard and custom orthogonal views and True type for Isometric, Dimetric and Trimetric views.

- *Cosmetic Thread Display*. Set either the High quality or Draft quality settings from the Model View PropertyManager. The setting in the Cosmetic Thread Display, if different, will override the Cosmetic thread display option that is set in the Document Properties, Detailing section. The selections are:

  - **High quality**. Displays precise line fonts and trimming in cosmetic threads. If a cosmetic thread is only partially visible, the High quality option will display only the visible portion.

  - **Draft quality**. Displays cosmetic threads with less detail. If a cosmetic thread is only partially visible, the Draft quality option will display the entire feature.

Tutorial: Projected View 13-1

Add a sheet to an existing drawing. Copy a drawing from Sheet1 to Sheet2. Insert a Projected View. Modify the Title box.

1. Open **Projected View 13-1** from the SOLIDWORKS 2018\Drawing folder. Add a Sheet to the drawing.

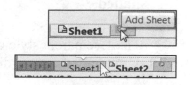

2. Click the **Add Sheet** icon in the bottom left corner of the drawing.

3. Return to Sheet1. Click the **Sheet1** tab in the bottom left corner of the drawing.

4. Copy **Drawing View1** from Sheet1 to Sheet2.

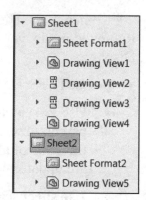

5. **Position** the view in the lower left corner of Sheet2. Drawing View5 is created.

6. Click the **Projected View** ⊞ tool from the View Layout tab. The Projected View PropertyManager is displayed.

7. Click a **position** directly to the right of Drawing View5. Drawing View6 is displayed.

8. Click **OK** ✔ from the Projected View PropertyManager.

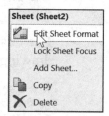

9. Right-click **Edit Sheet Format** in the Graphics window.

10. Double-click **Projected View 13-1** in the DWG NO. box.

11. Select **12** font from the Formatting dialog box.

12. Click **OK** ✔ from the Note PropertyManager.

13. Right-click in the **Graphics window**.

14. Click **Edit Sheet**. Return to Sheet1.

15. **Close** the drawing.

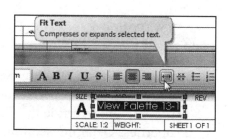

💡 Select the **Fit text** option from the Formatting dialog box to compress or expand selected text using control points.

Auxiliary View tool

The Auxiliary View  tool adds a view by unfolding a new view from a linear entity. Example: Edge, sketch entity, etc. An Auxiliary View is similar to a Projected View, but it is unfolded normal to a reference edge in an existing view. The Auxiliary View tool uses the Auxiliary View PropertyManager. The Auxiliary View PropertyManager provides the following selections:

- *Reference Configuration*. Provides the ability to change drawing view configurations.

- *Arrow*. The Arrow box provides the ability to display a view arrow or a set of arrows in the ANSI drafting standard indicating the direction of the projection.

  - **Label**. Only active when the Arrow box is selected. The first Label is A. Displays the entered text to be displayed with both the parent view and the auxiliary view.

- *Options*. Only available if the model was created with annotation views. The Options box provides the following selections.

  - **Annotation view(s)**. Select an annotation view, if the model was created with annotation views. The view will include annotations from the model.

- *Display State*. Only available for assemblies. Defines different combinations of settings for each component in an assembly and saves these settings in the Display State. The available settings are **Hide/Show state**, **Display Mode**, **Appearance** and **Transparency**.

- *Display Style*. Displays drawing views in the following modes. They are **Wireframe**, **Hidden Lines Visible**, **Hidden Lines Removed**, **Shaded With Edges** and **Shaded**.

The Use parent style check box option provides the ability to apply the same display style used from the parent.

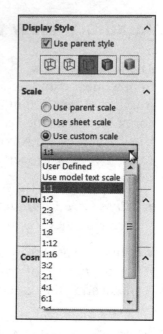

☀ Reference dimensions show measurements of the model, but they do not drive the model and you cannot change their values. However, when you change the model, the reference dimensions update accordingly. Reference dimensions are enclosed in parentheses by default.

- *Scale*. The Scale box provides the following sections:

  - **Use Parent scale**. Applies the same scale used for the parent view. If you modify the scale of a parent view, the scale of all child views that use the parent scale are updated.

  - **Use sheet scale**. Applies the same scale used for the drawing sheet.

  - **Use custom scale**. Create a custom scale for your drawing. There are two selections:

    - **User Defined**. Enter a scale value.

    - **Use model text scale**. Uses the model scale.

- *Dimension Type*. Sets the dimension type when you insert a drawing view. The Dimension Type box provides two dimension type selections:

  - **Projected**. Displays 2D dimensions.

  - **True**. Displays accurate model values.

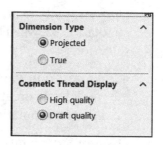

☀ SOLIDWORKS specifies Projected type dimensions for standard and custom orthogonal views and True type for Isometric, Dimetric and Trimetric views.

- *Cosmetic Thread Display*. Sets either the High quality or Draft quality settings from the Model View PropertyManager. The setting in the Cosmetic Thread Display, if different, will override the Cosmetic thread display option that was set in the Document Properties, Detailing section. The selections are:

  - **High quality**. Displays precise line fonts and trimming in cosmetic threads. If a cosmetic thread is only partially visible, the High quality option will display only the visible portion.

  - **Draft quality**. Displays cosmetic threads with less detail. If a cosmetic thread is only partially visible, the Draft quality option will display the entire feature.

Tutorial: Auxiliary View 13-1

Create an Auxilary view from an existing Front view. Reposition the view.

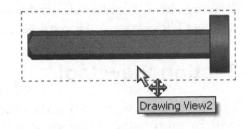

1. Open **Auxiliary View 13-1** from the SOLIDWORKS 2018\Drawing folder.

2. Click inside the **Front view** boundary of Drawing View2.

3. Click the **Auxiliary View**  drawing tool. The Auxiliary View PropertyManager is displayed.

4. Click the **right vertical edge** of Drawing View2 as illustrated. Hold the **Ctrl** key down.

5. Click a **position** above Drawing View2.

6. Release the **Ctrl** key. Flip the arrows if required.

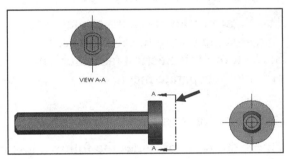

7. Drag the **View A-A** text below the Drawing View3 view boundary.

8. Drag the **A-A arrow** as illustrated.

9. Click **OK** ✔ from the Auxiliary View PropertyManager.

10. **Rebuild** the drawing.

11. **Close** the drawing.

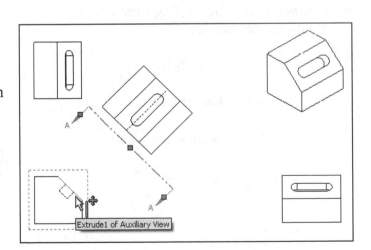

Tutorial: Auxiliary View 13-2

Create an Auxilary view from an existing Front view.

1. Open **Auxiliary View 13-2** from the SOLIDWORKS 2018\Drawing folder.

2. Click the **angled edge** in Drawing View1 as illustrated.

3. Click the **Auxiliary View** drawing tool.

4. Click a **position** up and to the right.

5. Drag the **A-A arrow** as illustrated.

6. Drag the **text** off the view.

7. Click **OK** ✔ from the Auxiliary View PropertyManager.

8. **Close** the drawing.

💡 You can set auxiliary views and section views to be orthographically aligned to the drawing sheet. Right-click the drawing view and click Align Drawing View ➤ Horizontal to Sheet Clockwise or Horizontal to Sheet Counterclockwise.

Section View tool

The Section View ↕ tool adds a section view by cutting the parent view with a section line. Use the section view sketch mode in conjunction with the Section tool user interface to create both Section views and Aligned Section views.

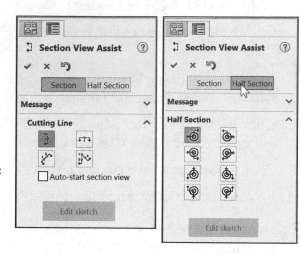

The section line can also include concentric arcs. The Section View tool uses the Section View PropertyManager.

The Section View PropertyManager provides the ability to select either the **Section** or **Half Section** tab. Each tab provides a separate menu.

Section View tool: Section tab

The Section tab provides the following selections:

- *Cutting Line*. The Cutting Line box provides the following selections:

    - **Vertical Cutting line** ↕.

    - **Horizontal Cutting line** ↔.

    - **Auxiliary Cutting line** ⟋.

    - **Aligned Cutting line** ⤧.

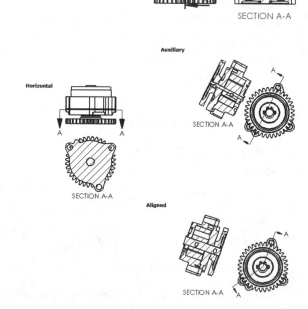

The Section View pop-up menu provides the ability to add offsets to the Section view. The following options are available in the pop-up menu:

- **Add Arc Offset** 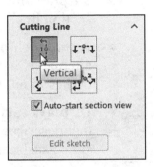. Select first point of arc on cutting line, then select second point of arc.

- **Add Single Offset** . Select first point of offset on cutting line, then select second point of offset.

- **Add Notch Offset** . Select first point of notch on cutting line, select second point on cutting line for width of notch, then select third point for depth of notch.

- **Step Back** . Previous step.

- **OK** . Adds the view.

- **Cancel** . Cancels the view.

- *Edit Sketch*.

  - **Edit Sketch** button. Select to create a custom cutting line.

Section View tool: Half Section tab

The Half Section tab provides the following Half Sections selections:

- **Topside Right** Half Section .

- **Topside Left** Half Section .

- **Bottomside Right** Half Section .

- **Bottomside Left** Half Section .

- **Leftside Down** Half Section .

- **Rightside Down** Half Section .

- **Leftside Up** Half Section .

- **Rightside Up** Half Section .

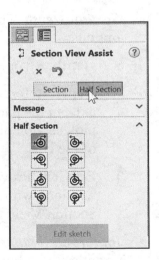

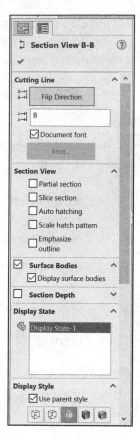

- *Section View*. The Section View box provides the following selections:

  - **Partial section**. If the section line does not completely cross the view, the Partial section option provides the ability to display a message which states, "The section line does not completely cut through the bounding box of the model in this view. Do you want this to be a partial section cut?" There are two selections:

    - **Yes**. The section view is displayed as a partial section view.

    - **No**. The section view is displayed with not cut.

  - **Display only cut faces(s)**. Displays the faces cut by the section line.

  - **Auto hatching**. Provides the ability to crosshatch patterns alternately between components in assemblies, or between bodies in multi-body parts and weldments.

  - **Displays surface bodies**. Surface body is a general term that describes connected zero-thickness geometries such as single surfaces, knit surfaces, trimmed and filleted surfaces, and so on.

- *Section Depth*. The Section Depth box provides the following options:

  - **Depth**. Enter Section depth.

  - **Depth Reference**. Select geometry, such as an edge or an axis, in the parent view for the Depth Reference. Drag the pink section plane in the Graphics window to set the depth of the cut.

  - **Preview**. Provides the ability to view how the section view will look with the section depth settings before you close the Section View PropertyManager.

- *Import annotation from*. The Import annotation from box provides the following options:

  - **Annotation views** box.

  - **Import annotations**. Selected by default.

  - **Design annotations**.

  - **DimXpert annotations**. Selected by default.

  - **Include items from hidden features**.

- ***Display State***. Only available for assemblies. Defines different combinations of settings for each component in an assembly and saves these settings in Display States. The available settings are **Hide/Show state**, **Display Mode**, **Appearance** and **Transparency**.

- ***Display Style***. Displays drawing views in the following style modes. They are **Wireframe**, **Hidden Lines Visible**, **Hidden Lines Removed**, **Shaded With Edges** and **Shaded**.

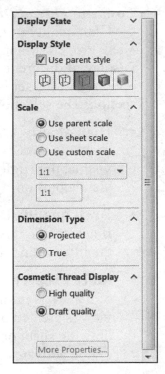

The Use parent style check box option provides the ability to apply the same display style used from the parent.

When in the Hidden Lines Visible or Hidden Lines Removed mode, you can select a style for Tangent Edge Display.

- ***Scale***. The Scale box provides the following sections:

    - **Use Parent scale**. Applies the same scale used for the parent view. If you modify the scale of a parent view, the scale of all child views that use the parent scale is updated.

    - **Use sheet scale**. Applies the same scale used for the drawing sheet.

    - **Use custom scale**. Creates a custom scale for your drawing. There are two selections:

        - **User Defined**. Enter a scale value.

        - **Use model text scale**. Uses the model scale.

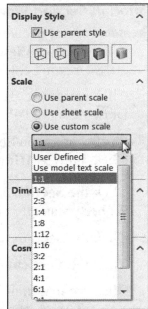

- ***Dimension Type***. Set the dimension type when you insert a drawing view. The Dimension Type box provides two dimension type selections:

    - **Projected**. Displays 2D dimensions.

    - **True**. Displays accurate model values.

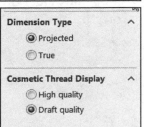

SOLIDWORKS specifies Projected type dimensions for standard and custom orthogonal views and True type for Isometric, Dimetric and Trimetric views.

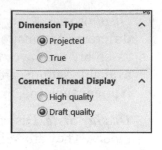

- ***Cosmetic Thread Display***. Provides the ability to set either the High quality or Draft quality settings from the Model View PropertyManager. The setting in the Cosmetic Thread Display, if different, will override the Cosmetic thread display option that was set in the Document Properties, Detailing section. The selections are:

  - **High quality**. Displays precise line fonts and trimming in cosmetic threads. If a cosmetic thread is only partially visible, the High quality option will display only the visible portion.

  - **Draft quality**. Displays cosmetic threads with less detail. If a cosmetic thread is only partially visible, the Draft quality option will display the entire feature.

  - **More Properties**. Provides the ability to modify the bill of materials information, show hidden edges, etc. after a view is created or by selecting an existing view from the drawing sheet or FeatureManager. The More Properties option activates the Drawing View Properties dialog box.

Tutorial: Section View 13-1

Create a vertical Section view and modify the drawing view scale.

1. Open **Section View 13-1** from the SOLIDWORKS 2018\Drawing folder.

2. Click inside the **Drawing View1** view boundary. The Drawing View1 PropertyManager is displayed.

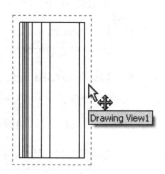

Display the origin on Sheet1.

3. Click **View** ➤ **Hide/Show** ➤ **Origins** from the Menu bar.

4. Click the **Section View** ⇱ drawing tool. The Section View PropertyManager is displayed.

5. Click the **Section** tab.

6. Click the **Vertical** Cutting Line button.

7. Click the **origin** as illustrated. Note: You can select the midpoint vs. the origin.

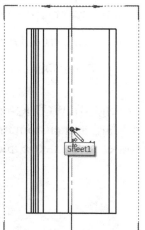

Place the Section view.

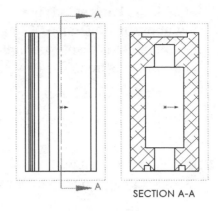

8. Click a **position** to the right of Drawing View1. The Section arrows point to the right. If required, click Flip direction.

9. Check **Auto hatching** from the Section View box.

10. Check the **Shaded With Edges** option from the Display Style box.

11. Click **OK** ✔ from the Section View A-A PropertyManager. Section View A-A is created and is displayed in the Drawing FeatureManager.

SECTION A-A

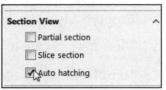

12. Click inside the **Drawing View1** view boundary. The Drawing View1 PropertyManager is displayed.

13. Modify the Scale to **1.5:1**.

14. Click **OK** ✔ from the Drawing View1 PropertyManager. Both drawing views are modified (parent and child).

15. **Close** the drawing.

Tutorial: Aligned Section View 13-1

Create an Aligned Section view.

1. Open **Aligned Section View 13-1** from the SOLIDWORKS 2018\Drawing folder.

2. Click inside the **Drawing View1** view boundary. The Drawing View1 PropertyManager is displayed.

3. Click the **Section View** ⇄ drawing tool. The Section View PropertyManager is displayed.

4. Click the **Half Section** tab.

5. Click the **Leftside Down** Half Section button.

6. Select the **center** of the drawing view as illustrated. The Section View dialog box is displayed.

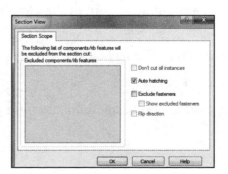

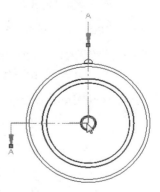

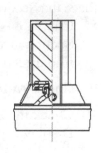

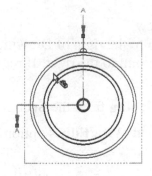

7. Click the **LENSCAP** in the Graphics window as illustrated. The LENSCAP component is displayed in the Excluded components box.

8. Check the **Auto hatching** box.

9. Click **OK** from the Section View dialog box.

Place the Section View.
10. Click a **position** above Drawing View1.

SECTION A-A
SCALE 1 : 4

11. Click **OK** ✔ from the Section View A-A PropertyManager.

12. Modify the **scale** to **1:4**.

13. **Rebuild** 🔳 the drawing.

14. **Move** the drawing views to fit the Sheet. View the drawing FeatureManager.

15. **Close** the drawing.

🔅 Use Section View Assist, previously called the Section View User Interface, to add offsets to existing section views. To edit existing section views with Section View Assist: Right-click the section view or its cutting line and click **Edit Cutting Line.**

Tutorial: Copy/Paste 13-1

Apply the Copy/Paste tool with the After selected sheet option.
1. Open **Copy-Paste 13-1** from the SOLIDWORKS 2018\Drawing folder.

2. Right-click the **Sheet1** tab at the bottom of the Graphics window. Click **Copy**.

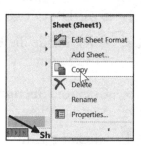

3. Right-click the **Sheet1 tab** at the bottom of the Graphics window.

4. Click **Paste**. The Insert Paste dialog box is displayed.

5. Check the **After selected sheet** box.

6. Click **OK**. View the results. You copied the drawing view from Sheet1 to Sheet2.

7. **Close** the model.

Detail View tool

The Detail View 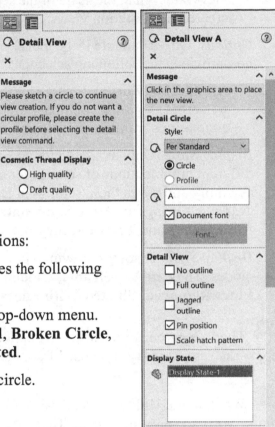 tool provides the ability to add a detail view to display a portion of a view, usually at an enlarged scale. Create a detail view in a drawing to display or highlight a portion of a view. This detail may be of an orthographic view, a non-planar (isometric) view, a section view, a crop view, an exploded assembly view, or another detail view.

The Detail View tool uses the Detail View PropertyManager. The Detail View PropertyManager provides the following selections:

- *Detail Circle*. The Detail Circle box provides the following selections:

  - **Style**. Select a display style from the drop-down menu. There are five selections: **Per Standard, Broken Circle, With Leader, No Leader** and **Connected**.

  - **Circle**. Selected by default. Displays a circle.

  - **Profile**. Displays a profile.

  - **Label**. Provides a letter associated with the section line and section view. The first Label is A by default.

To specify the label format, click **Options** ⚙ ➤ **Document Properties** ➤ **View Labels** from the Menu bar.

  - **Document font**. Uses the document font for the section line label.

  - **Font**. Un-check the Document font check box to use the Font button. Provides the ability to choose a font for the section line label other than the document's font.

- *Detail View*. The Detail View box provides the following selections:

  - **Full outline**. Displays the profile outline in the detail view.

  - **Pin position**. Selected by default. Keeps the detail view in the same relative position on the drawing sheet if you modify the scale of the view.

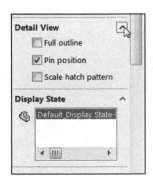

- **Scale hatch pattern**. Displays the hatch pattern based on the scale of the detail view rather than the scale of the section view. This option applies to detail views created from section views.

- *Options*. Only available if the model was created with annotation views. The Options box provides the following selection:

  - **Annotation view(s)**. Select an annotation view, if the model was created with annotation views. The view will include annotations from the model.

- *Display State*. Only available for assemblies. Defines the different combinations of settings for each component in an assembly and saves these settings in Display States. The available settings are **Hide/Show state**, **Display Mode**, **Appearance** and **Transparency**.

- *Display Style*. Displays drawing views in the following style modes. They are **Wireframe**, **Hidden Lines Visible**, **Hidden Lines Removed**, **Shaded With Edges** and **Shaded**.

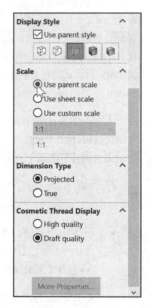

The Use parent style check box option provides the ability to apply the same display style used from the parent.

When in the Hidden Lines Visible or Hidden Lines Removed mode, you can select a style for Tangent Edge Display.

- *Scale*. The Scale box provides the following sections:

  - **Use parent scale**. Applies the same scale used for the parent view. If you modify the scale of a parent view, the scale of all child views that use the parent scale is updated.

  - **Use sheet scale**. Applies the same scale used for the drawing sheet.

  - **Use custom scale**. Creates a custom scale for your drawing. There are two selections:

    - **User Defined**. Enter a scale value.

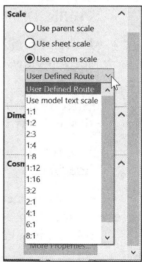

- *Dimension Type*. Sets the dimension type when you insert a drawing view. The Dimension Type box provides two dimension type selections:

  - **Projected**. Displays 2D dimensions.

  - **True**. Displays accurate model values.

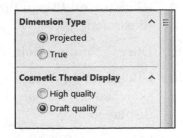

SOLIDWORKS specifies Projected type dimensions for standard and custom orthogonal views and True type for Isometric, Dimetric and Trimetric views.

- *Cosmetic Thread Display*. Sets either the High quality or Draft quality settings from the Model View PropertyManager. The setting in the Cosmetic Thread Display, if different, will override the Cosmetic thread display option that was set in the Document Properties, Detailing section. The selections are:

  - **High quality**. Displays precise line fonts and trimming in cosmetic threads. If a cosmetic thread is only partially visible, the High quality option will display only the visible portion.

  - **Draft quality**. Displays cosmetic threads with less detail. If a cosmetic thread is only partially visible, the Draft quality option will display the entire feature.

- *Save View as...* Drag the Manipulator on-screen to position the insertion point.

  - **More Properties**. The More Properties button provides the ability to modify the bill or materials information, show hidden edges, etc. after a view is created or by selecting an existing view from the drawing sheet or FeatureManager. The More Properties option activates the Drawing View Properties dialog box.

Tutorial: Detail View 13-1

Create a Detail view from an existing view.

1. Open **Detail View 13-1** from the SOLIDWORKS 2018\Drawing folder.

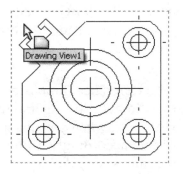

2. Click inside the **Drawing View1** view boundary. The Drawing View1 PropertyManager is displayed.

3. Click the **Detail View** drawing tool. The Circle Sketch tool is activated.

4. Click the **middle** of the Switch Grove in the Front view as illustrated.

5. Drag the **mouse pointer** outward.

6. Click a **position** just below the large circle to create the sketched circle. The Detail View PropertyManager is displayed.

7. Click a **position** to the left of DrawingView1.

8. Check the **Use custom scale** box.

9. Select **User Defined**.

10. Enter **3:1** in the Custom Scale text box.

11. Click **Hidden Lines Visible** from the Display Style box.

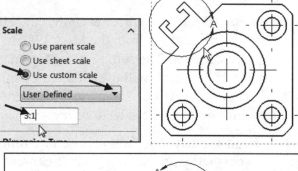

12. Click **OK** ✔ from the Detail View A PropertyManager.

13. Drag the **text** off the profile lines.

14. **Rebuild** the drawing.

15. **Close** the drawing.

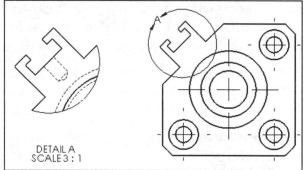

The Hole Wizard information from a model propagates to the drawing when the Hole Callout button is selected.

Use the Smart Dimension tool to input drawing dimensions. Typically, you create dimensions as you create each part feature, then insert those dimensions into the various drawing views. Changing a dimension in the model updates the drawing and changing an inserted dimension in a drawing changes the model.

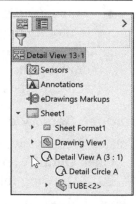

Reference dimensions show measurements of the model, but they do not drive the model and you cannot change their values. However, when you change the model, the reference dimensions update accordingly. Reference dimensions are enclosed in parentheses by default.

Broken-out Section tool

The Broken-out Section ⬚ tool provides the ability to add a broken-out section to an existing view exposing inner details of a model. A broken-out section is part of an existing drawing view, not a separate view. A closed profile, usually a spline, defines the broken-out section. Material is removed to a specified depth to expose inner details. The Broken-out Section tool uses the Broken-out Section PropertyManager. The Broken-out Section PropertyManager provides the following selections:

- *Depth*. The Depth box provides the following options:

  - **Depth Reference**. Displays the selected depth reference geometry, such as an edge or an axis from the Graphics window.

  - **Depth**. Enter a value for the depth.

  - **Preview**. Displays the broken-out section as you change the depth. When un-checked, the broken-out section is applied when you exit from the Broken-out Section PropertyManager.

  - **Auto hatching**. Only for assemblies. Automatically adjusts for neighboring components to alternate crosshatch patterns in 90 degree increments.

  - **Exclude fasteners**. Only for assemblies. Excludes fasteners from being sectioned. Fasteners include any item inserted from SOLIDWORKS Toolbox (nuts, bolts, washers, etc.).

💡 Use the 3D drawing view 🔖 tool to select an obscured edge for the depth. The 3D drawing view tool provides the ability to rotate a drawing view out of its plane so you can view components or edges obscured by other entities. 3D drawing view mode is not available for *Detail, Broken, Crop and Empty or Detached views.*

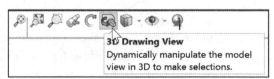

Tutorial: Broken-out Section View - 3D drawing View 13-1

Create a Broken-out Section view using the Spline Sketch tool and utilize the 3D drawing view tool.

1. Open **Broken-out Section View 13-1** from the SOLIDWORKS 2018\Drawing folder.

2. Click inside the **Drawing View1** view boundary. The Drawing View1 PropertyManager is displayed. Apply the 3D drawing view tool.

3. Click the **3D drawing view** 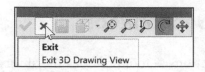 tool from the View toolbar.

4. Click the **Rotate** ↻ icon from the View toolbar.

5. **Rotate** the view with the middle mouse button. View the results.

6. Click the red **Exit** ✖ button to exit the 3D drawing view mode.

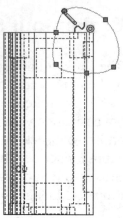

7. Click the **Broken-out Section** drawing tool. The Broken-out Section PropertyManager is displayed.

8. Sketch a **closed Spline** in the top right corner as illustrated.

9. Enter **.3**in for Depth.

10. Check **Preview**. View the Broken-out Section preview in the Graphics window.

11. Click **OK** ✔ from the Broken-out Section PropertyManager. View the Drawing FeatureManager.

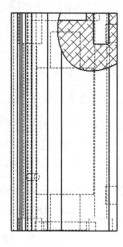

12. **Close** the drawing.

💡 You cannot create a broken-out section on a detail, section, or alternate position view. If you create a broken-out section on an exploded view, the view is no longer exploded.

Break tool

The Break 〽 tool provides the ability to add a break line to a selected view. Create a broken (or interrupted) view in a drawing. Use the Broken view to display the drawing view in a larger scale on a smaller size drawing sheet. Reference dimensions and model dimensions associated with the broken area reflect the actual model values. The Break tool uses the Broken View PropertyManager.

The Broken View PropertyManager provides the following selections:

- *Broken View Settings*. The Broken View Settings box provides the following options:

  - **Add vertical break line**. Selected by default. Adds a vertical break line.

  - **Add horizontal break line**. Adds a horizontal break line.

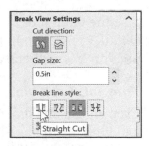

- **Gap size**. Sets the value of the space between the gap.

- **Break line style**. The Break line style option defines the break line type. Zig Zag Cut selected by default. There are four options to select from. They are **Straight Cut**, **Curve Cut**, **Zig Zag Cut**, **Small Zig Zag Cut and Jagged Cut**.

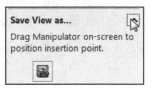

- *Save View as*. Drag the Manipulator on-screen to position the insertion point.

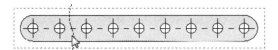

Tutorial: Break View 13-1

Insert a Broken view in an existing drawing view. Use the Break drawing tool. Note the available Break line options.

1. Open **Break View 13-1** from the SOLIDWORKS 2018\Drawing folder.

2. Click **inside** the Drawing View1 view boundary. The Drawing View1 PropertyManager is displayed.

3. Click the **Break** drawing tool. The Broken View PropertyManager is displayed.

4. Select **Add vertical break line** from the Broken View Settings box.

5. Select **Curve Cut** from the Break line style box.

6. Enter **.4**in for Gap.

7. Click the **two locations** as illustrated in the Graphics window.

8. Click **OK** from the Broken View PropertyManager. View the Drawing FeatureManager.

9. **Close** the drawing.

Crop tool

The Crop tool provides the ability to crop an existing drawing view. You cannot use the Crop tool on a Detail View, a view from which a Detail View has been created, or an exploded view.

Use the Crop tool to save steps. Example: instead of creating a Section View and then a Detail View, then hiding the unnecessary Section View, use the Crop tool to crop the Section View directly. The Crop tool does not use a PropertyManager.

Tutorial: Crop view 13-1

Crop an existing drawing view. Apply the Crop View drawing tool with the Spline Sketch tool.

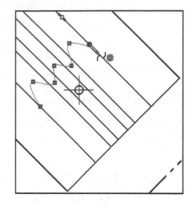

1. Open **Crop View 13-1** from the SOLIDWORKS 2018\Drawing folder.

2. Click inside the **Drawing View2** view boundary. The Drawing View2 PropertyManager is displayed.

3. Click the **Spline** ∿ Sketch tool. The Spline PropertyManager is displayed.

4. Sketch **7 control points** as illustrated.

5. Right-click **End Spline**.

6. Click the **Line** ⁄ Sketch tool. The Insert Line PropertyManager is displayed.

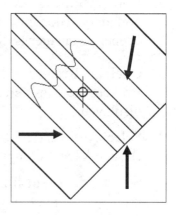

7. Sketch **three lines** as illustrated. The first line, the first point must display Endpoint interference with the first point of the spline. The second line is collinear with the bottom edge of the view. The third line, the last point must display Endpoint interference with the last point of the spline.

Deselect the active Line Sketch tool.

8. Right-click **Select** in the Graphics window.

9. Window-Select the **three lines** and the **Spline**. The selected sketch entities are displayed in the Properties PropertyManager. Delete any unwanted sketch entities.

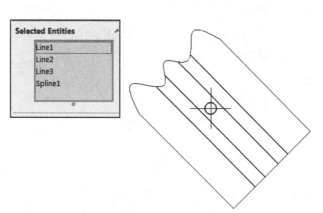

10. Click the **Crop View** ⊡ drawing tool. The selected view is cropped.

11. Click **OK** ✔ from the Properties PropertyManager. View the Drawing FeatureManager.

12. **Close** the drawing.

Alternate Position View tool

The Alternate Position View 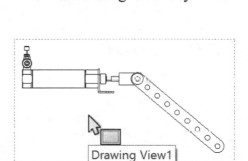 tool provides the ability to superimpose an existing drawing view precisely on another. The alternate position is displayed with phantom lines. Use the Alternate Position View to display the range of motion of an assembly. You can dimension between the primary view and the Alternate Position View. You cannot use the Alternate Position View tool with Broken, Section, or Detail views. The Alternate Position PropertyManager provides the following selections:

- *Configuration box*. The Configuration box provides the following selections:

  - **New configuration**. Selected by default. A default name is displayed in the configuration box. Accept the default name or type a name of your choice. Click OK from the Alternate Position PropertyManager. The assembly is displayed with the Move Component PropertyManager. See section on the Move Component PropertyManager for additional details.

  - **Existing configuration**. Provides the ability to select an existing assembly configuration that is displayed in the drop down arrow box. Click OK from the Alternate Position PropertyManager. The alternate position of the selected configuration is displayed in the drawing view. The Alternate Position view is complete.

Tutorial: Alternate Position View 13-1

Create an Alternate Position view in an existing assembly drawing view.

1. Open **Alternate Position View 13-1** from the SOLIDWORKS 2018\Drawing folder.

2. Click inside the **Drawing View1** view boundary.

3. Click the **Alternate Position View** 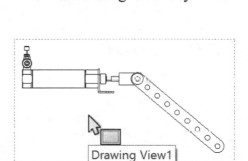 drawing tool. The Alternate Position PropertyManager is displayed.

4. Click **New configuration** from the Configuration box. Accept the default name.

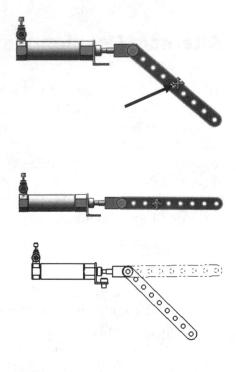

5. Click **OK** ✔ from the Alternate Position PropertyManager. If the assembly document is not already open, it opens automatically. The assembly is displayed with the Move Component PropertyManager. Free drag is selected by default in the Move box.

6. Click the **flatbar** in the Graphics window as illustrated.

7. Drag the **flatbar** upward to create the Alternate position.

8. Click **OK** ✔ from the Move Component PropertyManager. The Alternate Position view is displayed in the Graphics window.

9. **Close** the drawing.

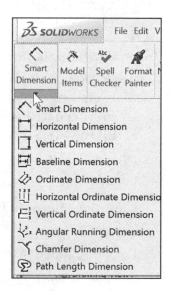

💡 To display an Exploded view of an assembly in a drawing, right-click inside the view boundary. Click Show in Exploded State. You first need an Exploded configuration of the assembly.

Annotation Toolbar

Add annotations to a drawing document using the Annotation toolbar from the CommandManager.

You can add most annotation types in a part or assembly document, and then insert them into a drawing document. However, there are some types, such as Center Marks and Area Hatch that you can only add in a drawing document.

Annotations behave like dimensions in a SOLIDWORKS document. You can add dimensions in a part or assembly document, then insert the dimensions into the drawing, or create dimensions directly in the drawing.

💡 The Smart Dimension drop-down menu provides access to the following options: *Smart Dimension, Horizontal Dimension, Vertical Dimension, Baseline Dimension, Ordinate Dimension, Horizontal Ordinate Dimension, Vertical Ordinate Dimension, Chamfer Dimension* and *Path Length Dimension*.

Smart Dimension tool

The Smart Dimension ✎ tool provides the ability to create a dimension for one or more selected entities in the drawing. The Smart Dimension tool uses the Dimension PropertyManager or the Autodimension PropertyManager. The Dimension PropertyManager provides the ability to select either the **DimXpert** or **Autodimension** tab. Each tab provides a separate menu. Note: The DimXpert tab is selected by default.

Smart Dimension tool: DimXpert tab

The DimXpert tab provides the following selections:

- *Dimension Assist Tools*. The Dimension Assist Tools box provides the following selections:

  - **Smart dimensioning**. Default setting. Provides the ability to create dimensions with the Smart Dimension tool. View Chapter 5 on Smart dimensions for detail command and PropertyManager information.

  - **Rapid dimensioning**. Provides the ability to enable or disable the rapid dimension manipulator. Select to enable; clear to disable. This setting persists across sessions.

  - **DimXpert**. Provides the ability to apply dimensions to fully define manufacturing features such as fillets, patterns, slots, etc. and locating dimensions. When selected, the following selections are available:

- *Style*. Provides the ability to apply default attributes to a dimension, add, delete, save or load a style.

- *Pattern Scheme*. The Pattern Scheme box provides the ability to apply a select a dimensioning scheme. The two pattern schemes are:

  - **Polar dimensioning**. Creates a scheme consisting of a callout and sometimes an angular locating dimension.

  - **Linear dimensioning**. Selected by default. Creates linear dimensions for a pattern.

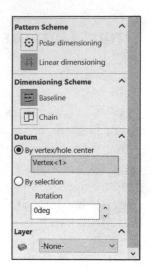

- **Dimensioning Scheme**. Baseline selected by default. The Dimensioning Scheme box provides the ability to select a dimensioning scheme. The two dimensioning schemes are: **Baseline** and **Chain**.

- **Datum**. The Datum box provides the following selections:

  - **By Vertex/hole center**. Measures the manufacturing features and the locating dimensions from the selected vertex in the Graphics window.

  - **By selection**. Measures the manufacturing features and the locating dimensions. Set the following selections:

  - **X**. Select the X edge to create a virtual point.

  - **Y**. Select the Y edge to create a virtual point.

  - **Rotation**. Set the angle or drag the origin to rotate if required.

- **Layer**. Provides the ability to select a drawing layer.

If the X and Y edges intersect, you must use the By vertex/hole center option.

Tutorial: Smart Dimension 13-1

Dimension a part drawing view using the Smart Dimension tool with the DimXpert option.

1. Open **Smart Dimension 13-1** from the SOLIDWORKS 2018\Drawing folder.

2. Click the **Smart Dimension** ⟍ tool. The Smart Dimension PropertyManager is displayed.

3. Click the **DimXpert** tab from the Dimension Assist Tools box. The Dimension PropertyManager is displayed.

4. Click **Linear dimensioning** for Pattern Scheme.

5. Click **Baseline** for Dimensioning Scheme.

6. Check the **By vertex/hole center** box.

7. Click the **Axis** of the left most hole. Axis<1> is displayed in the Datum box.

8. Click the **circular edge** of the left most hole.

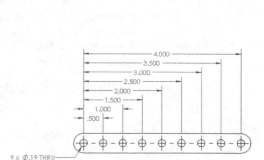

9. Click **OK** to the Warning message. The hole dimension is displayed.

10. Click **OK** ✔ from the Dimension PropertyManager. The dimensions are displayed on Drawing View1. **Close** the drawing.

Smart Dimension tool: AutoDimension tab

The Autodimension ✧ tool provides the ability to specify the following properties to insert reference dimensions into drawing views as baseline, chain, and ordinate dimensions.

The AutoDimension tool uses the Autodimension PropertyManager. The Autodimension PropertyManager provides the following selections:

- **Entities to Dimension**. The Entities to Dimension box provides two selections:

  - **All entities in view**. Selected by default. Dimensions all entities in the drawing view.

  - **Selected entities**. Displays the selected entities to dimension.

- **Horizontal Dimensions**. The Horizontal Dimensions box provides the following selections:

  - **Scheme**. Baseline selected by default. Sets the Horizontal Dimensioning Scheme and the entity used as the vertical point of origination (Datum - Vertical Model Edge, Model Vertex, Vertical Line or Point) for the dimensions. The Horizontal Dimensions Scheme provides three types of dimensions. They are **Baseline**, **Chain** and **Ordinate**.

By default the vertical point of origination for the horizontal dimensions is based on the first vertical entity relative to the geometric coordinates x0, y0. You can select other vertical model edges or points in the drawing view.

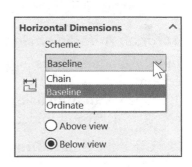

- **Dimension placement**. The Dimension placement section provides two options:

  - **Above view**. Locates the dimension above the drawing view.

  - **Below view**. Selected by default. Locates the dimension below the drawing view.

- ***Vertical Dimensions***. The Vertical Dimensions box provides the following selections:

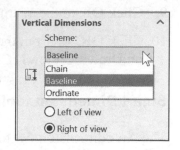

  - **Scheme**. Baseline selected by default. Sets the Horizontal Dimensioning Scheme and the entity used as the vertical point of origination (Datum - Vertical Model Edge, Model Vertex, Vertical Line or Point) for the dimensions. The Horizontal Dimensions Scheme provides three types of dimensions: **Baseline**, **Chain** and **Ordinate**.

  By default, the horizontal point of origination for the vertical dimensions (Datum - Horizontal Model Edge, Model Vertex, Horizontal Line or Point) is based on the first horizontal entity relative to the geometric coordinates x0,y0. You can select other horizontal model edges or points in the drawing view.

  - **Dimension placement**. Right of view selected by default. The Dimension placement section provides two selections:

    - **Left of view**. Locates the dimension to the left of the drawing view.

    - **Right of view**. Select by default. Locates the dimension to the right of the drawing view.

- ***Origin***. Sets the origin for the dimensions. Select a horizontal edge to set as the zero starting point for all dimensions.

  - **Apply**. Modifies the selected edge. Select a different edge. Click the Apply button.

Tutorial: Autodimension 13-1

Insert Autodimensions into an existing drawing view. Note your options.

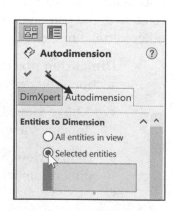

1. Open **Autodimension 13-1** from the SOLIDWORKS 2018\Drawing folder.

2. Click the **Smart Dimension** tool.

3. Click the **Autodimension** tab. The Autodimension PropertyManager is displayed.

4. Check the **Selected entities** box.

5. Click inside the **Drawing View1** view boundary.

6. Click the **left radius** of the flatbar as illustrated. Edge<3> is displayed in the PropertyManager.

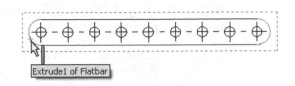

7. Check **Above view** in the Horizontal Dimensions box.

8. Click **OK** ✔ from the Autodimension PropertyManager. Dimensions are applied to the front view.

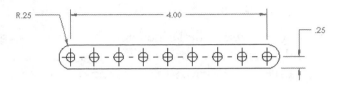

9. **Move** dimensions off the profile.

10. **Create** the needed dimension gaps as illustrated.

11. **Close** the drawing.

Model Items tool

The Model Items 🖉 tool provides the ability to import dimensions, annotations, and reference geometry from the referenced model into the selected view of the drawing. You can insert items into a selected feature, an assembly component, an assembly feature, a drawing view or all views. When inserting items into all drawing views, dimensions and annotations are displayed in the most appropriate view. Edit the inserted locations if required.

The Model Items tool uses the Model Items PropertyManager. The Model Items PropertyManager provides the following selections:

- *Source/Destination*. The Source/Destination box provides the following options:

  - **Source**. Selects the source of the model to insert the dimensions. There are four options:

    - **Entire model**. Inserts model items for the total model.

    - **Selected feature**. Inserts model items for the selected feature.

    - **Selected component**. Only available for assembly drawings. Inserts model items for the selected component.

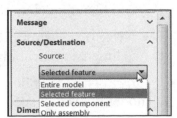

- **Only assembly**. Assembly drawing only. Inserts model items only for assembly features.

  - **Import items into all views**. Inserts model items into all drawing views on the selected sheet.

  - **Destination view(s)**. Lists the drawing views where the model items will be inserted.

- *Dimensions*. The Dimensions box provides various selections to insert the following model items if they exist: **Marked for drawing**, **Not marked for drawing**, **Instance/Revolution counts**, **Tolerance dimensions**, **Hole Wizard Profiles**, **Hole Wizard Locations** and **Hole callout**.

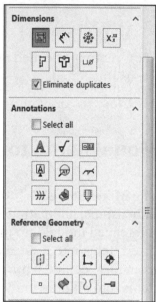

Hole Wizard information from the model propagates to the drawing when the Hole Wizard Profiles, Hole Wizard Locations and Hole callout buttons are selected.

  - **Eliminate duplicates**. Selected by default. Inserts only unique model items. Duplicates are not inserted.

- *Annotations*. The Annotations box provides selections to insert the following model items if they exist. If not, select the Select all option. The available selections are **Select all**, **Notes**, **Surface finish**, **Geometric tolerances**, **Datums**, **Datum targets**, **Weld Symbols**, **Caterpillar**, **End Treatment** and **Cosmetic thread**.

- *Reference Geometry*. The Reference Geometry box provides selections to insert the following model items if they exist. If not, select the Select all option. The available selections are **Select all**, **Planes**, **Axis**, **Origins**, **Center of Mass**, **Points**, **Surfaces**, **Curves** and **Routing points**.

- *Options*. The Options box provides the following selections:

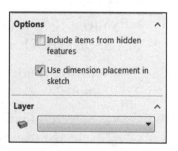

  - **Include items from hidden features**. Clear this option to prevent the insertion of annotations that belong to hidden model items. Performance is slower while hidden model items are filtered.

  - **Use dimension placement in sketch**. Uses the dimension location of the sketch.

- *Layer*. The Layer box provides the ability to insert the model items to a specified drawing layer using the drop-down menu.

Tutorial: Model Items view 13-1

Insert Annotations into a single drawing using the Model Items tool.

1. Open **Model Items 13-1** from the SOLIDWORKS 2018\Drawing folder.

2. Click the **Model Items** <span>tool from the Annotations tab in the CommandManager. The Model Items PropertyManager is displayed.

3. Select **Entire model** from the Source/Destination box. Import items into all views is checked by default.

4. Click **OK** ✔ from the Model Items PropertyManager. Dimensions are inserted into Drawing View1.

5. Click and drag each **dimension** off the model. View the results. Note: You can manually insert dimensions (Smart Dimension tool) and move dimensions from one drawing view to the other.

6. **Close** the drawing.

Note tool

The Note $\mathbf{A}$ tool adds a note to the selected drawing sheet. A note can be free floating or fixed. A note can be placed with a leader pointing to an item, face, edge, or vertex in your document. A note can contain simple text, symbols, parametric text, and hyperlinks. The leader can be straight, bent or multi-jog. Use the Note tool to create a note or to edit an existing note, balloon note or a revision symbol.

The Note tool uses the Note PropertyManager. The Note PropertyManager provides the following selections:

- *Style*. Define styles for dimensions and various annotations, (Notes, Geometric Tolerance Symbols, Surface Finish Symbols, Weld Symbols, etc.). The favorite box provides the following six selections. They are **Apply the default attributes to selected Notes**, **Add or Update a Favorite**, **Delete a Favorite**, **Save a Favorite** and **Load Favorites**.

Set notes in the Notes Area annotation view to behave as watermarks in models. You can display the watermarks under the model geometry or on top of the model geometry, with specified transparency.

The extensions for styles are Dimensions: .sldfvt, Notes: sldnotefvt, Geometric Tolerance Symbols: .sldgtolfvt and Weld Symbols: .sldweldfvt

- **Text Format**. The Text Format box provides the following format selections:

  - **Left Align**. Aligns the text to the left.

  - **Center Align**. Aligns the text in the center.

  - **Right Align**. Aligns the text to the right.

  - **Angle**. Enter the angle value. A positive angle rotates the note counterclockwise.

  - **Insert Hyperlink**. Includes a Hyperlink in the note.

  - **Link to Property**. Links a note to a document property.

  - **Add Symbol**. Adds a symbol from the Symbols dialog box.

  - **Lock/Unlock note**. Only available for drawings. Fixes the note to a selected location.

  - **Insert Geometric Tolerance**. Inserts a geometric tolerance symbol into the note. Use the Geometric Tolerance PropertyManager and the Properties dialog box.

  - **Insert Surface Finish Symbol**. Inserts a surface finish symbol into the selected note. Use the Surface Finish PropertyManager.

  - **Insert Datum Feature**. Inserts a datum feature symbol into the selected note. Use the Datum Feature PropertyManager.

  - **Flag Note Bank**. Flag notes are a method of cross-referencing one area or feature on a drawing to a list of notes, often called general notes.

  - **Manual view label**. Only for projected, detail, section, aligned section, and auxiliary views. Overrides the options in **Tools**, **Options**, **Document Properties**, View labels. When selected, you can edit the label text.

  - **Use document font**. Use the document font which is specified in the Tools, Options, Document Properties, Notes section of the Menu bar menu.

  - **Font**. Select a new font style, size, etc. from the Choose Font dialog box.

  - **All uppercase**. Sets the text of the note to display in uppercase.

  - **Block Attribute**. Only available when you are editing a note in a block.

  - **Attribute name**. Displays the text for notes with attributes imported from AutoCAD. There are three options:

    - **Read only, Invisible, Both**.

☀ The Attribute name and its properties are displayed in the Attributes editor from the Block PropertyManager.

- *Leader*. Provides the ability to select the required Leader type. Document dependent. The available types are:

  - **No Leader**.

  - **Auto Leader**. Selects leader or multi-jog depending on the location of the note.

  - **Leader**. Create a simple leader from the note to the drawing.

  - **Multi-jog Leader**. Create a leader from the note to the drawing with one or more bends.

  - **Spine Leader**. Create a leader with a spline.

  - **Straight Leader**. Leader with a straight line.

  - **Bent Leader**. Leader with a bent line.

  - **Underlined Leader**. Leader that is underlined.

  - **Attach Leader Top**. In multiline notes, attaches leader to top of note.

  - **Attach Leader Center**. In multiline notes, attaches leader to center of note.

  - **Attach Leader Bottom**. In multiline notes, attaches leader to bottom of note.

  - **Attach Leader Nearest**. Originates from the closest point.

  - **Nearest Leader**. Originates from the closest point.

    - Choose whether the leader is **Straight**, **Bent** or **Underlined**.

  - **To bounding box**. Select to position leader with bounding box instead of note content. The leaders associated with the note are vertically aligned based on the size of the bounding box instead of the text.

  - **Apply to all**. Select to apply a change to all of the arrowheads of the selected note. If the selected note has multiple leaders, and Auto Leader is not selected, you can use a different arrowhead style for each individual leader.

- *Leader Style*. Provides the ability to customize Leader style at the document level for line thickness and style.

  - **Use document display**. Selected by default.

  - **Leader Style**. Select a Leader Style from the drop-down menu.

  - **Leader Thickness**. Select a Leader thickness from the drop-down menu.

- *Border*. Provides the following style and size selections:

  - **Style**. Circular is selected by default. The Style option provides nine selections. Specify the style of the border from the drop-down arrow box. The Style selections are **None**, **Circular**, **Triangle**, **Hexagon**, **Box**, **Diamond**, **Pentagon**, **Flag - Five Sided** and **Flag - Triangle**.

  - **Size**. Tight Fit is selected by default. The Size option provides six selections. Specify the required size of the border from the drop-down arrow box. The Size selections are **Tight Fit**, **1 Character**, **2 Characters**, **3 Characters**, **4 Characters** and **5 Characters**.

  - **Padding**. A tight-fit border option for annotation notes and balloons where you can specify a distance to offset the border from the selected text or annotation note.

- *Parameters*. Provides the ability to input or view the X and Y coordinate of the Note on the drawing.

  - **X**. X coordinates.

  - **Y**. Y coordinates.

  - **Display on the screen**.

- *Layer*. The Layer box provides the ability to select a created Layer for the drawing view.

Tutorial: Note 13-1

Create a linked note in a drawing.

1. Open **Note 13-1** from the SOLIDWORKS 2018\Drawing folder.

2. Click the **Note** A tool from the Annotation tab. The Note PropertyManager is displayed.

3. Click **Left Align** from the Text Format box.

4. Click **Link to Property** from the Text Format box. The Link to Property dialog box is displayed.

5. Click the **Files Properties** button. The Summary Information dialog box is displayed. Enter **Link** for Keywords.

6. Click **OK** from the Summary Information dialog box.

7. Check the **Current document** box.

8. Select **SW-Keywords(Keywords)** from the drop-down menu.

9. Click **OK** from the Link to Property dialog box.

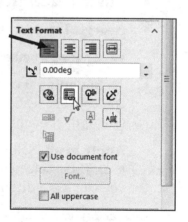

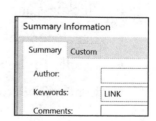

10. Check **Auto Leader** from the Leader box.

11. Click **Underlined Leader** from the Leader box. Remember, document dependent.

12. Select **Circular** from the Border drop-down menu.

13. Select **Tight Fit** from the Border drop-down menu.

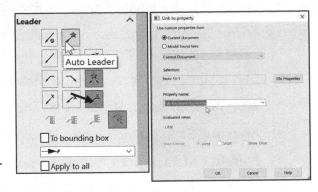

14. Click the **top horizontal** line of the flatbar in Drawing View1.

15. Click a **position** above the view as illustrated.

16. Click **OK** ✔ from the Note PropertyManager. View the results.

17. **Close** the drawing.

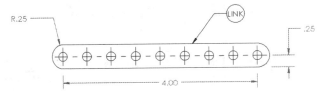

Displaying Annotation Notes in Uppercase. The All uppercase option displays note text and custom property values in uppercase letters in drawings, regardless of the case in the note edit field. You can set the option for individual annotation notes and balloons in the Text Format section of the Note PropertyManager.

Linear Note Pattern tool

The Linear Note Pattern ᴬᴬᴬ tool adds a note to the selected drawing sheet. A note can be free floating or fixed. A note can be placed with a leader pointing to an item, face, edge, or vertex in your document.

The Linear Note tool uses the Linear Note PropertyManager. The Linear Note PropertyManager provides the following selections:

- *Direction 1*. Provides the following options:

  - Sets distance between the pattern instances in Direction 1.

  - **Reverse Direction**. Reverses the Direction of the pattern.

  - **Spacing**. Sets distance between the pattern instances in Direction 1.

  - **Number of Instances**. Sets the number of pattern instances in Direction 1.

- **Angle**. Sets an angular direction from the horizontal (X axis).

- *Direction 2*. Setting Number of Instances in Direction 2 to a value greater than 1 activates Direction 2 settings. Direction 2 provides the following options:

  - Sets distance between the pattern instances in Direction 2.

  - **Reverse Direction**. Reverses the Direction of the pattern.

  - **Spacing**. Sets distance between the pattern instances in Direction 2.

  - **Number of Instances**. Sets the number of pattern instances in Direction 2.

  - **Angle**. Sets an angular direction from the horizontal (X axis).

- *Notes to Pattern*. Select the notes to pattern.

Circular Note Pattern tool

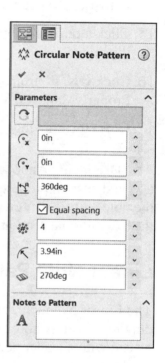

The Circular Note Pattern ⚶ tool adds a note to the selected drawing sheet. A note can be free floating or fixed. A note can be placed with a leader pointing to an item, face, edge, or vertex in your document.

The Circular Note tool uses the Circular Note PropertyManager. The Circular Note PropertyManager provides the following selections:

- *Parameters*. Provides the following options:

  - Sets distance between the pattern instances in Direction 1.

  - **Reverse Direction**. Reverses the Direction of the pattern.

  - **Center X**. Sets the pattern center along the X axis.

  - **Center Y**. Sets the pattern center along the Y axis.

  - **Spacing**. Sets the number of total degrees included in the pattern.

  - **Equal spacing**. Patterns equidistant instances.

  - **Number of Instances**. Sets the number of pattern instances.

  - **Radius**. Sets the pattern radius.

  - **Arc Angle**. Sets an angle measured from the center of the selected entities to the pattern center.

- *Notes to Pattern*. Select the notes to pattern.

Spell Checker tool

The Spell Checker ^{Abc}✓ tool checks for misspelled words in documents. This tool checks notes, dimensions with text, and drawing title blocks when you are in the Edit Sheet Format mode. The Spell Checker tool will not check words in a table and is only available in English using Microsoft Word 2000 or later.

The Spell Checker tool uses the Spelling Check PropertyManager. The Spelling Check PropertyManager provides the following selections:

- *Text*. The Text box displays the misspelled word. Edit the word if required or use the supplied words in the Suggestions box.

- *Suggestions*. The Suggestions box displays a list of possible replacements words associated with the misspelled word.

    - **Ignore**. Skips the instance of the misspelled word.

    - **Ignore All**. Skips all instances of the misspelled word.

    - **Change**. Changes the misspelled word to the highlighted word in the Suggestions box.

    - **Change All**. Changes all instances of the misspelled word to the highlighted word in the Suggestions box.

    - **Add**. Adds the currently selected word to the dictionary.

    - **Undo**. Undoes the last action.

- *Dictionary Language*. List the language used for the Microsoft dictionary. English is the default option.

- *Options*. The Options box provides the following selections:

    - **Check Notes**. Selected by default. Checks the spelling of notes.

    - **Check Dimensions**. Selected by default. Checks the spelling of dimensions.

    - **More Options**. Provides the ability to customize your spell checking options.

Format Painter tool

The Format Painter 🖌 tool provides the ability to copy visual properties from dimensions and annotations to other dimensions and annotations in the same document or another document. Format Painter is supported by parts, assemblies, and drawings. The Format tool uses the Format PropertyManager.

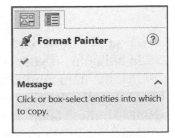

Tutorial: Format Painter 13-1

Apply the Format Painter tool to a Front drawing view.

1. Open the **Format Painter 13-1** drawing from the SOLIDWORKS 2018\Drawing folder.

2. Click the **Format Painter** ✐ tool from the Annotation tab. The Format Painter PropertyManager is displayed.

Select the source dimension for the format.

3. Click the **1.932** dimension as illustrated.

Select the destination dimension to paint with the source dimension format.

4. Click the **1.738** dimension in the view as illustrated. The destination dimension changes to the bilateral format and applies the same tolerance values.

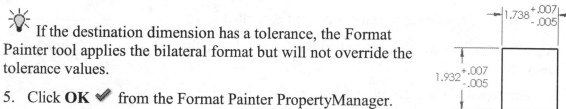

🔆 If the destination dimension has a tolerance, the Format Painter tool applies the bilateral format but will not override the tolerance values.

5. Click **OK** ✔ from the Format Painter PropertyManager.

6. **Close** the drawing.

Balloon tool

The Balloon 🎈 tool provides the ability to create a single balloon or multi balloons in an assembly or drawing document. The balloons label the parts in the assembly and relate them to item numbers on the bill of materials (BOM). The Balloon tool uses the Balloon PropertyManager. The Balloon PropertyManager provides the following selections:

🔆 You do not have to insert a BOM in order to add balloons. If your drawing does not have a BOM, the item numbers are the default values that SOLIDWORKS would use if you did have a BOM. If there is no BOM on the active sheet, but there is a BOM on another sheet, the numbers from that BOM are used.

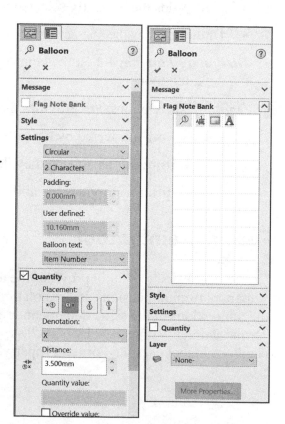

- *Flag Note Bank*. Select Flag Note Bank to include a flag note in the balloon. The following options are available: **Flag note number**, **Flag note zone location**, **Flag note sheet**, **Flag note text**.

- *Style*. Defines the balloon style.

- *Settings*. Select a style for the shape and border of the balloon from the list. The style None displays the balloon text with no border. The Balloon Settings box provides the following selections:

  - **Style**. Circular selected by default. Provides the ability to select a style for the shape and border of the balloon from the drop-down list. The available styles are **None**, **Circular**, **Triangle**, **Hexagon**, **Box**, **Diamond**, **Pentagon**, **Circular Split Line**, **Flag - Five Sided**, **Flag - Triangle**, **Underline**, **Square**, **Square Circle and Inspection**.

  - **Size**. 2 Characters is selected by default. Provides the ability to select a size of the balloon from the drop-down list. The available sizes are **Tight Fit**, **1 Character**, **2 Characters**, **3 Characters**, **4 Characters** and **5 Characters** and **User Defined**.

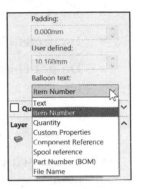

  - **Padding**. Provides the ability to add border spacing for annotation notes and balloons.

  - **Balloon text**. Item Number is selected by default. Provides the ability to select the type of text for the balloon, or for the upper section of a split balloon from the drop-down list. (Not available for auto balloons.) The available options are:

    - **Item Number**. Displays the item number from the Bill of Materials.

    - **Quantity**. Displays the quantity of this item in the assembly.

    - **Custom Properties**. Provides the ability to select a custom property.

    - **Component Reference.** Shows the text specified for **Component Reference** in the **Component Properties dialog box**. To link balloons to component reference values, the drawing must contain a BOM with a COMPONENT REFERENCE.

    - **Spool reference**. Provides the following options:

    - **Part Number (BOM)**. Shows the same information that is in the Part Number column of the BOM. If the BOM field is linked to the model, updates to the field in the BOM display in the balloon.

- **File Name**. Shows the file name of the associated part without the file extension.

- **Cut List Properties**. Allows you to select a cut list property.

- **Balloon text**. In assemblies, specifies where the balloon gets its numbering from:

  - **Assembly Structure**. Uses the numbering according to the component order in the FeatureManager design tree.

  - **Bill of Materials**. Uses the order displayed in the selected bill of Material.

- **Lower text**. If you select the Circular Split Line style, this box is available to specify the text for the lower section. You have the same options as in the Balloon text section.

- *Items Numbers*. When the drawing contains a bill of materials, use Item Numbers to set the balloon sequence. The following options are available:

  - **Start at**. Sets initial value.

  - **Increment**. Sets increment value.

  - **Do not change item numbers**. Keeps bill of materials item numbers as is.

  - **Follow assembly order**. Sets balloons and bill of materials to follow the assembly order in the FeatureManager design tree.

  - **Order sequentially**. Orders balloons and bill of materials items sequentially, starting with Start at and incrementing by Increment.

  - **First item**. When Order sequentially is selected, choose an item to be used as the first item.

- *Quantity*. Provides the following selections:

  - **Placement**. Displays the quantity to the **Left, Right, Top or Bottom of the balloons**.

  - **Denotation**. Lets you set a label for the quantity. **User Defined** lets you type a custom label.

  - **Quantity value**. Displays the quantity to appear in the balloon (read-only).

  - **Override value**. Lets you override a balloon's quantity. For example, you might insert a balloon in a drawing view that shows only two components, but the total number of assembly components is five. In one view a balloon can display a "2," and in the other view it can display a "3."

- *Layer*. The Layer box provides the ability to apply the balloon to a specified drawing layer that you created.

- **More Properties**. Provides the ability to customize your setting by using the Note PropertyManager.

Tutorial: Balloon 13-1

Manually insert two balloons in an assembly drawing.

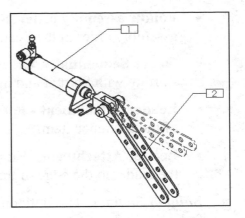

1. Open **Balloon 13-1** from the SOLIDWORKS 2018\Drawing folder.

2. Click the **Balloon** tool from the Annotation tab. The Balloon PropertyManager is displayed.

3. Select **Box** for Style. Select **2 Characters** for Size.

4. Select **Quantity** for Balloon text. Click the face of **Base-Extrude of NCJ2 Tubes<1>** as illustrated in the Graphics window.

5. Click a **position** above the tube for the first balloon. Balloon 1 is displayed.

6. Click the edge of **Extrude1 of Flatbar<1>** as illustrated in the Graphics window.

7. Click a **position** above the flatbar for the second balloon. View the results.

8. Click **OK** from the Balloon PropertyManager.

9. **Close** the drawing.

You can insert a balloon directly into a Note text simply by selecting the desired balloon from the active drawing sheet.

AutoBalloon tool

The AutoBalloon tool provides the ability to add balloons for all components in the selected view. The AutoBalloon tool uses the Auto Balloon PropertyManager. The Auto Balloon PropertyManager provides the following selections:

- *Style*. None selected by default. Provides the ability to select a style for the shape and border of the balloon.

- *Balloon Layout*. Square selected by default. Provides the following balloon selections: **Top**, **Bottom**, **Left**, **Right**, **Square**, and **Circular**.

- *Item Numbers*. When the drawing contains a bill of materials, use Item Numbers to set the balloon sequence.

  - **Ignore multiple instances**. Selected by default. Applies a balloon to only one instance for components with multiple instances.

- **Insert magnetic line (s)**. Attach balloons to magnetic lines. You can choose to space the balloons equally or not, and move the lines freely at any angle.

- **Do not change Item numbers**. Keeps bill of materials item numbers as is.

- **Follow assembly order**. Sets balloons and bill of materials to follow the assembly order in the FeatureManager design tree.

- **Order Sequentially**. Orders balloons and bill of materials items sequentially, starting with Start at and incrementing by Increment.

- **Leader Attachment - Faces**. Applies the leader to the face of the reference item.

- **Leader Attachment - Edges**. Selected by default. Applies the leader to the edge of the reference item.

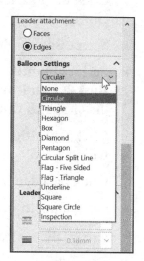

- *Balloon Settings*. The Balloon Settings box provides the following selections:

  - **Style**. Circular selected by default. Provides the following style options: **None, Circular, Triangle, Hexagon, Box, Diamond, Pentagon, Circular Split Line, Flag - Five Sided, Flag - Triangle, Underline, Square, Square Circle** and **Inspection**.

  - **Size**. 2 Characters selected by default. The Size box provides the following options from the drop-down list. They are **Tight Fit, 1 Character, 2 Characters, 3 Characters, 4 Characters, 5 Characters** and **User Defined**.

  - **Padding**. Provides the ability to add border spacing for annotation notes and balloons.

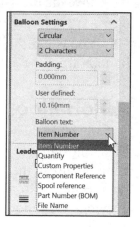

  - **Balloon text**. Item Number is selected by default. Provides the ability to select the type of text for the balloon, or for the upper section of a split balloon from the drop-down list. (Not available for auto balloons.) The available options are:

    - **Item Number**. Displays the item number from the Bill of Materials.

    - **Quantity**. Displays the quantity of this item in the assembly.

    - **Custom Properties**. Provides the ability to select a custom property.

    - **Component Reference**. Shows the text specified for **Component Reference** in the **Component Properties dialog box**. To link balloons to component reference values, the drawing must contain a BOM with a COMPONENT REFERENCE.

- **Spool reference**. Provides the following options:

- **Part Number (BOM)**. Shows the same information that is in the Part Number column of the BOM. If the BOM field is linked to the model, updates to the field in the BOM display in the balloon.

- **File Name**. Shows the file name of the associated part without the file extension.

- **Balloon text**. In assemblies, specifies where the balloon gets its numbering from:

  - **Assembly Structure**. Uses the numbering according to the component order in the FeatureManager design tree.

  - **Bill of Materials**. Uses the order displayed in the selected bill of Material.

  - **Lower text**. If you select the Circular Split Line style, this box is available to specify the text for the lower section. You have the same options as in the Balloon text section.

- *Leader Style*. Use document display selected by default. Provides the ability to apply the default document leader style set in the Document Properties section or to select a custom leader style for the drop-down menus.

- *Frame Style*. Use document display selected by default. Provides the ability to apply the default document Frame style set in the Document Properties section or to select a custom frame style from the drop-down menus.

- *Layer*. The Layer box provides the ability to apply the balloon to a specified drawing layer that you created.

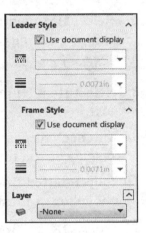

You can reattach any balloon (including dangling balloons within a stack) to associate it with any component within the same drawing view. To reattach a balloon, right-click the balloon and click **Reattach**.

☼ Padding is a tight-fit border option (Document properties) for annotation notes and balloons where you can specify a distance to offset the border from the selected text or annotation note.

Tutorial: AutoBalloon 13-1

Apply the AutoBalloon tool.

1. Open **AutoBalloon 13-1** from the SOLIDWORKS 2018\Drawing folder.

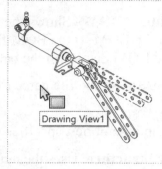

2. Click inside the **Drawing View1** view boundary. The Drawing View1 PropertyManager is displayed.

3. Click the **AutoBalloon** tool from the Annotation tab. The AutoBalloon PropertyManager is displayed.

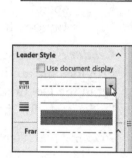

4. Select the **Layout Balloons to Top** button. View the results.

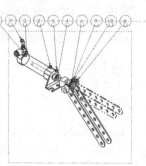

5. Select the **Layout Balloons to Square** button. View the results.

6. Select **Hexagon** from the Balloon Settings box.

7. Select **2 Characters** for Size.

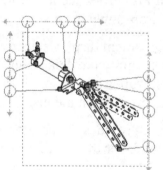

8. Select **Item Number** for Balloon text.

9. **Uncheck** the Use document display box in the Leader Style.

10. Select **dotted** from the Leader Style box as illustrated.

11. Click **OK** from the AutoBalloon PropertyManager. View the created balloons in the drawing.

12. Click and drag the **magnetic lines** to fit the balloons in the drawing view.

13. **Close** the drawing.

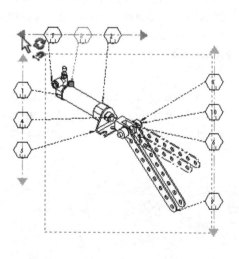

Use the **Insert magnetic line (s)** option to attach balloons to magnetic lines. You can choose to space the balloons equally or not, and move the lines freely at any angle.

Magnetic Line tool

The Magnetic Line  tool provides the ability to align balloons along a line at any angle. The Magnetic Line PropertyManager provides the following selections:

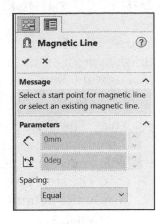

- **Length**. Sets the line length.

- **Angle**. Sets the angle of the line between 0 and 180 degrees from the horizontal.

- **Spacing**. Spacing between the balloons. There are two options:

  - **Equal**. Equally distributes the balloons between the outermost balloons.

  - **Free Drag**. Lets you place the balloons anywhere on the line.

In general, the Magnetic Line feature provides the ability to:

- Have multiple magnetic lines in a drawing view.

- Not to print the lines in the drawing.

- Maintain balloon alignment when the magnetic lines are not visible.

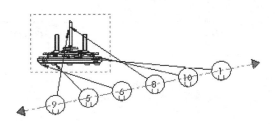

- Swap balloon position. Drag a balloon onto another balloon on a magnetic line; the balloons swap positions.

- Insert balloons before or after inserting magnetic lines.

- Insert magnetic lines automatically when you utilize the Auto Balloon command.

The Soft snaps option provides the ability to snap at 15 degree intervals to facilitate placing or changing the angle of the magnetic lines.

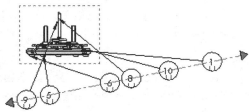

In SOLIDWORKS, inserted dimensions in the drawing are displayed in gray. Imported dimensions from the part are displayed in black.

Surface Finish tool

The Surface Finish ✓ tool provides the ability to add a surface finish symbol to the selected drawing view. The Surface Finish tool uses the Surface Finish PropertyManager. The Surface Finish PropertyManager provides the following selections:

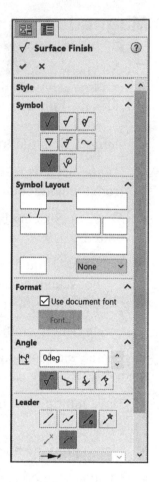

- *Style*. The Style box provides the ability to define styles for dimensions and various annotations (Notes, Geometric Tolerance Symbols, Surface Finish Symbols, Weld Symbols, etc). The style box provides the following selections. They are **Apply the default attributes to selected Surface Finish Symbols, Add or Update a Style, Delete a Style, Save a Style** and **Load Style**.

The extensions for styles are:

Dimensions:	.sldfvt
Notes:	.sldnotefvt
Geometric Tolerance Symbols:	.sldgtolfvt
Surface Finish Symbols:	.sldsffvt
Weld Symbols:	.sldweldfvt

- *Symbol*. Basic and Local are selected by default. The Symbol box provides the following selections for Surface Finish symbols. They are **Basic, Machining Required, Machining Prohibited, JIS Basic, JIS Machining Required, JIS Machining Prohibited Local** and **All Around**.

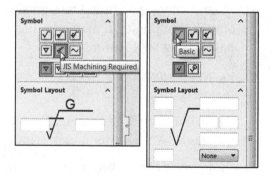

If you select JIS Basic or JIS Machining Required, several surface textures are available.

- *Symbol Layout*. For ANSI symbols and symbols using ISO and related standards prior to 2002, specify text for the predefined locations around the symbol:

  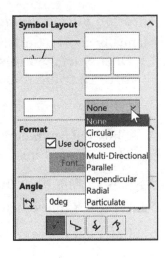

  - **Maximum Roughness, Minimum Roughness, Material Removal Allowance, Production Method/Treatment, Sampling Length, Other Roughness Values, Roughness Spacing** and **Lay Direction**.

  - For symbols using ISO and related standards per 2002, specify: **Manufacturing Method, Texture Requirement 1, Texture Requirement 2, Surface Lay And Orientation, Matching Allowance**.

  - For JIS symbols, specify: **Roughness Ra** and **Roughness Rz/Rmax**.

  - For GOST symbols, select: **Use for notation option**. The Use for notation option displays the surface finish symbol 0.5 times larger than the default size. **Add a default symbol**. The Add default symbol selection displays the default surface finish symbol in parentheses.

- *Format*. The Format box provides the following selections. They are:

  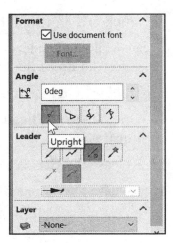

  - **Use document font**. Uses the same font style and type as in the document.

  - **Font**. Ability to customize the font for the document. Uncheck the Use document font check box.

- *Angle*. The Angle box provides the ability to set the angle value of rotation for the symbol. A positive angle rotates the note counterclockwise.

  - **Set a specified rotation**: There are four selections. They are **Upright**, **Rotated 90deg**, **Perpendicular** and **Perpendicular (Reverse)**.

- *Leader*. The Leader box provides the ability to select the Leader style and Arrow style. The available selections are **Leader, Multi-jog Leader, No Leader, Auto Leader, Straight Leader** and **Bent Leader**.

  - **Select an Arrow Style**. Provides the ability to select an Arrow Style from the drop down box.

- *Layer*. The Layer box provides the ability to apply the Surface finish to a specified drawing layer.

Tutorial: Surface Finish 13-1

Manually apply a Surface Finish to a drawing view.

1. Open **Surface Finish 13-1** from the SOLIDWORKS 2018\Drawing folder.

2. **Zoom in** on the Top view.

3. Click the **Surface Finish** ∇ tool from the Annotation tab. The Surface Finish PropertyManager is displayed.

4. Click **Basic** for Symbol.

5. Enter **0.8** micrometers for Maximum Roughness.

6. Enter **0.4** micrometers for Minimum Roughness.

7. Click **Leader** from the Leader box.

8. Click **Bent Leader** from the Leader box.

9. Click the **top horizontal edge** of the Top view for the arrowhead attachment.

10. Click a **position** for the Surface Finish symbol.

11. Click **OK** ✔ from the Surface Finish PropertyManager.

Create Multiple Leaders to the Surface finish symbol.

12. Click the **tip** of the arrowhead.

13. Hold the **Ctrl** key down.

14. Drag the **arrowhead** to the bottom edge of the Top view as illustrated.

15. Release the **Ctrl** key.

16. Release the **mouse button**. View the results.

17. **Close** the drawing.

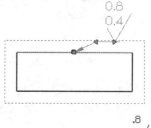

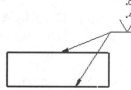

Weld Symbol tool

The Weld Symbol ↗ tool provides the ability to add a weld symbol on a selected entity, edge, face, etc. Create and apply the weld symbol by using the Properties dialog box under the ANSI Weld Symbol tab.

Weld beads use a simplified display. They are displayed as graphical representations in models. No geometry is created. The weld beads are lightweight and do not affect performance.

The dialog box is displayed for ISO, BSI, DIN, JIS and GB standards. Different dialog boxes are displayed for ANSI and GOST standards. The book addresses the ANSI dialog box. The dialog box displays numerous selections. Enter values and select the required symbols and options. A preview is displayed in the Graphics window.

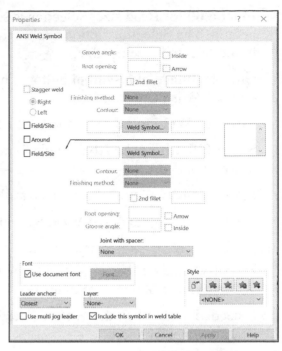

Click a face or edge where you want to locate a welded joint. If the weld symbol has a leader, click a location to place the leader first. Click to place the symbol.

If you selected a face or edge before you click the Weld Symbol tool from the Annotation toolbar, the leader is already placed; click once to place the symbol. Click as many times as necessary to place multiple weld symbols.

Construct weld symbols independently in a part, assembly, or drawing document. When you create or edit your weld symbol, you can:

- Add secondary weld fillet information to the weld symbol for certain types of weld, example: Square or Bevel.

- Choose a Leader anchor of None.

- Choose the text font for each weld symbol.

Tutorial: Weld Symbol 13-1

Manually insert a Weld Symbol into an assembly.

1. Open the **Weld Symbol 13-1** assembly from the SOLIDWORKS 2018\Drawing folder.

2. Click **Insert ≻ Assembly Feature ≻ Weld Bead** 🖉 from the Menu bar. The Weld Bead Type PropertyManager is displayed. Select the faces.

🔅 The Smart Weld Section ☞ tool provides the ability to let you drag the pointer over the faces where you want to apply a weld bead.

3. Click the **outside cylindrical face** of TUBE<1>. Face<1>@TUBE1-W-1 is displayed in the Contact Faces box.

4. Click the **top face** of PLATE<1>. Face<2>@PLATE1-W-1 is displayed in the Contact Faces box.

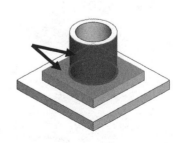

5. Enter **6.00** for Bead size. Selection is selected by default. The Selection option applies the weld bead to the selected face or edge.

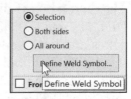

6. Click **Define Weld Symbol** button. The ANSI Weld dialog box is displayed.

7. Click the **Weld Symbol** button as illustrated.

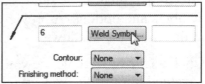

8. Select **Fillet** from the ANSI Weld Symbols drop down menu.

9. Select **Concave** for Surface Shape as illustrated.

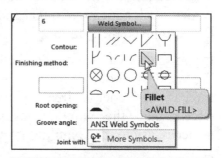

10. Click **OK** from the ANSI Weld Symbol dialog box.

11. Click **OK** ✔ from the Weld Bead PropertyManager. View the results. A Weld Folder is displayed in the FeatureManager.

12. If needed, right-click the **Weld** Folder in the FeatureManager.

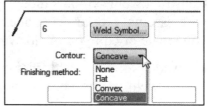

13. Click **Show Cosmetic Welds**. View the results in the Graphics window.

14. **Save** 💾 the document.

Edit the weld bead feature as you would any feature by right-clicking in the FeatureManager design tree and clicking Edit Feature.

🔅 View the additional options in the Weld Bead PropertyManager to create various gaps and pitch with the bead.

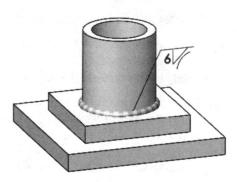

Geometric Tolerance tool

The SOLIDWORKS software supports the ANSI Y14.5 Geometric and True Position Tolerancing guidelines. These standards represent the drawing practices used by U.S. industry. The ASME Y14 practices supersede the American National Standards Institute ANSI standards.

The ASME Y14 Engineering Drawing and Related Documentation Practices are published by The American Society of Mechanical Engineers, New York, NY. References to the current ASME Y14 standards are used with permission.

ASME Y14 Standard Name:	American National Standard Engineering Drawing and Related Documentation:	Revision of the Standard:
ASME Y14.100M-1998	Engineering Drawing Practices	DOD-STD-100
ASME Y14.1-1995	Decimal Inch Drawing Sheet Size and Format	ANSI Y14.1
ASME Y14.1M-1995	Metric Drawing Sheet Size and Format	ANSI Y14.1M
ASME Y14.24M	Types and Applications of Engineering Drawings	ANSI Y14.24M
ASME Y14.2M(Reaffirmed 1998)	Line Conventions and Lettering	ANSI Y14.2M
ASME Y14.3M-1994	Multi-view and Sectional View Drawings	ANSI Y14.3
ASME Y14.41-2003	Digital Product Definition Data Practices	N/A
ASME Y14.5M –1994 (Reaffirmed 1999)	Dimensioning and Tolerancing	ANSI Y14.5-1982 (R1988)

The Geometric Tolerance tool uses the Properties dialog box and the Geometric Tolerance PropertyManager. As items are added, a preview is displayed. The Geometric Tolerance PropertyManager provides the following selections:

- *Style*. The Style box provides the ability to define styles, for dimensions and various annotations (Notes, Geometric Tolerance, Surface Finish, Weld Symbols, etc.). The style box provides the following selections: **Apply the default attributes to selected geometric tolerance**, **Add or Update a Style**, **Delete a Style**, **Save a Style** and **Load Style**.

- *Leader*. The Leader box provides the ability to select the Leader style and Arrow style. The available selections are **Leader, Straight Leader, Leader Left, Multi-jog Leader, Bend Leader, Leader Right, No Leader, Perpendicular Leader, Leader Nearest, Auto Leader** and **All Around Leader**.

- *Text*. Default text <Gtol>.

- *Leader Style*. Use document display selected by default. Provides the ability to apply the default document leader style set in the Document properties section or to select a custom leader style for the drop-down menu.

- *Frame Style*. Use document display selected by default. Provides the ability to apply the default document frame style set in the Document properties section or to select a custom frame style for the drop-down menus.

- *Angle*. The Angle box provides the ability to set the angle value of rotation for the symbol. A positive angle rotates the note counterclockwise.

- *Format*. The Format box provides the following selections. They are:

  - **Use document font**. Uses the same font style and type as in the document.

  - **Font**. Provides the ability to customize the font for the document. Uncheck the Use document font box.

- *Layer*. The Layer option provides the ability to apply the Geometric Tolerance to a specified drawing layer.

When you use the Auto Leader option, you must float over the entity to highlight the entity and to attach the leader. The leader is not displayed until you float over the entity with the mouse pointer.

Tutorial: Geometric Tolerance 13-1

Insert a Geometric tolerance. Add a Feature Control frame for the Ø22 hole dimension in the Front view.

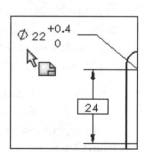

1. Open **Geometric Tolerance 13-1** from the SOLIDWORKS 2018\Drawing folder.

2. Click a position below the **Ø22 hole dimension** text in the Front view.

3. Click the **Geometric Tolerance** tool from the Annotation tab. The Geometric Tolerance Properties dialog box is displayed.

4. Click the **Symbol Library** Geometric Tolerance drop-down arrow.

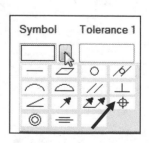

5. Click the **Position** icon as illustrated. The Feature Control Frame displays the Position symbol in the Preview box.

6. Click inside the **Tolerance 1** box.

7. Click the **Diameter Ø** button.

8. Enter **0.25** in the Tolerance 1 box.

9. Click the **M** button. The Maximum Material Condition Modifying Symbol is displayed.

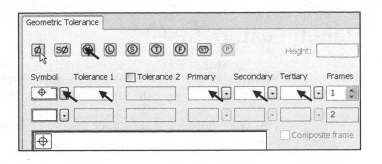

10. Click inside the **Primary** box.

11. Enter **A** in the Primary box.

12. Click inside the **Secondary** box.

13. Enter **B** in the Secondary box.

14. Click the **M** button.

15. Click inside the **Tertiary** box.

16. Enter **C** in the Tertiary box.

17. Click the **M** button.

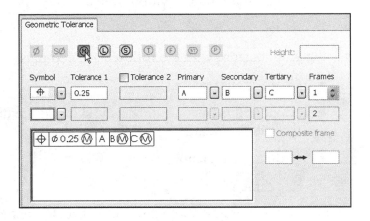

18. Click **OK** from the Properties dialog box.

Attach the Feature Control Frame.

19. Drag the Feature Control Frame to the **Ø22 hole dimension** text and position the mouse pointer on the Ø22.

20. Release the **mouse**. The Feature Control Frame is locked to the dimension text. **Close** the drawing.

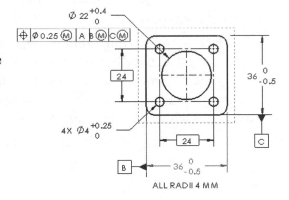

To move a dimension from one view to another view, apply the Shift key to click and drag the required dimension.

Click the dimension Palette rollover button to display the dimension palette. Use the dimension palette in the Graphics window to save mouse travel to the Dimension PropertyManager. Click on a dimension in a drawing view, and modify it directly from the dimension palette.

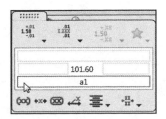

Datum Feature tool

The Datum Feature tool adds a Datum Feature symbol. Attach a datum feature symbol to the following items:

- In an assembly or part, on a reference plane, or on a planar model surface.

- In a drawing view, on a surface that is displayed as an edge, not a silhouette or on a section view surface.

- A geometric tolerance symbol frame.

- In a note.

- On a dimension, with the following exceptions:

  - Chamfer dimensions.

  - Angular dimensions, unless the symbols are displayed per the 1982 standard.

  - Ordinate dimensions.

  - Arc dimensions.

The Datum Feature symbol name automatically increments every time you select the symbol. The Datum Feature tool uses the Datum Feature PropertyManager. The Datum Feature PropertyManager provides the following selections:

- *Style*. None selected by default. The Style box provides the ability to define styles, for dimensions and various annotations. The style box provides the following selections: **Apply the default attributes to selected datum feature**, **Add or Update a Style**, **Delete a Style**, **Save a Style** and **Load Style**.

- *Label Settings*. The Label Settings box provides the following option:

    - **Label box**. Displays a Label to the leader line. The first label is A.

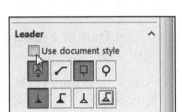

- *Leader*. The Leader box provides the following options:

    - **Use document style**. Selected by default. Uses the same style and type as in the document.

    - **Styles**. Each style of box has a different set of attachment styles. The available pre-set styles follow the ANSI, ISO, etc. standards. The book addresses the ANSI standard. The available styles for Square are **No Leader, Leader**, **Filled Triangle**, **Filled Triangle With Shoulder**, **Empty Triangle** and **Empty Triangle With Shoulder**. The styles for Round (GB) are **Perpendicular Vertical** and **Horizontal**.

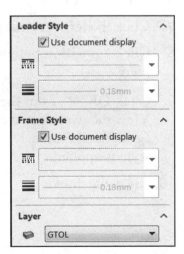

- *Text*. Default text <Gtol>.

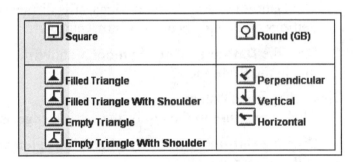

- *Leader Style*. Use document display selected by default. Provides the ability to apply the default document leader style set in the Document properties section or to select a custom leader style for the drop-down menu.

- *Frame Style*. Use document display selected by default. Provides the ability to apply the default document frame style set in the Document properties section or to select a custom frame style for the drop-down menu.

- *Layer*. The Layer option provides the ability to apply the Datum Feature to a specified drawing layer.

Tutorial: Datum Feature 13-1

Create two datum features. The method to insert a datum feature and geometric tolerance symbols is the same for a part or drawing. There is one exception. When a geometric tolerance frame is created in the drawing, anchor the frame to the corresponding dimension. In the part, position the geometric tolerance near the dimension. The frame is free to move.

1. Open **Datum Feature 13-1** from the SOLIDWORKS 2018\Drawing folder.

2. Click the **right vertical edge** of Drawing View2 as illustrated. The Drawing View2 PropertyManager is displayed.

Create the first Datum.

3. Click the **Datum Feature** tool from the Annotation tab. The Datum Feature PropertyManager is displayed. The Label Settings box displays an A. If not, enter A.

4. Drag the **Datum Feature Symbol A** above the top profile line in the Right view.

Create the second Datum.
5. Click a **position** to the left of the vertical edge as illustrated.

6. Drag the **Datum Feature Symbol B** to the left vertical edge in the Front view.

7. Click the **left vertical edge** in the Front view as illustrated.

8. Click a **position** below the bottom profile line in the Front view.

9. Click **OK** from the Datum Feature PropertyManager. View the two created datum features. Note: The default Leader and Frame style is applied.

10. **Close** the drawing.

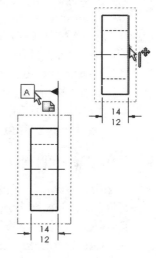

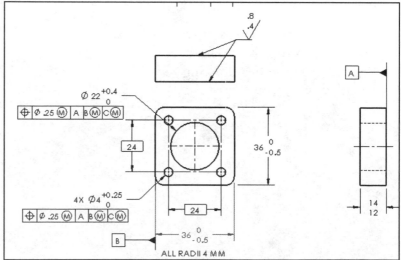

Datum Target tool

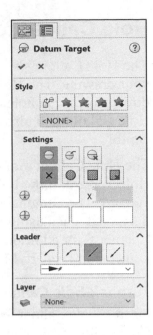

The Datum Target 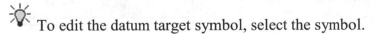 tool provides the ability to attach a datum target and symbol to a model face or edge in a document. The Datum Target tool uses the Datum Target PropertyManager. The Datum Target PropertyManager provides the following selections:

- *Style*. None selected by default. The Style box provides the ability to define styles for dimensions and various annotations. The style box provides the following selections. They are **Apply the default attributes to selected Datum Target, Add or Update a Style, Delete a Style, Save a Style** and **Load Style**.

- *Settings*. The Settings box provides the ability to select and attach a selected symbol to a model face or edge. The Setting box provides selections for Target symbols and Target area. The available Target symbols are **Target symbol, Target symbol with area size outside,** and **No target symbol**.

  - The available Target areas are **X target area, Circular target area, Rectangular target area** and **Do not display target area**.

  - **Target areas size box**. Provides the ability to specify the width and height for rectangles or the diameter for circles.

  - **Datum references**. Provides the ability to specify up to three references.

- *Leader*. The Leader box provides the ability to select the leader style and Arrow style. The available selections are **Bend, solid leader, Bend, dash leader, Straight, solid leader** and **Straight dash leader**.

- *Layer*. The Layer box provides the ability to apply the Datum Target to a specified drawing layer.

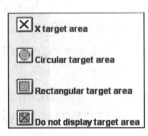

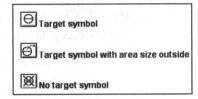

To move a datum target symbol, balloon or point, select the symbol and drag the item in the Graphics window to the new location.

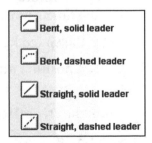

To edit the datum target symbol, select the symbol.

Hole Callout tool

The Hole Callout ⊔⌀ tool adds a Hole Callout to the selected view in your drawing. The Hole callout contains a diameter symbol and the dimension of the hole diameter. If the depth of the hole is known, the Hole Callout will also contain a depth symbol and the dimension of the depth.

The Hole Callout tool uses the information from the Hole Wizard, if the hole was created by the Hole Wizard. The Dimension PropertyManager is displayed when using the Hole Callout tool. See Chapter 5 for additional information on the Dimension PropertyManager.

Tutorial: Hole Callout 13-1

Insert a Hole Callout feature into a drawing view. The Hole Callout feature inserts a dimension with the correct hole annotation.

1. Open **Hole Callout 13-1** from the SOLIDWORKS 2018\Drawing folder.

2. **Zoom in** on Drawing View3.

Insert dimension with hole annotation.

3. Click the **Hole CallOut** ⊔⌀ tool from the Annotation tab.

4. Click the **circumference of the left port** on Drawing View3.

5. Click a **position** above the profile. The Dimension PropertyManager is displayed.

6. Enter **2X** before the <MOD-DIAM><DIM> text in the Dimension Text box as illustrated.

7. Press the **space** key.

8. Click a position after the <MOD-DIAM><DIM> text.

9. Click the **Deep/Depth** icon.

10. Enter **3.25**.

11. Click **OK** ✔ from the Dimension PropertyManager. View the results.

12. **Close** the drawing.

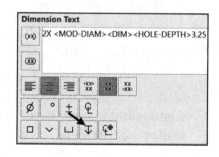

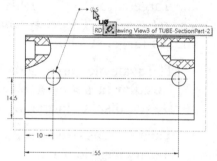

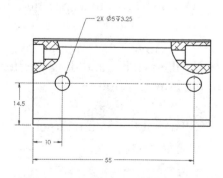

Revision Symbol tool

The Revision Symbol ⚠ tool
provides the ability to insert a
revision symbol into a drawing with a revision already in the
table. Insert a revision table into a drawing to track document
revisions. In addition to the functionality for all tables, you can
select Revision symbol shapes or an alphabetic or numeric
sequence. The latest revision is displayed under REV in the lower-
right corner of a sheet format.

The Revision Symbol tool uses the Revision Symbol
PropertyManager. The Revision Symbol PropertyManager uses
the same conventions as the Note PropertyManager. View the
Note tool section in this chapter and Annotation toolbar for
detailed PropertyManager information.

Tutorial: Revision Symbol 13-1

Insert a new revision into a drawing. Never delete an old revision.

1. Open **Revision Symbol 13-1** from the SOLIDWORKS
    2018\Drawing folder.

2. Right-click inside the **REVISIONS** table.

3. Click **Revisions, Add Revision**. The Revision Symbol ⓡ is
    displayed on the mouse pointer.

4. Click a **location** near the Ⓐ Symbol
    in the drawing sheet.

5. Click **OK** ✔ from the Revision
    Symbol PropertyManager.

6. Double-click inside the empty **Cell**
    under DESCRIPTION. The Pop-up
    menu is displayed.

7. Enter the following text: **ECO
    9443 RELEASE TO
    MANUFACTURING**. The
    Date is displayed in the
    REVISIONS table.

8. Double-click inside **Cell 4**
    under APPROVED.

9. Enter **DCP**.

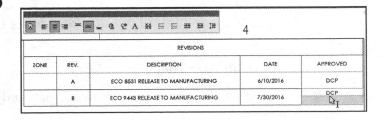

10. Click **outside** the REVISION table.

11. **Close** the drawing.

Revision Cloud tool

The Revision Cloud 🔲
tool provides the ability
to create cloud-like
shapes in a drawing.

The Revision Cloud tool
uses the Revision Cloud
PropertyManager. The following options are available:

TAG	X LOC	Y LOC	SIZE
A1	.88	1.17	Ø.129 THRU
B1	1.08	-.39	Ø.266 THRU
B2	1.85	1.27	Ø.266 THRU
C1	1.39	.57	Ø.781 THRU

- *Cloud Shape*. The Cloud Shape box provides the following
  options:

  - **Rectangle**.

  - **Ellipse**.

  - **Irregular Polygon**.

  - **Freehand**.

- *Maximum Arc Radius*. Select a radius.

- *Line Style*. Set style or thickness.

- *Layer*. Set to a layer on the drawing.

Area Hatch/Fill tool

The Area Hatch/Fill 🔲 tool provides the ability to add an area
hatch or fill pattern in a drawing. Apply a solid fill or a crosshatch
pattern to a closed sketch profile, model face, or to a region
bounded by a combination of model edges and sketch entities. You
can only apply an Area Hatch in drawings. Some characteristics of
Area Hatch include the following:

- If the Area Hatch is a solid fill, the default color of the fill is
  black.

- You can move an Area Hatch into a layer.

- You can select an Area Hatch in a Broken View only in its
  unbroken state.

- You cannot select an Area Hatch that crosses a break.

- Dimensions or annotations that belong to the drawing view are surrounded by a halo of space when they are on top of an area hatch or fill.

The Area Hatch/Fill tool uses the Area Hatch/Fill PropertyManager. The Area Hatch/Fill PropertyManager provides the following selections:

- *Properties*. The Properties box provides the following selections and displays the Area Hatch and Fill. The selections are:

  - **Hatch**. Selected by default. Applies a hatch pattern to the region. Illustrated in the Properties box. ANSI31 is selected by default.

  - **Solid**. Applies a black fill to the region.

  - **None**. Removes the area hatch or fill from the region.

  - **Pattern Scale**. Only available for the Hatch option.

  - **Hatch Pattern Angle**. Only available for the Hatch option.

- *Area to Hatch*. The Area to Hatch box provides the following selections:

  - **Region**. Applies the Area Hatch or Fill to a closed region bounded by model edges or sketch entities.

  - **Boundary**. Applies the Area Hatch or Fill to a combination of model edges and sketch entities that you select for Selected boundary or face.

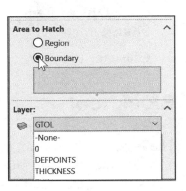

- *Layer*. The Layer box provides the ability to apply the Area Hatch/Fill to a specified drawing layer.

Tutorial: Area Hatch/Fill 13-1

Insert Area Hatch to a region of a drawing view. A hatch represents material type. Apply a hatch pattern for general steel.

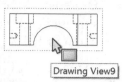

1. Open **Area Hatch 13-1** from the SOLIDWORKS 2018\Drawing folder.

2. Click inside the **Drawing View9** view boundary as illustrated.

3. Click the **Area Hatch/Fill** ◤ tool from the Annotation tab. The Area Hatch/Fill PropertyManager is displayed.

4. Check **Hatch** from the Properties box. Note the options in the PropertyManager.

5. Select **Steel** for Hatch Pattern from the drop-down menu. View your material options.

6. Select **4** for Hatch Pattern Scale. Accept the default angle. Region is checked by default.

7. Click on the **Drawing View** as illustrated. An area hatch is applied to the clamp in the selected region.

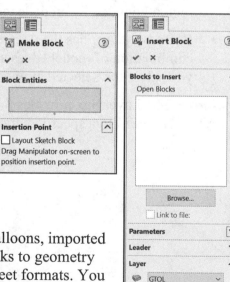

8. Click **OK** ✔ from the Area Hatch/Fill PropertyManager.

9. **Close** the drawing.

Block tool

The Block 🄰 tool from the Annotation toolbar provides the ability to either select the Make Block tool or the Insert Block tool. The Make Block tool uses the Make Block PropertyManager. The Insert Block tool uses the Insert Block PropertyManager.

You can make, save, edit, and insert blocks for drawing items that you use often, such as standard notes, title blocks, label positions, etc.

Blocks can include text, any type of sketch entity, balloons, imported entities and text, and area hatch. You can attach blocks to geometry or to drawing views, and you can insert them into sheet formats. You can also copy blocks between drawings and sketches, or insert blocks from the Design Library. View the Block section for additional information.

Center Mark tool

The Center Mark ⊕ tool provides the ability to locate center marks on circles or arcs in a drawing. Use the center mark lines as references for dimensioning. A few items to note about center marks are:

- Center marks are available as single marks, in Circular, Linear patterns or in a Slot center mark.

- The axis of the circle or arc must be normal to the drawing sheet.

- Center marks can be inserted automatically into new drawing views for holes or fillets if required using the option selection.

- Center marks propagate or insert automatically into patterns if the pattern is created from a feature and not a body or face.

- Center marks in Auxiliary Views are oriented to the viewing direction such that one of the lines of the center mark is parallel to the view arrow direction.

- Rotate center marks individually by specifying the rotation in degrees. Using the Rotate Drawing View dialog box.

The Center Mark tool uses the Center Mark PropertyManager. The Center Mark PropertyManager provides the following selections:

- *Style*. None selected by default. The Style box provides the ability to define styles for dimensions and various annotations. The style box provides the following selections: **Apply the default attributes to selected Center Mark, Add or Update a Style, Delete a Style, Save a Style** and **Load Style**.

- *Auto Insert*. Provides the following options: **For all holes, For all fillets** and **For all slots**.

- *Manual Insert Options*. The Options box provides the following selections:

  - **Single Center Mark**. Inserts a center mark into a single arc or circle. You can modify the Display Attributes and rotation Angle of the center mark.

  - **Linear Center Mark**. Inserts center marks into a Linear pattern of arcs and circles. You can select Connection lines and Display Attributes for Linear patterns.

- **Circular Center Mark**. Inserts center marks into a Circular pattern of arcs and circles. You can select Circular lines, Radial lines, Base center mark, and Display Attributes for Circular patterns.

- **Slot Center mark**. Selected by default. Insert center marks into slot centers in straight slots.

- **Slot Ends**. Insert center marks into slot ends in straight slots.

- **Arc Slot Centers**. Selected by default. Insert center marks into slot centers in arc slots.

- **Arc Slot Ends**. Insert center marks into slot ends in arc slots.

- *Display Attributes*. The Display Attributes box provides the following selections:

  - **Use document defaults**. Selected by default. Uses the default attributes set in the Documents Properties section.

  - **Mark size**. Sets a document value. Uncheck the Use document defaults option.

  - **Extended lines**. Displays the extended axis lines with a gap between the center mark and the extended lines. Uncheck the Use the document defaults option.

  - **Centerline font**. Displays the center mark lines in the centerline font. Uncheck the Use the document defaults option.

- *Angle*. The Angle section provides the following option:

  - **Angle box**. Not available for Linear or Circular pattern center marks. Sets the angle of the center mark. If you rotate the selected center mark, the rotation angle is displayed in the Angle box.

- *Layer*. The Layer box provides the ability to apply the center mark to a specified drawing layer.

Tutorial: Center Mark 13-1

Manually insert a Center Mark into a drawing view.

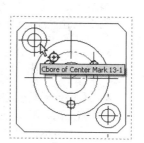

1. Open the **Center Mark 13-1** drawing from the SOLIDWORKS 2018\Drawing folder. Drawing View1 has two center marks.

2. Click inside the **Drawing View1** view boundary. The Drawing View1 PropertyManager is displayed.

Activate the Center Mark tool.

3. Click the **Center Mark** ⊕ tool from the Annotation tab. The Center Mark PropertyManager is displayed.

4. Click the **circumference** of the small top left circle as illustrated. A Center Mark is displayed. View your options and the available features.

5. Click **OK** ✔ from the Center Mark PropertyManager.

6. **Close** the drawing.

Centerline tool

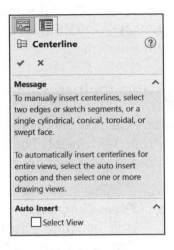

The Centerline tool provides the ability to insert centerlines into drawing views manually. The Centerline tool uses the Centerline PropertyManager. The Centerline PropertyManager provides the following two messages: 1.) Select two edges/sketch segments or single cylindrical/conical/toroidal face for Centerline insertion; 2.) To automatically insert centerlines for entire views, select the auto insert option and then select one or more drawing views.

☀ You can select either the tool or an entity first.

☀ To insert centerlines automatically into a drawing view, select **Centerlines** from the Document Properties, Detailing, Auto insert on view creation section.

Tutorial: Centerline 13-1

Insert a Centerline into a drawing view.
1. Open **Centerline 13-1** from the SOLIDWORKS 2018\Drawing folder.

2. Click inside the **Drawing View9** view boundary as illustrated.

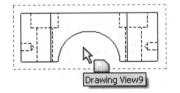

Activate the Centerline drawing tool.
3. Click the **Centerline** ⊞ tool from the Annotation tab. The Centerline PropertyManager is displayed.

4. Check the **Select View** box. Centerlines are displayed in the selected view. Click **OK** ✔ from the Centerline PropertyManager.

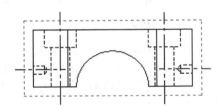

5. **Close** the drawing.

To modify a view scale, click inside the desired view. Click the Use custom size box. Enter the needed scale.

☀ Cosmetic thread callouts are editable, and therefore configurable, only when Standard is set to None in the Cosmetic Thread PropertyManager.

Consolidated Table Toolbar

The Consolidated Table toolbar provides the ability to add a table in a drawing. The toolbar is document and function dependent. See SOLIDWORKS help for additional information.

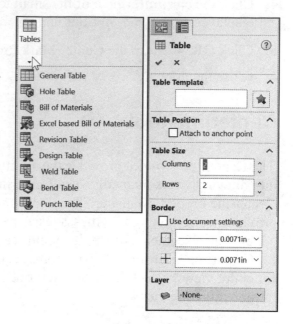

Table toolbar: General Table

The General Table option uses the Table PropertyManager. The Table PropertyManager provides the following selections:

- **_Table Template_**. Only available during table insertion. Display the selected standard or custom template.

- **_Table Position_**. Provides the ability to set the anchor corner to the table anchor. The available selections are **Top Left**, **Top Right**, **Bottom Left** and **Bottom Right**. Note: Top Left is the default.

- **_Table Size_**. The Table Size box provides the following selections:

  - **Columns**. Sets the number of columns in the table.

  - **Rows**. Sets the number of rows in the table.

- **_Border_**. The Border box provides the ability to address the Border and Grip of the table. The selections are:

  - **Use document settings**. Selected by default. Apply the default document settings in the Document Properties section.

  - **Border**. Thin is selected by default. Border provides the following selections: **Thin**, **Normal**, **Thick**, **Thick (2)**, **Thick (3)**, **Thick (4)**, **Thick (5)** and **Thick (6)**.

  - **Grid**. Thin is selected by default. Grid provides the following selections: **Thin**, **Normal**, **Thick**, **Thick (2)**, **Thick (3)**, **Thick (4)**, **Thick (5)** and **Thick (6)**.

- **_Layer_**. Select a layer type from the drop-down menu.

Table toolbar: Hole Table

The Hole Table tool uses the Hole Table PropertyManager. The Hole Table PropertyManager provides the following selections:

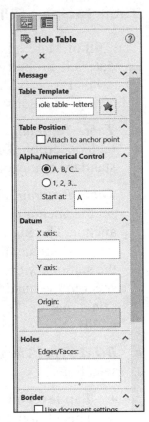

- *Table Template*. Only available during table insertion. Displays the selected standard or custom template.

- *Table Position*. Provides the ability to set the anchor corner to the table anchor. The available selections are **Top Left**, **Top Right**, **Bottom Left** and **Bottom Right**. Note: Top Left is the default.

- *Alpha/Numerical Control*. Specify the tags identifying the holes to be alphanumeric (**A, B, C...**) or numeric (**1, 2, 3...**). With alphanumeric tags, the letter prefix designates a specific hole size and the number designates an instance. Use **Start at** to specify a beginning letter or number.

- *Datum*. Provides the ability to select a vertex to define the origin with an X axis and a Y axis. The Datum box provides the following selections:

  - **X Axis Direction Reference**. Select a horizontal model edge from the Graphics window.

  - **Y Axis Direction Reference**. Select a vertical model edge from the Graphics window.

  - **Origin**. Select an Origin.

- *Holes*. The Holes box provides the ability to select individual hole edges, or select a model face to include all the holes in the face.

- *Border*. Use document settings selected by default. Provides the ability to select a custom Box or Grid border.

- *Layer*. None selected by default. The Layer box provides the ability to apply the hole table to a specified drawing layer.

  - **Next View**. Only available during table insertion. Provides the ability to set the Datum and Holes for another drawing view.

Table toolbar: Bill of Materials

The Bill of Materials option uses the Bill of Materials PropertyManager. The Bill of Materials PropertyManager provides the following selections:

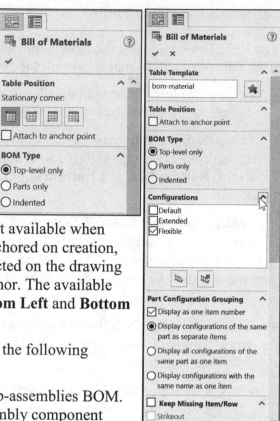

- *Table Template*. Only available during table insertion. Displays the selected standard or custom template.

- *Table Position*. Provides the ability to set the anchor corner to the table anchor. The Stationary corner option is not available when you first create the table. If the table is anchored on creation, the stationary corner is automatically selected on the drawing sheet in the quadrant nearest the table anchor. The available selections are **Top Left**, **Top Right**, **Bottom Left** and **Bottom Right**. Note: Top Left is the default.

- *BOM Type*. The BOM Type box provides the following selections:

  - **Top level only**. Creates a parts and sub-assemblies BOM. This option does not create a sub-assembly component BOM.

  - **Parts only**. Creates a sub-assembly components, "as individual items" BOM.

  - **Indented assemblies**. Lists the sub-assemblies. Lists the indents sub-assembly components below their sub-assemblies.

- *Configurations*. The Configurations box displays the list quantities in the Bill of Materials for all selected configurations.

- *Part Configuration Grouping*. The Part Configuration Grouping box provides the following selections:

  - **Display as one item number**. Only available when you first create a BOM. Uses the same item number for different configurations of a component in different top-level assembly configurations. Each unique component configuration can be present in only one of the assembly configurations in the BOM. Set the BOM Type to Top level only.

- **Display configurations of the same part as separate items**. List each component in the BOM if a component has multiple configurations.

- **Display all configurations of the same part as one item**. List each component in only one row in the BOM if a component has multiple configurations.

- **Display configurations with the same name as one item**. List one configuration name in one row of the BOM if more than one component has the same configuration name.

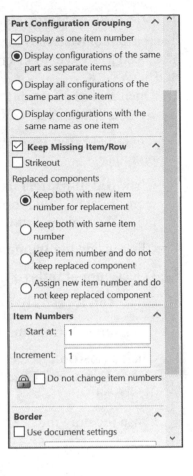

- *Keep Missing Items*. Keep components listed in the table if components have been deleted from the assembly since the Bill of Materials was created.

  - **Strikeout**. If missing components are still listed, the text for the items is displayed with the Strikeout option.

  - **Keep both with new item number for replacement**.

  - **Keep item number and do not keep replaced component**.

  - **Assign new item number and do not keep replace component**.

- *Item Numbers*. The Item Numbers box provides the following selections:

  - **Start**. The Start option provides a value for the beginning of the item number sequence. The sequence increases by a single digit.

  - **Increment**. The Increment option provides the ability to set the increment value for item numbers in the BOM.

  - **Do not change item numbers**.

- *Border*. The Border box provides the ability to address the Box and Grid of the table. The selections are:

  - **Border**. Thin is selected by default. Border provides the following selections: **Thin**, **Normal**, **Thick**, **Thick (2)**, **Thick (3)**, **Thick (4)**, **Thick (5)** and **Thick (6)**.

  - **Grid**. Thin is selected by default. Grid provides the following selections: **Thin**, **Normal**, **Thick**, **Thick (2)**, **Thick (3)**, **Thick (4)**, **Thick (5)** and **Thick (6)**.

- *Layer*. None selected by default. The Layer box provides the ability to apply the hole table to a specified drawing layer.

Tutorial: Bill of Materials 13.1

Insert a Bill of Materials in a drawing. Note: You need MS Excel to insert a BOM using SOLIDWORKS.

1. Open **Bill of Materials 13-1** from the SOLIDWORKS 2018\Drawing folder.

2. Click inside the **Isometric view** boundary.

3. Click **Bill of Materials** from the Consolidated Table toolbar (Annotation tab in the CommandManager).

4. Click inside the **Isometric view** boundary.

5. Select **bom-standard** from the Table Template.

6. Click **Top level only** from the BOM Type box.

7. Click **OK** ✔ from the Bill of Materials PropertyManager.

8. Click a **position** in the top left corner of Sheet1 to locate the Bill of Materials. The Bill of Materials PropertyManager is displayed.

9. **Close** the drawing.

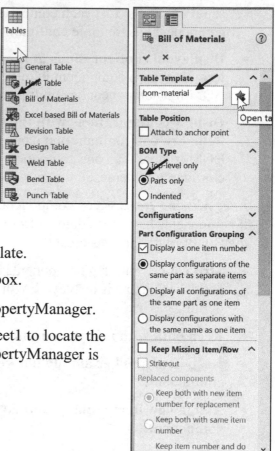

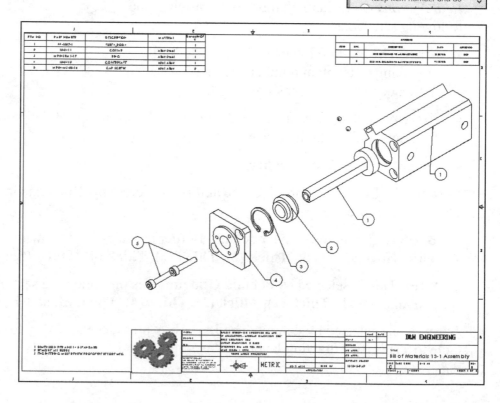

Table toolbar: Revision Table

The Revision Table 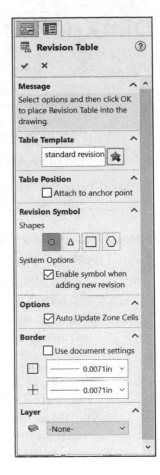 tool uses the Revision Table PropertyManager. The Revision Table PropertyManager provides the following selections:

- *Table Template*. Only available during table insertion. Displays the selected standard or custom template.

- *Table Anchor*. Provides the ability to set the anchor corner to the table anchor. The available selections are **Top Left**, **Top Right**, **Bottom Left** and **Bottom Right**.

- *Revision Symbol*. The Revision Symbol box provides the ability to select a border shape for revision symbols. The selections are **Circle**, **Triangle**, **Square** and **Hexagon**.

- *Options*. The Options box provides the following selection:

  - **Enable symbol when adding new revision**. Provides the ability to place revision symbols when you add a revision to the table.

- *Border*. The Border box provides the ability to address the Border and Grip of the table. The selections are:

  - **Border**. Thin is selected by default. Border provides the following selections: **Thin**, **Normal**, **Thick**, **Thick (2)**, **Thick (3)**, **Thick (4)**, **Thick (5)** and **Thick (6)**.

  - **Grid**. Thin is selected by default. Grid provides the following selections: **Thin**, **Normal**, **Thick**, **Thick (2)**, **Thick (3)**, **Thick (4)**, **Thick (5)** and **Thick (6)**.

- *Layer*. None selected by default. Select a layer type from the drop-down arrow.

Weld Table. Provides the ability to insert weld tables in drawings. Use weld tables to summarize weld specifications including Weld quantity, Weld size, Weld symbol, Weld length, and Custom weld bead properties such as weld material, process, mass, cost, and time.

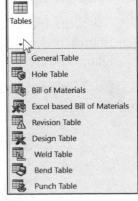

Bend Table. Provides the ability to specify the bend allowance or bend deduction values for a sheet metal part in a bend table. You can also specify K-Factor values in their own K-Factor bend tables. The bend table also contains values for bend radius, bend angle, and part thickness.

Punch Table. Contains the form features and library features used in sheet metal bodies. See SOLIDWORKS Help for additional information.

DimXpert Dimensions and Drawings

You can import dimensions and tolerances which you created using DimXpert for parts into drawings. If the DimXpert dimensions are not displayed in the drawing after import, follow the below procedure:

- In the FeatureManager design tree, right-click the **Annotations** folder.

- Select **Display Annotations** and **Show DimXpert Annotations.**

You can't use the Model Items PropertyManager to import DimXpert dimensions into a drawing.

Tutorial: DimXpert 13-1

Apply DimXpert: Plus and Minus option. Insert three drawing views using the View Palette with DimXpert Annotations.

1. Open the **DimXpert 13-1** part from the SOLIDWORKS 2018\Drawing folder. View the three Extrude features, Linear Pattern feature, and the Fillet feature.

2. Click the **DimXpertManager** ⊕ tab.

3. Click the **Auto Dimension Scheme** ⊕ tool from the DimXpertManager. The Auto Dimension PropertyManager is displayed. Prismatic and Plus and Minus is selected by default.

A key difference between the *Plus and Minus* option versus the *Geometric* option is how DimXpert controls the four-hole pattern, and how it applies tolerances to interrelate the datum features when in *Geometric* mode.

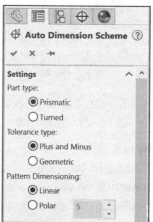

4. Check the **Prismatic** box.

5. Check the **Plus and Minus** box.

6. Click the **Linear** box.

7. Click the **back face** of the model. Plane1 is displayed in the Primary Datum box.

8. Click **inside** the Secondary Datum box.

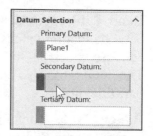

9. Click the **left face** of the model. Plane2 is displayed in the Secondary Datum box.

10. Click **inside** the Tertiary Datum box.

11. Click the **top face** of the model. The selected plane is displayed in the Tertiary Datum box.

12. Click **OK** ✔ from the Auto Dimension PropertyManager.

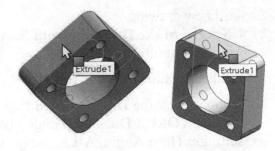

13. Display an **Isometric view**. View the dimensions. All features are displayed in green.

14. **Return** to the FeatureManager.

15. Right-click **Annotations** from the FeatureManager.

16. **Uncheck** Show DimXpert Annotations.

17. **View** the results in the Graphics window.

Display the DimXpert Annotations.
18. Right-click **Annotations** from the FeatureManager.

19. **Check** Show DimXpert Annotations.

20. **View** the results in the Graphics window.

💡 To modify the default DimXpert color, click **Options**, **Colors**. Under Color scheme settings, select **Annotations**, **DimXpert** and pick the new color.

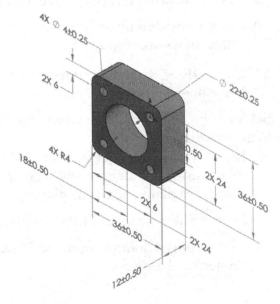

Create a new drawing.

21. Click the **Make Drawing from Part/Assembly** in the Menu bar toolbar.

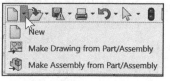

Accept the default Drawing Template.

22. Double-click the **Drawing** icon from the New SOLIDWORKS Document dialog box. Accept the default settings. Third Angle, A-Landscape is used in this tutorial.

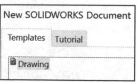

23. If required, click **OK** from the Sheet Format/Size dialog box.

Insert three drawing views using the View Palette.

24. Click the **View Palette** 📇 tab in the Task Pane. DimXpert 13-1 is displayed in the drop-down menu.

25. Check the **Import Annotations** box.

26. Check the **DimXpertAnnotations** box.

27. Check the **Include items from hidden features** box. Note: The (A) next to the drawing view informs the user that DimXpert Annotations are present.

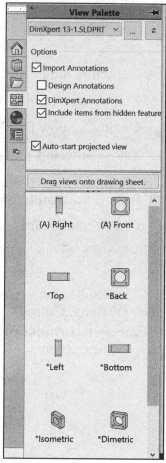

28. Click and drag the **(A) Front** view into Sheet1 in the lower left corner.

29. Click a **position** directly above the Front view.

30. Click a **position** directly to the right of the Front view. Three views are displayed.

31. Click **OK** ✔ from the Projected View PropertyManager. View the results.

Set Sheet Properties to fit the views and dimensions to the drawing sheet.

32. Right-click **Properties** in Sheet1. Do not select inside a drawing view. The Sheet Properties dialog box is displayed.

33. Set the **Sheet Scale** to fit the drawing.

34. Click **Apply Changes** from the Sheet Properties dialog box.

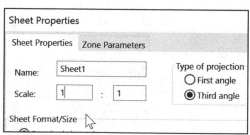

35. Click and drag **dimensions** off the model as illustrated.

36. **Close** the drawing. Do not save the drawing.

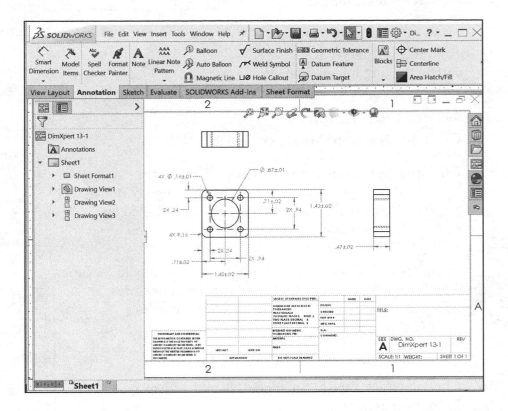

Tutorial: DimXpert 13-2

Apply DimXpert: Geometric option.
Insert three drawing views using the
View Palette with DimXpert
Annotations.

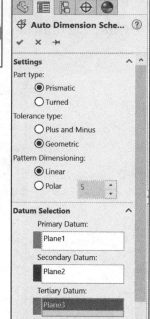

1. Open the **DimXpert 13-2** part
   from the SOLIDWORKS
   2018\Drawing folder. The Part
   FeatureManager is displayed.

Apply DimXpert to the part.

2. Click the **DimXpertManager** ⊕ tab.

3. Click the **Auto Dimension Scheme** tool from the
   DimXpertManager. The Auto Dimension PropertyManager is
   displayed. Prismatic and Plus and Minus is selected by
   default. In this section, select the Geometric option.

   DimXpert: Geometric option provides the ability to locate
   axial features with position and circular runout tolerances.
   Pockets and surfaces are located with surface profiles.

4. Click the **Prismatic** box.

5. Check the **Geometric** box. Click the **Linear** box.

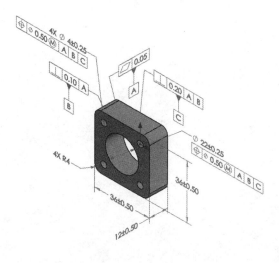

6. Click the **back face** of the model. Plane1 is displayed in the
   Primary Datum box.

7. Click **inside** the Secondary Datum box.

8. Click the **left face** of the model. Plane2 is displayed in the
   Secondary Datum box.

9. Click **inside** the Tertiary Datum box.

10. Click the **top face** of the model. Plane3 is
    displayed in the Tertiary Datum box.

11. Click **OK** ✔ from the Auto Dimension
    PropertyManager.

12. Display an **Isometric view**.

13. **View** the Datums, Feature Control Frames
    and Geometric tolerances.

14. **Return** to the FeatureManager.

Create a new drawing.

15. Click the **Make Drawing from Part/Assembly** in the Menu bar toolbar.

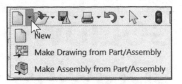

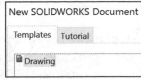

Accept the default Drawing Template.

16. Double-click the **Drawing** icon from the New SOLIDWORKS Document dialog box. Accept the default settings.

17. Click **OK** from the Sheet Format/Size dialog box.

Insert the Front, Top, and Right view from the View Palette.

18. Click the **View Palette** 🖼 tab in the Task Pane. DimXpert 13-2 is displayed in the drop-down menu.

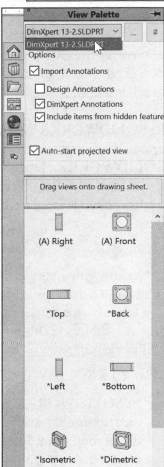

19. Check the **Import Annotations** box.

20. Check the **DimXpertAnnotations** box.

21. Check the **Include items from hidden features** box. The (A) next to the drawing view informs the user that DimXpert Annotations are present.

22. Click and drag the **(A) Front view** into Sheet1 in the lower left corner.

23. Click a **position** directly above the Front view.

24. Click a **position** directly to the right of the Front view.

25. Click **OK** ✔ from the Projected View PropertyManager.

Modify the Sheet Scale.

26. Right-click inside the **sheet boundary**. Do not click inside a view.

27. Click **Properties**. If needed, expand the drop-down menu.

28. Enter **1:1** for Scale.

29. Click **Apply Changes** from the Sheet Properties dialog box.

30. Click and drag **dimensions** off the model. View the drawing.

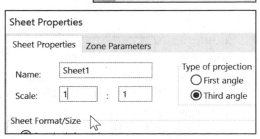

Close the drawing.

31. Click **File**, **Close**. Do not save the drawing.

SOLIDWORKS eDrawing

An eDrawing provides the power to create, view and share 3D models and 2D drawings. With the eDrawing Viewer, you can open SOLIDWORKS documents, modify, and then save as an eDrawing.

eDrawing files are an email enabled communications tool designed to dramatically improve sharing and interpreting 2D mechanical drawings.

In SOLIDWORKS® eDrawings® you can view and animate models and drawings and create documents convenient for sending to others who do not have SOLIDWORKS. SOLIDWORKS eDrawings® is installed automatically with SOLIDWORKS Professional and SOLIDWORKS Premium or with a Student Edition. You can view any SOLIDWORKS document (part, assembly or drawing) from SOLIDWORKS 97Plus and later.

eDrawings viewer is available for the iPad. Download the eDrawings viewer for iPad from your App store. Use eDrawings to view 3D ContentCentral. 3D ContentCentral is a free service for locating, configuring, downloading and requesting 2D and 3D parts and assemblies, 2D blocks, library features and macros. Start eDrawings on your iPad, and click http://www.3dcontentcentral.com/.

To open a SOLIDWORKS document in eDrawings:

- In SOLIDWORKS with an open part, assembly or drawing document, click **File ➢ Publish to eDrawings** from the Menu bar toolbar.

- You are prompted to specify configurations, display states, or drawing sheets (Current, All, or Selected) if the SOLIDWORKS document contains more than one. The SOLIDWORKS eDrawings Viewer opens and you can edit and save the file in eDrawings.

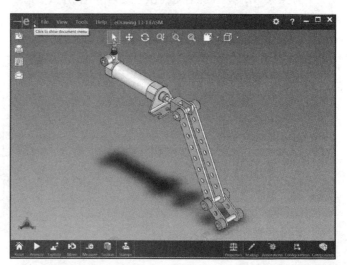

SOLIDWORKS Professional eDrawing is displayed in the book. Your SOLIDWORKS eDrawing dialog box may differ.

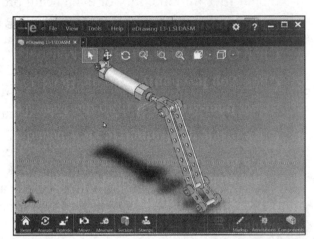

If you have SOLIDWORKS Professional, you can create an executable version of the document that contains the viewer to save the customer from downloading it.

Tutorial: eDrawing 13-1

Create a SOLIDWORKS eDrawing using an assembly.

1. Open **eDrawing 13-1** from the SOLIDWORKS 2018\eDrawing folder. The assembly is displayed. Create an eDrawing.

2. Click **File ➢ Publish to eDrawings** from the Menu bar toolbar. Accept the conditions if this is your first time using the SOLIDWORKS eDrawings. View the SOLIDWORKS eDrawings dialog box. Note: The screen shots reflect SOLIDWORKS Professional. Your options may vary.

3. Click the **Animate** button.

4. Click **Play**. **View** the results. Click the **Stop** button.

5. Click the **Reset** button to return to the original position.

6. Click the **Explode** button. Click **ExplView1**. **View** the results.

7. Click the **ExplView1** button again to collapse the assembly.

8. Click the **Mass Properties** button. **View** the results.

9. Click the **Mass Properties** button again to return to the original state.

10. Click the **Section** button. View the results and the options.

11. Click the **Reset** button to return to the original position.

Save the eDrawing.

12. Click **Save**. Accept the default name. View your options.

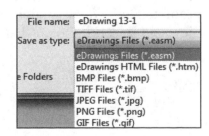

13. **Close** all models.

In the eDrawings software, you can view rotational exploded steps saved in assembly files created with SOLIDWORKS 2014 or later. You can also view exploded steps that rotate a component with or without linear translation in the eDrawings software.

SOLIDWORKS Detached Drawings

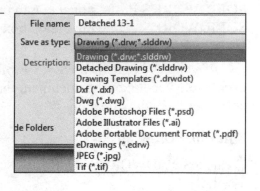

Detached drawings are designed so you can open and work in the drawing document without loading the model files into memory or even being present. This means you can send a Detached drawing to other SOLIDWORKS users without loading and sending the model files.

A Detached drawing provides the ability to save time when editing a drawing with complex parts or a very large assembly. A Detached drawing provides control over updating the drawing to the model. Members of the design team can work independently on the drawing, adding details and annotations, while other members edit the model.

When the drawing and the model are synchronized, all the details and dimensions added to the drawing update to any geometric or topological changes in the model.

You can save a regular drawing as a Detached drawing and vice versa. A Detached drawing cannot be a Lightweight drawing. In the FeatureManager design tree, the icons for Detached drawings display a broken link:

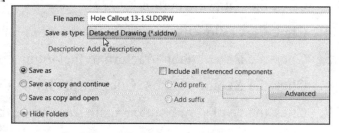

- Drawing icon for Detached drawing.

- Drawing view icon (model view) for Detached drawing with model not loaded.

If the referenced model is needed for an operation within a Detached drawing, you are prompted to load the model file. You can load the model manually by right-clicking a view and selecting Load Model.

Export Drawings to another Software Package

You may need to import your SOLIDWORKS data into other CAD software package for a quote or manufacturing purposes. When dealing with 2D drawings, the .dxf file format is the most common file format to use.

The DXF/DWG import wizard imports .dxf or .dwg files into SOLIDWORKS by guiding you through required steps. You have the option of importing to either a drawing or a part. If you import to a drawing, the layer, color, and line style information in the .dxf or .dwg file is also imported.

The DXF/DWG translator imports:

- AutoCAD® Mechanical annotations, known as proxy entities, (such as surface finish symbols or GTOL frames) and automatically drawn objects (such as cams and springs) when you import .dxf or .dwg files as SOLIDWORKS drawing documents. The translator converts these imported items to equivalent SOLIDWORKS objects, or creates them as blocks of primitive geometry, as appropriate.

- Associative and non-associative crosshatches as area hatches.

- XREFs in AutoCAD DWG files. If an imported block is an XREF, the symbol -> appears next to the block name in the FeatureManager design tree. If the XREF has a dangling definition, the symbol ->? appears.

- DWG files with multiple sheets.

- Mechanical Desktop® (MDT) parts, assemblies, and drawings.

See SOLIDWORKS Help for additional details.

Open a Drawing Document

When opening a SOLIDWORKS Drawing document the following options are available:

- *Mode*.

- *Configurations*.

- *Display State*.

- *Use Speedpak*.

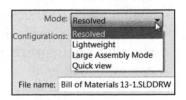

Mode

The Mode drop-down menu provides four options (document dependent). Resolved is selected by default.

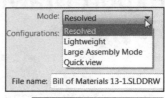

- **Resolved**. All individual components and sub-assemblies are loaded into the drawing.
- **Lightweight**. Only a subset of the data is loaded into memory. The remaining data of the model is loaded on an as-needed basis.

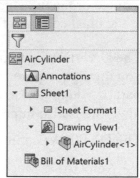

💡 A feather is displayed in the FeatureManager for a Lightweight drawing view. Right-click on the drawing view and click Set Lightweight to Resolved to resolve the drawing view.

- **Large Assembly Mode**. A collection of system settings that improves the performance of assemblies. You can turn on Large Assembly Mode at any time, or you can set a threshold for the number of components and have Large Assembly Mode turn on automatically when that threshold is reached.
- **Quickview**. Provides the ability to open a simplified representation of the drawing. For multi-sheet drawings, you can open one or more sheets in using the Select sheets to load dialog box.

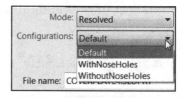

- **Configurations**. Provides the ability to open a drawing in the selected configuration.

- **Display State**. Provides the ability to open a drawing in the selected display state.

- **References**. Provides the ability to view the reference documents of the drawing.

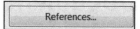

- **SpeedPak**. Creates a simplified configuration of a drawing without losing references. If you work with very large and complex drawings, using a SpeedPak configuration can significantly improve performance while working in the assembly and its drawing.

💡 You can create and update SpeedPak configurations of subassemblies from within the parent assembly. You can disable the SpeedPak graphics circle.

Insert a Center of Mass Point in a Drawing

Point at the Center of Mass

Add a Center of Mass (COM) point to parts, assemblies and drawings.

Add a COM point by clicking **Center of Mass** (Reference Geometry toolbar) or **Insert ➤ Reference Geometry ➤ Center of Mass** or checking the Create Center of Mass feature box in the Mass Properties dialog box.

In drawings of parts or assemblies that contain a COM point, you can show and reference the COM point. Center of mass is a selectable entity in drawings, and you can reference it to create dimensions. In a drawing, click **Insert ➤ Model Items**.

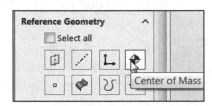

Under Reference Geometry, click **Center of Mass**.

The position of the COM point updates when the model's center of mass changes.

To view the center of mass in a drawing, click **View ➤ Hide/Show ➤ Center of mass** from the Main menu.

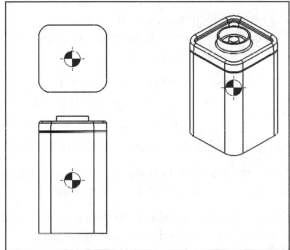

You can measure distances and add reference dimensions between the COM point and entities such as vertices, edges, and faces.

You cannot create driving dimensions from the COM point. However, you can create a Center of Mass Reference (COMR) point and use that point to define driving dimensions.

You can create Measurement sensors that reference COM and COMR points.

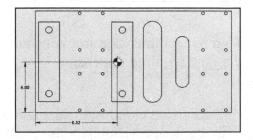

Summary

In this chapter, you learned about Drawings, Drawing Toolbar, Annotations Toolbar, eDrawings, Detached drawings, exporting a drawing and opening a SOLIDWORKS drawing document using various options.

You addressed Sheet Format, Size and Document Properties. You used the View Palette feature to insert drawing views into a drawing.

The View Palette shows drawing views placed on drawings. After placing a drawing view on a drawing, the view persists in the View Palette and is identified with a drawing icon.

You reviewed and addressed each tool in the Drawing Toolbar: *Model View, Projected View, Auxiliary View, Section View, Detail View, Standard 3 View, Broken-out Section, Break, Crop View and Alternate Position View.*

You reviewed and addressed each tool in the Annotations Toolbar: *Smart Dimension, Model Items, Spell Checker, Format Painter, Note, Linear Note Pattern, Circular Note Pattern, Balloon, AutoBalloon, Magnetic Line, Surface Finish, Weld Symbol, Hole Callout, Geometric Tolerance, Datum Feature, Datum Target, Area Hatch/Fill, Blocks, Center Mark, Centerline, Revision Symbol, Revision cloud and Tables.*

You Inserted DimXpert on a part, then applied the View Palette and inserted the DimXpert Annotation into drawing views.

You created an eDrawing, a detached drawing, exported drawings to another software package and learned about the various options in opening a drawing document.

You learned about creating a drawing with a Center of Mass point. You can add a Center of Mass (COM) point to a part, assembly or drawing.

Drawings consist of one or more views produced from a part or assembly. The part or assembly connected with the drawing must be saved before you can create the drawing. Drawing files have the .SLDDRW extension.

To move a dimension from one view to another view, apply the Shift key to click and drag the required dimension.

💡 Revision cloud is an annotation type that lets you create cloud-like shapes in a drawing. Use revision clouds to call attention to geometry changes. You can insert revision clouds in drawing views or on the drawing sheet.

💡 Use Section View Assist, previously called the Section View User Interface, to add offsets to existing section views. To edit existing section views with Section View Assist, Right-click the section view or its cutting line and click **Edit Cutting Line**.

In Chapter 14, explore Sheet Metal features and their associated tools.

CHAPTER 14: SHEET METAL FEATURES AND TOOLS

Chapter Objective

Chapter 14 provides a comprehensive understanding of the Sheet Metal features and their associated tools. On the completion of this chapter, you will be able to:

- Review and utilize the tools from the Sheet Metal toolbar:

  - Base-Flange/Tab, Convert to Sheet Metal, Lofted-Bend, Edge Flange, Miter Flange, Hem, Jog, Sketched Bend, Cross-Break, Closed Corners, Break-Corner/Corner-Trim, Extruded Cut, Simple Hole, Vent, Unfold, Fold, Flatten, No Bends, Insert Bends and Rip.

- Examine and utilize the Sheet Metal Library feature.

Sheet Metal

Sheet metal parts are generally used as enclosures for components or to provide support to other components. There are two methods to create a sheet metal part:

1. Build a part. Then convert the built part to a sheet metal part. There are a few instances where this method is to your advantage:

   - **Imported Solid Bodies**. If you import a sheet metal file with bends from another CAD system, the bends are already in the model. Using the Insert Bends tool from the Sheet Metal toolbar is your best option for converting the imported file to a SOLIDWORKS sheet metal part.

   - **Conical Bends**. Conical bends are not supported by specific sheet metal features, such as Base Flange, Edge Flange, etc. You must build the part using extrusions, revolves, etc., and then convert it to add bends to a conical sheet metal part.

2. Create the part as a sheet metal part using specific sheet metal features from the Sheet Metal toolbar. This method eliminates extra steps because you create a part as sheet metal from the initial design stage.

☀ The DXF/DWG PropertyManager in SOLIDWORKS exports sheet metal to .dxf or .dwg files. A preview shows what you are exporting and lets you remove unwanted entities, such as holes, cutouts and bend lines. The scale is 1:1 by default. Also automatically cleans up overlapping entities so a cleaner file is delivered to manufacturing.

Sheet Metal toolbar

The Sheet Metal toolbar provides the tools to create sheet metal parts. The available Sheet Metal tool options are dependent on the selected mode and selected part or assembly. The tools and menu options that are displayed in gray are called gray-out. The gray icon or text cannot be selected. Additional information is required for these options.

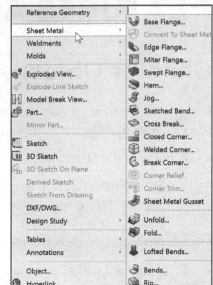

💡 Use the Swept Flange tool to create compound bends in sheet metal parts. See SOLIDWORKS help for additional information.

Base-Flange/Tab tool

The Base-Flange/Tab 🤚 tool provides the ability to add a base flange feature to a SOLIDWORKS sheet metal part. A Base-Flange/Tab feature is the first feature in a new sheet metal part.

The Base-Flange/Tab feature is created from a sketch. The sketch can be a single closed, a single open, or multiple enclosed profiles. The part is marked as a sheet metal part in the FeatureManager.

💡 The thickness and bend radius of the Base-Flange feature are the default values for the other sheet metal features.

The Base-Flange/Tab tool uses the Base Flange PropertyManager. The options on the Base Flange PropertyManager update according to the active sketch. Example, the Direction 1 and Direction 2 boxes are not displayed for a sketch with a single closed profile.

The Base Flange PropertyManager provides the following selections:

- **Direction1**. Blind is the default End Condition. Provides the ability to set the End Condition and depth if needed. View Chapter 6 on End Condition details.

- **Direction2**. Blind is the default End Condition. Provides the ability to set the End Condition and depth if needed.

- *Sheet Metal Gauges*. The Sheet Metal Gauges box provides the following selections:

  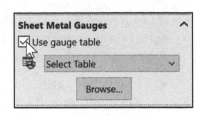

  - **Use gauge table**. Selected by default. Use the default gauge table.

  - **Select Table**. Select a Sheet Metal Gauge Table from the drop-down menu.

  - **Browse**. Browse to locate a Sheet Metal Gauge Table from the SW default folder.

- *Sheet Metal Parameters*. The Sheet Metal Parameters box provides the following selections:

  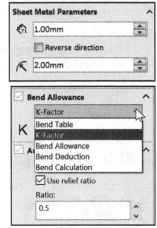

  - **Thickness**. Displays the selected value for the sheet metal thickness.

  - **Reverse direction**. Reverses the thicken sketch in the opposite direction if required.

  - **Bend Radius**. Displays the selected value for the bend radius.

- *Bend Allowance*. The Bend Allowance box provides the ability to select the following four types of bends:

  - **Bend Table**. Select a bend table from the drop-down menu.

    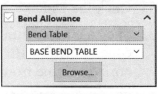

    - **Browse**. Browse to locate a bend table file.

  - **K-Factor**. Selected by default. Enter a K-Factor value. A K-Factor is a ratio that represents the location of the neutral sheet with respect to the thickness of the sheet metal part.

    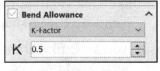

  - **Bend Allowance**. Enter a Bend Allowance value. A bend allowance is the arc length of the bend as measured along the neutral axis of the material.

    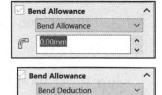

  - **Bend Deduction**. Enter a Bend Deduction value. A bend deduction is the difference between the bend allowance and twice the outside setback.

    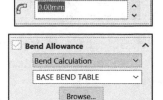

  - **Bend Calculation**. Select a table from the list, or click Browse to browse to a table.

- *Auto Relief*. The Auto Relief box provides the ability to select three relief types: **Rectangular**, **Tear** and **Obround** as illustrated.

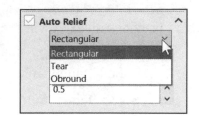

  - **Use relief ratio**. Selected by default for the Rectangular and Obround option. Uncheck the Use relief ration to enter Width and Depth.

Tutorial: Base/Flange 14-1

Create a Base Flange feature In-Context to a Sheet Metal assembly.

1. Open **BaseFlange 14-1** from the SOLIDWORKS 2018\SheetMetal folder.

2. Right-click **BRACKET1<1>** from the FeatureManager.

3. Click **Edit Part** from the Context toolbar. BRACKET<1> is displayed in blue.

4. Click **Front Plane** of CABINET from the FeatureManager.

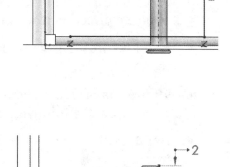

5. Click the **Line** ✏ Sketch tool.

6. Sketch a **horizontal line** collinear with the inside edge of the CABINET. Note: Sketch the horizontal line from left to right.

7. Sketch a **vertical line** and a **horizontal line** to complete the U-shaped profile as illustrated.

8. Add an **Equal Relation** between the top horizontal line and the bottom horizontal line.

9. Insert **dimensions** as illustrated.

10. Display an **Isometric** view.

11. Click **Base-Flange/Tab** 〰 from the Sheet Metal toolbar. The Base Flange PropertyManager is displayed.

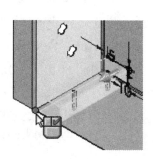

12. Select **Up to Vertex** for End Condition in Direction 1. Think about design intent.

13. Click the CABINET **Extruded1 front left vertex** as illustrated.
Vertex<1>@CABINET-1 is displayed in the Vertex box.

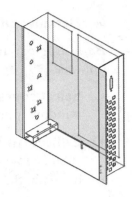

14. Enter **1.00**mm for Thickness in the Sheet Metal Parameters box.

15. Enter **2.00**mm for Bend Radius.

16. Enter **.45** for K factor. Click **OK** ✔ from the Base Flange PropertyManager.

Return to the assembly.

17. Click the **Exit Component** tool.

18. **Rebuild** the model. View the results.

19. **Close** the model.

Convert to Sheet Metal tool

The Convert to Sheet Metal tool provides the ability to convert a solid or surface body to a sheet metal part. The solid body can be an imported sheet metal part.

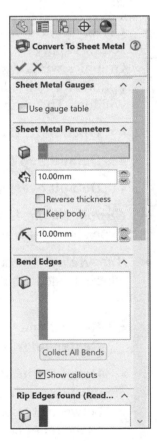

The Convert to Sheet Metal tool uses the Convert To Sheet Metal PropertyManager. The PropertyManager provides the following selections:

- *Use gauge table*. The sheet metal parameters (material thickness, bend radius, and bend calculation method) use the values stored in the gauge table unless you override them.

The Use gauge table option is only available the first time you use the Convert to Sheet Metal tool.

- *Sheet Metal Parameters*. The Sheet Metal Parameters box provides the following options:

  - **Select a fixed entity**. Select the face that remains in place when the part is flattened. You can select only one face.

  - **Sheet thickness**. Thickness of the sheet.

  - **Reverse Thickness**. Changes the direction in which the sheet thickness is applied.

  - **Keep body**. Keeps the solid body to use with multiple Convert to Sheet Metal features or designates that the entire body be consumed by the sheet metal feature.

- **Default radius for bends**. Bend radius.

- *Bend Edges*. The Bend Edges box provides the following options:

  - **Select edges/faces that represent bends**. In the graphics window, click an edge to add it to the list of bend edges.

  - **Collect All Bends**. When there are pre-existing bends, such as in an imported part, finds all of the appropriate bends in the part.

  - **Show callouts**. Displays callouts in the Graphics window for bend edges.

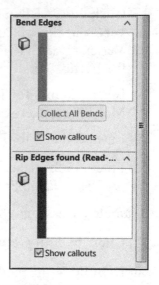

- *Rip Edges found (Read-only)*. When you select bend edges, the corresponding rip edges are selected automatically.

- *Rip Sketches*. Select a 2D or 3D sketch to define a required rip.

- *Corner Defaults*. When you set the Corner Defaults, the settings apply to all rips whose callouts say "Default" in the Graphics window. You can override these defaults by setting options for individual rips from callouts in the Graphics window. The available options are:

  - **Open butt, Overlap, Underlap**.

  - **Default gap for all rips**.

  - **Default overlap ratio for all rips**.

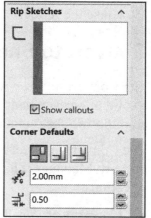

- *Custom Blend Allowance*. Select to set a Bend Allowance Type and a value for the bend allowance.

- *Auto Relief*. The software automatically adds relief cuts where needed when inserting bends. The available options are:

  - **Relief Type**. Select the type of relief cut to be added: **Rectangular**, **Obround** and **Tear**.

  - **Relief Ratio**. If you select Rectangular or Obround, you must set a relief ratio. The value of the relief ratio must be between .1 and 2. The higher the value, the larger the size of the relief cut added during the insertion of bends.

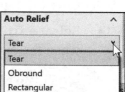

Use the Convert to Sheet Metal command with Solid or surface bodies that have No shells or fillets, either a shell or fillets, both a shell and fillets, or imported parts that are already in the form of a sheet metal part.

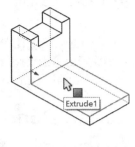

💡 When using the Convert to Sheet Metal tool, you can keep the solid body to use with multiple Convert to Sheet Metal features or specify that the entire body be consumed by the tool. See SOLIDWORKS Help for additional information and details.

Tutorial: Convert to Sheet Metal 14-1

Apply the Convert to Sheet Metal tool.

1. Open **Convert to Sheet Metal 14.1** from the SOLIDWORKS 2018\SheetMetal folder.

2. Click **Convert to Sheet Metal** 🗗 from the Sheet Metal toolbar. The Convert to Sheet Metal PropertyManager is displayed.

Select a face as the fixed face for the sheet metal part and set the sheet thickness and default bend radius. Under Bend Edges, select the model edges that will form bends.

3. Click the **top face** of Extrude1 as illustrated.

4. Enter **1mm** for Sheet Thickness.

5. Enter **1mm** for Default Radius for bend.

6. Select the **top 7 edges** of the model as illustrated. The selected edges are displayed in the Bend Edges dialog box. Note: If needed, open the solution folder to view the final model.

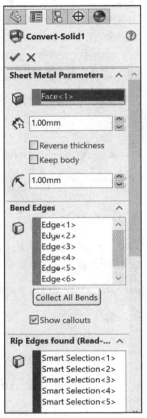

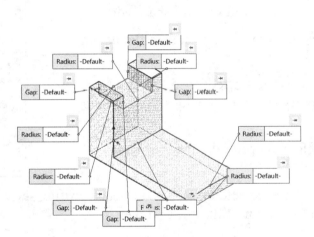

💡 To provide access to needed edges, display your model in the Hidden Lines Visible mode.

💡 Illustrations may vary depending on your SOLIDWORKS version and system setup.

Accept the default settings.

7. Click **OK** ✔ from the Convert to Sheet Metal PropertyManager. View the results in the Graphics window.

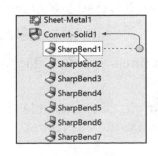

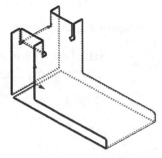

Display the model in a Flattened state.

8. Right-click **Flat-Pattern1** from the FeatureManager.

9. Click **Unsuppress**. View the results.

10. **Close** the model.

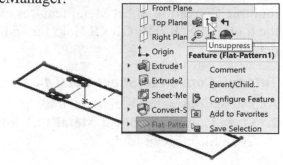

View SOLIDWORKS Help for additional information on the Convert to Sheet Metal feature.

Lofted Bend tool

The Lofted Bend 🛋 tool provides the ability to create a sheet metal part between two sketches using a loft feature. Lofted bends in sheet metal parts use two open-profile sketches that are connected by a loft. They are some additional items to note about the Lofted Bend feature:

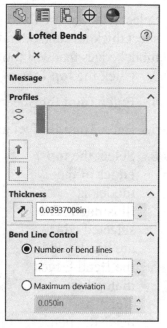

1. You cannot mirror a Lofted Bend feature.

2. Two sketches are required with the following restrictions:

   • An open profile is required without sharp edges.

   • An aligned profile is required to ensure flat pattern accuracy.

The Lofted Bend tool uses the Lofted Bend PropertyManager. The Lofted Bend PropertyManager provides the following selections:

• *Profiles*. The Profiles box provides the following selections:

   • **Profile**. Displays the selected sketches. For each profile, select the point from which you want the path of the loft to travel.

   • **Move Up**. Adjusts the order of the profiles up.

   • **Move Down**. Adjusts the order of the profiles down.

• *Thickness*. The Thickness box provides the ability to set the value for Thickness.

   • **Reverse Direction**. Reverses the direction of the lofted bend if required.

• *Bend Line Control*. The Bend Line Control box provides the following selections:

   • **Number of bend lines**. Sets the value for Setting to control coarseness of the flat pattern bend lines.

   • **Maximum deviation**. Only available if Maximum deviation is not selected. Sets the value for the maximum deviation.

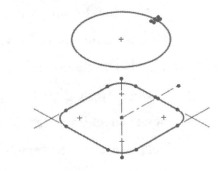

Decreasing the value of Maximum deviation increases the number of bend lines.

Tutorial: Lofted Bend 14-1

Create a Lofted Bend feature in a Sheet Metal part.

1. Open **Lofted Bend 14-1** from the SOLIDWORKS 2018\SheetMetal folder.

2. Click **Lofted Bend** 🔩 from the Sheet Metal toolbar. The Lofted Bend PropertyManager is displayed.

3. Click **both sketches** in the Graphics window as illustrated. The selected profiles are the traveled loft path.

4. Enter **.01**in for Thickness. The direction arrow points inward.

5. Click the **Reverse Direction** button if required.

6. Click **OK** ✔ from the Lofted Bend PropertyManager. The Lofted Bend1 feature is displayed in the FeatureManager. View the results.

7. **Close** the model.

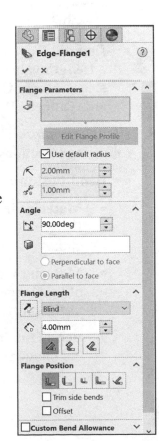

Edge Flange tool

The Edge Flange ![icon] tool provides the ability to add a wall to an edge of a sheet metal part. General edge flange characteristics include:

1. The thickness of the edge flange is linked to the thickness of the sheet metal part.

2. The sketch line of the profile must lie on the selected edge.

The Edge Flange tool uses the Edge-Flange PropertyManager. The Edge-Flange PropertyManager provides the following selections:

- *Flange Parameters*. The Flange Parameters box provides the following options:

    - **Select edges**. Displays the selected edges from the Graphics window.

    - **Edit Flange Profile**. Edits the sketch of the profile. The Profile Sketch dialog box is used.

- **Use default radius**. Uses the default radius. Un-check to enter bend radius.

- **Bend Radius**. Only available when the Use default radius option is un-checked. Enter a bend radius value.

- **Gap distance**. Sets the value of the gap distance.

- *Angle*. The Angle box provides the following selections:

  - **Flange Angle**. Sets the value of the Flange Angle.

  - **Select face**. Provides the ability to select a face from the Graphics window.

  - **Perpendicular to face**. Flange angle perpendicular to selected face.

  - **Parallel to face**. Flange angel parallel to selected face.

- *Flange Length*. The Flange Length box provides the following selections:

  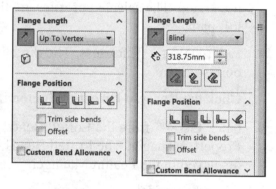

  - **Length End Condition**. Provides two selections:

    - **Blind**. Selected by default.

    - **Up To Vertex**. Select a Vertex from the Graphics window.

    - **Up To Edge And Merge**.

  - **Reverse Direction**. Reverses the direction of the edge flange if required.

  - **Length**. Only available if the Blind option is selected. Enter a **Length** value. Select the **origin** for the measurement. The selections are **Outer Virtual Sharp, Inner Virtual Sharp**, and **Tangent Bend**.

- *Flange Position*. The Flange Position box provides the following selections:

  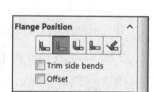

  - **Bend Position**. Select one of the following positions:

    - **Material Inside**. The top of the shaded preview of the flange coincides with the top of the fixed sheet metal entity.

    - **Material Outside**. The bottom of the shaded preview of the flange coincides with the top of the fixed sheet metal entity.

    - **Bend Outside**. The bottom of the shaded preview of the flange is offset by the bend radius.

- **Bend from Virtual Sharp**. This preserves the dimension to the original edge and will vary the bend material condition to automatically match with the flange's end condition.

- **Tangent to Bend**. Valid for all flange length options and for bends that are greater than 90º. The flange position will always be tangent to the side face attached to the selected edge, and the flange length will always maintain the exact length.

- **Trim side bends**. Not selected by default. Removes extra material.

- **Offset**. Not selected by default. Offsets the flange.

- **Offset End Condition**. Only available if the Offset option is checked. There are four End Conditions. They are **Blind, Up To Vertex, Up To Surface** and **Offset From Surface**.

- **Offset Distance**. Only available if the Offset option is checked. Enter the Offset distance value.

- *Custom Bend Allowance*. Provides the following four selections:

  - **Bend Table**. Selects a bend table from the drop down menu.

  - **Browse**. Browse to locate a bend table file.

  - **K-Factor**. Selected by default. Enter a K-Factor value. A K-Factor is a ratio that represents the location of the neutral sheet with respect to the thickness of the sheet metal part.

  - **Bend Allowance**. Enter a Bend Allowance value. A bend allowance is the arc length of the bend as measured along the neutral axis of the material.

  - **Bend Deduction**. Enter a Bend Deduction value. A bend deduction is the difference between the bend allowance and twice the outside setback Bend Allowance Type.

  - **Bend Calculation**. Calculate the developed length of sheet metal parts using bend calculation tables.

- *Custom Relief Type*. Provides the ability to select three relief types. They are **Rectangle, Tear** and **Obround**. The Rectangle custom relief type is selected by default. Note: If you want to add a Rectangle or Obround relief, you must specify the Relief ratio. The value of the Relief ratio must be between 0.05 and 2.0. The higher the value, the larger the size of the relief cut added during insertion of bends. See SOLIDWORKS help for additional information on Relief ratio.

  - **Use relief ratio**. Uncheck the Use relief ratio box. Enter **Width** and **Depth**.

  - **Tear**. Reliefs are the minimum size required to insert the bend and flatten the part. Two options are provided: **Rip** and **Extended**. Rip is selected by default.

Tutorial: Edge Flange 14-1

Create an Edge Flange feature on a Sheet metal part. Insert a front right flange.

1. Open **Edge Flange 14-1** from the SOLIDWORKS 2018\SheetMetal folder.

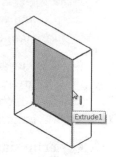

2. Click the **front right vertical** edge as illustrated.

3. Click **Edge Flange**  from the Sheet Metal toolbar. The Edge-Flange PropertyManager is displayed. Edge<1> is displayed in the Select edges box.

4. Select **Blind** for Length End Condition.

5. Enter **30**mm for Length. The direction arrow points to the right. If required, click the **Reverse Direction** button.

6. Click **Inner Virtual Sharp** for Flange Length.

7. Select **Material Outside** for Flange Position. Accept all other defaults.

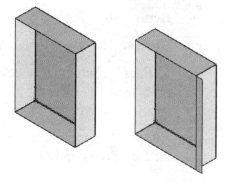

8. Click **OK** ✔ from the Edge-Flange PropertyManager. Edge-Flange1 is displayed in the FeatureManager.

9. **Close** the model.

Tutorial: Edge Flange 14-2

Create an Edge Flange feature on a Sheet metal part using the Edit Flange Profile option.

1. Open **Edge Flange 14-2** from the SOLIDWORKS 2018\SheetMetal folder.

2. Click the **top left edge** of the BRACKET.

3. Click **Edge Flange** 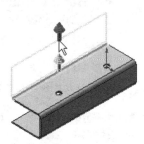 from the Sheet Metal toolbar. The PropertyManager is displayed. Edge<1> is displayed in the Select edges box.

4. Click a **position** above the BRACKET for direction. The direction arrow points upwards.

5. Enter **20**mm for Flange Length in the spin box.

6. Click the **Edit Flange Profile** button. Do not click the Finish button at this time.

7. Drag the **front left edge** of the edge flange towards the front hole as illustrated.

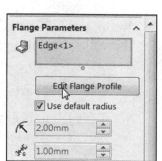

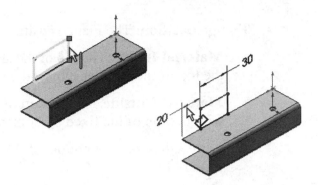

8. Drag the **back left edge** towards the front hole as illustrated. Add **dimensions** as illustrated.

9. Click **Finish** from the Profile Sketch box. Click **OK** ✔ from the PropertyManager. Edge-Flange1 is displayed in the FeatureManager.

10. **Close** the model.

Miter Flange tool

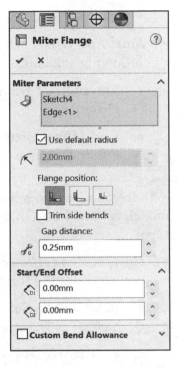

The Miter Flange ⬚ tool provides the ability to add a series of flanges to one or more edges of a sheet metal part. There are a few items to know when using the miter flange feature:

1. The sketch for a miter flange must adhere to the following requirements:

    •   The thickness of the miter flange is automatically linked to the thickness of the sheet metal part.

    •   The sketch for the miter flange can contain lines or arcs.

    •   The Miter Flange profile can contain more than one continuous line. Example: L-shaped profile

    •   The Sketch planc must be normal to the first created edge of the Miter Flange.

2. Specify an offset of the miter flange instead of creating a miter flange across the entire edge of a sheet metal part.

3. You can create a miter flange feature on a series of tangent or non-tangent edges.

The Miter Flange tool uses the Miter Flange PropertyManager. The Miter Flange PropertyManager provides the following selections:

•   *Miter Parameters*. The Miter Parameters box provides the following selections:

    •   **Along Edges**. Displays the selected edge.

    •   **Use default radius**. Uses the default bend radius for the miter flange.

    •   **Bend radius**. Only available when the Use default radius is un-checked. Set the bend radius value.

- **Flange position**. The Flange position option provides the following selections:

  - **Material Inside**. The top of the shaded preview of the flange coincides with the top of the fixed sheet metal entity.

  - **Material Outside**. The bottom of the shaded preview of the flange coincides with the top of the fixed sheet metal entity.

  - **Bend Outside**. The bottom of the shaded preview of the flange is offset by the bend radius.

  - **Trim side bends**. Removes extra material.

  - **Gap distance**. Sets the distance of the gap.

- *Start/End Offset*. The Start/End Offset box provides the following selections:

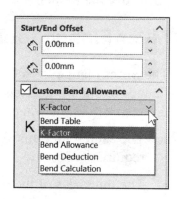

  - **Start Offset Distance**. Enter the value for Start Offset Distance.

  - **End Offset Distance**. Enter the value for the Start Offset Distance.

If you want the miter flange to span the entire edge of the model, set the Start Offset and End Offset Distance values to zero.

- *Custom Bend Allowance*. The Custom Bend Allowance box provides the following four selections:

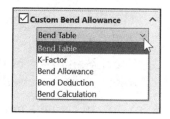

  - **Bend Table**. Select a bend table from the drop-down menu.

    - **Browse**. Browse to locate a bend table file.

  - **K-Factor**. Selected by default. Enter a K-Factor value. A K-Factor is a ratio that represents the location of the neutral sheet with respect to the thickness of the sheet metal part.

  - **Bend Allowance**. Enter a Bend Allowance value. A bend allowance is the arc length of the bend as measured along the neutral axis of the material.

  - **Bend Deduction**. Enter a Bend Deduction value. A bend deduction is the difference between the bend allowance and twice the outside setback Bend Allowance Type.

  - **Bend Calculation**. Select a table from the list, or click Browse to browse to a table.

Tutorial: Miter Flange 14-1

Create a Miter Flange feature in a Sheet metal part.

1. Open **Miter Flange 14-1** from the SOLIDWORKS 2018\SheetMetal folder.

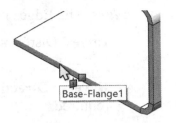

2. Click **Miter Flange**  from the Sheet Metal toolbar. The PropertyManager is displayed.

3. Click the **front inside edge** of the BRACKET as illustrated.

4. Click the **Line** Sketch tool.

5. Sketch a **small vertical line** on the front inside left corner as illustrated. The first point is coincident with the new sketch origin.

6. Add **dimensions** as illustrated. The line is 2mm.

7. **Exit** the Sketch. The Miter Flange PropertyManager is displayed.

8. Click the **Propagate** icon in the Graphics window to select inside tangent edges. The selected entities are displayed.

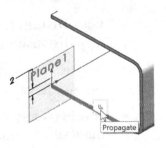

9. Uncheck **Use default radius**.

10. Enter **.50**mm for Bend Radius.

11. Click **Material Inside** for Flange position.

12. Check the **Trim side bends** box.

13. Enter **.50**mm for Rip Gap distance.

14. Click **OK** from the Miter Flange PropertyManager. The Miter Flange1 feature is displayed in the FeatureManager.

15. **Close** the model.

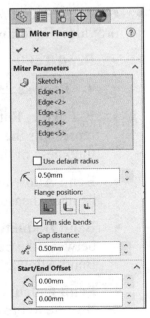

Hem tool

The Hem tool provides the ability to curl or hem the edge of a sheet metal part. There are a few items to know when using the Hem feature:

1. The selected edge to hem must be linear.

2. Mitered corners are automatically added to intersecting hems.

The Hem tool uses the Hem PropertyManager. The Hem PropertyManager provides the following options:

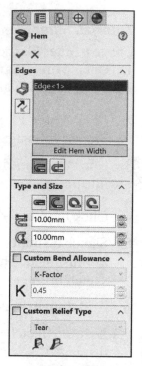

- *Edges*. The Edges box provides the following selections:

  - **Edges**. Displays the selected edges to hem from the Graphics window.

  - **Reverse Direction**. Reverse the direction of the hem if required.

  - **Material Inside**. Selected by default. The top of the shaded preview of the flange coincides with the top of the fixed sheet metal entity.

  - **Bend Outside**. The bottom of the shaded preview of the flange is offset by the bend radius.

- *Type and Size*. The Type and Size box provides the following selections:

  - **Closed**, **Open**, **Tear Drop** and **Rolled**.

  - **Length**. Sets the Length value.

  - **Gap Distance**. Only available for the Open option. Sets the Gap distance value.

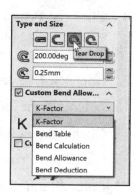

  - **Angle**. Only available for the Drop and Rolled options. Sets the Angle value.

  - **Radius**. Only available for the Drop and Rolled options. Sets the Radius value.

- *Custom Bend Allowance*. The Custom Bend Allowance box provides the following selections:

  - **Bend Table**. Select a bend table from the drop-down menu.

    - **Browse**. Browse to locate a bend table file.

  - **K-Factor**. Enter a K-Factor value. A K-Factor is a ratio that represents the location of the neutral sheet with respect to the thickness of the sheet metal part.

  - **Bend Allowance**. A bend allowance is the arc length of the bend as measured along the neutral axis of the material.

  - **Bend Deduction**. A bend deduction is the difference between the bend allowance and twice the outside setback Bend Allowance Type.

  - **Bend Calculation**. Select a table from the list, or click Browse to browse to a table.

Tutorial: Hem 14-1

Create an Open type Hem feature on a Sheet metal part.

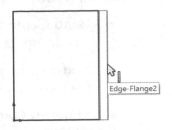

1. Open **Hem 14-1** from the SOLIDWORKS 2018\SheetMetal folder.

2. Display a **Front** view.

3. Click the **right vertical edge** of the right flange.

4. Click **Hem** ⬧ from the Sheet Metal toolbar. The Hem PropertyManager is displayed. Edge<1> is displayed in the Edges box.

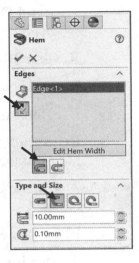

5. Display an **Isometric** view.

6. Click the **Reverse Direction** button.

7. Click **Open Hem** type.

8. Enter **10**mm for Hem Length.

9. Enter **.10**mm for Gap Distance.

10. Click **OK** ✔ from the Hem PropertyManager. The Hem1 feature is displayed in the FeatureManager.

11. **Close** the model.

Sketch Bend tool

The Sketch Bend 🖳 tool provides the ability to add a bend from a selected sketch in a sheet metal part. You can add a bend line to the sheet metal part while the part is in the folded state with a sketched bend feature. There are a few items to know when using the Sketch Bend feature:

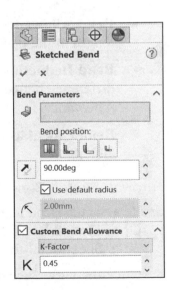

1. Only lines are allowed in the sketch.

2. You can have more than one line per sketch.

3. The bend line does not have to be the same length of the faces you are bending.

The Sketch Bend tool uses the Sketched Bend PropertyManager. The Sketched Bend PropertyManager provides the following selections:

- *Bend Parameters*. The Bend Parameters box provides the following selections:

  - **Fixed Faces**. Displays the selected face from the Graphics window. The Fixed face does not move as a result of the bend.

- **Bend position**. Select one of the four bend positions:

  - **Bend Centerline**. Selected by default. The bend line is located such that it equally splits the bend region in the flattened part.

  - **Material Inside**. The top of the shaded preview of the flange coincides with the top of the fixed sheet metal entity.

  - **Material Outside**. The bottom of the shaded preview of the flange coincides with the top of the fixed sheet metal entity.

  - **Bend Outside**. The bottom of the shaded preview of the flange is offset by the bend radius.

- **Reverse Direction**. Reverses the direction of the bend position if required.

- **Bend Angle**. 90deg selected by default. Sets the value of the bend angle.

- **Use default radius**. Selected by default. Uses the default bend radius.

- **Bend Radius**. Only available if the Use default radius is un-checked. Sets the radius of the bend.

- *Custom Blend Allowance*. The Custom Blend Allowance box provides the following selections:

  - **Bend Table**. Select a bend table from the drop down menu.

    - **Browse**. Browse to locate a bend table file.

  - **K-Factor**. Enter a K-Factor value. A K-Factor is a ratio that represents the location of the neutral sheet with respect to the thickness of the sheet metal part.

  - **Bend Allowance**. Enter a Bend Allowance value. A bend allowance is the arc length of the bend as measured along the neutral axis of the material.

  - **Bend Deduction**. Enter a Bend Deduction value. A bend deduction is the difference between the bend allowance and twice the outside setback Bend Allowance Type.

  - **Bend Calculation**. Select a table from the list, or click Browse to browse to a table.

Tutorial: Sketch Bend 14-1

Create a Sketch Bend feature in a Sheet Metal part.

1. Open **Sketch Bend 14-1** from the SOLIDWORKS 2018\SheetMetal folder.

2. Click the **right face** of Edge-Flange1.

3. Sketch a **line** on the planar face of the sheet metal part as illustrated.

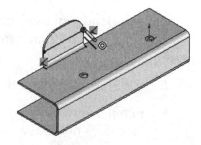

4. Click **Sketch Bend**  from the Sheet Metal toolbar. The Sketched Bend PropertyManager is displayed.

5. Click the **face** below the sketch line as illustrated in the Graphics window. The selected section does not move as the result of the bend.

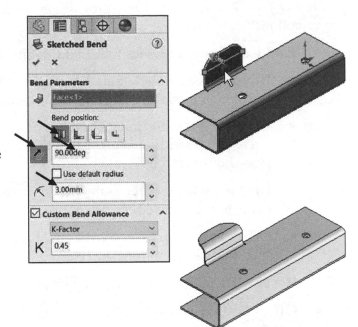

6. Click the **Reverse Direction** button. The direction arrow points towards the left.

7. Enter **90**deg for Angle.

8. Uncheck **Use default radius**.

9. Enter **3.0**mm for Radius.

10. Click **OK** ✅ from the Sketched Bend PropertyManager. The Sketched Bend1 feature is displayed in the FeatureManager.

11. **Close** the model.

Jog tool

The Jog  tool provides the ability to add two bends from a sketched line in a sheet metal part. There are a few items to know when using the Jog tool:

1. The line does not need to be vertical or horizontal.

2. Your sketch must contain at least one line.

3. The bend line does not have to be the same length of the faces that you are bending.

The Jog tool uses the Jog PropertyManager. The Jog PropertyManager provides the following selections:

- *Selections*. The Selections box provides the following options:

    - **Fixed Face**. Displays the selected fixed face for the Jog.

    - **Use default radius**. Selected by default. Uses the default Jog radius.

    - **Bend Radius**. Only available if the Use default radius is un-checked. Sets the radius of the Jog.

- *Jog Offset*. The Jog Offset box provides the following selections:

  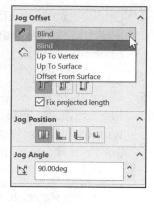

  - **End Condition**. Provides four End Condition selections:

    - **Blind**. Selected by default. Extends the feature from the Sketch plane from a specified distance.

    - **Up To Vertex**. Extends the feature from the Sketch plane to a plane that is parallel to the Sketch plane and passes through the selected vertex.

    - **Up To Surface**. Extends the feature from the Sketch plane to the selected surface.

    - **Offset From Surface**. Extends the feature from the Sketch plane to a specified distance from the selected surface.

  - **Reverse Direction**. Reverses the direction of the Jog if required.

  - **Offset Distance**. Displays the selected offset distance of the Jog.

  - **Dimension Position**: Outside Offset selected by default. There are three position types. They are **Outside Offset, Inside Offset** and **Overall Dimension**.

  - **Fix projected length**. Provides the ability for the selected face of the jog to stay the same length. The overall length of the tab is preserved.

- *Jog Position*. The Job Position box provides the following selections:

  - **Bend Centerline, Material Inside, Material Outside** and **Bend Outside**.

- *Jog Angle*. Displays the selected Jog Angle value.

- *Custom Blend Allowance*. The Custom Blend Allowance box provides the following selections:

  - **Bend Table**. Select a bend table from the drop down menu.

  - **Browse**. Browse to locate a bend table file.

  - **K-Factor**. Enter a K-Factor value. K-Factor is a ratio that represents the location of the neutral sheet with respect to the thickness of the sheet metal part.

  - **Bend Allowance**. Enter a Bend Allowance value. A bend allowance is the arc length of the bend as measured along the neutral axis of the material.

  - **Bend Deduction**. Enter a Bend Deduction value. A bend deduction is the difference between the bend allowance and twice the outside setback Bend Allowance Type.

  - **Bend Calculation**. Select a table from the list, or click Browse to browse to a table.

Tutorial: Jog 14-1

Insert a Jog feature in a Sheet Metal part.

1. Open **Jog 14-1** from the SOLIDWORKS 2018\SheetMetal folder.

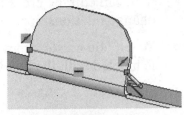

2. **Rotate** the model to view the back face of the tab.

3. Right-click the **back face** of the tab.

4. Click **Sketch**. Edge-Flange1 is highlighted in the FeatureManager.

5. Click the **Line** ✏ Sketch tool.

6. Sketch a **horizontal line** across the midpoints of the tab as illustrated.

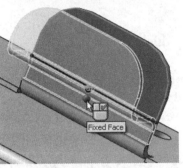

7. Click **Jog** 🗲 from the Sheet Metal toolbar. The Jog PropertyManager is displayed.

8. Click the **back face** of the tab below the horizontal line.

9. Enter **5**mm for Jog Offset Distance.

10. Display an **Isometric** view. The Jog direction arrow points to the left. If required, click the **Reverse Direction** button.

11. Click the **Outside Offset** button.

12. Enter **45**deg for Jog Angle.

13. Click **OK** ✔ from the Jog PropertyManager. The Jog1 feature is displayed in the FeatureManager.

14. **Close** the model.

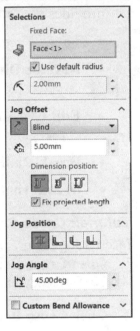

Cross-Break tool

In HVAC or duct work design, cross-breaks are used to stiffen sheet metal. The Cross-Break tool ⬦ provides the ability to insert a graphical representation of a cross break. There are a few items to know when using the Cross-Break feature:

- You can flatten a sheet metal part with a cross break.

- You can add edge or miter flanges to the edge of a face that contains a cross break.

- The dimensions of a part with a cross break are not changed.

- You can edit the cross break sketch to move corners and change relationships.

- When you create a part with a cross break, the flattened view is labeled with the bend direction, bend radius, and bend angle.

The corners of the cross break must be at the ends of the edge.

The Cross-Break tool uses the Cross Break PropertyManager. Use the Cross Break PropertyManager to add a graphical representation of a cross break to a sheet metal part. The Cross-Break PropertyManager provides the following selections:

- *Cross Break Parameters*. The Cross Break Parameters box provides the following selections:

    - **Faces**. Face on which the cross break is inserted.

    - **Reverse Direction**. Reverses the direction of the cross break in relation to the selected face.

    - **Edit Cross Profile**. Click to edit the profile sketch of the cross break to move corners and change relationships.

    - **Break Radius**. Radius to be used to create the cross break.

    - **Break Angle**. Angle to be used to create the cross break.

Tutorial: Cross-Break 14-1

Insert a Jog feature in a Sheet Metal part.

1. Open **Cross-Break 14-1** from the SOLIDWORKS 2018\SheetMetal folder.

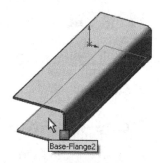

2. Click **Cross-Break** from the Sheet Metal toolbar. The Cross-Break PropertyManager is displayed.

3. Click the **bottom face** of Base-Flange2 as illustrated. The direction arrow points upwards.

Accept the default settings.

4. Click **OK** from the Cross-Break PropertyManager. The Cross Break1 feature is displayed in the FeatureManager. View the results in the Graphics window.

5. **Close** the model.

Consolidated Corner toolbar

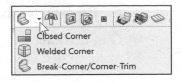

The Consolidated Corner toolbar provides access to the following sheet metals tools: **Closed Corner, Welded Corner** and **Break-Corner/Corner-Trim**.

 External corners cut material. Internal corners add material.

You can use the Closed Corner tool to add material between sheet metal features when there is a cut across flanges, with large radius bends, when the part is created using Insert Bends or Convert to Sheet Metal, with faces that are part of a larger corner.

Closed Corner tool

The Closed Corner 🔲 tool provides the ability to extend (add material) the face of a sheet metal part. Add closed corners between sheet metal flanges. The Closed Corner feature provides the following capabilities:

1. Close non-perpendicular corners.

2. Close or open the bend region.

3. Close multiple corners simultaneously by selecting the faces for all of the corners that you want to close.

4. Adjust the Gap distance. The distance between the two sections of material that were added by the Closed Corner feature.

5. Adjust the Overlap/underlap ratio. The ratio between the material that overlaps and the material that underlaps. A value of 1 indicates that the overlap and the underlap are equal.

The Closed Corner feature uses the Closed Corner PropertyManager. The Closed Corner PropertyManager provides the following selections:

- *Faces to Extend*. The Faces to Extend box provides the following selections:

  - **Faces to Extend**. Displays the selected planar faces from the Graphics window.

  - **Faces to Match**. SOLIDWORKS software attempts to find the Faces to match.

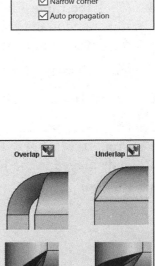

- **Corner type**. Overlap selected by default. Corner type provides three options:

  - **Butt**, **Overlap**, **Underlap**.

- **Gap distance**. Set a value for Gap distance. Distance between the two sections of material that were added by the Closed Corner feature.

- **Overlap/underlap ratio**. Only available with the Overlap or Underlap Corner type option. Sets the value. The value is the ratio between the material that overlaps and the material that underlaps.

- **Open bend region**. Preview is not displayed when selected.

- **Coplanar faces**. When cleared, all coplanar faces are selected automatically.

- **Narrow Corner**. Uses the algorithm for large bend radii to narrow the gap in the bend area.

- **Auto propagation**. Is automatically cleared if you clear or delete any selections under **Faces to Extend** or **Faces to Match**.

Tutorial: Close Corner 14-1

Create a Close Corner feature in a Sheet Metal part. Close off the open space between the adjacent angled edges.

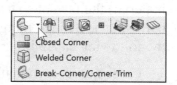

1. Open **Closed Corner 14-1** from the SOLIDWORKS 2018\SheetMetal folder.

2. Click the **Closed Corner** tool from the Sheet Metal toolbar. The Closed Corner PropertyManager is displayed.

3. **Zoom in** on the right front corner as illustrated.

4. Click the **Edge-Flange2 planar face** as illustrated in the Graphics window. Face<1> is displayed in the Faces to Extend box.

5. Click **Overlap** for Corner type.

6. Enter **.10**mm for Gap distance.

7. Click **OK** from the Closed Corner PropertyManager. The Closed Corner1 feature is displayed in the FeatureManager.

8. **Close** the model.

Welded Corner

The Welded Corner ⊞ tool provides the ability to add a weld bead to the corners of a folded sheet metal part, including miter flanges, edge flanges, and closed corners.

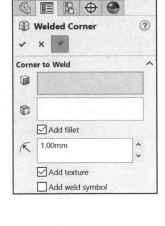

The Welded Corner uses the Welded Corner PropertyManager. The Welded Corner PropertyManager provides the following selections:

- *Corner to Weld*. Provides the ability to select a side face of a sheet metal corner to be welded. Display the selected face to apply the bead.

  - **Add Fillet**. Selected by default. Inserts a Fillet.

  - **Fillet Radius**. The fillet radius must be less than a tangent arc connecting the outer edges.

  - **Add Texture**. Selected by default. Adds a texture to the bead for display purposes.

  - **Add Weld Symbol**. Adds a weld symbol for the created weld.

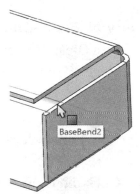

Tutorial: Welded Corner 14-1

Create a Welded Corner feature in a Sheet Metal part.
1. Open **Welded Corner 14-1** from the SOLIDWORKS 2018\SheetMetal folder.

2. Click the **Welded Corner** ⊞ tool from the Sheet Metal toolbar. The Welded Corner PropertyManager is displayed.

3. **Zoom in** on the right front side.

4. Click the flat face of **Edge-Flange1** as illustrated in the Graphics window. Face<1> is displayed.

5. Check the **Add fillet** box.

6. Enter **2.00**mm for fillet radius. The fillet radius must be less than a tangent arc connecting the outer edges.

7. Click **OK** ✔ from the PropertyManager.

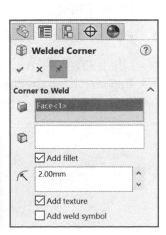

8. Click **OK** ✔ from the PropertyManager. Read the rebuild message. Click **OK** if required. View the FeatureManager. Welded Corner1 is displayed.

9. **Close** the model.

Break Corner/Corner-Trim tool

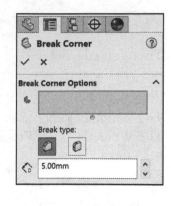

The Break-Corner/Corner Trim ⬚ tool uses the Break Corner PropertyManager. The Break Corner PropertyManager provides the following selections:

- **Break Corner Options**. The Break Corner Options box provides the following selections:

  - **Corner Edges and/or Flange Face**. Displays the selected corner edges and or Flange Face from the Graphics window.

  - **Break type**. Provides the following two break type selections: **Chamfer** and **Fillet**.

  - **Distance**. Only available for the Chamfer selection. Sets the value for the chamfer distance.

  - **Radius**. Only available for the Fillet option. Sets the value for the fillet radius.

Tutorial: Break-Corner/Corner Trim 14-1

Create a Break-Corner feature in a Sheet Metal part.

1. Open **Break-Corner 14-1** from the SOLIDWORKS 2018\SheetMetal folder.

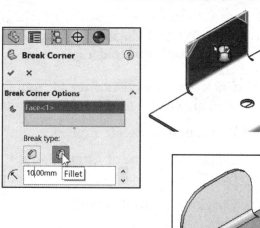

2. Click **Break-Corner/Corner Trim** ⬚ from the Sheet Metal toolbar. The Break Corner PropertyManager is displayed.

3. Click the **right face** of the tab as illustrated. Face<1> is displayed.

4. Click **Fillet** for Break type.

5. Enter **10**mm for Radius.

6. Click **OK** ✔ from the Break Corner PropertyManager. The Break-Corner1 feature is displayed in the FeatureManager.

7. **Close** the model.

Vent tool

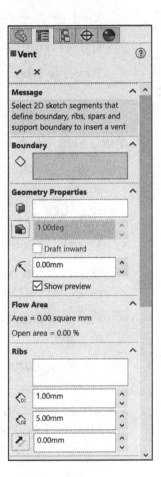

The Vent ▦ tool provides the ability to create various sketch elements for air flow in both plastic and sheet metal design. The Vent tool uses the Vent PropertyManager. The Vent PropertyManager provides the following selections:

- *Boundary*. The Boundary box provides the following option:

  - **Select 2D sketch segments**. Displays the selected sketch segment from the Graphics window to form a closed profile as the outer vent boundary.

- *Geometry Properties*. The Geometry Properties box provides the following selections:

  - **Select a face**. Displays the selected planar or non-planar face from the Graphics window for the vent. The entire vent sketch must fit on the selected face.

  - **Draft Angle**. Click Draft On/Off to apply draft to the boundary, fill-in boundary, plus all ribs and spars.

🔆 For vents on planar faces, draft is applied from the Sketch plane.

  - **Radius for the fillets**. Sets the fillet radius, which is applied to all intersections between the boundary, spars, ribs, and the fill-in boundary.

  - **Show preview**. Display a preview of the feature.

- *Flow Area*. The Flow Area box provides the following selections:

  - **Area**. Displays the total available area inside the boundary. This value remains fixed. The value is provided in square units.

  - **Open area**. Displays the open area inside boundary for air flow. This value updates as you add vent entities. Draft, fillets, ribs, spars, and the fill-in boundary reduce the open area value. The value is provided in percent.

- *Ribs*. The Ribs box provides the following selections:

  - **Select 2D sketch segments**. Displays the selected sketch segments for ribs from the Graphics window.

  - **Depth**. Sets the depth value of the ribs.

- **Width**. Sets the width value of the ribs.

- **Offset from Surface**. Sets the offsets value for all ribs from the surface.

- **Reverse Direction**. Reverses the direction of the offset if required.

You must create at least one rib before you can create spars.

- *Spars*. The Spars box provides the following selections:

  - **Select 2D sketch segments**. Displays the selected sketch segments for spars from the Graphics window.

  - **Depth**. Sets the depth value of the spars.

  - **Width**. Sets the width value of the spars.

  - **Offset from Surface**. Sets the offset value for all spars from the surface.

  - **Reverse Direction**. Reverses the direction of the offset if required.

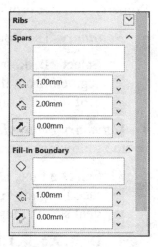

- *Fill-In Boundary*. The Fill-In Boundary box provides the selections:

  - **Select 2D sketch segments**. Select sketch entities that form a closed profile. At least one rib must intersect the fill-in boundary.

  - **Depth**. Sets the depth value of the support area.

  - **Offset**. Offsets the fill-in boundary from the surface.

  - **Reverse Direction**. Reverses the direction of the offset if required.

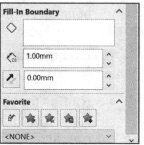

- *Favorite*. Manage a list of favorites that you can reuse in models.

Tutorial: Vent 14-1

Create a Vent feature for a Sheet metal part.

1. Open **Vent 14-1** from the SOLIDWORKS 2018\SheetMetal folder. Click **Vent** ⬚ from the Sheet Metal toolbar. The Vent PropertyManager is displayed.

2. Click the **large circle circumference** as illustrated. The 2D Sketch is a closed profile.

3. Click inside the **Ribs** box.

4. Click the **horizontal** sketch line.

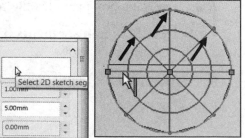

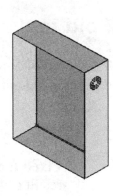

5. Click the **three remaining lines** as illustrated. The selected entities are displayed in the Ribs box. View the Flow area and Open area information.

6. Click **OK** ✔ from the Vent PropertyManager. Vent1 is created and is displayed in the FeatureManager. View the results.

7. **Close** the model.

Unfold tool

The Unfold 🖼 tool provides the ability to flatten one or more bends in a sheet metal part. The Unfold tool uses the Unfold PropertyManager. The Unfold PropertyManager provides the following selections:

- **Selections**. The Selections box provides the following options:

  - **Fixed face**. Displays the selected face that does not move as a result of the feature from the Graphics window.

  - **Bends to unfold**. Displays the selected bends from the Graphics window.

  - **Collect All Bends**. Selects all bends in the active part.

Tutorial: Unfold 14-1

Create an Unfold feature in a Sheet Metal part.

1. Open **Unfold 14-1** from the SOLIDWORKS 2018\SheetMetal folder.

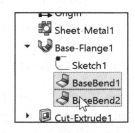

2. Click **Unfold** 🖼 from the Sheet Metal toolbar. The Unfold PropertyManager is displayed.

3. Click the **top face** of the bracket. Face<1> is displayed in the Fixed face box.

4. Click **BaseBend1** and **BaseBend2** from the Fly-out FeatureManager. BaseBend1 and BaseBend2 are displayed in the Bends to Unfold box.

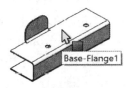

5. Click **OK** ✔ from the Unfold PropertyManager. The Unfold1 feature is displayed in the FeatureManager.

6. Display an **Isometric** view. View the results.

7. **Close** the model.

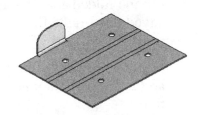

Fold tool

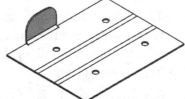

The Fold 🗃 tool provides the ability to bend one or more sections in a sheet metal part. The Fold tool uses the Fold PropertyManager.

The Fold PropertyManager provides the following selections:

- *Selections*. The Selections box provides the following options:

  - **Fixed face**. Displays the selected face that does not move as a result of the feature from the Graphics window.

  - **Bends to fold**. Displays the selected bends from the Graphics window.

  - **Collect All Bends**. Selects all bends in the active part.

Tutorial: Fold: 14-1

Create a Fold feature in a Sheet Metal part.
1. Open **Fold 14-1** from the SOLIDWORKS 2018\SheetMetal folder.

2. Click **Fold** 🗃 from the Sheet Metal toolbar. The Fold PropertyManager is displayed. Face<1> is displayed in the Fixed face box.

3. Click **BaseBend1** and **BaseBend2** from the Fly-out FeatureManager. BaseBend1 and BaseBend2 are displayed in the Bends to Fold box.

4. Click **OK** ✔ from the Fold PropertyManager. The Fold1 feature is displayed in the FeatureManager.

5. Display an **Isometric** view. View the results.

6. **Close** the model.

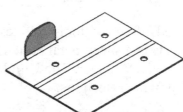

Flatten tool

The Flatten 🗃 tool provides the ability to display the Flat pattern for the existing sheet metal part. The Flatten feature is intended to be the last feature in the folded sheet metal part. All features before the Flatten feature in the FeatureManager design tree is displayed in both the folded and flattened sheet metal part. All features after the Flatten feature is displayed only in the flattened sheet metal part. Items to note about the Flat-pattern feature:

- New features in folded part. When Flat-Pattern1 is suppressed, all features that you add to the part automatically appear before this feature in the FeatureManager design tree.

- New features in flattened part. You flatten the entire sheet metal part by un-suppressing Flat-Pattern1. To add features to the flattened sheet metal part, you must first unsuppress Flat-Pattern1.

- Reorder features. You cannot reorder sheet metal features to go below Flat-Pattern1 in the FeatureManager design tree. So, you cannot order a cut with the Normal cut option underneath Flat-Pattern1.

- Modify parameters. You can modify the parameters of Flat-Pattern1 to control how the part bends, to enable or disable corner options, and to control the visibility of the bend region in the flattened sheet metal part.

- Sketches. You can transform sketches and their locating dimensions from a folded state to a flattened state and back again. The sketch and locating dimensions are retained.

The Flatten tool uses the Flat-Pattern PropertyManager. The Flat-Pattern PropertyManager provides the following selections:

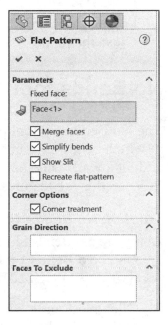

- *Parameters*. The Parameters box provides the following options:

  - **Fixed face**. Displays the selected fixed face from the Graphics window.

  - **Merge faces**. Merges faces that are planar and coincident in the flat pattern.

  - **Simplify bends**. Straighten curved edges in the selected flat pattern.

  - **Show Slit**. To show slits that are added for some corner relief features. When you create a rectangular or circular corner relief that is smaller than the bend area, a slit is added so that the part can still be bent. Enabling Show Slit makes the slit available in the flat pattern.

  - **Recreate flat-pattern**.

- *Corner Options*. The Corner Options box provides the following selection:

  - **Corner treatment**. Applies a smooth edge in the flat pattern.

- *Grain Direction*. Provides the ability to select an edge or line in the Graphics area.

Tutorial: Flatten 14-1

Display a flatten state for a Sheet Metal part.
1. Open **Flatten 14-1** from the SOLIDWORKS 2018\SheetMetal folder.

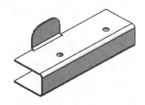

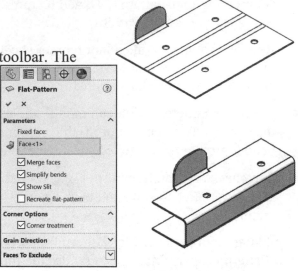

2. Click **Flatten** 📎 from the Sheet Metal toolbar. The Sheet Metal part is flattened.

Display the part in its fully formed state.

3. Click **Flatten** 📎 from the Sheet Metal toolbar. The part is returned to its fully formed state. View the Flat-Pattern PropertyManager.

4. Right-click **Flat-Pattern1** from the FeatureManager.

5. Click **Edit Feature**. View the Flat Pattern PropertyManager.

6. Click **OK** ✔ from the Flat-Pattern PropertyManager. View the results. **Close** the model.

No Bends tool

The No Bends 🖥 tool provides the ability to roll back all bends from a sheet metal part in which bends have been inserted. This provides the ability to create additions, such as adding a wall. This tool is only available in sheet metal parts with Flatten-Bends1 and Process-Bends1 features. The No Bends tool does not use a PropertyManager.

Tutorial: No Bends 14-1

Apply the No Bends SheetMetal tool.

1. Open **No Bends 14-1** from the SOLIDWORKS 2018\SheetMetal folder.

2. Click **No Bends** 🖥 from the Sheet Metal toolbar. The FeatureManager rolls back all bends from the sheet metal part. View the FeatureManager and the model in the Graphics window.

3. **Close** the model.

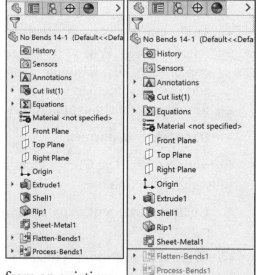

Insert Bends tool

The Inset Bends 📎 tool creates a sheet metal part from an existing part. You must create a solid body with uniform thickness when you use the Insert Bends tool.

💡 You can create a part directly out of sheet metal with the Base-Flange/Tab tool.

The Insert Bends tool uses the Bends PropertyManager. The Bends PropertyManager provides the following selections:

- **Bend Parameters**. The Bend Parameters box provides the following selections:

  - **Fixed Face or Edge**. Displays the selected fixed face or edge from the Graphics window.

  - **Bend Radius**. Sets the required bend radius value.

  - **Thickness**. Sets the required thickness value.

  - **Ignore beveled faces**. Excludes chamfers from being converted into sheet metal bends.

- **Bend Allowance**. The Bend Allowance box provides the following selections:

  - **Bend Table**. Select a bend table from the drop-down menu.

  - **Browse**. Browse to a bend table file.

  - **K-Factor**. Enter a K-Factor value. K-Factor is a ratio that represents the location of the neutral sheet with respect to the thickness of the sheet metal part.

  - **Bend Allowance**. Enter a Bend Allowance value. A bend allowance is the arc length of the bend as measured along the neutral axis of the material.

  - **Bend Deduction**. Enter a Bend Deduction value. A bend deduction is the difference between the bend allowance and twice the outside setback Bend Allowance Type.

- **Auto Relief**. Provides the ability to select three Relief types. The Rectangle relief type is selected by default. Note: If you want to add a Rectangle or Obround relief, you must specify the Relief ratio. See SOLIDWORKS help for additional information on Relief ratio.

- **Rip Parameters**. The Rip Parameters box provides the following selections:

  - **Edges to Rip**. Displays the selected edge to rip from the Graphics window.

  - **Change Direction**. Reverses the direction of the rip if required.

  - **Rip Gap**. Sets a rip gap distance value.

Rips are inserted in both directions by default. Each time you select the Change Direction option, the rip direction sequences from one direction, to the other direction, and then back to both directions.

Tutorial: Insert Bends 14-1

Insert bends in a Sheet Metal part.

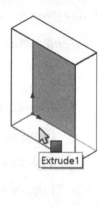

1. Open **Insert bends 14-1** from the SOLIDWORKS 2018\SheetMetal folder.

2. Click **Insert Bends** from the Sheet Metal toolbar. The Bends PropertyManager is displayed.

3. Click the **inside bottom face** to remain fixed. Face<1> is displayed in the Fixed Face box.

4. Enter **2.00**mm for Bend Radius.

5. Enter **.45** for K-Factor.

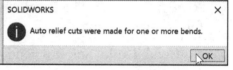

6. Select **Rectangular** for Auto Relief Type.

7. Enter **.5** for Relief Ratio.

8. Click **OK** ✓ from the Bends PropertyManager.

9. Click **OK** to the message, "Auto relief cuts were made for one or more bends."

10. Display an **Isometric** view. View the created feature in the FeatureManager and in the Graphics window.

11. **Close** the model.

Rip tool

The Rip tool provides the ability to create a gap between two edges in a sheet metal part. Create a rip feature:

- From linear sketch entities.

- Along selected internal or external model edges.

- By combining single linear sketch entities and model edges.

A Rip feature is commonly used to create sheet metal parts, but you can also add a rip feature to any part.

The Rip tool uses the Rip PropertyManager. The Rip PropertyManager provides the following selections:

- ***Rip Parameters***. The Rip Parameters box provides the following selections:

  - **Edges to Rip**. Displays the selected edges to rip from the Graphics window.

  - **Change Direction**. Reverses the direction of the rip if required.

  - **Rip Gap**. Sets a rip gap distance.

Rips are inserted in both directions by default. Each time you select the Change Direction option, the rip direction sequences from one direction, to the other direction, and then back to both directions.

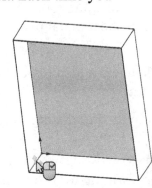

Tutorial: Rip 14-1

Create a Rip feature in a Sheet metal part.

1. Open **Rip 14-1** from the SOLIDWORKS 2018\SheetMetal folder.

2. Click **Rip** from the Sheet Metal toolbar. The Rip PropertyManager is displayed.

3. **Rotate** the part to view the inside edges.

4. Click the **inside lower left edge**.

5. Click the other **three inside edges**. The four selected edges are displayed in the Rip Parameters box.

6. Enter **.10mm** for Rip Gap.

7. Click **OK** from the Rip PropertyManager. Rip1 is created and is displayed in the FeatureManager.

8. **Close** the model.

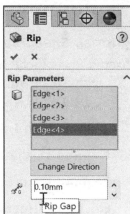

Sheet Metal Library Feature

Sheet metal manufacturers utilize dies and forms to create specialty cuts and shapes. The Design Library contains information on dies and forms. The SOLIDWORKS Design Library features folder contains examples of predefined sheet metal shapes.

Forming tools act as dies that bend, stretch, or otherwise form sheet metal to create form features such as louvers, lances, flanges, and ribs. The SOLIDWORKS software includes some sample forming tools. You can create forming tools and add them to sheet metal parts. When you create a forming tool:

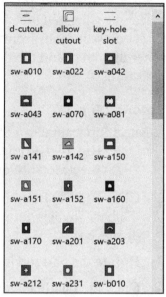

- The locating sketch is added to position the forming tool on the sheet metal part.

- The colors are applied to distinguish the Stopping Face from the Faces to Remove.

You can insert forming tools only from the Design Library and you can apply them only to sheet metal parts.

Tutorial: Sheet metal Library Feature 14-1

Insert a Sheet metal library feature.

1. Open **Sheet metal library feature 14.1** from the SOLIDWORKS 2018\SheetMetal folder.

2. Click **Design Library** from the Task Pane as illustrated.

3. **Expand** Design Library.

4. **Expand** features.

5. Click **Sheetmetal** from the features folder.

6. Click and drag the **d-cutout** Sheetmetal library feature to the right side of the Shell1 feature as illustrated. Accept the defaults.

7. Click **OK** from the d-cutout PropertyManager. The d-cutout feature is displayed in the FeatureManager and in the Graphics window.

8. **Close** the model.

Summary

In this chapter, you learned about Sheet Metal Features and the associated tools. You reviewed and addressed each tool in the Sheet Metal toolbar: *Base-Flange/Tab, Convert to Sheet Metal, Lofted-Bend, Edge Flange, Miter Flange, Hem, Jog, Sketched Bend, Cross-Break, Closed Corners, Break-Corner/Corner-Trim, Extruded Cut, Simple Hole, Vent, Unfold, Fold, Flatten, No Bends, Insert Bends* and *Rip*.

Use the Swept Flange tool to create compound bends in sheet metal parts. See SOLIDWORKS help for additional information.

When using the Convert to Sheet Metal tool, you can keep the solid body to use with multiple Convert to Sheet Metal features or specify that the entire body be consumed by the tool. See SOLIDWORKS Help for additional information and details.

In Chapter 15, explore PhotoView 360, the Measure tool and Mass Properties. Explore the various tools in PhotoView 360: Edit Appearance, Copy Appearance, Paste Appearance, Edit Scene, Edit Decal, Integrated Preview, Preview Window, Final Render, Options, Schedule Render and Recall Last Rendered Image.

The Measure tool measures distance, angle, radius, and size of and between lines, points, surfaces, and planes in sketches, 3D models, assemblies, or drawings.

When you select a vertex or sketch point, the x, y, and z coordinates are displayed.

Apply and understand the SOLIDWORKS Mass Properties tool. Apply standard and assign override values of Mass Properties.

Notes:

CHAPTER 15: PHOTOVIEW 360, MEASURE AND MASS PROPERTIES TOOL

Chapter Objective

Chapter 15 provides a general overview of SOLIDWORKS PhotoView 360, Measure and Mass Properties tool. On the completion of this chapter, you will be able to:

- Activate SOLIDWORKS PhotoView 360 and address a part or assembly with the following tools:

  - Edit Appearance, Copy Appearance, Paste Appearance, Edit Scene, Edit Decal, Integrated Preview, Preview Window, Final Render, Scene Illumination, Schedule Options, Schedule Render and Recall Last Render.

- Comprehend the SOLIDWORKS Measure tool with the Center of Mass point (COM) option.

- Apply standard values and assign override values of Mass Properties.

SOLIDWORKS PhotoView 360

Introduction

PhotoView 360 is a SOLIDWORKS Add-in that produces photo-realistic renderings of SOLIDWORKS models. The rendered image incorporates the appearances, lighting, scene, and decals included with the model.

PhotoView 360 is available with SOLIDWORKS Professional or SOLIDWORKS Premium. Drag an appearance from one entity and drop it onto another. Pasted appearances retain all the customizations made in the Edit Appearance PropertyManager.

Drag and drop multiple appearances with the Appearance Target pop-up toolbar pinned. Select the level (Face, Feature, Body, Part, etc.) you need. The Appearance Target toolbar is document dependent.

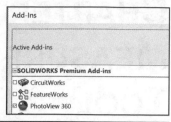

🔆 Illustrations may vary depending on your SOLIDWORKS version and system setup.

Active the Render tools for PhotoView 360. Select the PhotoView 360 menu either from the SOLIDWORKS Add-Ins tab or from Options, Add-Ins.

PhotoView 360 toolbar

Access the PhotoView 360 toolbar from the SOLIDWORKS Add-Ins. Click the PhotoView 360 icon. The Render Tools tab is displayed in the CommandManager. Click the Render Tools tab. The following tools are available:

- **Edit Appearance**. The Edit Appearance tool uses the Color PropertyManager and the Appearances, Scenes, and Decals tab in the Task Pane.

Drag and drop an Appearance, Scene or Decal on to your model (part or assembly) in the Graphics window.

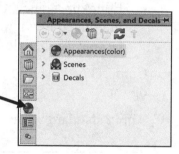

To control what is being affected by the Edit Appearance tool, utilize the Appearance Target pop-up toolbar to select *Face, Feature, Body, Part, Assembly, Appearance Filter* and the *Pin* tool. The Appearance Target toolbar is document dependent.

For an assembly, select either Apply at component level or Apply at part document level first from the color PropertyManager.

🔆 If the material is a blend of colors (for example automobile paint), Use the Color dialog box in the Color PropertyManager.

- **Copy Appearance**. Copy an appearance from one entity to another at a component or part document level.

- **Paste Appearance**. Pasted appearances retain all the customizations made in the Edit Appearance PropertyManager.

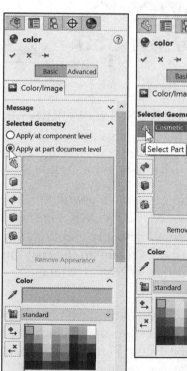

- *Edit Scene*. The Edit Scene tool uses the Edit Scene PropertyManager. The Edit Scene PropertyManager displays three tabs: *Basic*, *Advanced* and *PhotoView 360 Lighting*.

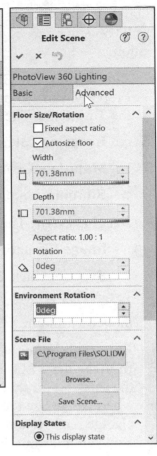

  - **Background**. The Background option provides the ability to use a background image with a scene so that what is visible behind the model is different from the reflections cast by the environment. For example, you might want a plain color behind the model while using the reflections from the courtyard scene.

  - **Environment**. The Environment option provides the ability to select any image to be spherically mapped as an environment for the scene.

  - **Floor**. The Floor option provides the following selections:

    - Floor reflections. Shows a reflection of the model on the floor.

    - Floor shadows. Shows shadows cast by the model on the floor.

    - Flatten floor (Background must be set to Use Environment). Flattens the floor of a spherical environment in scenes to improve the look of models that naturally rest on the ground or flat floors, particularly when performing view manipulations such as rotate or zoom.

    - Align floor with. Aligns the floor with a plane. Select one of XY, YZ, ZX, or Selected Plane.

Drag and drop a scene onto your model. As you rotate your model in the scene setting, you will view the model and scene and appearance from a camera perspective.

PhotoView provides various tools to render and view the model (part or assembly) quickly before you perform a full render.

To view a scene, you must first set the background appearance option.

- **Edit Decals**. The Edit Decals tool uses the Decals PropertyManager and the Appearances, Scenes, and Decals tab in the Task Pane. The Decals PropertyManager displays three tabs:

    o **Image**. Decal image and mask files.

    o **Mapping**. Geometry, Mapping, Size/Orientation, Rendering.

    o **Illumination**. Decal's response to illumination.

- **Decal Preview**. The decal is displayed in the window.

    o **Image file path**. The image file path is displayed.

    o **Browse**. Click **Browse** to choose another path and file.

- **Integrated Preview**. You can preview a rendering of the current model within the SOLIDWORKS graphics area.

- **Preview Window**. The Preview window is a separate window from the main SOLIDWORKS window. The window maintains the aspect ratio set in the PhotoView Options PropertyManager when you resize the window. Updates are interrupted when a change requires a rebuild. After the rebuild, the updates continue. You can interrupt the updates for performance reasons by clicking Pause.

Save renderings in progress directly from the PhotoView 360 Preview window. Use the continuous refinement settings in the PhotoView 360 Preview window to refine the rendering beyond the specified quality setting in the PhotoView options. The quality of the rendering increases over time. You can also specify that the PhotoView 360 Preview window create an image that uses the full specified output image size in the PhotoView options.

- **Final Render**. The Render Frame tool uses the Render dialog box. The dialog box opens and rendering begins.

- **Render Region**. Defines a render region to render.

- **Scene Illumination Proof Sheet**. PhotoView 360 proof sheets let you select lighting settings by viewing the effects of a range of settings in a series of proof sheets. Each image represents a different setting for your primary PhotoView 360 lighting.

Rendering brightness, background brightness, and scene reflectivity are controlled by the PhotoView 360 Scene Illumination Proof Sheet, a dialog box that lets you quickly see a wide range of variations of these parameters.

To open a PhotoView 360 Scene Illumination Proof Sheet, do one of the following:

- Click the **Render Tools** tab from the CommandManager, click **Scene Illumination Proof Sheet**.

- Click the DisplayManager tab. Click **View Scene, Lights**, and **Cameras**. Right-click **PhotoView 360 Lights** and click **Scene Illumination Proof Sheet**.

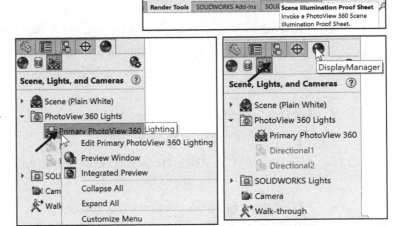

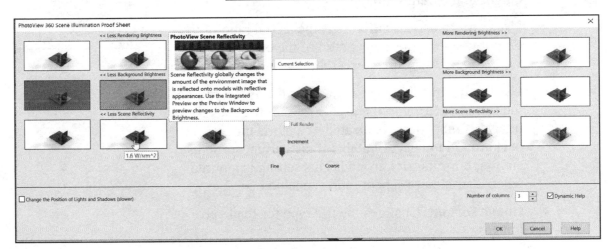

The proof sheet contains sets of thumbnail renderings. On the left, the thumbnails show the results of decreases in rendering brightness, background brightness, and scene reflectivity. On the right, they show increases in these values. By default, three columns of thumbnails appear on each side.

In the center, you can toggle between two larger thumbnails, one showing the original illumination and the other your selected illumination combination.

When you hover over a thumbnail, a tooltip shows the radiance for the setting.

- *Options*. The Options tool uses the PhotoView 360 Options PropertyManager. The PhotoView Options PropertyManager controls settings for PhotoView 360, including output image size and render quality.

  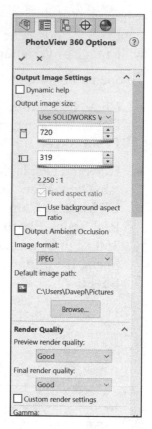

  - **Output Image Settings**. The available options are:

    - **Dynamic Help**. Displays pop-up tooltips for each property.

    - **Output image size - Preset image sizes**. Sets the size of the output image to a standard width and height. You can also select the settings assigned to the current camera or set custom values.

    - **Output image size - Image Width**. Sets the width of the output image, in pixels.

    - **Output image size - Image Height**. Sets the height of the output image, in pixels.

    - **Fixed aspect ratio**. Retains the current ratio of width to height in the output image.

    - **Use camera aspect ratio**. Sets the aspect ratio of the output image to the aspect ratio of the camera field of view. Available if the current view is through a camera.

    - **Use background aspect ratio**. Sets the aspect ratio of the final rendering to the aspect ratio of the background image. If this option is cleared, it distorts the background image. Available if the current scene uses an image for its background. This setting is ignored when Use camera aspect ratio is enabled.

    - **Image format**. Changes the file type for rendered images.

    - **Default image path**. Sets the default path for renderings that you schedule with the Task Scheduler.

  - **Render Quality**. The available options are:

    - **Preview render quality**. Sets the level of quality for the preview. Higher quality images require more time to render.

    - **Final render quality**. Sets the level of quality for the final render. Higher quality images require more time to render.

Typically, there is little difference between Best and Maximum. The Maximum setting is most effective when rendering interior scenes.

  - **Bloom**. The available options are:

- **Bloom**. Adds a bloom effect, a glow around very bright emissive or reflective objects in an image. The bloom is visible in the final rendering only, not in the preview.

- **Bloom setpoint**. Identifies the level of brightness or emissiveness to which the bloom effect is applied. Decreasing the percentage applies the effect to more items. Increasing it applies the effect to fewer items.

- **Bloom extent**. Sets the distance the bloom radiates from source.

- *Schedule Render*. The Schedule Render tool provides the ability to perform a rendering at a specified time and save it to a file.

- *Recall Last Render Image*. The Recall Last Render tool provides the ability to Recall the Last Render of the model.

A decal is a 2D image applied to a model. You can use decals to apply warning or instruction labels to models. You can also use decals to create model details that you can effectively represent with an image instead of with the model geometry, such as a car grill or a picture frame.

Tutorial: PhotoView 15-1

Open SOLIDWORKS. Apply PhotoView 360 to a SOLIDWORKS part.

1. **Start** a SOLIDWORKS session.

2. **Browse** to the SOLIDWORKS 2018\PhotoView 360 folder. Double-click **PhotoView 15-1**. The model is displayed in the Graphics window.

Activate the Render Tools for PhotoView 360.

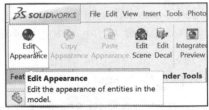

3. Click the **SOLIDWORKS Add-Ins** tab.

4. Click the **PhotoView 360** icon. The Render Tools tab is displayed in the CommandManager.

5. Click the **Render Tools** tab. View the available tools.

Apply the Edit Appearance tool to a part.

6. Click the **Edit Appearance** tool from the Render Tools tab in the CommandManager. The Color PropertyManager is displayed and the Appearances, Scenes, and Decal tab is selected in the Task Pane. View your options.

7. **Expand** the Appearances folder in the Task Pane.

8. **Expand** the Glass folder. Click **Gloss**.

9. Drag and drop **green glass** onto the side of the model as illustrated (Boss Battery). The Appearance Target pop-up toolbar is displayed.

10. Click the **Body** icon as illustrated from the Appearance Target pop-up toolbar. The model is displayed in green glass. The Part is a single body.

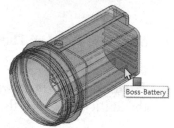

11. **Rotate** the model in the Graphics window to view the results. Click **OK** from the PropertyMananager.

To control what is being affected by the Edit Appearance tool, utilize the Appearance Target pop-up toolbar to select *Face, Feature, Body, Part, Assembly, Appearance Filter*, and the *Pin* tool. The Appearance Target toolbar is document dependent.

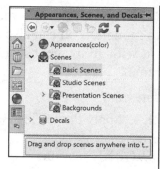

The Advanced tab in the PropertyManager provides the ability to access additional tabs and options.

Set an Environment for the part.

12. **Expand** the Scenes folder from the Task Pane as illustrated.

13. Click the **Basic Scenes** folder. View your options.

14. Drag and drop **Soft Box** into the Graphics window.

15. **Zoom out and rotate** the model in the Graphics window to view the new scene.

View your options in the Edit Scene PropertyManager.

16. Click the **Edit Scene**  tool. The Edit Scene PropertyManager is displayed. View your options.

Scene functionality is enhanced to allow full control of the scene that is visible behind the model. The DisplayManager lists the background and environment applied to the currently active model. The Edit Scene PropertyManager, available from the View Scene Pane in the DisplayManager, lets you size the floor, control the background or environment, and save custom scenes.

17. Click **Cancel** ✖ from the Edit Scene PropertyManager.

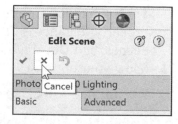

View the PhotoView 360 Options PropertyManager.

18. Click the **Options** 🔧 tool. The PhotoView 360 Options PropertyManager is displayed. View the available tools and options.

19. Click **Cancel** ✖ from the PhotoView 360 Options PropertyManager.

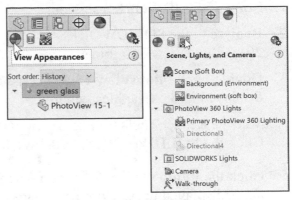

🔆 The DisplayManager is the central location for managing all aspects of lighting, including lighting controls that are available only when PhotoView 360 is Added in. The DisplayManager lists the lights applied to the currently active model. Light intensity is controlled with wattages.

Render the model.

20. Click the **Final Render** 🔴 tool.

21. Click **Continue without Camera or Perspective View**. The Final Render dialog box is displayed.

22. **View** the final render of the model in the Graphics window.

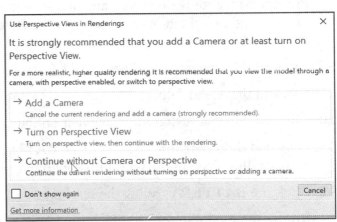

Save the Image.

23. Click **Save Preview Image** from the Render dialog box.

24. **Browse** to the needed folder and save the Image.

25. **Close** the Dialog box.

26. **Close** the Preview window.

27. **Close** the model.

🔆 In PhotoView 360, you can view PhotoWorks decals which were applied in SOLIDWORKS and that are visible when the part or assembly was saved.

Tutorial: PhotoView 15-2

Open SOLIDWORKS. Apply PhotoView 360 to an assembly and apply a decal.

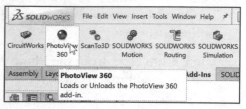

1. **Start** a SOLIDWORKS session.

2. **Browse** to the SOLIDWORKS 2018\PhotoView 360 folder.

3. Double-click **PhotoView 15-2**. The assembly is displayed in the Graphics window.

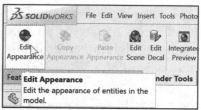

Activate the Render Tools for PhotoView 360.

4. Click the **SOLIDWORKS Add-Ins** tab.

5. Click the **PhotoView 360** ⊙ icon. The Render Tools tab is displayed in the CommandManager.

6. Click the **Render Tools** tab. View the available tools.

Apply the Edit Appearance tool to an Assembly.

7. Click the **Edit Appearance** ⊛ tool from the Render Tools tab in the CommandManager. The Color PropertyManager is displayed and the Appearances, Scenes, and Decal tab is selected in the Task Pane. View your options.

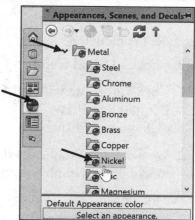

8. **Expand** the Appearances folder in the Task Pane.

9. **Expand** the Metal folder.

10. Click **Nickel** as illustrated. View your options.

11. Drag and drop **Duranickl(R)** onto the BATTERY body as illustrated in the Graphics window.

12. Click the **BATTERY**. View the results.

13. **Rotate** the model in the Graphics window to view the results. If needed, click **OK** ✔ from the **Duranickl(R)** PropertyManager.

14. **Display** an Isometric view.

Apply a Decal to the Assembly. Use the Mapping tab option.

15. Click the **Decals** ⊟ folder from the Task Pane.

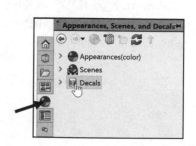

16. Drag and drop the **SOLIDWORKS** decal on the assembly in the Graphics window as illustrated. The Decals PropertyManager is displayed and the decal is displayed on the assembly.

17. Click the **Mapping** tab.

18. Select **Cylindrical** for Mapping type.

19. Select **XY** for direction.

20. Enter **80.00deg** About Axis.

21. Check **Fixed aspect ratio**.

22. Enter **5in** for Width.

23. Click inside the **Height** box.

24. Click **OK** ✔ from the Decals PropertyManager. View the decal on the model.

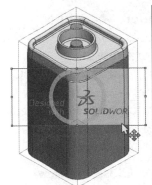

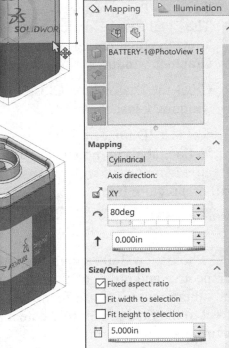

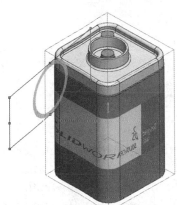

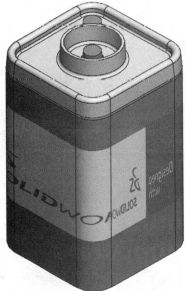

Render the Assembly.

25. Click the **Final Render**  tool from either the Render Tools tab in the CommandManager or from the PhotoView 360 Menu Bar toolbar. Click **Turn on Perspective View** if needed. The Final Render dialog box is displayed. View the final render of the model in the Graphics window.

26. **Close** the Dialog box. Close **Preview** window. **Close** the model.

A decal is a 2D image applied to a model. You can use decals to apply warning or instruction labels to models. You can also use decals to create model details that you can effectively represent with an image instead of with the model geometry, such as a car grill or a picture frame.

As an exercise, apply an Environment to the model.

PhotoView 360 camera supports additional effects; namely, you can switch between perspective and orthogonal views, create renderings where part of the image is in focus and other parts are not in focus through the Depth-of-field controls, and add a bloom effect to the final render to create a glow for emissive appearances or areas of very bright environment reflections.

Measure tool

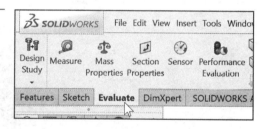

The Measure tool measures **distance**, **angle**, **radius**, and **size** of and between lines, points, surfaces and planes in **sketches**, **models**, **assemblies** or **drawings**.

When you select a vertex or sketch point, the x, y, and z coordinates are displayed.

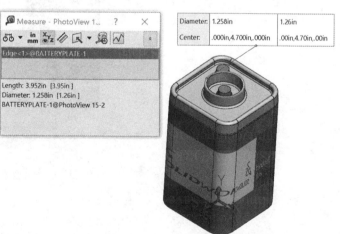

The Measure tool uses the Measure dialog box. The Measure dialog box provides the following selections:

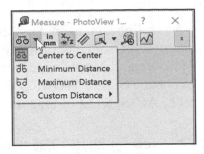

- *Arc/Circle Measurements*. Specify the distance to display when arcs or circles are selected. The options are:

  - **Center to Center.**

  - **Minimum Distance.**

  - **Maximum Distance.**

  - **Custom Distance.**

- *Units/Precision*. Provides access to the Measure Units/Precision dialog box. Select Use document settings or Use Custom Settings. Custom Settings provides access to Length unit, Scientific Notation, Decimal places, and Use Dual Units.

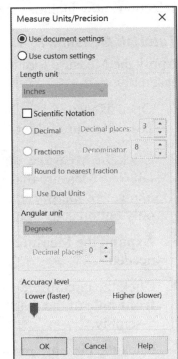

- *Show XYZ Measurements*. Select to display dX, dY, and dZ measurements between selected entities in the Graphics window. Clear to display only the minimum distance between selected entities.

- *Point to Point*. Enables point-to-point mode to measure the distance between any two points on the model. Selected by default.

- *Projected On*. Displays the distance between your selected entities as projected on one of the following. The options are:

  - **None.** Projection and Normal are not calculated. This is the default.

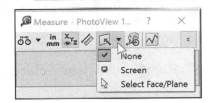

  - **Screen.** Displays in the Graphics window.

  - **Select Face/Plane.** Select a face or plane in the Graphics window or the FeatureManager design tree.

- *Measure History*. Opens the Measurement History dialog box, where you can view all measurements made during the current session of SOLIDWORKS.

- *Create Sensor*. Opens the Sensor PropertyManager, where you can set the software to alert you if the measurement value changes.

🔅 Define a coordinate system for a part or assembly. Use the coordinate system with the Measure and Mass Properties tools, and for exporting SOLIDWORKS documents to IGES, STL, ACIS, STEP, Parasolid, VRML and VDA.

🔅 New for 2018 - The input box that lists selections now includes six items.

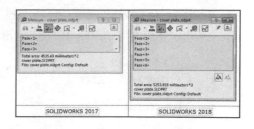

Tutorial: Measure 15-1

Apply the Measure tool to verify a design of an assembly.

1. Open **Measure 15-1** from the SOLIDWORKS 2018\Measure folder.

2. Click the **Measure** 🔎 tool from the Evaluate tab. The Measure dialog box is displayed. The *Show XYZ Measurements* button should be the only button activated.

3. **Expand** the Measure dialog box if needed.

4. Click the circumference of the **right front hole** of the f718b-30-b5-g-1 assembly.

5. Click the circumference of the **right back hole** of the f718b-30-b5-g-1 assembly. The distance 114.3mm is displayed.

6. Right-click **Clear Selections** in the display box.

7. Click the **front shaft** as illustrated. View the results in the Measure Dialog box.

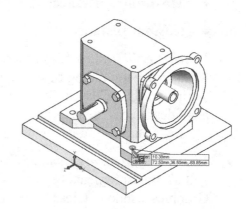

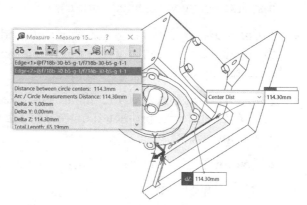

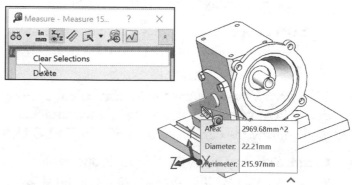

Apply Custom settings and dual dimensions.

8. Click **Units/Precision** from the Measure dialog box.

9. Check the **Use custom settings** box. Accept the default length units (Millimeters).

10. Select **3** Decimal places.

11. Check the **Use Dual Units** box.

Select the second length unit.

12. Select **inches**.

13. Click **OK**. View the updated results.

14. **Close** the Measure dialog box.

15. **Close** the model.

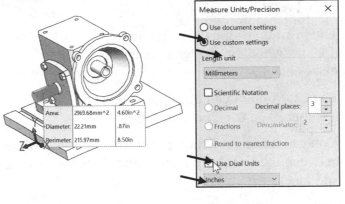

☼ Use the Measure Units/Precision dialog box to modify units of measure used by the Measure tool. To modify units of measure: Click **Units/Precision** in the Measure dialog box, select options, then click OK.

Tutorial: Center of Mass Point in an assembly 15-2

Insert a COM point. Apply the Measure tool in the assembly to validate the location of the Center of Mass.

1. Open **Center of Mass Point 15-2** from the SOLIDWORKS 2018\Measure\COM folder.

2. Click **Insert ▷ Reference Geometry ▷ ⊕ Center of Mass**. The **center of mass** of the model is displayed in the Graphs window and in the FeatureManager design tree just below the origin.

3. Click the **Measure** 🔎 tool from the Evaluate tab. The Measure dialog box is displayed. The *Show XYZ Measurements* button should be the only button activated.

4. **Expand** the Measure dialog box if needed.

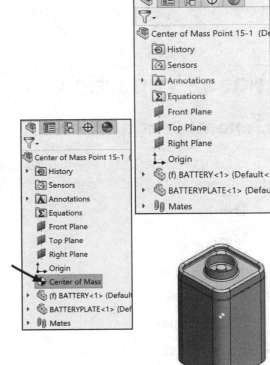

5. Click the **top edge** of the BATTERY assembly.

6. Click the **Center of Mass**. View the results.

7. **Close** the model.

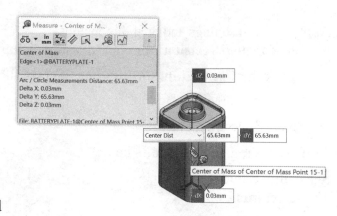

The position of the COM point updates when the model's center of mass changes.

The COM point can be suppressed and unsuppressed for configurations.

You can measure distances and add reference dimensions between the COM point and entities such as vertices, edges, and faces.

If you want to display a reference point where the CG was located at some particular point in the FeatureManager, you can insert a Center of Mass Reference Point.

A COMRP is a reference point created at the current center of mass of the part. It remains at the coordinates where you create it even if the COM point moves due to changes in the geometry of the part.

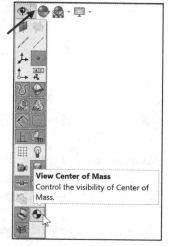

Mass Properties

General Introduction

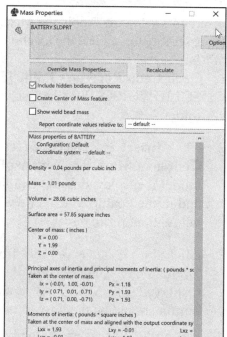

The SOLIDWORKS Mass properties tool provides information on the following:

- Density.

- Mass.

- Volume.

- Surface area.

- Center of mass.

- Principal axes of inertia.

- Moments of inertia and products of inertia.

☼ When you do not apply any material to a part, the density of water is used as the default for mass and inertia calculations.

Apply Material

Always apply material to a part or component. To apply material:

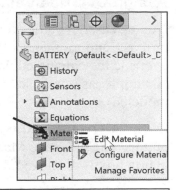

- Right-click in the **Materials** folder.

- Click **Edit Material** to add a material.

- Select Material **Category**.

- Select Material **type**.

- **Apply** the Material to the model.

View the updated Material folder in the FeatureManager.

Calculate Mass Properties

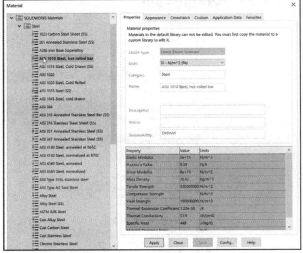

View the calculated mass properties of a part or assembly. Click the Mass Properties button from the Evaluate tab.

View the Mass Properties dialog box. View the provided information.

Click the Options button in the Mass Properties dialog box. View the Mass/Section Property Options dialog box.

Modify your decimal places, units, etc. if needed to view the mass properties in the Mass Properties dialog box.

In SOLIDWORKS, you can create numerous Coordinate systems. Select the correct Coordinate system to view relative values.

Assign Override Values

You can assign values for mass, center of mass, and moments of inertia to override the calculated values (Example: as purchased vs. as modeled).

Click the **Override Mass Properties** button. The Override Mass Properties dialog box is displayed.

Enter the **override values**.

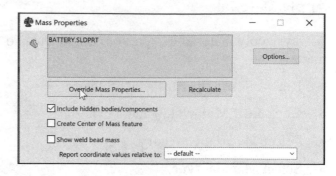

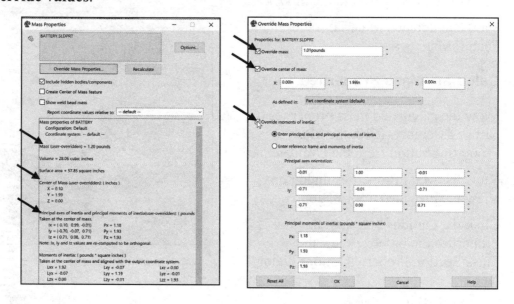

Click **OK**. View the results in the Mass Properties dialog box.

Summary

In this chapter, you were exposed to a general overview of SOLIDWORKS PhotoView 360 and the Measure tool.

PhotoView 360 is a SOLIDWORKS add-in that produces photo-realistic renderings of SOLIDWORKS models. The rendered image incorporates the appearances, lighting, scene, and decals included with the model. PhotoView 360 is available with SOLIDWORKS Professional or SOLIDWORKS Premium.

The Appearance Target palette appears when you add a new appearance to a model so you can add the appearance at the face, feature, body, part, or component level. You can pin the palette, which allows you to add or paste appearances quickly.

Scenes and their lighting schemes are closely connected. Note that lighting works differently in SOLIDWORKS compared with PhotoView.

In addition to controls for point, spot, and directional lights, you can control Scene Illumination when PhotoView is added in. These controls are available in the Scene PropertyManager, on the Illumination tab.

The Measure  tool measures distance, angle, radius, and size of and between lines, points, surfaces, and planes in sketches, 3D models, assemblies or drawings.

When you select a vertex or sketch point, the x, y, and z coordinates are displayed.

You reviewed and applied the SOLIDWORKS Mass Properties tool. You applied standard and assign override values of Mass Properties.

The Appearance filter relates to the Appearances set in SOLIDWORKS.

If the material is a blend of colors (for example automobile paint), use the Color dialog box in the Color PropertyManager.

Chapter 16 provides a general understanding and overview of saving a part, assembly or drawing document in SOLIDWORKS. It also shows how to use the Pack and Go tool, and address PDFs in SOLIDWORKS, SOLIDWORKS Toolbox, SOLIDWORKS Design Library, SOLIDWORKS Explorer and SOLIDWORKS Parts Reviewer.

Notes:

CHAPTER 16: SAVING, PACK AND GO, PDFS, TOOLBOX, DESIGN LIBRARY, SOLIDWORKS EXPLORER & PARTS REVIEWER

Chapter Objective

Chapter 16 provides a general understanding and overview of saving a part, assembly or drawing document in SOLIDWORKS, how to use the Pack and Go tool, and address PDFs in SOLIDWORKS, SOLIDWORKS Toolbox, SOLIDWORKS Design Library, SOLIDWORKS Explorer and SOLIDWORKS Parts Reviewer. On the completion of this chapter, you will be able to:

- Save a SOLIDWORKS Part Document using:

  - Save, Save as, Save as and continue, and Save as copy and open.

- Save a SOLIDWORKS Assembly or Drawing Document using:

  - Advanced button.

  - Pack and Go.

- Export a SOLIDWORKS part, assembly or drawing document as an Adobe® Portable Document Format (PDF) file and part and assembly documents as a 3D PDF or U3D file.

- Understand the 3D Views tab.

- Understand and utilize the SOLIDWORKS Toolbox.

- Comprehend and use the SOLIDWORKS Design Library.

- Know and apply SOLIDWORKS Explorer:

  - Rename and Save components.

- Knowledge of the SOLIDWORKS Parts Reviewer.

Save a Part Document

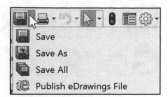

SOLIDWORKS provides various Save options to save a part, assembly or drawing document. The Save command saves the active document.

To save a part document, click Save 🖫 from the Main menu toolbar or click File, Save, or press Ctrl + S.

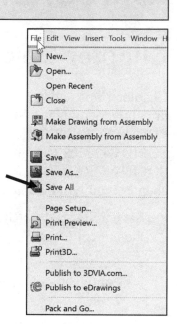

When you save a new part document, the Save As dialog box opens so that you can enter a filename or accept a default filename.

The Save As dialog box also provides various save options:

- **Save as**. Saves the active document to disk with a new name or saves it in a different format for export to another application.

- **Save as copy and continue**. Saves the document to a new file name without replacing the active document.

- **Save as copy and open**. Saves the document to a new file name that becomes the active document. The original document remains open. References to the original document are not automatically assigned to the copy.

- **Save All**. Saves all the files open in SOLIDWORKS that have been modified since they were last saved.

Save an Assembly or Drawing Document

In an assembly or drawing, there are references, toolbox components, studies, etc. that need to be saved in the same folder as the main document.

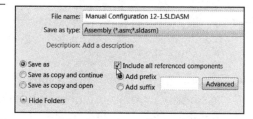

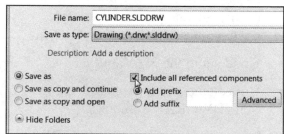

For an assembly or drawing, The Save As dialog box provides various save options:

- *Include all referenced components*. Copies all referenced components to the new location, adding a prefix or suffix to the component names, as specified.

- *Advanced*. Displays a list of the documents referenced by the currently selected assembly or drawing. You can edit the locations of the listed files. Click the Browse button to edit the location. You can also include Toolbox parts, broken references, Nested view, studies and more. See SOLIDWORKS help for additional information.

The Pack and Go feature provides the ability to take an assembly or drawing document and copy all of the related documents into a single folder location or zip file. We will address this in the next section.

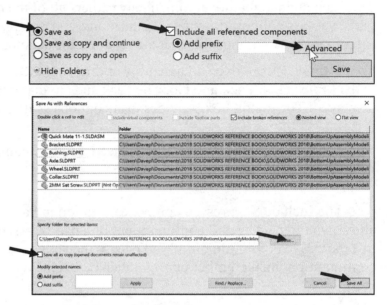

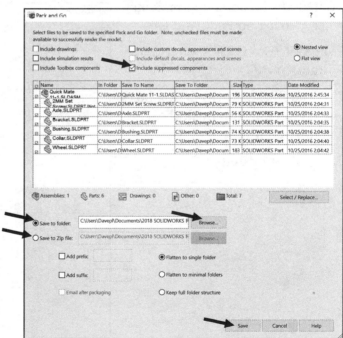

Pack and Go

SOLIDWORKS provides a helpful feature called Pack and Go. The Pack and Go feature provides the ability to take an assembly or drawing document and copy all of the related documents into a single folder location or zip file.

The Pack and Go feature automatically gathers all of the related parts, sub-assemblies, drawings, Simulations, and the Design Library toolbox components that are required for the model.

To use this feature - open an assembly or drawing document and under "File" click Pack and Go. The Pack and Go dialog box is displayed. The dialog box provides many options as illustrated.

You can *include* or *exclude* referenced Toolbox components when setting up Pack and Go.

The Include Suppressed components option in Pack and Go allows you to package all configurations of a part or assembly or just the active configuration. If you choose the active configuration, then all other configurations are removed from the copy.

To include Toolbox components when you set up Pack and Go:

1. In SOLIDWORKS, click **File ➢ Pack and Go**.

2. In the Pack and Go dialog box, either **check** or **uncheck** the **Toolbox components** box.

3. Set other **options**.

4. Click **Save**.

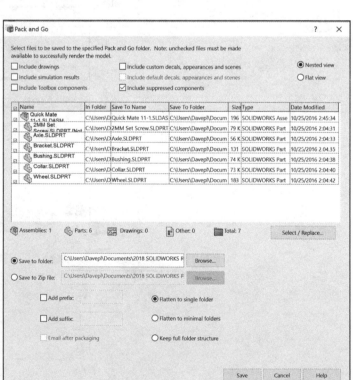

Pack and Go dialog box:

- Select files to include in the **Pack and Go** folder or zip file.

- To hide columns, right-click the column header and select **Hide Column**.

- To rename a copied file, double-click the file under **Save To Name**. The file name extension cannot be changed.

Options:

- *Include drawings*. Adds associated drawings in the model folders or in SOLIDWORKS or SOLIDWORKS Explorer search paths.

- *Include simulation results*. Adds SOLIDWORKS Simulation reports associated with the model, not necessarily in the same folder as the model.

- *Include Toolbox components*. Adds Toolbox components to the folder.

- *Include custom decals, appearances and scenes*. Adds custom decals, appearances and scenes associated with the model, not necessarily in the same folder as the model.

- *Include suppressed components*. Allows you to package all configurations of a part or assembly or just the active configuration. If you choose the active configuration, then all other configurations are removed from the copy.

- *Nested view*. Default setting. Preserve folder structure. The model and all its references are saved in one folder for Flat view.

- *Flat view*. Align all names at the left margin.

- *Select/Replace*. Search a column and select or clear items or replace text based on search criteria.

- *Save to folder*. Specify path and folder name.

- *Save to Zip file*. Specify path and file name.

- *Add prefix* or *Add suffix*. Add the specified prefix or suffix to all file names.

- *Flatten to single folder*. Default setting. Save the model and all its references in one folder. When cleared, the folder structure is preserved.

- *Email after packaging*. Available when **Save to Zip file** is selected. When you click **Send**, an email message opens with the zip file attached.

Tutorial: Pack and Go 16-1

Apply the Pack and Go feature to an assembly document. Create a Zip file of an assembly.

1. Open **Pack and Go 16-1** from the SOLIDWORKS 2018\Pack and Go folder. The assembly is displayed. Apply the Pack and Go feature.

2. Click **File ➢ Pack and Go** from the Menu bar toolbar. View the Pack and Go dialog box. View your options.

3. Check the **Save to Zip file** box.

4. Click the **Browse** button.

5. Select the **folder location** for the zip file of the assembly.

6. Enter the name of the zip file. Example: **Test Zip file**. Click **Save** from the Save As dialog box.

7. Click **Save** from the Pack and Go dialog box. View the location for the Zip file.

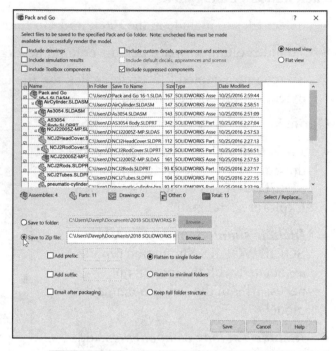

Check the **Email after packaging** box in the Pack and Go dialog box to email the assembly directly to a customer or colleague.

8. **Close** all models.

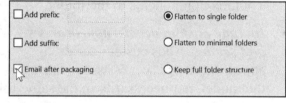

Illustrations may vary depending on your SOLIDWORKS version and system setup.

If the Toolbox components are not breaking their reference, uncheck the **Make this folder the default search location for Toolbox components** under System Options, Hole Wizard/Toolbox as illustrated.

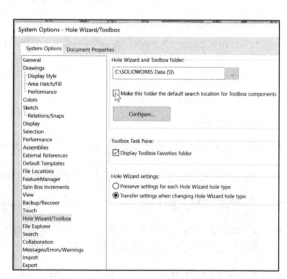

Tutorial: Pack and Go 16-2

Apply the Pack and Go feature to an assembly document and its related drawing and components. Also include the Toolbox components. Create a zip file of the assembly and assembly drawing.

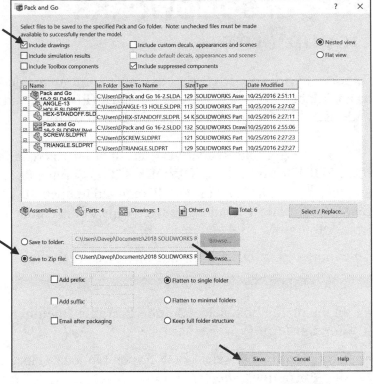

1. Open **Pack and Go 16-2** from the SOLIDWORKS 2018\Pack and Go folder. The assembly is displayed. Apply the Pack and Go feature.

2. Click **File ➤ Pack and Go** from the Menu bar toolbar. View the Pack and Go dialog box. View your options.

3. Check the **Include drawings** box. The Assembly drawing - Pack and Go 16-2.SLDDRW is added to the dialog box. You have options to add or subtract components, drawings, and simulation results.

4. Check the **Save to Zip file** box.

5. Click the **Browse** button.

6. Select the **folder location** for the zip file of the assembly.

7. Enter the **name** of the zip file.

8. Click **Save** from the Save As dialog box.

9. Click **Save** from the Pack and Go dialog box.

10. **View** the location for the Zip file.

11. **Close** all models.

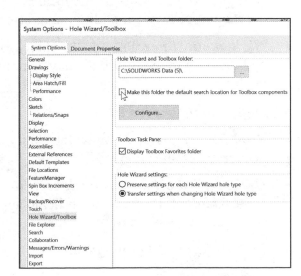

☼ If the Toolbox components are not breaking their reference, uncheck the **Make this folder the default search location for Toolbox components** under System Options, Hole Wizard/Toolbox as illustrated.

PDFs of SOLIDWORKS Documents

Export a SOLIDWORKS part, assembly, or drawing document as an Adobe® Portable Document Format (PDF) file and part and assembly documents as a 3D PDF or U3D file.

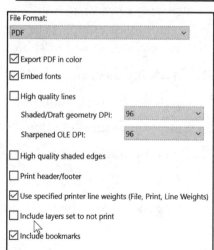

☀ Adobe Acrobat Reader version 5.0 or later is required to open PDF files from SOLIDWORKS 2006 Service Pack 2.1 and later. Version 7.0.7 or later is required to open 3D PDF files. You can download an updated copy of the Adobe Reader from http://www.adobe.com.

To export a SOLIDWORKS document as a PDF file:

- Click **File ≻ Save As**.

- In the dialog box, select **Adobe Portable Document Format (*.pdf)** in Save as type.

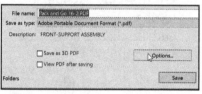

- Click **Options** to select PDF Export Options, select or clear options, then click **OK**.

- Type the **file name** in File name, and click **Save**.

The Export Options dialog box provides the following tools as illustrated: **Export PDF in color**. Provide the ability to match the colors or gray scale of the document.

- *Embed fonts*. To reduce the size of the file, PDF files use existing fonts on the local PC to generate the view. See help for additional information.

- *High quality lines*. Provide the ability to drawings and the drawing views to display them in high quality.

- *High quality shaded edges*.

- *Print header/footer*. Provide the ability to print a header or footer in the PDF.

- *Use specified printer line weights (File, Print, Line Weights)*. Provide the ability to apply the specified line weight to the PDF.

- *Include layers set to not print*.

- *Include bookmarks*.

3D Views tab

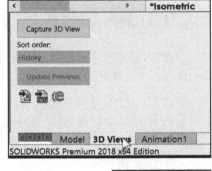

The 3D Views tab is visible to all users. Previously, only MBD users could access the 3D View tab. All users can now access the 3D Views tab when a model includes 3D Views, and can activate any of the views. You must have an MBD license to capture, edit, or publish 3D Views.

To support model based definition (MBD) create 3D drawing views of your parts and assemblies that contain the model settings needed for review and manufacturing. The output you create lets users navigate back to those settings as they evaluate the design.

Use the tools on the SOLIDWORKS MBD CommandManager to set up your model with selected configurations, including explodes and abbreviated views, annotations, display states, zoom level, view orientation and section views. You then capture those settings so that you and other users can return to them at any time using the 3D view palette.

To access the 3D View palette, click the 3D Views tab at the bottom of the SOLIDWORKS window or the SOLIDWORKS MBD tab in the CommandManager. The Capture 3D View button opens the Capture 3D View PropertyManager, where you specify the 3D view name, and the configuration, display state, and annotation view to capture. See SOLIDWORKS help for additional information.

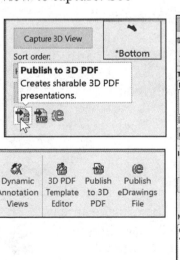

Publish to 3D PDF

The Publish to 3D PDF icon provides the ability to create sharable 3D PDF presentations. Use the Publish to 3D PDF PropertyManager for additional control.

SOLIDWORKS Toolbox

SOLIDWORKS Toolbox is an add-in that requires SOLIDWORKS Professional, Premium or an Educational version.

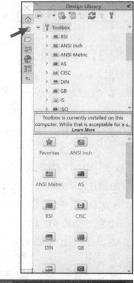

SOLIDWORKS Toolbox includes a library of standard parts that is fully integrated with SOLIDWORKS. Toolbox users select a standard and type of part and drag and drop a Toolbox component into an assembly.

SOLIDWORKS Toolbox supports various international drafting standards: ANSI, AS, BSI, GB, etc. as illustrated.

Check with your system administrator before you modify the SOLIDWORKS Toolbox. In a network environment, there may be restrictions. With the proper permission you can place the toolbox components in a central location on your network and streamline Toolbox to include only parts that comply with your corporate manufacturing standards. You can also control access to the Toolbox library to prevent users from changing the Toolbox components, specify how component files are handled, and assign part numbers and other custom properties to Toolbox components.

💡 SOLIDWORKS Toolbox is integrated with the Hole Wizard feature. Add-in SOLIDWORKS Toolbox to utilize the Hole Wizard feature or click SOLIDWORKS Toolbox directly from the SOLIDWORKS Add-ins tab.

Using the SOLIDWORKS Toolbox

SOLIDWORKS Toolbox has two components:

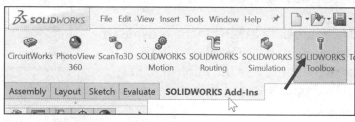

- Toolbox Library.
- Toolbox Utilities.

Before using the SOLIDWORKS Toolbox, make sure it is properly configured according to your company or group policy.

In many cases, a group will configure and maintain a central Toolbox. Check with the SOLIDWORKS administrator for guidance on how to specify a Toolbox during SOLIDWORKS installation.

There are a few key things to understand when you use the SOLIDWORKS Toolbox:

- Fasteners are merely a representation--they do not include accurate thread details.

- Gears are merely a representation--they are not true involute gears--but are a representation of a gear and should never be used for machining purposes and should not be included in a FEA if you require accurate information on stress concentrations in the components.

By default, the SOLIDWORKS Toolbox includes more than 2000 component types of different sizes for 12 tool standards, as well as other industry-specific content, resulting in millions of components. See SOLIDWORKS Help for additional information.

SOLIDWORKS Design Library

The SOLIDWORKS Design Library tab in the Task Pane provides a central location for reusable elements: parts, assemblies, sketches, annotation favorites, blocks, library features and DXF/DWG files.

The Design Library does not recognize non-reusable elements such as SOLIDWORKS drawings, text files, or other non-SOLIDWORKS files.

The Design Library tab contains the following standard folders: *Design Library*, *Toolbox*, *3D ContentCentral* and *SOLIDWORKS Content*.

The following tools are available on the Design Library tab:

- *Add to Library* . Adds content to the Design Library.

- *Add File Location* . Adds an existing folder to the Design Library.

- *Create New Folder* . Creates a new folder on disk and in the Design Library.

- *Refresh* . Refreshes the view of the Design Library tab.

- *Up one Level* . Displays the previous level.

- *Configure Toolbox* . Provides the ability to configure the Toolbox using the Toolbox dialog box.

Specify folders for the Design Library by clicking **Options ➢ System Options ➢ File locations**.

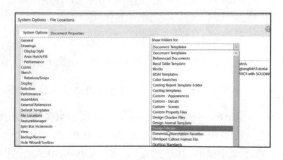

Using the SOLIDWORKS Design Library

You can drag and drop copies of parts, assemblies, features, annotations, etc. from:

- The Design Library into your Graphics window.

- The Graphics window into the lower pane of the Design Library.

- Folder to folder in the Design Library.

- Microsoft Internet Explorer and Windows Explorer into the Design Library.

When you click and drag items into the Design Library, the Add to Library PropertyManager is displayed with a default file name, the default file type, and the selected folder for the Design Library folder.

Tutorial: Assembly Design Library 16-1

Create a new file location in the Design Library. Creating a new design library file allows a user to easily bring in commonly used parts or assemblies into a model.

1. Open **Assembly Design Library 16-1** from the SOLIDWORKS 2018\Design Library-1 folder.

2. Click the **Design Library** 🎯 tab from the Task Pane.

3. Click the **Add File Location** 🎯 icon.

4. Browse to the **SOLIDWORKS 2018\Design Library-1** folder.

5. Click **OK** from the Choose Folder dialog box. The Design Library-1 folder is added to the Design Library.

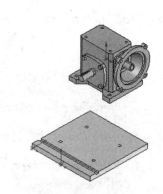

6. Click the **Design Library-1** folder from the Design Library.

7. Click and drag the **f718b-30-b5-g-1** assembly into the Graphics window above the plate.

8. Click **Cancel** ✖ from the Insert Component PropertyManager. View the Assembly Design Library FeatureManager. The part is added to the assembly.

9. Display an **Isometric** view.

10. **Close** all models.

🔆 View the below link on a video to create a new folder location in the design library. In this case, molecules were created using elements from the newly created design folder. Creating a design library folder allows a user to easily bring in commonly used parts or assemblies into a model. Also shown is how to use pack and go to make sure all of your parts are in one location. This is important if a model is going to be shared and you want to be sure that all the parts are grouped together.

Link:
http://blogs.SOLIDWORKS.com /teacher/2013/03/creating-a-new-folder-location-in-the-design-library.html

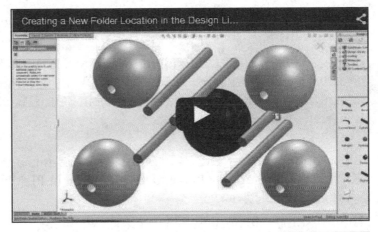

🔆 View the model in the SOLIDWORKS 2018\Design Library-1\Molecules folder.

Add a Design Library tab

Work directly from the Design Library. Add a Design Library tab to reference a folder.

Tutorial: Assembly Design Library 16-2

Create a new file location for the Design Library.

1. Click **Options** ⚙, **File Locations** from the Menu bar toolbar.

2. Select **Design Library** from the drop down menu.

3. Click the **Add** button.

4. **Browse** to the folder that you want to add to the Design Library.

5. Click **OK** from the Browse of Folder dialog box. Click **OK**. View the new tab in the Design Library.

SOLIDWORKS Explorer

SOLIDWORKS Explorer  is a file management tool designed to help you perform such tasks as *renaming*, *replacing* and *copying* SOLIDWORKS files. You can display a document's references, search for documents using a variety of criteria, and list all the places where a document is used.

🔅 Renamed files are still available to those documents that reference them.

Execute SOLIDWORKS Explorer within SOLIDWORKS or directly from the desktop. The SOLIDWORKS documents remain closed while manipulating names in SOLIDWORKS Explorer.

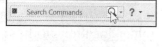

🔅 Apply SOLIDWORKS Explorer with or without the SOLIDWORKS application and with or without PDMWorks Workgroup added in.

The first time you open SOLIDWORKS Explorer, it is displayed in its collapsed view, with only the SOLIDWORKS Search box visible.

In the collapsed view, you can:

1.) Search for text strings and tags in all indexed documents, including SOLIDWORKS and Microsoft Office documents.

2.) Click the **Expand**  icon to display the File Explorer pane.

3.) Right-click near the outer edge and select **Close** or **Minimize**.

4.) Leave the collapsed view open on your desktop. It becomes transparent while you use other applications.

5.) Insert tags in documents to use as search criteria. Remember, tags provide the ability to associate keywords with documents to make it easier to search for them.

When you perform a search or click the **Expand** ⊞ icon, the view expands to display the File Explorer pane, which has two tabs:

- *File Explorer*. The File Explorer tab displays the folders and documents on your computer. If PDM Works Workgroup Contributor is added in, the vault view is displayed at the bottom of the pane.

- *Results*. Displays results of the searches.

🔆 Activate SOLIDWORKS Explorer from the SOLIDWORKS Tools menu or from the Windows Start menu. The application is located on the SOLIDWORKS installation DVDs or on the SOLIDWORKS website.

Tutorial: SOLIDWORKS Explorer 16-1

Activate SOLIDWORKS Explorer. Rename an assembly.

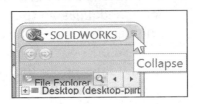

1. Click **Tools ➢ SOLIDWORKS Applications ➢ SOLIDWORKS Explorer** from the Menu bar menu. SOLIDWORKS Explorer is displayed.

2. If needed click **Accept**. Click **Cancel** from the SOLIDWORKS Workgroup Login dialog box. The SOLIDWORKS Search icon is displayed.

3. Click **Expand** ⊞ as illustrated if this is your first time using SOLIDWORKS Explorer.

Display the MGPM12-1010-11-1 assembly.

4. Click the **MGPM12-1010-11-1** assembly from the SOLIDWORKS 2018\SOLIDWORKS Explorer folder. The MGPM12-1010 assembly is displayed in the main window. Note: If you double-click MGPM12-1010-11-1, the assembly opens in the SOLIDWORKS Graphics window. Do not open the assembly at this time.

5. Click the **Info** tab.

6. **View** the provided information.

7. Click the **Properties** tab. View the provided information.

8. Click the **References** tab. View the provided information. Explore the other tabs.

Depending on the type of items selected, different tabs are displayed in the expanded view, where you can perform data management tasks. Each tab can contain several columns of information.

Display assembly references with the References tab.

9. Click the **MGPM12-1010-11-1** assembly as illustrated. The pop-up toolbar is displayed.

When you select a document, a pop-up toolbar helps you perform additional tasks. The pop-up toolbar provides the following tools:

- *Open a document*. Opens the selected document.

- *SOLIDWORKS Pack and Go*. Gathers all related files for a model design (parts, assemblies, drawings, references, design tables, Design Binder content, PhotoWorks content and Simulation/COSMOS results) into a folder or zip file. See SOLIDWORKS Help for additional information. This is a very useful tool to take an entire assembly and save it, keeping all of its relations in another file structure.

- *Rename*. Renames one or more selected documents and updates all the references.

- *Replace*. Replaces a selected part or assembly document and updates its references.

- *Move*. Moves a file and updates it when used.

Rename the document with SOLIDWORKS Explorer. SOLIDWORKS Explorer updates file references and requires less work. Microsoft Windows Explorer does not update file references. You are required to locate individual file references.

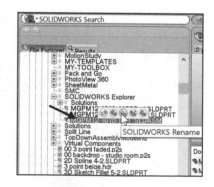

Rename the MGPM12-1010-11-1 assembly.

10. Click the **Rename** icon in the pop-up toolbar. The Rename Document dialog box is displayed.

11. Enter **GUIDE-CYLINDER-12MM** for new name.

12. Click **OK** from the Rename Document dialog box. View the results.

13. Return to the original name. Enter **MGPM12-1010-11-1**.

14. **Close** SOLIDWORKS Explorer and return to SOLIDWORKS.

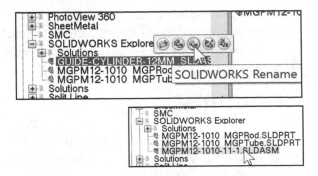

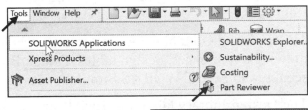

SOLIDWORKS Part Review

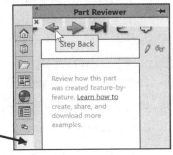

Learn best practices from creating various types of sketches and features with parts in SOLIDWORKS. Use the rollback bar in the FeatureManager to view the feature, sketch and design intent.

SOLIDWORKS Part Review is an automatic way to review "play" how a part was created and to view the feature, sketch and design intent in the FeatureManager. Comments are displayed with the sketch or feature.

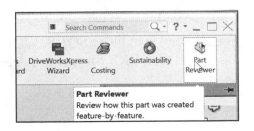

Click Part Reviewer from the Evaluate tab in the CommandManager. View the Part Reviewer dialog box in the Task Pane. View the welcome screen with a hyperlink. The hyperlink provides example SOLIDWORKS Part Reviewer files. **Note**: The authored by is defined by the Author filed within the File/Properties dialog.

Use the play controls to walk through the model feature-by-feature or play the entire part automatically. Note the comments appear as the feature and dimensions are displayed.

Part Reviewer sample parts are located in the SOLIDWORKS Forum. These sample parts contain comments about their features to teach you how they were created and why they were used. You can also create your own parts and add your own comments.

Summary

In this chapter, you obtained a general understanding and overview of how to save a part or an assembly document in SOLIDWORKS, applied the Pack and Go tool, generated PDFs of SOLIDWORKS documents, and learned about the SOLIDWORKS Toolbox, SOLIDWORKS Design Library, SOLIDWORKS Explorer and SOLIDWORKS Parts Reviewer.

The Pack and Go tool gathers all related files for a model design (parts, assemblies, drawings, references, design tables, Design Binder content, decals, appearances, and scenes, and SOLIDWORKS Simulation results) into a folder or zip file.

The Design Library tab in the Task Pane provides a central location for reusable elements such as parts, assemblies and sketches. It does not recognize non-reusable elements such as SOLIDWORKS drawings, text files, or other non-SOLIDWORKS files.

Export SOLIDWORKS part, assembly and drawing documents as Adobe® Portable Document Format (PDF) files and part and assembly documents as 3D PDF or U3D files.

Use Part Reviewer to review how parts are created feature-by-feature. Part Reviewer can help you learn best practices for creating various types of parts.

Part Reviewer sample parts are located in the SOLIDWORKS Forum. These sample parts contain comments about their features to teach you how they were created and why they were used. You can also create your own parts and add your own comments.

In Chapter 17, explore SOLIDWORKS Simulation and the key components that are in the SOLIDWORKS Simulation Associate - Finite Element Analysis (CSWSA-FEA) exam.

CHAPTER 17: SOLIDWORKS SIMULATION

Chapter Objective

Chapter 17 provides a general overview of SOLIDWORKS Simulation and the type of questions that are on the SOLIDWORKS Simulation Associate - Finite Element Analysis (CSWSA-FEA) exam. On the completion of this chapter, you will be able to:

- Recognize the power of SOLIDWORKS Simulation.

- Apply SOLIDWORKS Simulation to:

  - Define a Static Analysis Study.

  - Apply Material to a part model.

  - Work with a Solid and Sheet Metal model.

  - Define Solid, Shell and Beam elements.

  - Define Standard and Advanced Fixtures and External loads.

  - Define Local and Global coordinate systems.

  - Understand the axial forces, shear forces, bending moments and factor of safety.

  - Work with Multi-body parts as different solid bodies.

  - Select different solvers as directed to optimize problems.

  - Determine if the result is valid.

  - Ability to use SOLIDWORKS Simulation Help and the on-line Tutorials.

  - Understand the type of problems and questions that are on the CSWSA-FEA exam.

Basic FEA Concepts

SOLIDWORKS Simulation uses the Finite Element Method (FEM). FEM is a numerical technique for analyzing engineering designs. FEM is accepted as the standard analysis method due to its generality and suitability for computer implementation.

FEM divides the model into many small pieces of simple shapes called elements effectively replacing a complex problem by many simple problems that need to be solved simultaneously.

Elements share common points called nodes. The process of dividing the model into small pieces is called meshing.

The behavior of each element is well-known under all possible support and load scenarios. The finite element method uses elements with different shapes.

The response at any point in an element is interpolated from the response at the element nodes. Each node is fully described by a number of parameters depending on the analysis type and the element used.

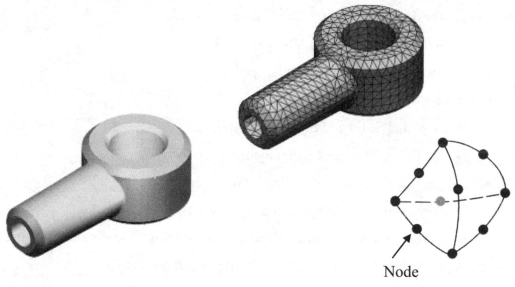

Node

Tetrahedral

For example, the temperature of a node fully describes its response in thermal analysis. For structural analyses, the response of a node is described, in general, by three translations and three rotations. These are called degrees of freedom (DOFs). Analysis using the FEM is called a Finite Element Analysis (FEA).

SOLIDWORKS Simulation formulates the equations governing the behavior of each element taking into consideration its connectivity to other elements. These equations relate the response to known material properties, restraints, and loads.

Next, SOLIDWORKS Simulation organizes the equations into a large set of simultaneous algebraic equations and solves for the unknowns.

In stress analysis, for example, the solver finds the displacements at each node and then the program calculates strains and finally stresses.

Static studies calculate displacements, reaction forces, strains, stresses, and factor of safety distribution. Material fails at locations where stresses exceed a certain level. Factor of safety calculations are based on one of the following failure criterion:

- **Maximum von Mises Stress**.

- **Maximum shear stress (Tresca)**.

- **Mohr-Coulomb stress**.

- **Maximum Normal stress**.

- **Automatic** (Automatically selects the most appropriate failure criterion across all element types).

Static studies can help avoid failure due to high stresses. A factor of safety less than unity indicates material failure. Large factors of safety in a contiguous region indicate low stresses and that you can probably remove some material from this region.

Simulation Advisor

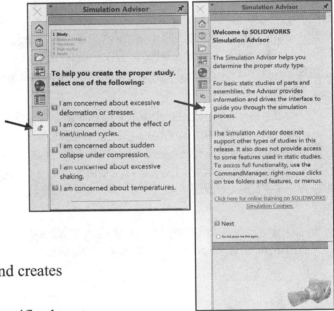

Simulation Advisor is a set of tools that guide you through the analysis process. By answering a series of questions, these tools collect the necessary data to help you perform your analysis. Simulation Advisor includes:

- **Study Advisor**. Recommends study types and outputs to expect. Helps you define sensors and creates studies automatically.

- **Bodies and Materials Advisor**. Specifies how to treat bodies within a part or an assembly and apply materials to components.

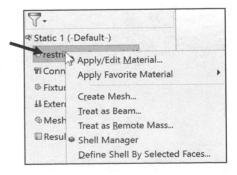

Apply material in SOLIDWORKS Simulation. Right-click on the part icon in the study. Click Apply/Edit Material.

Interactions Advisor. Defines internal interactions between bodies in the model as well as external interactions between the model and the environment. Interactions can include loads, fixtures, connectors, and contacts.

- **Mesh and Run Advisor**. Helps you specify the mesh and run the study.

- **Results Advisor**. Provides tips for interpreting and viewing the output of the simulation. Also, helps determine if frequency or buckling might be areas of concern.

While taking the CSWSA-FEA exam, the Simulation Advisor and or SOLIDWORKS Simulation Help topics may provide required information to answer the exam questions.

Simulation Advisor works with the SOLIDWORKS Simulation interface by starting the appropriate PropertyManager and linking to online help topics for additional information. Simulation Advisor leads you through the analysis workflow from determining the study type through analyzing the simulation output.

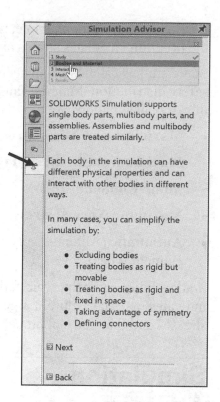

The purpose of this chapter is not to educate a new or intermediate user on SOLIDWORKS Simulation, but to cover and to inform you on the types of questions, layout and what to expect when taking the CSWSA-FEA exam.

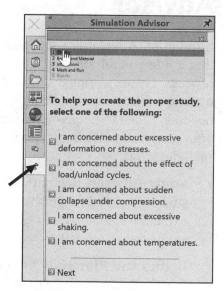

Illustrations and values may vary slightly depending on your SOLIDWORKS release.

SOLIDWORKS Simulation Help & Tutorials

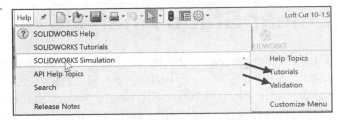

SOLIDWORKS Simulation is an Add-in. Use SOLIDWORKS Simulation during the CSWSA-FEA exam to discover information during the exam. Utilize the Contents and Search tabs to locate subject matter.

Review the SOLIDWORKS Simulation Tutorials on Static parts and assemblies. Understand your options and setup parameters. Questions in these areas will be on the exam.

Review the SOLIDWORKS Simulation Validation, Verification Problems, Static section and the SOLIDWORKS Simulation, Verification, NAFEMS Benchmarks, Linear Static section.

Access SOLIDWORKS Simulation directly from the SOLIDWORKS Add-Ins tab in the CommandManager.

A model needs to be open to obtain access to SOLIDWORKS Simulation Help or SOLIDWORKS Simulation Tutorials.

Linear Static Analysis

This section provides the basic theoretical information required for a Static analysis using SOLIDWORKS Simulation. The CSWSA-FEA only covers Static analysis. SOLIDWORKS Simulation and SOLIDWORKS Simulation Professional cover the following topics:

General Simulation.

- *Static Analysis.*
    - o Use 2D Simplification.
    - o Import Study Features.
- *Frequency Analysis.*

Design Insight.

- *Topology Study.*
- *Design Study.*

Advanced Simulation.

- *Thermal.*
- *Buckling.*
- *Fatigue.*
- *NonLinear.*
- *Linear Dynamic.*

Specialized Simulation.

- *Design Study.*
- *Drop Test.*
- *Pressure Vessel Design.*

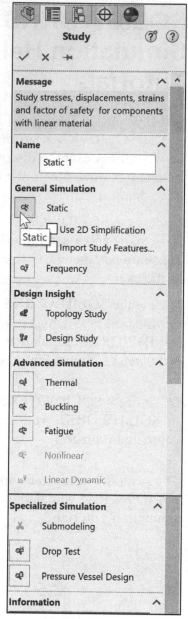

Linear Static Analysis

When loads are applied to a body, the body deforms and the effect of loads is transmitted throughout the body. The external loads induce internal forces and reactions to render the body into a state of equilibrium.

Linear Static analysis calculates displacements, strains, stresses, and reaction forces under the effect of applied loads.

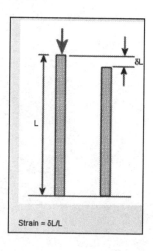

Strain = δL/L

Strain ε is the ratio of change, δ L, to the original length, L. Strain, is a dimensionless quantity. Stress σ is defined in terms of Force per unit Area.

Linear Static analysis makes the following assumptions:

- **Static Assumption**. All loads are applied *slowly* and gradually until they reach their full magnitudes. After reaching their full magnitudes, loads *remain constant* (time-invariant). This assumption neglects inertial and damping forces.

  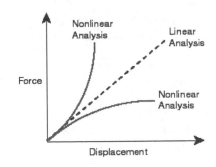

  Time-variant loads that induce considerable inertial and/or damping forces and require dynamic analysis.

  Dynamic loads change with time and in many cases induce considerable inertial and damping forces that cannot be neglected.

- **Linearity Assumption**. The relationship between loads and induced responses is linear. For example, if you double the loads, the response of the model (displacements, strains, and stresses) will also double. Apply the linearity assumption if:

  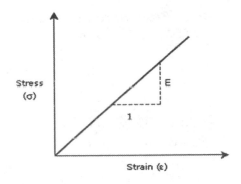

  - All materials in the model comply with Hooke's law; that is, stress is directly proportional to strain.
  - The induced displacements are small enough to ignore the change in stiffness caused by loading.
  - Boundary conditions do not vary during the application of loads. Loads must be constant in magnitude, direction, and distribution. They should not change while the model is deforming.

🔆 In SOLIDWORKS Simulation for Static analysis, all displacements are small relative to the model geometry (unless the Large Displacements option is activated).

SOLIDWORKS Simulation assumes that the normals to contact areas do not change direction during loading. Hence, it applies the full load in one step. This approach may lead to inaccurate results or convergence difficulties in cases where these assumptions are not valid.

- Elastic Modulus, E, is the stress required to cause one unit of strain. The material behaves linearly at low stresses.

- Elastic Modulus (Young's Modulus) is the slope defined as stress divided by strain. E = modulus of elasticity (Pa (N/m²), N/mm², psi).

- Stress σ is proportional to strain in a Linear Elastic Material. Units: (Pa (N/m²), N/mm², psi).

You must be able to work in SI and English units for the CSWSA-FEA exam within the same problem. For example, you apply a Force in Newtons and then you determine displacement in inches.

Different materials have different stress property levels. Mathematical equations derived from Elasticity theory and Strength of Materials are utilized to solve for displacement and stress. These analytical equations solve for displacement and stress for simple cross sections.

General Procedure to Perform a Linear Static Analysis

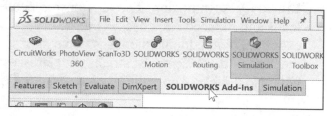

- Complete a Linear Static study by performing the following steps:

- Add-in SOLIDWORKS Simulation.

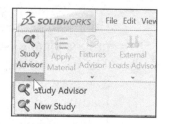

- Select the Simulation tab.

- Create a new Study. To create a new study, expand the Study Advisor and select New Study.

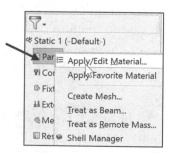

- Define Material. To define a material, right-click on the model icon in the Simulation study tree and select Apply/Edit Material.

- Define Restraints/Fixtures/Connections.

- Define External Loads. Right-click the External Loads icon in the Simulation study tree and select from the list.

- For assemblies and multi-body parts, use component contact and contact sets to simulate the behavior of the model.

- Mesh the model and Run the study. Select Mesh type and parameters. The Mesh PropertyManager lets you mesh models for solid, shell, and mixed mesh studies.

- View the results. In viewing the results after running a study, you can generate plots, lists, graphs, and reports depending on the study and result types.

- Double-click an icon in a results folder to display the associated plot.

- To define a new plot, right-click the Results folder, and select the desired option. You can plot displacements, stresses, strains, and deformation.

- To assess failure based on a yield criterion, right-click the Results folder, and select Define Factor of Safety Plot.

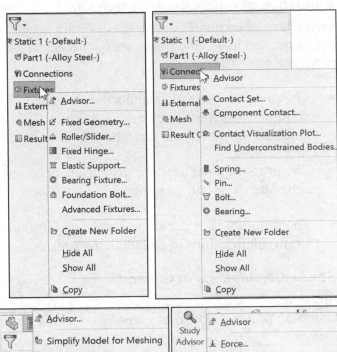

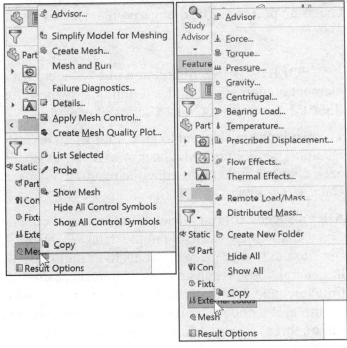

Sequence of Calculations in General

Given a meshed model with a set of displacement restraints and loads, the linear static analysis program proceeds as follows:

- The program constructs and solves a system of linear simultaneous finite element equilibrium equations to calculate displacement components at each node.

- The program then uses the displacement results to calculate the strain components.

- The program uses the strain results and the stress-strain relationships to calculate the stresses.

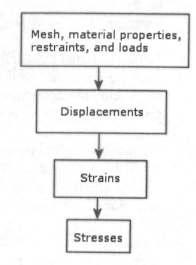

Stress Calculations in General

Stress results are first calculated at special points, called Gaussian points or Quadrature points, located inside each element.

☀ SOLIDWORKS Simulation utilizes a tetrahedral element containing 10 nodes (High quality mesh for a Solid). Each node contains a series of equations.

These points are selected to give optimal numerical results. The program calculates stresses at the nodes of each element by extrapolating the results available at the Gaussian points.

Node

Tetrahedral

After a successful run, nodal stress results at each node of every element are available in the database. Nodes common to two or more elements have multiple results. In general, these results are not identical because the finite element method is an approximate method. For example, if a node is common to three elements, there can be three slightly different values for every stress component at that node.

Overview of the Yield or Inflection Point in a Stress-Strain curve

When viewing stress results, you can ask for element stresses or nodal stresses. To calculate element stresses, the program averages the corresponding nodal stresses for each element.

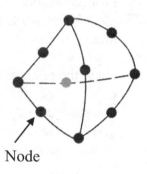

Stresss versus Strain Plot
Linearly Elastic Material

To calculate nodal stresses, the program averages the corresponding results from all elements sharing that node.

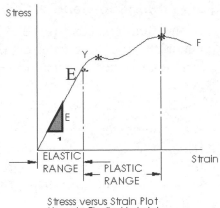

Stresss versus Strain Plot
Linearly Elastic Material

- The material remains in the Elastic Range until it reaches the elastic limit.

- The point E is the elastic limit. The material begins Plastic deformation.

- Yield Stress is the stress level at which the material ceases to behave elastically.

- The point Y is called the Yield Point. The material begins to deform at a faster rate. In the Plastic Range the material behaves non-linearly.

- The point U is called the ultimate tensile strength. Point U is the maximum value of the non-linear curve. Point U represents the maximum tensile stress a material can handle before a fracture or failure.

- Point F represents where the material will fracture.

- Designers utilize maximum and minimum stress calculations to determine if a part is safe. Simulation reports a recommended Factor of Safety during the analysis.

- The Simulation Factor of Safety is a ratio between the material strength and the calculated stress.

Brittle materials do not have a specific yield point and hence it is not recommended to use the yield strength to define the limit stress for the criterion.

Material Properties in General

Before running a study, you must define all material properties required by the associated analysis type and the specified material model. A material model describes the behavior of the material and determines the required material properties. Linear isotropic and orthotropic material models are available for all structural and thermal studies. Other material models are available for nonlinear stress studies. The von Mises plasticity model is available for drop test studies. Material properties can be specified as function of temperature.

- For solid assemblies, each component can have a different material.

- For shell models, each shell can have a different material and thickness.

- For beam models, each beam can have a different material.

- For mixed mesh models, you must define the required material properties for solid and shell separately.

Connections in General

A connection replaces a piece of hardware or fastener by simulating its effect on the rest of the model. Connections include Bolts, Springs, Flexible Support, Bearings, Bonding - Weld/Adhesives, Welds, etc.

The automatic detection tool in SOLIDWORKS Simulation defines Contact Sets. Sometimes additional contact sets and types need to be defined. The SOLIDWORKS Simulation Study Advisor can help.

For example, the behavior of an adhesive depends on its strength and thickness. You can select the Type manually.

Fixtures: Adequate restraints to prevent the body from rigid body motion. If your model is not adequately constrained, check the Use soft springs to stabilize the model option in the Static dialog box.

When importing loads from SOLIDWORKS Motion, make sure that Use inertial relief option is checked. These options are available for the Direct Sparse solver and FFEPlus solver.

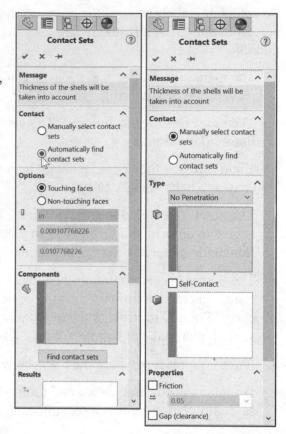

☀ See SOLIDWORKS Simulation Help for additional information.

Restraint Types

The Fixture PropertyManager provides the ability to prescribe zero or non-zero displacements on vertices, edges, or faces for use with static, frequency, buckling, dynamic and nonlinear studies. This section will only address standard restraint types, namely Fixed Geometry and Immovable (No translation).

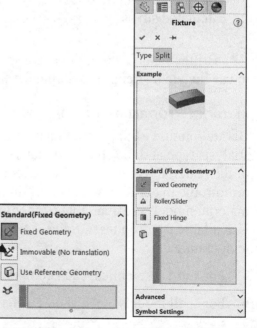

☀ The Immovable option is displayed for Sheet Metal parts.

Fixed: For solids, this restraint type sets all translational degrees of freedom to zero. For shells and beams, it sets the translational and the rotational degrees of freedom to zero. For truss joints, it sets the translational degrees of freedom to zero. When using this restraint type, no reference geometry is needed.

View the illustrated table for the attributes and input needed for this restraint.

Attribute	Value
DOFs restrained for solid meshes	3 translations
DOFs restrained for shells and beams	3 translations and 3 rotations
DOFs restrained for truss joints	3 translations
3D symbol (the arrows are for translations and the discs are for rotations)	
Selectable entities	Vertices, edges, faces and beam joints
Selectable reference entity	N/A
Translations	N/A
Rotations	N/A

Immovable (No translation): This restraint type sets all translational degrees of freedom to zero. It is the same for shells, beams and trusses. No reference geometry is used.

To access the immovable restraint, right-click on Fixtures in the Simulation study tree and select Fixed Geometry. Under Standard, select Immovable (No translation). View the illustrated table for the attributes and input needed for this restraint.

Attribute	Value
DOFs restrained for shell meshes	3 translations
DOFs restrained for beam and truss meshes	3 translations
3D symbol	
Selectable entities	Vertices, edges, faces and beam joints
Selectable reference entity	N/A
Translations	N/A
Rotations	N/A

There are differences for Shells and Beams between Immovable (No translation) and Fixed restraint types.

The Immovable option is not available for Solids.

The Fixture PropertyManager allows you to prescribe zero or non-zero displacements on vertices, edges or faces for use with static, frequency, buckling, dynamic and nonlinear studies.

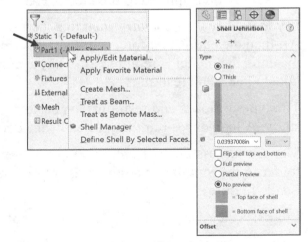

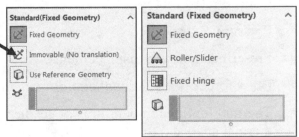

The Fixed Geometry Fixture allows for additional Advanced options: Symmetry, Circular Symmetry, User Reference Geometry, On Flat Faces, On Cylindrical Faces, On Spherical Faces.

Attributes of each option are available in SOLIDWORKS Simulation Help.

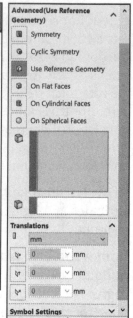

Loads and Restraints in General

Loads and restraints are necessary to define the service environment of the model. The results of analysis directly depend on the specified loads and restraints.

Loads and restraints are applied to geometric entities as features that are fully associative to geometry and automatically adjust to geometric changes.

For example, if you apply a pressure P to a face of area A_1, the equivalent force applied to the face is PA_1.

If you modify the geometry such that the area of the face changes to A_2, then the equivalent force automatically changes to PA_2. Re-meshing the model is required after any change in geometry to update loads and restraints.

The types of loads and restraints available depend on the type of the study. A load or restraint is applied by the corresponding Property

Manager accessible by right-clicking the Fixtures or External Loads folder in the Simulation study tree, or by clicking Simulation, Loads/Fixture.

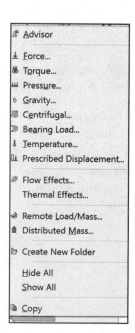

Loads: At least one of the following types of loading is required:

- Concentrated force.

- Pressure.

- Prescribed nonzero displacements.

- Body forces (gravitational and/or centrifugal).

- Thermal (define temperatures or get the temperature profile from thermal analysis).

- Imported loads from SOLIDWORKS Motion.

- Imported temperature and pressure from Flow Simulation.

In a linear static thermal stress analysis for an assembly, it is possible to input different temperature boundary conditions for different parts.

Under the External Loads folder you can define Remote Load/Mass and Distributed Mass. In the Remote Load/Mass you define Load, Load/Mass or Displacement. Input values are required for the Remote Location and the Force.

By default, the Location is set to x=0, y=0, z=0. The Force is set to $F_x=0$, $F_y=0$, $F_z=0$. The Force requires you to first select the direction and then enter the value.

Meshing in General

Meshing splits continuous mathematical models into finite elements. The types of elements created by this process depend on the type of geometry meshed. SOLIDWORKS Simulation offers three types of elements:

- Solid elements - solid geometry.

- Shell elements - surface geometry.

- Beam elements - wire frame geometry.

In CAD terminology, "Solid" denotes the type of geometry. In FEA terminology, "Solid" denotes the type of element used to mesh the solid CAD Geometry.

Meshing Types

Meshing splits continuous mathematical models into finite elements. Finite element analysis looks at the model as a network of interconnected elements.

Meshing is a crucial step in design analysis. SOLIDWORKS Simulation automatically creates a mixed mesh of:

- **Solid**: The Solid mesh is appropriate for bulky or complex 3D models. In meshing a part or an assembly with solid elements, Simulation generates one of the following types of elements based on the active mesh options for the study:

  - Draft quality mesh. The automatic mesher generates linear tetrahedral solid elements **(4 nodes)**.

  - High quality mesh. The automatic mesher generates parabolic tetrahedral solid elements **(10 nodes)**.

Linear elements are also called first-order, or lower-order elements. Parabolic elements are also called second-order, or higher-order elements.

A linear tetrahedral element is defined by four corner nodes connected by six straight edges.

Linear solid element

A parabolic tetrahedral element assigns 10 nodes to each solid element: four corner nodes and one node at the middle of each edge (a total of six mid-side nodes).

In general, for the same mesh density (number of elements), parabolic elements yield better results than linear elements because:

- They represent curved boundaries more accurately.

- They produce better mathematical approximations. However, parabolic elements require greater computational resources than linear elements.

Parabolic solid element

For structural problems, each node in a solid element has three degrees of freedom that represent the translations in three orthogonal directions.

SOLIDWORKS Simulation uses the X, Y, and Z directions of the global Cartesian coordinate system in formulating the problem.

- **Shell**: Shell elements are suitable for thin parts (sheet metal models). Shell elements are 2D elements capable of resisting membrane and bending loads. When using shell elements, Simulation generates one of the following types of elements depending on the active meshing options for the study:

  - Draft quality mesh. The automatic mesher generates linear triangular shell elements **(3 nodes)**.

  - High quality mesh. The automatic mesher generates parabolic triangular shell elements **(6 nodes)**.

Linear triangular element

Parabolic triangular element

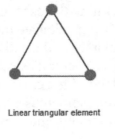

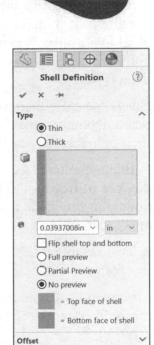

A linear triangular shell element is defined by three corner nodes connected by three straight edges.

A parabolic triangular element is defined by three corner nodes, three mid-side nodes, and three parabolic edges.

The Shell Definition PropertyManager is used to define the thickness of thin and thick shell elements. The program automatically extracts and assigns the thickness of the sheet metal to the shell. You cannot modify the thickness. You can select between the thin shell and thick shell formulations. You can also define a shell as a composite for static, frequency, and buckling studies. In general use thin shells when the thickness-to-span ratio is less than 0.05.

 Surface models can only be meshed with shell elements.

- **Beam or Truss**: Beam or Truss elements are suitable for extruded or revolved objects and structural members with constant cross-sections. Beam elements can resist bending, shear, and torsional loads. The typical frame shown is modeled with beams elements to transfer the load to the supports. Modeling such frames with truss elements fails since there is no mechanism to transfer the applied horizontal load to the supports.

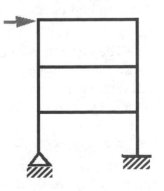

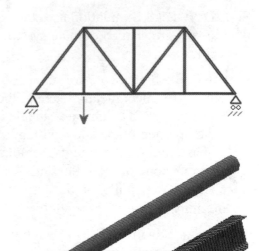

A truss is a special beam element that can resist axial deformation only.

Beam elements require defining the exact cross section so that the program can calculate the moments of inertia, neutral axes and the distances from the extreme fibers to the neutral axes. The stresses vary within the plane of the cross-section and along the beam. A beam element is a line element defined by two end points and a cross-section.

Consider a 3D beam with cross-sectional area (A) and the associated mesh. Beam elements can be displayed on actual beam geometry or as hollow cylinders regardless of their actual cross-section.

Beam elements are capable of resisting axial, bending, shear, and torsional loads. Trusses resist axial loads only. When used with weldments, the software defines cross-sectional properties and detects joints.

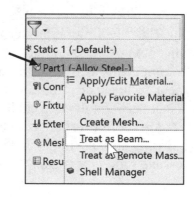

Mesh on cyclinders and beam geometry

Beam and truss members can be displayed on actual beam geometry or as hollow cylinders regardless of their actual cross-sectional shape.

A Beam element has 3 nodes (one at each end) with 6 degrees of freedom (3 translational and 3 rotational) per node plus one node to define the orientation of the beam cross section.

A Truss element has 2 nodes with 3 translational degrees of freedom per node.

The accuracy of the solution depends on the quality of the mesh. In general, the finer the mesh the better the accuracy.

A compatible mesh is a mesh where elements on touching bodies have overlaying nodes.

The curvature-based mesher supports multi-threaded surface and volume meshing for assembly and multi-body part documents. The standard mesher supports only multi-threaded volume meshing.

It is possible to mesh a part or assembly with a combination of solids, shells and beam elements (mixed mesh) in SOLIDWORKS Simulation.

SOLIDWORKS Simulation Meshing Tips

SOLIDWORKS Simulation Help lists the following Meshing tips that you should know for the CSWSA-FEA exam.

- When you mesh a study, the SOLIDWORKS Simulation meshes all unsuppressed solids, shells and beams:

- Use Solid mesh for bulky objects.

- Use Shell elements for thin objects like sheet metals.

- Use Beam or Truss elements for extruded or revolved objects with constant cross-sections.

- Simplify structural beams to optimize performance in Simulation to be modeled with beam elements. The size of the problem and the resources required are dramatically reduced in this case. For the beam formulation to produce acceptable results, the length of the beam should be 10 times larger than the largest dimension of its cross section.

- Compatible meshing (a mesh where elements on touching bodies have overlaying nodes) is more accurate than incompatible meshing in the interface region. Requesting compatible meshing can cause mesh failure in some cases. Requesting incompatible meshing can result in successful results. You can request compatible meshing and select Re-mesh failed parts with incompatible mesh so that the software uses incompatible meshing only for bodies that fail to mesh.

- Check for interferences between bodies when using a compatible mesh with the curvature-based mesher. If you specify a bonded contact condition between bodies, they should be touching. If interferences are detected, meshing stops, and you can access the Interference Detection PropertyManager to view the interfering parts. Make sure to resolve all interferences before you mesh again.

- If meshing fails, use the Failure Diagnostics tool to locate the cause of mesh failure. Try the proposed options to solve the problem. You can also try different element size, define mesh control, or activate Enable automatic looping for solids.

- The SOLIDWORKS Simplify utility lets you suppress features that meet a specified simplification factor. In the Simulation study tree, right-click Mesh and select Simplify Model for Meshing. This displays the Simplify utility.

- It is good practice to check mesh options before meshing. For example, the Automatic transition can result in generating an unnecessarily large number of elements for models with many small features. The high quality mesh is recommended for most cases. The Automatic looping can help solve meshing problems automatically, but you can adjust its settings for a particular model. The Curvature-based mesher automatically uses smaller element sizes in regions with high curvature.

- To improve results in important areas, use mesh control to set a smaller element size. When meshing an assembly with a wide range of component sizes, default meshing results in a relatively coarse mesh for small components. Component mesh control offers an easy way to give more importance to the selected small components. Use this option to identify important small components.

- For assemblies, check component interference. To detect interference in an assembly, click Tools, Interference Detection. Interference is allowed only when using shrink fit. The Treat coincidence as interference and Include multi-body part interferences options allow you to detect touching areas. These are the only areas affected by the global and component contact settings.

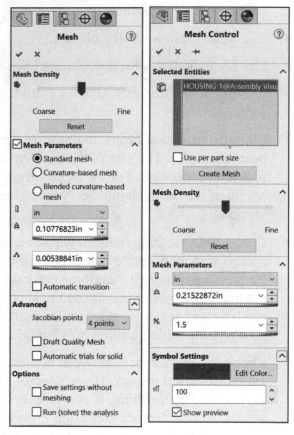

Use the mesh and displacement plots to calculate the distance between two nodes using SOLIDWORKS Simulation.

The Global element size parameter provides the ability to set the global average element size. SOLIDWORKS suggests a default value based on the model volume and surface area. This option is only available for a standard mesh.

The Ratio value in Mesh Control provides the geometric growth ratio from one layer of elements to the next. To access Mesh Control, right-click the Mesh folder in the Simulation study tree and click Apply Mesh Control.

Running the Study

When you run a study, Simulation calculates the results based on the specified input for materials, restraints, loads and mesh.

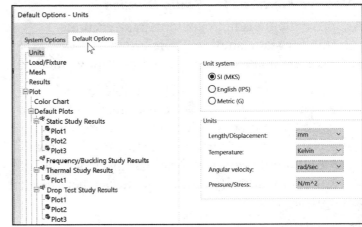

Set the default plots that you want to see in your Simulation Study tree under Simulation, Options from the Main menu.

When you run one or multiple studies, they run as background processes.

In viewing the results after running a study, you can generate plots, lists, graphs, and reports depending on the study and result types.

☀ Run multiple studies (batches) either by using the SOLIDWORKS Task Scheduler or the Run all Studies command.

☀ If you modify the study (force, material, etc.) you only need to re-run the study to update the results. You do not need to re-mesh unless you modified contact conditions.

Displacement Plot - Output of Linear Static Analysis

The Displacement Plot PropertyManager allows you to plot displacement and reaction force results for static, nonlinear, dynamic, drop test studies, or mode shapes for bucking and frequency studies. By default, directions X, Y and Z refer to the global coordinate system.

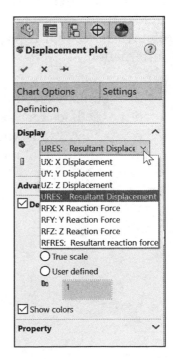

If you choose a reference geometry, these directions refer to the selected reference entity. Displacement components are:

UX = Displacement in the X-direction.

UY = Displacement in the Y-direction.

UZ = Displacement in the Z-direction.

URES = Resultant displacement.

RFX = Reaction force in the X-direction.

RFY = Reaction force in the Y-direction.

RFZ = Reaction force in the Z-axis.

RFRES = Resultant reaction force.

The Probe function allows you to query a plot and view the values of plotted quantities at defined nodes or centers of elements. When you probe a mesh plot, Simulation displays the node or element number and the global coordinates of the node. When you probe a result plot, SOLIDWORKS Simulation displays the node or element number, the value of the plotted result, and the global coordinates of the node or center of the element. For example, in a nodal stress plot, the node number, the stress value, and the global x, y, and z coordinates appear.

Adaptive Methods for Static Studies

Adaptive methods help you obtain an accurate solution for static studies. There are two types of adaptive methods: h-adaptive and p-adaptive method.

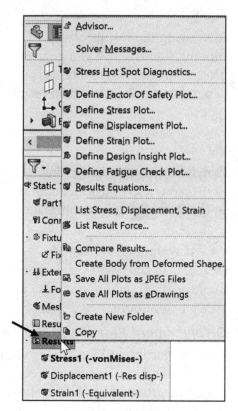

The concept of the h-method (available for solid part and assembly documents) is to use smaller elements (increase the number of elements) in regions with high relative errors to improve accuracy of the results. After running the study and estimating errors, the software automatically refines the mesh where needed to improve results.

The p-adaptive method (available for solid part and assembly documents) increases the polynomial order of elements with high relative errors. The p-method does not change the mesh. It changes the order of the polynomials used to approximate the displacement field using a unified polynomial order for all elements. See SOLIDWORKS Help for additional information.

A complete understanding of p-adaptive and h-adaptive methods requires advanced study; for the CSWSA- FEA exam, you require general knowledge of these two different methods.

Sample Exam Questions

These questions are examples of what to expect on the certification exam. The multiple choice questions should serve as a check for your knowledge of the exam materials.

1. What is the Modulus of Elasticity?

- The slope of the Deflection-Stress curve
- The slope of the Stress-Strain curve in its linear section
- The slope of the Force-Deflection curve in its linear section
- The first inflection point of a Strain curve

2. What is Stress?

- A measure of power
- A measure of strain
- A measure of material strength
- A measure of the average amount of force exerted per unit area

3. Which of the following assumptions are true for a static analysis in SOLIDWORKS Simulation with small displacements?

- Inertia effects are negligible and loads are applied slowly
- The model is not fully elastic. If loads are removed, the model will not return to its original position
- Results are proportional to loads
- All the displacements are small relative to the model geometry

4. What is Yield Stress?

- The stress level beyond which the material becomes plastic
- The stress level beyond which the material breaks
- The strain level above the stress level which the material breaks
- The stress level beyond the melting point of the material

5. A high quality Shell element has _____ nodes.

- 4
- 5
- 6
- 8

6. Stress σ is proportional to _____ in a Linear Elastic Material.

- Strain
- Stress
- Force
- Pressure

7. The Elastic Modulus (Young's Modulus) is the slope defined as _____ divided by _____.

- Strain, Stress
- Stress, Strain
- Stress, Force
- Force, Area

8. Linear static analysis assumes that the relationship between loads and the induced response is _____.

- Flat
- Linear
- Doubles per area
- Translational

9. In SOLIDWORKS Simulation, the Factor of Safety (FOS) calculations are based on one of the following failure criterion.

- Maximum von Mises Stress
- Maximum shear stress (Tresca)
- Mohr-Coulomb stress
- Maximum Normal stress

10. The Yield point is the point where the material begins to deform at a faster rate than at the elastic limit. The material behaves _____ in the Plastic Range.

- Flatly

- Linearly

- Non-Linearly

- Like a liquid

☼ The purpose of this chapter is not to educate a new or intermediate user on SOLIDWORKS Simulation, but to cover and to inform you on the types of questions, layout and what to expect when taking the CSWSA-FEA exam.

11. What are the Degrees of Freedom (DOFs) restrained for a Solid?

- None

- 3 Translations

- 3 Translations and 3 Rotations

- 3 Rotations

12. What are the Degrees of Freedom (DOFs) restrained for Truss joints?

- None

- 3 Translations

- 3 Translations and 3 Rotations

- 3 Rotations

13. What are the Degrees of Freedom (DOFs) restrained for Shells and Beams?

- None

- 3 Translations

- 3 Translations and 3 Rotations

- 3 Rotations

14. Which statements are true for Material Properties using SOLIDWORKS Simulation?

- For solid assemblies, each component can have a different material.

- For shell models, each shell cannot have a different material and thickness.

- For shell models, the material of the part is used for all shells.

- For beam models, each beam cannot have a different material.

15. A Beam element has _____nodes (one at each end) with _____degrees of freedom per node plus_____ node to define the orientation of the beam cross section.

- 6, 3, 1

- 3, 3, 1

- 3, 6, 1

- None of the above

16. A Truss element has _____ nodes with _____ translational degrees of freedom per node.

- 2, 3

- 3, 3

- 6, 6

- 2, 2

17. In general the finer the mesh the better the accuracy of the results.

- True

- False

18. How does SOLIDWORKS Simulation automatically treat a Sheet metal part with uniform thickness?

- Shell

- Solid

- Beam

- Mixed Mesh

19. Use the mesh and displacement plots to calculate the distance between two _____ using SOLIDWORKS Simulation.

- Nodes

- Elements

- Bodies

- Surfaces

20. Surface models can only be meshed with _____ elements.

- Shell
- Beam
- Mixed Mesh
- Solid

21. The shell mesh is generated on the surface (located at the mid-surface of the shell).

- True
- False

22. In general, use Thin shells when the thickness-to-span ratio is less than _____.

- 0.05
- 0.5
- 1
- 2

23. The model (a rectangular plate) has a length to thickness ratio of less than 5. You extracted its mid-surface to use it in SOLIDWORKS Simulation. You should use a _____.

- Thin Shell element formulation
- Thick Shell element formulation
- Thick or Thin Shell element formulation, it does not matter
- Beam Shell element formulation

24. The model, a rectangular sheet metal part, uses SOLIDWORKS Simulation. You should use a:

- Thin Shell element formulation
- Thick Shell element formulation
- Thick or Thin Shell element formulation, it does not matter
- Beam Shell element formulation

25. The Global element size parameter provides the ability to set the global average element size. SOLIDWORKS Simulation suggests a default value based on the model volume and _____ area. This option is only available for a standard mesh.

- Force

- Pressure

- Surface

- None of the above

26. A remote load applied on a face with a Force component and no Moment can result in: Note: Remember (DOFs restrain).

- A Force and Moment of the face

- A Force on the face only

- A Moment on the face only

- A Pressure and Force on the face

27. There are _____ DOFs restrain for a Solid element.

- 3

- 1

- 6

- None

28. There are _____ DOFs restrain for a Beam element.

- 3

- 1

- 6

- None

29. What best describes the difference(s) between a Fixed and Immovable (No translation) boundary condition in SOLIDWORKS Simulation?

- There are no differences

- There are no difference(s) for Shells but it is different for Solids

- There are no difference(s) for Solids but it is different for Shells and Beams

- There are only differences(s) for a Static Study.

30. Can a non-uniform pressure or force be applied on a face using SOLIDWORKS Simulation?

- No

- Yes, but the variation must be along a single direction only

- Yes. The non-uniform pressure distribution is defined by a reference coordinate system and the associated coefficients of a second order polynomial.

- Yes, but the variation must be linear

31. You are performing an analysis on your model. You select five faces, 3 edges and 2 vertices and apply a force of 20lbf. What is the total force applied to the model using SOLIDWORKS Simulation?

- 100lbf

- 1600lbf

- 180lbf

- 200lbf

32. Yield strength is typically determined at _____ strain.
- 0.1%

- 0.2%

- 0.02%

- 0.002%

33. There are four key assumptions made in Linear Static Analysis: 1. Effects of inertia and damping are neglected, 2. The response of the system is directly proportional to the applied loads, 3. Loads are applied slowly and gradually, and_____ .

- Displacements are very small. The highest stress is in the linear range of the stress-strain curve.

- There are no loads

- Material is not elastic

- Loads are applied quickly

34. How many degrees of freedom does a physical structure have?

- Zero

- Three - Rotations only

- Three - Translations only

- Six - Three translations and three rotational

35. Brittle material has little tendency to deform (or strain) before fracture and does not have a specific yield point. It is not recommended to apply the yield strength analysis as a failure criterion on brittle material. Which of the following failure theories is appropriate for brittle materials?

- Mohr-Columb stress criterion

- Maximum shear stress criterion

- Maximum von Mises stress criterion

- Minimum shear stress criterion

36. You are performing an analysis on your model. You select three faces and apply a force of 40lbf. What is the total force applied to the model using SOLIDWORKS Simulation?

- 40lbf

- 20lbf

- 120lbf

- Additional information is required

37. A material is orthotropic if its mechanical or thermal properties are not unique and independent in three mutually perpendicular directions.

- True

- False

38. An increase in the number of elements in a mesh for a part will.

- Decrease calculation accuracy and time

- Increase calculation accuracy and time

- Have no effect on the calculation

- Change the FOS below 1

39. SOLIDWORKS Simulation uses the von Mises Yield Criterion to calculate the Factor of Safety of many ductile materials. According to the criterion:

- Material yields when the von Mises stress in the model equals the yield strength of the material.

- Material yields when the von Mises stress in the model is 5 times greater than the minimum tensile strength of the material.

- Material yields when the von Mises stress in the model is 3 times greater than the FOS of the material.

- None of the above.

40. SOLIDWORKS Simulation calculates structural failure on:

- Buckling

- Fatigue

- Creep

- Material yield

41. Apply a uniform total force of 200lbf on two faces of a model. The two faces have different areas. How do you apply the load using SOLIDWORKS Simulation for a Linear Static Study?

- Select the two faces and input a normal to direction force of 200lbf on each face.

- Select the two faces and a reference plane. Apply 100lbf on each face.

- Apply equal force to the two faces. The force on each face is the total force divided by the total area of the two faces.

- None of the above.

42. Maximum and Minimum value indicators are displayed on Stress and Displacement plots in SOLIDWORKS Simulation for a Linear Static Study.

- True

- False

43. What SOLIDWORKS Simulation tool should you use to determine the result values at specific locations (nodes) in a model using SOLIDWORKS Simulation?

- Section tool

- Probe tool

- Clipping tool

- Surface tool

44. What criteria are best suited to check the failure of ductile materials in SOLIDWORKS Simulation?

- Maximum von Mises Strain and Maximum Shear Strain criterion

- Maximum von Mises Stress and Maximum Shear Stress criterion

- Maximum Mohr-Coulomb Stress and Maximum Mohr-Coulomb Shear Strain criterion

- Mohr-Coulomb Stress and Maximum Normal Stress criterion.

45. Set the scale factor for plots_____ to avoid any misinterpretation of the results, after performing a Static analysis with gap/contact elements.

- Equal to 0

- Equal to 1

- Less than 1

- To the Maximum displacement value for the model

46. It is possible to mesh _____ with a combination of Solids, Shells and Beam elements in SOLIDWORKS Simulation.

- Parts and Assemblies

- Only Parts

- Only Assemblies

- None of the above

47. SOLIDWORKS Simulation supports multi-body parts. Which of the following is a true statement?

- You can employ different mesh controls to each Solid body

- You can classify Contact conditions between multiple Solid bodies

- You can classify a different material for each Solid body

- All of the above are correct

48. Which statement best describes a Compatible mesh?

- A mesh where only one type of element is used

- A mesh where elements on touching bodies have overlaying nodes

- A mesh where only a Shell or Solid element is used

- A mesh where only a single Solid element is used

49. The Ratio value in Mesh Control provides the geometric growth ratio from one layer of elements to the next.

- True

- False

50. The structures displayed in the following illustration are best analyzed using:

- Shell elements

- Solid elements

- Beam elements

- A mixture of Beam and Shell elements

51. The structure displayed in the following illustration is best analyzed using:

- Shell elements
- Solid elements
- Beam elements
- A mixture of Beam and Shell elements

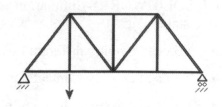

52. The structure displayed in the following illustration is best analyzed using:

- Shell elements
- Solid elements
- Beam elements
- A mixture of Beam and Shell elements

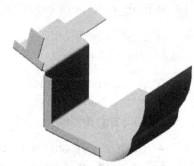

Sheet metal model

53. The structure displayed in the following illustration is best analyzed using:

- Shell elements
- Solid elements
- Beam elements
- A mixture of Beam and Shell elements

54. Surface models can only be meshed with _____ elements.

- Shell elements
- Solid elements
- Beam elements
- A mixture of Beam and Shell elements

55. Use the _____ and _____ plots to calculate the distance between two nodes using SOLIDWORKS Simulation.

- Mesh and Displacement

- Displacement and FOS

- Resultant Displacement and FOS

- None of the above

56. You can simplify a large assembly in a Static Study by using the _____ or _____ options in your study.

- Make Rigid, Fix

- Shell element, Solid element

- Shell element, Compound element

- Make Rigid, Load element

57. A force "F" applied in a static analysis produces a resultant displacement URES. If the force is now 2x F and the mesh is not changed, then URES will:

- Double if there are no contacts specified and there are large displacements in the structure

- Be divided by 2 if contacts are specified

- The analysis must be run again to find out

- Double if there is not a source of nonlinearity in the study (like contacts or large displacement options)

58. To compute thermal stresses on a model with a uniform temperature distribution, what type/types of study/studies are required?

- Static only

- Thermal only

- Both Static and Thermal

- None of these answers is correct

59. In an h-adaptive method, use smaller elements in mesh regions with high errors to improve the accuracy of results.

- True
- False

60. In a p-adaptive method, use elements with a higher order polynomial in mesh regions with high errors to improve the accuracy of results.

- True
- False

61. Where will the maximum stress be in the illustration?

- A
- B
- C
- D

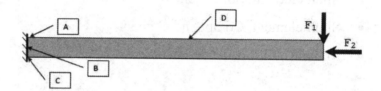

The purpose of this chapter is not to educate a new or intermediate user on SOLIDWORKS Simulation, but to cover and to inform you on the types of questions, layout and what to expect when taking the CSWSA-FEA exam.

 The CSWSA-FEA only covers Linear Static analysis.

FEA Modeling Section

Tutorial FEA Model 7-1

An exam question in this category could read:

In the figure displayed, what is the vertical displacement in the Global Y direction in (inches) at the location of the dot? Calculate your answer to 3 decimal places.

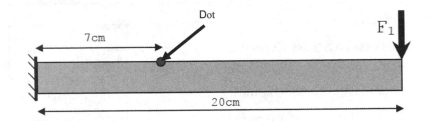

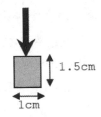

Given Information:

Material: Alloy Steel (SS) from the SOLIDWORKS Simulation Library.

Elastic modulus = $2.1e11$ N/m^2

Poisson's ratio = 0.28

F_1 = 200lbf

Use the default high quality element size to mesh.

Use the models from the CSWSA-FEA Model folder for this section.

Let's start.

1. **Open** Model 7-1 from the CSWSA-FEA Model folder.

Think about the problem. Think about the model. The bar that you opened was created on the Front Plane.

The upper left corner of the rectangle is located at the origin. This simplifies the actual deformation of the part. The height dimension references zero in the Global Y direction.

Split Lines were created to provide the ability to locate the needed Joints in this problem. To add the force at the center of the beam, a split line in the shape of a small circle on the top face of the right end of the beam is used.

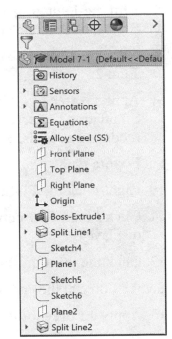

The proper model setup is very important to obtain the correct mesh and to obtain the correct final results.

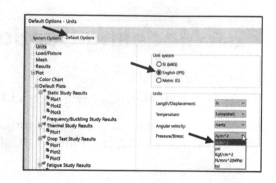

Set Simulation Options and start a Static SOLIDWORKS Simulation Study.

2. **Add-In** SOLIDWORKS Simulation.

3. Click **Simulation, Options** from the Main menu. Click the **Default Options** tab.

Set the Unit system and Mesh quality.

4. Select **English (IPS)** and **Pressure/Stress (N/m²)** as illustrated.

5. Click the **Mesh folder**. Select **High** for Mesh quality.

6. **Create** a new Static Study. Accept the default name (Static #). Treat the model as a **Beam**.

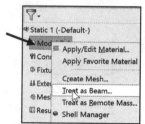

7. **Edit** the Joint groups folder from the Study Simulation tree. Split Line 2 is selected by default.

8. Click the **Calculate** button and accept the Results: Joint 1, Joint 2.

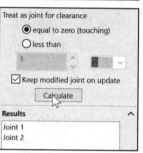

9. Apply **Material** - Alloy Steel (SS) from the SOLIDWORKS Simulation Library. A check mark is displayed next to the model name in the Study Simulation tree.

Set Fixture type.

10. Right-click the **Fixtures** folder.

11. Click **Fixed Geometry**. The Fixture PropertyManager is displayed.

12. Click the **joint on the left side** of the beam.

Set the Force at the end of the beam.

13. Apply a **Force** for the External load.

14. Click the **Joints** option.

15. Click the **joint on the right side** of the beam.

16. Click the **end face** of the beam as the plane for direction.

17. Enter **200**lbf Along Plane Direction1. The arrow is displayed downwards; reverse direction if needed.

Mesh and Run the model.

18. **Mesh and Run** the model. Use the standard default setting for the mesh.

19. Double-click the **Stress1** folder. View the results.

20. Double-click the **Displacement1** folder. View the results. If needed, Right-click the Displacement1 folder, and click Edit Definition. Set Chart Options to floating.

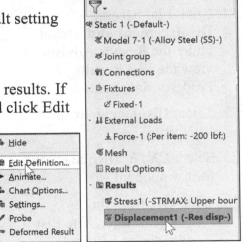

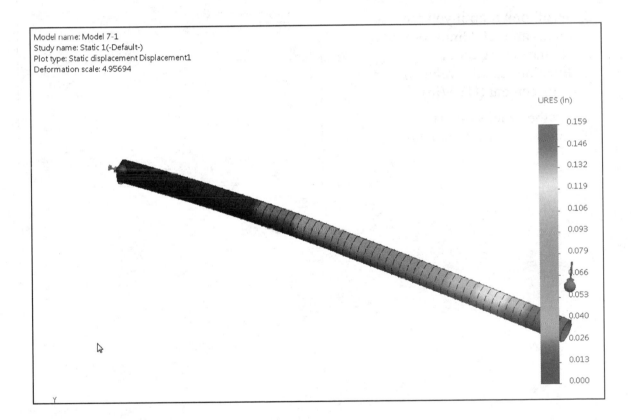

Illustrations may vary slightly depending on your SOLIDWORKS version.

Locate the displacement at 7cm.

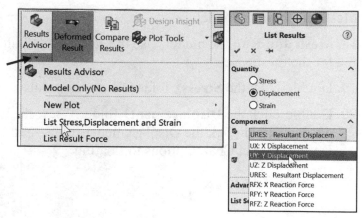

21. Click **List Stress, Displacement and Strain** under the Results Advisor to view the List Results PropertyManager.

22. Select **Displacement** for Quantity.

23. Select **UY: Y Displacement** for Component.

24. Click **OK** from the PropertyManager. The List Results dialog box is displayed. View the results.

25. **Scroll down** until you see values around 70mm for the distance along the X direction. See the value of displacement (UY) (in).

To find the exact value at 70mm, use linear interpolation.

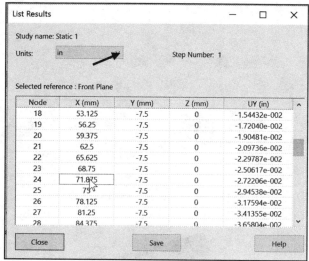

This method is shown below and uses the values greater and less than the optimal one to find the actual displacement.

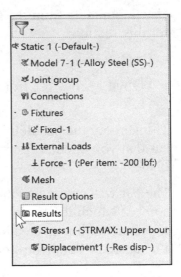

$$\frac{X_U - X_L}{UY_U - UY_L} = \frac{X_U - X_O}{UY_U - UY_O}$$

$$\frac{71.875 - 68.75}{-0.0272206 - -0.0250617} = \frac{71.875 - 70}{-0.02718 - UY_O}$$

$$UY_o = 0.0258$$

At the distance of 7cm (70mm) the displacement is found to be approximately 0.026in.

The correct answer is **B**.

A = 0.034in

B = 0.026in

C = 0.043in

D = 0.021in

The purpose of this chapter is not to educate a new or intermediate user on SOLIDWORKS Simulation, but to cover and to inform you on the types of questions, layout and what to expect when taking the CSWSA-FEA exam.

Tutorial FEA Model 7-2

Below is a second way to address the first problem (Tutorial FEA Model 7-1) with a different model, without using Split lines, using the Study Advisor and the Probe tool.

In the figure displayed, what is the vertical displacement in the Global Y direction in (inches) at the location of the red dot? Report your answer to 3 decimal places.

Given Information:

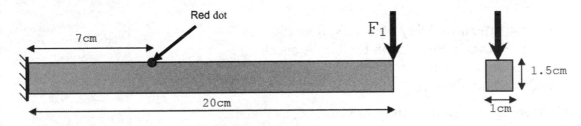

Material: Alloy Steel (SS) from the SOLIDWORKS Simulation Library.

Elastic modulus = 2.1e11 N/m²

Poisson's ratio = 0.28

F₁ = 200lbf

Use the default high quality element size to mesh.

Use the models from the CSWSA-FEA Model folder for this section.

Let's start.

1. **Open** Model 7-2 from the CSWSA-FEA Model folder.

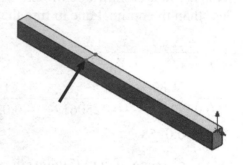

Think about the problem. Think about the model. The bar that you opened is created on the Front Plane.

The bar was created so that the origin is the point at which the force is applied. A construction line is created across the part at a distance of 7cm from the end that is to be fixed.

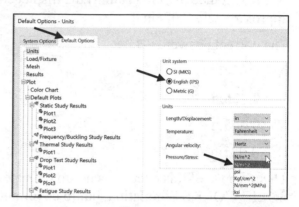

Set Simulation Options and start a Static SOLIDWORKS Simulation Study.

2. **Add-In** SOLIDWORKS Simulation.

3. Click **Simulation**, **Options** from the Main menu. Click the **Default Options** tab.

Set the Unit system and Mesh quality.

4. Select **English (IPS)** and **Pressure/Stress (N/m²)**.

5. Click the **Mesh folder**. Select **High** for Mesh quality.

6. **Create** a new Static Study. Accept the default name (Static #).

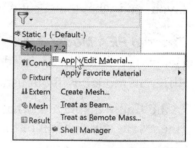

7. Apply **Material** - Alloy Steel (SS) from the SOLIDWORKS Simulation Library. A check mark is displayed next to the model name in the Study Simulation tree.

Set Fixture type. Use the Study Advisor. Set as Fixed Geometry.

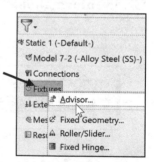

8. Right-click the **Fixtures** folder from the Simulation study tree.

9. Click **Advisor**.

10. Click **Add a fixture**.

11. Select the **proper face (left end)** as illustrated.

12. Set as **Fixed Geometry**.

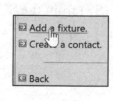

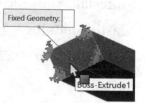

Set the Force.

13. Right-click the **External loads** folder from the Simulation study tree. Select **Remote Load/Mass**. The Remote Loads/Mass PropertyManager is displayed.

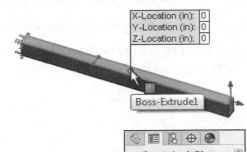

14. Click **Load (Direct transfer)** for Type.

15. Select the **top face of the part** - for Faces for Remote Load.

16. Set the Reference Coordinate system to **Global**.

17. Leave all of the location boxes at **zero** value.

18. Select the **Force** check box.

19. Select **lbf** for units.

20. Enter **200lbf in Y-Direction**. Use the X-, Y-, and Z-direction boxes to direct the load; reverse direction as necessary.

Mesh and Run the model.

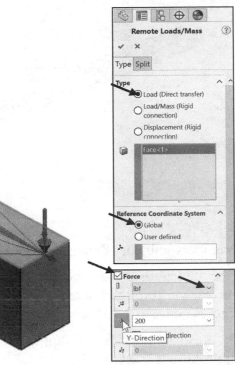

21. **Mesh and Run** the model. Use the standard default setting for the mesh.

22. Double-click the **Stress1** folder.

23. **View** the results.

24. Double-click the **Displacement1** folder.

25. **View** the results. If needed, Right-click the Displacement1 folder, and click Edit Definition. Set Chart Options to floating.

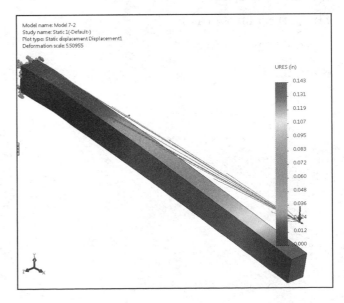

Locate the displacement at 7cm using the Probe tool.

26. Click the **Displacement1** Results folder if needed.

27. Click **Probe** from the Plot Tools drop-down menu.

28. Select **two points**: one on each side of the construction line as illustrated in the Results table.

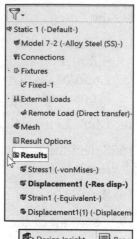

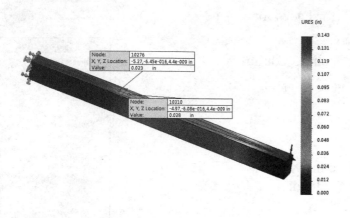

Numbers will vary depending on the selection location.

Use the length and displacement values of the selected points to find the answer through linear interpolation. The prescribed distance of 7cm is equal to 2.755in.

At the distance of 7cm (70mm) the displacement is found to be approximately 0.026in.

The correct answer is **B**.

A = 0.034in

B = 0.026in

C = 0.043in

D = 0.021in

Tutorial FEA Model 7-3

Proper model setup is very important to obtain the correct mesh and to obtain the correct final results. In the last two examples, you needed to manually apply linear interpolation to locate your final answer.

How can you eliminate the need to manually apply linear interpolation for your final answer? Create a Sensor or use Split Lines at the 7cm point.

Below is the Tutorial FEA Model 7-2 problem. Let's use the same model, but apply a Split line at the 7cm point before we begin the Linear Static study.

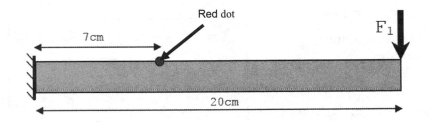
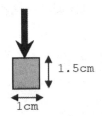

In the figure displayed, what is the vertical displacement in the Global Y direction in (inches) at the location of the red dot? Calculate your answer to 3 decimal places.

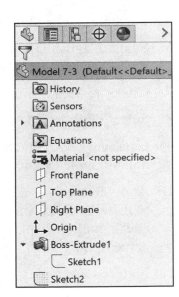

Given Information:

Material: Alloy Steel (SS) from the SOLIDWORKS Simulation Library.

Elastic modulus = 2.1e11 N/m^2

Poisson's ratio = 0.28

F$_1$ = 200lbf

Use the default high quality element size to mesh.

Let's start.

1. **Open** Model 7-3 from the CSWSA-FEA Model Folder.

Think about the problem. This is the same model that you opened in the second example. How can you eliminate the need to manually apply linear interpolation for your final answer?

Address this in the initial setup of the provided model.

Create a Projected Split line at 70mm.

Apply the Probe tool and select the Split Line for the exact point.

2. Create a **Split line feature** with a new sketch (**Insert, Curve, Split Line**) at 70mm.

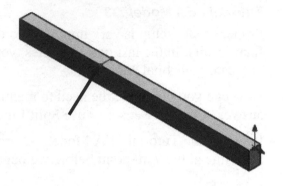

Set Simulation Options and start a Static SOLIDWORKS Simulation Study.

3. **Add-In** SOLIDWORKS Simulation.

4. Click **Simulation, Options** from the Main menu.

5. Click the **Default Options** tab.

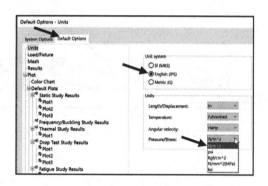

Set the Unit system and Mesh quality.

6. Select **English (IPS)** and **Pressure/Stress (N/m²)** as illustrated.

7. Click the **Mesh folder**.

8. Select **High** for Mesh quality.

9. **Create** a new Static Study. Accept the default name (Static #).

10. Apply **Material** - Alloy Steel (SS) from the SOLIDWORKS Simulation Library. A check mark is displayed next to the model name in the Study Simulation tree.

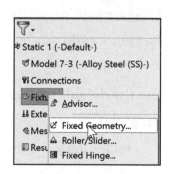

Set Fixture type. Set as Fixed Geometry.

11. Right-click **Fixtures** folder from the Simulation study tree.

12. Click **Fixed Geometry**.

13. Select the **proper face (left end)** as illustrated.

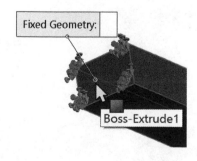

Set the External Load.

14. Right-click the **External Loads** folder from the Simulation study tree.

15. Click **Remote Load/Mass**. The Remote Loads/Mass PropertyManager is displayed.

16. Click **Load (Direct transfer)** for Type.

17. Select the **top face of the part** - for Faces for Remote Load. Note: Click **on both sides** of the Split line.

18. Set the Reference Coordinate system to **Global**.

19. Leave all of the location boxes at **zero** value.

20. Select the **Force** check box. Click the **Y-Direction** box.

21. Enter **200**. Use the X-, Y- and Z-direction boxes to direct the load; reverse direction as necessary.

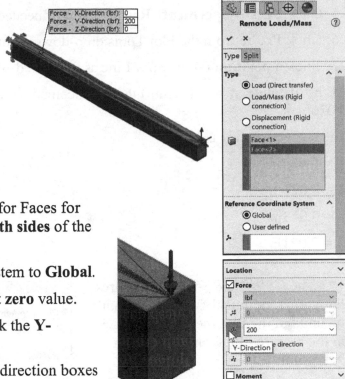

Mesh and Run the model.

22. **Mesh and Run** the model. Use the standard default setting for the mesh.

23. Double-click the **Stress1** folder. View the results.

24. Double-click the **Displacement1** folder. View the results. If needed, Right-click the Displacement1 folder, and click Edit Definition. Set Chart Options to floating.

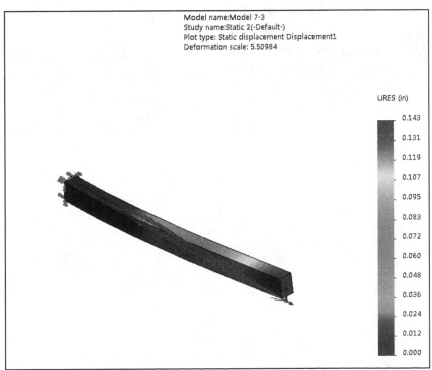

Model name:Model 7-3
Study name:Static 2(-Default-)
Plot type: Static displacement Displacement1
Deformation scale: 5.50984

URES (in)

0.143
0.131
0.119
0.107
0.095
0.083
0.072
0.060
0.048
0.036
0.024
0.012
0.000

Locate the displacement at 7cm using the Probe tool.

25. Click the **Displacement1** Results folder if needed.

26. Click **Probe** from the Plot Tools drop-down menu.

27. Click a **position** on the Split Line as illustrated. View the results.

At the distance of 7cm (70mm) the displacement is found to be 0.026in.

The correct answer is **B**.

A = 0.034in

B = 0.026in

C = 0.043in

D = 0.021in

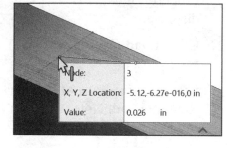

The CSWSA-FEA exam requires that you work quickly. You can modify the units in the Plot menu. Right-click on the required Plot, and select Edit Definition.

Each hands-on problem in the CSWSA-FEA exam requires a single answer. To save time, set the Chart value to floating and the number of decimal places required. Right-click on the required Plot, and select Chart Options.

Decimal places required change often. Verify units and decimal places.

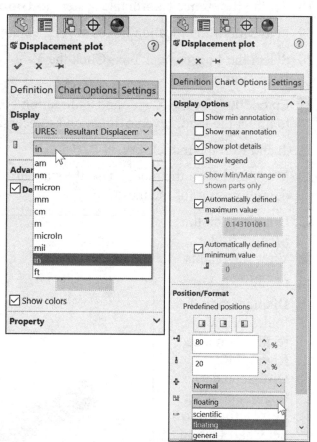

Tutorial FEA Model 7-4

An exam question in the Solid category could read:

In the figure displayed, what is the maximum resultant displacement in millimeters on the annular face of the model? The three holes are fixed.

Calculate your answer to 3 decimal places.

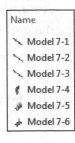

Given Information:

Material: Alloy Steel (SS).

A normal force F_1 is applied to the annular face.
$F_1 = 3000$lbf.

The three holes are **fixed**.

Use the default high quality element size to mesh.

Use the models from the CSWSA-FEA Model folder for this section.

Let's start.

1. **Open** Model 7-4 from the CSWSA-FEA Model folder. Think about the problem. Think about the model.

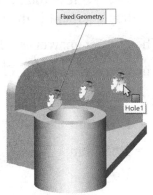

Set Simulation Options and start a Static SOLIDWORKS Simulation Study.

2. **Add-In** SOLIDWORKS Simulation.

3. Click **Simulation**, **Options** from the Main menu.

4. Click the **Default Options** tab.

Set the Unit system and Mesh quality.

5. Select **English (IPS)** and **Pressure/Stress (psi)**.

6. Click the **Mesh folder**.

7. Select **High** for Mesh quality.

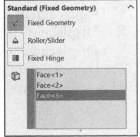

8. **Create** a new Static Study. Accept the default name (Static #).

9. Apply **Material - Alloy Steel (SS)** from the SOLIDWORKS Simulation Library. A check mark is displayed next to the model name in the Study Simulation tree.

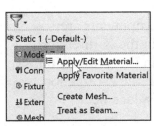
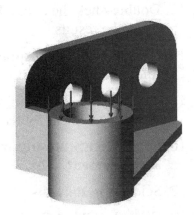

10. Apply Fixed Geometry **Fixtures**. Select the three cylindrical faces of the hole pattern as illustrated.

11. Apply an External Load of **3000lbf** normal to on the annular face of the model as illustrated.

Mesh and Run the model.

12. **Mesh** and **Run** the model. Use the default setting for the mesh.

Create a Displacement Plot for the maximum resultant displacement in millimeters.

13. Double-click the **Stress1** folder. View the results.

14. Double-click the **Displacement1** folder. View the results.

15. Right-click **Edit Definition** from the Displacement folder.

16. Select **URES, Resultant Displacement**.

17. Select **mm** for units.

18. Click the **Chart Options** tab.

19. Select **Show max annotation**. The maximum displacement is displayed: 1.655mm.

The correct answer is **±1%** of this value.

The correct answer is **C**.

A = 1.112mm

B = 1.014mm

C = 1.655mm

D = 1.734mm

 Displacement components are:
UX = Displacement in the X-direction,
UY = Displacement in the Y-direction,
UZ = Displacement in the Z-direction,
URES = Resultant displacement.

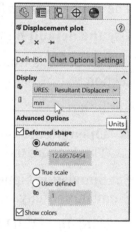

After you calculate displacement or other parameters in a Simulation Study, the CSWSA-FEA exam will also deliver a series of successive questions to test the understanding of the results.

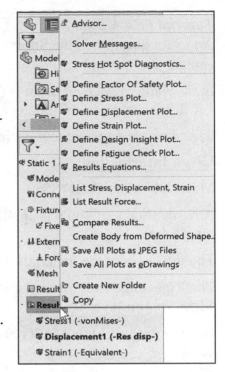

In the first question you calculate Displacement; in the second question you calculate Resultant Force. Do not re-mesh and re-run. Create the required parameter in the Results folder.

In the third question, you are asked to determine if the results are valid or invalid. If the materials yield strength was passed, then the results are invalid.

Use the Define Factor of Safety Plot to determine if your results are valid or invalid. The CSWSA-FEA exam requires you to apply Finite Element Method theory and review displacement values, factory of safety, mesh refinement, and material properties such as yield strength.

The following are some statements you will encounter in the CSWSA-FEA exam:

- The results are invalid because the material's yield strength was passed.

- The results area is invalid because the displacement was more than ½ the plate's thickness.

- The results are valid as they are, even if mesh refinement was better.

- The results are invalid because for sure a dynamic study is required.

In general use Thin Shells when thickness to span ratio < 0.05.

Displacement components are UX = Displacement in the X-direction, UY = Displacement in the Y-direction, UZ = Displacement in the Z-direction, URES = Resultant displacement.

Tutorial FEA Model 7-5

An exam question in the Sheet Metal category could read:

In the figure displayed, what is the maximum *UX* displacement in millimeters?

Calculate your answer to 4 decimal places.

Given Information:

Material: Alloy Steel (SS).

A normal force of **400N** is applied to the **inside** face.

The thickness of the materials is **0.15in**.

The Right edge as illustrated and circular edges are **immovable**.

Use the default high quality element size to mesh.

Let's start.

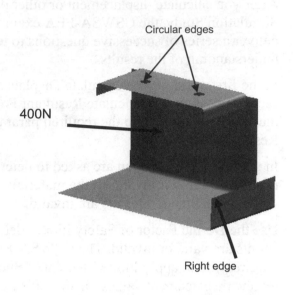

You need to define the thickness. The Fixture option Immovable is added for shell models. Thin models created with no Sheet Metal feature require you to define the use of Shell elements in SOLIDWORKS Simulation.

Use the models from the CSWSA-FEA Model folder for this section. Models created with the Sheet Metal feature automatically create Shell elements in SOLIDWORKS Simulation.

1. **Open** Model 7-5 from the CSWSA-FEA Model folder. Think about the problem. Think about the model.

Set Simulation Options and start a Static SOLIDWORKS Simulation Study.

2. **Add-In** SOLIDWORKS Simulation.

3. Click **Simulation, Options** from the Main menu.

4. Click the **Default Options** tab.

Set the Unit system and Mesh quality.

5. Select **SI (MKS)** and **Pressure/Stress (N/m²)**.

6. Click the **Mesh folder**.

7. Select **High** for Mesh quality.

8. **Create** a new Static Study. Accept the default name (Static #).

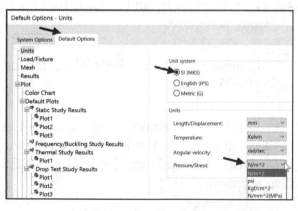

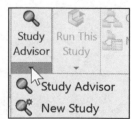

9. Apply **Material -** Alloy Steel (SS) from the SOLIDWORKS
    Simulation Library. A check mark is displayed next to the
    model name in the Study Simulation tree.

Define the Shell thickness.

10. Right-click **Model 7-5** in the Study tree.

11. Click **Edit Definition** from the drop-down menu. The Shell
    Definition PropertyManager is displayed.

12. Enter **0.15in** for **Thin** Type.

In general use Thin Shells
when thickness to span ratio
< 0.05.

Apply Fix Geometry Fixtures.

13. Click the **Immovable** option.

14. Select the **edge** and the **two
    circular edges** of the model
    as illustrated.

Apply an External Load (Force)
of 400N.

15. Apply an **External Load
    (Force)** of **400N** normal to
    the inside face as
    illustrated.

Mesh and Run the model.

16. **Mesh** and **Run** the model.
    Use the standard default
    setting for the mesh.
    Review the Results.

View the Results.

17. Double-click the **Stress1**
    folder. View the results.

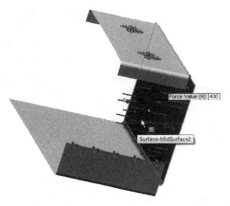

18. Double-click the
    Displacement1 folder.
    View the results.

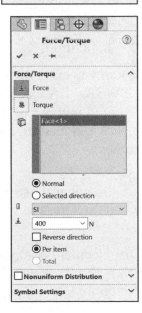

Calculate Displacement in X.

19. Right-click **Displacement1** from the study tree.

20. Click **Edit Definition**. The Displacement plot PropertyManager is displayed.

21. Select **UX: X Displacement** from the display drop-down menu.

22. Select **mm**.

Select Chart Options.

23. Click the **Chart Options** tab. The Chart Options PropertyManager is displayed.

24. Select **4 decimal places**, millimeter display and floating.

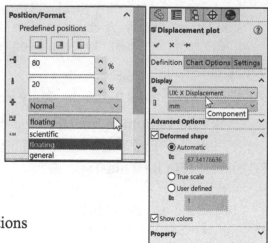

The maximum displacement in X, *UX* = 0.4370mm.

The correct answer is **±1%** of this value.

The correct answer is **D**.

A = 0.4000mm

B = 0.4120mm

C = 1.655mm

D = 0.4370mm

When you enter a value in the CSWSA-FEA exam, include the required number of decimal places and leading and trailing zeroes.

In general use Thin Shells when thickness to span ratio < 0.05.

Displacement components are UX = Displacement in the X-direction, UY = Displacement in the Y-direction, UZ = Displacement in the Z-direction, URES = Resultant displacement.

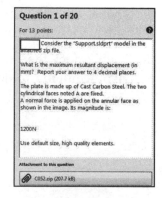

Screen shot from the exam

Definitions:

The following are a few key definitions for the exam:

Axisymmetry: Having symmetry around an axis.

Brittle: A material is brittle if, when subjected to stress, it breaks without significant deformation (strain). Brittle materials, such as concrete and carbon fiber, are characterized by failure at small strains. They often fail while still behaving in a linear elastic manner, and thus do not have a defined yield point. Because strains are low, there is negligible difference between the engineering stress and the true stress. Testing of several identical specimens will result in different failure stresses; this is due to the Weibull modulus of the brittle material.

Compatible meshing: A mesh where elements on touching bodies have overlaying nodes.

Cyclic Symmetry: To define the number of sectors and the axis of symmetry in a cyclic symmetric structure for use in a cyclic symmetry calculation.

Deflection: is a term to describe the magnitude to which a structural element bends under a load.

Deformation: is the change in geometry created when stress is applied (in the form of force loading, gravitational field, acceleration, thermal expansion, etc.). Deformation is expressed by the displacement field of the material.

Distributed Mass Load: Distributes a specified mass value on the selected faces for use with static, frequency, buckling, and linear dynamic studies. Use this functionality to simulate the effect of components that are suppressed or not included in the modeling when their mass can be assumed to be uniformly distributed on the specified faces. The distributed mass is assumed to lie directly on the selected faces, so rotational effects are not considered.

Ductile Material: In materials science, ductility is a solid material's ability to deform under tensile stress; this is often characterized by the material's ability to be stretched into a wire. Stress vs. Strain curve typical of aluminum.

Maximum Normal Stress criterion: The maximum normal stress criterion also known as Coulomb's criterion is based on the Maximum normal stress theory. According to this theory failure occurs when the maximum principal stress reaches the ultimate strength of the material for simple tension.

This criterion is used for brittle materials. It assumes that the ultimate strength of the material in tension and compression is the same. This assumption is not valid in all cases. For example, cracks decrease the strength of the material in tension considerably while their effect is far less small in compression because the cracks tend to close.

Brittle materials do not have a specific yield point and hence it is not recommended to use the yield strength to define the limit stress for this criterion.

This theory predicts failure to occur when:

$$\sigma_1 \geq \sigma_{limit}$$

where σ_1 is the maximum principal stress.

Hence:

Factor of safety = $\sigma_{limit} / \sigma_1$

Maximum Shear Stress criterion: The maximum shear stress criterion, also known as Tresca yield criterion, is based on the Maximum Shear stress theory.

This theory predicts failure of a material to occur when the absolute maximum shear stress (τ_{max}) reaches the stress that causes the material to yield in a simple tension test. The Maximum shear stress criterion is used for ductile materials.

$$\tau_{max} \geq \sigma_{limit} / 2$$

τ_{max} is the greatest of τ_{12}, τ_{23} and τ_{13}

Where:

$$\tau_{12} = (\sigma_1 - \sigma_2)/2; \tau_{23} = (\sigma_2 - \sigma_3)/2; \tau_{13} = (\sigma_1 - \sigma_3)/2$$

Hence:

Factor of safety (FOS) = $\sigma_{limit} / (2 * \tau_{max})$

Maximum von Mises Stress criterion: The maximum von Mises stress criterion is based on the von Mises-Hencky theory, also known as the Shear-energy theory or the Maximum distortion energy theory.

In terms of the principal stresses s1, s2, and s3, the von Mises stress is expressed as:

$$\sigma_{vonMises} = \{[(s1 - s2)2 + (s2 - s3)2 + (s1 - s3)2]/2\}(1/2)$$

The theory states that a ductile material starts to yield at a location when the von Mises stress becomes equal to the stress limit. In most cases, the yield strength is used as the stress limit. However, the software allows you to use the ultimate tensile or set your own stress limit.

$$\sigma_{vonMises} \geq \sigma_{limit}$$

Yield strength is a temperature-dependent property. This specified value of the yield strength should consider the temperature of the component. The factor of safety at a location is calculated from:

Factor of Safety (FOS) = $\sigma_{limit} / \sigma_{vonMises}$

Modulus of Elasticity or Young's Modulus: The Elastic Modulus (Young's Modulus) is the slope defined as stress divided by strain. E = modulus of elasticity (Pa (N/m^2), N/mm^2, psi). The Modulus of Elasticity can be used to determine the stress-strain relationship in the linear-elastic portion of the stress-strain curve. The linear-elastic region is either below the yield point, or if a yield point is not easily identified on the stress-strain plot it is defined to be between 0 and 0.2% strain, and is defined as the region of strain in which no yielding (permanent deformation) occurs.

Force is the action of one body on another. A force tends to move a body in the direction of its action.

Mohr-Coulomb: The Mohr-Coulomb stress criterion is based on the Mohr-Coulomb theory, also known as the Internal Friction theory. This criterion is used for brittle materials with different tensile and compressive properties. Brittle materials do not have a specific yield point and hence it is not recommended to use the yield strength to define the limit stress for this criterion.

Mohr-Coulomb Stress criterion: The Mohr-Coulomb stress criterion is based on the Mohr-Coulomb theory, also known as the Internal Friction theory. This criterion is used for brittle materials with different tensile and compressive properties. Brittle materials do not have a specific yield point and hence it is not recommended to use the yield strength to define the limit stress for this criterion.

This theory predicts failure to occur when:

$$\sigma_1 \geq \sigma_{TensileLimit} \quad \text{if } \sigma_1 > 0 \text{ and } \sigma_3 > 0$$

$$\sigma_3 \geq -\sigma_{CompressiveLimit} \quad \text{if } \sigma_1 < 0 \text{ and } \sigma_3 < 0$$

$$\sigma_1 / \sigma_{TensileLimit} + \sigma_3 / -\sigma_{CompressiveLimit} \geq 1 \text{ if } \sigma_1 \geq 0 \text{ and } \sigma_3 \leq 0$$

The factor of safety is given by:

Factor of Safety (FOS) = $\{\sigma_1 / \sigma_{TensileLimit} + \sigma_3 / -\sigma_{CompressiveLimit}\}^{(-1)}$

Stress: Stress is defined in terms of Force per unit Area:

$$Stress = \frac{f}{A}.$$

Stress vs. Strain diagram: Many materials display linear elastic behavior, defined by a linear stress-strain relationship, as shown in the figure up to point 2, in which deformations are completely recoverable upon removal of the load; that is, a specimen loaded elastically in tension will elongate, but will return to its original shape and size when unloaded. Beyond this linear region, for ductile materials such as steel, deformations are plastic. A plastically deformed specimen will not return to its original size and shape

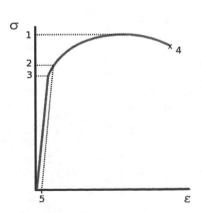

Stress vs. Strain curve typical of aluminum

when unloaded. Note that there will be elastic recovery of a portion of the deformation. For many applications, plastic deformation is unacceptable and is used as the design limitation.

1 - Ultimate Strength

2 - Yield Strength

3 - Proportional Limit Stress

4 - Rupture

5 - Offset Strain (usually 0.2%)

Tensile strength: Ultimate tensile strength (UTS), often shortened to tensile strength (TS) or ultimate strength, is the maximum stress that a material can withstand while being stretched or pulled before necking, which is when the specimen's cross-section starts to significantly contract. Tensile strength is the opposite of compressive strength and the values can be quite different.

Yield Stress: The stress level beyond which the material becomes plastic.

Yield Strength: is the lowest stress that produces a permanent deformation in a material. In some materials, like aluminum alloys, the point of yielding is difficult to identify, thus it is usually defined as the stress required to cause 0.2% plastic strain. This is called a 0.2% proof stress.

Young's Modulus, or the "Modulus of Elasticity": The Elastic Modulus (Young's Modulus) is the slope defined as stress divided by strain. E = modulus of elasticity (Pa (N/m2), N/mm2, psi). The Modulus of Elasticity can be used to determine the stress-strain relationship in the linear-elastic portion of the stress-strain curve. The linear-elastic region is either below the yield point, or if a yield point is not easily identified on the stress-strain plot it is defined to be between 0 and 0.2% strain, and is defined as the region of strain in which no yielding (permanent deformation) occurs.

CHAPTER 18: INTELLIGENT MODELING TECHNIQUES

Chapter Objective

Understand some of the available tools in SOLIDWORKS to perform intelligent modeling. Intelligent modeling is incorporating design intent into the definition of the sketch, feature, part, assembly or drawing document. Intelligent modeling is most commonly addressed through design intent using a:

- Sketch:
  - Geometric relations.
  - Fully defined Sketch tool.
  - SketchXpert.
  - Equations:
    - Explicit Dimension Driven.
    - Parametric Driven Curve.
  - Curves:
    - Curve Through XYZ Points.
    - Projected Composite.
- Feature:
  - End Conditions:
    - Blind, Through All, through All - Both, Up to Next, Up to Vertex, Up to Surface, Offset from Surface, Up to Body and Mid Plane.
  - Along a Vector.
  - FeatureXpert (Constant Radius).
  - Symmetry:
    - Mirror.
- Plane
- Assembly:
  - Symmetry (Mirror/Pattern).
  - Assembly Visualization.
  - SOLIDWORKS Sustainability.

       o   MateXpert.

- Drawing:

       o   DimXpert (Slots, Pockets, Machined features, etc.).

Design Intent

What is design intent? All designs are created for a purpose. Design intent is the intellectual arrangement of features and dimensions of a design. Design intent governs the relationship between sketches in a feature, features in a part and parts in an assembly or drawing document.

The SOLIDWORKS definition of design intent is the process in which the model is developed to accept future modifications. Models behave differently when design changes occur.

Design for change. Utilize geometry for symmetry, reuse common features, and reuse common parts. Build change into the following areas that you create: sketch, feature, part, assembly and drawing.

When editing or repairing geometric relations, it is considered best practice to edit the relation vs. deleting it.

Sketch

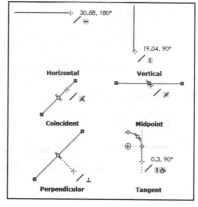

In SOLIDWORKS, relations between sketch entities and model geometry, in either 2D or 3D sketches, are an important means of building in design intent. In this chapter we will address 2D sketches.

Apply design intent in a sketch as the profile is created. A profile is determined from the Sketch Entities. Example: Rectangle, Circle, Arc, Point, Slot etc.

Develop design intent as you sketch with Geometric relations. Sketch relations are geometric constraints between sketch entities or between a sketch entity and a plane, axis, edge, or vertex. Relations can be added automatically or manually.

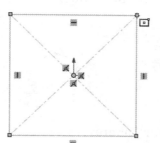

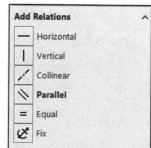

As you sketch, allow SOLIDWORKS to automatically add relations. Automatic relations rely on Inferencing, Pointer display, Sketch snaps and Quick Snaps.

After you sketch, manually add relations using the Add Relations tool, or edit existing relations using the Display/Delete Relations tool.

Fully Defined Sketch

Sketches are generally in one of the following states:

*   Under defined.

*   Fully defined.

*   Over defined.

Although you can create features using sketches that are not fully defined, it is a good idea to always fully define sketches for production models. Sketches are parametric, and if they are fully defined, changes are predictable. However, sketches in drawings, although they follow the same conventions as sketches in parts, do not need to be fully defined since they are not the basis of features.

SOLIDWORKS provides a tool to help the user fully define a sketch. The Fully Defined Sketch tool provides the ability to calculate which dimensions and relations are required to fully define under defined sketches or selected sketch entities. You can access the Fully Define Sketch tool at any point and with any combination of dimensions and relations already added. See Chapter 5 for additional information.

☼ Your sketch should include some dimensions and relations before you use the Fully Define Sketch tool.

The Fully Define Sketch tool uses the Fully Define Sketch PropertyManager. The Fully Define Sketch PropertyManager provides the following selections:

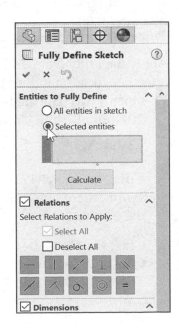

*   ***Entities to Fully Define***. The Entities to Fully Define box provides the following options:

    *   **All entities in sketch**. Fully defines the sketch by applying combinations of relations and dimensions.

    *   **Selected entities**. Provides the ability to select sketch entities.

        *   **Entities to Fully Define**. Only available when the Selected entities box is checked. Applies relations and dimensions to the specified sketch entities.

    *   **Calculate**. Analyzes the sketch and generates the appropriate relations and dimensions.

- *Relations*. The Relations box provides the following selections:

  - **Select All**. Includes all relations in the results.

  - **Deselect All**. Omits all relations in the results.

  - **Individual relations**. Include or exclude needed relations. The available relations are **Horizontal, Vertical, Collinear, Perpendicular, Parallel, Midpoint, Coincident, Tangent, Concentric** and **Equal radius/Length**.

- *Dimensions*. The Dimensions box provides the following selections:

  - **Horizontal Dimensions**. Displays the selected Horizontal Dimensions Scheme and the entity used as the Datum - Vertical Model Edge, Model Vertex, Vertical Line or Point for the dimensions. The available options are **Baseline, Chain** and **Ordinate**.

  - **Vertical Dimensions**. Displays the selected Vertical Dimensions Scheme and the entity used as the Datum - Horizontal Model Edge, Model Vertex, Horizontal Line or Point for the dimensions. The available options are **Baseline, Chain** and **Ordinate**.

  - **Dimension**. Below sketch and Left of sketch is selected by default. Locates the dimension. There are four selections: **Above sketch, Below the sketch, Right of sketch** and **Left of sketch**.

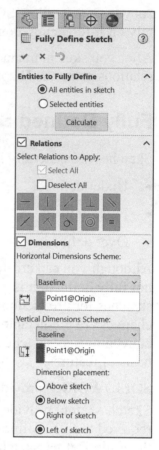

Tutorial: Fully Defined Sketch 18-1

Apply the Fully Defined Sketch tool. Modify the dimension reference location in the sketch profile with control points.

1. Open **Fully Defined Sketch 18-1** from the SOLIDWORKS 2018\Intelligent Modeling folder. View the FeatureManager. Sketch1 is under defined.

2. **Edit** Sketch1. The two circles are equal and symmetrical about the y axis. The rectangle is centered at the origin.

3. Click the **Fully Define Sketch** tool from the Consolidated Display/Delete Relations drop-down menu. The Fully Defined Sketch PropertyManager is displayed.

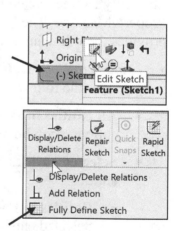

4. The All entities in the sketch is selected by default. Click **Calculate**. View the results. Sketch1 is fully defined to the origin.

5. Click **OK** ✔ from the PropertyManager. Drag all dimensions off the profile.

6. Modify the **vertical dimension to 50mm** and the **diameter dimension to 25mm** as illustrated. SOLIDWORKS suggests a dimension scheme to create a fully defined sketch.

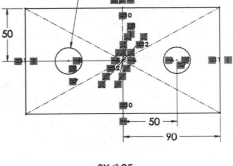

7. Click **View**, **Hide/Show**, uncheck **Sketch Relations** from the Main menu.

Modify the dimension reference location in the sketch profile with control points.

8. Click the **90**mm dimension in the Graphic window.

9. Click and drag the **left control point** as illustrated.

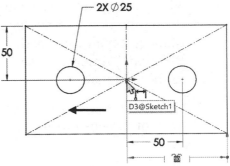

10. Release the mouse pointer on the **left vertical line** of the profile. View the new dimension reference location of the profile.

11. Repeat the same procedure for the horizontal **50**mm dimension. Select the new reference location to the left hole diameter as illustrated.

12. **Close** the model. View the results.

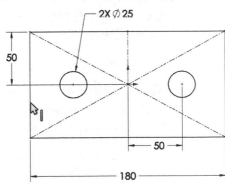

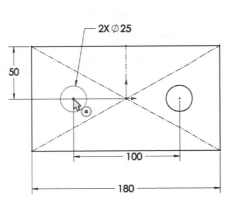

SketchXpert

SketchXpert resolves conflicts in over defined sketches and proposes possible solution sets. Color codes are displayed in the SOLIDWORKS Graphics window to represent the sketch states. The SketchXpert tool uses the SketchXpert PropertyManager. The SketchXpert PropertyManager provides the following selections:

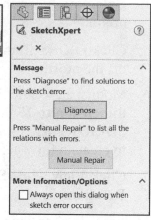

- *Message*. The Message box provides access to the following selections:

  - **Diagnose**. The Diagnose button generates a list of solutions for the sketch. The generated solutions are displayed in the Results section of the SketchXpert PropertyManager.

  - **Manual Repair**. The Manual Repair button generates a list of all relations and dimensions in the sketch. The Manual Repair information is displayed in the Conflicting Relations/Dimensions section of the SketchXpert PropertyManager.

- *More Information/Options*. Provides information on the relations or dimensions that would be deleted to solve the sketch.

  - **Always open this dialog when sketch error occurs**. Selected by default. Opens the dialog box when a sketch error is detected.

- *Results*. The Results box provides the following selections:

  - **Left or Right arrows**. Provides the ability to cycle through the solutions. As you select a solution, the solution is highlighted in the Graphics window.

  - **Accept**. Applies the selected solution. The sketch is no longer over-defined.

- *More Information/Options*. The More Information/Options box provides the following selections:

  - **Diagnose**. The Diagnose box displays a list of the valid generated solutions.

  - **Always open this dialog when sketch error occurs**. Selected by default. Opens the dialog box when a sketch error is detected.

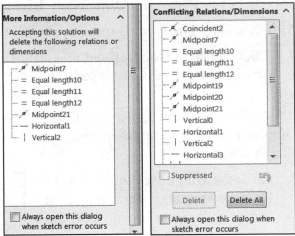

- *Conflicting Relations/Dimensions*. The Conflicting Relations/Dimensions box provides the ability to select a displayed conflicting relation or dimension. The selected item is highlighted in the Graphics window. The options include:

  - **Suppressed**. Suppresses the relation or dimension.

  - **Delete**. Removes the selected relation or dimension.

  - **Delete All**. Removes all relations and dimensions.

  - **Always open this dialog when sketch error occurs**. Selected by default. Opens the dialog box when a sketch error is dctcctcd.

Tutorial: SketchXpert 18-1

Create an over defined sketch. Apply SketchXpert to sclcct a solution.

1. Open **SketchXpert 18-1** from the SOLIDWORKS 2018\Intelligent Modeling folder.

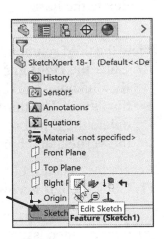

2. **Edit** ✏️ Sketch1. Sketch1 is fully defined. The rectangle has a midpoint relation to the origin, and an equal relation with all four sides. The top horizontal line is dimensioned.

Insert a dimension to create an over defined sketch.

3. Click **Smart Dimension** ✨.

4. Add a dimension to the **left vertical line**. This makes the sketch over defined. The Make Dimension Driven dialog box is displayed.

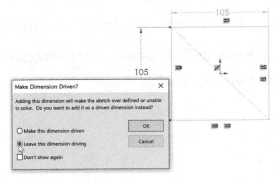

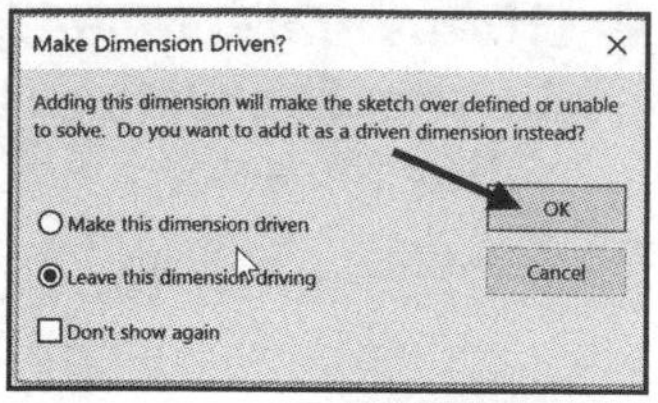

5. Check the **Leave this dimension driving** box option.
6. Click **OK**. The Over Defined warning is displayed.
7. Click the **red Over Defined** message. The SketchXpert PropertyManager is displayed.

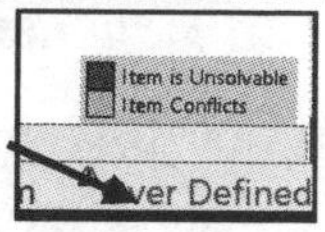

Color codes are displayed in the Graphics window to represent the sketch states.

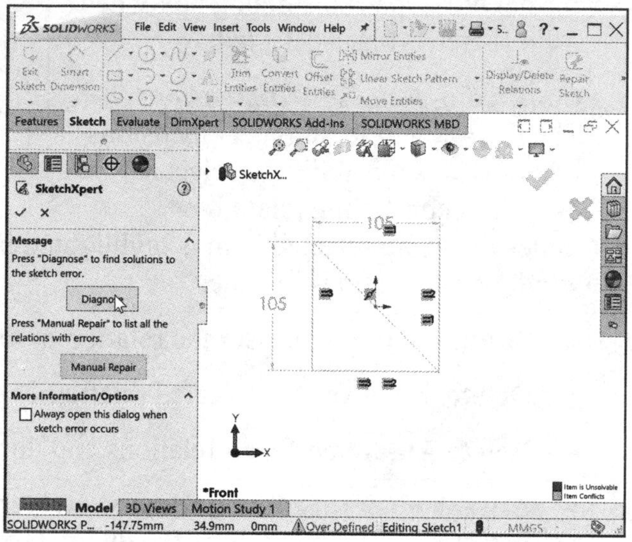

8. Click the **Diagnose** button. The Diagnose button generates a list of solutions for your sketch. You can either accept the first solution or click the Right arrow key in the Results box to view the section solution. The first solution is to delete the vertical dimension of 105mm.

View the second solution.

9. Click the **right arrow** key in the Results box. The second solution is displayed. The second solution is to delete the horizontal dimension of 105mm.

View the third solution.

10. Click the **right arrow** key in the Results box. The third solution is displayed. The third solution is to delete the Equal relation between the vertical and horizontal lines.

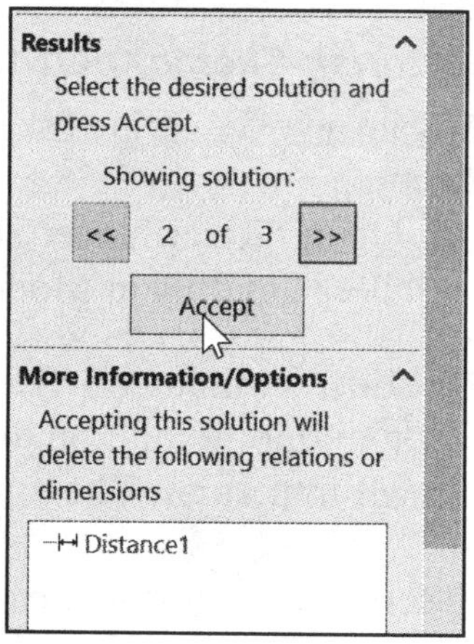

Accept the Second solution.

11. Click the **left arrow** key to obtain the second solution.
12. Click the **Accept** button. The SketchXpert tool resolves the over defined issue. A message is displayed.
13. Click **OK** from the SketchXpert PropertyManager.
14. **Rebuild** the model. **View** the results.
15. **Close** the model.

Equations

Dimension driven by equations

You want to design a hinge that you can modify easily to make similar sizes. You need an efficient way to create multi sizes. Equations create a mathematical relation between dimensions. You can use equations to locate entities instead of setting explicit dimensions.

In the example below, you set one screw hole location to be one half the height of the hinge, and the other screw hole to be one third the length of the hinge. If you change the height or length hinge dimension, the screw holes maintain this mathematical relation. Below I used the Mirror feature with the equation to quickly copy the existing hole feature across the Front plane.

Add comments to equations to document your design intent. Place a single quote (') at the end of the equation, then enter the comment. Anything after the single quote is ignored when the equation is evaluated. Example: "D2@Sketch1" = "D1@Sketch1" / 2 'height is 1/2 width. You can also use comment syntax to prevent an equation from being evaluated. Place a single quote (') at the beginning of the equation. The equation is then interpreted as a comment, and it is ignored. See SOLIDWORKS Help for additional information.

Tutorial: Equation 18-1

Close all parts, assemblies and drawings. Create two driven equations. Insert two Countersink holes. To position each hole on the hinge, one dimension is fixed, and the other is driven by an equation.

1. Open the **Hinge 18-1** part from the SOLIDWORKS 2018\Intelligent Modeling folder.

Add two Countersink holes using the Hole Wizard.

2. Click **Hole Wizard** from the Features toolbar. The Hole Specification PropertyManager is displayed.

3. Click **Countersink** for Hole Type as illustrated.

4. Select **ANSI Metric** for Standard. Accept the default Type.

5. Select **Flat Head Screw - ANSI B18.6.7M**.

6. Select **M8** for Size from the drop-down menu.

7. Select **Through All** for End Condition.

Place the location of the holes.

8. Click the **Positions** tab in the Hole Specification PropertyManager. Display a **Right** view.

9. Click the **face** of the model. The Point Sketch tool is displayed. Click to **place the holes** as illustrated.

10. **De-select** the Point Sketch tool.

Dimension the two holes.

11. Click **Smart Dimension**.

12. **Dimension** the holes as illustrated.

13. Click **OK** from the Hole Position PropertyManager.

14. Click **OK** from the Hole Specification PropertyManager. View the new feature in the FeatureManager.

Add equations to control the locations of the Countersink holes. First, add an equation to control the location of one of the points. Create an equation that sets the distance between the point and the bottom edge to one half the height of the hinge.

15. Right-click **Sketch4**. Click **Edit Sketch**.

16. Double-click **Extruded-Thin1** from the FeatureManager. Dimensions are displayed in the Graphics window.

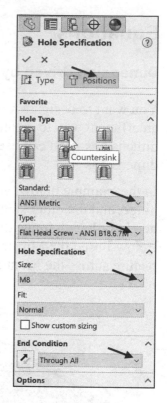
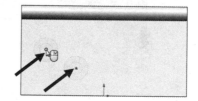
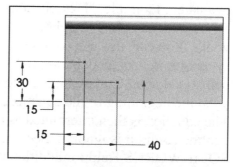
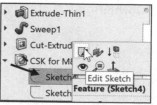

17. Click **Tools ➤ Equations** from the Menu bar. The Equations, Global Variables, and Dimensions dialog box is displayed. Click the **Equations View** Σ icon as illustrated.

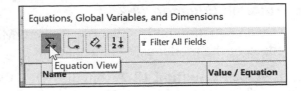

18. Click **inside** the first empty cell under Equations.

19. Click the **30mm** dimension in the Graphics window. An = sign is displayed in the Value/Equation cell.

20. Click the **60mm** dimension in the Graphics window.

21. Enter **/2** in the dialog box to complete the dimension. A green check mark is displayed.

22. Click inside the **Value / Equation** box. This equation sets the distance between the point and the bottom edge to one-half the height of the hinge.

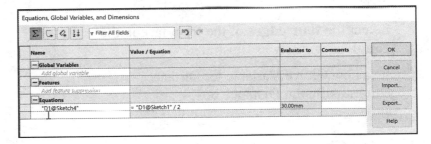

Create the second equation. Add an equation to control the location of the other point.

23. Click **inside** the first empty cell under Equations.

24. Click the **40mm** dimension from the Graphics window. An = sign is displayed in the Value/Equation cell.

25. Click the **120mm** dimension for the base.

26. Enter **/3** in the dialog box to complete the dimension. A green check mark is displayed.

27. Click inside the **Evaluate to** box. View the active equations. This sets the distance between the points and the side edge to one-third the length of the hinge. Click **OK** from the Equations dialog box.

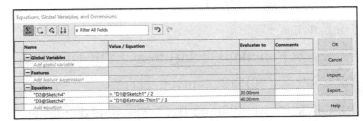

28. If needed, **Exit** the Sketch.

29. **Close** the model.

Tutorial: Equation 18-2

Use an equation in the Chamfer PropertyManager. Bevel the edges of the model using an equation.

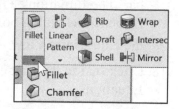

1. Open **Equation 18-2** from the SOLIDWORKS 2018 folder. The FeatureManager is displayed.

2. Right-click the **Equations** folder in the FeatureManager.

3. Click **Manage Equations**. View the Equations dialog box.

4. Click **OK** to close the Equations dialog box.

5. Click **Chamfer** 🟡 from the Features toolbar.

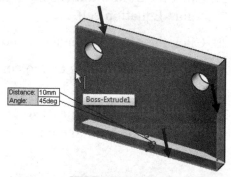

6. Click the **Angle Distance** Chamfer Type.

7. Click inside the **Edges and Faces or Vertex** box.

8. Select the **four edges** of the front face in the Graphics window as illustrated.

Create a new Global Variable for Distance.

9. Click **inside** the Distance box.

10. Enter =.

11. Select **Chamfer** as illustrated.

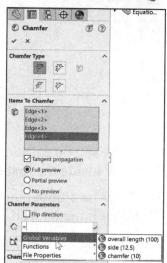

12. Click **OK** ✔ in the input field. The field displays an Equations 🌐 button.

13. Click the **Equations** button to toggle the display between the equation and the value.

14. Modify the Distance value from **10mm** to **6mm**.

15. Create a new Global Variable for Angle. Click **inside** the Angle box.

16. Enter =.

17. Select **Functions > sin()** from the fly-out menu.

18. Enter sin**(90)*10**. Units: **degrees**.

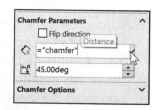

19. Click ✔ in the input field. The field displays an Equations Σ button.

20. Right-click the **Equations** folder in the FeatureManager.

21. Click **Manage Equations.** View the Equations dialog box. The Global Variable "chamfer" and the angle equation are listed in the Equations dialog box.

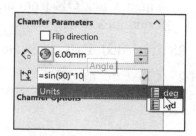

22. **Return** to the FeatureManager. View the results. **Close** the model.

💡 View the Equations folder for additional information.

Equation Driven Curve

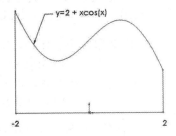

SOLIDWORKS provides the ability to address Explicit and Parametric equation types. When you create equation driven curves, the values you use must be in radians. You cannot use global variables directly for equations driven curves. However, you can create a global variable and associate it with a dimension, then use the dimension in the equation for the curve.

Explicit Equation Driven Curve

Explicit provides the ability to define x values for the start and endpoints of the range. Y values are calculated along the range of x values. Mathematical equations can be inserted into a sketch. Utilize parentheses to manage the order of operations. For example, calculate the volume of a solid bounded by a curve as illustrated. The region bounded by the equation y = 2+(x*cos(x)), and the x-axis, over the interval x = -2 to x = 2, is revolved about the x-axis to generate a solid.

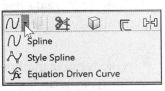

Tutorial: Explicit Equation Driven Curve 18-1

Create an Explicit Equation Driven Curve on the Front plane. Revolve the curve. Calculate the volume of the solid.

1. Create a **New** 🗋 part. Use the default ANSI, IPS Part template.

Create a 2D Sketch on the Front Plane.

2. Right-click **Sketch** from the Front Plane in the FeatureManager. Front Plane is your Sketch plane.

3. Click the **Equation Driven Curve Sketch** 🔏 tool from the Consolidated drop-down menu. Enter the **Equation y_x** as illustrated. Enter the **parameters x_1, x_2** that defines the lower and upper bound of the equation as illustrated.

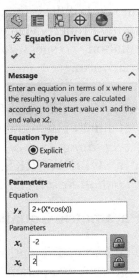

4. Click **OK** ✔ from the PropertyManager. View the curve in the Graphics window. Size the curve in the Graphics window. The Sketch is under defined.

5. Insert **three lines** to close the profile as illustrated. Insert a **Coincident** relation between the origin and the front left vertex as illustrated. Insert **dimensions** to fully define the sketch.

Create a Revolved feature.

6. Click the **horizontal line**.

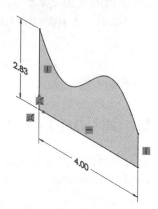

7. Click the **Revolved Boss/Base** 🍥 tool from the Feature toolbar. 360 degrees is the default.

8. Click **OK** ✔ from the PropertyManager. View the results in the Graphics window. Revolve1 is displayed. Utilize the Section tool parallel with the Right plane to view how each cross section is a circle.

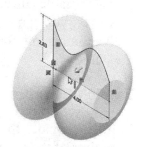

9. Apply **Brass** for material.

10. Calculate the **volume** of the part using the Mass Properties tool. View the results. Also note the surface area and the Center of mass.

11. **Close** the model.

🔆 You can create parametric (in addition to explicit) equation-driven curves in both 2D and 3D sketches.

🔆 Use regular mathematical notation and order of operations to write an equation. x_1 and x_2 are for the beginning and end of the curve. Use the transform options at the bottom of the PropertyManager to move the entire curve in x-, y- or rotation. To specify $x = f(y)$ instead of $y = f(x)$, use a 90 degree transform.

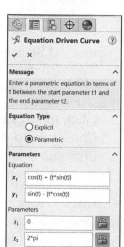

🔆 View the .mp4 files located in the book to better understand the potential of the Equation Driven Curve tool. The first one: *Calculating Area of a region bounded by two curves (secx)^2 and sin x* in SOLIDWORKS. The second one: *Determine the Volume of a Function Revolved Around the x Axis* in SOLIDWORKS.

Parametric Equation Driven Curve

The Parametric option of Equation Driven Curve can be utilized to represent two parameters, x- and y-, in terms of a third variable, t.

In the illustration below, a string is wound about a fixed circle of radius 1, and is then unwound while being held taut in the x-y- plane. The end point of the string, P, traces an involute of the circle.

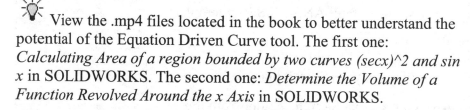

The initial point (1,0) is on the x- axis and the string is tangent to the circle at Q. The angle t is measured in radians from the positive x- axis.

Use the Equation Driven Curve, Parametric option to illustrate how the parametric equations $x = \cos t + t \sin t$ and $y = \sin t - t \cos t$ represent the involute of the circle from $t = 0$ to $t = 2\pi$.

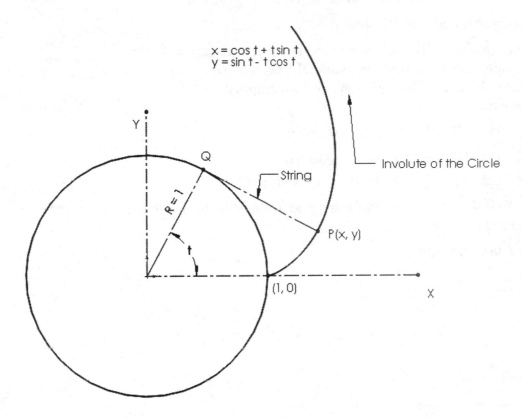

$x = \cos t + t \sin t$
$y = \sin t - t \cos t$

Involute of the Circle

String

Q

Y

R = 1

t

P(x, y)

(1, 0)

X

Rename a feature or sketch for clarity. Slowly click the feature or sketch name twice and enter the new name when the old one is highlighted.

Tutorial: Parametric Equation Driven Curve 18-1

Close all parts, assemblies and drawings. Create a Parametric Equation Driven Curve.

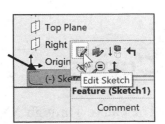

1. Open the **Parametric Equation 18.1** part from the SOLIDWORKS 2018\Intelligent Modeling folder. The circle of radius 1 is displayed.

2. **Edit** Sketch1 from the FeatureManager.

Activate the Equation Driven Curve tool.

3. Click **Tools ➤ Sketch Entities ➤ Equation Driven Curve** from the Menu bar. The Equation Driven Curve PropertyManager is displayed.

4. Click **Parametric** in the Equation Type dialog box.

5. Enter the parametric equations for x_t and y_t as illustrated.

6. Enter t_1 and t_2 as illustrated.

You can also utilize sketch parameters in your equation. For example, the diameter dimension DIA2 is represented as "D2@Sketch2." The equation can be multiplied by the diameter.

X = "D2@Sketch2"*(cos(t)+t*(sin(t))).

Y = "D2@Sketch2"*(sin(t)-t*(cos(t))).

7. Click **OK** ✔ from the PropertyManager.

8. **Exit** the Sketch. **Rebuild** the model. View the results in the Graphics window.

9. **Close** the model.

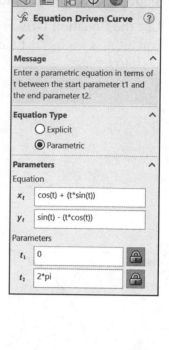

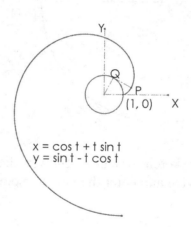

$$x = \cos t + t \sin t$$
$$y = \sin t - t \cos t$$

Curves

The Curve Through XYZ Points ℧ feature provides the ability to either type in (using the Curve File dialog box) or click Browse and import a text file with x-, y-, z-, coordinates for points on a curve.

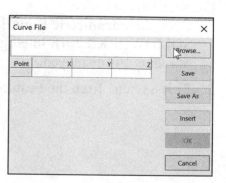

A text file can be generated by any program which creates columns of numbers. The Curve 𝒰 feature reacts like a default spline that is fully defined.

It is highly recommended that you insert a few extra points on either end of the curve to set end conditions or tangency in the Curve File dialog box.

Imported files can have an extension of either *.sldcrv or *.text. The imported data x-, y-, z- must be separated by a space, a comma, or a tab.

The National Advisory Committee for Aeronautics (NACA) developed airfoil shapes for aircraft wings. The shape of the aerofoil is defined by parameters in the numerical code that can be entered into equations to generate accurate cross-sections.

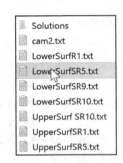

Tutorial: Curve Through XYZ Points 18-1

Create a curve using the Curve Through XYZ Points tool and the Composite curve tool. Import the x-, y-, z- data for an NACA aerofoil file for various cross sections.

1. Create a **New** ▯ part. Use the default ANSI, MMGS Part template.

Browse to select the imported x-, y-, z- data.

2. Click **Insert ➤ Curve ➤ Curve Through XYZ Points** from the Menu bar. The Curve File dialog box is displayed.

3. **Browse** to the SOLIDWORKS 2018\Intelligent Modeling folder.

4. Select **Text Files (*.txt)** for file type.

5. Double-click **LowerSurfSR5.txt**.

6. **View** the results in the Curve File dialog box.

7. Click **OK** from the Curve File dialog box. Curve1 is displayed in the FeatureManager.

8. **Repeat** the above procedure to create Curve2 from the UpperSurfSR5.txt file. Both the Lower and Upper curves are on the same Z plane. They are separate entities.

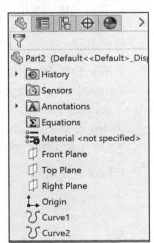

Point	X	Y	Z
1	0in	0in	187.81in
2	2.91in	-3.32in	187.81in
3	5.81in	-4.53in	187.81in
4	11.62in	-5.79in	187.81in
5	17.43in	-6.37in	187.81in
6	23.24in	-6.65in	187.81in
7	34.86in	-6.69in	187.81in
8	46.48in	-6.37in	187.81in
9	58.1in	-5.81in	187.81in
10	69.72in	-5.25in	187.81in
11	92.96in	-4.18in	187.81in

Use the Composite curve tool to join the Lower and Upper curves.

9. Click **Insert** ➤ **Curve** ➤ **Composite** from the Menu bar.

10. Select **Curve1** and **Curve2** from the fly-out FeatureManager. Both curves are displayed in the Entities to Join box.

11. Click **OK** ✔ from the PropertyManager. View the results in the FeatureManager.

12. **Close** the model.

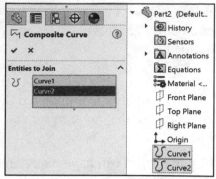

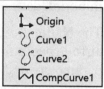

Tutorial: Curve Through XYZ points 18-2

Create a curve using the Curve Through XYZ Points tool. Import the x-, y-, z- data from a program. Verify that the first and last points in the curve file are the same for a closed profile.

1. Open **Curve Through XYZ points 18-2** from the SOLIDWORKS 2018\Intelligent Modeling folder.

2. Click the **Curve Through XYZ Points** ℧ tool from the Features CommandManager. The Curve File dialog box is displayed.

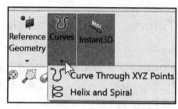

Import the curve data.

3. Click **Browse** from the Curve File dialog box.

4. **Browse** to the SOLIDWORKS 2018\Intelligent Modeling folder.

5. Set file type to **Text Files**.

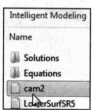

6. Double-click **cam2.text**. View the data in the Curve File dialog box. View the sketch in the Graphics window. Review the data points in the dialog box.

7. Click **OK** from the Curve File dialog box. Curve1 is displayed in the FeatureManager. You can now use this curve to create a fully defined sketch and then apply a feature. Let's view a sample result using this curve.

8. **Close** the model.

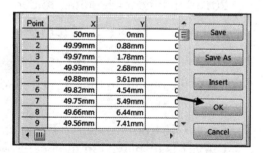

9. Open **Curve Through XYZ points 18.3**
    from the SOLIDWORKS 2018\Intelligent
    Modeling folder to view the final model
    using the created curve in Sketch1.

10. **Close** the model.

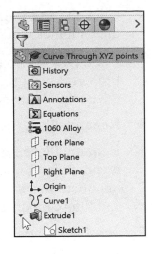

Tutorial: Projected Composite Curves 18-1

Create a plane and sketch for two composite
curves. Note: Before you create a Loft feature,
you must insert a plane for each curve and add
additional construction geometry to the sketch.

1. Open the **Projected Composite Curves
    18-1** part from the SOLIDWORKS
    2018\Intelligent Modeling folder. Two
    NACA aerofoil profiles are displayed
    created from x-, y-, z- data.

Create a reference plane through each point of
the composite curve.

2. Click **Insert ➢ Reference
    Geometry ➢ Plane** from
    the Menu bar. The Plane
    PropertyManager is
    displayed.

3. Click **Front Plane** for
    First Reference from the
    fly-out FeatureManager.

4. Click the **end point** of
    CompCurve1 in the
    Graphics window as
    illustrated for Second
    Reference. The Plane is
    fully defined.

5. Click **OK** ✅ from the
    Plane PropertyManager.
    Plane1 is displayed in the FeatureManager.

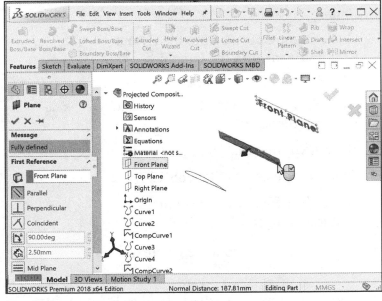

Create a Reference Plane through CompCurve 2, parallel to Plane1.

6. Click **Insert ➢ Reference Geometry ➢ Plane** from the Menu
    bar. Plane1 is the First Reference.

7. Click the **end point** of CompCurve2 in the Graphics window as illustrated.

8. Click **OK** ✔ from the Plane PropertyManager. Plane2 is displayed in the FeatureManager.

9. **Rename** Plane1 to PlaneSR5.

10. **Rename** Plane2 to PlaneSR9. When preparing multiple cross sections it is important to name the planes for clarity and future modification in the design.

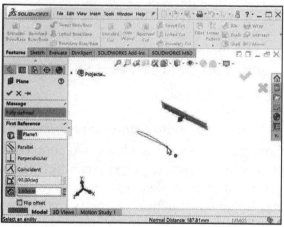

Convert CompCurve1 to Plane SR5 and CompCurve2 to Plane SR9.

11. Right-click **PlaneSR5** from the FeatureManager.

12. Click **Sketch** ⬚ from the Context toolbar.

13. Click the **Convert Entities** ⬡ Sketch tool.

14. Click **CompCurve1** from the fly-out FeatureManager in the Graphics window.

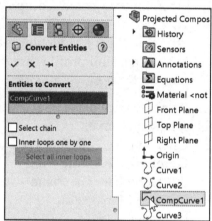

15. Click **OK** ✔ from the Convert Entities PropertyManager.

16. Click **Exit Sketch**.

17. **Rename** Sketch1 to SketchSR5.

18. Right-click **PlaneSR9** from the FeatureManager.

19. Click **Sketch** ⬚ from the Context toolbar.

20. Click the **Convert Entities** ⬡ Sketch tool.

21. Click **CompCurve2** from the fly-out FeatureManager in the Graphics window.

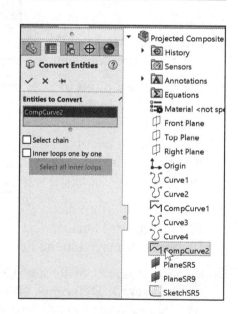

22. Click **OK** ✔ from the Convert Entities PropertyManager.

23. Click **Exit Sketch**.

24. **Rename** Sketch2 to SketchSR9.

25. Display an **Isometric** view.

26. **Show** Front Plane. View the results in the Graphics window.

27. **Close** the model.

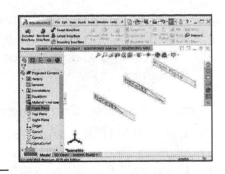

To save time in developing a series of curves, sketches can be copied and rotated.

Feature - End Conditions

Build design intent into a feature by addressing End Conditions (Blind, Through All, Up to Next, Up to Vertex, Up to Surface, Offset from Surface, Up to Body and Mid Plane) symmetry, feature selection and the order of feature creation.

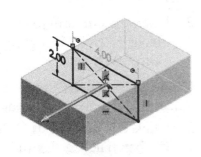

Example A: The Extruded Base feature remains symmetric about the Front Plane. Utilize the Mid Plane End Condition option in Direction 1. Modify the depth, and the feature remains symmetric about the Front Plane.

Example B: Create 34 teeth in the model. Do you create each tooth separately using the Extruded Cut feature? No.

Create a single tooth and then apply the Circular Pattern feature. Modify the Circular Pattern from 32 to 24 teeth. Think about Design Intent when you apply an End Condition in a feature during modeling.

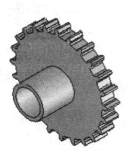

The basic End Conditions are:

- *Blind* - Extends the feature from the selected sketch plane for a specified distance - default End Condition.
- *Through All* - Extends the feature from the selected sketch plane through all existing geometry.
- *Up to Next* - Extends the feature from the selected sketch plane to the next surface that intercepts the entire profile. The intercepting surface must be on the same part.
- *Up to Vertex* - Extends the feature from the selected sketch plane to a plane that is parallel to the sketch plane and passing through the specified vertex.
- *Up to Surface* - Extends the feature from the selected sketch plane to the selected surface.
- *Offset from Surface* - Extends the feature from the selected sketch plane to a specified distance from the selected surface.

- *Up to Body* - Extends the feature up to the selected body. Use this option with assemblies, mold parts, or multi-body parts.
- *Mid Plane* - Extends the feature from the selected sketch plane equally in both directions.

Tutorial: Feature - End Conditions 18-1

Create Extruded Cut features using various End Condition options. Think about Design Intent of the model.

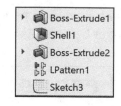

1. Open **End Condition 18-1** from the SOLIDWORKS 2018\Intelligent Modeling folder. The FeatureManager displays two Extrude features, a Shell feature and a Linear Pattern feature.

2. Click the circumference of the **front most circle**. Sketch3 is highlighted in the FeatureManager.

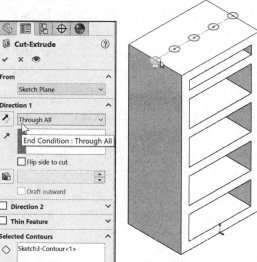

Create an Extruded Cut feature using the Selected Contours and Through All option.

3. Click **Extruded Cut** from the Features toolbar. The Cut-Extrude PropertyManager is displayed.

4. **Expand** the Selected Contours box. Sketch3-Contour<1> is displayed in the Selected Contours box (front most circle). The direction arrow points downward—if not, click the Reverse Direction icon.

5. Select **Through All** for End Condition in Direction 1. Only the first circle of your sketch is extruded.

6. Click **OK** from the Cut-Extruded PropertyManager. Cut-Extrude1 is displayed in the FeatureManager.

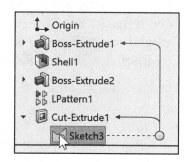

Create an Extruded Cut feature using the Selected Contours and the Up To Next option. Think about Design Intent.

7. Click **Sketch3** from the FeatureManager. Note the icon. This is a Selected Contour sketch.

8. Click **Extruded Cut** from the Features toolbar. The Cut-Extrude PropertyManager is displayed.

9. **Expand** the Selected Contours box.

10. Delete **Sketch3** from the Selected Contours box.

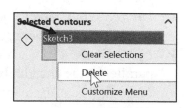

11. Click the **circumference of the second circle** as illustrated. Sketch3-Contour<1> is displayed in the Selected Contours box.

12. Select **Up To Next** for End Condition in Direction 1. Only the second circle of your sketch is extruded.

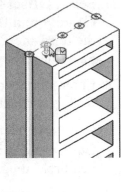

13. Click **OK** ✔ from the Cut-Extrude PropertyManager. Cut-Extrude2 is displayed in the FeatureManager.

14. **Rotate** your model and view the created feature.

Create an Extruded Cut feature using the Selected Contours and the Up To Vertex option.

15. Click **Sketch3** from the FeatureManager. Note the icon. It's a shared selected contour sketch icon.

16. Click the **Extruded Cut** ▥ Features tool. The Cut-Extrude PropertyManager is displayed. **Expand** the Selected Contours box.

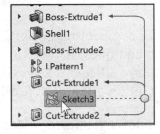

17. Delete **Sketch3** from the Selected Contours box.

18. Click the **circumference of the third circle** from the left. Sketch3-Contour<1> is displayed in the Selected Contours box.

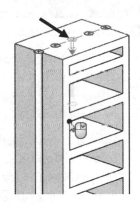

19. Select **Up To Vertex** for End Condition in Direction 1. Only the third circle of your sketch is extruded.

20. Click the **Vertex** point as illustrated.

21. Click **OK** ✔ from the PropertyManager. Cut-Extrude3 is displayed in the FeatureManager. The third circle has an Extruded Cut feature through the top two shelves.

Create an Extruded Cut feature using the Selected Contours and the Offset From Surface option.

22. Click **Sketch3** from the FeatureManager.

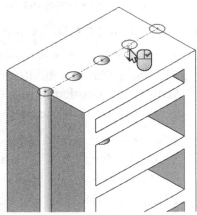

23. Click the **Extruded Cut** ▥ Features tool.

24. **Expand** the Selected Contours box.

25. Delete **Sketch3** from the Selected Contours box.

26. Click the circumference of the **fourth circle** as illustrated. Sketch3-Contour<1> is displayed in the Selected Contours box.

27. Select **Offset From Surface** for End
   Condition in Direction 1. Click the **face** of
   the third shelf. Face<1> is displayed in the
   Face/Plane box in Direction1.

28. Enter **60**mm for Offset Distance. Click the
   Reverse offset box.

29. Click **OK** ✔ from the PropertyManager.
   Cut-Extrude4 is displayed in the
   FeatureManager.

30. Display an **Isometric** view. View the
   created features. **Close** the model.

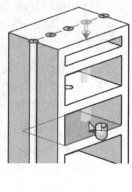

Along a Vector

In engineering design, vectors are utilized in modeling
techniques. In an extrusion, the sketch profile is extruded
perpendicular to the sketch plane. The Direction of Extrusion
Option allows the profile to be extruded normal to the vector.

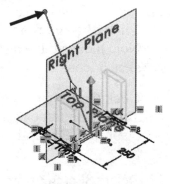

Tutorial: Along a Vector 18-1

Utilize a vector, created in a separate sketch, and the
Direction of Extrusion to modify the extruded feature.

1. Open **Along Vector 18-1** from the SOLIDWORKS
   2018\Intelligent Modeling folder. View the model in the
   Graphics window. The current sketch profile is extruded normal
   to the Top Sketch Plane.

2. Edit **Boss-Extrude1** from the
   FeatureManager. The Boss-Extrude1
   PropertyManager is displayed. Select **Up to
   Vertex** for End Condition in Direction 1.

3. Click the **top endpoint** of Sketch Vector.

4. Click inside the **Direction of Extrusion** box.

5. Click the **Sketch Vector** from the
   Graphics window as illustrated. The
   feature is extruded along the Sketch
   Vector normal to the Sketch Profile.

6. Click **OK** ✔ from the Boss-Extrude1
   PropertyManager. View the results in
   the Graphics window.

7. **Close** the part.

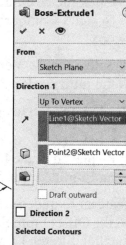

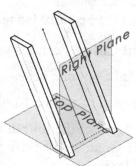

FeatureXpert (Constant Radius)

FeatureXpert manages the interaction between fillet and draft features when features fail. The FeatureXpert, manages fillet and draft features for you so you can concentrate on your design. See Xperts Overview in SOLIDWORKS Help for additional information.

When you add or make changes to constant radius fillets and neutral plane drafts that cause rebuild errors, the What's Wrong dialog appears with a description of the error. Click FeatureXpert in the dialog to run the FeatureXpert to attempt to fix the error.

The FeatureXpert can change the feature order in the FeatureManager design tree or adjust the tangent properties so a part successfully rebuilds. The FeatureXpert can also, to a lesser extent, repair reference planes that have lost references.

Supported features:

- Constant radius fillets.

- Neutral plane drafts.

- Reference planes.

Unsupported items:

- Other types of fillets or draft features.

- Mirror or pattern features. When mirror or pattern features contain a fillet or draft feature, the FeatureXpert cannot manipulate those features in the mirrored or patterned copies.

- Library features. Fillet or draft features in a library feature are ignored by the FeatureXpert and the entire Library feature is treated as one rigid feature.

- Configurations and Design Tables. The FeatureXpert is not available for parts that contain these items.

Utilize symmetry, feature order and reusing common features to build design intent into a part. Example A: Feature order. Is the entire part symmetric? Feature order affects the part.

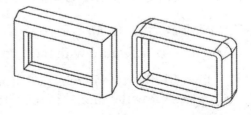

Apply the Shell feature before the Fillet feature, and the inside corners remain perpendicular.

Symmetry

An object is symmetrical when it has the same exact shape on opposite sides of a dividing line (or plane) or about a center or axis. The simplest type of Symmetry is a "Mirror" as we discussed above in this chapter.

Symmetry can be important when creating a 2D sketch, a 3D feature or an assembly. Symmetry is important because:

- Mirrored shapes have symmetry where points on opposite sides of the dividing line (or mirror line) are the same distance away from the mirror line.

- For a 2D mirrored shape, the axis of symmetry is the mirror line.

- For a 3D mirrored shape, the symmetry is about a plane.

Molded symmetrical parts are often made using a mold with two halves, one on each side of the axis of symmetry.

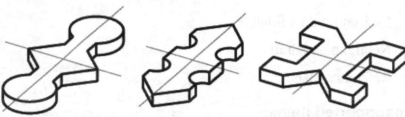

The axis or line where two mold parts join is called a parting line.

When items are removed from a mold, sometimes a small ridge of material is left on the object. Have you ever noticed a parting line on a molded object such as your toothbrush or a screwdriver handle?

Parting line

Bodies to mirror

When a model contains single point entities, the pattern and mirror features may create disjointed geometry and the feature will fail. For example, a cone contains a single point at its origin. To mirror the cone about a plane would create disjointed geometry. To resolve this issue, utilize Bodies to Pattern option.

Tutorial: Bodies to Mirror 18-1

Create a Mirror feature with the Top plane and utilize the Body to Mirror option.

1. Open **Bodies to Mirror 18-1** from the SOLIDWORKS 2018\Intelligent Modeling folder. View the model in the Graphics window.

2. Select **Mirror** ⊮⊣ from the Features toolbar. The Mirror PropertyManager is displayed.

3. **Expand** the Bodies to Mirror dialog box.

4. Click the face of the **Cone** in the Graphics window. Note the icon feedback symbol.

5. Click inside the **Mirror Face/Plane** dialog box.

6. Click **Top Plane** from the fly-out FeatureManager.

7. Uncheck the **Merge solids** box. (You cannot merge these two cones at a single point.)

8. Click **OK** ✔ from the Mirror PropertyManager. View the results in the Graphics window. Mirror1 is displayed in the FeatureManager.

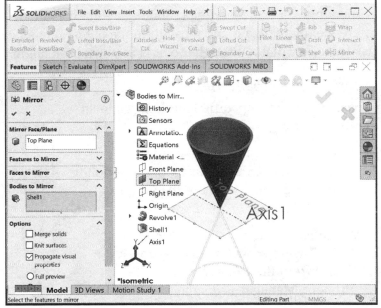

Planes

Certain types of models are better suited to design automation than others. When setting up models for automation, consider how they fit into an assembly and how the parts might change when automated.

Create planes so that sketches and features can be referenced to them (coincident, up to surface, etc.). This provides the ability to dimension for the plane to be changed and the extrusion extended with it. When placed in an assembly, other components can be mated to a plane so that they move with consideration to the parts altered.

Incorporating planes into the design process prepares the model for future changes. As geometry becomes more complex, additional sketch geometry (construction lines, circles) or reference geometry (axis, planes) may be required to construct planes.

For example, a sketched line coincident with the silhouette edge of the cone provides a reference to create a plane through to the outer cone's face. An axis, a plane and an angle creates a new plane through the axis at a specified angle.

Tutorial: Angle Plane 18-1

Create a plane at a specified angle through an axis.

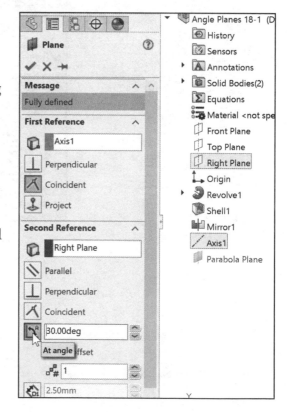

1. Open **Angle Planes 18-1** from the SOLIDWORKS 2018\Intelligent Modeling folder. View the model in the Graphics window.

Create an Angle Plane.

2. Click **Insert ➤ Reference Geometry ➤ Plane** from the Menu bar. The Plane PropertyManager is displayed.

3. Click **Axis1** in the Graphics window. Axis1 is displayed in the First Reference box.

4. Click **Right Plane** from the fly-out FeatureManager. Right Plane is displayed in the Second Reference box.

Set the angle of the Plane.

5. Click the **At angle** box.

6. Enter **30** for Angle. The Plane is fully defined.

7. Click **OK** ✔ from the Plane PropertyManager. Plane1 is displayed in the FeatureManager.

8. **Close** the Model.

As geometry becomes more complex in mechanical design, planes and conical geometry can be combined to create circular, elliptical, parabolic and hyperbolic sketches with the Intersection Curve Sketch Curve.

Conic Sections and Planes

Conic sections are paths traveled by planets, satellites, electrons and other bodies whose motions are driven by inverse-square forces. Once the path of a moving body is known, information about its velocity and forces can be derived. Using planes to section cones creates circular, elliptical, parabolic, hyperbolic and other cross sections that are utilized in engineering design.

The Intersection Curve tool provides the ability to open a sketch and create a sketched curve at the following kinds of intersections: a plane and a surface or a model face, two surfaces, a surface and a model face, a plane and the entire part, and a surface and the entire part.

Tutorial: Conic Section 18-1

Apply the Intersection Curve Sketch tool. Create a sketched curve.

1. Open **Conic Section 18-1** from the SOLIDWORKS 2018\Intelligent Modeling folder. View the model in the Graphics window.

Create a Sketch.
2. Right-click **Plane1 Circle** from the FeatureManager.

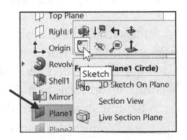

3. Click **Sketch** 🖉 from the Context toolbar. The Sketch toolbar is displayed in the CommandManager.

4. Click **Tools** ➤ **Sketch Tools** ➤ **Intersection Curve** from the Menu bar. The Intersection Curve PropertyManager is displayed.

5. Click the **inside Shell face** as illustrated. Face<1> is displayed in the Selected Entities dialog box.

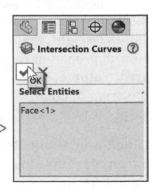

6. Click **OK** ✔ from the Intersection Curve PropertyManager. Sketch2 is created.

7. Click **OK** ✔.

8. Click **Exit Sketch**.

9. **Hide** Plane1 Circle. The intersection of the cone with the plane creates a circle.

10. **Close** the model. You can use the resulting sketched intersection curve in the same way that you use any sketched curve.

Assembly

Utilizing symmetry, reusing common parts and using the Mate relation between parts builds the design intent into an assembly. For example: Reuse geometry in an assembly. The assembly contains a linear pattern of holes. Insert one screw into the first hole. Utilize the Component Pattern feature to copy the machine screw to the other holes.

Assembly Visualization

In an assembly, the designer selects material based on cost, performance, physical properties, manufacturing processes, sustainability, etc. The SOLIDWORKS Assembly Visualization tool includes a set of predefined columns to help troubleshoot assembly performance. You can view the open and rebuild times for the components, and the total number of graphics triangles for all instances of components. The Assembly Visualization tool provides the ability to rank components based on the default values (**weight, mass, density, volume, etc.**) or their custom properties (**cost, sustainability, density, surface area, volume, etc.**) or an equation and activate a spectrum of colors that reflects the relative values of the properties for each component.

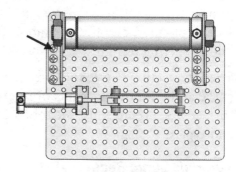

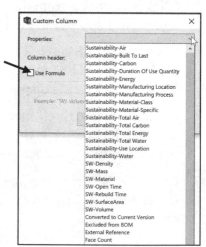

Hide/Show Value Bar ▤ **icon**. Available for numeric properties. Turns the value bars off and on. When the value bars are on, the component with the highest value displays the longest bar. You can set the length of the bars to be calculated relative to the highest-value component or relative to the entire assembly.

Flat Nested View **icon**. Nested view, where subassemblies are indented. Flat view, where subassembly structures are ignored (similar to a parts-only BOM).

Grouped/Ungrouped View **icon**. Groups multiple instances of a component into a single line item in the list. Grouped View is useful when listing values for properties that are identical for every instance of the component. Ungrouped views list each instance of a component individually. Ungrouped View is useful when listing values for instance-specific properties, such as Fully mated, which might be different for different instances of the component.

Performance Analysis **icon**. Provides additional information on the open, display, and rebuild performance of models in an assembly.

Filter **icon**. Filters the list by text and by component show/hide state.

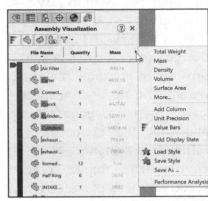

The Assembly Visualization tab in the FeatureManager design tree panel contains a list of all components in the assembly, sorted initially by file name. There are three default columns:

- File name.

- Quantity.

- Mass.

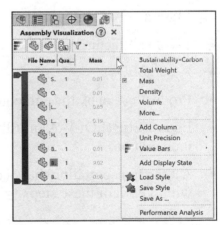

Mass is the default column. View your options as illustrated other than Mass.

Tutorial: Assembly Visualization 18-1

Apply the Assembly Visualization tool to an assembly. View your options.

1. Open **Assembly Visualization 18-1** from the SOLIDWORKS 2018\Intelligent Modeling folder. View the assembly in the Graphics window. Material was assigned to each component in the assembly.

2. Click the **Assembly Visualization** tool from the Evaluate Tab in the CommandManager. The Assembly Visualization PropertyManager is displayed.

3. Click the **expand arrow** to the right of Mass as illustrated to display the default visualization properties. Mass is selected by default. **View** the More selection.

4. **Explore** the available tabs and SOLIDWORKS Help for additional information.

5. **Close** the model.

SOLIDWORKS Sustainability

With SOLIDWORKS Sustainability you can determine Life Cycle Assessment (LCA) properties for a part or assembly. By integrating Life Cycle Assessment (LCA) into the design process, you can see how decisions about material, manufacturing, and location (where parts are manufactured and where they are used) influence a design's environmental impact. You specify various parameters that SOLIDWORKS Sustainability uses to perform a comprehensive evaluation of all the steps in a design's life. LCA includes:

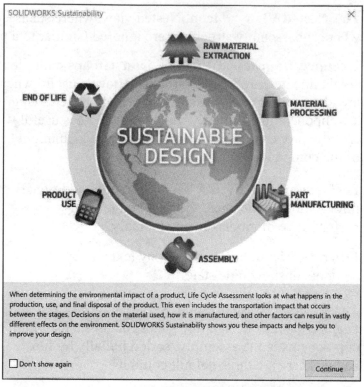

When determining the environmental impact of a product, Life Cycle Assessment looks at what happens in the production, use, and final disposal of the product. This even includes the transportation impact that occurs between the stages. Decisions on the material used, how it is manufactured, and other factors can result in vastly different effects on the environment. SOLIDWORKS Sustainability shows you these impacts and helps you to improve your design.

- Ore extraction from the earth.

- Material processing.

- Part manufacturing.

- Assembly.

- Product usage by the end consumer.

- End of Life (EOL) - Landfill, recycling, and incineration.

- All the transportation that occurs between and within each of these steps.

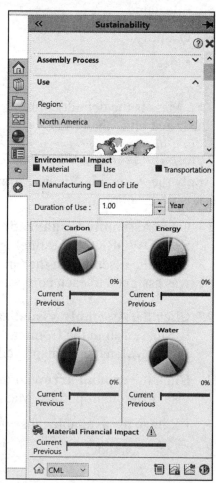

By combining SOLIDWORKS Sustainability and Assembly Visualization, you can determine the components in the assembly that contain the greatest carbon footprint and review materials with similar properties to reduce CO_2 emissions.

🔆 You need SOLIDWORKS Professional or SOLIDWORKS Premium to access SOLIDWORKS Sustainability for an assembly.

MateXpert

The MateXpert is a tool that provides the ability to identify mate problems in an assembly. You can examine the details of mates that are not satisfied, and identify groups of mates which over define the assembly. If the introduction of a component leads to multiple mate errors, it may be easier to delete the component, review the design intent, reinsert the component and then apply new mates. See SOLIDWORKS Help for additional information.

Drawings

Utilize dimensions, tolerance and notes in parts and assemblies to build the design intent into a drawing.

Example A: Tolerance and material in the drawing. Insert an inside diameter tolerance +.000/-.002 into the Pipe part. The tolerance propagates to the drawing.

Define the Custom Property Material in the Part. The Material Custom Property propagates to your drawing.

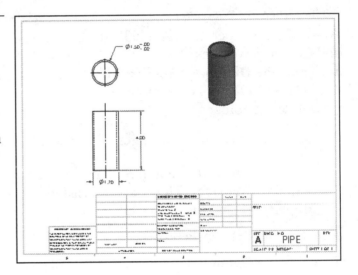

DimXpert

DimXpert for parts is a set of tools you use to apply dimensions and tolerances to parts according to the requirements of the ASME Y14.41-2009 standard.

DimXpert dimensions show up in a different color to help identify them from model dims and reference dims. DimXpert dims are the dimensions that are used when calculating tolerance stack-up using TolAnalyst.

DimXpert applies dimensions in drawings so that manufacturing features (patterns, slots, pockets, etc.) are fully-defined.

DimXpert for parts and drawings automatically recognize manufacturing features. What are manufacturing features? Manufacturing features are *not SOLIDWORKS features*. Manufacturing features are defined in 1.1.12 of the ASME Y14.5M-1994 Dimensioning and Tolerancing standard as "The general term applied to a physical portion of a part, such as a surface, hole or slot."

The DimXpertManager provides the following selections:

Auto Dimension Scheme ⌗, **Basic Location Dimension** ⌶, **Basic Size Dimension** ⌐ **Show Tolerance Status** ⊹, **Copy Scheme** ⊕ and **TolAnalyst Study** ⛭.

☼ Care is required to apply DimXpert correctly on complex surfaces or with some existing models. See SOLIDWORKS help for additional information on DimXpert.

Tutorial: DimXpert 18-1

Apply the DimXpert tool. Apply Prismatic and Geometric options.

1. Open the **DimXpert 18-1** part from the SOLIDWORKS 2018\Intelligent Modeling folder.

2. Click the **DimXpertManager** ⊕ tab as illustrated.

3. Click the **Auto Dimension Scheme** ⌗ tab from the DimXpertManager. The Auto Dimension Scheme PropertyManager is displayed.

4. Click the **Prismatic** box.

5. Click the **Geometric** box.

6. Click the **Linear** box.

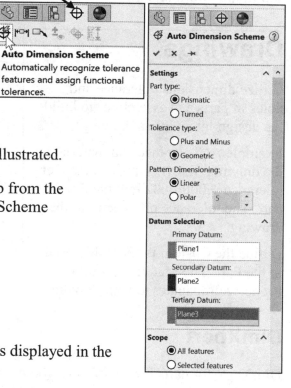

Select the Primary Datum.
7. Click the **front face** of the model. Plane1 is displayed in the Primary Datum box.

Select the Secondary Datum.
8. Click **inside** the Secondary Datum box.

9. Click the **left face** of the model. Plane2 is displayed in the Secondary Datum box.

Select the Tertiary Datum.
10. Click **inside** the Tertiary Datum box.

11. Click the **bottom face** of the model. Plane3 is displayed in the Tertiary Datum box. Accept the default options.

12. Click **OK** ✔ from the Auto Dimension
    Scheme PropertyManager. View the
    results in the Graphics window and in the
    DimXpertManager.

13. **Close** all models. Do not save the
    updates.

🔆 Right-click Delete to delete the Auto
Dimension Scheme in the DimXpert
PropertyManager.

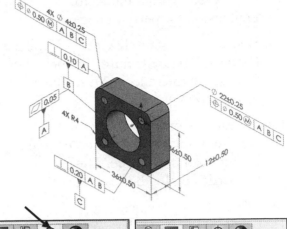

Tutorial: DimXpert 18-2

Apply DimXpert: Geometric option.
Edit a Feature Control Frame.

1. Open the **DimXpert 18-2** part
   from the SOLIDWORKS
   2018\Intelligent Modeling folder.

2. Click the **DimXpertManager** ⊕
   tab as illustrated.

3. Click the **Auto Dimension Scheme** ⊕ tool from the
   DimXpertManager. The Auto Dimension PropertyManager
   is displayed.

4. Click the **Prismatic box**.

5. Click the **Geometric** box.

6. Click the **Linear** box.

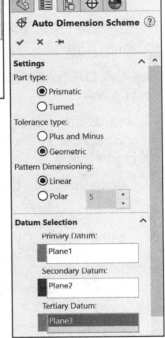

🔆 DimXpert: Geometric option provides the ability to locate
axial features with position and circular runout tolerances.
Pockets and surfaces are located with surface profiles.

Select the Primary Datum.
7. Click the **back face** of the model. Plane1 is displayed in the
   Primary Datum box.

Select the Secondary Datum.
8. Click **inside** the Secondary Datum box.

9. Click the **left face** of the model. Plane2 is displayed in the
   Secondary Datum box.

Select the Tertiary Datum.
10. Click **inside** the Tertiary Datum box.

11. Click the **top face** of the model. Plane3 is displayed in the Tertiary Datum box.

12. Click **OK** ✔ from the Auto
    Dimension PropertyManager.

13. Display an **Isometric** view. View the
    Datums, Feature Control Frames, and
    Geometric tolerances. All features are
    displayed in green.

Edit a Feature Control Frame.

14. **Double-click** the illustrated Position
    Feature Control Frame. The
    Properties dialog box is displayed.

Modify the 0.50 tolerance.

15. Click **inside** the Tolerance 1 box.

16. **Delete** the existing text.

17. Enter **0.25**.

18. Click **OK** from the Properties dialog
    box.

19. **Repeat** the above procedure for the
    second Position Feature Control
    Frame. View the results.

20. **Close** the model. Do not save the
    model.

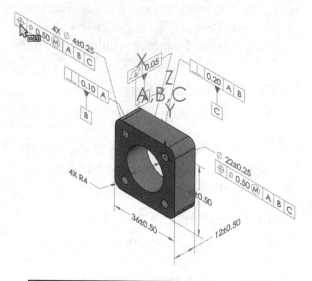

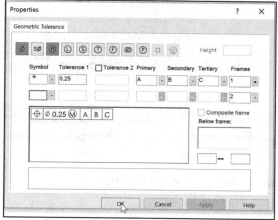

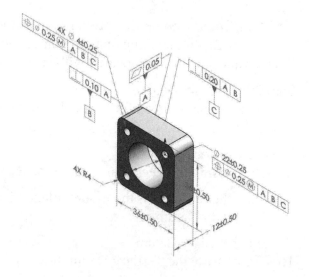

Summary

In this chapter, you performed short step-by-step tutorials to understanding some of the available tools in SOLIDWORKS to perform intelligent modeling. Intelligent modeling is incorporating design intent into the definition of the Sketch, Feature, Part and Assembly or Drawing document.

All designs are created for a purpose. Design intent is the intellectual arrangement of features and dimensions of a design. Design intent governs the relationship between sketches in a feature, features in a part and parts in an assembly or drawing document.

The SOLIDWORKS definition of design intent is the process in which the model is developed to accept future modifications. Models behave differently when design changes occur.

You create Explicit Dimension Driven and Parametric Equation Driven Curves and generated curves using the Curve Through XYZ Points tool for a NACA aerofoil and a data text file.

You also applied SketchXpert, DimXpert and FeatureXpert.

Notes:

Chapter 19

Additive Manufacturing - 3D Printing

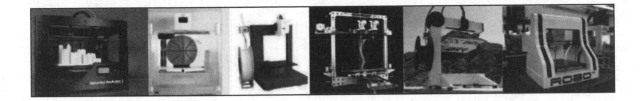

Below are the desired outcomes and usage competencies based on the completion of Chapter 19.

Desired Outcomes:	Usage Competencies:
• Knowledge of Additive Manufacturing.	• Discuss the advantages and disadvantages of Additive Manufacturing.
• Identify key features of a low cost $500 - $3,000 3D printer.	• Describe the differences between a Cartesian and Delta printer.
• Create an STL (STereoLithography) file.	• Explain 3D printer technology: STereoLithography (STL), Fused Filament Fabrication (FFF), Fused Deposition Model (FDM), and Digital Light Process (DLP).
• 3D Print directly from SOLIDWORKS.	
• Select the proper Slicer engine and print parameters.	• Recognize common 3D printer terminology: Raft, Skirt, Brim, Support, Touching Buildplate, Heated vs. Non-Heated build area.
• Create a successful print.	• Choose the proper Slicer and print parameters.
	• Select the proper filament.
	• Understand part orientation during the print cycle.
	• Define general 3D printing tips.
	• Address fit tolerance for interlocking parts.

Notes:

Chapter 19 - Additive Manufacturing - 3D Printing

Chapter Objective

Provide a basic understanding between the differences of Additive vs. Subtractive Manufacturing. Comprehend 3D printer terminology along with a working knowledge of preparing, saving, and printing a 3D CAD model on a low cost ($500 - $3,000) printer.

On the completion of this chapter, you will be able to:

- Discuss Additive vs Subtractive Manufacturing.

- Determine the differences between a Cartesian and Delta printer.

- Create a STereoLithography (STL) file in SOLIDWORKS.

- 3D print directly from SOLIDWORKS using an Add-In.

  o Save an STL file to G-code.

- Discuss printer hardware.

- Select the correct filament type:

  o PLA (Polylactic acid), ABS (Acrylonitrile butadiene styrene) or Nylon.

- Prepare the G-code.

  o Address model setup, print orientation, extruder temperature, bed temperature, support type, layer height, infill, and number of shells.

- Comprehend the following 3D printer terminology:

  o (STereoLithography) file - STL.

  o Fused Filament Fabrication - FFF.

  o Fused Deposition Model - FDM.

  o Digital Light Process - DLP.

  o Dissolvable Support System - DDS.

  o Fast Layer Deposition - FLD.

  o Raft, Skirt and Brim.

  o Support and Touching Buildplate.

  o Slicer.

  o G-code.

- Address fit tolerance for interlocking parts.

- Define general 3D Printing tips.

Additive vs. Subtractive Manufacturing

In April 2012, *The Economist* published an article on 3D printing. In the article they stated that this was the "beginning of a third industrial revolution, offering the potential to revolutionize how the world makes just about everything."

Avi Reichental, President and CEO of 3D Systems, stated, "With 3D printing, complexity is free. The printer doesn't care if it makes the most rudimentary shape or the most complex shape, and that is completely turning design and manufacturing on its head as we know it."

Over the past five years, companies are now using 3D printing to evaluate more concepts in less time to improve decisions early in product development. As the design process moves forward, technical decisions are iteratively tested at every step to guide decisions big and small, to achieve improved performance, lower manufacturing costs, delivering higher quality and more successful product introductions. In pre-production, 3D printing is enabling faster first article production to support marketing and sales functions, and early adopter customers. And in final production processes, 3D printing is enabling higher productivity, increased flexibility, reduced warehouse and other logistics costs, economical customization, improved quality, reduced product weight, and greater efficiency in a growing number of industries.

Technology for 3D printing continues to advance in three key areas: **printers** and **printing methods**, **design software**, and **materials** used in printing.

Already, 3D printing is being used in the medical industry to help save lives and in some space exploration efforts. But how will 3D printing affect the average, middle-class person in the future? Low cost 3D printers are addressing this consumer market.

Additive manufacturing is the process of joining materials to create an object from a 3D model, usually adding layer over layer.

Subtractive manufacturing relies upon the removal of material to create something. The blacksmith hammered away at heated metal to create a product. Today, a Computer Numerical Control CNC machine cuts and drills and otherwise removes material from a larger initial block to create a product.

Additive manufacturing, sometimes known as *rapid prototyping*, can be slower than Subtractive manufacturing. Both take skill in creating the G-code and understanding the machine limitations.

A few advantages of Additive manufacturing:

- Lower cost (different entry levels) into the manufacturing environment.

- Lowers the barriers (space, power, safety, and training) to traditional subtractive manufacturing.

- Reduce part count in an assembly from traditional subtractive manufacturing (complex parts vs. assemblies).

- Build complex features, shapes, and objects.

- Reduce prototyping time.

- Faster development cycle.

- Quicker customer feedback.

- Faster product to market.

- Quicker product customization and configuration.

- Parallel verticals: develop and prototype at the same time.

- Open source slicing engines: (Slic3r, Skeinforge, Netfabb, KISSkice, Cura, etc.).

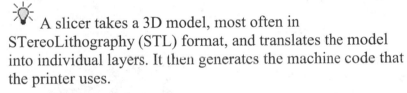

A slicer takes a 3D model, most often in STereoLithography (STL) format, and translates the model into individual layers. It then generates the machine code that the printer uses.

A slicer program allows the user to calibrate printer settings: filament type, part orientation, extruder speed, extruder temperature, bed temperature, cooling fan rate, raft, support, percent infill, infill pattern type, etc.

3D printers are controlled either through a small on-board control screen, an external memory device (USB, Sims card, etc.) or through a computer interface. User interface/control software allows a user to generate the needed machine code (G-code, .gcode) file from the computer to the 3D printer.

A few disadvantages of Additive manufacturing:

- Slow build rates. Many printers lay down material at a speed of one to five cubic inches per hour. Depending on the part needed, other manufacturing processes may be significantly faster.

- Requires post-processing. The surface finish and dimensional accuracy may be lower quality than other manufacturing methods.

- Poor mechanical properties. Layering and multiple interfaces can cause defects in the product.

- Frequent calibration is required. Without frequent calibration, prints may not be the correct dimensions, they may not stick to the build surface, and a variety of other not-so-wanted effects can occur.

- Limited component size/small build volume. In most cases, polymer products are about 1 cubic yard in size; metal parts may only be one cubic foot. While larger machines are available, they come at a cost.

Cartesian Printer vs. Delta Printer

Cartesian and Delta style 3D printers are the most common styles of desktop 3D printers currently available. What are the main differences? To understand their unique abilities one first needs to understand the basic technology of a Fused Deposition Model printer.

A Fused Filament Fabrication (FFF) printer takes a filament of a material (usually plastic) and extrudes it through a print-head-like nozzle. This material is then laid down in thin layers to form a 3D object on a platform, built from the bottom up.

Cartesian 3D printers are named after the dimensional coordinate system - the X, Y, and Z-axis - which is used to determine where and how to move in three dimensions.

Cartesian 3D printers typically have a print bed which moves only in the Z-axis. The extruder sits on the X-axis and Y-axis, where it can move in four directions on a gantry. This principle can be seen in action on popular models from Ultimaker, Sindoh, and MakerBot.

Controlling a linear Cartesian system is relatively simple, which is why most low-cost 3D printers on the market today use this type of design. The Cartesian coordinate system has long been used for tools like plotters, CNC milling machines, and 2D printers.

Delta 3D printers also work within the Cartesian plane, but use a very different approach to getting the print-nozzle where it needs to be.

Identifying characteristics start with the circular print bed. The extruder is suspended above the print bed by three arms in a triangular configuration, thus the name "Delta."

These arms move up and down independently to the print-nozzle keeping it precisely located throughout the print. Delta printers have the advantage in the ability to make taller objects due to the height of the printer and the tall arms.

Rather than using simple Cartesian geometry to calculate where the print-nozzle should go, Delta printers estimate the head position using trigonometric functions.

Create an STL file in SOLIDWORKS

STL (STereoLithography) is a file format native to the Stereolithography CAD software created by 3D Systems. STL has several after-the-fact backronyms such as "Standard Triangle Language" and "Standard Tessellation Language."

An STL file describes only the surface geometry of a three-dimensional object without any representation of color, texture, or other common CAD model attributes. The STL format specifies both ASCII and Binary representations.

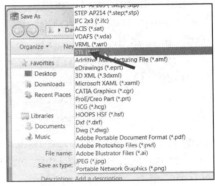

Binary files are more common, since they are more compact. An STL file describes a raw unstructured triangulated (point cloud) surface by the unit normal and vertices (ordered by the right-hand rule) of the triangles using a three-dimensional Cartesian coordinate system.

STL Save options allow you to control the number and size of the triangles by setting the various parameters in the CAD software.

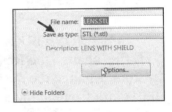

To save a SOLIDWORKS model as a STL file, click **File**, **Save As** from the Main menu or **Save As** from the Main menu toolbar. The Save As dialog is displayed. Select **STL(*.stl)** as the Save as type.

Click the **Options** button. The Export Options dialog box is displayed. View your options. For most parts, utilize the default setting.

Close the Export Options dialog box. Click **OK**. View the generate point cloud of the part. Click **Yes**. The STL file is now ready to be imported into your 3D printer software.

3D Print Directly from SOLIDWORKS

Export part and assembly files to STL (.stl), 3D Manufacturing Format (.3mf), or Additive Manufacturing File Format (.amf) format.

The 3MF (3D Manufacturing Format) and AMF (Additive Manufacturing File) formats provide additional model information over the .stl file format. Therefore, it requires less post-processing to define data such as the position of your model relative to the selected 3D printer, orientation, color, materials, etc.

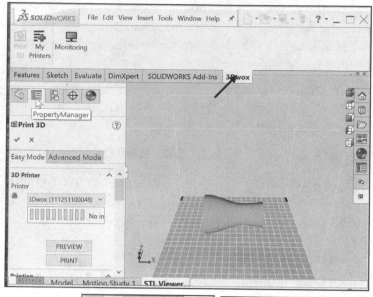

During the print process, users can monitor the print time, needed filament, and estimated time to completion in the SOLIDWORKS graphics window.

In SOLIDWORKS for a smoother STL file, change the Resolution to Custom. Change the deviation to 0.0005in (0.01mm). Change the angle to 5. Smaller deviations and angles produce a smoother file but increase the file size and print time.

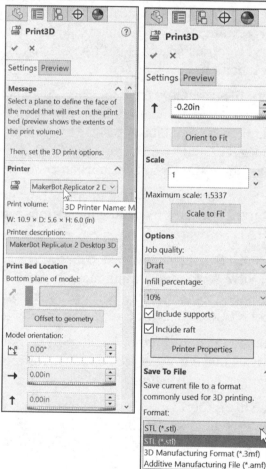

Print Material

As printed parts cool (PLA, ABS and Nylon), various areas of the object cool at different rates. Depending on the model being printed and the filament material, this effect can lead to warping, curling and or layer delamination.

There are many materials that are being explored for 3D printing; however, you will find that the two dominant plastics are ABS (Acrylonitrile butadiene styrene) and PLA (Polylactic acid). Both ABS and PLA are known as thermoplastics; that is, they become soft and moldable when heated and return to a solid when cooled.

This process can be repeated again and again. Their ability to melt and be processed again is what has made them so prevalent in society and is why most of the plastics you interact with on a daily basis are thermoplastics.

For a material to prove viable for 3D printing, it should pass four general criteria:

- Low cost and easily obtainable.

- Controllable extrusion.

- Second extrusion (two or more heads) and trace-binding during the 3D printing process.

- End use application.

HIPS (High Impact Polystyrene) filament is a soluble support material used with dual head extruders.

ABS - Storage

ABS, PLA, and Nylon do best if, before use or when stored long term, they arc sealed off from the atmosphere to prevent the absorption of moisture from the air. This does not mean your material will be ruined by a week of sitting on a bench in the shop, but long term exposure to a humid environment can have detrimental effects, both to the printing process and to the quality of finished parts.

Moisture laden ABS will tend to bubble and spurt from the tip of the extruder nozzle when printing, reducing the visual quality of the part, part accuracy, strength and introducing the risk of a stripping or clogging in the nozzle. ABS can be easily dried using a source of hot (preferably dry) air such as a food dehydrator.

ABS - Part Accuracy

For most, the single greatest hurdle for accurate parts in ABS is a curling upwards off the surface in direct contact with the 3D printer's print bed. A combination of heating the print surface and ensuring it is smooth, flat and clean helps to eliminate this issue. Additionally, some find various solutions can be useful when applied beforehand to the print surface. Example: A mixture of ABS dissolved in Acetone or a shot of hairspray on the build plate. Keep these solutions away from heat.

For fine features on parts involving sharp corners, such as gears, there may be a slight rounding of the corner. A fan, to provide a small amount of active cooling around the nozzle, can improve corners but one does also run the risk of introducing too much cooling and reducing adhesion between layers, eventually leading to cracks in the finished part.

PLA - Storage

PLA responds somewhat differently to moisture. In addition to bubbles or spurting at the nozzle, you may see discoloration and a reduction in 3D printed part properties. PLA (Polylactic acid) is a bioplastic derived from corn and is biodegradable and can react with water at high temperatures and undergo de-polymerization.

PLA can be dried using something as simple as a food dehydrator. It is important to note that this can alter the crystallinity ratio in the PLA and will lead to changes in extrusion temperature and other extrusion characteristics. For many 3D printers, moisture is not a major concern.

PLA - Part Accuracy

Compared to ABS and Nylon, PLA demonstrates much less part warping. For this reason it is possible to successfully print without a heated bed and use more commonly available "Blue" painter's tape as a print surface.

PLA undergoes more of a phase-change when heated and becomes much more liquid. If actively cooled, sharper details can be seen on printed corners without the risk of cracking or warping. The increased flow can also lead to stronger binding between layers, improving the strength of the printed part.

Nylon - Storage

Nylon is very hygroscopic, more so than PLA or ABS. Nylon can absorb more than 10% of its weight in water in less than 24 hours. Successful 3D printing with nylon requires dry filament. When you print with nylon that isn't dry, the water in the filament explodes causing air bubbles during printing that prevents good layer adhesion and greatly weakens the part. It also ruins the surface finish. To dry nylon, place it in an oven at 50-60C for 6-8 hours. After drying, store in an airtight container, preferably with desiccant.

Type 6, 6 Nylon requires a higher extruder temperature than ABS. Type 6, 6 Nylon melts at 255-265C (490-510F), type 6 nylon melts at 210-220C (410-428F) and polypropylene melts at 160-175C (320-347F).

Nylon - Part Accuracy

Compared to ABS and PLA, Nylon and ABS warp approximately the same. PLA demonstrates much less part warping. A heated plate (70 - 80C) is required for Nylon printing. It's best to apply a PVA (Polyvinyl acetate) Elmer's or Scotch based glue stick to the bed for adhesion in a cross-hatch pattern. Remember, less is more when applying the glue stick to the plate.

💡 Nylon is a flexible, stronger and more durable alternative to PLA and ABS. With Nylon, do not use layer cooling fans and avoid drafty or cool rooms for best results.

💡 **Flex PLA** - Common flexible filaments are polyester-based (non-toxic) with a low melting point. Print temp: 120C. It is highly recommended to drastically lower your printing speed to around 20% from standard PLA. To take best advantage of the filament's properties, print it with 10% infill or less. Most flexible filament adheres well to acrylic; it does not adhere well to painter's tape and glass.

Build Plates

PLA in general has a lower shrinkage factor than ABS and Nylon. They are common materials which can warp, curl, delaminate and potentially destroy the print. The following equipment and procedures deal with these potential areas.

Non-Heated

People have experimented with different build surfaces such as steel, glass and various kinds of plastic. However, from experience, both ABS and PLA material seem to stick fairly well to any of the above build surfaces when used with a thin even layer of Polymide Kapton tape or a good quality blue painter's tape. It is not recommended to print ABS or Nylon with a non-heated build plate. Using a raft is always a good idea until you fully understand your system limitations.

💡 Some users apply a small amount of acetone to the build plate while rubbing an old print on the plate. This applies a base layer of material to improve adhesion.

Polymide Kapton tape used in 3D printing.

Heated

A heated build plate or platform helps keep the lowest levels of a print warm as the higher layers are printed. This allows the overall print to cool more evenly. A heated build plate helps tremendously with most ABS prints and large PLA prints. You need a heated build plate for Nylon. Heated build plate temperatures should range between 90 - 95C for ABS, 60 - 65C for PLA and 70 - 80 for Nylon. Polymide Kapton tape works the best due to its temperature range capabilities for ABS and PLA. For Nylon, it is recommended to use a PVA (Elmer's or Scotch) based glue stick to the bed for adhesion in a cross-hatch pattern. Remember, less is more when applying the glue stick on the plate.

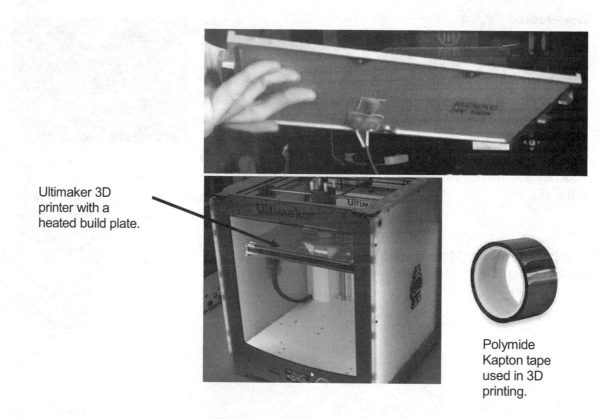

Ultimaker 3D printer with a heated build plate.

Polymide Kapton tape used in 3D printing.

Some users apply a small amount of acetone to the build plate while rubbing an old print to the plate. This applies a base layer of material to improve adhesion.

💡 Most PSUs (power supply units) supplied with non-heated build plate printers need to be upgraded to handle a heated plate upgrade.

Clean

Print materials stick better to a clean build plate. Clean your build plate after every build. Remove any build up material (hairspray, glue, old print residue, etc.). Replace needed tape (blue, Kapton, etc.).

Level

Level the build plate after every build. Do not print with a badly calibrated printer.

A level printing surface means every thin trace of support will be laid down at the intended height. A slope can cause some traces to be laid down too high from the print bed and prevent them from adhering.

Rule of thumb: For a low cost 3D printer, place the print surface (Z axis) approximately 0.1mm away from the nozzle at six points around the print bed.

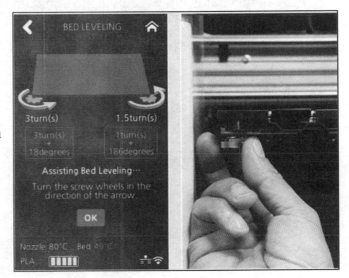

Control Temperature

Eliminate all drafts (air conditioner, heater, windows) and control air flow that may cause a temperature gradient within the build area. Changes in temperature during a build cycle can cause curling, cracking and layer delamination, especially on long thin parts. Below are five fully enclosed 3D printers: A modified MakerBot Replicator 2, an Ultimaker 2 Go, a bq witbox, a Sindoh DP200 3DWox and an XYZ da Vinci.

☀ PLA in general is more forgiving to temperature fluctuations and moisture than ABS and Nylon during a build cycle.

Filament Storage

Below is a filament roll. The filament roll is packed with a package of desiccant. A desiccant is a hygroscopic substance that induces or sustains a state of dryness (desiccation) in its vicinity. Do not open filament packages before you need them. Moisture is your enemy. The filament will swell with moisture and can be an issue (size and temperature) in the extruder head during the print cycle.

Darker colors in general (PLA and ABS) often require higher extruder temperatures. Extruding at higher temperatures may help prevent future delamination if you do not have a heated build plate or control build area.

Desiccant package

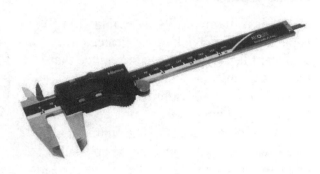

☀ There are set defaults that come in many low cost printers. The default setting for the diameter of the filament in the Replicator G software is 1.85mm. If you are having build/extruder problems, use a pair of calibers to measure the diameter of the filament. Most calibers will have measurements down to .01mm. To measure your filament, take an average reading of at least three readings and re-set the default filament setting if needed.

Never focus too much on one single issue. These machines are complex, and trouble often arises from multiple reasons. A slipping filament may not only be caused by a bad gear drive, but also by an obstructed nozzle, a wrong feed value, a too low (or too high) temperature or a combination of all these.

Prepare the Model

There are many available open source software packages for preparing your model in an STL file format. To turn the .STL file into code for your printer, you'll need a slicing program. The MakerWare software (slicing program) for the MakerBot Replicator 2 Desktop and the Sindoh DP200 3DWox is displayed in this section.

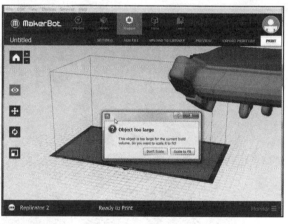

The first action is to add/insert or load your STL file using the slicing program in the build area.

Depending on your printer "slicing program," you may or may not receive a message indicating your object is too large for the current build volume.

If the object is too large, then you will need to scale it down or to redesign it into separate parts.

Use caution when scaling if you require fasteners or a minimum wall thickness.

You should center the part and have it lay flat on the build plate.

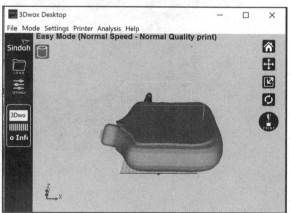

If you are printing more than one part, space them evenly on the plate or position them for a single build.

⌥ In SOLIDWORKS, lay the parts out in an assembly. Save the assembly as a part. Save the part file as an STL file. Open the STL in the printer using your slicer.

Consideration should be used when printing an assembly. If the print takes 20 hours, and a failure happens after 19 hours, you just wasted a lot of time versus printing each part individually.

Example 1: Part Orientation

Part orientation is very important on build strength and the amount of raft and support material required for the build. Part orientations can also be related to warping, curling, and delamination.

If maximizing strength is an issue, select the part orientation on the build plate so that the "grain" of the print is oriented to maximize the strength of the part.

Example 1: First Orientation - Vertical

In the first orientation (vertical), due to the number of holes and slots, additional support material is required (with minimum raft material) to print the model.

Removing this material in these geometrics can be very time consuming.

Example 1: Second Orientation - Horizontal

In the second orientation (horizontal), additional raft material is used, and the support material is reduced.

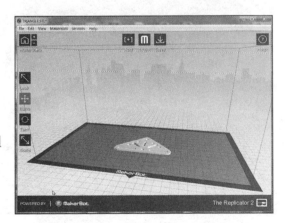

Second orientation - horizontal

The raft material can be easily removed with a pair of needle-nose pliers and no support material clean-up is required for the holes and slots. Note: In some cases, raft material is not needed.

Example 2: Part Orientation

The lens part is orientated in a vertical position with the large face flat on the build plate. This reduces the required support material and ensures proper contact with the build plate.

The needed support material is created mainly internal to the part to print the CBORE feature. Note: There is some outside support material on the top section of the part.

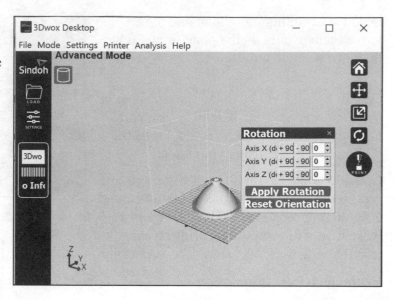

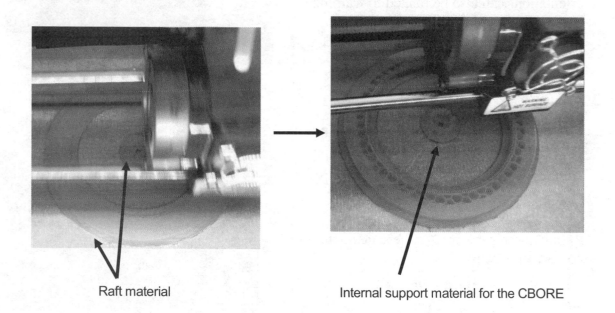

Raft material   Internal support material for the CBORE

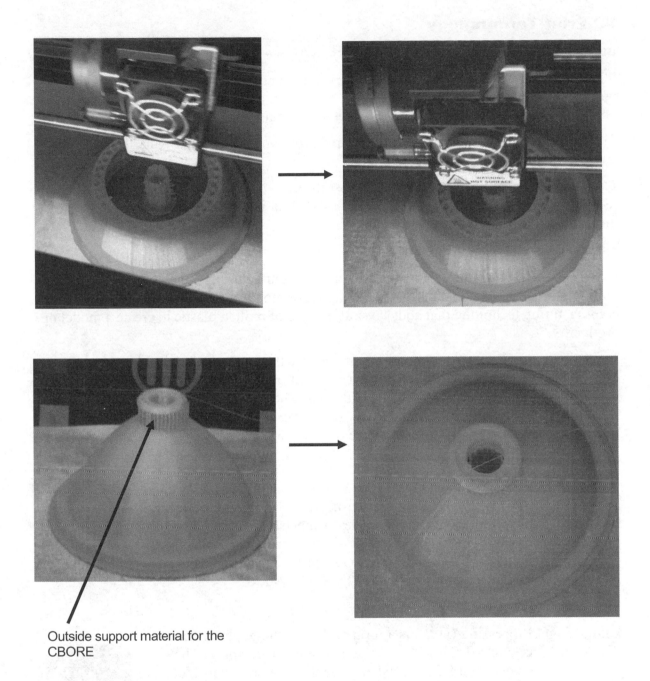

Outside support material for the
CBORE

3D Print Terminology

Stereolithography (SL or SLA)

Stereolithography (SL or SLA) is one of several methods used to create 3D printed objects. This is a liquid-based process that consists in the curing or solidification of a photosensitive polymer when an ultraviolet laser makes contact with the resin.

The process was patented as a means of rapid prototyping in 1986 by Charles Hull, co-founder of 3D Systems, Inc.

The process starts with a model in a CAD software and then it is translated to a STL (STereoLithography) file in which the pieces are "cut in slices" containing the information for each layer.

Fused Filament Fabrication (FFF)

Fused Filament Fabrication (FFF) is an additive manufacturing technology used for building three-dimensional products, prototypes or models. It is a rapid prototyping and manufacturing technique that adds layer after layer of molten plastic to create a model or product.

Typically, the FFF mechanism consists of a nozzle that emits material and deposits it onto a moving table. The FFF machine takes input from G-code and starts moving the nozzle and build plate to the needed coordinates. The material is heated in the nozzle to form liquid, which solidifies immediately when it is deposited onto the layer surface. The nozzle works layer-by-layer until the product is finished.

Fused Deposition Fabrication (FDM)

Fused Deposition Modeling (FDM) is an additive manufacturing technology that builds parts up layer-by-layer by heating and extruding thermoplastic filament. Ideal for building durable components with complex geometries in nearly any shape and size, FDM is the only 3D printing process that uses materials like ABS, PC-ISO polycarbonate, and ULTEM 9085. The actual term "Fused Deposition Modeling" and its abbreviation "FDM" are trademarked by Stratasys Ltd.

Digital Light Process (DLP)

Digital Light Processing (DLP) is a form of Stereolithography (SL, or SLA) that is used in Additive Manufacture or rapid prototyping. The main difference between DLP and SLA is the use of a projector light rather than a laser to cure photo-sensitive polymer resin. A DLP 3D printer projects the image of the object's cross section onto the surface of the resin. The exposed resin hardens while the machine's build platform descends, setting the stage for a new layer of fresh resin to be coated to the object and cured by light. Once a complete object is formed, additional post processing may be required such as removal of support material, chemical bath, and UV curing.

An example of a DLP 3D printer is the Formlabs Form1 printer as shown.

Raft, Skirt and Brim

A Raft is a horizontal latticework of filament located underneath the part. A Raft is used to help the part stick to the build plate (heated or non-heated).

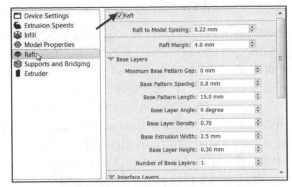

Rafts are also used to help stabilize thin tall parts with small build plate footprints.

When the print is complete, remove the part from the build plate. Peel the raft away from the part. If needed, use a scraper or spatula.

The Raft Settings dialog box illustrates numerous settings and options.

A Skirt is a layer of filament that surrounds the part with a 3-4mm offset. The layer does not connect the part directly to the build plate. The Skirt primes the extruder and establishes a smooth flow of filament. In some slicers, the skirt is added automatically when you select None for Bed Adhesion Type.

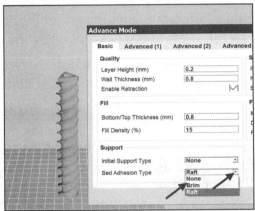

A Brim is basically like a Skirt for the part. A Brim has a zero offset from the part. It is a layer of filament that is laid down around the base of the part to increase its surface area. A Brim, however, does not extend underneath the part, which is the key difference between a brim and a raft.

Proper part orientation for thin parts will make the removal of the raft easier.

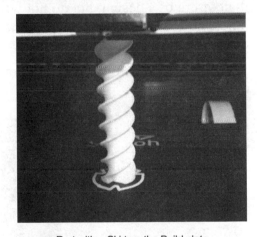

Part with a Skirt on the Build plate

Part with a Raft on the Build plate

Support and Touching Build plate

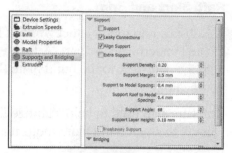

Printing with support is required when material must be deposited on a layer where there is no or insufficient material on the previous layer. This includes steep overhanging surfaces, straight overhangs, and fully suspended islands. Learning to print objects on a 3D Printer that require support structures will dramatically expand the potential of your printer and give you the confidence to undertake printing tasks that perhaps you had previously avoided.

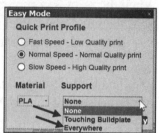

You can remove supports after building the object, but be careful in the initial orientation of the part.

If the part has numerous holes, sharp edges, steep angles or thin bodies, additional support will be added and can make it difficult to remove cleanly.

In some slicers, there is a Touching Buildplate option. This option provides supports only where the part touches the build plate. This reduces build time, clean up and support build material. Use this option when you have overhangs and tricky angles toward the bottom of a design, but do not wish to plug up holes, hollow spaces, or arches in the rest of the design.

See the below illustration of a support material example.

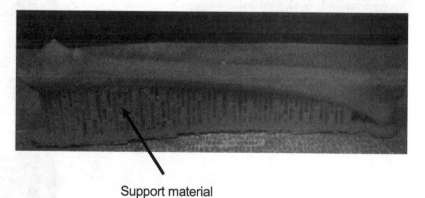

Support material

My favorite tools to remove support and raft material are a good pair of needle nose pliers, angle tweezers and a small flat head screwdriver.

Slicer Engine

All 3D printers use a slicer engine. Slicing generates the G-code necessary to feed into your printer. Slicing is the process of turning your 3D model into a toolpath for your 3D printer. Most people call it slicing because the first thing the slicing engine does is cut your 3D model into thin horizontal layers. Open source slicing engines include Slic3r, Skeinforge, Netfabb, KISSkice, Cura, etc. to name a few.

G-code

There are different ways to prepare G-code for a printer. One is to use a slicer engine. These programs take a CAD model, slice it into layers, and output the G-code required for each layer. Slicers are an easy way to go from a 3D model to a printed part, but the user sacrifices some flexibility when using them. Another option for G-code generation is to use me-code, a lower level open source library. Me-code libraries give precise control over the toolpath providing a solution for a complex print that is not suitable for slicing. If you need to run a few test lines while calibrating your 3D printer, then the final option is to write your own G-code.

Infill

Infill is the internal structure of your object, which can be as sparse or as substantial as you would like it to be. A higher percentage will result in a more solid object, so 100% infill will make your object completely solid, while 0% infill will give you something completely hollow. The higher the infill percentage, the more material and longer the print time required. In general, use a 10% - 15% infill. 100% infill is very rarely used.

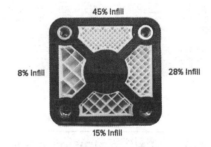

Infill Pattern/Shape

When using any infill percentage, a pattern is used to create a strong and durable structure inside the print. A few standard patterns are Rectilinear, Honeycomb, Circular and Triangular.

Shells/Parameters

Shells are the outer layers of a print which make the walls of an object, prior to the various infill levels being printed within.

The Number of Shells Display shows examples of cubes printed with 1, 2, 3, 4, 5, 10, 15, 20, 25, and 30 shells. The number of shells affect stability and translucency of the model.

Picture from 3D Printing.com
Duncan Smith

More shells result in a stronger object but longer printing time and more material.

Shells are also referred to as perimeters in some software and documents.

💡 Do not use additional shells on fine featured models, such as small text. It will obscure the detail.

View the different
infill and number of
shells between the
two models.

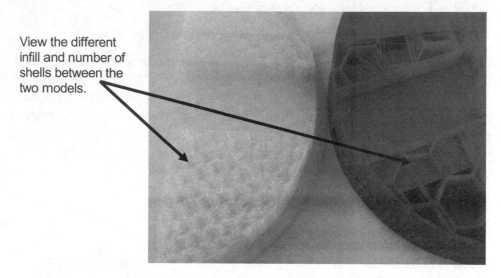

Layer Height

Traditionally, extrusion-based 3D printers have emphasized the layer height (such as 0.1mm or 0.3mm) as the main indicator of accuracy or quality. One reason for this may be that 3D prints made with this technology typically have visible ridges between different layers.

Layer height has a significant effect on the resolution of organic (curvy) parts. In general use 0.3mm for fast draft quality, 0.25mm for medium draft quality, 0.2mm for standard quality, and 0.1mm for high quality.

💡 (1micron = 1μm = 0.001mm).

Influence of Percent Infill

How much infill is needed for your print? It depends. Infill refers to the structure that is printed inside the model. It is extruded in a designated percent and pattern. The designated percentage and pattern is set in the slicer.

The percent infill and pattern type (Rectilinear, Honeycomb, Circular), to name a few, influence material usage, strength, print weight, print time and sometimes flatness and decorative properties. As a general rule, the larger the percentage infill, the stronger the print, but the longer it takes to print. In most cases, 10% - 15% infill is sufficient. 100% infill is very rarely used.

Strength corresponds to the maximum stress the print can take before breaking.

Remove the Model from the Build Plate

If you have a non-heated build plate with a sheet of good quality blue painter's tape, it is not too difficult to remove the part. Utilize a flat edge tool (a thin steel spatula), gently work under the part, and lift the raft or the part directly on the tape. You may rip the tape when taking the part off. Replace the tape. Re-level your build plate after every build.

If you have a heated build plate your temperatures can range between 60 - 90C (Kapton tape) depending on the material, so be careful. Again, as above, utilize a flat edge tool (a thin steel spatula), gently work under the part, and lift the part directly from the plate. Re-level your build plate after every build.

Know the Printer's Limitations

There are features that are too small to be printed in plastic on a desktop 3D printer. An important, but often overlooked, variable in what your printer can achieve is thread width.

Thread width is determined by the diameter of your printer's nozzle. Most printers have a 0.4mm or 0.5mm nozzle. Practically, this means that a circle drawn by a 3D printer is always two thread widths deep: 0.8mm thick with a 0.4mm nozzle to 1mm thick for a 0.5mm nozzle. A good rule of thumb is "The smallest feature you can create is double the thread width."

Tolerance for Interlocking Parts

For objects with multiple interlocking parts, design in your fit tolerance. Getting tolerances correct can be difficult. Typical minimum wall thickness for a model is roughly 1mm, and printed part accuracy is ±0.2mm. It is recommended to use a ±0.2mm offset for tight fit (press fit parts, connectors) and use a ±0.4mm offset for loose fit (hinges, box lids).

Keep in mind the underlying problems with low cost 3D printers. The same 3D model will produce prints with varying part tolerances when printed on different types and brands of 3D printers. Users need to check the specifications from the 3D printer and filament manufacturer and then adjust the model geometries accordingly. Also, with any 3D printer, make sure the printer is calibrated to ensure the resolution will be the same across multiple prints.

Test the fit yourself with the particular model to determine the right tolerance for the items you are creating and material you are using.

General Printing Tips

Reduce Infill

When printing a model, you can choose to print it hollow or completely solid, or some percentage between 0 and 100. Infill is a settable variable in most Slicer engines. The material inside the part exerts a force on the entire printed part as it cools. More material increases cost of the part and build time.

Parts with a lower percentage of Infill should have a lower internal force between layers and can reduce the chance of curling, cracking, and layer delamination along with a low build cost and time.

Control build area temperature

For a consistent quality build, control your build area and environment temperature. Eliminate all drafts and control air flow that may cause a temperature gradient within the build area.

Changes in temperature during a build cycle can cause curling, cracking and layer delamination, especially on long thin parts. From 1000s of hours in 3D printing experience, having a top cover, sides and a heated build plate along with a consistent room temperature provides the best and repeatable builds.

When troubleshooting issues with your printer, it is always best to know if the nozzle and heated bed are achieving the desired temperatures. A thermocouple and a thermometer come in handy. I prefer a Type-K Thermocouple connected to a multi-meter. But a non-contact IR or Laser-based thermometer also works well.

One important thing to remember is that IR and Laser units are not 100% accurate when it comes to shiny reflective surfaces. A Type-K thermocouple can be taped to the nozzle or heated bed using Kapton tape, for a very accurate temperature measurement.

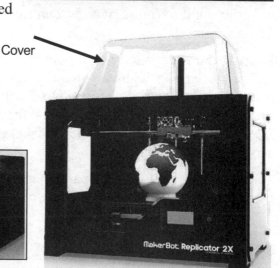

Cover

Add Pads

Sometimes, when you are printing a large flat object, such as a simple box container or a very long thin feature, you may view warping at the corners or extremities. One way to address this is to create small pads to your part during the modeling process. Create the model for the print. Think before you print and know your printer limitations. The pads can be any size and shape, but generally, diameter 10mm cylinders that are 1-2 layers thick work well. After the part is printed, remove them.

Makerbot Image

Unique Shape or a Large Part

If you need to make parts larger than your build area or create parts that have intricate projections, here are a few suggestions:

- Fuse smaller sections together using acetone (if using ABS). Glue if using PLA.

- Design smaller parts to be attached together (without hardware).

- Design smaller parts to be screwed together (with hardware).

Safe Zone Rule

Parts may have a safe zone. The safe zone is called "self-supporting" and no support material is required to build the part.

The safe zone can range between 30° to 150°. If the part's features are below 30° or greater than 150°, it should have support material during the build cycle. This is only a rule. Are there other factors to consider? Yes. They are layer thickness, extrusion speed, material type, length of the overhang along with the general model design of features.

💡 Design your part for your printer. Use various modeling techniques (ribs, fillets, pads, etc.) during the design process to eliminate or to minimize the need for supports and clean up.

Wall Thickness

In 3D printing, wall thickness refers to the distance between one surface of your model and the opposite sheer surface. A model made in stereolithography with a minimal wall thickness of 1mm provides you with a strong solid surface.

💡 If you scale your model with the 3D printer software, the wall thickness also scales proportionally. Make modifications in the CAD model to ensure the minimal wall thickness of 1mm.

Extruder Temperature

When working with a new roll of filament for the first time, I generally start out printing at about 200 - 210C (heated plate) and 225 - 230C (non-heated plate) for PLA and then adjust the temperature up or down by a few degrees until I obtain the quality of the print and the strength of the part to be in good balance with each other.

If the temperature is too high, you will see more strings between the separate parts of your print and you may notice that the extruder leaks out material while moving between separate areas of the print.

If the temperature is too cold, you will either see that the filament is not sticking to the (non-heated build plate) or the previous layer and you are getting a rough surface. You will get a part that is not strong and can be pulled apart easily.

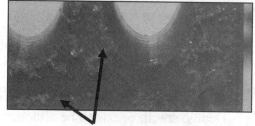

Too cold – rough surface

The extruder temperature for PLA ranges between 200C - 215C (heated bed 60C - 65C), 225C - 230C (non-heated bed). For ABS they are 225C - 240 (heated bed 90C - 95C).

Various factors can influence the optimal extruding temperature:

- Moisture content.

- Temperature of the printing environment.

- Color.

- Glow in the dark ability.

- Elevation (from sea level) of the printing environment.

First Layer Not Sticking

One of the toughest aspects of 3D printing is to get your prints to stick to the build surface or bed platform. Investigate the following:

- Make sure the bed is level.

- Make sure the bed is hot enough. Do not use blue painter's tape with a heated plate.

- Make sure that the ambient temperature of the print environment isn't too hot or cold (or else adjust accordingly).

- Make sure you put the adhesive on the bed.

- Make sure the print head is close enough to make a nice squished first layer.

- Make sure you run the extruder enough before your print starts so there is filament going onto the bed during the entire first layer.

- Kapton/PET tape is a great way to print ABS. It makes a great shiny bottom layer and the heated bed ensures that your parts stay nice and flat.

Level Build Platform

An unleveled build platform will cause many headaches during a print. You can quickly check the platform by performing the business card test: use a single business card to judge the height of your extruder nozzle over the build platform. Achieve a consistent slight resistance when you position the business card between the tip of the extruder and the platform for all leveling positions.

Minimize Internal Support

Design your part for your printer. Use various modeling techniques (ribs, fillets, pads, etc.) during the design process to eliminate or to minimize the need for support and final clean up.

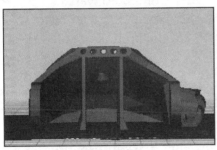 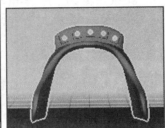 

Water Tight Mesh

A water-tight mesh is achieved by having closed edges creating a solid volume. If you were to fill your geometry with water, would you see a leak? You may have to clean up any internal geometry that could have been left behind accidentally from Booleans.

Clearance

If you are creating separate or interlocking parts, make sure there is a large enough distance between tight areas. 3D Printing production, such as Selective Laser Sintering (SLS), makes moving parts without assembly a possibility that was not there before. Take advantage of this strength by creating enough clearance that the model's pieces do not fuse together or trap support material inside.

In General

- Keep your software and firmware up to date.

- Think before you print. Design your model for your printer.

- Understand the printer's limitations. Adjust one thing at a time between prints and keep notes about the settings and effect on the print. Label test prints with a Sharpie and take photographs.

- Always level and clean the build plate before a build.

- Select the correct material (filament) for the build and application. Materials are still an area of active exploration. Use ABS, PLA and Nylon, each of which has specific printing requirements.

- Select the correct part orientation on the build plate. Many of these devices require some attention when printing. 3D printers fabricate objects from thin layers of material; there is a grain to the structure of the printed parts, much like there is grain in wood. Try to print in an orientation so that the "grain" of the print is oriented to maximize the strength of the part.

- Make sure that the bottom surface of each part of the print is firmly secured to the build platform, both for the success of the print and for the surface finish on the bottom face of the object.

- Select the correct Slicing engine settings. Models go through two software processes on their way to becoming finished prints: slicing and sending. Slicing divides a model into printable layers and plots the toolpaths to fill them in. The printer client then sends these movements to the hardware and provides a control interface for its other functions.

- Control the environment for the filament and build area (temperature, humidity, etc.).

- If in doubt, create your first build with a raft and support.

- 3D printers cannot print perfectly, so there will always be a bit of variation between the model and the output. If you make a round hole that's 10mm in diameter and a 10mm diameter drive shaft to fill that hole, it won't have enough clearance to fit into the hole. The best current commercially available printers have a resolution of 0.1mm. This means there could be up to 0.2mm difference between the actual surface of your print and the 3D model. Models also shrink a little after being printed. The best thing to do after a print has been completed is to cool down your machine entirely before switching it off. But, if you are in a rush, make sure to at least get it under 100C before switching it off.

Summary

With 1000s of hours using multiple low-cost 3D printers, learning about additive print technology is a great experience. Students face a multitude of obstacles with their first 3D prints. Understanding what went wrong and knowing the capabilities of the 3D printer produces positive results.

Never focus too much on one single issue. These machines are complex, and trouble often arises from multiple reasons. A slipping filament may not only be caused by a bad gear drive, but also by an obstructed nozzle, a wrong feed value, a too low (or too high) temperature or a combination of all these.

As printed parts cool (PLA, ABS and Nylon), various areas of the object cool at different rates. Depending on the model being printed and the filament material, this effect can lead to warping, curling and or layer delamination.

Design your part for your printer. Use various modeling techniques (ribs, fillets, pads, etc.) during the design process to eliminate or to minimize the need for supports and clean up.

Avi Reichental, President and CEO, 3D Systems stated, "With 3D printing, complexity is free. The printer doesn't care if it makes the most rudimentary shape or the most complex shape, and that is completely turning design and manufacturing on its head as we know it."

Additive manufacturing, sometimes known as ***rapid prototyping***, can be slower than Subtractive manufacturing. Both take skill in creating the G-code and understanding the machine limitations.

Notes:

Appendix

SOLIDWORKS Keyboard Shortcuts

Below are some of the pre-defined keyboard shortcuts in SOLIDWORKS:

Action:	Key Combination:
Model Views	
Rotate the model horizontally or vertically	**Arrow** keys
Rotate the model horizontally or vertically 90 degrees	**Shift** + **Arrow** keys
Rotate the model clockwise or counterclockwise	**Alt** + left of right **Arrow** keys
Pan the model	**Ctrl** + **Arrow** keys
Magnifying glass	**g**
Zoom in	**Shift + z**
Zoom out	**z**
Zoom to fit	**f**
Previous view	**Ctrl + Shift + z**
View Orientation	
View Orientation menu	**Spacebar**
Front view	**Ctrl + 1**
Back view	**Ctrl + 2**
Left view	**Ctrl + 3**
Right view	**Ctrl + 4**
Top view	**Ctrl + 5**
Bottom view	**Ctrl + 6**
Isometric view	**Ctrl + 7**
NormalTo view	**Ctrl + 8**
Selection Filters	
Filter edges	**e**
Filter vertices	**v**
Filter faces	**x**
Toggle Selection Filter toolbar	**F5**
Toggle selection filters on/off	**F6**
File menu items	
New SOLIDWORKS document	**Ctrl + n**
Open document	**Ctrl + o**
Open From Web Folder	**Ctrl + w**
Make Drawing from Part	**Ctrl + d**
Make Assembly from Part	**Ctrl + a**
Save	**Ctrl +s**
Print	**Ctrl + p**
Additional items	
Access online help inside of PropertyManager or dialog box	**F1**
Rename an item in the FeatureManager design tree	**F2**

Action:	Key Combination:
Rebuild the model	Ctrl + b
Force rebuild - Rebuild the model and all its features	Ctrl + q
Redraw the screen	Ctrl + r
Cycle between open SOLIDWORKS document	Ctrl + Tab
Line to arc/arc to line in the Sketch	a
Undo	Ctrl + z
Redo	Ctrl + y
Cut	Ctrl + x
Copy	Ctrl + c
Paste	Ctrl + v
Delete	Delete
Next window	Ctrl + F6
Close window	Ctrl + F4
View previous tools	s
Selects all text inside an Annotations text box	Ctrl + a

In a sketch, the **Esc** key un-selects geometry items currently selected in the Properties box and Add Relations box.

In the model, the **Esc** key closes the PropertyManager and cancels the selections.

Use the **g** key to activate the Magnifying glass tool. Use the Magnifying glass tool to inspect a model and make selections without changing the overall view.

Use the **s** key to view/access previous command tools in the Graphics window.

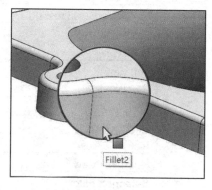

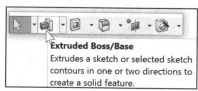

Extruded Boss/Base
Extrudes a sketch or selected sketch contours in one or two directions to create a solid feature.

Modeling - Best Practices

Best practices are simply ways of bringing about better results in easier, more reliable ways. The Modeling - Best Practice list is a set of rules helpful for new users and users who are trying to experiment with the limits of the software.

These rules are not inflexible, but conservative starting places; they are concepts that you can default to, but that can be broken if you have good reason. The following is a list of suggested best practices:

- Create a folder structure (parts, drawings, assemblies, simulations, etc.). Organize into project or file folders.

- Construct sound document templates. The document template provides the foundation that all models are built on. This is especially important if working with other SOLIDWORKS users on the same project; it will ensure consistency across the project.

- Generate unique part filenames. SOLIDWORKS assemblies and drawings may pick up incorrect references if you use parts with identical names.

- Apply Custom Properties. Custom Properties is a great way to enter text-based information into the SOLIDWORKS parts. Users can view this information from outside the file by using applications such as Windows Explorer, SOLIDWORKS Explorer, and Product Data Management (PDM) applications.

- Understand part orientation. When you create a new part or assembly, the three default Planes (Front, Right and Top) are aligned with specific views. The plane you select for the Base sketch determines the orientation.

- Learn to sketch using automatic relations.

- Limit your usage of the Fixed constraint.

- Add geometric relations, then dimensions in a 2D sketch. This keeps the part from having too many unnecessary dimensions. This also helps to show the design intent of the model. Dimension what geometry you intend to modify or adjust.

- Fully define all sketches in the model. However, there are times when this is not practical, generally when using the Spline tool to create a freeform shape.

- When possible, make relations to sketches or stable reference geometry; such as the Origin or standard planes, instead of edges or faces. Sketches are far more stable than faces, edges, or model vertices, which change their internal ID at the slightest change and may disappear entirely with fillets, chamfers, split lines, and so on.

- Do not dimension to edges created by fillets or other cosmetic or temporary features.

- Apply names to sketches, features, dimensions, and mates that help to make their function clear.

- When possible, use feature fillets and feature patterns rather than sketch fillets and sketch patterns.

- Apply the Shell feature before the Fillet feature, and the inside corners remain perpendicular.

- Apply cosmetic fillets and chamfers last in the modeling procedure.

- Combine fillets into as few fillet features as possible. This enables you to control fillets that need to be controlled separately, such as fillets to be removed and simplified configurations.

- Create a simplified configuration when building very complex parts or working with large assemblies.

- Use symmetry during the modeling process. Utilize feature patterns and mirroring when possible. Think End Conditions.

- Use global variables and equations to control commonly applied dimensions (design intent).

- Add comments to equations to document your design intent. Place a single quote (') at the end of the equation, then enter the comment. Anything after the single quote is ignored when the equation is evaluated.

- Avoid redundant mates. Although SOLIDWORKS allows some redundant mates (all except distance and angle), these mates take longer to solve and make the mating scheme harder to understand and diagnose if problems occur.

- Fix modeling errors in the part or assembly when they occur. Errors cause rebuild time to increase, and if you wait until additional errors exist, troubleshooting will be more difficult.

- Create a Library of Standardized notes and parts.

- Utilize the Rollback bar. Troubleshoot feature and sketch errors from the top of the design tree.

- Determine the static and dynamic behavior of mates in each sub-assembly before creating the top level assembly.

- Plan the assembly and sub-assemblies in an assembly layout diagram. Group components together to form smaller sub-assemblies.

- When you create an assembly document, the base component should be fixed, fully defined or mated to an axis about the assembly origin.

- In an assembly, group fasteners into a folder at the bottom of the FeatureManager. Suppress fasteners and their assembly patterns to save rebuild time and file size.

- When comparing mass, volume and other properties with assembly visualization, utilize similar units.

- Use limit mates sparingly because they take longer to solve and whenever possible, mate all components to one or two fixed components or references. Long chains of components take longer to solve and are more prone to mate errors.

Helpful On-line Information

The SOLIDWORKS URL http://www.SOLIDWORKS.com contains information on Local Resellers, Solution Partners, Certifications, SOLIDWORKS users groups and more.

Access 3D ContentCentral using the Task Pane to obtain engineering electronic catalog model and part information.

Use the SOLIDWORKS Resources tab in the Task Pane to obtain access to Customer Portals, Discussion Forums, User Groups, Manufacturers, Solution Partners, Labs and more.

Helpful on-line SOLIDWORKS information is available from the following URLs:

- http://www.swugn.org/

List of all SOLIDWORKS User groups.

- https://www.solidworks.com/sw/education/certification-programs-cad-students.htm

The SOLIDWORKS Academic Certification Programs.

- http://www.solidworks.com/sw/industries/education/engineering-education-software.htm

The SOLIDWORKS Education Program.

- https://solidworks.virtualtester.com/#home_button

The SOLIDWORKS Certification Center - Virtual tester site.

*On-line tutorials are for educational purposes only. Tutorials are copyrighted by their respective owners.

SOLIDWORKS Document Types

SOLIDWORKS has three main document file types: Part, Assembly and Drawing, but there are many additional supporting types that you may want to know. Below is a brief list of these supporting file types:

Design Documents	Description
.sldprt	SOLIDWORKS Part document
.slddrw	SOLIDWORKS Drawing document
.sldasm	SOLIDWORKS Assembly document

Templates and Formats	Description
.asmdot	Assembly Template
.asmprp	Assembly Template Custom Properties tab
.drwdot	Drawing Template
.drwprp	Drawing Template Custom Properties tab
.prtdot	Part Template
.prtprp	Part Template Custom Properties tab
.sldtbt	General Table Template
.slddrt	Drawing Sheet Template
.sldbombt	Bill of Materials Template (Table-based)
.sldholtbt	Hole Table Template
.sldrevbt	Revision Table Template
.sldwldbt	Weldment Cutlist Template
.xls	Bill of Materials Template (Excel-based)

Library Files	Description
.sldlfp	Library Part file
.sldblk	Blocks

Other	Description
.sldstd	Drafting standard
.sldmat	Material Database
.sldclr	Color Palette File
.xls	Sheet metal gauge table

Chapter 17: Answer Key

1. What is the Modulus of Elasticity?

- The slope of the Deflection-Stress curve

- **The slope of the Stress-Strain curve in its linear section**

- The slope of the Force-Deflection curve in its linear section

- The first inflection point of a Strain curve

2. What is Stress?

- A measure of power

- A measure of strain

- A measure of material strength

- **A measure of the average amount of force exerted per unit area**

3. Which of the following assumptions are true for a static analysis in SOLIDWORKS Simulation with small displacements?

- **Inertia effects are negligible and loads are applied slowly**

- The model is not fully elastic. If loads are removed, the model will not return to its original position

- **Results are proportional to loads**

- **All the displacements are small relative to the model geometry**

4. What is Yield Stress?

- **The stress level beyond which the material becomes plastic**

- The stress level beyond which the material breaks

- The strain level above the stress level which the material breaks

- The stress level beyond the melting point of the material

5. A high quality Shell element has _____ nodes.

- 4
- 5
- **6**
- 8

6. Stress σ is proportional to _____ in a Linear Elastic Material.

- **Strain**
- Stress
- Force
- Pressure

7. The Elastic Modulus (Young's Modulus) is the slope defined as _____ divided by
_____.

- Strain, Stress
- **Stress, Strain**
- Stress, Force
- Force, Area

8. Linear static analysis assumes that the relationship between loads and the induced
response is _____.

- Flat
- **Linear**
- Doubles per area
- Translational

9. In SOLIDWORKS Simulation, the Factor of Safety (FOS) calculations are based on
one of the following failure criterion.

- **Maximum von Mises Stress**
- **Maximum shear stress (Tresca)**
- **Mohr-Coulomb stress**
- **Maximum Normal stress**

10. The Yield point is the point where the material begins to deform at a faster rate than at the elastic limit. The material behaves _____ in the Plastic Range.

- Flatly
- Linearly
- **Non-Linearly**
- Like a liquid

11. What are the Degrees of Freedom (DOFs) restrained for a Solid?

- None
- **3 Translations**
- 3 Translations and 3 Rotations
- 3 Rotations

12. What are the Degrees of Freedom (DOFs) restrained for Truss joints?

- None
- **3 Translations**
- 3 Translations and 3 Rotations
- 3 Rotations

13. What are the Degrees of Freedom (DOFs) restrained for Shells and Beams?

- None
- 3 Translations
- **3 Translations and 3 Rotations**
- 3 Rotations

14. Which statements are true for Material Properties using SOLIDWORKS Simulation?

- **For solid assemblies, each component can have a different material**
- For shell models, each shell cannot have a different material and thickness
- **For shell models, the material of the part is used for all shells**
- For beam models, each beam cannot have a different material

15. A Beam element has _____nodes (one at each end) with _____degrees of freedom per node plus_____ node to define the orientation of the beam cross section.

- 6, 3, 1
- 3, 3, 1
- **3, 6, 1**
- None of the above

16. A Truss element has _____ nodes with _____ translational degrees of freedom per node.

- **2, 3**
- 3, 3
- 6, 6
- 2, 2

17. In general, the finer the mesh the better the accuracy of the results.

- **True**
- False

18. How does SOLIDWORKS Simulation automatically treat a Sheet metal part with uniform thickness?

- **Shell**
- Solid
- Beam
- Mixed Mesh

19. Use the mesh and displacement plots to calculate the distance between two _____ using SOLIDWORKS Simulation.

- **Nodes**
- Elements
- Bodies
- Surfaces

20. Surface models can only be meshed with _____ elements.

- **Shell**
- Beam
- Mixed Mesh
- Solid

21. The shell mesh is generated on the surface (located at the mid-surface of the shell).

- **True**
- False

22. In general, use Thin shells when the thickness-to-span ratio is less than _____.

- **0.05**
- .5
- 1
- 2

23. The model (a rectangular plate) has a length to thickness ratio of less than 5. You extracted its mid-surface to use it in SOLIDWORKS Simulation. You should use a _____.

- Thin Shell element formulation
- **Thick Shell element formulation**
- Thick or Thin Shell element formulation, it does not matter
- Beam Shell element formulation

24. The model, a rectangular sheet metal part, uses SOLIDWORKS Simulation. You should use a:

- **Thin Shell element formulation**
- Thick Shell element formulation
- Thick or Thin Shell element formulation, it does not matter
- Beam Shell element formulation

25. The Global element size parameter provides the ability to set the global average element size. SOLIDWORKS Simulation suggests a default value based on the model volume and _____ area. This option is only available for a standard mesh.

- Force

- Pressure

- **Surface**

- None of the above

26. A remote load applied on a face with a Force component and no Moment can result in: Note: Remember (DOFs restrain).

- **A Force and Moment of the face**

- A Force on the face only

- A Moment on the face only

- A Pressure and Force on the face

27. There are _____ DOFs restrain for a Solid element.

- **3**

- 1

- 6

- None

28. There are _____ DOFs restrain for a Beam element.

- 3

- 1

- **6**

- None

29. What best describes the difference(s) between a Fixed and Immovable (No translation) boundary condition in SOLIDWORKS Simulation?

- There are no differences

- There are no difference(s) for Shells but it is different for Solids

- **There is no difference(s) for Solids but it is different for Shells and Beams**

- There are only differences(s) for a Static Study

30. Can a non-uniform pressure of force be applied on a face using SOLIDWORKS Simulation?

- No

- Yes, but the variation must be along a single direction only

- **Yes. The non-uniform pressure distribution is defined by a reference coordinate system and the associated coefficients of a second order polynomial**

- Yes, but the variation must be linear

31. You are performing an analysis on your model. You select five faces, 3 edges and 2 vertices and apply a force of 20lbf. What is the total force applied to the model using SOLIDWORKS Simulation?

- 100lbf

- 1600lbf

- 180lbf

- **200lbf**

32. Yield strength is typically determined at _____ strain.

- 0.1%

- **0.2%**

- 0.02%

- 0.002%

33. There are four key assumptions made in Linear Static Analysis: 1. Effects of inertia and damping are neglected, 2. The response of the system is directly proportional to the applied loads, 3. Loads are applied slowly and gradually, and_____ .

- **Displacements are very small. The highest stress is in the linear range of the stress-strain curve**

- There are no loads

- Material is not elastic

- Loads are applied quickly

34. How many degrees of freedom does a physical structure have?

- Zero

- Three - Rotations only

- Three - Translations only

- **Six - Three translations and three rotational**

35. Brittle material has little tendency to deform (or strain) before fracture and does not have a specific yield point. It is not recommended to apply the yield strength analysis as a failure criterion on brittle material. Which of the following failure theories is appropriate for brittle materials?

- **Mohr-Columb stress criterion**

- Maximum shear stress criterion

- Maximum von Mises stress criterion

- Minimum shear stress criterion

36. You are performing an analysis on your model. You select three faces and apply a force of 40lb. What is the total force applied to the model using SOLIDWORKS Simulation?

- 40lb

- 20lb

- **120lb**

- Additional information is required

37. A material is orthotropic if its mechanical or thermal properties are not unique and independent in three mutually perpendicular directions.

- True

- **False**

38. An increase in the number of elements in a mesh for a part will:

- Decrease calculation accuracy and time

- **Increase calculation accuracy and time**

- Have no effect on the calculation

- Change the FOS below 1

39. SOLIDWORKS Simulation uses the von Mises Yield Criterion to calculate the Factor of Safety of many ductile materials. According to the criterion:

- **Material yields when the von Mises stress in the model equals the yield strength of the material**

- Material yields when the von Mises stress in the model is 5 times greater than the minimum tensile strength of the material

- Material yields when the von Mises stress in the model is 3 times greater than the FOS of the material

- None of the above

40. SOLIDWORKS Simulation calculates structural failure on:

- Buckling

- Fatigue

- Creep

- **Material yield**

41. Apply a uniform total force of 200lb on two faces of a model. The two faces have different areas. How do you apply the load using SOLIDWORKS Simulation for a Linear Static Study?

- Select the two faces and input a normal to direction force of 200lb on each face

- Select the two faces and a reference plane. Apply 100lb on each face

- **Apply equal force to the two faces. The force on each face is the total force divided by the total area of the two faces**

- None of the above

42. Maximum and Minimum value indicators are displayed on Stress and Displacement plots in SOLIDWORKS Simulation for a Linear Static Study.

- **True**

- False

43. What SOLIDWORKS Simulation tool should you use to determine the result values at specific locations (nodes) in a model using SOLIDWORKS Simulation?

- Section tool
- **Probe tool**
- Clipping tool
- Surface tool

44. What criteria are best suited to check the failure of ductile materials in SOLIDWORKS Simulation?

- Maximum von Mises Strain and Maximum Shear Strain criterion
- **Maximum von Misses Stress and Maximum Shear Stress criterion**
- Maximum Mohr-Coulomb Stress and Maximum Mohr-Coulomb Shear Strain criterion
- Mohr-Coulomb Stress and Maximum Normal Stress criterion

45. Set the scale factor for plots_____ to avoid any misinterpretation of the results, after performing a Static analysis with gap/contact elements.

- Equal to 0
- **Equal to 1**
- Less than 1
- To the Maximum displacement value for the model

46. It is possible to mesh _____ with a combination of Solids, Shells and Beam elements in SOLIDWORKS Simulation.

- **Parts and Assemblies**
- Only Parts
- Only Assemblies
- None of the above

47. SOLIDWORKS Simulation supports multi-body parts. Which of the following is a true statement?

- You can employ different mesh controls to each Solid body

- You can classify Contact conditions between multiple Solid bodies

- You can classify a different material for each Solid body

- **All of the above are correct**

48. Which statement best describes a Compatible mesh?

- A mesh where only one type of element is used

- **A mesh where elements on touching bodies have overlaying nodes**

- A mesh where only a Shell or Solid element is used

- A mesh where only a single Solid element is used

49. The Ratio value in Mesh Control provides the geometric growth ratio from one layer of elements to the next.

- **True**

- False

50. The structures displayed in the following illustration are best analyzed using:

- Shell elements

- Solid elements

- **Beam elements**

- A mixture of Beam and Shell elements

51. The structure displayed in the following illustration is best analyzed using:

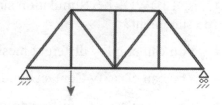

- Shell elements
- Solid elements
- **Beam elements**
- A mixture of Beam and Shell elements

52. The structure displayed in the following illustration is best analyzed using:

- **Shell elements**
- Solid elements
- Beam elements
- A mixture of Beam and Shell elements

Sheet metal model

53. The structure displayed in the following illustration is best analyzed using:

- Shell elements
- **Solid elements**
- Beam elements
- A mixture of Beam and Shell elements

54. Surface models can only be meshed with _____ elements.

- **Shell elements**
- Solid elements
- Beam elements
- A mixture of Beam and Shell elements

55. Use the _____ and _____ plots to calculate the distance between two nodes using SOLIDWORKS Simulation.

- **Mesh and Displacement**
- Displacement and FOS
- Resultant Displacement and FOS
- None of the above

56. You can simplify a large assembly in a Static Study by using the _____ or _____ options in your study.

- **Make Rigid, Fix**
- Shell element, Solid element
- Shell element, Compound element
- Make Rigid, Load element

57. A force "F" applied in a static analysis produces a resultant displacement URES. If the force is now 2x F and the mesh is not changed, then URES will:

- Double if there are no contact specified and there are large displacements in the structure
- Be divided by 2 if contacts are specified
- The analysis must be run again to find out
- **Double if there is no source of nonlinearity in the study (like contacts or large displacement options)**

58. To compute thermal stresses on a model with a uniform temperature distribution, what type/types of study/studies are required?

- **Static only**
- Thermal only
- Both Static and Thermal
- None of these answers is correct

59. In an h-adaptive method, use smaller elements in mesh regions with high errors to improve the accuracy of results.

- **True**
- False

60. In a p-adaptive method, use elements with a higher order polynomial in mesh regions with high errors to improve the accuracy of results.

- **True**
- False

61. Where will the maximum stress be in the illustration?

- A
- B
- C
- D

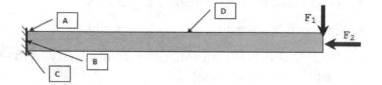

GLOSSARY

Alphabet of Lines: Each line on a technical drawing has a definite meaning and is drawn in a certain way. The line conventions recommended by the American National Standards Institute (ANSI) are presented in this text.

Alternate Position View: A drawing view superimposed in phantom lines on the original view. Utilized to show range of motion of an assembly.

Anchor Point: The origin of the Bill of Material in a sheet format.

Annotation: An annotation is a text note or a symbol that adds specific information and design intent to a part, assembly, or drawing. Annotations in a drawing include specific note, hole callout, surface finish symbol, datum feature symbol, datum target, geometric tolerance symbol, weld symbol, balloon, and stacked balloon, center mark, centerline marks, area hatch and block.

ANSI: American National Standards Institute.

Area Hatch: Apply a crosshatch pattern or solid fill to a model face, to a closed sketch profile, or to a region bounded by a combination of model edges and sketch entities. Area hatch can be applied only in drawings.

ASME: American Society of Mechanical Engineering, publisher of ASME Y14 Engineering Drawing and Documentation Practices that controls drawing, dimensioning and tolerancing.

Assembly: An assembly is a document in which parts, features and other assemblies (sub-assemblies) are put together. A part in an assembly is called a component. Adding a component to an assembly creates a link between the assembly and the component. When SOLIDWORKS opens the assembly, it finds the component file to show it in the assembly. Changes in the component are automatically reflected in the assembly. The filename extension for a SOLIDWORKS assembly file name is *.sldasm.

Attachment Point: An attachment point is the end of a leader that attaches to an edge, vertex, or face in a drawing sheet.

AutoDimension: The Autodimension tool provides the ability to insert reference dimensions into drawing views such as baseline, chain, and ordinate dimensions.

Auxiliary View: An Auxiliary View is similar to a Projected View, but it is unfolded normal to a reference edge in an existing view.

AWS: American Welding Society, publisher of AWS A2.4, Standard Location of Elements of a Welding Symbol.

Axonometric Projection: A type of parallel projection, more specifically a type of orthographic projection, used to create a pictorial drawing of an object, where the object is rotated along one or more of its axes relative to the plane of projection.

Balloon: A balloon labels the parts in the assembly and relates them to item numbers on the bill of materials (BOM) added in the drawing. The balloon item number corresponds to the order in the Feature Tree. The order controls the initial BOM Item Number.

Baseline Dimensions: Dimensions referenced from the same edge or vertex in a drawing view.

Bill of Materials: A table inserted into a drawing to keep a record of the parts and materials used in an assembly.

Block: A symbol in the drawing that combines geometry into a single entity.

BOM: Abbreviation for Bill of Materials.

Broken-out Section: A broken-out section exposes inner details of a drawing view by removing material from a closed profile. In an assembly, the Broken-out Section displays multiple components.

CAD: The use of computer technology for the design of objects, real or virtual. CAD often involves more than just shapes.

Cartesian Coordinate System: Specifies each point uniquely in a plane by a pair of numerical coordinates, which are the signed distances from the point to two fixed perpendicular directed lines measured in the same unit of length. Each reference line is called a coordinate axis or just axis of the system, and the point where they meet is its origin.

Cell: Area to enter a value in an EXCEL spreadsheet, identified by a Row and Column.

Center Mark: A cross that marks the center of a circle or arc.

Centerline: An axis of symmetry in a sketch or drawing displayed in a phantom font.

CommandManager: The CommandManager is a Context-sensitive toolbar that dynamically updates based on the toolbar you want to access. By default, it has toolbars embedded in it based on the document type. When you click a tab below the Command Manager, it updates to display that toolbar. For example, if you click the Sketch tab, the Sketch toolbar is displayed.

Component: A part or sub-assembly within an assembly.

ConfigurationManager: The ConfigurationManager is located on the left side of the SOLIDWORKS window and provides the means to create, select and view multiple configurations of parts and assemblies in an active document. You can split the

ConfigurationManager and either display two ConfigurationManager instances, or combine the ConfigurationManager with the FeatureManager design tree, PropertyManager or third party applications that use the panel.

Configurations: Variations of a part or assembly that control dimensions, display and state of a model.

Coordinate System: SOLIDWORKS uses a coordinate system with origins. A part document contains an original origin. Whenever you select a plane or face and open a sketch, an origin is created in alignment with the plane or face. An origin can be used as an anchor for the sketch entities, and it helps orient perspective of the axes. A three-dimensional reference triad orients you to the X, Y, and Z directions in part and assembly documents.

Copy and Paste: Utilize copy/paste to copy views from one sheet to another sheet in a drawing or between different drawings.

Cosmetic Thread: An annotation that represents threads.

Crosshatch: A pattern (or fill) applied to drawing views such as section views and broken-out sections.

Cursor Feedback: The system feedback symbol indicates what you are selecting or what the system is expecting you to select. As you move the mouse pointer across your model, system feedback is provided.

Datum Feature: An annotation that represents the primary, secondary and other reference planes of a model utilized in manufacturing.

Depth: The horizontal (front to back) distance between two features in frontal planes. Depth is often identified in the shop as the thickness of a part or feature.

Design Table: An Excel spreadsheet that is used to create multiple configurations in a part or assembly document.

Detail View: A portion of a larger view, usually at a larger scale than the original view. Create a detail view in a drawing to display a portion of a view, usually at an enlarged scale. This detail may be of an orthographic view, a non-planar (isometric) view, a section view, a crop view, an exploded assembly view or another detail view.

Detailing: Detailing refers to the SOLIDWORKS module used to insert, add and modify dimensions and notes in an engineering drawing.

Dimension Line: A line that references dimension text to extension lines indicating the feature being measured.

Dimension Tolerance: Controls the dimension tolerance values and the display of non-integer dimensions. The tolerance types are *None, Basic, Bilateral, Limit, Symmetric, MIN, MAX, Fit, Fit with tolerance* or *Fit (tolerance only)*.

Dimension: A value indicating the size of the 2D sketch entity or 3D feature. Dimensions in a SOLIDWORKS drawing are associated with the model, and changes in the model are reflected in the drawing, if you DO NOT USE DimXpert.

Dimensioning Standard - Metric: - ASME standards for the use of metric dimensioning required all the dimensions to be expressed in millimeters (mm). The (mm) is not needed on each dimension, but it is used when a dimension is used in a notation. No trailing zeroes are used. The Metric or International System of Units (S.I.) unit system in drafting is also known as the Millimeter, Gram Second (MMGS) unit system.

Dimensioning Standard - U.S: - ASME standard for U.S. dimensioning use the decimal inch value. When the decimal inch system is used, a zero is not used to the left of the decimal point for values less than one inch, and trailing zeroes are used. The U.S. unit system is also known as the Inch, Pound, Second (IPS) unit system.

DimXpert for Parts: A set of tools that applies dimensions and tolerances to parts according to the requirements of the ASME Y.14.41-2009 standard.

DimXpertManager: The DimXpertManager lists the tolerance features defined by DimXpert for a part. It also displays DimXpert tools that you use to insert dimensions and tolerances into a part. You can import these dimensions and tolerances into drawings. DimXpert is not associative.

Document: In SOLIDWORKS, each part, assembly, and drawing is referred to as a document, and each document is displayed in a separate window.

Drawing Sheet: A page in a drawing document.

Drawing Template: A document that is the foundation of a new drawing. The drawing template contains document properties and user-defined parameters such as sheet format. The extension for the drawing template filename is .DRWDOT.

Drawing: A 2D representation of a 3D part or assembly. The extension for a SOLIDWORKS drawing file name is .SLDDRW. Drawing refers to the SOLIDWORKS module used to insert, add, and modify views in an engineering drawing.

Edit Sheet Format: The drawing sheet contains two modes. Utilize the Edit Sheet Format command to add or modify notes and Title block information. Edit in the Edit Sheet Format mode.

Edit Sheet: The drawing sheet contains two modes. Utilize the Edit Sheet command to insert views and dimensions.

eDrawing: A compressed document that does not require the referenced part or assembly. eDrawings are animated to display multiple views in a drawing.

Empty View: An Empty View creates a blank view not tied to a part or assembly document.

Engineering Graphics: Translates ideas from design layouts, specifications, rough sketches, and calculations of engineers & architects into working drawings, maps, plans and illustrations which are used in making products.

Equation: Creates a mathematical relation between sketch dimensions, using dimension names as variables, or between feature parameters, such as the depth of an extruded feature or the instance count in a pattern.

Exploded view: A configuration in an assembly that displays its components separated from one another.

Export: The process to save a SOLIDWORKS document in another format for use in other CAD/CAM, rapid prototyping, web or graphics software applications.

Extension Line: The line extending from the profile line indicating the point from which a dimension is measured.

Extruded Cut Feature: Projects a sketch perpendicular to a Sketch plane to remove material from a part.

Face: A selectable area (planar or otherwise) of a model or surface with boundaries that help define the shape of the model or surface. For example, a rectangular solid has six faces.

Family Cell: A named empty cell in a Design Table that indicates the start of the evaluated parameters and configuration names. Locate Comments in a Design Table to the left or above the Family Cell.

Fasteners: Includes Bolts and nuts (threaded), Set screws (threaded), Washers, Keys, and Pins to name a few. Fasteners are not a permanent means of assembly such as welding or adhesives.

Feature: Features are geometry building blocks. Features add or remove material. Features are created from 2D or 3D sketched profiles or from edges and faces of existing geometry.

FeatureManager: The FeatureManager design tree located on the left side of the SOLIDWORKS window provides an outline view of the active part, assembly, or drawing. This makes it easy to see how the model or assembly was constructed or to examine the various sheets and views in a drawing. The FeatureManager and the Graphics window are dynamically linked. You can select features, sketches, drawing views and construction geometry in either pane.

First Angle Projection: In First Angle Projection the Top view is looking at the bottom of the part. First Angle Projection is used in Europe and most of the world. However, America and Australia use a method known as Third Angle Projection.

Fully defined: A sketch where all lines and curves in the sketch, and their positions, are described by dimensions or relations, or both, and cannot be moved. Fully defined sketch entities are shown in black.

Foreshortened radius: Helpful when the centerpoint of a radius is outside of the drawing or interferes with another drawing view: Broken Leader.

Foreshortening: The way things appear to get smaller in both height and depth as they recede into the distance.

French curve: A template made out of plastic, metal or wood composed of many different curves. It is used in manual drafting to draw smooth curves of varying radii.

Fully Defined: A sketch where all lines and curves in the sketch, and their positions, are described by dimensions or relations, or both, and cannot be moved. Fully defined sketch entities are displayed in black.

Geometric Tolerance: A set of standard symbols that specify the geometric characteristics and dimensional requirements of a feature.

Glass Box method: A traditional method of placing an object in an *imaginary glass box* to view the six principle views.

Global Coordinate System: Directional input refers by default to the Global coordinate system (X-, Y- and Z-), which is based on Plane1 with its origin located at the origin of the part or assembly.

Graphics Window: The area in the SOLIDWORKS window where the part, assembly, or drawing is displayed.

Grid: A system of fixed horizontal and vertical divisions.

Handle: An arrow, square or circle that you drag to adjust the size or position of an entity such as a view or dimension.

Heads-up View Toolbar: A transparent toolbar located at the top of the Graphic window.

Height: The vertical distance between two or more lines or surfaces (features) which are in horizontal planes.

Hidden Lines Removed (HLR): A view mode. All edges of the model that are not visible from the current view angle are removed from the display.

Hidden Lines Visible (HLV): A view mode. All edges of the model that are not visible from the current view angle are shown gray or dashed.

Hole Callouts: Hole callouts are available in drawings. If you modify a hole dimension in the model, the callout updates automatically in the drawing if you did not use DimXpert.

Hole Table: A table in a drawing document that displays the positions of selected holes from a specified origin datum. The tool labels each hole with a tag. The tag corresponds to a row in the table.

Import: The ability to open files from other software applications into a SOLIDWORKS document. The A-size sheet format was created as an AutoCAD file and imported into SOLIDWORKS.

Isometric Projection: A form of graphical projection, more specifically, a form of axonometric projection. It is a method of visually representing three-dimensional objects in two dimensions, in which the three coordinate axes appear equally foreshortened and the angles between any two of them arc 120º.

Layers: Simplifies a drawing by combining dimensions, annotations, geometry and components. Properties such as display, line style and thickness are assigned to a named layer.

Leader: A solid line created from an annotation to the referenced feature.

Line Format: A series of tools that controls Line Thickness, Line Style, Color, Layer and other properties.

Local (Reference) Coordinate System: Coordinate system other than the Global coordinate system. You can specify restraints and loads in any desired direction.

Lock Sheet Focus: Adds sketch entities and annotations to the selected sheet. Double-click the sheet to activate Lock Sheet Focus. To unlock a sheet, right-click and select Unlock Sheet Focus or double click inside the sheet boundary.

Lock View Position: Secures the view at its current position in the sheet. Right-click in the drawing view to Lock View Position. To unlock a view position, right-click and select Unlock View Position.

Mass Properties: The physical properties of a model based upon geometry and material.

Menus: Menus provide access to the commands that the SOLIDWORKS software offers. Menus are Context-sensitive and can be customized through a dialog box.

Model Item: Provides the ability to insert dimensions, annotations, and reference geometry from a model document (part or assembly) into a drawing.

Model View: A specific view of a part or assembly. Standard named views are listed in the view orientation dialog box such as isometric or front. Named views can be user-defined names for a specific view.

Model: 3D solid geometry in a part or assembly document. If a part or assembly document contains multiple configurations, each configuration is a separate model.

Motion Studies: Graphical simulations of motion and visual properties with assembly models. Analogous to a configuration, they do not actually change the original assembly model or its properties. They display the model as it changes based on simulation elements you add.

Mouse Buttons: The left, middle, and right mouse buttons have distinct meanings in SOLIDWORKS. Use the middle mouse button to rotate and Zoom in/out on the part or assembly document.

Oblique Projection: A simple type of graphical projection used for producing pictorial, two-dimensional images of three-dimensional objects.

OLE (Object Linking and Embedding): A Windows file format. A company logo or EXCEL spreadsheet placed inside a SOLIDWORKS document are examples of OLE files.

Ordinate Dimensions: Chain of dimensions referenced from a zero ordinate in a drawing or sketch.

Origin: The model origin is displayed in blue and represents the (0,0,0) coordinate of the model. When a sketch is active, a sketch origin is displayed in red and represents the (0,0,0) coordinate of the sketch. Dimensions and relations can be added to the model origin but not to a sketch origin.

Orthographic Projection: A means of representing a three-dimensional object in two dimensions. It is a form of parallel projection, where the view direction is orthogonal to the projection plane, resulting in every plane of the scene appearing in affine transformation on the viewing surface.

Parametric Note: A Note annotation that links text to a feature dimension or property value.

Parent View: A Parent view is an existing view in which other views are dependent on.

Part Dimension: Used in creating a part, they are sometimes called construction dimensions.

Part: A 3D object that consist of one or more features. A part inserted into an assembly is called a component. Insert part views, feature dimensions and annotations into 2D drawing. The extension for a SOLIDWORKS part filename is .SLDPRT.

Perspective Projection: The two most characteristic features of perspective are that objects are drawn smaller as their distance from the observer increases, and they are foreshortened: the size of an object's dimensions along the line of sight is relatively shorter than dimensions across the line of sight.

Plane: To create a sketch, choose a plane. Planes are flat and infinite. Planes are represented on the screen with visible edges.

Precedence of Line Types: When obtaining orthographic views, it is common for one type of line to overlap another type. When this occurs, drawing conventions have established an order of precedence.

Precision: Controls the number of decimal places displayed in a dimension.

Projected View: Projected views are created for Orthogonal views using one of the following tools: Standard 3 View, Model View or the Projected View tool from the View Layout toolbar.

Properties: Variables shared between documents through linked notes.

PropertyManager: Most sketch, feature, and drawing tools in SOLIDWORKS open a PropertyManager located on the left side of the SOLIDWORKS window. The PropertyManager displays the properties of the entity or feature so you specify the properties without a dialog box covering the Graphics window.

RealView: Provides a simplified way to display models in a photo-realistic setting using a library of appearances and scenes. RealView requires graphics card support and is memory intensive.

Rebuild: A tool that updates (or regenerates) the document with any changes made since the last time the model was rebuilt. Rebuild is typically used after changing a model dimension.

Reference Dimension: Dimensions added to a drawing document are called Reference dimensions, and are driven; you cannot edit the value of reference dimensions to modify the model. However, the values of reference dimensions change when the model dimensions change.

Relation: A relation is a geometric constraint between sketch entities or between a sketch entity and a plane, axis, edge or vertex.

Relative view: The Relative View defines an Orthographic view based on two orthogonal faces or places in the model.

Revision Table: The Revision Table lists the Engineering Change Orders (ECO), in a table form, issued over the life of the model and the drawing. The current Revision letter or number is placed in the Title block of the Drawing.

Right-Hand Rule: Is a common mnemonic for understanding notation conventions for vectors in 3 dimensions.

Rollback: Suppresses all items below the rollback bar.

Scale: A relative term meaning "size" in relationship to some system of measurement.

Section Line: A line or centerline sketched in a drawing view to create a section view.

Section Scope: Specifies the components to be left uncut when you create an assembly drawing section view.

Section View: You create a section view in a drawing by cutting the parent view with a cutting, or section line. The section view can be a straight cut section or an offset section defined by a stepped section line. The section line can also include concentric arcs. Create a Section View in a drawing by cutting the Parent view with a section line.

Sheet Format: A document that contains the following: page size and orientation, standard text, borders, logos, and Title block information. Customize the Sheet format to save time. The extension for the Sheet format filename is .SLDDRT.

Sheet Properties: Sheet Properties display properties of the selected sheet. Sheet Properties define the following: Name of the Sheet, Sheet Scale, Type of Projection (First angle or Third angle), Sheet Format, Sheet Size, View label, and Datum label.

Sheet: A page in a drawing document.

Silhouette Edge: A curve representing the extent of a cylindrical or curved face when viewed from the side.

Sketch: The name to describe a 2D profile is called a sketch. 2D sketches are created on flat faces and planes within the model. Typical geometry types are lines, arcs, corner rectangles, circles, polygons, and ellipses.

Spline: A sketched 2D or 3D curve defined by a set of control points.

Stacked Balloon: A group of balloons with only one leader. The balloons can be stacked vertically (up or down) or horizontally (left or right).

Standard views: The three orthographic projection views, Front, Top and Right positioned on the drawing according to First angle or Third angle projection.

Suppress: Removes an entity from the display and from any calculations in which it is involved. You can suppress features, assembly components, and so on. Suppressing an entity does not delete the entity; you can unsuppress the entity to restore it.

Surface Finish: An annotation that represents the texture of a part.

System Feedback: Feedback is provided by a symbol attached to the cursor arrow indicating your selection. As the cursor floats across the model, feedback is provided in the form of symbols riding next to the cursor.

System Options: System Options are stored in the registry of the computer. System Options are not part of the document. Changes to the System Options affect all current and future documents. There are hundreds of Systems Options.

Tangent Edge: The transition edge between rounded or filleted faces in hidden lines visible or hidden lines removed modes in drawings.

Task Pane: The Task Pane is displayed when you open the SOLIDWORKS software. It contains the following tabs: SOLIDWORKS Resources, Design Library, File Explorer, Search, View Palette, Document Recovery and RealView/PhotoWorks.

Templates: Templates are part, drawing and assembly documents that include user-defined parameters and are the basis for new documents.

Third Angle Projection: In Third angle projection the Top View is looking at the Top of the part. First Angle Projection is used in Europe and most of the world. America and Australia use the Third Angle Projection method.

Thread Class or Fit: Classes of fit are tolerance standards; they set a plus or minus figure that is applied to the pitch diameter of bolts or nuts. The classes of fit used with almost all bolts sized in inches are specified by the ANSI/ASME Unified Screw Thread standards (which differ from the previous American National standards).

Thread Lead: The distance advanced parallel to the axis when the screw is turned one revolution. For a single thread, lead is equal to the pitch; for a double thread, lead is twice the pitch.

Tolerance: The permissible range of variation in a dimension of an object. Tolerance may be specified as a factor or percentage of the nominal value, a maximum deviation from a nominal value, an explicit range of allowed values, be specified by a note or published standard with this information, or be implied by the numeric accuracy of the nominal value.

Toolbars: The toolbar menus provide shortcuts enabling you to access the most frequently used commands. Toolbars are Context-sensitive and can be customized through a dialog box.

T-Square: A technical drawing instrument, primarily a guide for drawing horizontal lines on a drafting table. It is used to guide the triangle that draws vertical lines. Its name comes from the general shape of the instrument where the horizontal member of the T slides on the side of the drafting table. Common lengths are 18", 24", 30", 36" and 42".

Under-defined: A sketch is under defined when there are not enough dimensions and relations to prevent entities from moving or changing size.

Units: Used in the measurement of physical quantities. Decimal inch dimensioning and Millimeter dimensioning are the two types of common units specified for engineering parts and drawings.

Vertex: A point at which two or more lines or edges intersect. Vertices can be selected for sketching, dimensioning, and many other operations.

View Palette: Use the View Palette, located in the Task Pane, to insert drawing views. It contains images of standard views, annotation views, section views, and flat patterns (sheet metal parts) of the selected model. You can drag views onto the drawing sheet to create a drawing view.

Weld Bead: An assembly feature that represents a weld between multiple parts.

Weld Finish: A weld symbol representing the parameters you specify.

Weld Symbol: An annotation in the part or drawing that represents the parameters of the weld.

Width: The horizontal distance between surfaces in profile planes. In the machine shop, the terms length and width are used interchangeably.

Zebra Stripes: Simulate the reflection of long strips of light on a very shiny surface. They allow you to see small changes in a surface that may be hard to see with a standard display.

Index